Collins

Student Book, **Higher 2**

Delivering the Edexcel specification

NEW GCSE MATHS
Edexcel Modular

Fully supports the 2010 GCSE Specification

Brian Speed • Keith Gordon • Kevin Evans • Trevor Senior • Chris Pearce

CONTENTS

Remember you will also need the Core content in Book 1 to complete your Unit 3 exam.

UNIT 3: Number, Algebra, Geometry 2

INTRODUCTION

Welcome to Collins New GCSE Maths for Edexcel Modular Higher Book 2. This book covers the content required for Unit 3. You will also need to remind yourself of the 'Core' material that you covered in Edexcel Modular Student Book 1, which will also be assessed in your Unit 3 exam.

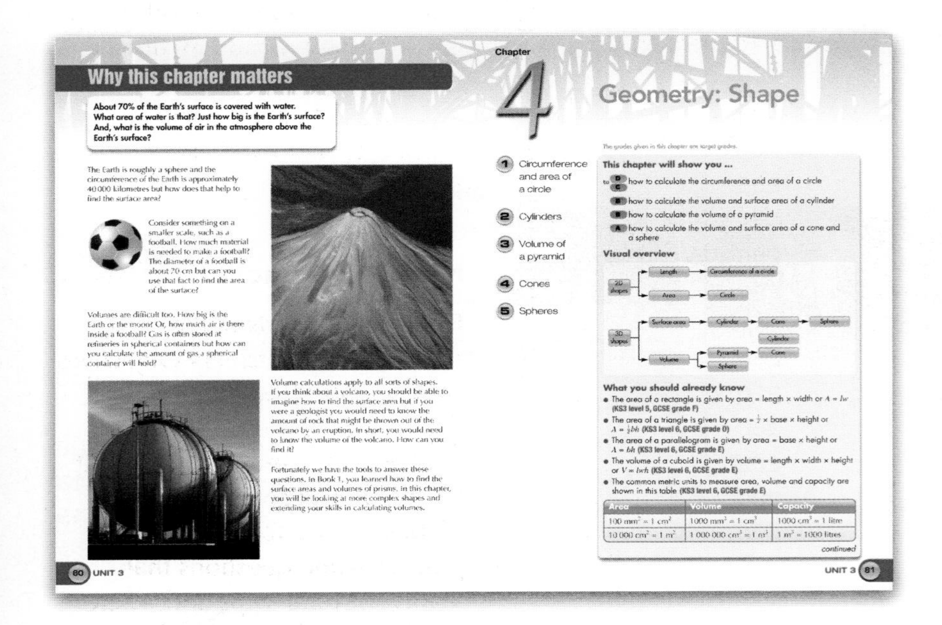

Why this chapter matters

About 70% of the Earth's surface is covered with water. What area of water is that? Just how big is the Earth's surface? And, what is the volume of air in the atmosphere above the Earth's surface?

The Earth is roughly a sphere and the circumference of the Earth is approximately 40 000 kilometres but how does that help to find the surface area?

Consider something on a smaller scale, such as a football. How much material is needed to make a football? The diameter of a football is about 70 cm but can you use that fact to find the area of the surface?

Volumes are difficult too. How big is the Earth or the moon? Or, how much air is there inside a football? Gas is often stored at refineries in spherical containers but how can you calculate the amount of gas a spherical container will hold?

Volume calculations apply to all sorts of shapes. If you think about a volcano, you should be able to imagine how to find the surface area but if you were a geologist you would need to know the amount of rock that might be thrown out of the volcano by an eruption. In short, you would need to know the volume of the volcano. How can you find it?

Fortunately we have the tools to answer these questions. In Book 1, you learned how to find the surface areas and volumes of prisms. In this chapter, you will be looking at more complex shapes and extending your skills in calculating volumes.

80 UNIT 3

Chapter 4

Geometry: Shape

1 Circumference and area of a circle
2 Cylinders
3 Volume of a pyramid
4 Cones
5 Spheres

The grades given in this chapter are target grades.

This chapter will show you ...

- D to C how to calculate the circumference and area of a circle
- how to calculate the volume and surface area of a cylinder
- how to calculate the volume of a pyramid
- A how to calculate the volume and surface area of a cone and a sphere

Visual overview

What you should already know

- The area of a rectangle is given by area = length × width or $A = lw$ **(KS3 level 5, GCSE grade F)**
- The area of a triangle is given by area = $\frac{1}{2}$ × base × height or $A = \frac{1}{2}bh$ **(KS3 level 6, GCSE grade D)**
- The area of a parallelogram is given by area = base × height or $A = bh$ **(KS3 level 6, GCSE grade E)**
- The volume of a cuboid is given by volume = length × width × height or $V = lwh$ **(KS3 level 6, GCSE grade E)**
- The common metric units to measure area, volume and capacity are shown in this table **(KS3 level 6, GCSE grade E)**

Area	Volume	Capacity
$100\ mm^2 = 1\ cm^2$	$1000\ mm^3 = 1\ cm^3$	$1000\ cm^3 = 1$ litre
$10\ 000\ cm^2 = 1\ m^2$	$1\ 000\ 000\ cm^3 = 1\ m^3$	$1\ m^3 = 1000$ litres

continued

UNIT 3 81

Why this chapter matters

Find out why each chapter is important through the history of maths, seeing how maths links to other subjects and cultures, and how maths is related to real life.

Chapter overviews

Look ahead to see what maths you will be doing and how you can build on what you already know.

Colour-coded grades

Know what target grade you are working at and track your progress with the colour-coded grade panels at the side of the page.

Use of calculators

Questions where you must or could use your calculator are marked with icon.

Explanations involving calculators are based on *CASIO fx–83ES*.

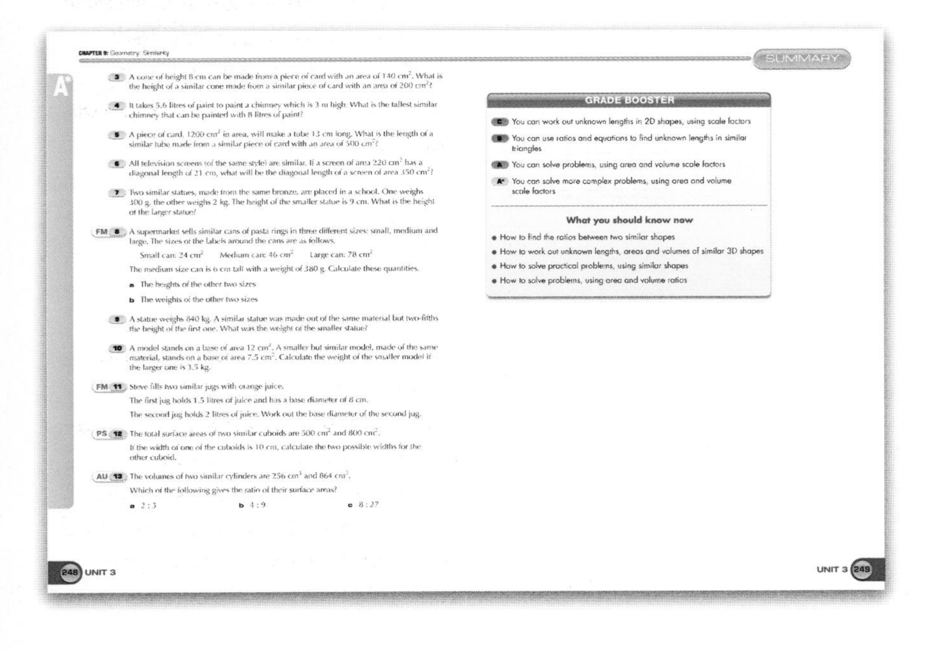

CHAPTER 9: Geometry: Similarity

A*

3 A cone of height 8 cm can be made from a piece of card with an area of $140\ cm^2$. What is the height of a similar cone made from a similar piece of card with an area of $200\ cm^2$?

4 It takes 5.6 litres of paint to paint a chimney which is 3 m high. What is the tallest similar chimney that can be painted with 8 litres of paint?

5 A piece of card, $1200\ cm^2$ in area, will make a tube 13 cm long. What is the length of a similar tube made from a similar piece of card with an area of $500\ cm^2$?

6 All television screens (of the same style) are similar. If a screen of area $220\ cm^2$ has a diagonal length of 21 cm, what will be the diagonal length of a screen of area $350\ cm^2$?

7 Two similar statues, made from the same bronze, are placed in a school. One weighs 300 g, the other weighs 2 kg. The height of the smaller statue is 9 cm. What is the height of the larger statue?

FM 8 A supermarket sells similar cans of pasta rings in three different sizes: small, medium and large. The sizes of the labels around the cans are as follows.

Small can: $24\ cm^2$ Medium can: $46\ cm^2$ Large can: $78\ cm^2$

The medium size can is 6 cm tall with a weight of 380 g. Calculate these quantities.

a The heights of the other two sizes

b The weights of the other two sizes

9 A statue weighs 840 kg. A similar statue was made out of the same material but two-fifths the height of the first one. What was the weight of the smaller statue?

10 A model stands on a base of area $12\ cm^2$. A smaller but similar model, made of the same material, stands on a base of area $7.5\ cm^2$. Calculate the weight of the smaller model if the larger one is 3.5 kg.

FM 11 Steve fills two similar jugs with orange juice.

The first jug holds 1.5 litres of juice and has a base diameter of 8 cm.

The second jug holds 2 litres of juice. Work out the base diameter of the second jug.

PS 12 The total surface areas of two similar cuboids are $500\ cm^2$ and $800\ cm^2$.

If the width of one of the cuboids is 10 cm, calculate the two possible widths for the other cuboid.

AU 13 The volumes of two similar cylinders are $256\ cm^3$ and $864\ cm^3$.

Which of the following gives the ratio of their surface areas?

a 2 : 3 **b** 4 : 9 **c** 8 : 27

248 UNIT 3

SUMMARY

GRADE BOOSTER

- C You can work out unknown lengths in 2D shapes, using scale factors
- B You can use ratios and equations to find unknown lengths in similar triangles
- A You can solve problems, using area and volume scale factors
- A* You can solve more complex problems, using area and volume scale factors

What you should know now

- How to find the ratios between two similar shapes
- How to work out unknown lengths, areas and volumes of similar 3D shapes
- How to solve practical problems, using similar shapes
- How to solve problems, using area and volume ratios

UNIT 3 249

Grade booster

Review what you have learnt and how to get to the next grade with the Grade booster at the end of each chapter.

Worked examples

Understand the topic before you start the exercise by reading the examples in blue boxes. These take you through questions step by step.

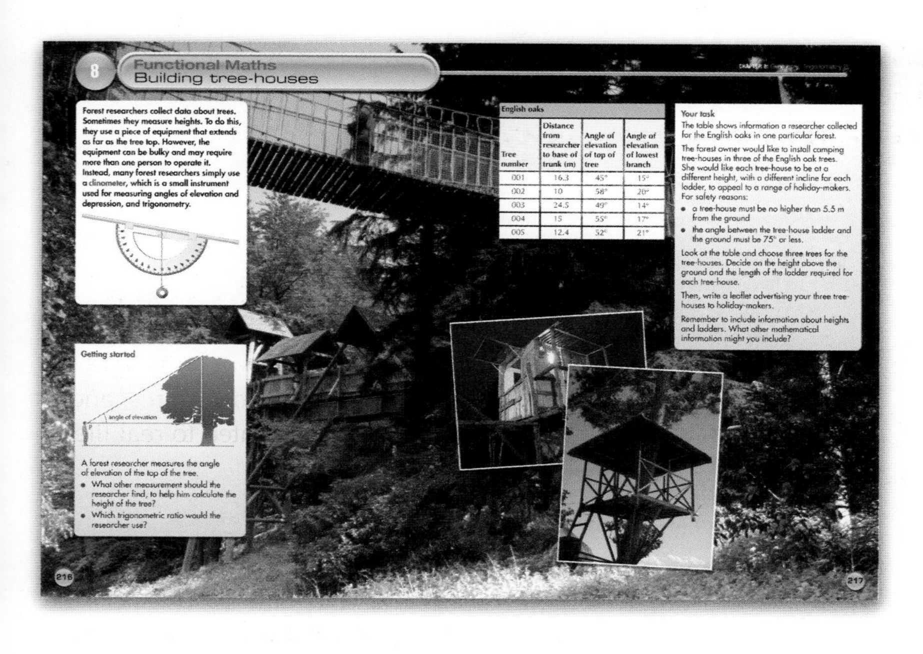

8 Functional Maths
Building tree-houses

Forest researchers collect data about trees. Sometimes they measure heights. To do this, they use a piece of equipment that extends as far as the tree top. However, the equipment can be bulky and may require more than one person to operate it. Instead, many forest researchers simply use a clinometer, which is a small instrument used for measuring angles of elevation and depression, and trigonometry.

Getting started

angle of elevation

A forest researcher measures the angle of elevation of the top of the tree.

- What other measurement should the researcher find, to help him calculate the height of the tree?
- Which trigonometric ratio would the researcher use?

English oaks

Tree number	Distance from researcher to base of trunk (m)	Angle of elevation of top of tree	Angle of elevation of lowest branch
001	16.3	45°	15°
002	10	58°	20°
003	24.5	49°	14°
004	15	55°	17°
005	12.4	52°	21°

Your task

The table shows information a researcher collected for the English oaks in one particular forest.

The forest owner would like to install camping tree-houses in three of the English oak trees. She would like each tree-house to be at a different height, with a different incline for each ladder, to appeal to a range of holiday-makers. For safety reasons:

- a tree-house must be no higher than 5.5 m from the ground
- the angle between the tree-house ladder and the ground must be 75° or less.

Look at the table and choose three trees for the tree-houses. Decide on the height above the ground and the length of the ladder required for each tree-house.

Then, write a leaflet advertising your three tree-houses to holiday-makers.

Remember to include information about heights and ladders. What other mathematical information might you include?

216 217

Functional maths

Practise functional maths skills to see how people use maths in everyday life. Look out for practice questions marked FM. There are also extra functional-maths and problem-solving activities at the end of every chapter to build and apply your skills.

New Assessment Objectives

Practise new parts of the curriculum (Assessment Objectives AO2 and AO3) with questions that assess your understanding marked AU and questions that test if you can solve problems marked PS. You will also practise some questions that involve several steps and where you have to choose which method to use; these also test AO2. There are also plenty of straightforward questions (AO1) that test if you can do the maths.

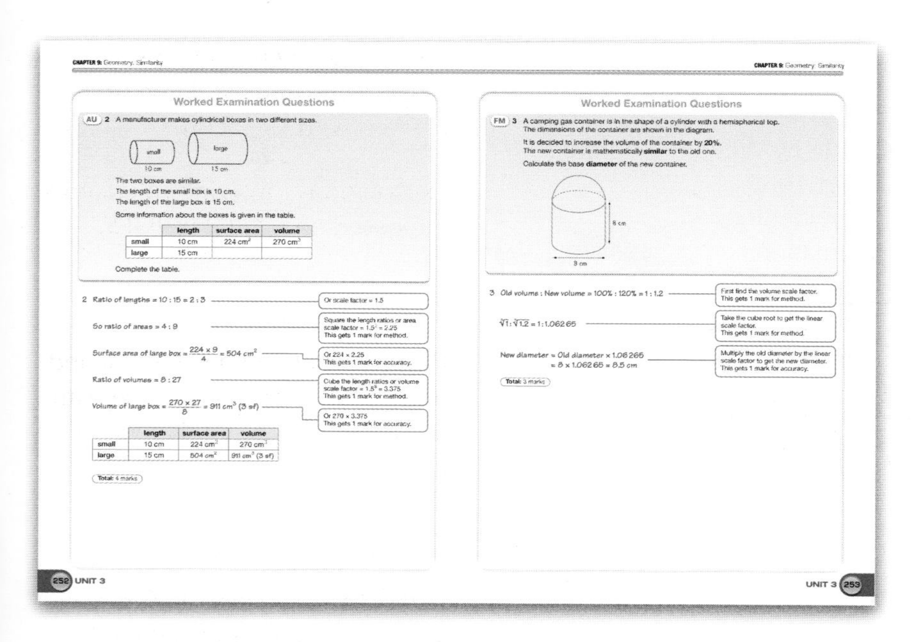

CHAPTER 9: Geometry: Similarity

Worked Examination Questions

AU 2 A manufacturer makes cylindrical boxes in two different sizes.

small 10 cm large 15 cm

The two boxes are similar.
The length of the small box is 10 cm.
The length of the large box is 15 cm.
Some information about the boxes is given in the table.

	length	surface area	volume
small	10 cm	224 cm²	270 cm³
large	15 cm		

Complete the table.

2 Ratio of lengths = 10 : 15 = 2 : 3 — Or scale factor = 1.5

So ratio of areas = 4 : 9 — Square the length ratios or area scale factor = 1.5^2 = 2.25. This gets 1 mark for method.

Surface area of large box = $\frac{224 \times 9}{4}$ = 504 cm² — Or 224 × 2.25. This gets 1 mark for accuracy.

Ratio of volumes = 8 : 27 — Cube the length ratios or volume scale factor = 1.5^3 = 3.375. This gets 1 mark for method.

Volume of large box = $\frac{270 \times 27}{8}$ = 911 cm³ (3 sf) — Or 270 × 3.375. This gets 1 mark for accuracy.

	length	surface area	volume
small	10 cm	224 cm²	270 cm³
large	15 cm	504 cm²	911 cm³ (3 sf)

Total: 4 marks

252 UNIT 3

CHAPTER 9: Geometry: Similarity

Worked Examination Questions

FM 3 A camping gas container is in the shape of a cylinder with a hemispherical top.
The dimensions of the container are shown in the diagram.
It is decided to increase the volume of the container by **20%**.
The new container is mathematically **similar** to the old one.
Calculate the base **diameter** of the new container.

8 cm
8 cm

3 Old volume : New volume = 100% : 120% = 1 : 1.2 — First find the volume scale factor. This gets 1 mark for method.

$\sqrt[3]{1} : \sqrt[3]{1.2}$ = 1 : 1.062 65 — Take the cube root to get the linear scale factor. This gets 1 mark for method.

New diameter = Old diameter × 1.062 65
= 8 × 1.062 65 = 8.5 cm — Multiply the old diameter by the linear scale factor to get the new diameter. This gets 1 mark for accuracy.

Total: 3 marks

UNIT 3 253

Exam practice

Prepare for your exams with past exam questions and detailed worked exam questions with examiner comments to help you score maximum marks.

Quality of Written Communication (QWC)

Practise using accurate mathematical vocabulary and writing logical answers to questions to ensure you get your QWC (Quality of Written Communication) marks in the exams. The Glossary and worked exam questions will help you with this.

Why this chapter matters

Technology is increasingly important in our lives. It helps us do many things more efficiently than we could without it.

Modern **calculators** take away the need to perform long calculations by hand. They can help to improve accuracy – but a calculator is only as good as the person using it. If you press the buttons in the wrong order when doing a calculation then you will get the wrong answer. That is why learning to use a calculator effectively is important.

The earliest known calculating device was a **tally stick**, which was a stick with notches cut into it so that small numbers could be recorded.

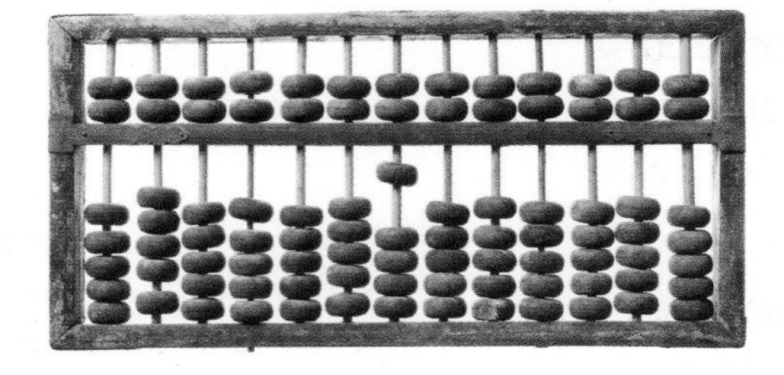

In about 2000 BC the **abacus** was invented in Eygpt.

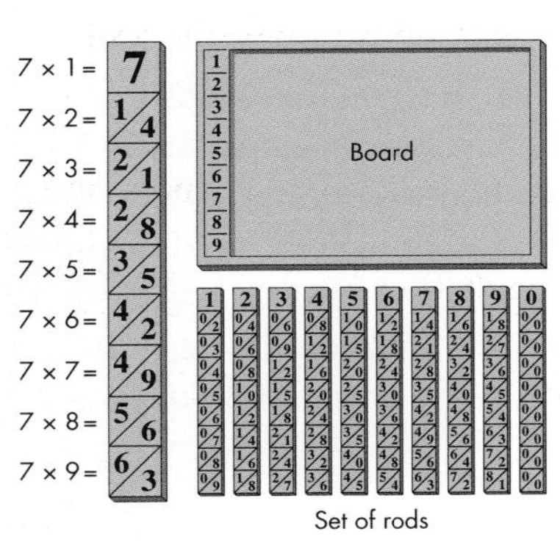

Abacuses are still used widely in China today and they were used widely for almost 3500 years, until John Napier devised a calculating aid called **Napier's bones**.

These led to the invention of the **slide rule** by William Oughtred in 1622. These stayed in use until the mid-1960s. Engineers working on the first ever moon landings used slide rules to do some of their calculations.

In the mid-16th century the first **mechanical calculating machines** were produced. These were based on a series of cogs and gears and so were too expensive to be widely used.

The first **electronic computers** were produced in the mid-20th century. Once the transistor was perfected, the power increased and the cost and size decreased until the point where the average scientific calculator that students use in schools has more computing power than the first craft that went into space.

Chapter

Number: Using a calculator

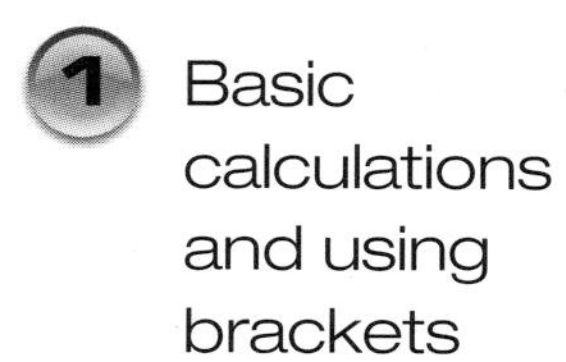

1 Basic calculations and using brackets

2 Adding and subtracting fractions with a calculator

3 Multiplying and dividing fractions with a calculator

The grades given in this chapter are target grades.

This chapter will show you ...

to D C how to use a calculator effectively

Visual overview

Basic calculations (+, −, ×, ÷) → Inputting fractions → Calculating with fractions

What you should already know

- How to add, subtract, multiply and divide with whole numbers and decimals **(KS3 level 5, GCSE grade E)**
- How to simplify fractions and decimals **(KS3 level 5, GCSE grade E)**
- How to convert improper fractions to mixed numbers or decimals and vice versa **(KS3 level 6, GCSE grade E)**
- The rules of BIDMAS/BODMAS with decimals **(KS3 level 5, GCSE grade E)**
- How to add and subtract fractions and decimals **(KS3 level 6, GCSE grade D)**

Quick check

1 Complete these calculations. Do not use a calculator.

a $48 + 89$ **b** $102 - 37$ **c** 23×7

d $336 \div 8$ **e** $3.6 + 2.9$ **f** $8.4 - 3.8$

g 3×4.5 **h** $7.8 \div 6$

2 a Convert these mixed numbers into improper fractions.

i $2\frac{2}{5}$ **ii** $3\frac{1}{4}$ **iii** $1\frac{7}{9}$

b Convert these improper fractions into mixed numbers.

i $\frac{11}{6}$ **ii** $\frac{7}{3}$ **iii** $\frac{23}{7}$

3 Work these out without using a calculator.

a $2 + 3 \times 4$ **b** $(2 + 3) \times 4$

c $6 + 4 - 3^2$ **d** $6 + (4 - 3)^2$

4 Work these out without using a calculator.

a $\frac{2}{3} + \frac{3}{4}$ **b** $\frac{1}{5} + \frac{2}{7}$

c $\frac{4}{5} - \frac{1}{4}$ **d** $2\frac{1}{3} - 1\frac{2}{5}$

1.1 Basic calculations and using brackets

This section will show you how to:

- use some of the important keys, including the bracket keys, to do calculations on a calculator

Key words
brackets
equals
function key
key
shift key

Most of the calculations in this unit are carried out to find the final answer of an algebraic problem or a geometric problem. The examples are intended to demonstrate how to use some of the **function keys** on the calculator. Remember that some functions will need the **shift key** SHIFT to make them work. When you have **keyed** in the calculation, press the **equals** key = to give the answer.

Some calculators display answers to fraction calculations as fractions. There is always a key to convert this to a decimal. In examinations, an answer given as a fraction or a decimal will always be acceptable unless the question asks you to round to a given accuracy.

Most scientific calculators can be set up to display the answers in the format you want.

EXAMPLE 1

These three angles are on a straight line.

To find the size of angle **a**, subtract the angles 68° and 49° from 180°.

a 49° 68°

You can do the calculation in two ways.

180 – 68 – 49 or 180 – (68 + 49)

You will learn more about angles in Chapters 5 and 10.

Try keying each calculation into your calculator.

180 – 68 – 49

1 8 0 – 6 8 – 4 9 =

The display will show 63.

180 – (68 + 49)

1 8 0 – (6 8 + 4 9) =

Again, the display should show 63.

It is important that you can do this both ways.

You must use the correct calculation or use **brackets** to combine parts of the calculation.

A common error is to work out 180 – 68 + 49, which will give the wrong answer.

EXAMPLE 2

Work out the area of this trapezium, where $a = 12.3$, $b = 16.8$ and $h = 2.4$.

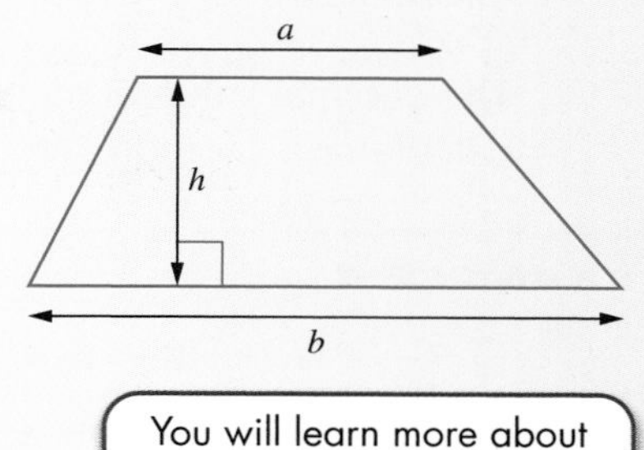

To work out the area of the trapezium, you use the formula:

$$A = \tfrac{1}{2}(a + b)h$$

Remember, you should always substitute into a formula before working it out.

$$A = \tfrac{1}{2}(12.3 + 16.8) \times 2.4$$

You will learn more about shapes in Chapter 4.

Between the brackets and the numbers at each end there is an assumed multiplication sign, so the calculation is:

$$\tfrac{1}{2} \times (12.3 + 16.8) \times 2.4$$

Be careful, $\frac{1}{2}$ can be keyed in lots of different ways:

- As a division

  [1] [÷] [2] [=]

  The display should show 0.5.

- Using the fraction key and the arrows

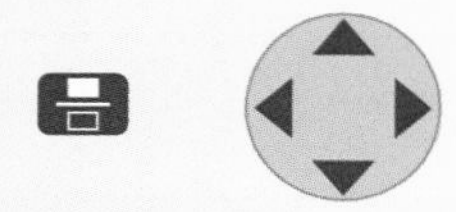

  [fraction] [1] ▼ [2] [=]

  The display should show $\frac{1}{2}$.

Keying in the full calculation, using the fraction key:

[fraction] [1] ▼ [2] ▶ [×] [(] [1] [2] [•] [3] [+] [1] [6] [•] [8] [)] [×] [2] [•] [4] [=]

The display should show 34.92 or $\frac{873}{25}$.

Your calculator has a power key [$x^{\square}$] and a cube key [x^3].

EXAMPLE 3

Find the value of $4.5^3 - 2 \times 4.5$.

Try keying in:

[4] [•] [5] [x^3] [3] [−] [2] [x] [4] [•] [5] [=]

The display should show 82.125 or $\frac{657}{8}$.

Most calculations involving circles will involve the number π (pronounced 'pi'), which has its own calculator button **π**.

You will learn more about circles in Chapter 4.

The decimal value of π goes on for ever. It has an approximate value of 3.14 but the value in a calculator is far more accurate and may be displayed as 3.1415926535 or π.

EXAMPLE 4

Work out: **a** $\pi \times 3.2^2$ **b** $2 \times \pi \times 4.9$

Give your answers to 1 decimal place.

a Try keying in:

π **×** **3** **•** **2** **x²** **=**

The display should show 32.16990877 or $\frac{256}{25}$ π. (Convert this to a decimal.)

This is 32.2 to 1 decimal place.

b Try keying in:

2 **×** **π** **x** **4** **•** **9** **=**

The display should show 30.78760801 or $\frac{49}{5}$ π. (Convert this to a decimal.)

This is 30.8 to 1 decimal place.

EXERCISE 1A

Use your calculator to work out the following.

Try to key in the calculation in as one continuous set, without writing down any intermediate values.

1 Work these out.

a $(10 - 2) \times 180 \div 10$

b $180 - (360 \div 5)$

2 Work these out.

a $\frac{1}{2} \times (4.6 + 6.8) \times 2.2$

b $\frac{1}{2} \times (2.3 + 9.9) \times 4.5$

3 Work out the following and give your answers to 1 decimal place.

a $\pi \times 8.5$ **b** $2 \times \pi \times 3.9$ **c** $\pi \times 6.8^2$ **d** $\pi \times 0.7^2$

FM 4 At Sovereign garage, Jon bought 21 litres of petrol for £21.52.

At the Bridge garage he paid £15.41 for 15 litres.

At which garage is the petrol cheaper?

D

D

AU **5** A teacher asked her class to work out $\frac{2.3 + 8.9}{3.8 - 1.7}$.

Abby keyed in:

(2 • 3 + 8 • 9) ÷ 3 • 8 − 1 • 7 =

Bobby keyed in:

2 • 3 + 8 • 9 ÷ 3 • 8 − 1 • 7 =

Col keyed in:

(2 • 3 + 8 • 9) ÷ (3 • 8 − 1 • 7) =

Donna keyed in:

2 • 3 + 8 • 9 ÷ (3 • 8 − 1 • 7) =

They each rounded their answers to 3 decimal places.

Work out the answer each of them found.

Who had the correct answer?

PS **6** Show that a speed of 31 metres per seconds is approximately 70 miles per hour.

You will need to know that 1 mile ≈ 1610 metres.

7 Work the value of each of these, if $a = 3.4$, $b = 5.6$ and $c = 8.8$.

a abc **b** $2(ab + ac + bc)$

C

8 Work out the following giving your answer to 2 decimal places.

a $\sqrt{(3.2^2 - 1.6^2)}$ **b** $\sqrt{(4.8^2 + 3.6^2)}$

9 Work these out.

a $7.8^3 + 3 \times 7.8$ **b** $5.45^3 - 2 \times 5.45 - 40$

1.2 Adding and subtracting fractions with a calculator

This section will show you how to:
- use a calculator to add and subtract fractions

Key words

fraction	mixed number
improper fraction	proper fraction
key	shift key

In this lesson, questions requiring calculation of **fractions** are set in a context linked to other topics, such as algebra or geometry.

You will recall from Unit 2 that a fraction with the numerator bigger than the denominator is an **improper fraction** or a *top-heavy fraction*.

You will also recall that a **mixed number** is made up of a whole number and a **proper fraction**.

For example:

$\frac{14}{5} = 2\frac{4}{5}$ and $3\frac{2}{7} = \frac{23}{7}$

Using a calculator with improper fractions

Check that your calculator has a fraction key. Remember, for some functions, you may need to use the **shift key** SHIFT.

To **key** in a fraction, press ⊟.

Input the fraction so that it looks like this:

$\frac{9}{5}$ or 9⌟5

Now press the equals key = so that the fraction displays in the answer part of the screen.

Pressing shift and the key S⇔D will convert the fraction to a mixed number.

1⌟4⌟5

This is the mixed number $1\frac{4}{5}$.

Pressing the equals sign again will convert the mixed number back to an improper fraction.

- Can you see a way of converting an improper fraction to a mixed number without using a calculator?
- Test your idea. Then use your calculator to check it.

Using a calculator to convert mixed numbers to improper fractions

To input a mixed number, press the shift key first and then the fraction key ⊟.

Pressing the equals sign will convert the mixed number to an improper fraction.

- Now key in at least 10 improper fractions and convert them to mixed numbers.
- Remember to press the equals sign to change the mixed numbers back to improper fractions.
- Now input at least 10 mixed numbers and convert them to improper fractions.
- Look at your results. Can you see a way of converting a mixed number to an improper fraction without using a calculator?
- Test your idea. Then use your calculator to check it.

EXAMPLE 5

A water tank is half full. One-third of the capacity of the full tank is poured out.

What fraction of the tank is now full of water?

The calculation is $\frac{1}{2} - \frac{1}{3}$.

Keying in the calculation gives:

[fraction] 1 ▼ 2 ▶ − [fraction] 1 ▼ 3 ▶ =

The display should show $\frac{1}{6}$.

The tank is now one-sixth full of water.

EXAMPLE 6

Work out the perimeter of a rectangle $1\frac{1}{2}$ cm long and $3\frac{2}{3}$ cm wide.

To work out the perimeter of this rectangle, you can use the formula:

$$P = 2l + 2w$$

where $l = 1\frac{1}{2}$ cm and $w = 3\frac{2}{3}$ cm.

$$P = 2 \times 1\frac{1}{2} + 2 \times 3\frac{2}{3}$$

Keying in the calculation gives:

2 × SHIFT [fraction] 1 ▶ 1 ▼ 2 ▶ + 2 × SHIFT [fraction] 3 ▶ 2 ▼ 3 ▶ =

The display should show $10\frac{1}{3}$.

So the perimeter is $10\frac{1}{3}$ cm.

EXERCISE 1B

1 Use your calculator to work these out. Give your answers as mixed numbers.

Try to key in the calculation as one continuous set, without writing down any intermediate values.

a $4\frac{3}{4} + 1\frac{4}{5}$ **b** $3\frac{5}{6} + 4\frac{7}{10}$ **c** $7\frac{4}{5} + 8\frac{9}{20}$ **d** $9\frac{3}{8} + 2\frac{9}{25}$

e $6\frac{7}{20} + 1\frac{3}{16}$ **f** $2\frac{5}{8} + 3\frac{9}{16} + 5\frac{3}{5}$ **g** $6\frac{9}{20} - 3\frac{1}{12}$ **h** $4\frac{3}{4} - 2\frac{7}{48}$

i $8\frac{11}{32} - 5\frac{1}{6}$ **j** $12\frac{4}{5} + 3\frac{9}{16} - 8\frac{2}{3}$ **k** $9\frac{7}{16} + 5\frac{3}{8} - 7\frac{1}{20}$ **l** $10\frac{3}{4} + 6\frac{2}{9} - 12\frac{3}{11}$

2 A water tank is three-quarters full. Two-thirds of a full tank is poured out.

What fraction of the tank is now full of water?

D

D

3

a What is the distance between Wickersley and Redbrook, using these roads?

b How much further is it to Redbrook than to Wickersley?

C

4 Here is a calculation.

$\frac{3}{25} + \frac{7}{10}$

Imagine that you are trying to explain to someone how to use a calculator to do this.

Write down what you would say.

FM 5 There are the same number of boys and girls in a school.

Because of snow $\frac{4}{5}$ of the boys are absent and $\frac{5}{12}$ of the girls are absent.

What fraction of the students are present?

PS 6 **a** Use your calculator to work out $\frac{18}{37} - \frac{23}{43}$.

b Explain how your answer tells you that $\frac{23}{43}$ is greater than $\frac{18}{37}$.

7 Jon is working out $\frac{9}{32} + \frac{5}{7}$ without using a calculator.

He adds the numerators and the denominators to get an answer of $\frac{14}{39}$ which is not correct.

a Use a calculator to work out the correct answer.

b Work out $\frac{14}{39} - \frac{9}{32}$ on your calculator.

c Work out $\frac{14}{39} - \frac{5}{7}$ on your calculator.

d Explain why your answers to parts **b** and **c** show that $\frac{14}{39}$ is a fraction between $\frac{9}{32}$ and $\frac{5}{7}$.

AU 8 **a** Choose two other fractions to add together.

Write down the incorrect answer that Jon would get.

Repeat the steps of question **7** for these fractions.

b Is Jon's answer between your two fractions?

AU 9 To work out the perimeter of a rectangle the following formula is used.

$P = 2l + 2w$

Work out the perimeter when $l = 5\frac{1}{8}$ cm and $w = 4\frac{1}{3}$ cm.

PS 10 A shape is rotated 90° clockwise and then a further 60° clockwise.

What fraction of a turn is needed to return it to its original position.

Give both possible answers.

1.3 Multiplying and dividing fractions with a calculator

This section will show you how to:
- use a calculator to multiply and divide fractions

Key words
fraction
key
shift key

In this lesson, questions requiring calculation of **fractions** will be set in a context linked to other topics such as algebra or geometry. Remember, for some functions, you may need to use the **shift key** SHIFT.

EXAMPLE 7

Work out the area of a rectangle of length $3\frac{1}{2}$ m and width $2\frac{2}{3}$ m.

The formula for the area of a rectangle is:

area = length × width

Keying in the calculation, where length = $3\frac{1}{2}$ and width = $2\frac{2}{3}$ gives:

SHIFT ⊟ 3 ▶ 1 ▼ 2 → × SHIFT ⊟ 2 ▶ 2 ▼ 3 ▶ =

The display should show $9\frac{1}{3}$.

The area is $9\frac{1}{3}$ cm^2.

EXAMPLE 8

Work out the average speed of a bus that travels $20\frac{1}{4}$ miles in $\frac{3}{4}$ hour.

The formula for the average speed is:

$$\text{average speed} = \frac{\text{distance}}{\text{time}}$$

Use this formula to work the average speed of the bus, where distance is $20\frac{1}{4}$ and time is $\frac{3}{4}$.

Keying in the calculation gives:

SHIFT ⊟ 2 0 ▶ 1 ▼ 4 ÷ ⊟ 3 ▼ 4 ▶ =

The display should show 27.

The average speed is 27 mph.

EXERCISE 1C

D

1 Use your calculator to work these out. Give your answers as fractions.

Try to key in the calculation as one continuous set, without writing down any intermediate values.

a $\frac{3}{4} \times \frac{4}{5}$ **b** $\frac{5}{6} \times \frac{7}{10}$ **c** $\frac{4}{5} \times \frac{9}{20}$ **d** $\frac{3}{8} \times \frac{9}{25}$

e $\frac{7}{20} \times \frac{3}{16}$ **f** $\frac{5}{8} \times \frac{9}{16} \times \frac{3}{5}$ **g** $\frac{9}{20} \div \frac{1}{12}$ **h** $\frac{3}{4} \div \frac{7}{48}$

i $\frac{11}{32} \div \frac{1}{6}$ **j** $\frac{4}{5} \times \frac{9}{16} \div \frac{2}{3}$ **k** $\frac{7}{16} \times \frac{3}{8} \div \frac{1}{20}$ **l** $\frac{3}{4} \times \frac{2}{9} \div \frac{3}{11}$

2 The formula for the area of a rectangle is: area = length × width

Use this formula to work the area of a rectangle of length $\frac{2}{3}$ m and width $\frac{1}{4}$ m.

3 Some steps are each $\frac{1}{5}$ m high. How many steps are needed to climb 3 m?

AU **4** **a** Use your calculator to work out $\frac{3}{4} \times \frac{9}{16}$ **b** Write down the answer to $\frac{9}{4} \times \frac{3}{16}$

AU **5** **a** Use your calculator to work out $\frac{2}{3} \div \frac{5}{6}$. **b** Use your calculator to work out $\frac{2}{3} \times \frac{6}{5}$.

c Use your calculator to work out $\frac{4}{7} \div \frac{3}{4}$. **d** Write down the answer to $\frac{4}{7} \times \frac{4}{3}$.

6 Use your calculator to work these out. Give your answers as mixed numbers.

Try to key in the calculation as one continuous set, without writing down any intermediate values.

a $4\frac{3}{4} \times 1\frac{4}{5}$ **b** $3\frac{5}{6} \times 4\frac{7}{10}$ **c** $7\frac{4}{5} \times 8\frac{9}{20}$ **d** $9\frac{3}{8} \times 2\frac{9}{25}$

e $6\frac{7}{20} \times 1\frac{3}{16}$ **f** $2\frac{5}{8} \times 3\frac{9}{16} \times 5\frac{3}{5}$ **g** $6\frac{9}{20} \div 3\frac{1}{12}$ **h** $4\frac{3}{4} \div 2\frac{7}{48}$

i $8\frac{11}{32} \div 5\frac{1}{6}$ **j** $12\frac{4}{5} \times 3\frac{9}{16} \div 8\frac{2}{3}$ **k** $9\frac{7}{16} \times 5\frac{3}{8} \div 7\frac{1}{20}$ **l** $10\frac{3}{4} \times 6\frac{2}{9} \div 12\frac{3}{11}$

C

7 The formula for the area of a rectangle is: area = length × width

Use this formula to work the area of a rectangle of length $5\frac{2}{3}$ metres and width $3\frac{1}{4}$ metres.

8 The volume of a cuboid is $26\frac{3}{4}$ cm^3. It is cut into eight equal pieces.

Work out the volume of one of the pieces.

9 The formula for the distance travelled is: distance = average speed × time taken

Work out how far a car travelling at an average speed of $36\frac{1}{4}$ mph will travel in $2\frac{1}{2}$ hours.

10 Glasses are filled from litre bottles of water.

Each glass holds $\frac{1}{2}$ pint. 1 litre = $1\frac{3}{4}$ pints

How many litre bottles are needed to fill 10 glasses?

PS FM **11** The ribbon on a roll is $3\frac{1}{2}$ m long. Joe wants to cut pieces of ribbon that are each $\frac{1}{6}$ m long.

He needs 50 piece.

How many rolls will he need?

GRADE BOOSTER

- **D** You can use BIDMAS or BODMAS to carry out operations in the correct order
- **D** You can use a calculator to add, subtract, multiply and divide fractions
- **C** You can use a calculator to add, subtract, multiply and divide mixed numbers

What you should know now

- How to use a calculator effectively, including the brackets and fraction keys

1 Use your calculator to work out the exact value of

$$\frac{15.6}{1.18 + 2.07}$$ *(2 marks)*

Edexcel, March 2008, Paper 10 (Calculator), Question 4

2 Use your calculator to work out

$$(3.4 + 2.1)^2 \times 5.7$$

Write down all the figures on your calculator display. *(2 marks)*

Edexcel, June 2008, Paper 10 (Calculator), Question 1

3 Use your calculator to work out

$$\frac{22.4 \times 14.5}{8.5 + 3.2}$$

Write down all the figures on your calculator display. *(2 marks)*

Edexcel, June 2008, Paper 15 (Calculator), Question 3

4 **a** Use your calculator to work out

$$\frac{1000}{7.3^2 - 16.3}$$

Write down all the figures on your calculator display. *(2 marks)*

b Write your answer to part a correct to 1 decimal place. *(1 mark)*

Edexcel, June 2008, Paper 11 Section A (Calculator), Question 2

5 Use your calculator to work out

$$\sqrt{12.63 + 18^2}$$

Write down all the figures on your calculator display. *(2 marks)*

Edexcel, November 2008, Paper 10 (Calculator), Question 1

6 Use a calculator to work out

$$\sqrt{\frac{21.6 \times 15.8}{3.8}}$$

Write down all the figures on your calculator display. *(2 marks)*

Edexcel, November 2008, Paper 15 (Calculator), Question 2

7 **a** Use your calculator to work out $\frac{26.4 + 8.2}{\sqrt{5.76}}$ as a decimal.

Write down all the figures on your calculator display. *(2 marks)*

b Write your answer to part **a** correct to 2 decimal places. *(1 mark)*

Edexcel, March 2009, Paper 10 (Calculator), Question 3

8 Use your calculator to work out $\frac{\sqrt{13.2 - 6.8}}{3.25 + 4.9}$

Write down all the figures on your calculator display. *(2 marks)*

Edexcel, June 2007, Paper 11 Section A (Calculator), Question 5

Worked Examination Questions

AU **1** The perimeter of a rectangle is $32\frac{1}{2}$ cm.
Work out a pair of possible values for the length and the width of the rectangle.

1 Perimeter is 2 × length + 2 × width
Length + width = $32\frac{1}{2} \div 2$

This gets 1 mark for method.

Length + width = $16\frac{1}{4}$ cm

This gets a mark for an accurate calculation.

Possible length and width are:
Length = 10 cm
Width = $6\frac{1}{4}$ cm

Any two values with a sum of $16\frac{1}{4}$ would score the final mark.

Total: 3 marks

FM **2** A driver is travelling 200 miles.
He sets off at 10 am.
He stops for a 20-minute break.
His average speed when travelling is $42\frac{1}{2}$ mph.
He wants to arrive before 3 pm.
Is he successful?

2 Time travelling = distance ÷ average speed
= $200 \div 42\frac{1}{2}$

This gets 1 mark for method.

= $4\frac{12}{17}$ or 4.7058...

20 minutes = $\frac{1}{3}$ hour or 0.33

This is the next step and gets 1 mark for method.

$4\frac{12}{17} + \frac{1}{3}$ or 4.7058... + 0.33...
= $5\frac{2}{51}$ hours 5.039 hours

This gets 1 mark for accuracy.

10 am to 3 pm is 5 hours so he arrives after 3 pm

A statement giving the correct conclusion from correct working would get 1 mark for quality of written communication (QWC).

Total: 4 marks

1 Functional Maths
Setting up your own business

You have been asked by your Business Studies teacher to set up a jewellery stall selling beaded jewellery at an upcoming Young Enterprise fair. There will be 50 stalls at this fair (many of which will be selling jewellery) and it is expected that there will be 500 attendees.

You will be competing against every other stall to sell your products to the attendees, either as one-off purchases or as bulk orders. In order to be successful in this you must carefully plan the design, cost and price of your jewellery, to ensure that people will buy your products and that you make a profit.

Getting started

Answer these questions to begin thinking about how beads can be used to make a piece of jewellery.

1 How many 6 mm beads are needed to make a bracelet?

2 How many 8 mm beads are needed to make an anklet?

3 How many 10 mm beads are needed to make a short necklace?

4 How many 10 mm beads are needed to make a long necklace?

5 You are asked to make a bracelet with beads of two different lengths. You decide to use 6 mm red beads and 8 mm blue beads. How many would you need if you used them alternately?

How to make beaded jewellery

Beads are sold in different sizes and wire is sold in different thicknesses, called the gauge. To make a piece of jewellery the beads are threaded onto the wire.

Step 1 Choose a gauge of wire and cut the length required.

Step 2 Put a fastening on one end.

Step 3 Thread on beads of different sizes in a pattern.

Step 4 Put a fastener on the other end.

Your jewellery is now complete.

Cost of materials

Here are the costs of the raw materials that you will need to make your jewellery.

6 mm beads	10p each
8 mm beads	12p each
10 mm beads	15p each
24-gauge wire	10p per centimetre
20-gauge wire	8p per centimetre
Fasteners for both ends: 30p per item of jewellery	

Advice

For bracelets and anklets use 20-gauge wire.

For necklaces use 24-gauge wire.

Beads are available in three lengths: 6 mm, 8 mm and 10 mm.

Beads are available in three colours, green, blue and red.

Standard lengths for bracelets and necklaces	
Bracelet	17 cm
Anklet	23 cm
Short necklace	39 cm
Long necklace	46 cm

Your task

With a partner, draw up a business plan for your jewellery stall, to ensure you produce high quality beaded jewellery that will turn a good profit. In your plan, you should include:

- an outline of who you expect to buy your jewellery (your 'target market')
- a design for at least one set of jewellery that will appeal to your target market
- a list of all the materials you will need
- the cost of your designs
- a fair price at which to sell your jewellery
- a discounting plan for bulk orders, or if you must reduce your prices on the day
- an expected profit.

Use all the information given on these pages to create your business plan.

Be sure to justify your plan, using appropriate mathematics and describing the calculations that you have done.

Present your business plan as a report to the Young Enterprise committee.

Why this chapter matters

The number system we use today is not the first one that was ever developed. Over thousands of years, lots of different systems have been used, in many different civilisations. However, most were based on the number 10. Can you think of a good reason for this?

The Egyptian number system (from about 3000BC) used different symbols to represent 1, 10, 100, 1000 and so on.

So the number 23 would be written: ∩∩ |||

Decimal number	Egyptian symbol	
1 =	𓏺	staff
10 =	𓎆	heel bone
100 =	𓍢	piece of rope
1000 =	𓆼	flower
10 000 =	𓂭	pointing finger
100 000 =	𓆐	tadpole
1 000 000 =	𓁨	man

The Chinese number system used sticks.

𝍠	𝍡	𝍢	𝍣	𝍤	𝍥	𝍦	𝍧	𝍨
1	2	3	4	5	6	7	8	9

𝍩	𝍪	𝍫	𝍬	𝍭	𝍮	𝍯	𝍰	𝍱
10	20	30	40	50	60	70	80	90

So, the number 23 would be written: 𝍪𝍢

The Roman number system is still used today. You will see it on clocks and sometimes for the numbers of the first few pages in a book.

I	V	X	L	C	D	M
1	5	10	50	100	500	1000

So, the number 23 would be written: XXIII

Interestingly, although the representation of 4 in Roman numerals is IV, on clocks it generally appears as IIII. Why would this be?

The system we use today is called the **Hindu-Arabic system** and uses the symbols 0, 1, 2, 3, 4, 5, 6, 7, 8 and 9. This has been widely used since about 900AD.

Here you can see the Roman numerals for 1–12 on a clock face.

Roman numerals also appear on Big Ben.

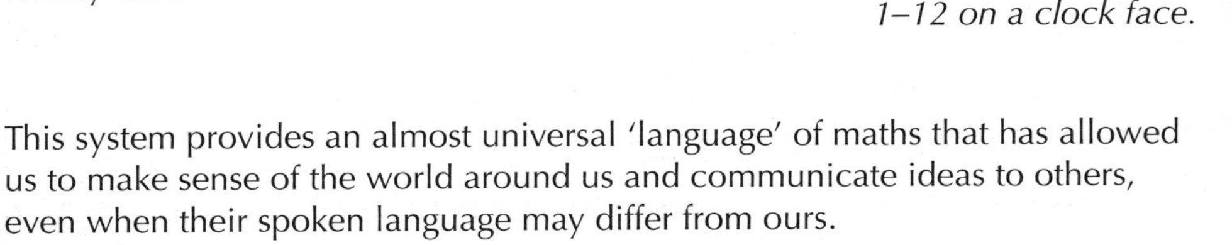

This system provides an almost universal 'language' of maths that has allowed us to make sense of the world around us and communicate ideas to others, even when their spoken language may differ from ours.

Chapter

Number: Decimals, percentages and powers

The grades given in this chapter are target grades.

1. Multiplication and division with decimals
2. Compound interest and repeated percentage change
3. Reverse percentage (working out the original quantity)
4. Powers (indices)
5. Reciprocals and rational numbers
6. Standard form

This chapter will show you ...

- **D** how to calculate with decimals
- **D to C** how to recognise rational numbers, terminating decimals and recurring decimals
- **D to A** how to calculate with powers (indices)
- **C to B** how to calculate compound interest
- **C to B** how to solve problems involving repeated percentage change
- **C to A*** how to work out a reciprocal of a rational number
- **C to A** how to calculate the original value after a percentage increase or decrease
- **B** how to write numbers in standard form and how to calculate with standard form
- **B to A*** how to convert fractions to terminating decimals and recurring decimals, and vice versa

Visual overview

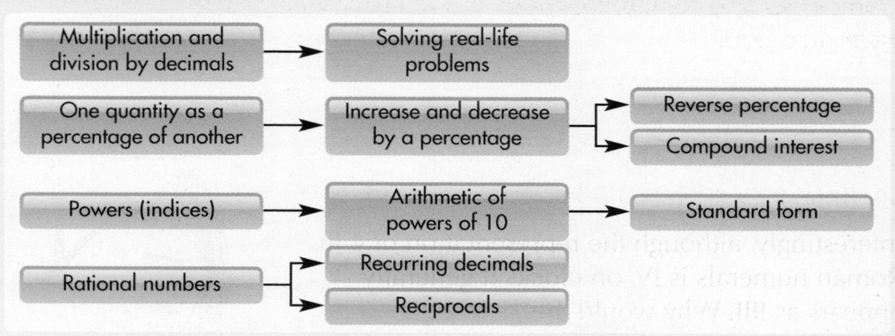

What you should already know

- How to work out simple percentages, such as 10%, of quantities **(KS3 level 4, GCSE grade F)**
- The meaning of square root and cube root **(KS3 level 5, GCSE grade E)**

Quick check

1 Work out the following:
a 23×145 **b** $984 \div 24$ **c** $(16 + 9)^2$

2 Work out the following:
a $2 + 3 \times 5$ **b** $(2 + 3) \times 5$ **c** $2 + 32 - 6$

3 What is 10% of **a** £250 **b** £39.00 **c** £4.70?

2.1 Multiplication and division with decimals

This section will show you how to:
- multiply a decimal number by another decimal number
- divide by decimals by changing the calculation to division by an integer

Key words
decimal places
decimal point
integer

Multiplying two decimal numbers together

Follow these steps to multiply one decimal number by another decimal number.

- First, complete the whole calculation as if the **decimal points** were not there.
- Then, count the total number of **decimal places** (dp) in the two decimal numbers. This gives the number of decimal places in the answer.

EXAMPLE 1

Work out: 3.42×2.7

Ignoring the decimal points gives the following calculation.

$$\begin{array}{r} 342 \\ \times \quad 27 \\ \hline 2394 \\ 6840 \\ \hline 9234 \\ \hline \end{array}$$

Now, 3.42 has two decimal places (.42) and 2.7 has one decimal place (.7). So, the total number of decimal places in the answer is three.

So $3.42 \times 2.7 = 9.234$

Dividing by a decimal

EXAMPLE 2

Work out the following. **a** $42 \div 0.2$ **b** $19.8 \div 0.55$

a The calculation is $42 \div 0.2$ which can be rewritten as $420 \div 2$. In this case both values have been multiplied by 10 to make the divisor into a whole number or **integer**. This is then a straightforward division to which the answer is 210.

Another way to view this is as a fraction problem.

$$\frac{42}{0.2} = \frac{42}{0.2} \times \frac{10}{10} = \frac{420}{2} = \frac{210}{1} = 210$$

b $19.8 \div 0.55 = 198 \div 5.5 = 1980 \div 55$

This then becomes a long division problem.

This has been solved by the method of repeated subtraction.

$$\begin{array}{rr} 1980 & \\ -\ 1100 & 20 \times 55 \\ \hline 880 & \\ -\ 440 & 8 \times 55 \\ \hline 440 & \\ -\ 440 & 8 \times 55 \\ \hline 0 & 36 \times 55 \end{array}$$

So $19.8 \div 0.55 = 36$

EXERCISE 2A

D

1 Work out each of these.

a 0.14×0.2 **b** 0.3×0.3 **c** 0.24×0.8 **d** 5.82×0.52

e 5.8×1.23 **f** 5.6×9.1 **g** 0.875×3.5 **h** 9.12×5.1

2 For each of the following:

i estimate the answer by first rounding each number to the nearest whole number

ii calculate the exact answer and then calculate the difference between this and your answers to part **i**.

a 4.8×7.3 **b** 2.4×7.6 **c** 15.3×3.9 **d** 20.1×8.6

e 4.35×2.8 **f** 8.13×3.2 **g** 7.82×5.2 **h** 19.8×7.1

AU 3 **a** Use any method to work out: 26×22

b Use your answer to part **a** to work out the following.

i 2.6×2.2 **ii** 1.3×1.1 **iii** 2.6×8.8

AU 4 Lee is trying to work out the answer to 8.6×4.7. His answer is 40.24.

a Without working it out, can you tell whether his answer is correct?

b Tracy says the answer is 46.42.
Without working out the answer can you tell whether her answer is correct?
In each part, show how you decided.

PS 5 Here are three calculations. $26.66 \div 3.1$ $17.15 \div 3.5$ $55.04 \div 8.6$

Which has the largest answer? Show how you know.

6 Work out each of these.

a $3.6 \div 0.2$ **b** $56 \div 0.4$ **c** $0.42 \div 0.3$ **d** $8.4 \div 0.7$ **e** $4.26 \div 0.2$

f $3.45 \div 0.5$ **g** $83.7 \div 0.03$ **h** $0.968 \div 0.08$ **i** $7.56 \div 0.4$

7 Work out each of these.

a $67.2 \div 0.24$ **b** $6.36 \div 0.53$ **c** $0.936 \div 5.2$ **d** $162 \div 0.36$ **e** $2.17 \div 3.5$

f $98.8 \div 0.26$ **g** $0.468 \div 1.8$ **h** $132 \div 0.55$ **i** $0.984 \div 0.082$

8 A pile of paper is 6 cm high. Each sheet is 0.008 cm thick. How many sheets are in the pile of paper?

9 Doris buys a big bag of safety pins. The bag weighs 180 g. Each safety pin weighs 0.6 g. How many safety pins are in the bag?

AU 10 **a** Use any method to work out: $81 \div 3$

b Use your answer to part **a** to work these out.

i $8.1 \div 0.3$ **ii** $0.81 \div 30$ **iii** $0.081 \div 0.3$

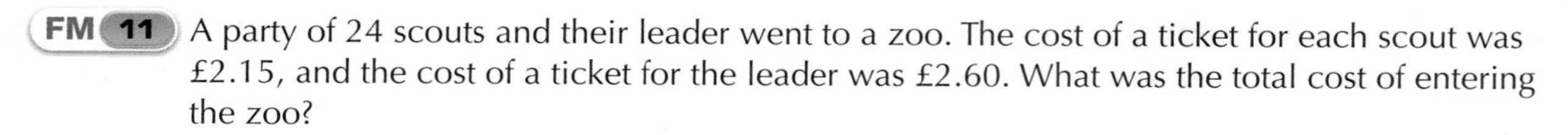

D

FM 11 A party of 24 scouts and their leader went to a zoo. The cost of a ticket for each scout was £2.15, and the cost of a ticket for the leader was £2.60. What was the total cost of entering the zoo?

PS 12 Mark went shopping.
He went into three stores and bought one item from each store.

Music Store	
CDs	£5.98
DVDs	£7.99

Clothes Store	
Shirt	£12.50
Jeans	£32.00

Book Store	
Magazine	£2.25
Pen	£3.98

In total he spent £43.97. What did he buy?

2.2 Compound interest and repeated percentage change

This section will show you how to:

- calculate compound interest
- solve problems involving repeated percentage change

Key words
annual rate
compound interest
multiplier
principal

Banks and building societies usually pay **compound interest** on savings accounts.

When compound interest is used, the interest earned each year is added to the original amount (**principal**) and the new total then earns interest at the **annual rate** in the following year. This pattern is then repeated each year while the money is in the account.

The most efficient way to calculate the total amount in the account after several years is to use a **multiplier**.

EXAMPLE 3

Elizabeth invests £400 in a savings account. The account pays compound interest at 6% each year. How much will she have in the account after three years?

The amount in the account increases by 6% each year, so the multiplier is 1.06.

After 1 year she will have £400 × 1.06 = £424
After 2 years she will have £424 × 1.06 = £449.44
After 3 years she will have £449.44 × 1.06 = £476.41 (rounded)

If you calculate the differences, you can see that the amount of interest increases each year (£24, £25.44 and £26.97).

From this example, you should see that you could have used £400 × $(1.06)^3$ to find the amount after three years. That is, you could have used the following formula for calculating the total amount due at any time:

total amount = P × multiplier raised to the power n = $P \times (1 + \frac{r}{100})^n$

or $A = P(1 + r)^n$ where P is the original amount invested, r is the percentage interest rate, giving a multiple of $(1 + r)$, and n is the number of years for which the money is invested.

So, in Example 3, P = £400, $r = 0.06$ and $n = 3$,

and the total amount = £400 × $(1.06)^3$

Using your calculator

You may have noticed that you can do the above calculation on your calculator without having to write down all the intermediate steps.

To add on the 6% each time, just multiply by 1.06 each time. So you can do the calculation as:

4 0 0 × 1 • 0 6 × 1 • 0 6 × 1 • 0 6 =

or

4 0 0 × 1 • 0 6 $x^{\blacksquare}$ 3 =

or

4 0 0 × 1 0 6 % $x^{\blacksquare}$ 3 =

You need to find the method with which you are comfortable and which you understand.

The idea of compound interest does not only concern money. It can be about, for example, the growth in population, increases in salaries, or increases in body weight or height. Also, the idea can involve regular reduction by a fixed percentage: for example, car depreciation, population losses and even water losses. The next exercise shows the extent to which compound interest ideas are used.

EXERCISE 2B

1 A baby octopus increases its body weight by 5% each day for the first month of its life. In a safe ocean zoo, a baby octopus was born weighing 10 g.

a What was its weight after:

i 1 day **ii** 2 days **iii** 4 days **iv** 1 week?

b After how many days will the octopus first weigh over 15 g?

2 A certain type of conifer hedging increases in height by 17% each year for the first 20 years. When I bought some of this hedging, it was all about 50 cm tall. How long will it take to grow 3 m tall?

3 The manager of a small family business offered his staff an annual pay increase of 4% for every year they stayed with the firm.

a Gareth started work at the business on a salary of £12 200. What salary will he be on after 4 years?

PS **b** Julie started work at the business on a salary of £9350. How many years will it be until she is earning a salary of over £20 000?

C

C

4 Scientists have been studying the shores of Scotland and estimate that due to pollution the seal population of those shores will decline at the rate of 15% each year. In 2006 they counted about 3000 seals on those shores.

a If nothing is done about pollution, how many seals did they expect to be there in:

i 2007 **ii** 2008 **iii** 2011?

PS **b** How long will it take for the seal population to be less than 1000?

5 I am told that if I buy a new car its value will depreciate at the rate of 20% each year. If I bought a car in 2009 priced at £8500, what would be the value of the car in:

a 2010

b 2011

c 2013?

6 At the peak of a drought during the summer, a reservoir in Derbyshire was losing water at the rate of 8% each day. On 1 August this reservoir held 2.1 million litres of water.

a At this rate of losing water, how much would have been in the reservoir on the following days?

i 2 August **ii** 4 August **iii** 8 August

FM **b** The danger point is when the water drops below 1 million litres. When would this have been if things had continued as they were?

7 The population of a small country, Yebon, was only 46 000 in 2001, but it steadily increased by about 13% each year during the 2000s.

a Calculate the population in:

i 2002 **ii** 2006 **iii** 2010.

PS **b** If the population keeps growing at this rate, when will it be half a million?

B

PS **8** How long will it take to accumulate one million pounds in the following situations?

a An investment of £100 000 at a rate of 12% compound interest

b An investment of £50 000 at a rate of 16% compound interest

PS **9** An oak tree is 60 cm tall. It grows at a rate of 8% per year. A conifer is 50 cm tall. It grows at a rate of 15% per year. How many years does it take before the conifer is taller than the oak?

PS **10** A tree increases in height by 18% per year. When it is 1 year old, it is 8 cm tall. How long will it take the tree to grow to 10 m?

PS **11** Show that a 10% increase followed by a 10% increase is equivalent to a 21% increase overall.

Here are two advertisements for savings accounts.

Which account is worth more after 2 years?

You **must** show your working.

Bradley Bank

Invest £1000 for two years and earn 3.2% interest overall.

Monastery Building Society

Invest £1000. Interest rate 1.3% compound per annum. Bonus of 0.5% on balance after 2 years.

PS 13 A fish weighs 3 kg and increases in weight by 10% each month. A crab weighs 6 kg but decreases in weight by 10% each month. After how many months will the fish weigh more than the crab?

FM 14 There is a bread shortage.

Each week during the shortage a shop increases its price of bread by 20% of the price the week before.

After how many weeks would the price of the bread have doubled?

PS 15 In a survey exactly 35% of the people surveyed wanted a new supermarket.

What is the least number that could have been surveyed?

2.3 Reverse percentage (working out the original quantity)

This section will show you how to:

- calculate the original amount, given the final amount, after a known percentage increase or decrease

Key words

final amount
multiplier
original amount
unitary method

Reverse percentage questions involve working backwards from the **final amount** to find the **original amount** when you know, or can work out, the final amount as a percentage of the original amount.

Method 1: The unitary method

The **unitary method** has three steps.

Step 1: Equate the final percentage to the final value.

Step 2: Use this to calculate the value of 1%.

Step 3: Multiply by 100 to work out 100% (the original value).

EXAMPLE 4

In a factory, 70 workers were given a pay rise. This was 20% of all the workers. How many workers are there altogether?

20% represents 70 workers.

Divide by 20.
1% represents 70 ÷ 20 workers. (There is no need to work out this calculation yet.)

Multiply by 100.
100% represents all the workers: 70 ÷ 20 × 100 = 350

So there are 350 workers altogether.

EXAMPLE 5

The price of a car increased by 6% to £9116. Work out the price before the increase.

106% represents £9116.

Divide by 106.
1% represents £9116 ÷ 106

Multiply by 100.
100% represents original price: £9116 ÷ 106 × 100 = £8600

So the price before the increase was £8600.

Method 2: The multiplier method

The **multiplier** method involves fewer steps.

Step 1: Write down the multiplier.

Step 2: Divide the final value by the multiplier to give the original value.

EXAMPLE 6

In a sale the price of a freezer is reduced by 12%. The sale price is £220. What was the price before the sale?

A decrease of 12% gives a multiplier of 0.88.

Dividing the sale price by the multiplier gives £220 ÷ 0.88 = £250

So the price before the sale was £250.

EXERCISE 2C

C

1 Find what 100% represents in these situations.

- **a** 40% represents 320 g
- **b** 14% represents 35 m
- **c** 45% represents 27 cm
- **d** 4% represents £123
- **e** 2.5% represents £5
- **f** 8.5% represents £34

B

2 On a gruelling army training session, only 28 youngsters survived the whole day. This represented 35% of the original group. How large was the original group?

3 VAT is a government tax added to goods and services. With VAT at 17.5%, what is the pre-VAT price of the following priced goods?

T-shirt	£9.87	Tights	£1.41
Shorts	£6.11	Sweater	£12.62
Trainers	£29.14	Boots	£38.07

4 Howard spends £200 a month on food. This represents 24% of his monthly take-home pay. How much is his monthly take-home pay?

5 Tina's weekly pay is increased by 5% to £315. What was Tina's pay before the increase?

6 The number of workers in a factory fell by 5% to 228. How many workers were there originally?

7 In a sale the price of a TV is reduced to £325.50. This is a 7% reduction on the original price. What was the original price?

8 If 38% of plastic bottles in a production line are blue and the remaining 7750 plastic bottles are brown, how many plastic bottles are blue?

9 I received £3.85 back from HM Revenue and Customs, which represented the 17.5% VAT on a piece of equipment. How much did I pay for this equipment in the first place?

A

FM 10 A company is in financial trouble. The workers are asked to take a 10% pay cut for each of the next two years.

- **a** Rob works out that his pay in two years' time will be £1296 per month. How much is his pay now?
- **b** Instead he offers to take an immediate pay cut of 14% and have his pay frozen at that level for two years. Has he made the correct decision?

AU 11 The population in a village is 30% of the size of the population in a neighbouring town.

- **a** If both populations double, what is the population of the village as a percentage of the population of the town?
- **b** If the population of the village stays the same but the population of the town doubles, what is the population of the village as a percentage of the population of the town?

A

PS 12 A man's salary was increased by 5% in one year and reduced by 5% in the next year. Is his final salary greater or less than the original one and by how many per cent?

PS 13 A woman's salary increased by 5% in one year and then increased the following year by 5% again.

Her new salary was £19 845.

How much was the increase, in pounds, in the first year?

PS 14 A quick way of estimating the pre-VAT price of an item with VAT added is to divide by 6 and then multiply by 5. For example, if an item costs £360 including VAT, it cost approximately $(360 \div 6) \times 5 =$ £300 before VAT. Show that this gives an estimate to within £5 of the pre-VAT price for items costing up to £280.

PS 15 After a 6% increase followed by an 8% increase, the monthly salary of a chef was £1431. What was the original monthly salary?

PS 16 Cassie invests some money at 4% interest per annum for five years. After five years, she had £1520.82 in the bank. How much did she invest originally?

AU 17 A teacher asked her class to work out the original price of a cooker for which, after a 12% increase, the price was £291.20.

This is Baz's answer: 12% of 291.20 = £34.94

Original price = $291.20 - 34.94 = 256.26 \approx$ £260

When the teacher read out the answer Baz ticked his work as correct.

What errors has he made?

2.4 Powers (indices)

This section will show you how to:
- use powers (also known as indices)

Key words
cube, index, indices, power, reciprocal, square

Powers are a convenient way of writing repetitive multiplications. (Powers are also called **indices**, singular **index**.)

The power tells you the number of times a number is multiplied by itself. For example:

$4^6 = 4 \times 4 \times 4 \times 4 \times 4 \times 4$ six lots of 4 multiplied together

$6^4 = 6 \times 6 \times 6 \times 6$ four lots of 6 multiplied together

$7^3 = 7 \times 7 \times 7$

$12^2 = 12 \times 12$

You are expected to know **square** numbers (power 2) up to $15^2 = 225$

You should also know the **cubes** of numbers (power 3):

$1^3 = 1$, $2^3 = 8$, $3^3 = 27$, $4^3 = 64$, $5^3 = 125$ and $10^3 = 1000$

EXAMPLE 7

a What is the value of:

i 7 squared **ii** 5 cubed?

b Write each of these numbers out in full.

i 4^6 **ii** 6^4 **iii** 7^3 **iv** 12^2

c Write the following multiplications using powers.

i $3 \times 3 \times 3 \times 3 \times 3 \times 3 \times 3 \times 3$

ii $13 \times 13 \times 13 \times 13 \times 13$

iii $7 \times 7 \times 7 \times 7$

iv $5 \times 5 \times 5 \times 5 \times 5 \times 5 \times 5$

a The value of 7 squared is $7^2 = 7 \times 7 = 49$

The value of 5 cubed is $5^3 = 5 \times 5 \times 5 = 125$

b **i** $4^6 = 4 \times 4 \times 4 \times 4 \times 4 \times 4$ **ii** $6^4 = 6 \times 6 \times 6 \times 6$

iii $7^3 = 7 \times 7 \times 7$ **iv** $12^2 = 12 \times 12$

c **i** $3 \times 3 \times 3 \times 3 \times 3 \times 3 \times 3 \times 3 = 3^8$

ii $13 \times 13 \times 13 \times 13 \times 13 = 13^5$

iii $7 \times 7 \times 7 \times 7 = 7^4$

iv $5 \times 5 \times 5 \times 5 \times 5 \times 5 \times 5 = 5^7$

Working out powers on your calculator

The power button on your calculator will probably look like this $x^{\square}$.

To work out 5^7 on your calculator use the power key.

5^7 = 5 $x^{\square}$ 7 = 78 125

Two special powers

Power 1

Any number to the power 1 is the same as the number itself. This is always true so normally you do not write the power 1.

For example: $5^1 = 5$ $32^1 = 32$ $(-8)^1 = -8$

Power zero

Any number to the power 0 is equal to 1.

For example: $5^0 = 1$ $32^0 = 1$ $(-8)^0 = 1$

You can check these results on your calculator.

EXERCISE 2D

D

1 Write these expressions using index notation. Do not work them out yet.

a $2 \times 2 \times 2 \times 2$
b $3 \times 3 \times 3 \times 3 \times 3$
c 7×7
d $5 \times 5 \times 5$
e $10 \times 10 \times 10 \times 10 \times 10 \times 10 \times 10$
f $6 \times 6 \times 6 \times 6$
g 4
h $1 \times 1 \times 1 \times 1 \times 1 \times 1 \times 1$
i $0.5 \times 0.5 \times 0.5 \times 0.5$
j $100 \times 100 \times 100$

2 Write these power terms out in full. Do not work them out yet.

a 3^4 b 9^3 c 6^2 d 10^5 e 2^{10}
f 8^1 g 0.1^3 h 2.5^2 i 0.7^3 j 1000^2

3 Using the power key on your calculator (or another method), work out the values of the power terms in question **1**.

4 Using the power key on your calculator (or another method), work out the values of the power terms in question **2**.

FM 5 A storage container is in the shape of a cube. The length of the container is 5 m.

To work out the volume of a cube, use the formula:

volume = $(\text{length of edge})^3$

Work out the total storage space in the container.

AU 6 Write each number as a power of a different number.

The first one has been done for you.

a $32 = 2^5$ b 100 c 8 d 25

C

7 Without using a calculator, work out the values of these power terms.

a 2^0 b 4^1 c 5^0 d 1^9 e 1^{235}

8 The answers to question **7**, parts **d** and **e**, should tell you something special about powers of 1. What is it?

PS 9 Write the answer to question **1**, part **j** as a power of 10.

PS 10 Write the answer to question **2**, part **j** as a power of 10.

11 Using your calculator, or otherwise, work out the values of these power terms.

a $(-1)^0$ b $(-1)^1$ c $(-1)^2$ d $(-1)^4$ e $(-1)^5$

C

PS 12 Using your answers to question **11**, write down the answers to these power terms.

a $(-1)^8$ **b** $(-1)^{11}$ **c** $(-1)^{99}$

d $(-1)^{80}$ **e** $(-1)^{126}$

PS 13 The number 16 777 216 is a power of 2.

It is also a power of 4, a power of 8 and a power of 16.

Write the number using each of the powers.

Negative powers (or negative indices)

A negative index is a convenient way of writing the **reciprocal** of a number or term. (That is, one divided by that number or term.) For example,

$$x^{-a} = \frac{1}{x^a}$$

Here are some other examples.

$5^{-2} = \frac{1}{5^2}$ $3^{-1} = \frac{1}{3}$ $5x^{-2} = \frac{5}{x^2}$

EXAMPLE 8

Rewrite the following in the form 2^n.

a 8 **b** $\frac{1}{4}$ **c** -32 **d** $-\frac{1}{64}$

a $8 = 2 \times 2 \times 2 = 2^3$ **b** $\frac{1}{4} = \frac{1}{2^2} = 2^{-2}$

c $-32 = -2^5$ **d** $-\frac{1}{64} = -\frac{1}{2^6} = -2^{-6}$

EXERCISE 2E

1 Write down each of these in fraction form.

a 5^{-3} **b** 6^{-1} **c** 10^{-5} **d** 3^{-2}

e 8^{-2} **f** 9^{-1} **g** w^{-2} **h** t^{-1}

i x^{-m} **j** $4m^{-3}$

2 Write down each of these in negative index form.

a $\frac{1}{3^2}$ **b** $\frac{1}{5}$ **c** $\frac{1}{10^3}$

d $\frac{1}{m}$ **e** $\frac{1}{t^n}$

HINTS AND TIPS

If you move a power from top to bottom, or vice versa, the sign changes. Negative power means the reciprocal: it does not mean the answer is negative.

B

A

3 Change each of the following expressions into an index form of the type shown.

a All of the form 2^n

i 16 **ii** $\frac{1}{2}$ **iii** $\frac{1}{16}$ **iv** -8

b All of the form 10^n

i 1000 **ii** $\frac{1}{10}$ **iii** $\frac{1}{100}$ **iv** 1 million

c All of the form 5^n

i 125 **ii** $\frac{1}{5}$ **iii** $\frac{1}{25}$ **iv** $\frac{1}{625}$

d All of the form 3^n

i 9 **ii** $\frac{1}{27}$ **iii** $\frac{1}{81}$ **iv** -243

4 Rewrite each of the following expressions in fraction form.

a $5x^{-3}$ **b** $6t^{-1}$ **c** $7m^{-2}$ **d** $4q^{-4}$ **e** $10y^{-5}$

f $\frac{1}{2}x^{-3}$ **g** $\frac{1}{2}m^{-1}$ **h** $\frac{3}{4}t^{-4}$ **i** $\frac{4}{5}y^{-3}$ **j** $\frac{7}{8}x^{-5}$

5 Write each fraction in index form.

a $\frac{7}{x^3}$ **b** $\frac{10}{p}$ **c** $\frac{5}{t^2}$ **d** $\frac{8}{m^5}$ **e** $\frac{3}{y}$

6 Find the value of each of the following.

a $x = 5$

i x^2 **ii** x^{-3} **iii** $4x^{-1}$

b $t = 4$

i t^3 **ii** t^{-2} **iii** $5t^{-4}$

c $m = 2$

i m^3 **ii** m^{-5} **iii** $9m^{-1}$

d $w = 10$

i w^6 **ii** w^{-3} **iii** $25w^{-2}$

PS 7 Two different numbers can be written in the form 2^n.

The sum of the numbers is 40.

What is the difference of the numbers?

AU 8 x and y are integers.

$$x^2 - y^3 = 0$$

Work out possible values of x and y.

AU 9 You are given that $8^7 = 2\,097\,152$

Write down the value of 8^{-7}.

PS 10 Put these in order from smallest to largest:

x^5 x^{-5} x^0

a when x is greater than 1 **b** when x is between 0 and 1 **c** when $x = -10$

Rules for multiplying and dividing numbers in index form

When you *multiply* powers of the same number or variable, you *add* the indices. For example,

$3^4 \times 3^5 = 3^{(4+5)} = 3^9$

$2^3 \times 2^4 \times 2^5 = 2^{12}$

$10^4 \times 10^{-2} = 10^2$

$10^{-3} \times 10^{-1} = 10^{-4}$

$a^x \times a^y = a^{(x+y)}$

When you *divide* powers of the same number or variable, you *subtract* the indices. For example,

$a^4 \div a^3 = a^{(4-3)} = a^1 = a$

$b^4 \div b^7 = b^{-3}$

$10^4 \div 10^{-2} = 10^6$

$10^{-2} \div 10^{-4} = 10^2$

$a^x \div a^y = a^{(x-y)}$

When you *raise* a power to a further power, you *multiply* the indices. For example,

$(a^2)^3 = a^{2 \times 3} = a^6$

$(a^{-2})^4 = a^{-8}$

$(a^2)^6 = a^{12}$

$(a^x)^y = a^{xy}$

Here are some examples of different kinds of expressions using powers.

$2a^2 \times 3a^4 = (2 \times 3) \times (a^2 \times a^4) = 6 \times a^6 = 6a^6$

$4a^2b^3 \times 2ab^2 = (4 \times 2) \times (a^2 \times a) \times (b^3 \times b^2) = 8a^3b^5$

$12a^5 \div 3a^2 = (12 \div 3) \times (a^5 \div a^2) = 4a^3$

$(2a^2)^3 = (2)^3 \times (a^2)^3 = 8 \times a^6 = 8a^6$

EXERCISE 2F

C

1 Write these as single powers of 5.

a $5^2 \times 5^2$ **b** 5×5^2 **c** $5^{-2} \times 5^4$ **d** $5^6 \times 5^{-3}$ **e** $5^{-2} \times 5^{-3}$

2 Write these as single powers of 6.

a $6^5 \div 6^2$ **b** $6^4 \div 6^4$ **c** $6^4 \div 6^{-2}$ **d** $6^{-3} \div 6^4$ **e** $6^{-3} \div 6^{-5}$

C

3 Simplify these and write them as single powers of a.

a $a^2 \times a$ **b** $a^3 \times a^2$ **c** $a^4 \times a^3$

d $a^6 \div a^2$ **e** $a^3 \div a$ **f** $a^5 \div a^4$

AU **4** **a** $a^x \times a^y = a^{10}$

Write down a possible pair of values of x and y.

b $a^x \div a^y = a^{10}$

Write down a possible pair of values of x and y.

B

5 Write these as single powers of 4.

a $(4^2)^3$ **b** $(4^3)^5$ **c** $(4^1)^6$

d $(4^3)^{-2}$ **e** $(4^{-2})^{-3}$ **f** $(4^7)^0$

6 Simplify these expressions.

a $2a^2 \times 3a^3$ **b** $3a^4 \times 3a^{-2}$ **c** $(2a^2)^3$

d $-2a^2 \times 3a^2$ **e** $-4a^3 \times -2a^5$ **f** $-2a^4 \times 5a^{-7}$

7 Simplify these expressions.

a $6a^3 \div 2a^2$ **b** $12a^5 \div 3a^2$

c $15a^5 \div 5a$ **d** $18a^{-2} \div 3a^{-1}$

e $24a^5 \div 6a^{-2}$ **f** $30a \div 6a^5$

HINTS AND TIPS

Deal with numbers and indices separately and do not confuse the rules.
For example: $12a^5 \div 4a^2$
$= (12 \div 4) \times (a^5 \div a^2)$

8 Simplify these expressions.

a $2a^2b^3 \times 4a^3b$ **b** $5a^2b^4 \times 2ab^{-3}$ **c** $6a^2b^3 \times 5a^{-4}b^{-5}$

d $12a^2b^4 \div 6ab$ **e** $24a^{-3}b^4 \div 3a^2b^{-3}$

9 Simplify these expressions.

a $\dfrac{6a^4b^3}{2ab}$ **b** $\dfrac{2a^2bc^2 \times 6abc^3}{4ab^2c}$ **c** $\dfrac{3abc \times 4a^3b^2c \times 6c^2}{9a^2bc}$

FM **10** Write down **two** possible:

a multiplication questions with an answer of $12x^2y^5$

b division questions with an answer of $12x^2y^5$.

A

PS **11** a, b and c are three different positive integers.

What is the smallest possible value of a^2b^3c?

FM PS **12** Use the general rule for dividing powers of the same number, $\dfrac{a^x}{a^y} = a^{x-y}$, to prove that any number raised to the power zero is 1.

Indices of the form $\frac{1}{n}$

Consider the problem $7^x \times 7^x = 7$. This can be written as:

$$7^{(x+x)} = 7$$

$$7^{2x} = 7^1 \Rightarrow 2x = 1 \Rightarrow x = \tfrac{1}{2}$$

If you now substitute $x = \frac{1}{2}$ back into the original equation, you see that:

$$7^{\frac{1}{2}} \times 7^{\frac{1}{2}} = 7$$

This makes $7^{\frac{1}{2}}$ the same as $\sqrt{7}$.

You can similarly show that $7^{\frac{1}{3}}$ is the same as $\sqrt[3]{7}$. And that, generally,

$$x^{\frac{1}{n}} = \sqrt[n]{x} \text{ (}n\text{th root of } x\text{)}$$

So in summary:

Power $\frac{1}{2}$ is the same as positive square root.

Power $\frac{1}{3}$ is the same as cube root.

Power $\frac{1}{n}$ is the same as nth root.

For example,

$49^{\frac{1}{2}} = \sqrt{49} = 7$ $\quad 8^{\frac{1}{3}} = \sqrt[3]{8} = 2$ $\quad 10\,000^{\frac{1}{4}} = \sqrt[4]{10\,000} = 1036^{-\frac{1}{2}} = \frac{1}{\sqrt{36}} = \frac{1}{6}$

EXERCISE 2G

1 Evaluate the following.

a $25^{\frac{1}{2}}$	**b** $100^{\frac{1}{2}}$	**c** $64^{\frac{1}{2}}$	**d** $81^{\frac{1}{2}}$	**e** $625^{\frac{1}{2}}$
f $27^{\frac{1}{3}}$	**g** $64^{\frac{1}{3}}$	**h** $1000^{\frac{1}{3}}$	**i** $125^{\frac{1}{3}}$	**j** $512^{\frac{1}{3}}$
k $144^{\frac{1}{2}}$	**l** $400^{\frac{1}{2}}$	**m** $625^{\frac{1}{4}}$	**n** $81^{\frac{1}{4}}$	**o** $100\,000^{\frac{1}{5}}$
p $729^{\frac{1}{6}}$	**q** $32^{\frac{1}{5}}$	**r** $1024^{\frac{1}{10}}$	**s** $1296^{\frac{1}{4}}$	**t** $216^{\frac{1}{3}}$
u $16^{-\frac{1}{2}}$	**v** $8^{-\frac{1}{3}}$	**w** $81^{-\frac{1}{4}}$	**x** $3125^{-\frac{1}{5}}$	**y** $1\,000\,000^{-\frac{1}{6}}$

2 Evaluate the following.

a $\left(\frac{25}{36}\right)^{\frac{1}{2}}$	**b** $\left(\frac{100}{36}\right)^{\frac{1}{2}}$	**c** $\left(\frac{64}{81}\right)^{\frac{1}{2}}$	**d** $\left(\frac{81}{25}\right)^{\frac{1}{2}}$	**e** $\left(\frac{25}{64}\right)^{\frac{1}{2}}$
f $\left(\frac{27}{125}\right)^{\frac{1}{3}}$	**g** $\left(\frac{8}{512}\right)^{\frac{1}{3}}$	**h** $\left(\frac{1000}{64}\right)^{\frac{1}{3}}$	**i** $\left(\frac{64}{125}\right)^{\frac{1}{3}}$	**j** $\left(\frac{512}{343}\right)^{\frac{1}{3}}$

3 Use the general rule for raising a power to another power to prove that $x^{\frac{1}{n}}$ is equivalent to $\sqrt[n]{x}$.

PS 4 Which of these is the odd one out?

$16^{-\frac{1}{4}}$ $\quad 64^{-\frac{1}{2}}$ $\quad 8^{-\frac{1}{3}}$

Show how you decided.

AU 5 Imagine that you are the teacher.

Write down how you would teach the class that $27^{-\frac{1}{3}}$ is equal to $\frac{1}{3}$.

PS 6 $x^{-\frac{2}{3}} = y^{\frac{1}{3}}$

Find values for x and y that make this equation work.

Indices of the form $\frac{a}{b}$

Here are two examples of this form.

$t^{\frac{2}{3}} = t^{\frac{1}{3}} \times t^{\frac{1}{3}} = (\sqrt[3]{t})^2$ $\qquad$ $81^{\frac{3}{4}} = (\sqrt[4]{81})^3 = 3^3 = 27$

EXAMPLE 9

Evaluate the following. **a** $16^{-\frac{1}{4}}$ **b** $32^{-\frac{4}{5}}$

When dealing with the negative index remember that it means reciprocal.

Do problems like these one step at a time.

Step 1: Rewrite the calculation as a fraction by dealing with the negative power.

Step 2: Take the root of the base number given by the denominator of the fraction.

Step 3: Raise the result to the power given by the numerator of the fraction.

Step 4: Write out the answer as a fraction.

a **Step 1:** $16^{-\frac{1}{4}} = \left(\frac{1}{16}\right)^{\frac{1}{4}}$ **Step 2:** $16^{\frac{1}{4}} = \sqrt[4]{16} = 2$ **Step 3:** $2^1 = 2$ **Step 4:** $16^{-\frac{1}{4}} = \frac{1}{2}$

b **Step 1:** $32^{-\frac{4}{5}} = \left(\frac{1}{32}\right)^{\frac{4}{5}}$ **Step 2:** $32^{\frac{1}{5}} = \sqrt[5]{32} = 2$ **Step 3:** $2^4 = 16$ **Step 4:** $32^{-\frac{4}{5}} = \frac{1}{16}$

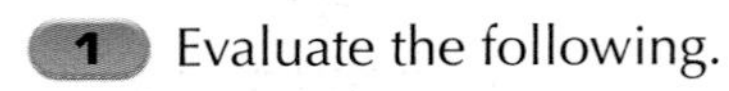

EXERCISE 2H

1 Evaluate the following.

a $32^{\frac{4}{5}}$ **b** $125^{\frac{2}{3}}$ **c** $1296^{\frac{3}{4}}$ **d** $243^{\frac{4}{5}}$

2 Rewrite the following in index form.

a $\sqrt[3]{t^2}$ **b** $\sqrt[4]{m^3}$ **c** $\sqrt[5]{k^2}$ **d** $\sqrt{x^3}$

3 Evaluate the following.

a $8^{\frac{2}{3}}$ **b** $27^{\frac{2}{3}}$ **c** $16^{\frac{3}{2}}$ **d** $625^{\frac{5}{4}}$

4 Evaluate the following.

a $25^{-\frac{1}{2}}$ **b** $36^{-\frac{1}{2}}$ **c** $16^{-\frac{1}{4}}$ **d** $81^{-\frac{1}{4}}$

e $16^{-\frac{1}{2}}$ **f** $8^{-\frac{1}{3}}$ **g** $32^{-\frac{1}{5}}$ **h** $27^{-\frac{1}{3}}$

5 Evaluate the following.

a $25^{-\frac{3}{2}}$ **b** $36^{-\frac{3}{2}}$

c $16^{-\frac{3}{4}}$ **d** $81^{-\frac{3}{4}}$

e $64^{-\frac{4}{3}}$ **f** $8^{-\frac{2}{3}}$

g $32^{-\frac{2}{5}}$ **h** $27^{-\frac{2}{3}}$

6 Evaluate the following.

a $100^{-\frac{5}{2}}$ **b** $144^{-\frac{1}{2}}$

c $125^{-\frac{2}{3}}$ **d** $9^{-\frac{3}{2}}$

e $4^{-\frac{5}{2}}$ **f** $64^{-\frac{5}{6}}$

g $27^{-\frac{4}{3}}$ **h** $169^{-\frac{1}{2}}$

PS 7 Which of these is the odd one out?

$16^{-\frac{3}{4}}$ $64^{-\frac{1}{2}}$ $8^{-\frac{2}{3}}$

Show how you decided.

AU 8 Imagine that you are the teacher.

Write down how you would teach the class that $27^{-\frac{2}{3}}$ is equal to $\frac{1}{9}$.

2.5 Reciprocals and rational numbers

This section will show you how to:

- recognise rational numbers, reciprocals, terminating decimals and recurring decimals
- convert terminal decimals to fractions
- convert fractions to recurring decimals
- find reciprocals of numbers or fractions

Key words
rational number
reciprocal
recurring decimal
terminating decimal

Rational numbers

A **rational number** is a number that can be written as a fraction, for example, $\frac{1}{4}$ or $\frac{10}{3}$.

When a fraction is converted to a decimal it will either be:

- a **terminating decimal** or
- a **recurring decimal**.

A terminating decimal has a finite number of digits. For example, $\frac{1}{4} = 0.25$, $\frac{1}{8} = 0.125$.

A recurring decimal has a digit, or block of digits, that repeat. For example, $\frac{1}{3} = 0.3333\ \ldots$, $\frac{2}{11} = 0.181818\ \ldots$

Recurring digits can be shown by putting a dot over the first and last digit of the group that repeats.

0.3333 … becomes $0.\dot{3}$

0.181818 … becomes $0.\dot{1}\dot{8}$

0.123123123 … becomes $0.\dot{1}2\dot{3}$

0.58333 … becomes $0.58\dot{3}$

0.6181818 … becomes $0.6\dot{1}\dot{8}$

0.4123123123 … become $0.4\dot{1}2\dot{3}$

Converting fractions into recurring decimals

A fraction that does not convert to a terminating decimal will give a recurring decimal. You may already know that $\frac{1}{3} = 0.333\ldots = 0.\dot{3}$

This means that the 3s go on forever and the decimal never ends.

To convert the fraction, you can usually use a calculator to divide the numerator by the denominator. Note that calculators round the last digit so it may not always be a true recurring decimal in the display. Use a calculator to check the following recurring decimals.

$$\frac{2}{11} = 0.181\,818\ldots = 0.\dot{1}\dot{8}$$

$$\frac{4}{15} = 0.2666\ldots = 0.2\dot{6}$$

$$\frac{8}{13} = 0.615\,384\,615\,384\,6\ldots = 0.\dot{6}15\,38\dot{4}$$

Converting terminal decimals into fractions

To convert a terminating decimal to a fraction, take the decimal number as the numerator. Then the denominator is 10, 100, 1000 …, depending on the number of decimal places. Because a terminating decimal has a specific number of decimal places, you can use place value to work out exactly where the numerator and the denominator end. For example:

- $0.7 = \frac{7}{10}$
- $0.23 = \frac{23}{100}$
- $0.045 = \frac{45}{1000} = \frac{9}{200}$
- $2.34 = \frac{234}{100} = \frac{117}{50} = 2\frac{17}{50}$
- $0.625 = \frac{625}{1000} = \frac{5}{8}$

Converting recurring decimals into fractions

To convert a recurring decimal to a fraction you have to use the algebraic method shown in the examples below.

EXAMPLE 10

Convert $0.\dot{7}$ to a fraction.

Let x be the fraction. Then:

$$x = 0.777\,777\,777\ldots \quad (1)$$

Multiply (1) by 10 $\quad 10x = 7.777\,777\,777\ldots \quad (2)$

Subtract (2) – (1) $\quad 9x = 7$

$$\Rightarrow x = \frac{7}{9}$$

EXAMPLE 11

Convert $0.\dot{5}6\dot{4}$ to a fraction.

Let x be the fraction. Then:

$$x = 0.564\,564\,564\ldots \quad (1)$$

Multiply (1) by 1000 $\quad 1000x = 564.564\,564\,564\ldots \quad (2)$

Subtract (2) – (1) $\quad 999x = 564$

$$\Rightarrow x = \frac{564}{999} = \frac{188}{333}$$

As a general rule, multiply by 10 if one digit recurs, multiply by 100 if two digits recur, multiply by 1000 if three digits recur, and so on.

Finding reciprocals of numbers or fractions

The **reciprocal** of any number is 1 divided by the number.

For example, the reciprocal of 2 is $1 \div 2 = \frac{1}{2} = 0.5$

The reciprocal of 0.25 is $1 \div 0.25 = 4$

You can find the reciprocal of a fraction by inverting it.

For example, the reciprocal of $\frac{2}{3}$ is $\frac{3}{2}$.

The reciprocal of $\frac{7}{4}$ is $\frac{4}{7}$.

EXERCISE 2I

1 Work out each of these fractions as a decimal. Give them as terminating decimals or recurring decimals as appropriate.

a $\frac{1}{2}$ **b** $\frac{1}{3}$ **c** $\frac{1}{4}$ **d** $\frac{1}{5}$ **e** $\frac{1}{6}$ **f** $\frac{1}{7}$ **g** $\frac{1}{8}$ **h** $\frac{1}{9}$ **i** $\frac{1}{10}$ **j** $\frac{1}{13}$

C

PS **2** There are several patterns to be found in recurring decimals. For example,

$\frac{1}{7} = 0.142\ 857\ 142\ 857\ 142\ 857\ 142\ 857\ldots$

$\frac{2}{7} = 0.285\ 714\ 285\ 714\ 285\ 714\ 285\ 714\ldots$

$\frac{3}{7} = 0.428\ 571\ 428\ 571\ 428\ 571\ 428\ 571\ldots$

and so on.

a Write down the decimals for $\frac{4}{7}, \frac{5}{7}, \frac{6}{7}$ to 24 decimal places.

b What do you notice?

3 Work out the ninths, $\frac{1}{9}, \frac{2}{9}, \frac{3}{9}$ and so on, up to $\frac{8}{9}$, as recurring decimals.

Describe any patterns that you notice.

4 Work out the elevenths, $\frac{1}{11}, \frac{2}{11}, \frac{3}{11}$ and so on, up to $\frac{10}{11}$, as recurring decimals.

Describe any patterns that you notice.

5 Write each of these fractions as a decimal. Use your results to write the list in order of size, smallest first.

$\frac{4}{9}$ $\frac{5}{11}$ $\frac{3}{7}$ $\frac{9}{22}$ $\frac{16}{37}$ $\frac{6}{13}$

6 Write each of the following as a fraction with a denominator of 120. Use your results to put them in order of size, smallest first.

$\frac{19}{60}$ $\frac{7}{24}$ $\frac{3}{10}$ $\frac{2}{5}$ $\frac{5}{12}$

7 Convert each of these terminating decimals to a fraction.

a 0.125 **b** 0.34 **c** 0.725 **d** 0.3125

e 0.89 **f** 0.05 **g** 2.35 **h** 0.218 75

8 Use a calculator to work out the reciprocal of each of the following.

a 12 **b** 16 **c** 20 **d** 25 **e** 50

9 Write down the reciprocal of each of the following fractions.

a $\frac{3}{4}$ **b** $\frac{5}{6}$ **c** $\frac{2}{5}$ **d** $\frac{7}{10}$ **e** $\frac{11}{20}$ **f** $\frac{4}{15}$

10 **a** Write the fractions and their reciprocals from question **9** as decimals. Write them as terminating decimals or recurring decimals as appropriate.

b Is it always true that a terminating decimal has a reciprocal that is a recurring decimal?

AU **11** Explain why zero has no reciprocal.

AU **12** **a** Work out the reciprocal of the reciprocal of 10.

b Work out the reciprocal of the reciprocal of 2.

c What do you notice?

AU 13 x and y are two positive numbers.

If x is less than y, which statement is true?

The reciprocal of x is less than the reciprocal of y.

The reciprocal of x is greater than the reciprocal of y.

It is impossible to tell.

Give an example to support your answer.

AU 14 Explain why a number multiplied by its reciprocal is equal to 1. Use examples to show that this is true for negative numbers.

15 $x = 0.242\ 424\ \ldots$

a What is $100x$?

b By subtracting the original value from your answer to part **a**, work out the value of $99x$.

c What is x as a fraction?

16 Convert each of these recurring decimals to a fraction.

a $0.\dot{8}$ **b** $0.\dot{3}\dot{4}$ **c** $0.\dot{4}\dot{5}$

d $0.\dot{5}6\dot{7}$ **e** $0.\dot{4}$ **f** $0.0\dot{4}$

g $0.1\dot{4}$ **h** $0.0\dot{4}\dot{5}$ **i** $2.\dot{7}$

j $7.\dot{6}\dot{3}$ **k** $3.\dot{3}$ **l** $2.\dot{0}\dot{6}$

PS 17 **a** $\frac{1}{7}$ is a recurring decimal. $(\frac{1}{7})^2 = \frac{1}{49}$ is also a recurring decimal.

Is it true that when you square any fraction that is a recurring decimal, you get another fraction that is also a recurring decimal? Try this with at least four numerical examples before you make a decision.

b $\frac{1}{4}$ is a terminating decimal. $(\frac{1}{4})^2 = \frac{1}{16}$ is also a terminating decimal.

Is it true that when you square any fraction that is a terminating decimal, you get another fraction that is also a terminating decimal? Try this with at least four numerical examples before you make a decision.

c What type of fraction do you get when you multiply a fraction that gives a recurring decimal by another fraction that gives a terminating decimal? Try this with at least four numerical examples before you make a decision.

PS 18 **a** Convert the recurring decimal $0.\dot{9}$ to a fraction.

b Prove that $0.4\dot{9}$ is equal to 0.5

2.6 Standard form

This section will show you how to:

- change a number into standard form
- calculate using numbers in standard form

Key words
powers
standard form
standard index form

Arithmetic of powers of 10

Multiplying

You are now going to look at **powers** of 10.

How many zeros does a million have? What is a million as a power of 10? This table shows some of the pattern of the powers of 10.

Number	0.001	0.01	0.1	1	10	100	1000	10 000	100 000
Powers	10^{-3}	10^{-2}	10^{-1}	10^0	10^1	10^2	10^3	10^4	10^5

What pattern is there in the top row? What pattern is there to the powers in the bottom row?

To multiply by any power of 10, you simply move the digits according to these two rules.

- When the index is *positive*, move the digits to the *left* by the same number of places as the value of the index.
- When the index is *negative*, move the digits to the *right* by the same number of places as the value of the index.

EXAMPLE 12

Write these as ordinary numbers.

a 12.356×10^2 **b** 3.45×10^1

c 753.4×10^{-2} **d** 6789×10^{-1}

a $12.356 \times 10^2 = 1235.6$ **b** $3.45 \times 10^1 = 34.5$

c $753.4 \times 10^{-2} = 7.534$ **d** $6789 \times 10^{-1} = 678.9$

In certain cases, you have to insert the 'hidden' zeros.

EXAMPLE 13

Write these as ordinary numbers.

a 75×10^4 **b** 2.04×10^5

c 6.78×10^{-3} **d** 0.897×10^{-4}

a $75 \times 10^4 = 750\,000$ **b** $2.04 \times 10^5 = 204\,000$

c $6.78 \times 10^{-3} = 0.006\,78$ **d** $0.897 \times 10^{-4} = 0.000\,0897$

Dividing

To divide by any power of 10, you simply move the digits according to these two rules.

- When the index is *positive*, move the digits to the *right* by the same number of places as the value of the index.
- When the index is *negative*, move the digits to the *left* by the same number of places as the value of the index.

EXAMPLE 14

Write these as ordinary numbers.

a $712.35 \div 10^2$ **b** $38.45 \div 10^1$

c $3.463 \div 10^{-2}$ **d** $6.789 \div 10^{-1}$

a $712.35 \div 10^2 = 7.1235$ **b** $38.45 \div 10^1 = 3.845$

c $3.463 \div 10^{-2} = 346.3$ **d** $6.789 \div 10^{-1} = 67.89$

In certain cases, you have to insert the 'hidden' zeros.

EXAMPLE 15

Write these as ordinary numbers.

a $75 \div 10^4$ **b** $2.04 \div 10^5$

c $6.78 \div 10^{-3}$ **d** $0.08 \div 10^{-4}$

a $75 \div 10^4 = 0.0075$ **b** $2.04 \div 10^5 = 0.000\,0204$

c $6.78 \div 10^{-3} = 6780$ **d** $0.08 \div 10^{-4} = 800$

When doing the next exercise, remember:

$10\,000 = 10 \times 10 \times 10 \times 10 = 10^4$

$1000 = 10 \times 10 \times 10 = 10^3$

$100 = 10 \times 10 = 10^2$

$10 = 10 = 10^1$

$1 = 10^0$

$0.1 = 1 \div 10 = 10^{-1}$

$0.01 = 1 \div 100 = 10^{-2}$

$0.001 = 1 \div 1000 = 10^{-3}$

EXERCISE 2J

1 Write down the value of each of the following.

a 3.1×10 **b** 3.1×100 **c** 3.1×1000 **d** $3.1 \times 10\,000$

2 Write down the value of each of the following.

a 6.5×10 **b** 6.5×10^2 **c** 6.5×10^3 **d** 6.5×10^4

D

D

3 Write down the value of each of the following.

a $3.1 \div 10$ b $3.1 \div 100$ c $3.1 \div 1000$ d $3.1 \div 10\,000$

4 Write down the value of each of the following.

a $6.5 \div 10$ b $6.5 \div 10^2$ c $6.5 \div 10^3$ d $6.5 \div 10^4$

5 Evaluate the following.

a 2.5×100 b 3.45×10 c 4.67×1000 d 34.6×10

e 20.789×10 f 56.78×1000 g 2.46×10^2 h 0.076×10

i 0.999×10^6 j 234.56×10^2 k 98.7654×10^3 l 43.23×10^6

m $0.003\,457\,8 \times 10^5$ n 0.0006×10^7 o $0.005\,67 \times 10^4$ p 56.0045×10^4

6 Evaluate the following.

a $2.5 \div 100$ b $3.45 \div 10$ c $4.67 \div 1000$ d $34.6 \div 10$

e $20.789 \div 100$ f $56.78 \div 1000$ g $2.46 \div 10^2$ h $0.076 \div 10$

i $0.999 \div 10^6$ j $234.56 \div 10^2$ k $98.7654 \div 10^3$ l $43.23 \div 10^6$

m $0.003\,4578 \div 10^5$ n $0.0006 \div 10^7$ o $0.005\,67 \div 10^4$ p $56.0045 \div 10^4$

7 Without using a calculator, work out the following.

a 2.3×10^2 b 5.789×10^5 c 4.79×10^3 d 5.7×10^7

e 2.16×10^2 f 1.05×10^4 g 3.2×10^{-4} h 9.87×10^3

8 Which of these statements is true about the numbers in question **7**?

a The first part is always a number between 1 and 10.

b There is always a multiplication sign in the middle of the expression.

c There is always a power of 10 at the end.

d Calculator displays sometimes show numbers in this form.

C

AU PS **9** The mass of Mars is 6.4×10^{23} kg.

The mass of Venus is 4.9×10^{24} kg.

Without working out the answers, explain how you can tell which planet is the heavier.

B

PS **10** A number is between one million and 10 million. It is written in the form 4.7×10^n.

What is the value of n?

Standard form

Standard form is also known as **standard index form**.

Standard form is a way of writing very large and very small numbers using powers of 10. In this form, a number is given a value between 1 and 10 multiplied by a power of 10. That is,

$a \times 10^n$ where $1 \leqslant a < 10$, and n is a whole number.

Follow through these examples to see how numbers are written in standard form.

$52 = 5.2 \times 10 = \mathbf{5.2} \times \mathbf{10^1}$

$73 = 7.3 \times 10 = \mathbf{7.3} \times \mathbf{10^1}$

$625 = 6.25 \times 100 = \mathbf{6.25} \times \mathbf{10^2}$ The numbers in bold are in standard form.

$389 = 3.89 \times 100 = \mathbf{3.89} \times \mathbf{10^2}$

$3147 = 3.147 \times 1000 = \mathbf{3.147} \times \mathbf{10^3}$

When writing a number in this way, you must always follow two rules.

- The first part must be a number between 1 and 10 (1 is allowed but 10 isn't).
- The second part must be a whole-number (negative or positive) power of 10. Note that you would *not normally* write the power 1.

Standard form on a calculator

A number such as 123 000 000 000 is obviously difficult to key into a calculator. Instead, you enter it in standard form (assuming you are using a scientific calculator):

$123\,000\,000\,000 = 1.23 \times 10^{11}$

The key strokes to enter this into your calculator will be:

1 • 2 3 ×10ˣ 1 1

Your calculator display will display the number either as an ordinary number, if there is enough space, or in standard form.

Standard form of numbers less than 1

These numbers are written in standard form. Make sure that you understand how they are formed.

a $0.4 = 4 \times 10^{-1}$ **b** $0.05 = 5 \times 10^{-2}$ **c** $0.007 = 7 \times 10^{-3}$

d $0.123 = 1.23 \times 10^{-1}$ **e** $0.007\,65 = 7.65 \times 10^{-3}$ **f** $0.9804 = 9.804 \times 10^{-1}$

g $0.0098 = 9.8 \times 10^{-3}$ **h** $0.000\,0078 = 7.8 \times 10^{-6}$

On a calculator you would enter 1.23×10^{-6}, for example, as:

1 • 2 3 ×10ˣ (−) 6

Try entering some of the numbers **a** to **h** (above) into your calculator for practice.

EXERCISE 2K

D

1 Write down the value of each of the following.

a 3.1×0.1 **b** 3.1×0.01 **c** 3.1×0.001 **d** 3.1×0.0001

2 Write down the value of each of the following.

a 6.5×10^{-1} **b** 6.5×10^{-2} **c** 6.5×10^{-3} **d** 6.5×10^{-4}

PS **3** **a** What is the largest number you can enter into your calculator?

b What is the smallest number you can enter into your calculator?

4 Work out the value of each of the following.

a $3.1 \div 0.1$ **b** $3.1 \div 0.01$ **c** $3.1 \div 0.001$ **d** $3.1 \div 0.0001$

5 Work out the value of each of the following.

a $6.5 \div 10^{-1}$ **b** $6.5 \div 10^{-2}$ **c** $6.5 \div 10^{-3}$ **d** $6.5 \div 10^{-4}$

C

6 Write these numbers out in full.

a 2.5×10^{2} **b** 3.45×10 **c** 4.67×10^{-3} **d** 3.46×10

e 2.0789×10^{-2} **f** 5.678×10^{3} **g** 2.46×10^{2} **h** 7.6×10^{3}

i 8.97×10^{5} **j** 8.65×10^{-3} **k** 6×10^{7} **l** 5.67×10^{-4}

7 Write these numbers in standard form.

a 250 **b** 0.345 **c** 46 700

d 3 400 000 000 **e** 20 780 000 000 **f** 0.000 567 8

g 2460 **h** 0.076 **i** 0.000 76

j 0.999 **k** 234.56 **l** 98.7654

m 0.0006 **n** 0.005 67 **o** 56.0045

In questions **8** to **10**, write the numbers given in each question in standard form.

8 One year, 27 797 runners completed the New York marathon.

9 The largest number of dominoes ever toppled by one person is 281 581, although 30 people set up and toppled 1 382 101.

10 The asteroid *Phaethon* comes within 12 980 000 miles of the Sun, whilst the asteroid *Pholus*, at its furthest point, is a distance of 2997 million miles from the Earth. The closest an asteroid ever came to Earth was 93 000 miles from the planet.

B

AU **11** How many times bigger is 3.2×10^{6} than 3.2×10^{4}?

12 The speed of sound (Mach 1) is 1236 kilometres per hour or about 1 mile in 5 seconds.

A plane travelling at Mach 2 would be travelling at twice the speed of sound.

How many miles would a plane travelling at Mach 3 cover in 1 minute?

B

Calculating with standard form

Calculations involving very large or very small numbers can be done more easily using standard form.

In these examples, you will see how to work out the area of a pixel on a computer screen, and how long it takes light to reach the Earth from a distant star.

EXAMPLE 16

A pixel on a computer screen is 2×10^{-2} cm long by 7×10^{-3} cm wide.

What is the area of the pixel?

The area is given by length times width.

$$\text{Area} = 2 \times 10^{-2} \text{ cm} \times 7 \times 10^{-3} \text{ cm}$$
$$= (2 \times 7) \times (10^{-2} \times 10^{-3}) \text{ cm}^2 = 14 \times 10^{-5} \text{ cm}^2$$

Note that you multiply the numbers and add the powers of 10. (You should not need to use a calculator to do this calculation.) The answer is not in standard form as the first part is not between 1 and 10, so you have to change it to standard form.

$$14 = 1.4 \times 10^1$$

So area $= 14 \times 10^{-5} \text{ cm}^2 = 1.4 \times 10^1 \times 10^{-5} \text{ cm}^2 = 1.4 \times 10^{-4} \text{ cm}^2$

EXERCISE 2L

B

1 These numbers are not in standard form. Write them in standard form.

a 56.7×10^2 **b** 0.06×10^4 **c** 34.6×10^{-2}

d 0.07×10^{-2} **e** 56×10 **f** $2 \times 3 \times 10^5$

g $2 \times 10^2 \times 35$ **h** 160×10^{-2} **i** 23 million

j 0.0003×10^{-2} **k** 25.6×10^5 **l** $16 \times 10^2 \times 3 \times 10^{-1}$

m $2 \times 10^4 \times 56 \times 10^{-4}$ **n** $(18 \times 10^2) \div (3 \times 10^3)$ **o** $(56 \times 10^3) \div (2 \times 10^{-2})$

2 Work out the following. Give your answers in standard form.

a $2 \times 10^4 \times 5.4 \times 10^3$ **b** $1.6 \times 10^2 \times 3 \times 10^4$ **c** $2 \times 10^4 \times 6 \times 10^4$

d $2 \times 10^{-4} \times 5.4 \times 10^3$ **e** $1.6 \times 10^{-2} \times 4 \times 10^4$ **f** $2 \times 10^4 \times 6 \times 10^{-4}$

g $7.2 \times 10^{-3} \times 4 \times 10^2$ **h** $(5 \times 10^3)^2$ **i** $(2 \times 10^{-2})^3$

3 Work out the following. Give your answers in standard form, rounding to an appropriate degree of accuracy where necessary.

a $2.1 \times 10^4 \times 5.4 \times 10^3$ **b** $1.6 \times 10^3 \times 3.8 \times 10^3$ **c** $2.4 \times 10^4 \times 6.6 \times 10^4$

d $7.3 \times 10^{-6} \times 5.4 \times 10^3$ **e** $(3.1 \times 10^4)^2$ **f** $(6.8 \times 10^{-4})^2$

g $5.7 \times 10 \times 3.7 \times 10$ **h** $1.9 \times 10^{-2} \times 1.9 \times 10^9$ **i** $5.9 \times 10^3 \times 2.5 \times 10^{-2}$

j $5.2 \times 10^3 \times 2.2 \times 10^2 \times 3.1 \times 10^3$ **k** $1.8 \times 10^2 \times 3.6 \times 10^3 \times 2.4 \times 10^{-2}$

4 Work out the following. Give your answers in standard form.

a $(5.4 \times 10^4) \div (2 \times 10^3)$ **b** $(4.8 \times 10^2) \div (3 \times 10^4)$ **c** $(1.2 \times 10^4) \div (6 \times 10^4)$

d $(2 \times 10^{-4}) \div (5 \times 10^3)$ **e** $(1.8 \times 10^4) \div (9 \times 10^{-2})$ **f** $\sqrt{(36 \times 10^{-4})}$

g $(5.4 \times 10^{-3}) \div (2.7 \times 10^2)$ **h** $(1.8 \times 10^6) \div (3.6 \times 10^3)$ **i** $(5.6 \times 10^3) \div (2.8 \times 10^2)$

5 Work out the following. Give your answers in standard form, rounding to an appropriate degree of accuracy where necessary.

a $(2.7 \times 10^4) \div (5 \times 10^2)$ **b** $(2.3 \times 10^4) \div (8 \times 10^6)$ **c** $(3.2 \times 10^{-1}) \div (2.8 \times 10^{-1})$

d $(2.6 \times 10^{-6}) \div (4.1 \times 10^3)$ **e** $\sqrt{(8 \times 10^4)}$ **f** $\sqrt{(30 \times 10^{-4})}$

g $5.3 \times 10^3 \times 2.3 \times 10^2 \div (2.5 \times 10^3)$ **h** $1.8 \times 10^2 \times 3.1 \times 10^3 \div (6.5 \times 10^{-2})$

6 A typical adult has about 20 000 000 000 000 red corpuscles. Each red corpuscle has a mass of about 0.000 000 000 1 g. Write both of these numbers in standard form and work out the total mass of red corpuscles in a typical adult.

PS 7 A man puts one grain of rice on the first square of a chess board, two on the second square, four on the third, eight on the fourth and so on.

a How many grains of rice will he put on the 64th square of the board?

b How many grains of rice will there be altogether?

Give your answers in standard form.

HINTS AND TIPS

Compare powers of 2 with the running totals. By the fourth square you have 15 grains altogether, and $2^4 = 16$.

8 The surface area of the Earth is approximately 2×10^8 square miles. The area of the Earth's surface that is covered by water is approximately 1.4×10^8 square miles.

a Calculate the area of the Earth's surface *not* covered by water. Give your answer in standard form.

b What percentage of the Earth's surface is not covered by water?

9 The Moon is a sphere with a radius of 1.080×10^3 miles. The formula for working out the surface area of a sphere is:

surface area = $4\pi r^2$

Calculate the surface area of the Moon.

B

10 Evaluate $\frac{E}{M}$ when $E = 1.5 \times 10^3$ and $M = 3 \times 10^{-2}$, giving your answer in standard form.

11 Work out the value of $\frac{3.2 \times 10^7}{1.4 \times 10^2}$ giving your answer in standard form, correct to 2 significant figures.

12 In 2009, British Airways carried 33 million passengers. Of these, 70% passed through Heathrow Airport. On average, each passenger carried 19.7 kg of luggage. Calculate the total mass of the luggage carried by these passengers.

FM 13 In 2009 the world population was approximately 6.77×10^9. In 2010 the world population is approximately 6.85×10^9.

By how much did the population rise? Give your answer as an ordinary number.

A

PS 14 Here are four numbers written in standard form.

1.6×10^4 $\quad$ 4.8×10^6 $\quad$ 3.2×10^2 $\quad$ 6.4×10^3

a Work out the smallest answer when two of these numbers are multiplied together.

b Work out the largest answer when two of these numbers are added together.

Give your answers in standard form.

FM 15 Many people withdraw money from their banks by using hole-in-the-wall machines. Each day there are eight million withdrawals from 32 000 machines. What is the average number of withdrawals per machine?

PS 16 The mass of Saturn is 5.686×10^{26} tonnes. The mass of the Earth is 6.04×10^{21} tonnes. How many times heavier is Saturn than the Earth? Give your answer in standard form to a suitable degree of accuracy.

AU 17 A number, when written in standard form, is greater than 100 million and less than 1000 million.

Write down a possible value of the number, in standard form.

GRADE BOOSTER

- **D** You can add, subtract, multiply and divide decimal numbers
- **C** You can calculate percentage increases and decreases
- **C** You can work out compound interest problems
- **C** You know how to use the rules of indices for negative and fractional values
- **B** You can solve complex problems involving percentage increases and percentage decreases
- **B** You can do reverse percentage problems
- **B** You can write numbers in standard or index form
- **B** You can calculate with numbers written in index form
- **A** You can convert recurring decimals to fractions

What you should know now

- How multiply and divide by decimals with up to two decimal places
- How to find the reciprocals of rational numbers
- You can work out compound interest problems
- How to manipulate indices, both integer (positive and negative) and fractional
- You can solve complex problems involving percentage increases and percentage decreases
- How to do reverse percentage problems
- How to write numbers in standard form
- How to solve problems, using numbers in standard form
- How to convert recurring decimals into fractions

1 Use a calculator to work out:

$$\sqrt{\frac{21.6 \times 15.8}{3.8}}$$

a Write down all the figures on your calculator display. *(2 marks)*

b Give your answer to part a to 3 significant figures. *(1 mark)*

Edexcel, November 2008, Paper 4, Question 2

2 Use your calculator to work out the value of $\sqrt{7.08^2 - 6.57^2}$

a Write down all the figures on your calculator display. *(2 marks)*

b Write your answer to part **a** correct to 2 significant figures. *(1 mark)*

Edexcel, March 2008, Paper 11, Question 3

3 Find the value of

$$\frac{212 \times 7.88}{0.365}$$

4 A year ago, Donna weighed 51.5 kg. Donna now weighs $8\frac{1}{2}$% less.

Work out how much Donna now weighs. Give your answer to an appropriate degree of accuracy. *(2 marks)*

Edexcel, November 2005, Paper 6 Higher, Question 2

5 There are 75 penguins at a zoo.

a There are 15 baby penguins. What percentage of the penguins are babies?

b The number of penguins increases by 40% each year. Calculate the number of penguins in the zoo after 2 years.

6 Zoe invests £6000 in a savings account that pays 3.5% compound interest per year. How much does she have in the account after 6 years?

7 Leila's savings account earns 10% per year compound interest.

a She invests £1500 in the savings account. How much will she have in her account after 2 years?

b Lewis has the same kind of account as his sister. After earning interest for one year, he has £858 in his account. How much money did Lewis invest?

8 Jack invests £3000 for 2 years at 4% per annum compound interest.

Work out the value of the investment at the end of 2 years. *(3 marks)*

Edexcel, November 2008, Paper 15, Question 6

9 Toby invested £4500 for 2 years in a savings account.

He was paid 4% per annum compound interest.

a How much did Toby have in his savings account after 2 years? *(3 marks)*

Jaspar invested £2400 for n years in a savings account.

He was paid 7.5% per annum compound interest.

At the end of the n years he had £3445.51 in the savings account.

b Work out the value of n. *(2 marks)*

Edexcel, June 2009, Paper 4, Question 19

10 $x = \sqrt{\dfrac{p+q}{pq}}$

$p = 4 \times 10^8 \quad q = 3 \times 10^6$

Find the value of x. Give your answer in standard form correct to 2 significant figures.

Edexcel, March 2005, Paper 13 Higher, Question 3

11 **a** Work out $3^6 \div 3^2$ *(1 mark)*

b Write down the value of $36^{\frac{1}{2}}$ *(1 mark)*

c $3^n = \frac{1}{9}$

Find the value of n. *(1 mark)*

Edexcel, June 2007, Paper 11, Question 7

12 **a** Write down the exact value of 3^{-2} *(1 mark)*

b Simplify $\dfrac{7^2 \times 7^4}{7^3}$ *(2 marks)*

Edexcel, May 2008, Paper 14, Question 14

13 **a** Simplify $t^6 \times t^2$ *(1 mark)*

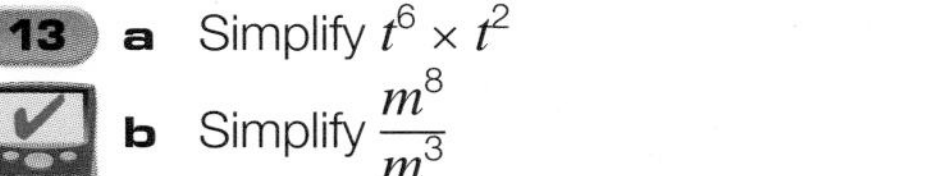

b Simplify $\dfrac{m^8}{m^3}$ *(1 mark)*

c Simplify $(2x)^3$ *(2 marks)*

d Simplify $3a^2h \times 4a^5h^4$ *(2 marks)*

Edexcel, June 2009, Paper 4, Questi

14 The cost of bananas increased by 25% one week but then fell the following week back to the original price.

By what percentage did the cost of bananas fall in the following week?

15 **a** Write the number 45 000 in standard form. *(1 mark)*

b Write 6×10^{-2} as an ordinary number. *(1 mark)*

Edexcel, November 2008, Paper 11, Question 7

16 The number of atoms in one kilogram of helium is 1.51×10^{26}

Calculate the number of atoms in 20 kilograms of helium.

Give your answer in standard form. *(2 marks)*

Edexcel, June 2008, Paper 4, Question 17

17 **a** Simplify

i $t^4 \times t^3$ **ii** $\frac{m}{m^6}$ **iii** $(3k^2m^2) \times (4k^3m)$

18 Simplify **a** $\frac{8x^2y \times 3xy^3}{6x^2y^2}$ **b** $(2m^4p^2)^3$

Worked Examination Questions

FM **1** Kelly bought a television set. After a reduction of 15% in a sale, the one she bought cost her £319.60. What was the original price of the television set?

1 Multiplier is 0.85 or 85% is equivalent to £319.60

A 15% reduction is a multiplier of 0.85, or realising the sale price is 85% scores 1 mark for method.

£319.60 ÷ 0.85, or 100% is equivalent to £319.60 ÷ 85 × 100

Showing the correct calculation that will lead to the correct answer scores 1 mark for method.

= £376

£376 gets 1 mark for accuracy.

Total: 3 marks

PS **2** A plant in a greenhouse is 10 cm high. It increases its height by 15% each day. How many days does it take to double in height?

2 1.15 is the multiplier

Recognising the multiplier scores 1 mark.

$10 \times 1.15 = 11.5$ (one day)

10×1.15^2 (or 11.5×1.15) $= 13.225$

Getting to this stage scores 1 mark for method.

$10 \times 1.15^3 = 15.2$

$10 \times 1.15^4 = 17.5$

$10 \times 1.15^5 = 20.1$ and therefore it takes 5 days to double its height

Reaching the correct solution gets 1 mark for accuracy.

Total: 3 marks

2 Functional Maths
Oil crisis

Throughout the world, people are concerned about our energy supplies. At present, we rely greatly on fossil fuels such as oil for our energy. This is a cause for concern for two main reasons: burning fossil fuels causes damage to the environment and supplies of fossil fuels are now running out. Understanding the figures related to energy use throughout the world is vital in finding a solution to the energy crisis.

Getting started

Use your knowledge of science and current affairs to bring together what you know about energy resources. The questions below may help you.

- Which sources of energy do we use in the UK?
- Where does energy come from?
- Which forms of energy are considered to be 'green'?

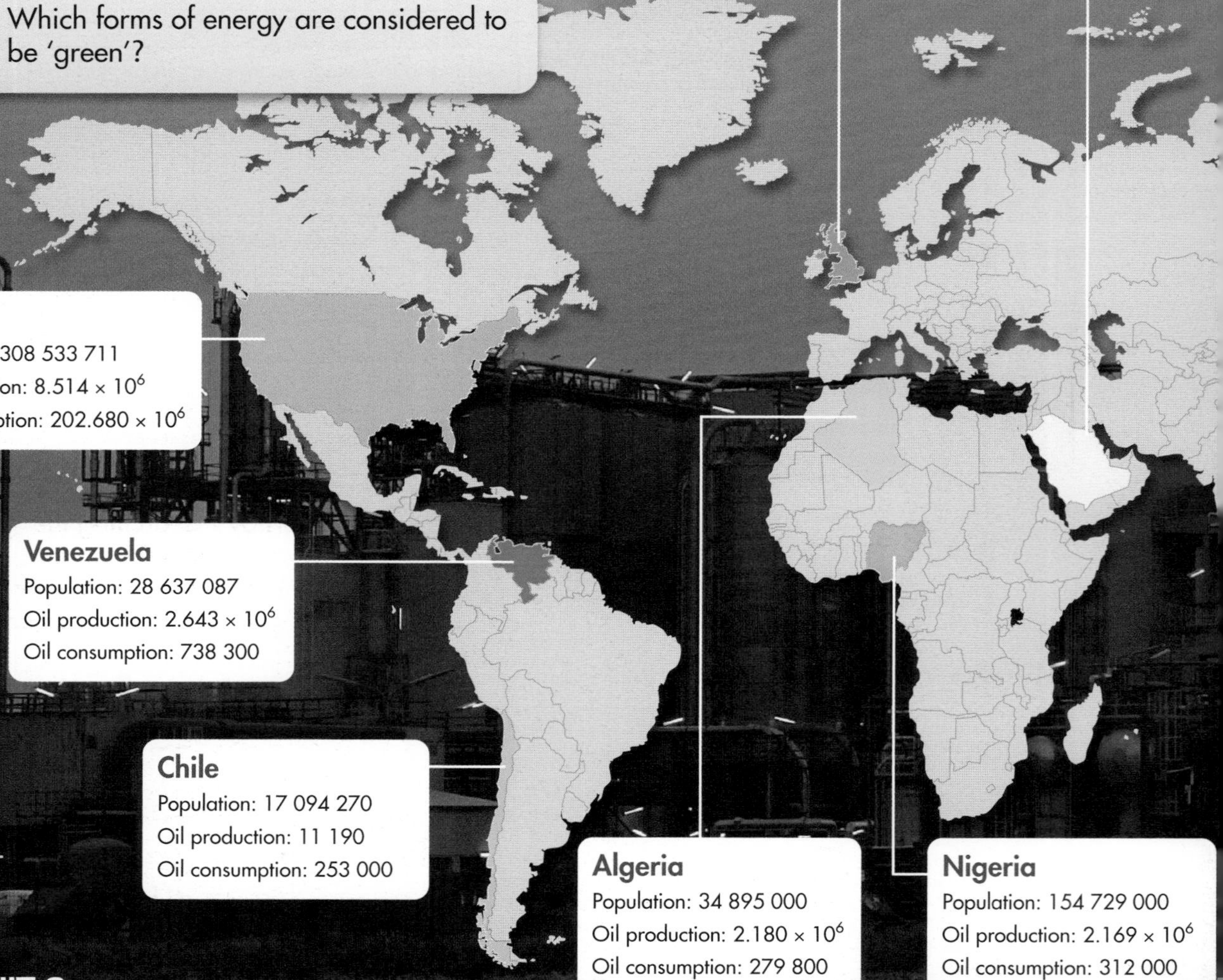

UK
Population: 62 041 708
Oil production: 1.584×10^6
Oil consumption: 1.765×10^6

Saudi Arabia
Population: 25 721 00
Oil production: 10.7×10^6
Oil consumption: 2.34×10^6

USA
Population: 308 533 711
Oil production: 8.514×10^6
Oil consumption: 202.680×10^6

Venezuela
Population: 28 637 087
Oil production: 2.643×10^6
Oil consumption: 738 300

Chile
Population: 17 094 270
Oil production: 11 190
Oil consumption: 253 000

Algeria
Population: 34 895 000
Oil production: 2.180×10^6
Oil consumption: 279 800

Nigeria
Population: 154 729 000
Oil production: 2.169×10^6
Oil consumption: 312 000

The energy that we get from oil can also be generated from more environmentally-friendly, sustainable sources.

Your task

A scientist is researching the production and consumption of oil in the world, in order to inform people about the global energy crisis.

She looks at 10 oil-producing countries and, for each country, finds the most recent figures on the country's population, and its oil production and consumption, measured in barrels per day.

The scientist asks you to write a report about the production and consumption of oil throughout the world for a national newspaper.

Your report should use evidence to explain:

- how many barrels of oil are produced per person, per year
- in which year the world is 'greener'
- which country is the 'best' consumer of oil
- which country is the 'worst' consumer of oil.

Year	World population	World oil production, barrels per day
1984	4.77×10^9	5.45×10^7
1989	5.19×10^9	5.99×10^7
1994	5.61×10^9	6.10×10^7
1999	6.01×10^9	6.58×10^7
2004	6.38×10^9	7.25×10^7
2009	6.79×10^9	8.49×10^7

Oil production and oil consumption figures on this map are measured in barrels.

Japan
Population: 127 530 000
Oil production: 133 100
Oil consumption: 5.007×10^6

Indonesia
Population: 234 181 400
Oil production: 1.051×10^6
Oil consumption: 1.219×10^6

Australia
Population: 22 125 030
Oil production: 86 400
Oil consumption: 966 200

Extension

Develop your report by researching the 'green', sustainable energy resources that many countries produce and consume. How do the figures related to 'green' energy compare to those related to oil?

Why this chapter matters

We use proportion and speed as part of our everyday lives to help when dealing with facts or to compare two or more pieces of information.

Proportion is often used to compare sizes while speed is used to compare distances with the time taken to travel them.

Speed

What do you think of as a high speed?

On 16 August 2009 Usain Bolt set a new world record for the 100 m sprint of 9.58 seconds. This is an average speed of 23.3 mph.

The sailfish is the fastest fish and can swim at 68 mph.

The cheetah is the fastest land animal and can run at 75 mph.

The fastest bird is the swift, which can fly at 106 mph.

Proportion facts

- Russia is the largest country. Vatican City is the smallest country. The area of Russia is nearly 39 million times bigger than the area of Vatican City.
- Monaco has the most people per square mile. Mongolia has the least people per square mile.
- Japan has the highest life expectancy. Sierra Leone has the lowest life expectancy. On average, people in Japan live over twice as long as people in Sierra Leone.
- Taiwan has the most mobile phones per 100 people (106.5). This is approximately four times more than in Thailand (26.04).
- About one-seventh of England is green-belt land.

This chapter is about comparing pieces of information. You can compare the speeds of Usain Bolt, the sailfish, the cheetah and the swift by answering questions such as: How much faster is a sailfish than Usain Bolt?

Chapter 3

Number: Compound measures

1 Limits of accuracy

2 Speed, time and distance

3 Direct proportion problems

4 Density

The grades given in this chapter are target grades.

This chapter will show you ...

- D how to solve problems involving direct proportion
- D how to calculate speed
- C how to find the limits of numbers rounded to certain accuracies
- B how to calculate density
- B to A* how to use limits of accuracy in calculations

Visual overview

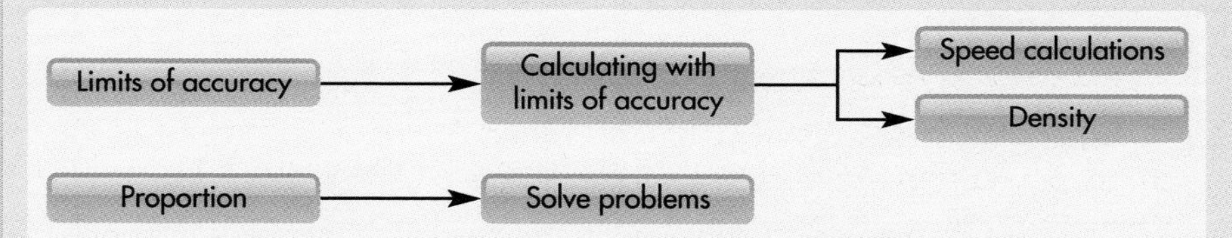

What you should already know

- How to round numbers to the nearest 10, 100 or 1000 **(KS3 level 4, GCSE grade G)**
- How to round numbers to a given number of decimal places **(KS3 level 5, GCSE grade F)**
- How to round numbers to a given number of significant figures **(KS3 level 6–7, GCSE grade E)**

Quick check

1 Round 6374 to:

a the nearest 10
b the nearest 100
c the nearest 1000.

2 Round 2.389 to:

a one decimal place
b two decimal places.

3 Round 47.28 to:

a one significant figure
b three significant figures.

3.1 Limits of accuracy

This section will show you how to:

- find the limits of accuracy of numbers that have been rounded to different degrees of accuracy

Key words
continuous data
discrete data
limits of accuracy
lower bound
upper bound

Any recorded measurements have usually been rounded.

The true value will be somewhere between the **lower bound** and the **upper bound**.

The lower and upper bounds are sometimes known as the **limits of accuracy**.

Discrete data

Discrete data can only take certain values within a given range; amounts of money and numbers of people are examples of discrete data.

EXAMPLE 1

A coach is carrying 50 people, to the nearest 10.

What are the minimum and maximum numbers of people on the coach?

45 is the lowest whole number that rounds to 50 to the nearest 10.
54 is the highest whole number that rounds to 50 to the nearest 10.

So minimum = 45 and maximum = 54

The limits are 45 $\leqslant$ number of people $\leqslant$ 54

Remember that you can only have a whole number of people.

Continuous data

Continuous data can take any value, within a given range; length and mass are examples of continuous data.

Upper and lower bounds

A journey of 26 miles measured to the nearest mile could actually be as long as 26.4999999… miles or as short as 25.5 miles. It could not be 26.5 miles, as this would round up to 27 miles. However, 26.499 999 9… is virtually the same as 26.5.

You overcome this difficulty by saying that 26.5 is the upper bound of the measured value and 25.5 is its lower bound. You can therefore write:

25.5 miles $\leqslant$ actual distance $<$ 26.5 miles

which states that the actual distance is *greater than or equal to* 25.5 miles but *less than* 26.5 miles.

When stating the upper bound, follow the accepted practice, as demonstrated here, which eliminates the difficulties of using recurring decimals.

A mathematical peculiarity

Let: $x = 0.999\ 999\ldots$ (1)

Multiply by 10: $10x = 9.999\ 999\ldots$ (2)

Subtract (1) from (2): $9x = 9$

Divide by 9: $x = 1$

So, we have: $0.\dot{9} = 1$

Hence, it is valid to give the upper bound without using recurring decimals.

EXAMPLE 2

A stick of wood measures 32 cm, to the nearest centimetre.

What are the lower and upper limits of the actual length of the stick?

The lower limit is 31.5 cm as this is the lowest value that rounds to 32 cm to the nearest centimetre.

The upper limit is 32.499 999 999… cm as this is the highest value that rounds to 32 cm to the nearest centimetre as 32.5 cm would round to 33 cm.

However, you say that 32.5 cm is the upper bound. So you write:

31.5 cm $\leqslant$ length of stick $<$ 32.5 cm

Note the use of the strict inequality ($<$) for the upper bound.

EXAMPLE 3

A time of 53.7 seconds is accurate to 1 decimal place.

What are the limits of accuracy?

The smallest possible value is 53.65 seconds.

The largest possible value is 53.749 999 999… but 53.75 seconds is the upper bound.

So the limits of accuracy are 53.65 seconds $\leqslant$ time $<$ 53.75 seconds.

EXAMPLE 4

A skip has a mass of 220 kg measured to 3 significant figures. What are the limits of accuracy of the mass of the skip?

The smallest possible value is 219.5 kg.

The largest possible value is 220.499 999 99… kg but 220.5 kg is the upper bound.

So the limits of accuracy are 219.5 kg $\leqslant$ mass of skip $<$ 220.5 kg.

EXERCISE 3A

C

1 Write down the limits of accuracy of the following.

a 7 cm measured to the nearest centimetre

b 120 g measured to the nearest 10 g

c 3400 km measured to the nearest 100 km

d 50 mph measured to the nearest miles per hour

e £6 given to the nearest pound

f 16.8 cm to the nearest tenth of a centimetre

g 16 kg to the nearest kilogram

h A football crowd of 14 500 given to the nearest 100

i 55 miles given to the nearest mile

j 55 miles given to the nearest 5 miles

B

2 Write down the limits of accuracy for each of the following values, which are rounded to the given degree of accuracy.

a 6 cm (1 significant figure)

b 17 kg (2 significant figures)

c 32 min (2 significant figures)

d 238 km (3 significant figures)

e 7.3 m (1 decimal place)

f 25.8 kg (1 decimal place)

g 3.4 h (1 decimal place)

h 87 g (2 significant figures)

i 4.23 mm (2 decimal places)

j 2.19 kg (2 decimal places)

k 12.67 min (2 decimal places)

l 25 m (2 significant figures)

m 40 cm (1 significant figure)

n 600 g (2 significant figures)

o 30 min (1 significant figure)

p 1000 m (2 significant figures)

q 4.0 m (1 decimal place)

r 7.04 kg (2 decimal places)

s 12.0 s (1 decimal place)

t 7.00 m (2 decimal places)

3 Write down the lower and upper bounds of each of these values, rounded to the accuracy stated.

a 8 m (1 significant figure)

b 26 kg (2 significant figures)

c 25 min (2 significant figures)

d 85 g (2 significant figures)

e 2.40 m (2 decimal places)

f 0.2 kg (1 decimal place)

g 0.06 s (2 decimal places)

h 300 g (1 significant figure)

i 0.7 m (1 decimal place)

j 366 d (3 significant figures)

k 170 weeks (2 significant figures)

l 210 g (2 significant figures)

PS **4** A bus has 53 seats of which 37 are occupied.

The driver estimates that at the next bus stop 20 people, to the nearest 10, will get on and no one will get off.

If he is correct, is it possible they will all get a seat?

B

FM **5** A chain is 30 m long, to the nearest metre.

A chain is needed to fasten a boat to a harbour wall, a distance that is also 30 m, to the nearest metre.

Which statement is definitely true? Explain your decision.

A: The chain will be long enough.

B: The chain will not be long enough.

C: It is impossible to tell whether or not the chain is long enough.

AU **6** A bag contains 2.5 kg of soil, to the nearest 100 g.

What is the least amount of soil in the bag?

Give your answer in kilograms and grams.

7 Billy has 40 identical marbles. Each marble has a mass of 65 g (to the nearest gram).

a What is the greatest possible mass of one marble?

b What is the least possible mass of one marble?

c What is the greatest possible mass of all the marbles?

d What is the least possible mass of all the marbles?

3.2 Speed, time and distance

This section will show you how to:

- recognise the relationship between speed, distance and time
- calculate average speed from distance and time
- calculate distance travelled from the speed and the time
- calculate the time taken on a journey from the speed and the distance

Key words

average
distance
speed
time

The relationship between **speed**, **time** and **distance** can be expressed in three ways:

$$\textit{speed} = \frac{\textit{distance}}{\textit{time}} \qquad \textit{distance} = \textit{speed} \times \textit{time} \qquad \textit{time} = \frac{\textit{distance}}{\textit{speed}}$$

In problems relating to speed, you usually mean **average** speed, as it would be unusual to maintain one exact speed for the whole of a journey.

The diagram will help you remember the relationships between distance (D), time (T) and speed (S).

D
S T

$D = S \times T \qquad S = \frac{D}{T} \qquad T = \frac{D}{S}$

EXAMPLE 5

Paula drove a distance of 270 miles in 5 hours. What was her average speed?

Paula's average speed $= \frac{\text{distance she drove}}{\text{time she took}} = \frac{270}{5} = 54$ miles per hour (mph)

EXAMPLE 6

Edith drove from Sheffield to Peebles in $3\frac{1}{2}$ hours at an average speed of 60 mph. How far is it from Sheffield to Peebles?

Since: *distance = speed × time*

the distance from Sheffield to Peebles is given by

$60 \times 3.5 = 210$ miles

Note: You need to change the time to a decimal number and use 3.5 (not 3.30).

EXAMPLE 7

Sean is going to drive from Newcastle upon Tyne to Nottingham, a distance of 190 miles. He estimates that he will drive at an average speed of 50 mph. How long will it take him?

Sean's time $= \frac{\text{distance he covers}}{\text{his average speed}} = \frac{190}{50} = 3.8$ hours

Change the 0.8 hour to minutes by multiplying by 60, to give 48 minutes.

So, the time for Sean's journey will be 3 hours 48 minutes.

Remember: When you calculate a time and get a decimal answer, as in Example 7, *do not mistake* the decimal part for minutes. You must either:

- leave the time as a decimal number and give the unit as hours, or
- change the decimal part to minutes by multiplying it by 60 (1 hour = 60 minutes) and give the answer in hours and minutes.

EXERCISE 3B 

HINTS AND TIPS

Remember to convert time to a decimal if you are using a calculator, for example, 8 hours 30 minutes is 8.5 hours.

1 A cyclist travels a distance of 90 miles in 5 hours. What was her average speed?

2 How far along a motorway would you travel if you drove at 70 mph for 4 hours?

3 I drive to Bude in Cornwall from Sheffield in about 6 hours. The distance from Sheffield to Bude is 315 miles. What is my average speed?

4 The distance from Leeds to London is 210 miles. The train travels at an average speed of 90 mph. If I catch the 9.30 am train in London, at what time should I expect to arrive in Leeds?

5 How long will an athlete take to run 2000 m at an average speed of 4 m per second?

6 Copy and complete the following table.

	Distance travelled	Time taken	Average speed
a	150 miles	2 hours	
b	260 miles		40 mph
c		5 hours	35 mph
d		3 hours	80 km/h
e	544 km	8 hours 30 minutes	
f		3 hours 15 minutes	100 km/h
g	215 km		50 km/h

7 Eliot drove from Sheffield to Inverness, a distance of 410 miles, in 7 hours 45 minutes.

a Change the time 7 hours 45 minutes to a decimal.

b What was the average speed of the journey? Round your answer to 1 decimal place.

8 Colin drives home from his son's house in 2 hours 15 minutes. He says that he drives at an average speed of 44 mph.

a Change the 2 hours 15 minutes to a decimal.

b How far is it from Colin's home to his son's house?

9 The distance between Paris and Le Mans is 200 km. The express train between Paris and Le Mans travels at an average speed of 160 km/h.

a Calculate the time taken for the journey from Paris to Le Mans, giving your answer as a decimal number of hours.

b Change your answer to part a to hours and minutes.

FM 10 The distance between Sheffield and Land's End is 420 miles.

a What is the average speed of a journey from Sheffield to Land's End that takes 8 hours 45 minutes?

b If Sam covered the distance at an average speed of 63 mph, how long would it take him?

D

C

C

FM **11** A train travels at 50 km/h for 2 hours, then slows down to do the last 30 minutes of its journey at 40 km/h.

a What is the total distance of this journey?

b What is the average speed of the train over the whole journey?

FM **12** Jade runs and walks the 3 miles from home to work each day. She runs the first 2 miles at a speed of 8 mph, then walks the next mile at a steady 4 mph.

a How long does it take Jade to get to work?

b What is her average speed?

13 Change the following speeds to metres per second.

a 36 km/h **b** 12 km/h **c** 60 km/h

d 150 km/h **e** 75 km/h

HINTS AND TIPS

Remember that there are 3600 seconds in an hour and 1000 metres in a kilometre. So to change from km/h to m/s multiply by 1000 and divide by 3600.

14 Change the following speeds to kilometres per hour.

a 25 m/s **b** 12 m/s **c** 4 m/s **d** 30 m/s **e** 0.5 m/s

AU **15** A train travels at an average speed of 18 m/s.

a Express its average speed in km/h.

b Find the approximate time the train would take to travel 500 m.

c The train set off at 7.30 on a 40 km journey. At approximately what time will it reach its destination?

HINTS AND TIPS

To change from m/s to km/h multiply by 3600 and divide by 1000.

16 A cyclist is travelling at an average speed of 24 km/h.

a What is this speed in metres per second?

b What distance does he travel in 2 hours 45 minutes?

c How long does it take him to travel 2 km?

d How far does he travel in 20 seconds?

HINTS AND TIPS

To convert a decimal fraction of an hour to minutes, just multiply by 60.

PS **17** How much longer does it take to travel 100 miles at 65 mph than at 70 mph?

18 It takes me 20 minutes to walk from home to the bus station.

I catch the bus from the bus station to work each morning. My bus journey is 10 miles and usually takes 30 minutes. I can catch a bus at 20 minutes past the hour or 10 minutes to the hour. When I get off the bus it takes me 5 minutes to walk to work.

FM **a** What is the average speed of my bus?

PS **b** I have to be at work for 08.30. What time is the latest I can leave home to be at work on time?

3.3 Direct proportion problems

This section will show you how to:

- recognise and solve problems, using direct proportion

Key words
direct proportion
unitary method
unit cost

Suppose you buy 12 items that each cost the same. The total amount you spend is 12 times the cost of one item.

That is, the total cost is said to be in **direct proportion** to the number of items bought. The cost of a single item (the **unit cost**) is the constant factor that links the two quantities.

Direct proportion is not only concerned with costs. Any two related quantities can be in direct proportion to each other.

The best way to solve all problems involving direct proportion is to start by finding the single unit value. This method is called the **unitary method**, because it involves referring to a single *unit* value.

Remember: Before solving a direct proportion problem, think carefully about it to make sure that you know how to find the required single unit value.

EXAMPLE 8

If eight pens cost £2.64, what is the cost of five pens?

First, you need to find the cost of *one* pen. This is £2.64 ÷ 8 = £0.33

So, the cost of five pens is £0.33 × 5 = £1.65

EXAMPLE 9

Eight loaves of bread will make packed lunches for 18 people. How many packed lunches can be made from 20 loaves?

First, find how many lunches one loaf will make.

One loaf will make 18 ÷ 8 = 2.25 lunches

So, 20 loaves will make 2.25 × 20 = 45 lunches

EXERCISE 3C

1 If 30 matches weigh 45 g, what would 40 matches weigh?

2 Five bars of chocolate cost £2.90. Find the cost of nine bars.

3 Eight men can chop down 18 trees in a day. How many trees can 20 men chop down in a day?

D

D

4 Find the cost of 48 eggs when 15 eggs can be bought for £2.10.

5 Seventy maths textbooks cost £875.

a How much will 25 maths textbooks cost?

b How many maths textbooks can you buy for £100?

> **HINTS AND TIPS**
>
> **Remember** to work out the value of one unit each time. Always check that answers are sensible.

FM **6** A lorry uses 80 litres of diesel fuel on a trip of 280 miles.

a How much diesel would the same lorry use on a trip of 196 miles?

b How far would the lorry get on a full tank of 100 litres of diesel?

FM **7** During the winter, I find that 200 kg of coal keeps my open fire burning for 12 weeks.

a If I want an open fire all through the winter (18 weeks), how much coal will I need?

b Last year I bought 150 kg of coal. For how many weeks did I have an open fire?

8 It takes a photocopier 16 seconds to produce 12 copies. How long will it take to produce 30 copies?

9 A recipe for 12 biscuits uses:

200 g margarine   400 g sugar   500 g flour   300 g ground rice

a What quantities are needed for:

i 6 biscuits   **ii** 9 biscuits   **iii** 15 biscuits?

PS **b** What is the maximum number of biscuits I could make if I had just 1 kg of each ingredient?

AU **10** Peter the butcher sells sausages in pack of 6 for £2.30.

Paul the butcher sells sausages in packs of 10 for £3.50.

I have £10 to spend on sausages. If I want to buy as many sausages as possible from one shop, which shop should I use? Show your working.

C

PS **11** A shredding machine can shred 20 sheets of paper in 14 seconds. The bin has room for 1000 sheets of shredded paper.

How long will it take to fill the bin if the machine has to stop for 3 minutes after every 200 sheets to prevent overheating?

FM **12** Here is a recipe for making Yorkshire pudding.

Adjust this recipe to use it for two people.
Justify any decision you make.

> **Yorkshire pudding recipe** (Serves 8)
>
> 125 g plain flour
> 235 ml whole milk
> 2 eggs
> 3 g salt
> 45 ml beef dripping or lard

FM **13** An aircraft has two fuel tanks, one in each wing.

The tanks each hold 40 litres when full.

The left tank is quarter full. The right tank is half full.

How much fuel is needed so that both tanks are three-quarters full?

3.4 Density

This section will show you how to:
- solve problems involving density

Key words
density
mass
volume

Density is the **mass** of a substance per unit **volume**, usually expressed in grams per cm^3. The relationship between the three quantities is:

$$density = \frac{mass}{volume}$$

You can remember this with a triangle similar to that for distance, speed and time.

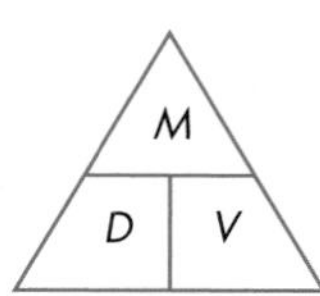

mass = *density* × *volume*
density = *mass* ÷ *volume*
volume = *mass* ÷ *density*

Note: Density is defined in terms of mass. The common metric units for mass are grams and kilograms. Try not to mix up mass with weight. The common metric unit for weight is the newton (N). You may have learnt about the difference between mass and weight in science.

EXAMPLE 10

A piece of metal has a mass of 30 g and a volume of 4 cm^3. What is the density of the metal?

$$\text{Density} = \frac{\text{mass}}{\text{volume}} = \frac{30}{4} = 7.5 \text{ g/cm}^3$$

EXAMPLE 11

What is the mass of a piece of rock which has a volume of 34 cm^3 and a density of 2.25 g/cm^3?

Mass = volume × density = 34 × 2.25 = 76.5 g

EXERCISE 3D

B

1 Find the density of a piece of wood with a mass of 6 g and a volume of 8 cm^3.

2 Calculate the density of a metal if 12 cm^3 of it has a mass of 100 g.

3 Calculate the mass of a piece of plastic, 20 cm^3 in volume, if its density is 1.6 g/cm^3.

4 Calculate the volume of a piece of wood which has a mass of 102 g and a density of 0.85 g/cm^3.

5 Find the mass of a marble model, 56 cm^3 in volume, if the density of marble is 2.8 g/cm^3.

6 Calculate the volume of a liquid with a mass of 4 kg and a density of 1.25 g/cm^3.

7 Find the density of the material of a pebble which has a mass of 34 g and a volume of 12.5 cm^3.

8 It is estimated that the statue of Queen Victoria in Endcliffe Park, Sheffield, has a volume of about 4 m^3.

The density of the material used to make the statue is 9.2 g/cm^3. What is the estimated mass of the statue?

9 I bought a 50 kg bag of coal, and estimated the total volume of coal to be about 28 000 cm^3.

What is the density of coal, in g/cm^3?

10 A 1 kg bag of sugar has a volume of about 625 cm^3. What is the density of sugar in g/cm^3?

PS 11 Two statues look identical and both appear to be made out of gold. One of them is a fake.

The density of gold is 19.3 g/cm^3.

The statues each have a volume of approximately 200 cm^3.

The first statue has a mass of 5.2 kg.

The second statue has a mass of 3.8 kg.

Which one is the fake?

AU 12 A piece of metal has a mass of 345 g and a volume of 15 cm^3.

A different piece of metal has a mass of 400 g and a density of 25 g/cm^3.

Which piece of metal has the bigger volume and by how much?

FM 13 Two pieces of scrap metal are melted down to make a single piece of metal.

The first piece has a mass of 1.5 tonnes and a density of 7000 kg/m^3.

The second piece has a mass of 1 tonne and a density of 8000 kg/m^3.

Work out the total volume of the new piece.

GRADE BOOSTER

- **D** You can calculate average speeds from data
- **D** You can calculate distance from speed and time
- **D** You can calculate time from speed and distance
- **D** You can solve problems involving direct proportion
- **B** You can solve problems involving density
- **B** You can find the measures of accuracy for numbers given to whole number accuracies
- **A** You can find measures of accuracy for numbers given to decimal place or significant figure accuracies
- **A*** You can calculate the limits of compound measures

What you should know now

- How to find the limits of numbers given to various accuracies
- How to find the limits of compound measures by combining the appropriate limits of the variables involved
- The relationships between speed, time and distance
- How to solve problems involving direct proportion
- How to work out the density of materials

1 Here are the ingredients for making cheese pie for 6 people.

> Cheese pie for **6** people
> 180 g flour
> 240 g cheese
> 80 g butter
> 4 eggs
> 160 ml milk

Bill makes a cheese pie for 3 people.

a Work our how much flour he needs. *(2 marks)*

Jenny makes a cheese pie for 15 people.

b Work out how much milk she needs. *(2 marks)*

Edexcel, November 2008, Paper 4, Question 1

2 Julie buys 19 identical calculators.

The total cost is £143.64

Work out the total cost of 31 of these calculators. *(3 marks)*

Edexcel, June 2009, Paper 4, Question 5

3 Stuart drives 180 km in 2 hours 15 minutes.

Work out Stuart's average speed. *(3 marks)*

Edexcel, November 2008, Paper 11, Question 3

4 Joe travelled 60 miles in 1 hour 30 minutes.

Work out Joe's average speed.

Give your answer in miles per hour. *(2 marks)*

Edexcel, June 2007, Paper 11, Question 2

5 The length of a line is 63 centimetres, correct to the nearest centimetre.

a Write down the **least** possible length of the line. *(1 mark)*

b Write down the **greatest** possible length of the line. *(1 mark)*

Edexcel, May 2009, Paper 3, Question 11

6 Bob measures the length of his book.

The length of the book is 22 cm correct to the nearest centimetre.

a Write down the maximum possible length it could be. *(1 mark)*

b Write down the minimum possible length it could be. *(1 mark)*

Edexcel, June 2007, Paper 11, Question 4

7 Mandy completes a journey in two stages. In stage 1 of her journey, she drives at an average speed of 96 km/h and takes 2 hours 15 minutes.

a How far does Mandy travel in stage 1 of her journey?

b Altogether, Mandy drives 370 km and takes a total time of 4 hours. What is her average speed, in km/h, in stage 2 of her journey?

8 A ball is thrown vertically upwards with a speed V metres per second.

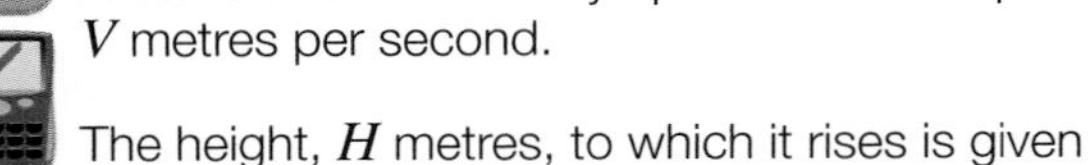

The height, H metres, to which it rises is given by $H = \frac{V^2}{2g}$

where g m/s^2 is the acceleration due to gravity.

$V = 24.4$ correct to 3 significant figures.

$g = 9.8$ correct to 2 significant figures.

a Write down the lower bound of g. *(1 mark)*

b Calculate the upper bound of H.

Give your answer correct to 3 significant figures. *(2 marks)*

Edexcel, November 2008, Paper 4, Question 23

9 Katy drove for 238 miles, correct to the nearest mile.

She used 27.3 litres of petrol, to the nearest tenth of a litre.

$$\text{Petrol consumption} = \frac{\text{Number of miles travelled}}{\text{Number of litres of petrol used}}$$

Work out the upper bound for the petrol consumption for Katy's journey.

Give your answer correct to 2 decimal places. *(3 marks)*

Edexcel, June 2008, Paper 4, Question 22

10 Jerry measures a piece of wood as 60 cm correct to the nearest centimetre.

a Write down the minimum possible length of the piece of wood.

b Write down the maximum possible length of the piece of wood.

Edexcel, March 2005, Paper 13A Higher, Question 1

11 $x = 1.8$ measured to 1 decimal place,
$y = 4.0$ measured to 2 significant figures,
$z = 2.56$ measured to 3 significant figures.

Calculate the upper limit of $\frac{x^2 + y}{z}$.

12 A girl runs 60 metres in a time of 8.0 seconds. The distance is measured to the nearest metre and the time is measured to 2 significant figures.

What is the least possible speed?

13 The volume of a gold bar is 100 cm^3. The density of gold is 19.3 grams per cm^3.

Work out the mass of the gold bar.

Worked Examination Questions

1 To be on time, a train must complete a journey of 210 miles in 3 hours.

a Calculate the average speed of the train for the whole journey when it is on time.

b The train averages a speed of 56 mph over the first 98 miles of the journey. Calculate the average speed for the remainder of the journey so that the train arrives on time.

1 a Average speed = distance ÷ time

$210 \div 3 = 70$ mph

This gets 1 mark for method and 1 mark for the correct answer.

Note that one question in the examination will ask you to state the units of your answer. This is often done with a speed question.

b $98 \div 56 = 1.75$ which is 1 hour and 45 minutes.

First find out how long the train took to do the first 98 miles.

98 ÷ 56 gets 1 mark for method.
1.75 or 1 hour 45 minutes gets 1 mark for accuracy.

$(210 - 98) \div (3 - 1.75) = 112 \div 1.25$

$= 89.6$ mph

Now work out the distance still to be travelled (112 miles) and the time left (1 hour 15 minutes = 1.25 hours). Divide distance by time to get the average speed.

(210 – 98) ÷ (3 – 1.75) gets 1 mark for method even if there is an error in one of the figures.
89.6 or an answer following an arithmatic error gets 1 mark for accuracy.

Total: 6 marks

PS **2** Jonathan is comparing two ways to travel from his flat in London to his parents' house in Doncaster.

Tube, train and taxi

It takes 35 minutes to get to the railway station by tube in London. A train journey from London to Doncaster takes 1 hour 40 minutes. From Doncaster it is 15 miles by taxi at an average speed of 20 mph.

Car

The car journey is 160 miles at an average speed of 50 mph. Which is the slower journey, tube, train and taxi or car?

2 Time = distance ÷ speed = $\frac{15}{20}$

Work out the time taken by taxi.
This gets 1 mark for method.

= 0.75 hour (or 45 minutes)

This gets 1 mark for accuracy.

Total time = 35 minutes + 1 hour 40 minutes + 45 minutes

Work out the **total time** for tube, train and taxi.

= 3 hours

This is required to compare with the car.
This gets 1 mark.

Time = distance ÷ speed = $160 \div 50$

Work out the time taken by car.
This gets 1 mark for method and 1 mark for accuracy.

= 3.2 hours (or 3 hours 12 minutes)

Car is 12 minutes slower.

State the conclusion following from your results. This gets 1 mark.

Total: 6 marks

Worked Examination Questions

3 The magnification of a lens is given by the formula

$$m = \frac{v}{u}$$

In an experiment, u is measured as 8.5 cm and v is measured as 14.0 cm, both correct to the nearest 0.1 cm. Find the least possible value of m. You must show full details of your calculation.

3 $8.45 \leqslant u < 8.55$ — Write down the limits of both variables. This gets 1 mark each.

$13.95 \leqslant v < 14.05$ — As the calculation is a division the least value will be given by least u ÷ greatest v. This gets 1 mark for method.

Least m = least u ÷ greatest v = 8.45 ÷ 14.05

= 0.6014234875 — This gets 1 mark for accuracy.

= 0.601 or 0.6 — It is good practice to round to a suitable degree of accuracy.

Total: 4 marks

4 A long rod with a square cross-section is made with a side of 5 cm. A circular hole is drilled with a radius of 3.6 cm. All measurements are to the nearest $\frac{1}{10}$ cm. Will the rod fit into the circle?

This is a using and applying maths question. You need to have a strategy to solve it.
Step 1: Find the largest possible diagonal of the square using Pythagoras' theorem.
Step 2: Work out the smallest possible diameter of the circle.
Step 3: Compare the values to see if the diagonal is smaller than the diameter.

4 Limits of side of square

$4.95 < \text{side} < 5.05$

Limits of radius

$3.55 < \text{radius} < 3.65$

Always start by writing down the limits of the variables in the question. These get 1 mark each.

Largest diagonal = $\sqrt{(5.05^2 + 5.05^2)}$ — Evidence of using Pythagoras' theorem gives 1 method mark.

= 7.141 778 49

= 7.142 (4 significant figures) — Work out the square root but do not round to less than 4 sf. This gets 1 mark for method and 1 mark for accuracy.

Smallest diameter = 2 × 3.55 = 7.1 — Work out the smallest diameter. This gets 1 mark for method.

As 7.142 > 7.1, the rod may not fit in the circle. — Obtaining 7.1 and giving the correct conclusion gets 1 mark for accuracy.

Total: 7 marks

Functional Maths
Organising your birthday dinner

To celebrate your birthday, you have decided to hold a large dinner party for your friends and family.

You have invited 15 people in total and so far you have had responses from eight people, all of whom can attend.

As the dinner party will be a big event for you, you want to begin preparing for it straight away. You decide to plan for several different themes, so that you can work out the price of the ingredients for each menu for your dinner and work out how much money you will need to spend.

Chilli pasta Serves 2

175 g pasta
40 ml olive oil
2 onions
4 cloves garlic
1 red jalapeno pepper
185 g yellow peppers, roasted
Basil

Ragu Bolognese Serves 16

450 g minced beef
450 g minced pork
14 kg pasta
30 ml olive oil
225 g chicken liver
2 onions
4 garlic cloves
150 g streaky bacon
400 g chopped tomatoes
200 g tomato purée
400 ml red wine
Basil

Wild mushroom tart Serves 4

275 g puff pastry
25 g butter
300 g wild mushrooms
25 g cheese
1 clove garlic
1 egg

Steak and kidney pudding Serves 6

450 g diced beef
150 g kidney
30 ml beef dripping
2 onions
40 g plain flour
Thyme
Bay leaves
Parsley
1 pint brown beef stock
175 g self-raising flour
75 g suet

Getting started

- Find the cost of 175 g of pasta, if you can buy 500 g for 70p.
- The butcher sells sausages in packs of eight for £2.50. How much would you pay for three sausages, if he will sell them individually?
- How many 300-g portions will 1.3 kg provide?
- Which is cheaper, 200 g of tomatoes for £1.90 or 350 g for £2.50?

Prices of ingredients	Quantity	Cost
Beef dripping	500 g	55p
Beef stock	12 cubes	98p
Minced beef	500 g	£2.89
Diced beef	1 kg	£5.20
Kidney	400 g	£1.20
Minced pork	500 g	£2.00
Chicken liver	400 g	99p
Streaky bacon	300 g	£2.00
Suet	200 g	65p
Puff pastry	500 g	79p
Butter	250 g	85p
Eggs	6	91p
Pasta	500 g	70p
Onion	1kg	75p
Jalapeno pepper	200 g jar	£1.25
Yellow pepper	each	80p
Plain flour	1.5 kg	43p
Self-raising flour	1.5 kg	43p
Cheese	1 kg	£5.12
Chopped tomatoes	400 g tin	55p
Tomato purée	200 g	33p
Wild mushrooms	125 g	£1.69
Olive oil	500 ml	£1.88
Garlic (8 cloves per bulb)	3 bulbs	89p
Thyme	16 g	68p
Bay leaves	6 g	88p
Parsley	180 g	79p
Basil	115 g	£2.20
Red wine	750 ml	£3.99

Your task

You have not chosen the main course of your dinner yet. To help you decide, you have chosen four of your favourite recipes.

You now need to look at the lists of ingredients and the price list, then work out what you will need for each one, and how much it will cost.

1. First, work on the assumption that no more than the eight guests, who have already confirmed that they are coming, will be able to attend your party.

   Work out the ingredients you will need for each recipe, and the cost of these ingredients (in total and per portion).
2. To make sure that you are able to cater for more guests, work out the ingredients and costs for each recipe for:
   - 10 guests
   - 15 guests.
3. You decide to set a budget for the ingredients for your main course and you choose £75 as a starting point.

   Evaluate how realistic this budget is, considering the number of guests that could attend and the level of choice you would like to offer, taking account that some will eat meat and some will be vegetarians.

   Remember: you must include yourself when you are working out the ingredients that you must buy and the potential cost.

Extension

Your friends and family will need to travel to get to your dinner party. Your friend Sam will set out at 4.30pm and travel 35 miles to your house; your cousin Charlie will set out at 4.00pm and travel 65 miles. If they both travel at the same speed, who will reach your house first? How fast will they each have to travel for both to reach your house at the same time?

Why this chapter matters

About 70% of the Earth's surface is covered with water. What area of water is that? Just how big is the Earth's surface? And, what is the volume of air in the atmosphere above the Earth's surface?

The Earth is roughly a sphere and the circumference of the Earth is approximately 40 000 kilometres but how does that help to find the surface area?

Consider something on a smaller scale, such as a football. How much material is needed to make a football? The diameter of a football is about 70 cm but can you use that fact to find the area of the surface?

Volumes are difficult too. How big is the Earth or the moon? Or, how much air is there inside a football? Gas is often stored at refineries in spherical containers but how can you calculate the amount of gas a spherical container will hold?

Volume calculations apply to all sorts of shapes. If you think about a volcano, you should be able to imagine how to find the surface area but if you were a geologist you would need to know the amount of rock that might be thrown out of the volcano by an eruption. In short, you would need to know the volume of the volcano. How can you find it?

Fortunately we have the tools to answer these questions. In Book 1, you learned how to find the surface areas and volumes of prisms. In this chapter, you will be looking at more complex shapes and extending your skills in calculating volumes.

Chapter

Geometry: Shape

1. Circumference and area of a circle
2. Cylinders
3. Volume of a pyramid
4. Cones
5. Spheres

You will also need to revisit calculating the length of an arc and area of a sector (Book 1, lesson 8.2) and finding the volume of a prism (Book 1, lesson 8.3) to complete Unit 3.

The grades given in this chapter are target grades.

This chapter will show you ...

- D to C how to calculate the circumference and area of a circle
- B how to calculate the volume and surface area of a cylinder
- B how to calculate the volume of a pyramid
- A how to calculate the volume and surface area of a cone and a sphere

Visual overview

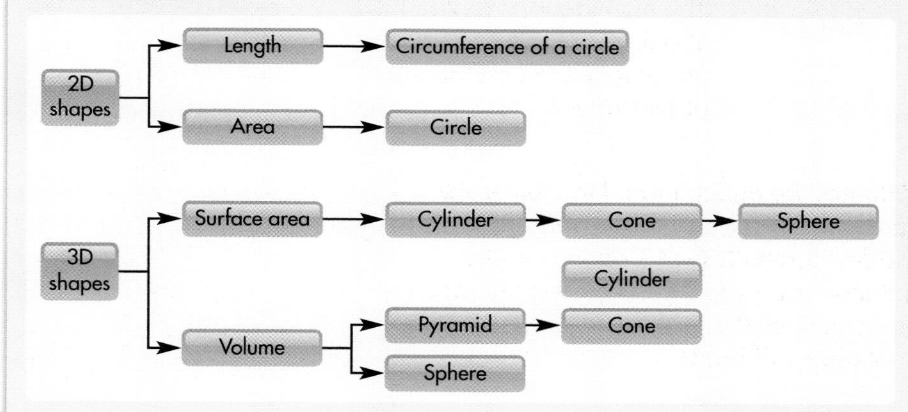

What you should already know

- The area of a rectangle is given by area = length × width or $A = lw$ **(KS3 level 5, GCSE grade F)**
- The area of a triangle is given by area = $\frac{1}{2}$ × base × height or $A = \frac{1}{2}bh$ **(KS3 level 6, GCSE grade D)**
- The area of a parallelogram is given by area = base × height or $A = bh$ **(KS3 level 6, GCSE grade E)**
- The volume of a cuboid is given by volume = length × width × height or $V = lwh$ **(KS3 level 6, GCSE grade E)**
- The common metric units to measure area, volume and capacity are shown in this table **(KS3 level 6, GCSE grade E)**

Area	Volume	Capacity
100 mm^2 = 1 cm^2	1000 mm^3 = 1 cm^3	1000 cm^3 = 1 litre
10 000 cm^2 = 1 m^2	1 000 000 cm^3 = 1 m^3	1 m^3 = 1000 litres

continued

Quick check

1 Find the areas of the following shapes.

a

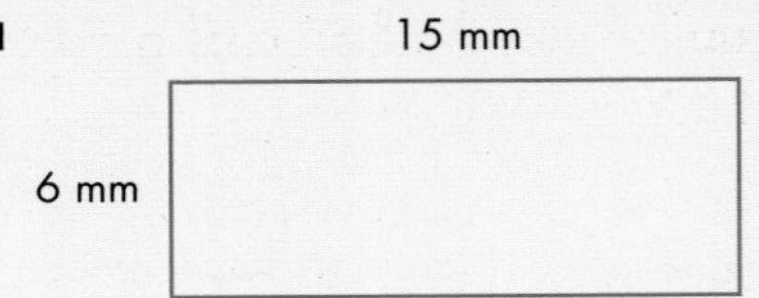

b

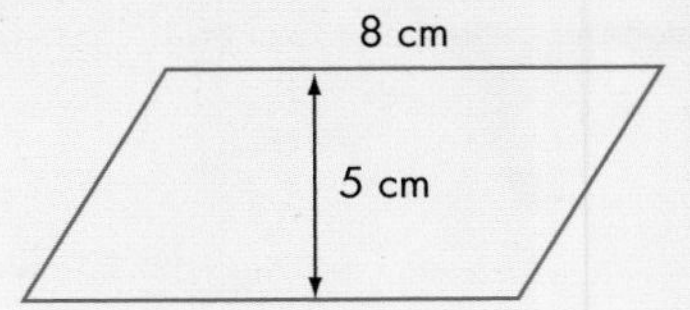

c

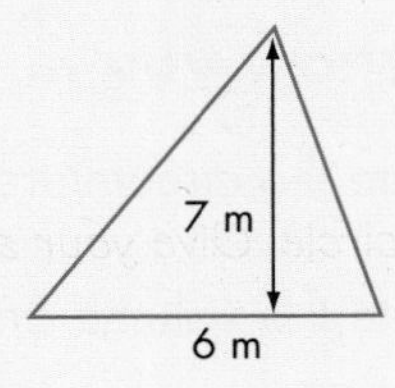

2 Find the volume of this cuboid.

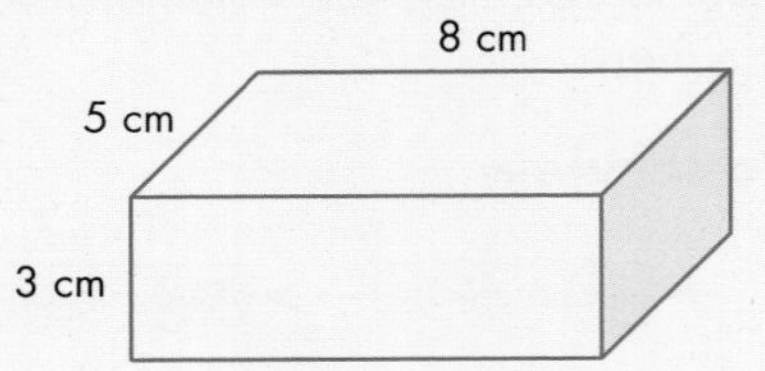

If you need to revise circle calculations, you should work through Exercise 4A.

4.1 Circumference and area of a circle

This section will show you how to:
- calculate the circumference and area of a circle

Key words
π
area
circumference

EXAMPLE 1

Calculate the **circumference** of the circle. Give your answer to 3 significant figures.

$C = \pi d$
$= \pi \times 5.6$ cm
$= 17.6$ cm (to 3 significant figures)

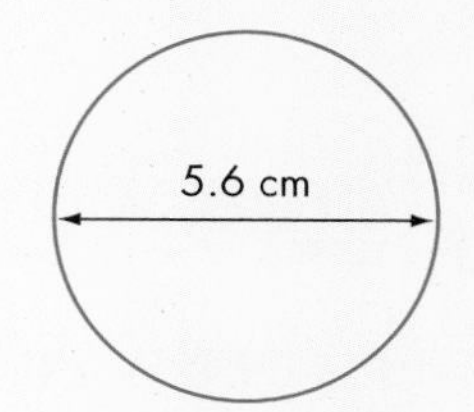

EXAMPLE 2

Calculate the **area** of the circle. Give your answer in terms of π.

$A = \pi r^2$
$= \pi \times 6^2$ m^2
$= 36\pi$ m^2

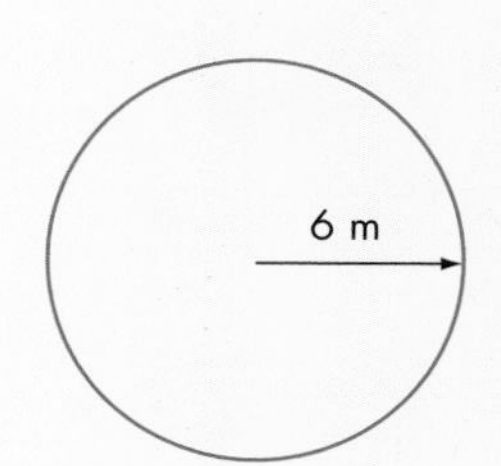

EXERCISE 4A

1 Copy and complete the following table for each circle. Give your answers to 3 significant figures.

	Radius	Diameter	Circumference	Area
a	4.0 cm			
b	2.6 m			
c		12.0 cm		
d		3.2 m		

2 Find the circumference of each of the following circles. Give your answers in terms of π.

a Diameter 5 cm **b** Radius 4 cm **c** Radius 9 m **d** Diameter 12 cm

3 Find the area of each of the following circles. Give your answers in terms of π.

a Radius 5 cm **b** Diameter 12 cm **c** Radius 10 cm **d** Diameter 1 m

D

AU 4 A rope is wrapped eight times around a capstan (a cylindrical post), the diameter of which is 35 cm. How long is the rope?

PS 5 The roller used on a cricket pitch has a radius of 70 cm.

A cricket pitch has a length of 20 m. How many complete revolutions does the roller make when rolling the pitch?

6 The diameter of each of the following coins is as follows.

1p: 2.0 cm, 2p: 2.6 cm, 5p: 1.7 cm, 10p: 2.4 cm

Calculate the area of one face of each coin. Give your answers to 1 decimal place.

7 The distance around the outside of a large pipe is 2.6 m. What is the diameter of the pipe?

AU 8 What is the total perimeter of a semicircle of diameter 15 cm?

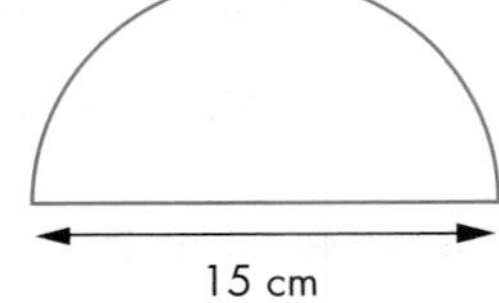

FM 9 A restaurant sells two sizes of pizza. The diameters are 24 cm and 30 cm. The restaurant claims that the larger size is 50% bigger.

Your friend disagrees and wants to complain to the local trading standards officer. What would you advise? Give a reason for your answer.

10 Calculate the area of each of these shapes, giving your answers in terms of π.

a

12 cm

b

4 cm

AU 11 Calculate the area of the shaded part of the diagram, giving your answer in terms of π.

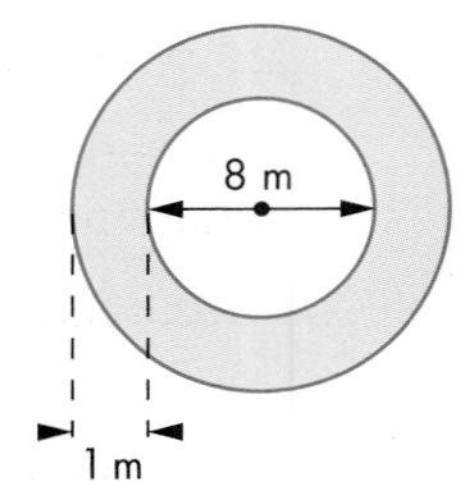

12 This is the plan of a large pond with a gravel path all around it. What area needs to be covered with gravel?

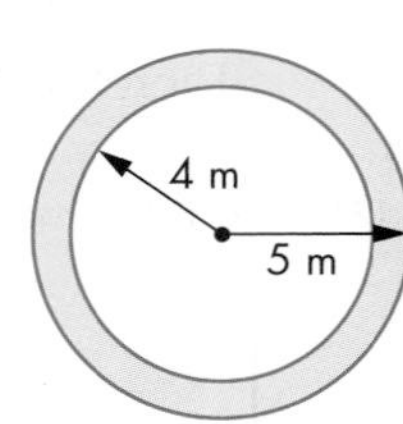

13 A tree in Sequoia National Park in USA is considered to be the largest in the world. It has a circumference at the base of 31.3 m. Would the base of the tree fit inside your classroom?

4.2 Cylinders

This section will show you how to:
- calculate the volume and surface area of a cylinder

Key words
cylinder
surface area
volume

Volume

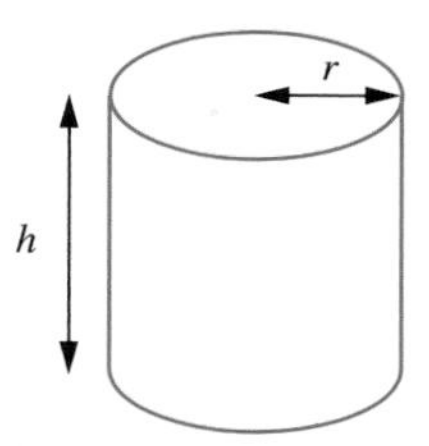

Since a **cylinder** is an example of a prism, its **volume** is found by multiplying the area of one of its circular ends by the height.

That is, volume = $\pi r^2 h$

where r is the radius of the cylinder and h is its height or length.

EXAMPLE 3

What is the volume of a cylinder having a radius of 5 cm and a height of 12 cm?

Volume = area of circular base × height
$= \pi r^2 h$
$= \pi \times 5^2 \times 12 \text{ cm}^3$
$= 942 \text{ cm}^3$ (3 significant figures)

Surface area

The total **surface area** of a cylinder is made up of the area of its curved surface plus the area of its two circular ends.

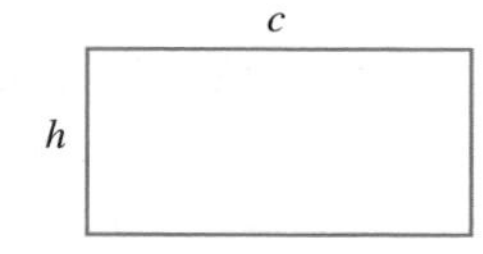

The curved surface area, when opened out, is a rectangle with length equal to the circumference of the circular end.

curved surface area = circumference of end × height of cylinder
$= 2\pi rh$ **or** πdh

area of one end $= \pi r^2$

Therefore, total surface area = $2\pi rh + 2\pi r^2$ **or** $\pi dh + 2\pi r^2$

EXAMPLE 4

What is the total surface area of a cylinder with a radius of 15 cm and a height of 2.5 m?

First, you must change the dimensions to a *common unit*. Use centimetres in this case.

Total surface area = $\pi dh + 2\pi r^2$
$= \pi \times 30 \times 250 + 2 \times \pi \times 15^2 \text{ cm}^2$
$= 23\,562 + 1414 \text{ cm}^2$
$= 24\,976 \text{ cm}^2$
$= 25\,000 \text{ cm}^2$ (3 significant figures)

EXERCISE 4B

1 For the cylinders below find:

i the volume **ii** the total surface area.

Give your answers to 3 significant figures.

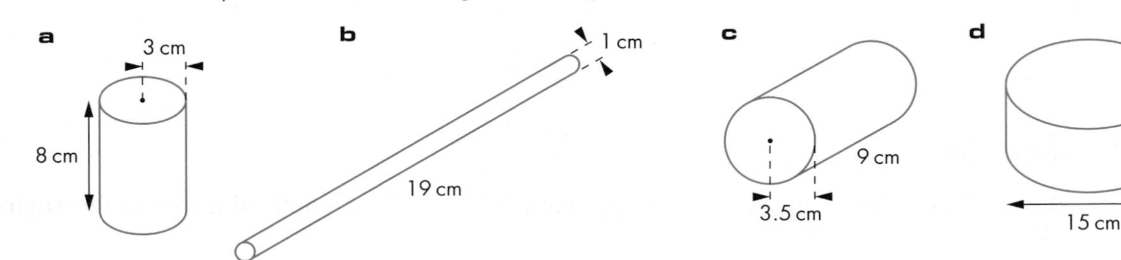

2 For each of these cylinder dimensions find:

i the volume **ii** the curved surface area.

Give your answers in terms of π.

a Base radius 3 cm and height 8 cm **b** Base diameter 8 cm and height 7 cm

c Base diameter 12 cm and height 5 cm **d** Base radius of 10 m and length 6 m

AU 3 The diameter of a marble, cylindrical column is 60 cm and its height is 4.2 m. The cost of making this column is quoted as £67.50 per cubic metre. What is the estimated total cost of making the column?

AU 4 Find the mass of a solid iron cylinder 55 cm high with a base diameter of 60 cm. The density of iron is 7.9 g/cm^3.

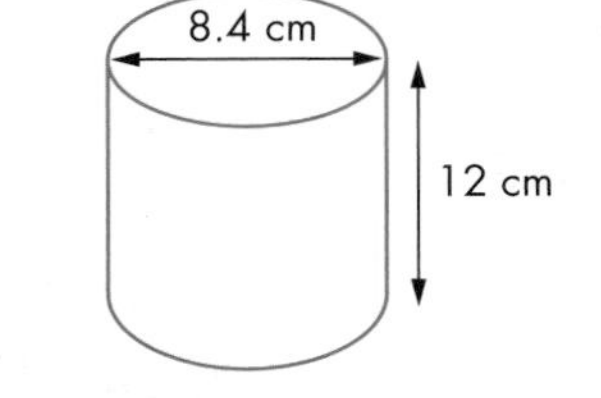

5 A solid cylinder has a diameter of 8.4 cm and a height of 12.0 cm. Calculate the volume of the cylinder.

FM 6 A cylindrical food can has a height of 10.5 cm and a diameter of 7.4 cm.

What can you say about the size of the paper label around the can?

7 A cylindrical container is 65 cm in diameter. Water is poured into the container until it is 1 m deep. How much water is in the container? Give your answer in litres.

FM 8 A drinks manufacturer wishes to market a new drink in a can. The quantity in each can must be 330 ml.

Suggest a suitable height and diameter for the can.

You might like to look at the dimensions of a real drinks can.

9 A cylindrical can of soup has a diameter of 7 cm and a height of 9.5 cm. It is full of soup, which weighs 625 g. What is the density of the soup?

AU 10 A metal bar, 1 m long and with a diameter of 6 cm, has a mass of 22 kg. What is the density of the metal from which the bar is made?

PS 11 Wire is commonly made by putting hot metal through a hole in a plate.

What length of wire of diameter 1 mm can be made from a 1 cm cube of metal?

FM 12 The engine size of a car is measured in litres. This tells you the total capacity of the cylinders in which the pistons move up and down. For example, in a 1.6 litre engine with four cylinders, each cylinder will have a capacity of 0.4 litres.

Cylinders of a particular size can be long and thin or short and fat; they will give the engine different running characteristics.

In a racing car, the diameter can be approximately twice the length. This means the engine will run at very high revs.

Suggest possible dimensions for a 0.4 litre racing car cylinder.

4.3 Volume of a pyramid

This section will show you how to:

- calculate the volume of a pyramid

Key words

apex
frustum
pyramid
volume

A **pyramid** is a 3D shape with a base from which triangular faces rise to a common vertex, called the **apex**. The base can be any polygon, but is usually a triangle, a rectangle or a square.

The **volume** of a pyramid is given by:

volume = $\frac{1}{3}$ × base area × vertical height

$$V = \tfrac{1}{3}Ah$$

where A is the base area and h is the vertical height.

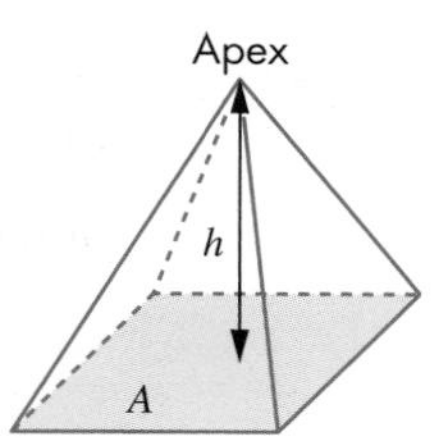

EXAMPLE 5

Calculate the volume of the pyramid on the right.

Base area = 5 × 4 = 20 cm²

Volume = $\frac{1}{3}$ × 20 × 6 = 40 cm³

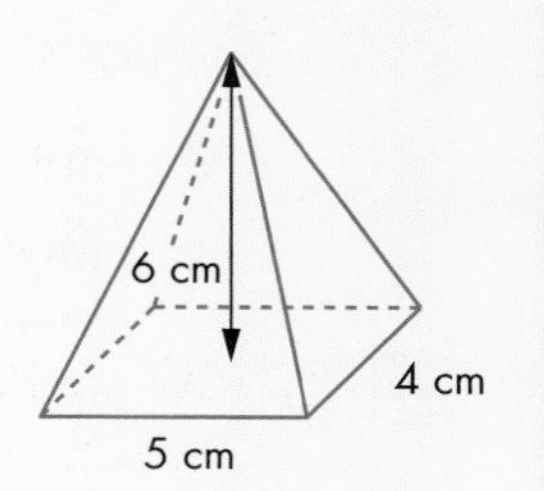

EXAMPLE 6

A pyramid, with a square base of side 8 cm, has a volume of 320 cm^3. What is the vertical height of the pyramid?

Let h be the vertical height of the pyramid. Then,

$$\text{volume} = \tfrac{1}{3} \times 64 \times h = 320 \text{ cm}^3$$

$$\frac{64h}{3} = 320 \text{ cm}^3$$

$$h = \frac{960}{64} \text{ cm}$$

$$h = 15 \text{ cm}$$

EXERCISE 4C

B

1 Calculate the volume of each of these pyramids, all with rectangular bases.

a

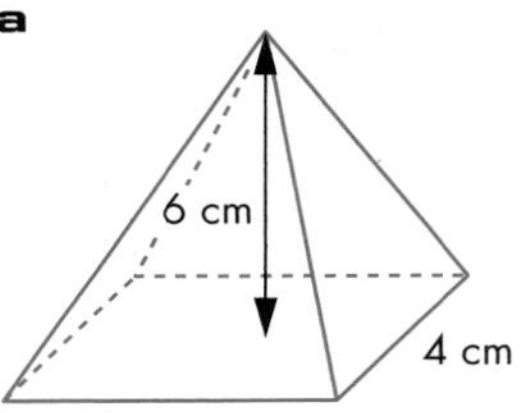

b

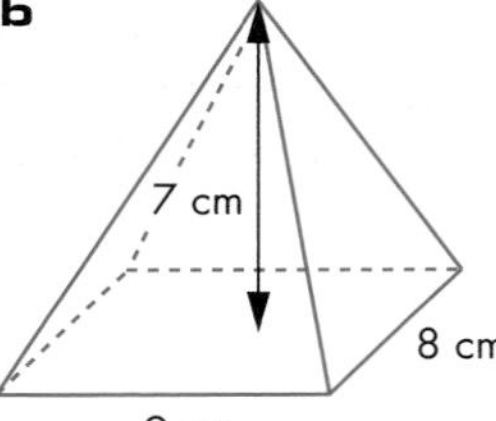

c

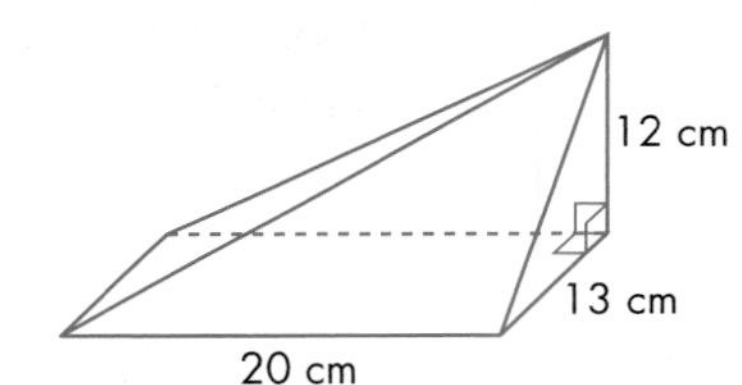

d

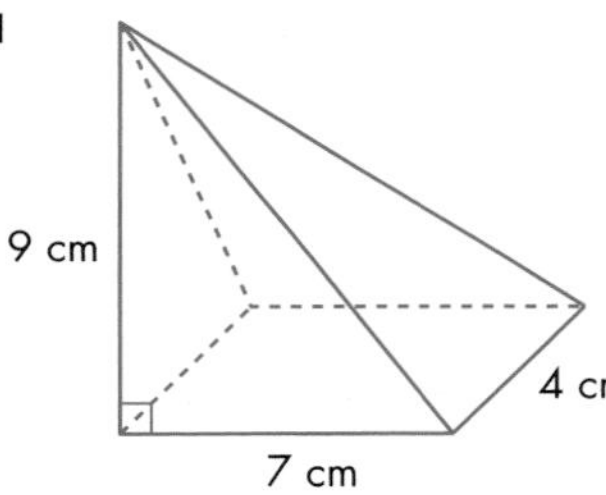

e

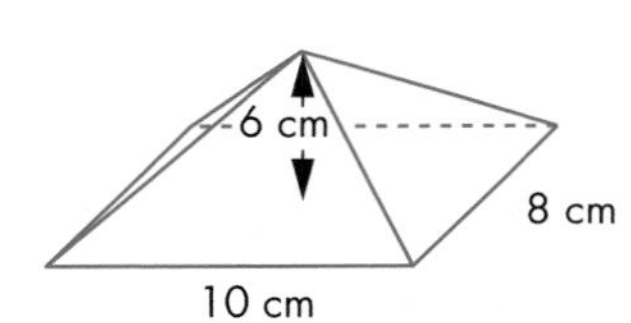

2 Calculate the volume of a pyramid having a square base of side 9 cm and a vertical height of 10 cm.

AU 3 Suppose you have six pyramids which have a height that is half the side of the square base.

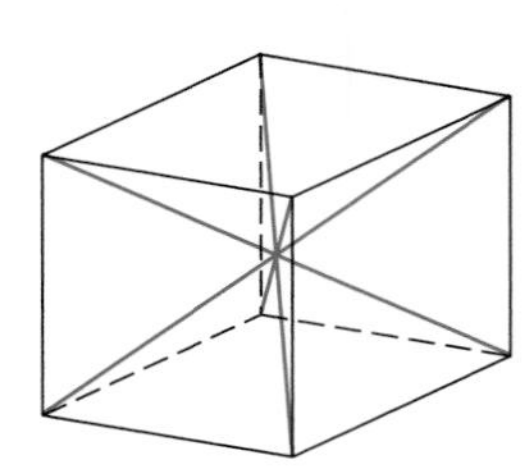

a Explain how they can fit together to make a cube.

b How does this show that the formula for the volume of a pyramid is correct?

4 The glass pyramid outside the Louvre Museum in Paris was built in the 1980s. It is 20.6 m tall and the base is a square of side 35 m. The design was very controversial.

Suppose that instead of a pyramid, the building was a conventional shape with the same square base, a flat roof and the same volume.

How high would it have been?

5 Calculate the volume of each of these shapes.

a

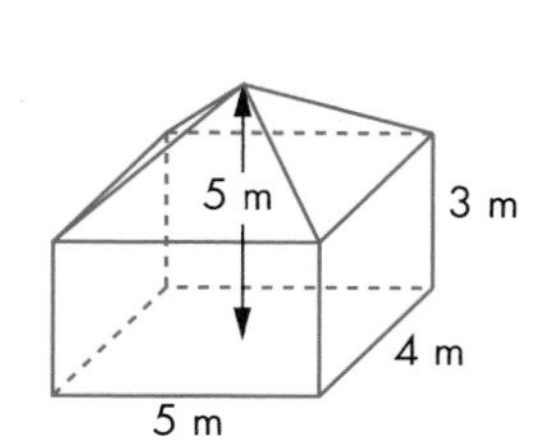

b

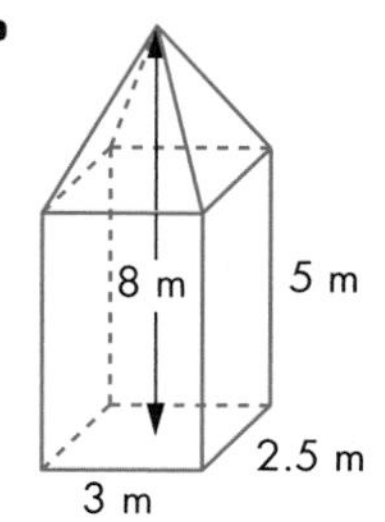

c

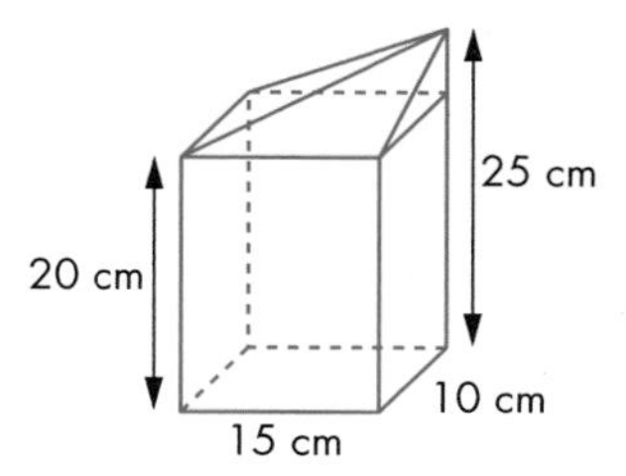

AU 6 What is the mass of a solid pyramid having a square base of side 4 cm, a height of 3 cm and a density of 13 g/cm^3? (1 cm^3 has a mass of 13 g.)

PS 7 A crystal is in the form of two square-based pyramids joined at their bases (see diagram). The crystal has a mass of 31.5 g. What is the mass of 1 cm^3 of the substance?

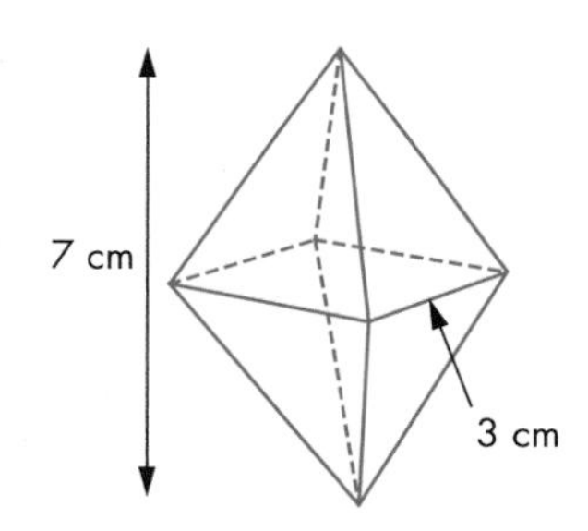

8 A pyramid has a square base of side 6.4 cm.

Its volume is 81.3 cm^3.

Calculate the height of the pyramid.

PS 9 A pyramid has the same volume as a cube of side 10.0 cm.

The height of the pyramid is the same as the side of the square base.

Calculate the height of the pyramid.

10 The pyramid in the diagram has its top 5 cm cut off as shown. The shape which is left is called a **frustum**. Calculate the volume of the frustum.

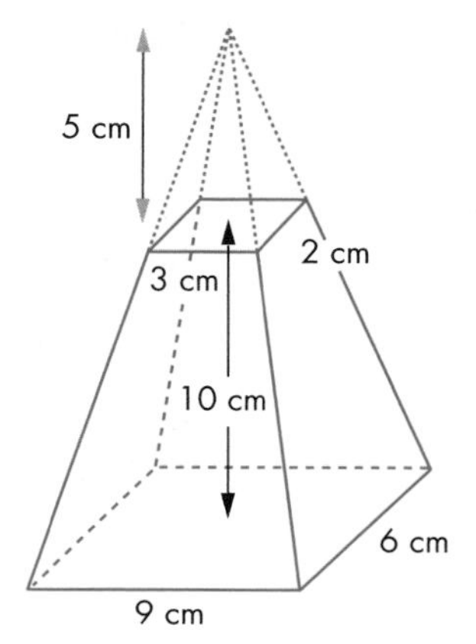

4.4 Cones

This section will show you how to:
- calculate the volume and surface area of a cone

Key words
slant height
surface area
vertical height
volume

A cone can be treated as a pyramid with a circular base. Therefore, the formula for the **volume** of a cone is the same as that for a pyramid.

volume = $\frac{1}{3}$ × base area × vertical height

$$V = \frac{1}{3}\pi r^2 h$$

where r is the radius of the base and h is the **vertical height** of the cone.

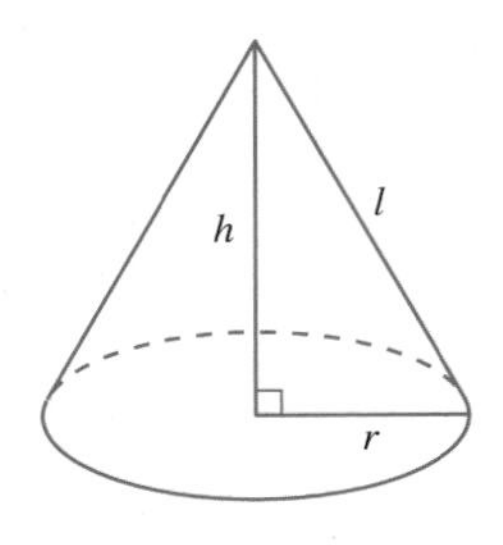

The curved **surface area** of a cone is given by:

curved surface area = π × radius × slant height

$$S = \pi r l$$

where l is the **slant height** of the cone.

So the total surface area of a cone is given by the curved surface area plus the area of its circular base.

$$A = \pi r l + \pi r^2$$

EXAMPLE 7

For the cone in the diagram, calculate:

i its volume

ii its total surface area.

Give your answers in terms of π.

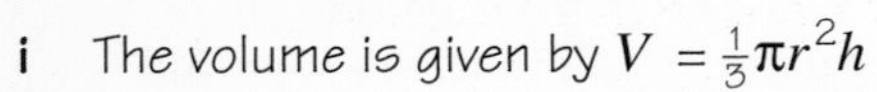

i The volume is given by $V = \frac{1}{3}\pi r^2 h$

$= \frac{1}{3} \times \pi \times 36 \times 8 = 96\pi \text{ cm}^3$

ii The total surface area is given by $A = \pi r l + \pi r^2$

$= \pi \times 6 \times 10 + \pi \times 36 = 96\pi \text{ cm}^2$

EXERCISE 4D 

1 For each cone, calculate:

i its volume **ii** its total surface area.

Give your answers to 3 significant figures.

a

b

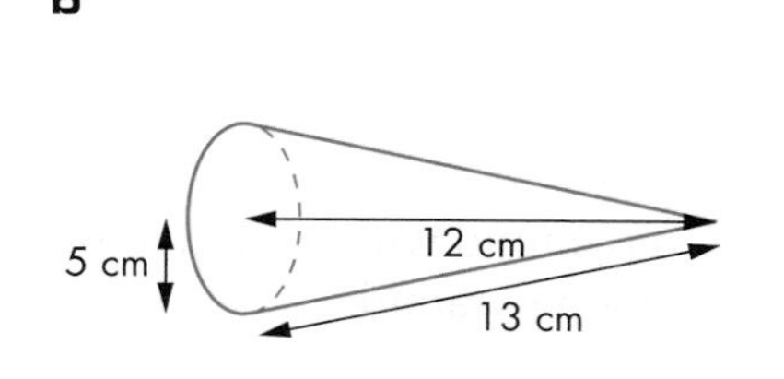

c

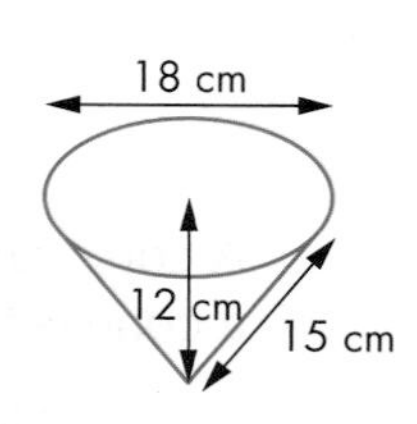

2 A solid cone, base radius 6 cm and vertical height 8 cm, is made of metal whose density is 3.1 g/cm^3. Find the mass of the cone.

3 Find the total surface area of a cone whose base radius is 3 cm and slant height is 5 cm. Give your answer in terms of π.

4 Calculate the volume of each of these shapes. Give your answers in terms of π.

a

b

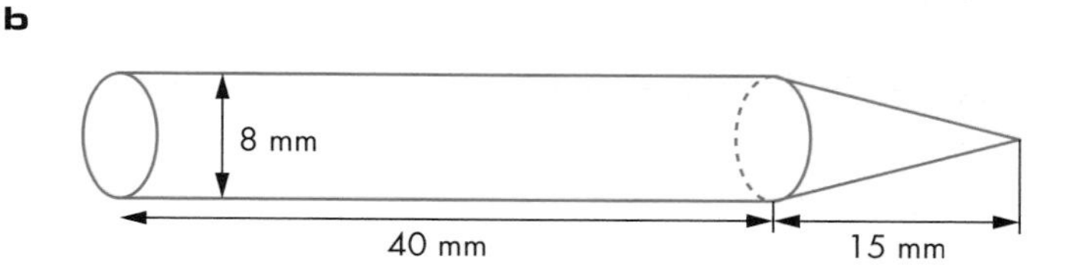

5 You could work with a partner on this question.

A sector of a circle, as in the diagram, can be made into a cone (without a base) by sticking the two straight edges together.

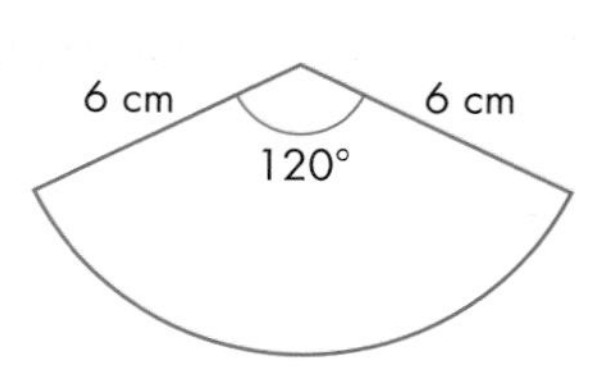

a What would be the diameter of the base of the cone in this case?

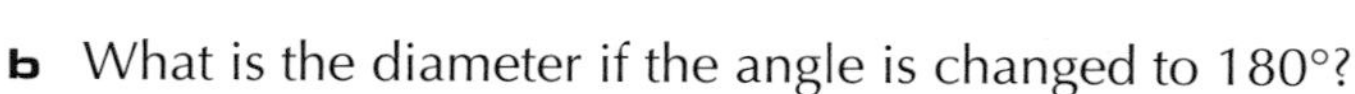

b What is the diameter if the angle is changed to 180°?

c Investigate other angles.

6 A cone has the dimensions shown in the diagram.

Calculate the total surface area, leaving your answer in terms of π.

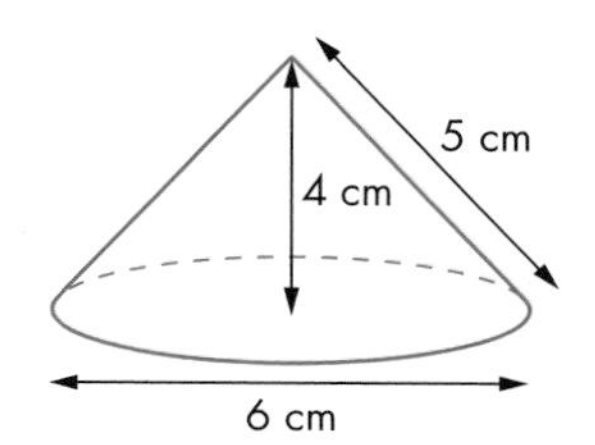

PS 7 If the slant height of a cone is equal to the base diameter, show that the area of the curved surface is twice the area of the base.

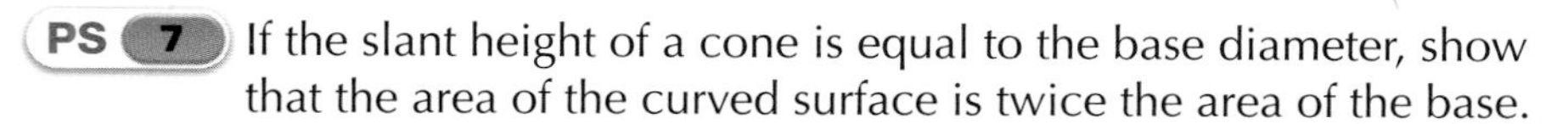

A*

8 The model shown on the right is made from aluminium.
What is the mass of the model, given that the density of aluminium is 2.7 g/cm^3?

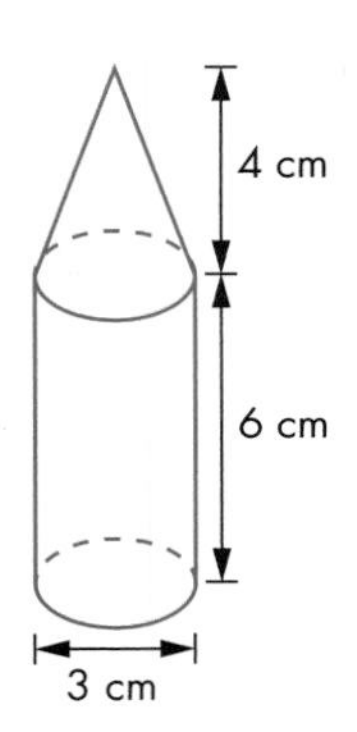

PS 9 A container in the shape of a cone, base radius 10 cm and vertical height 19 cm, is full of water. The water is poured into an empty cylinder of radius 15 cm. How high is the water in the cylinder?

4.5 Spheres

This section will show you how to:
- calculate the volume and surface area of a sphere

Key words
sphere
surface area
volume

The **volume** of a **sphere**, radius r, is given by:

$V = \frac{4}{3}\pi r^3$

Its **surface area** is given by:

$A = 4\pi r^2$

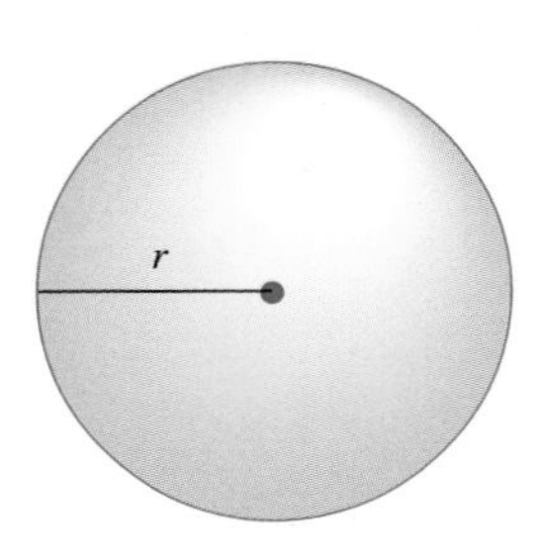

EXAMPLE 8

For a sphere of radius of 8 cm, calculate **i** its volume and **ii** its surface area.

i The volume is given by:

$V = \frac{4}{3}\pi r^3$

$= \frac{4}{3} \times \pi \times 8^3 = \frac{2048}{3} \times \pi = 2140 \text{ cm}^3$ (3 significant figures)

ii The surface area is given by:

$A = 4\pi r^2$

$= 4 \times \pi \times 8^2 = 256 \times \pi = 804 \text{ cm}^2$ (3 significant figures)

EXERCISE 4E

1 Calculate the volume of each of these spheres. Give your answers in terms of π.

a Radius 3 cm **b** Radius 6 cm **c** Diameter 20 cm

2 Calculate the surface area of each of these spheres. Give your answers in terms of π.

a Radius 3 cm **b** Radius 5 cm **c** Diameter 14 cm

3 Calculate the volume and the surface area of a sphere with a diameter of 50 cm.

4 A sphere fits exactly into an open cubical box of side 25 cm. Calculate the following.

a The surface area of the sphere **b** The volume of the sphere

AU **5** A metal sphere of radius 15 cm is melted down and recast into a solid cylinder of radius 6 cm. Calculate the height of the cylinder.

PS **6** Lead has a density of 11.35 g/cm^3. (This means that 1 cm^3 of lead has a mass of 11.35 g.) Calculate the maximum number of shot (spherical lead pellets) of radius 1.5 mm which can be made from 1 kg of lead.

FM **7** The standard (size 5) football must be between 68 cm and 70 cm in circumference and weigh between 410 g and 450 g. They are usually made from 32 panels: 12 regular pentagons and 20 regular hexagons.

a Will a maker of footballs be more interested in the surface area or the volume of the ball? Why?

b What variation in the surface area of a football is allowed?

AU **8** A sphere has a radius of 5.0 cm.

A cone has a base radius of 8.0 cm.

The sphere and the cone have the same volume.

Calculate the height of the cone.

PS **9** A sphere of diameter 10 cm is carved out of a wooden block in the shape of a cube of side 10 cm.

What percentage of the wood is wasted?

AU **10** A manufacturer is making cylindrical packaging for a sphere as shown. The curved surface of the cylinder is made from card.

Show that the area of the card is the same as the surface area of the sphere.

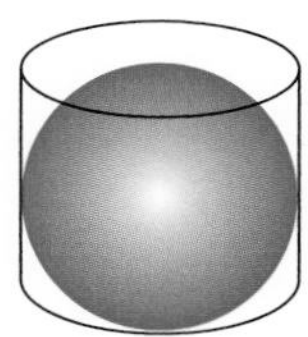

A

GRADE BOOSTER

D You can calculate the circumference and area of a circle

C You can calculate the volume of cylinders

B You can calculate the surface area of cylinders, cones and spheres

B You can calculate the volume of pyramids, cones and spheres

A You can calculate volume and surface area of compound 3D shapes

---

What you should know now

- The volume of a cylinder is given by $V = \pi r^2 h$, where r is the radius and h is the height or length of the cylinder
- The curved surface area of a cylinder is given by $S = 2\pi rh$, where r is the radius and h is the height or length of the cylinder
- The volume of a pyramid is given by $V = \frac{1}{3}Ah$, where A is the area of the base and h is the vertical height of the pyramid
- The volume of a cone is given by $V = \frac{1}{3}\pi r^2 h$, where r is the base radius and h is the vertical height of the cone
- The curved surface area of a cone is given by $S = \pi rl$, where r is the base radius and l is the slant height of the cone
- The volume of a sphere is given by $V = \frac{1}{3}\pi r^3$, where r is its radius
- The surface area of a sphere is given by $A = 4\pi r^2$, where r is its radius

1 A semi-circular protractor has a diameter of 10 cm.

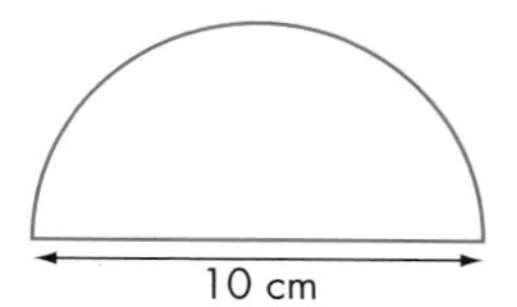

Calculate its perimeter. Give your answer in terms of π.

2 A solid cylinder has a radius of 6 cm and a height of 20 cm.

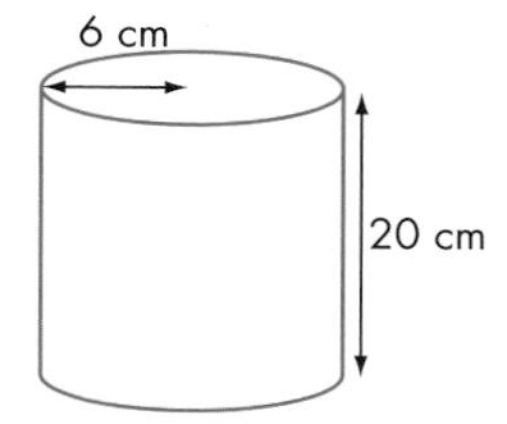

a Calculate the volume of the cylinder. Give your answer correct to 3 significant figures.

The cylinder is made of a material that has a density of 1.5 g/cm^3.

b Calculate the mass of the cylinder. Give your answer correct to 3 significant figures.

3 A solid cylinder has a radius of 4 cm and a height of 10 cm.

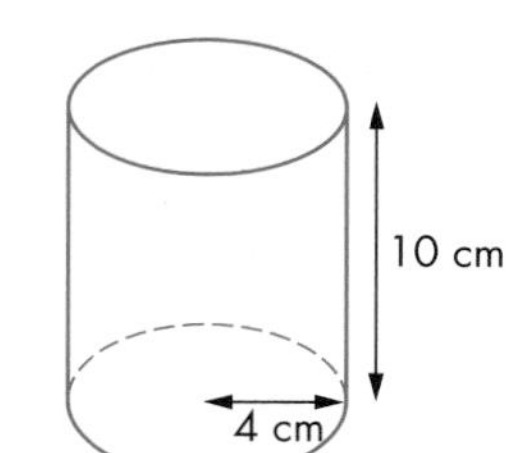

a Work out the volume of the cylinder.

Give your answer correct to 3 significant figures. *(2 marks)*

The cylinder is made from wood.

The density of the wood is 0.6 grams per cm^3.

b Work out the mass of the cylinder.

Give your answer correct to 3 significant figures. *(2 marks)*

Edexcel, June 2008, Paper 4, Question 13

4 A solid cone has base radius 6 cm and slant height 10 cm. Calculate the total surface area of the cone. Give your answer in terms of π.

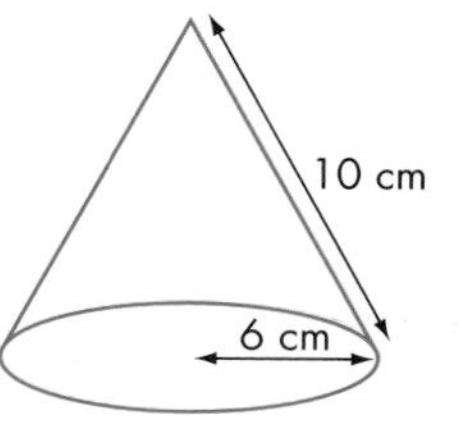

5 Not drawn accurately

$2x$ cm, x cm, h cm, x cm

A cylinder has base radius x cm and height $2x$ cm.

A cone has base radius x cm and height h cm.

The volume of the cylinder and the volume of the cone are equal.

Find h in terms of x.

Give your answer in its simplest form. *(3 marks)*

Edexcel, May 2008, Paper 3, Question 26

6 The diagram shows a storage tank.

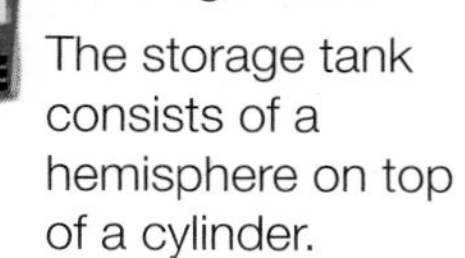

The storage tank consists of a hemisphere on top of a cylinder.

The height of the cylinder is 30 metres.

The radius of the cylinder is 3 metres.

The radius of the hemisphere is 3 metres.

3 m, 3 m, 30 m, 3 m — Not drawn accurately

a Calculate the total volume of the storage tank.

Give your answer correct to 3 significant figures. *(3 marks)*

A sphere has a volume of 500 m^3.

b Calculate the radius of the sphere.

Give your answer correct to 3 significant figures. *(3 marks)*

Edexcel, November 2008, Paper 4, Question 24

Worked Examination Questions

AU **1** The diagram shows a pepper pot. The pot consists of a cylinder and a hemisphere. The cylinder has a diameter of 5 cm and a height of 7 cm.

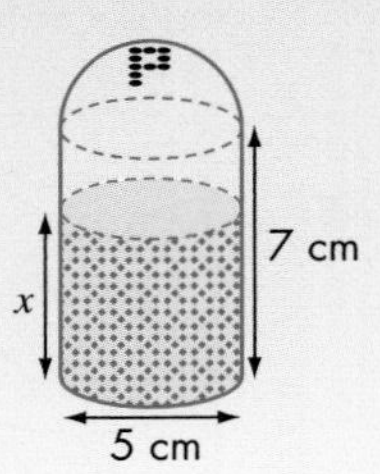

The pepper takes up half the total volume of the pot.
Find the depth of pepper in the pot marked x in the diagram.

1 *Volume of pepper pot*

$= \pi r^2 h + \frac{2}{3}\pi r^3$ — This gets 1 method mark for setting up equation.

$= \pi \times 2.5^2 \times 7 + \frac{2}{3} \times \pi \times 2.5^3\ \text{cm}^3$ — This gets 1 method mark for correct substitutions.

$= 170.2\ \text{cm}^3$ — This gets 1 accuracy mark for correct answers.

So volume of pepper $= 85.1\ \text{cm}^3$

Therefore,

$\pi r^2 x = 85.1\ \text{cm}^3$ — This gets 1 method mark for setting up equation.

and $x = \dfrac{85.1}{\pi \times 2.5^2} = 4.3$ *cm (1 decimal place)* — This gets 1 method mark for correct rearrangement and 1 accuracy mark for correct answer.

Total: 6 marks

FM **2** Aluminium craft wire is available in different diameters. One manufacturer sells a 5 metre coil of wire with a diameter of 1.6 mm for £5.00.

What volume of aluminium is that?

2 *Volume* $= \pi \times r^2 \times h = \pi \times 0.8^2 \times 5000$ — Find the volume of a cylinder with a length of 5 m and a diameter of 1.6 mm. This gets 1 mark for method. The units need to be the same. In millimetres, the length is 5000mm. This gets 1 mark for method. This gets 1 mark for accuracy for correct substitution.

$= 10\,000\ \text{mm}^3$ *to 2 sf* — This gets 1 mark for accuracy and rounding answer to 2 sf. The answer could also be given in cm^3 (10 or 10.1). Rounding off to 3 sf would also be acceptable (10 100).

Total: 4 marks

Worked Examination Questions

PS **3** Three balls of diameter 8.2 cm just fits inside a cylindrical container.

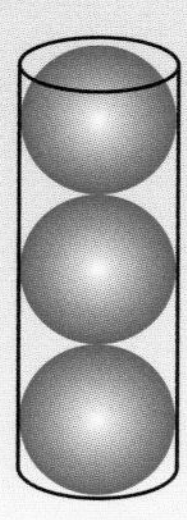

What is the internal volume of the container?

3 The diameter of the cylinder is 8.2 cm and the height is 24.6 cm. — This gets 1 mark for method.

The internal volume $= \pi \times 4.12 \times 24.6$

$= 1299.13\ldots$

$= 1300\ \text{cm}^3$ to 2 sf

This gets 1 mark for method, 1 mark for accuracy and 1 mark for sensible rounding.

Total: 4 marks

4

Functional Maths
Organising a harvest

Farmers have to do mathematical calculations almost every day. For example, an arable farmer may need to know how much seed to buy, how much water is required to irrigate the field each day, how much wheat they expect to grow and how much storage space they need to store wheat once it is harvested.

Farming can be filled with uncertainties, including changes in weather, crop disease and changes in consumption. It is therefore important that farmers correctly calculate variables that are within their control, to minimise the impact of changes that are outside their control.

Important information about wheat crops (yield data)

- A 1 kg bag of seeds holds 26 500 seeds
- A 1 kg bag of seeds costs 50p
- I want to plant 60 bags of seed
- I need to plant 100 seeds in each square metre (m^2) of field
- I need to irrigate each square metre of the field with 5 litres of water each day
- I expect to harvest 0.7 kg of wheat from each square metre of the field
- Every cubic metre (m^3) of storage will hold 800 kg of wheat

Grain storage

Wheat is stored in large containers called silos. These are usually big cylinders but can also be various other shapes.

Your task

Rufus, a crop farmer, is going to grow his first field of wheat next summer. Using all the information that he has gathered, help him to plan for his wheat crop. You should consider:

- the size of the field that he will require
- how much seed he will need
- how much water he will need to irrigate the crops, per day, and how it will be stored
- how he will store the seeds and wheat
- how much profit he could make if grain is sold at £92.25 per tonne.

Getting started

Think about these points to help you create your plan.

- What different shapes and sizes of field could the crops be grown in?
- If Rufus needs a container to hold one day's irrigation water, what size cylinder would he need? How would this change if he chose a cuboid? What other shapes and sizes of container could he use?
- What shapes and sizes could the silos be?

Handy hints

Remember: 1000 litres = 1 m^3
1000 kg = 1 tonne

Why this chapter matters

How can you find the height of a mountain?

How do you draw an accurate map?

How can computers take an image and make it rotate so that you can view it from different directions?

How were sailors able to navigate before GPS? And how does GPS work? How can music be produced electronically?

The answer is by looking at the angles and sides of triangles and the connections between them. This important branch of mathematics is called trigonometry and has a huge range of applications in science, engineering, electronics and everyday life. This chapter gives a brief introduction to the subject and introduces some important new mathematical tools.

The first major work of trigonometry that still survives is called the *Almagest*. It was a written by an astronomer called Ptolemy who lived in Alexandria, Egypt, over 1800 years ago. In it, there are tables of numbers, called 'trigonometric ratios', used in making calculations about the positions of stars and planets. Trigonometry also helped Ptolemy to make a map of the known world at that time. Today we no longer need to look up tables of values of trigonometric ratios because they are programmed into calculators and computers.

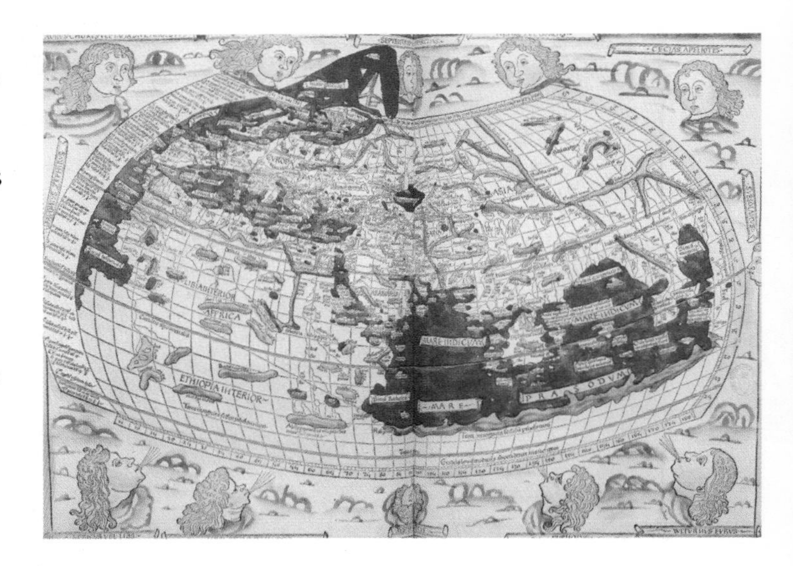

Nearer our own time, the 19th century French mathematician Jean Fourier showed how all musical sounds can be broken down into a combination of pure tones that can be described by trigonometry. His work is what makes it possible to imitate the sound of any instrument electronically – a clear example of how mathematics has a profound influence on other areas of life, including art.

Chapter 5

Geometry: Pythagoras' theorem and trigonometry

The grades given in this chapter are target grades.

1 Pythagoras' theorem

2 Finding a shorter side

3 Applying Pythagoras' theorem to real-life situations

4 Pythagoras' theorem in three dimensions

5 Trigonometric ratios

6 Calculating angles

7 Using the sine and cosine functions

8 Using the tangent function

9 Which ratio to use

10 Solving problems using trigonometry 1

11 Solving problems using trigonometry 2

This chapter will show you ...

- **C** how to use Pythagoras' theorem in right-angled triangles
- **C** how to solve problems using Pythagoras' theorem
- **B** how to use Pythagoras' theorem in three dimensions
- **B** how to use trigonometric ratios in right-angled triangles
- **A** how to use trigonometry to solve problems

Visual overview

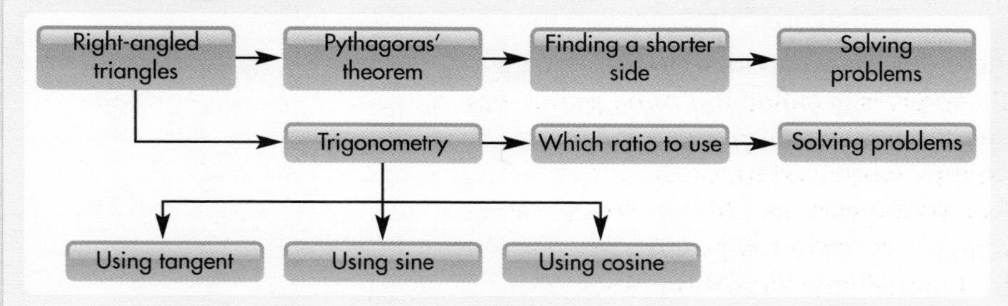

What you should already know

- how to find the square and square root of a number **(KS3 level 5, GCSE grade F)**
- how to round numbers to a suitable degree of accuracy **(KS3 level 6, GCSE grade E)**

Quick check

Use your calculator to evaluate the following, giving your answers to one decimal place.

1 2.3^2

2 15.7^2

3 0.78^2

4 $\sqrt{8}$

5 $\sqrt{260}$

6 $\sqrt{0.5}$

5.1 Pythagoras' theorem

This section will show you how to:

- calculate the length of the hypotenuse in a right-angled triangle

Key words
hypotenuse
Pythagoras' theorem

Pythagoras, who was a philosopher as well as a mathematician, was born in 580BC, on the island of Samos in Greece. He later moved to Crotona (Italy), where he established the Pythagorean Brotherhood, which was a secret society devoted to politics, mathematics and astronomy. It is said that when he discovered his famous theorem, he was so full of joy that he showed his gratitude to the gods by sacrificing a hundred oxen.

Consider squares being drawn on each side of a right-angled triangle, with sides 3 cm, 4 cm and 5 cm.

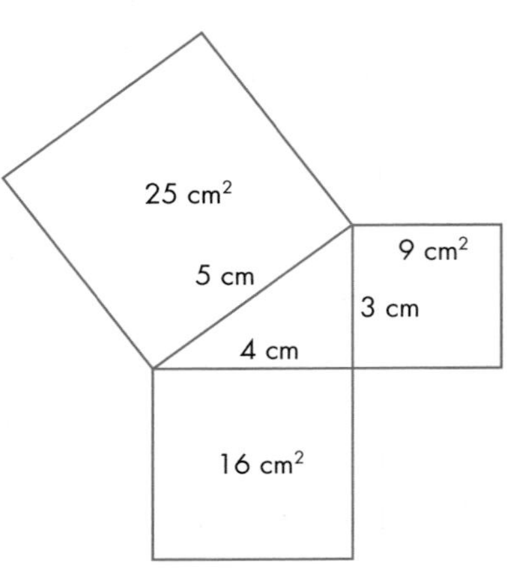

The longest side is called the **hypotenuse** and is always opposite the right angle.

Pythagoras' theorem can then be stated as follows:

For any right-angled triangle, the area of the square drawn on the hypotenuse is equal to the sum of the areas of the squares drawn on the other two sides.

The form in which most of your parents would have learnt the theorem when they were at school – and which is still in use today – is as follows:

In any right-angled triangle, the square of the hypotenuse is equal to the sum of the squares of the other two sides.

Pythagoras' theorem is more usually written as a formula:

$$c^2 = a^2 + b^2$$

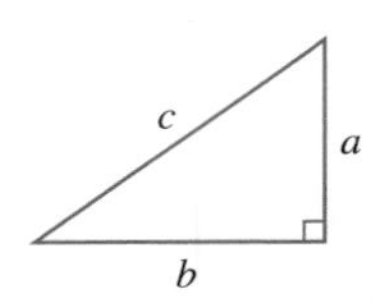

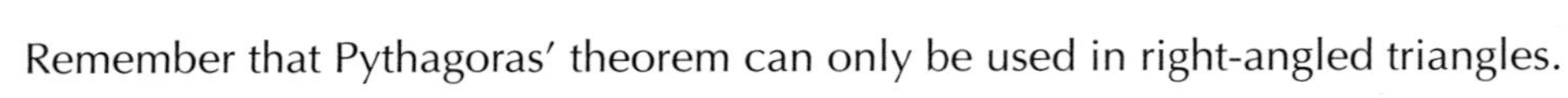
Remember that Pythagoras' theorem can only be used in right-angled triangles.

Finding the hypotenuse

EXAMPLE 1

Find the length of the hypotenuse, marked x on the diagram.

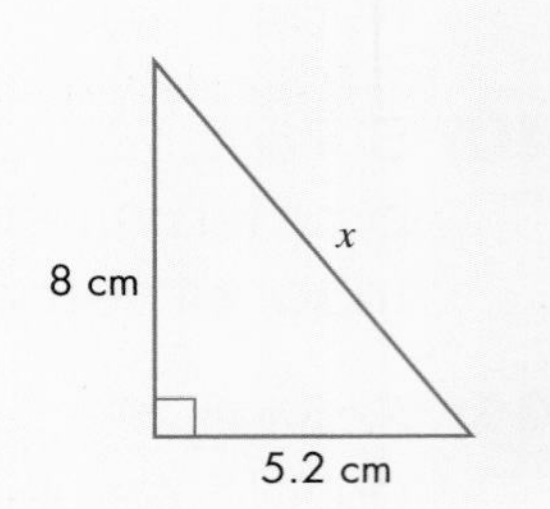

Using Pythagoras' theorem gives:

$x^2 = 8^2 + 5.2^2\ \text{cm}^2$
$= 64 + 27.04\ \text{cm}^2$
$= 91.04\ \text{cm}^2$

So $x = \sqrt{91.04} = 9.5$ cm (1 decimal place)

EXERCISE 5A

For each of the following triangles, calculate the length of the hypotenuse, x, giving your answers to 1 decimal place.

> **HINTS AND TIPS**
>
> In these examples you are finding the hypotenuse. The squares of the two short sides are added in every case.

C

1

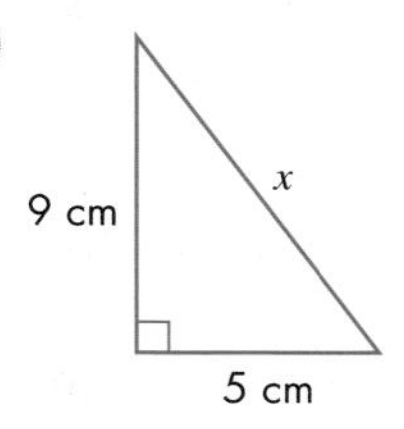

2

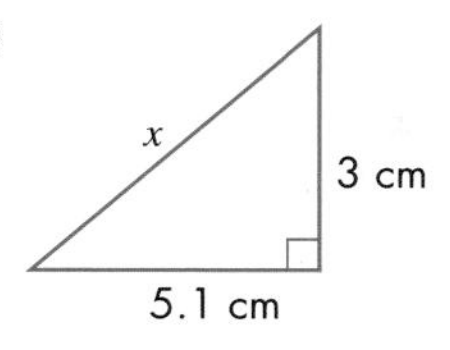

3

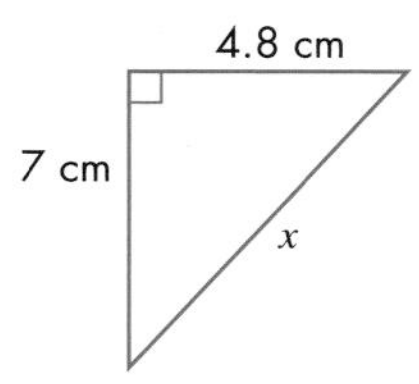

4

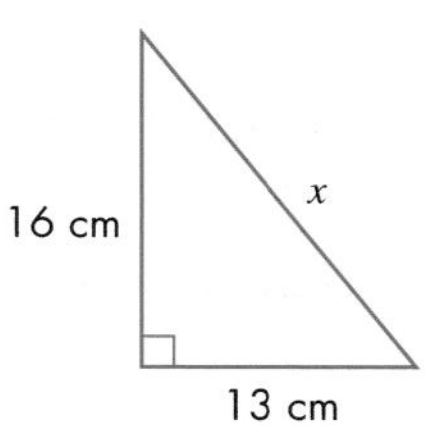

5

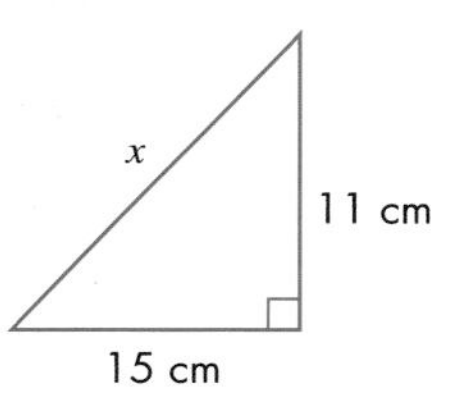

6

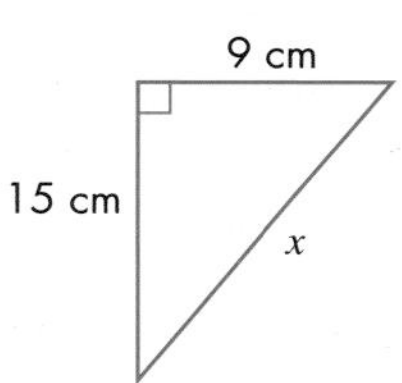

7

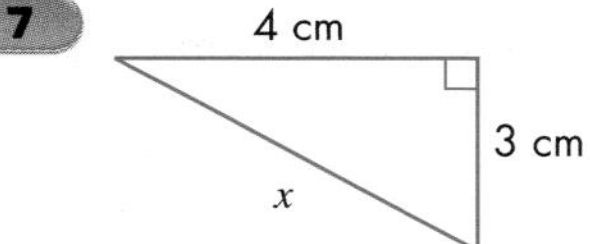

8

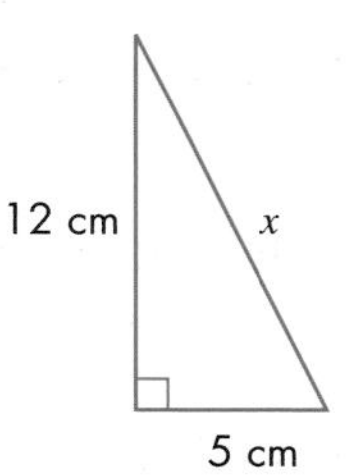

9

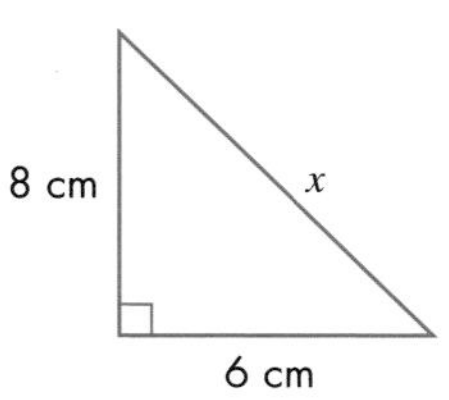

PS 10 How does this diagram show that Pythagoras' theorem is true?

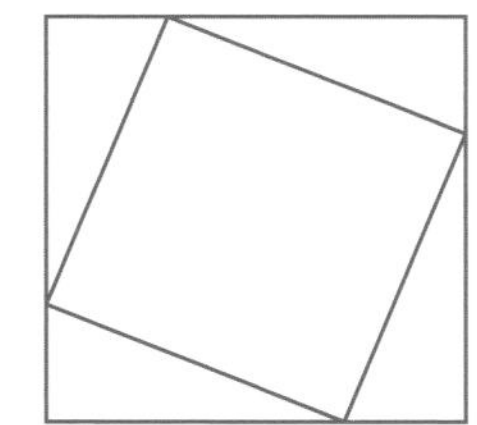

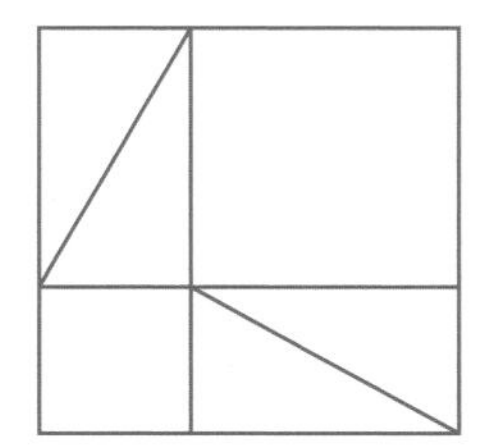

5.2 Finding a shorter side

This section will show you how to:

- calculate the length of a shorter side in a right-angled triangle

Key words
Pythagoras' theorem

By rearranging the formula for **Pythagoras' theorem**, the length of one of the shorter sides can easily be calculated.

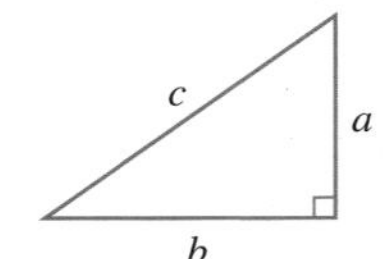

$$c^2 = a^2 + b^2$$

So, $a^2 = c^2 - b^2$ or $b^2 = c^2 - a^2$

EXAMPLE 2

Find the length x.

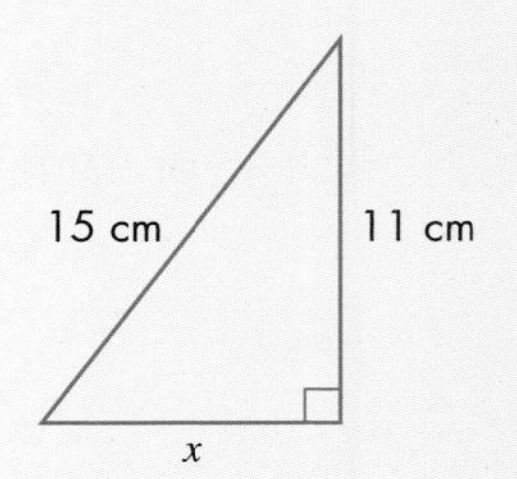

x is one of the shorter sides.

So using Pythagoras' theorem gives:

$$\begin{aligned} x^2 &= 15^2 - 11^2 \text{ cm}^2 \\ &= 225 - 121 \text{ cm}^2 \\ &= 104 \text{ cm}^2 \end{aligned}$$

So $x = \sqrt{104} = 10.2$ cm (1 decimal place)

EXERCISE 5B

1 For each of the following triangles, calculate the length x, giving your answers to 1 decimal place.

a

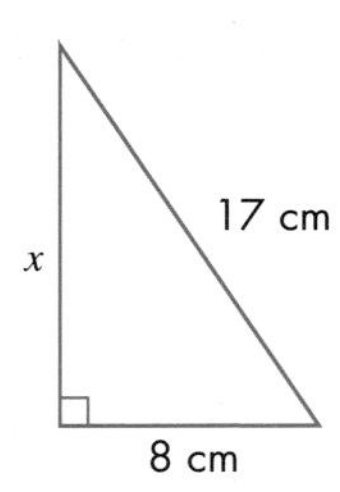

b

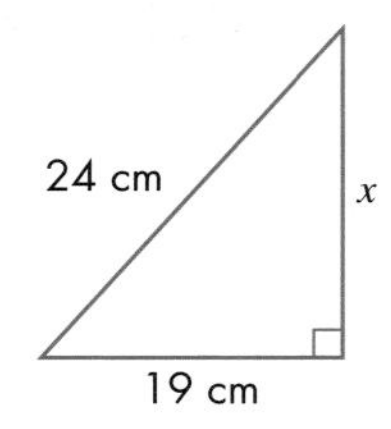

c

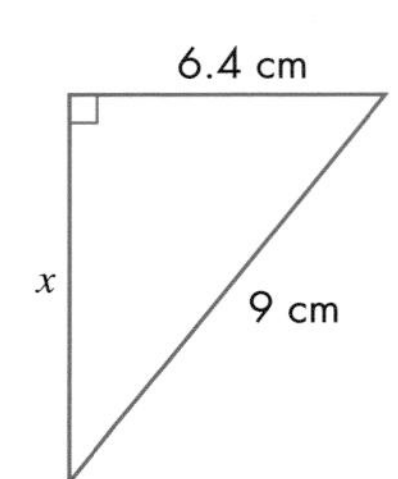

d

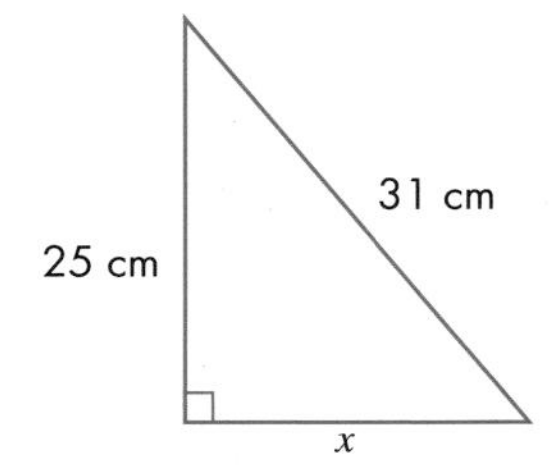

HINTS AND TIPS

In these examples you are finding a short side. The square of the other short side is subtracted from the square of the hypotenuse in every case.

C

2 For each of the following triangles, calculate the length x, giving your answers to 1 decimal place.

a

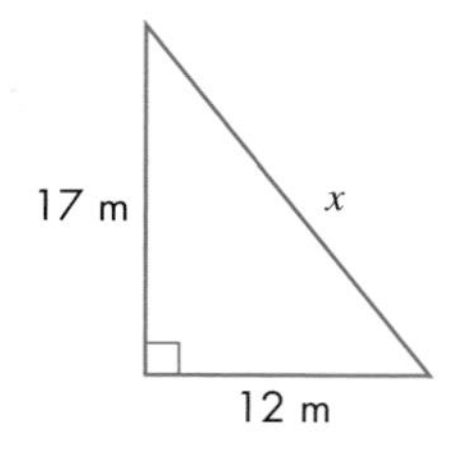

b

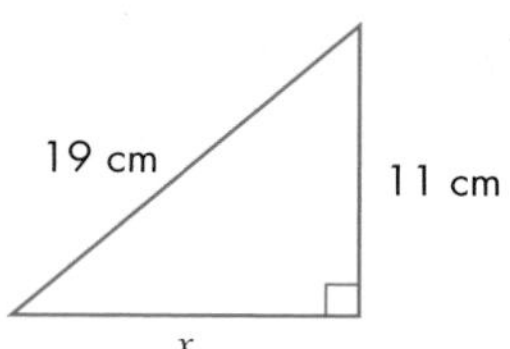

c

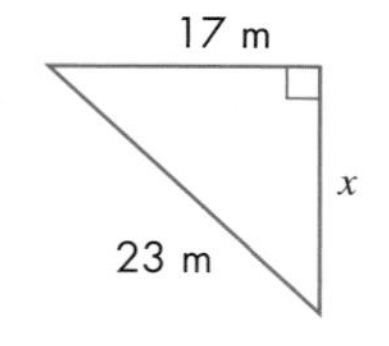

d

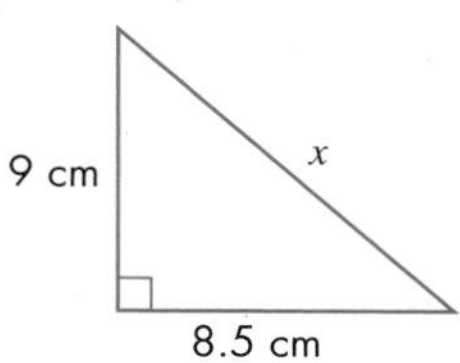

HINTS AND TIPS

These examples are a mixture. Make sure you combine the squares of the sides correctly. If you get it wrong in an examination, you get no marks.

3 For each of the following triangles, find the length marked x.

a

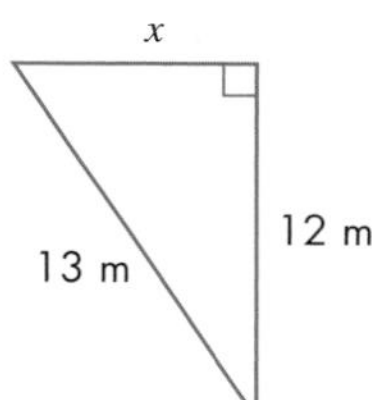

b

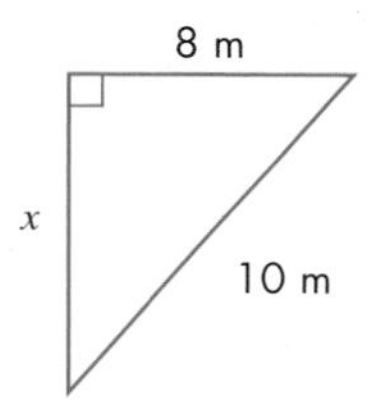

c

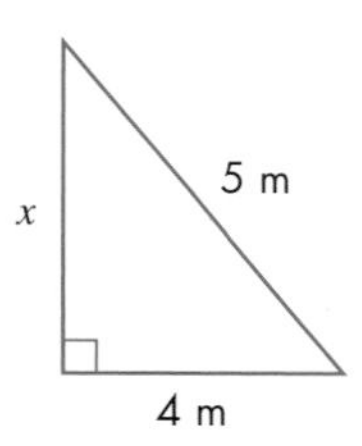

d

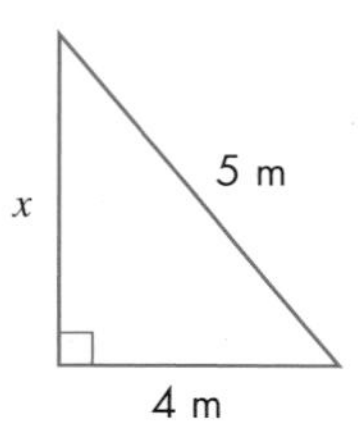

PS 4 In question **3** you found sets of three whole numbers which satisfy $a^2 + b^2 = c^2$.

Can you find any more?

5 Calculate the value of x.

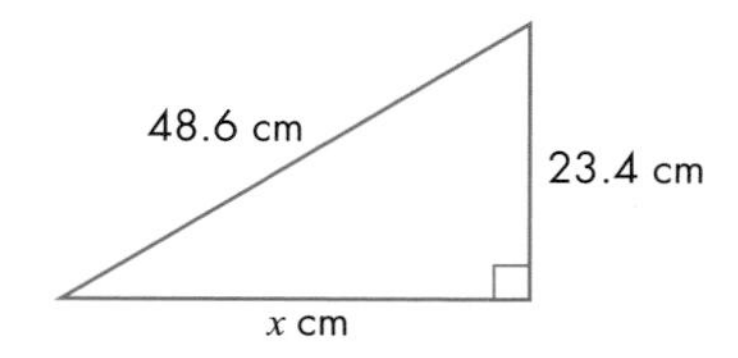

5.3 Applying Pythagoras' theorem to real-life situations

This section will show you how to:
- solve problems using Pythagoras' theorem

Key words
isosceles triangle
Pythagoras' theorem

Pythagoras' theorem can be used to solve certain practical problems. When a problem involves two lengths only, follow these steps.

- Draw a diagram for the problem that includes a right-angled triangle.
- Look at the diagram and decide which side has to be found: the hypotenuse or one of the shorter sides. Label the unknown side x.
- If it's the hypotenuse, then square both numbers, add the squares and take the square root of the sum.
- If it's one of the shorter sides, then square both numbers, subtract the squares and take the square root of the difference.
- Finally, round the answer to a suitable degree of accuracy.

EXAMPLE 3

A plane leaves Manchester airport heading due east. It flies 160 km before turning due north. It then flies a further 280 km and lands. What is the distance of the return flight if the plane flies straight back to Manchester airport?

First, sketch the situation.

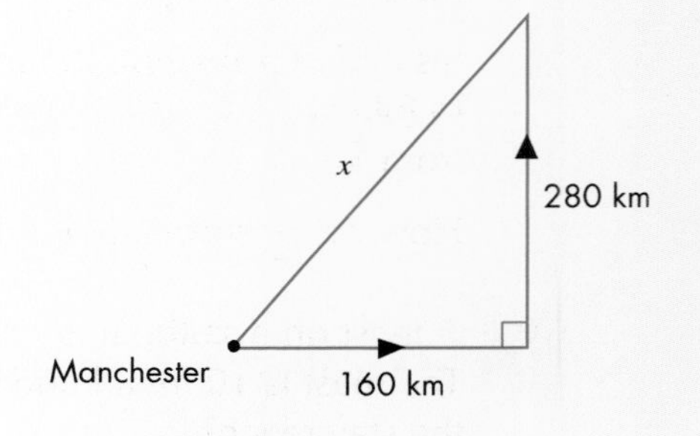

Using Pythagoras' theorem gives:

$$x^2 = 160^2 + 280^2 \text{ km}^2$$
$$= 25\,600 + 78\,400 \text{ km}^2$$
$$= 104\,000 \text{ km}^2$$

So $x = \sqrt{104104000} = 322$ km
(3 significant figures)

Remember the following tips when solving problems.

- Always sketch the right-angled triangle you need. Sometimes, the triangle is already drawn for you but some problems involve other lines and triangles that may confuse you. So identify which right-angled triangle you need and sketch it separately.
- Label the triangle with necessary information, such as the length of its sides, taken from the question. Label the unknown side x.
- Set out your solution as in Example 3. Avoid short cuts, since they often cause errors. You gain marks in your examination for clearly showing how you are applying Pythagoras' theorem to the problem.
- Round your answer to a suitable degree of accuracy.

EXERCISE 5C

FM 1 A ladder, 12 m long, leans against a wall. The ladder reaches 10 m up the wall. The ladder is safe if the foot of the ladder is about 2.5 m away from the wall. Is this ladder safe?

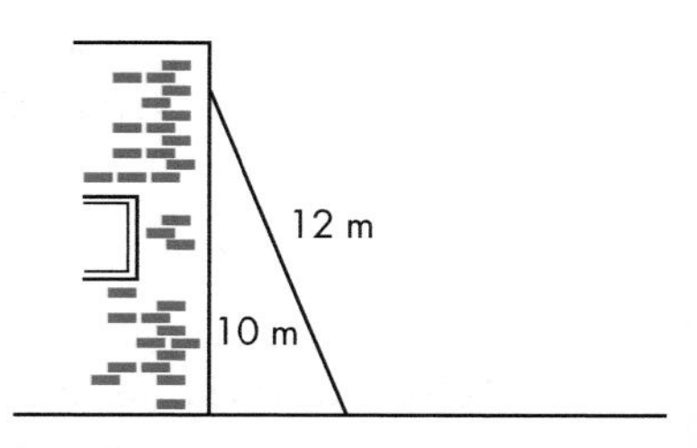

2 A model football pitch is 2 m long and 0.5 m wide. How long is the diagonal?

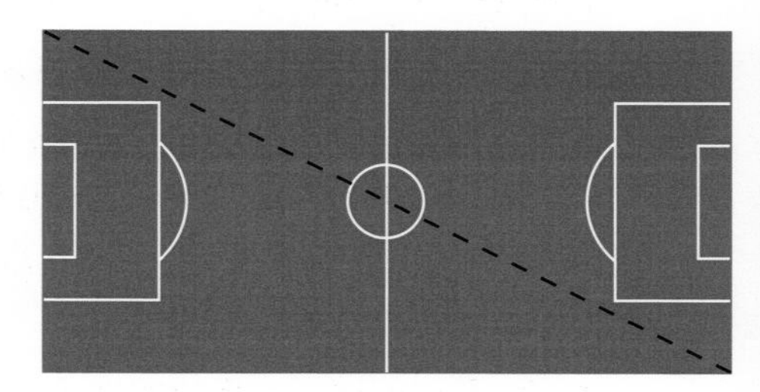

PS 3 How long is the diagonal of a square with a side of 8 m?

AU 4 A ship going from a port to a lighthouse steams 15 km east and 12 km north. The journey takes 1 hour. How much time would be saved by travelling directly to the lighthouse in a straight line?

FM 5 Some pedestrians want to get from point X on one road to point Y on another. The two roads meet at right angles.

Instead of following the roads, they decide to follow a footpath which goes directly from X to Y.

How much shorter is this route?

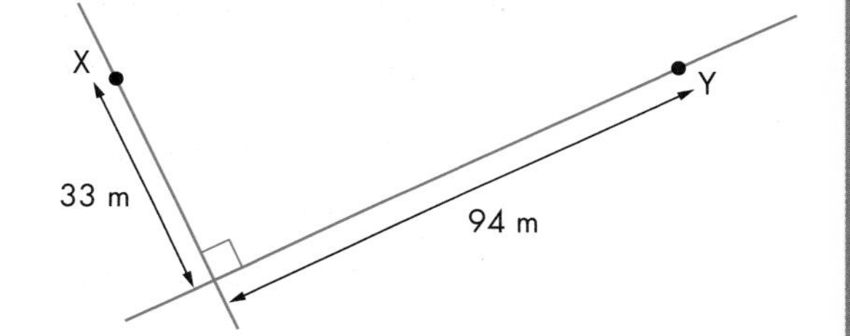

6 A mast on a sailboat is strengthened by a wire (called a stay), as shown on the diagram. The mast is 10 m tall and the stay is 11 m long. How far from the base of the mast does the stay reach?

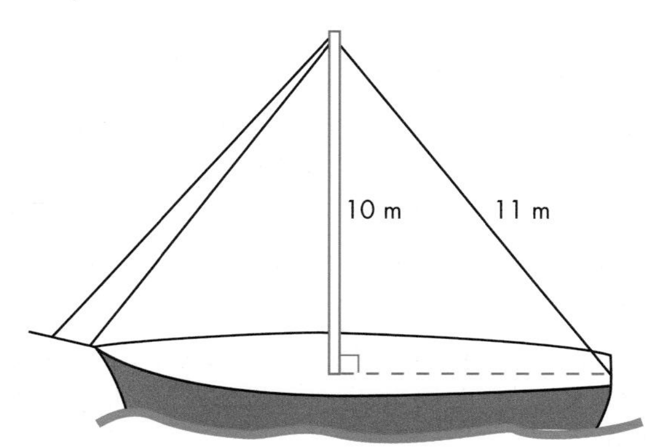

C

C

FM 7 A ladder, 4 m long, is put up against a wall.

a How far up the wall will it reach when the foot of the ladder is 1 m away from the wall?

b When it reaches 3.6 m up the wall, how far is the foot of the ladder away from the wall?

AU 8 A pole, 8 m high, is supported by metal wires, each 8.6 m long, attached to the top of the pole. How far from the foot of the pole are the wires fixed to the ground?

AU 9 A and B are two points on a coordinate grid. They have coordinates (13, 6) and (1, 1). How long is the line that joins them?

FM 10 The regulation for safe use of ladders states that: *the foot of a 5 m ladder must be placed between 1.2 m and 1.3 m from the foot of the wall.*

a What is the maximum height the ladder can safely reach up the wall?

b What is the minimum height the ladder can safely reach up the wall?

AU 11 Is the triangle with sides 7 cm, 24 cm and 25 cm a right-angled triangle? Give a reason for your answer.

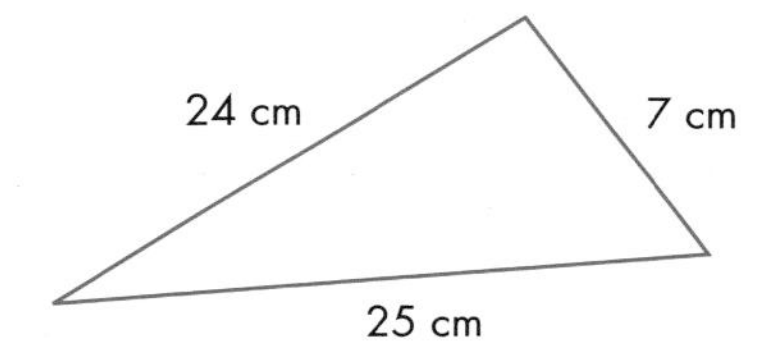

PS 12 A 4 m long ladder is leaning against a wall. The foot of the ladder is 1 m from the wall. The foot of the ladder is not securely held and slips 20 cm further away from the wall.

How far does the top of the ladder move down the wall?

PS 13 The diagonal of a rectangle is 10 cm. What can you say about the perimeter of the rectangle?

Pythagoras' theorem and isosceles triangles

This section shows you how to to use Pythagoras' theorem in isosceles triangles.

Every **isosceles triangle** has a line of symmetry that divides the triangle into two congruent right-angled triangles. So when you are faced with a problem involving an isosceles triangle, be aware that you are quite likely to have to split that triangle down the middle to create a right-angled triangle which will help you to solve the problem.

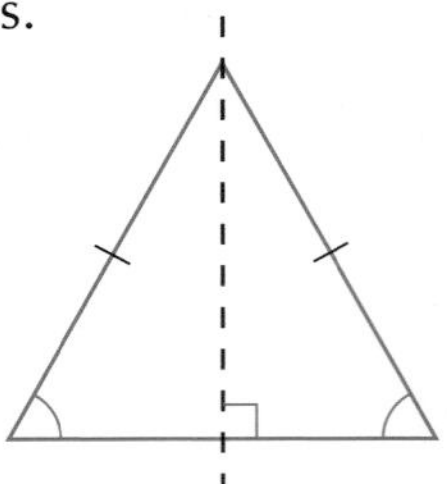

EXAMPLE 4

Calculate the area of this triangle.

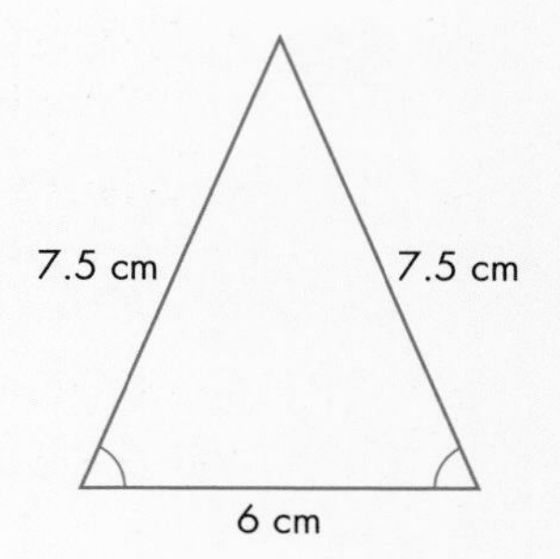

It is an isosceles triangle and you need to calculate its height to find its area.

First split the triangle into two right-angled triangles to find its height.

Let the height be x.

Then, using Pythagoras' theorem,

$$x^2 = 7.5^2 - 3^2 \text{ cm}^2$$
$$= 56.25 - 9 \text{ cm}^2$$
$$= 47.25 \text{ cm}^2$$

So $x = \sqrt{47.25}$ cm

$x = 6.87$ cm

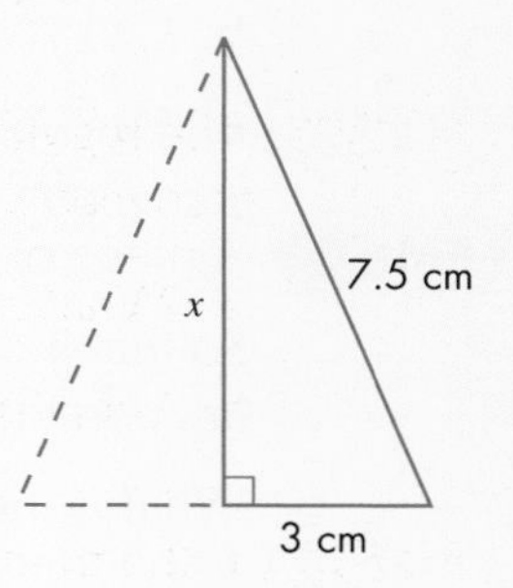

Keep the accurate figure in the calculator memory.

The area of the triangle is $\frac{1}{2} \times 6 \times 6.87 \text{ cm}^2$ (from the calculator memory), which is 20.6 cm^2 (1 decimal place).

EXERCISE 5D

1 Calculate the areas of these isosceles triangles.

a

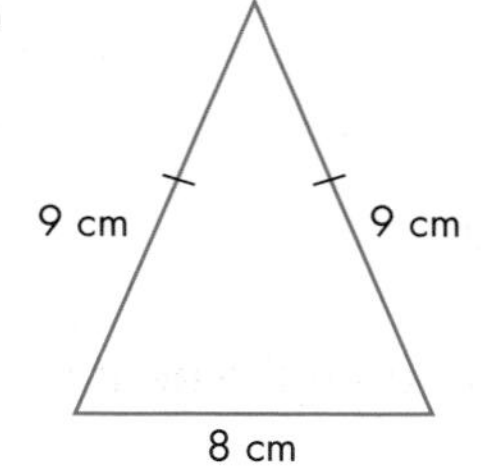

b

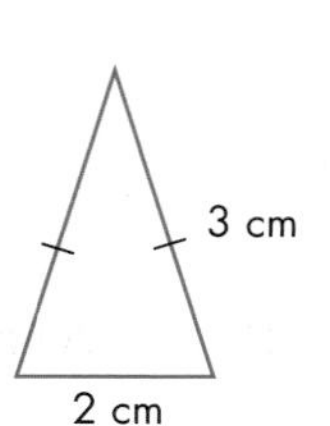

c

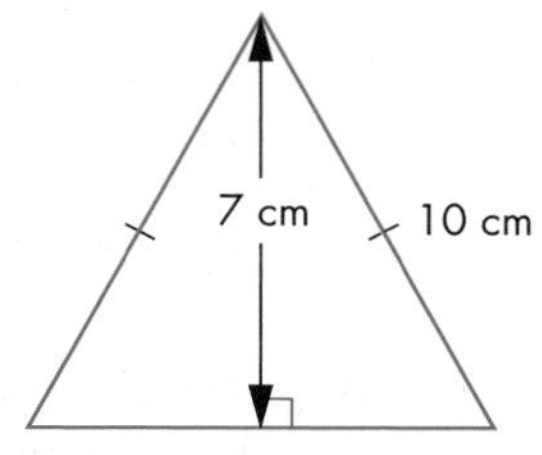

2 Calculate the area of an isosceles triangle whose sides are 8 cm, 8 cm and 6 cm.

PS 3 Calculate the area of an equilateral triangle of side 6 cm.

PS 4 An isosceles triangle has sides of 5 cm and 6 cm.

a Sketch the two different isosceles triangles that fit this data.

b Which of the two triangles has the greater area?

5 **a** Sketch a regular hexagon, showing all its lines of symmetry.

b Calculate the area of the hexagon if its side is 8 cm.

B

B

PS 6 Calculate the area of a hexagon of side 10 cm.

PS 7 These isosceles triangles have the same perimeter.

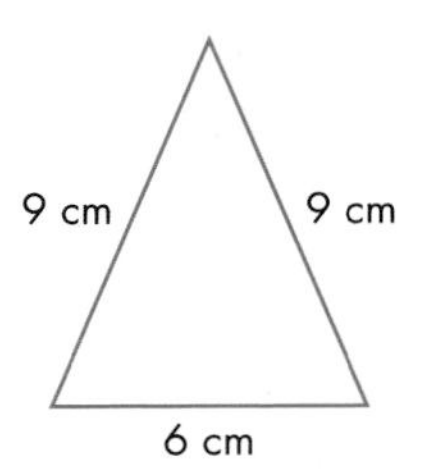

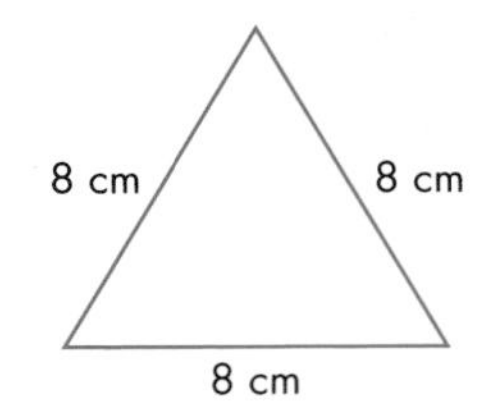

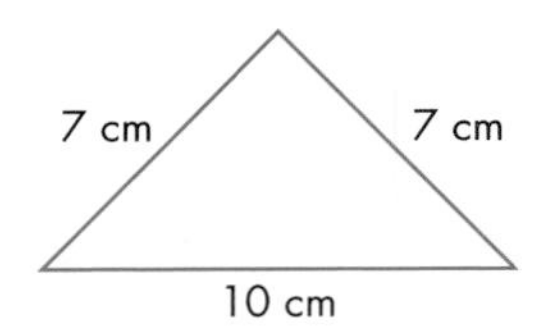

a Do the three triangles have the same area?

b Can you find an isosceles triangle with the same perimeter but a larger area?

c Can you generalise your findings?

FM 8 A piece of land is in the shape of an isosceles triangle with sides 6.5 m, 6.5 m and 7.4 m.

So that it can be sown with the correct quantity of grass seed to make a lawn, you have been asked to calculate the area.

What is the area of the land?

9 The diagram shows an isosceles triangle ABC.

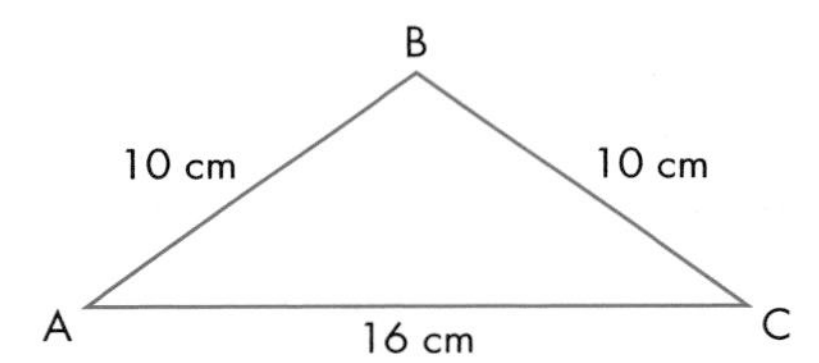

Calculate the area of triangle ABC.

State the units of your answer.

10 Calculate the lengths marked x in these isosceles triangles.

a

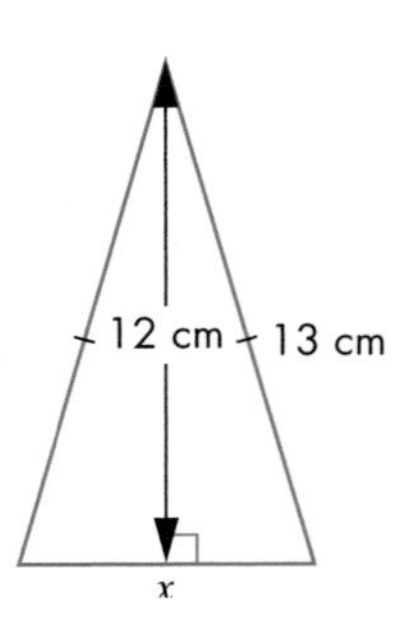

b

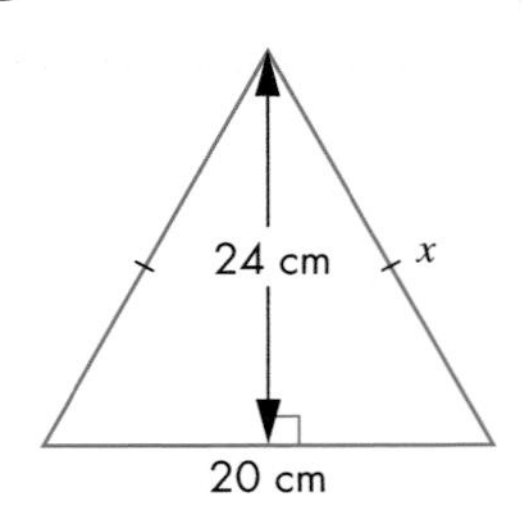

AU **c**

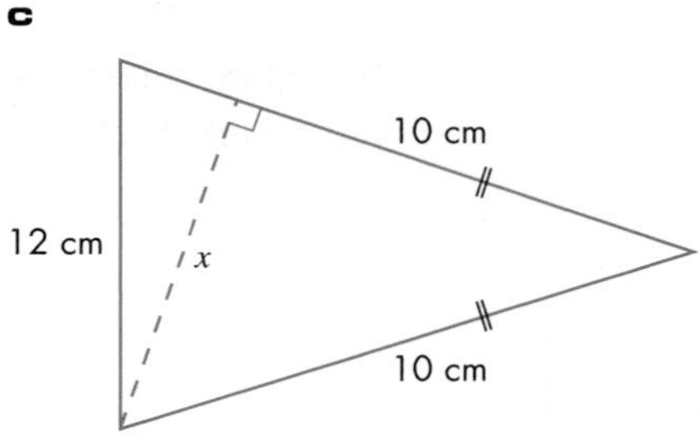

HINTS AND TIPS

Find the area first.

5.4 Pythagoras' theorem in three dimensions

This section will show you how to:

- use Pythagoras' theorem in problems involving three dimensions

Key words
3D
Pythagoras' theorem

This section shows you how to solve problems in **3D** using **Pythagoras' theorem**.

In your GCSE examinations, there may be questions which involve applying Pythagoras' theorem in 3D situations. Such questions are usually accompanied by clearly-labelled diagrams, which will help you to identify the lengths needed for your solutions.

You deal with these 3D problems in exactly the same way as 2D problems.

- Identify the right-angled triangle you need.
- Redraw this triangle and label it with the given lengths and the length to be found, usually x or y.
- From your diagram, decide whether it is the hypotenuse or one of the shorter sides which has to be found.
- Solve the problem, rounding to a suitable degree of accuracy.

EXAMPLE 5

What is the longest piece of straight wire that can be stored in this box measuring 30 cm by 15 cm by 20 cm?

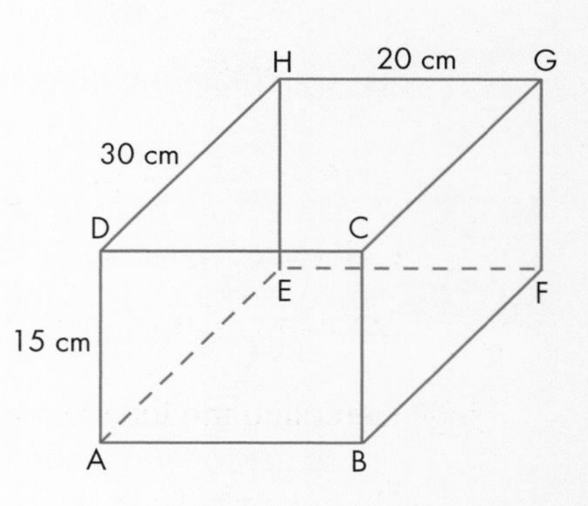

The longest distance across this box is any one of the diagonals AG, DF, CE or HB.

Let us take AG.

First, identify a right-angled triangle containing AG and draw it.

This gives a triangle AFG, which contains two lengths you do not know, AG and AF.

Let AG = x and AF = y

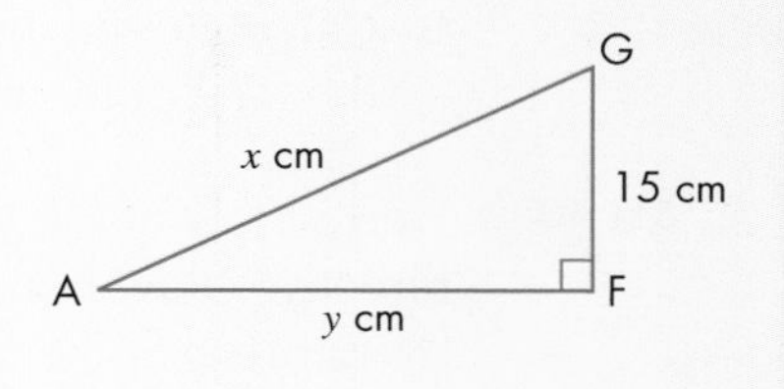

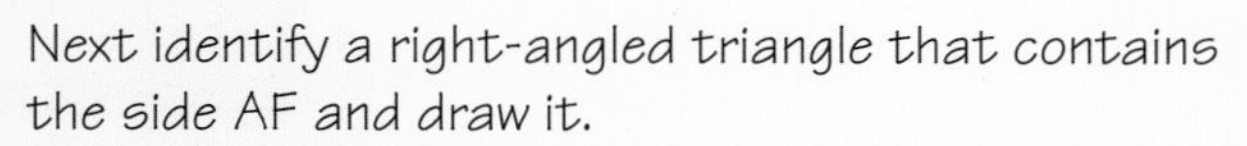

Next identify a right-angled triangle that contains the side AF and draw it.

This gives a triangle ABF. You can now find AF.

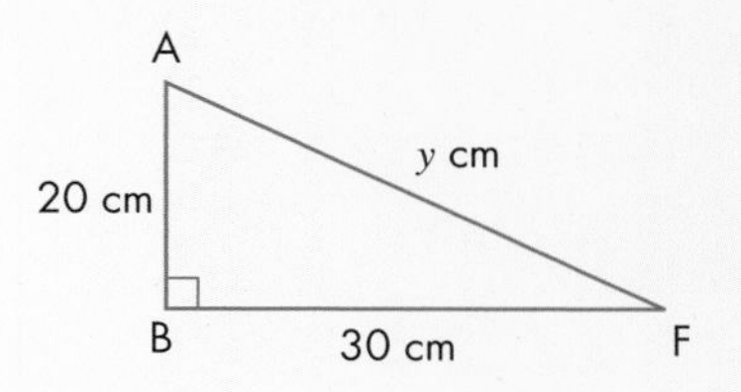

By Pythagoras' theorem,

$y^2 = 30^2 + 20^2 \text{ cm}^2$

$y^2 = 1300 \text{ cm}^2$ (there is no need to find y)

EXAMPLE 5 continued

Now find AG using triangle AFG.

By Pythagoras' theorem,

$x^2 = y^2 + 15^2$ cm^2

$x^2 = 1300 + 225 = 1525$ cm^2

So $x = 39.1$ cm (1 decimal place)

So, the longest straight wire that can be stored in the box is 39.1 cm.

Note that in any cuboid with sides a, b and c, the length of a diagonal is given by:

$\sqrt{(a^2 + b^2 + c^2)}$

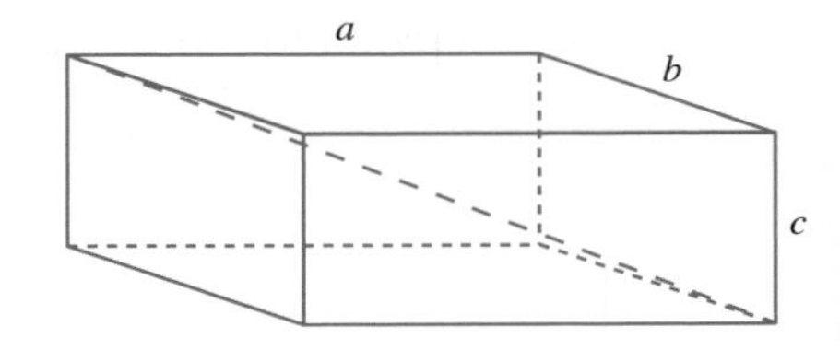

EXERCISE 5E

A

1 A box measures 8 cm by 12 cm by 5 cm.

a Calculate the lengths of the following.

i AC **ii** BG **iii** BE

b Calculate the diagonal distance BH.

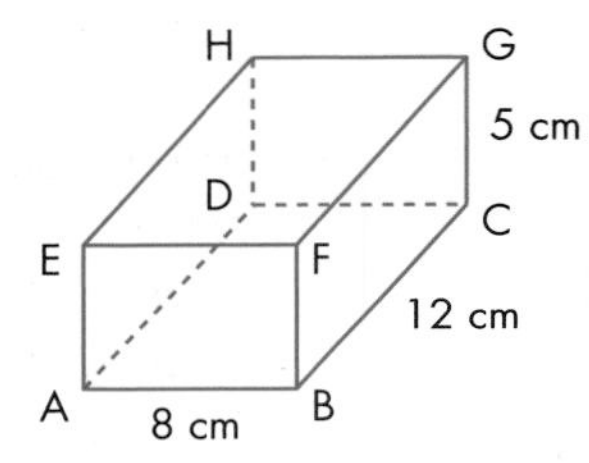

AU 2 A garage is 5 m long, 3 m wide and 3 m high. Can a 7 m long pole be stored in it?

AU 3 Spike, a spider, is at the corner S of the wedge shown in the diagram. Fred, a fly, is at the corner F of the same wedge.

a Calculate the shortest distance Spike would have to travel to get to Fred if she used the edges of the wedge.

b Calculate the distance Spike would have to travel across the face of the wedge to get directly to Fred.

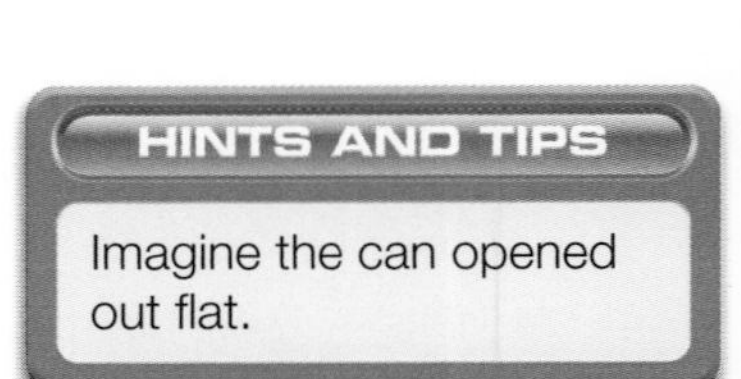

AU 4 Fred is now at the top of a baked-beans can and Spike is directly below him on the base of the can. To catch Fred by surprise, Spike takes a diagonal route round the can. How far does Spike travel?

HINTS AND TIPS

Imagine the can opened out flat.

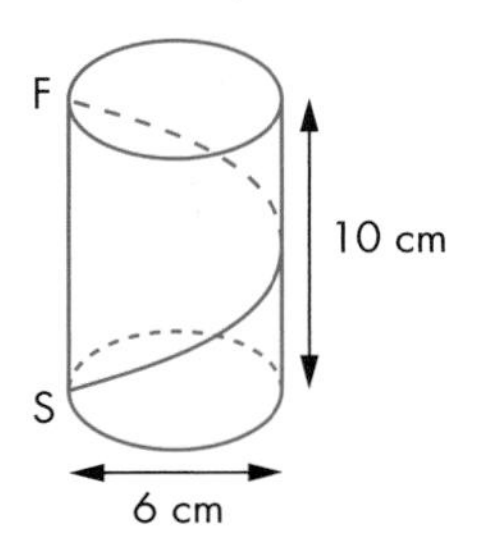

FM 5 A corridor is 3 m wide and turns through a right angle, as in the diagram.

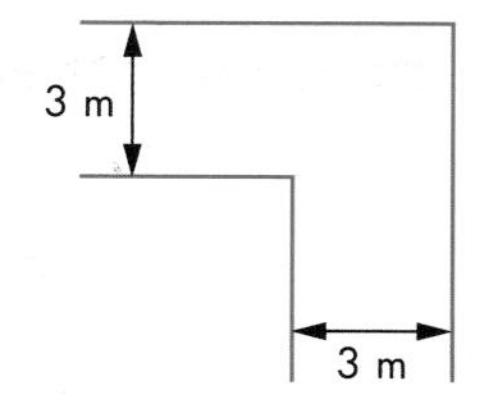

a What is the longest pole that can be carried along the corridor horizontally?

b If the corridor is 3 m high, what is the longest pole that can be carried along in any direction?

PS 6 If each side of a cube is 10 cm long, how far will it be from one corner of the cube to the opposite one?

AU 7 A pyramid has a square base of side 20 cm and each sloping edge is 25 cm long.

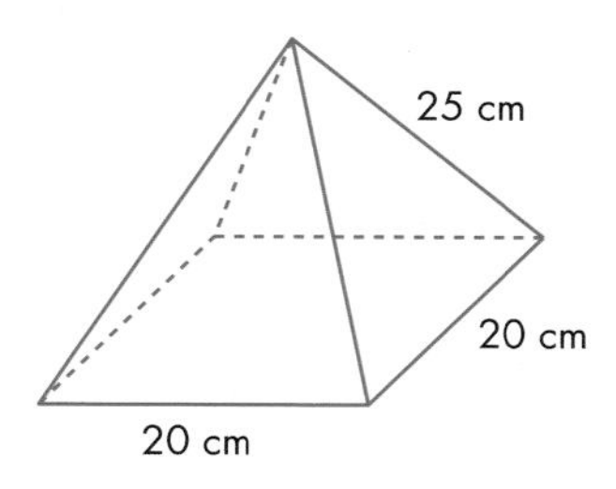

How high is the pyramid?

AU 8 The diagram shows a square-based pyramid with base length 8 cm and sloping edges 9 cm. M is the midpoint of the side AB, X is the midpoint of the base, and E is directly above X.

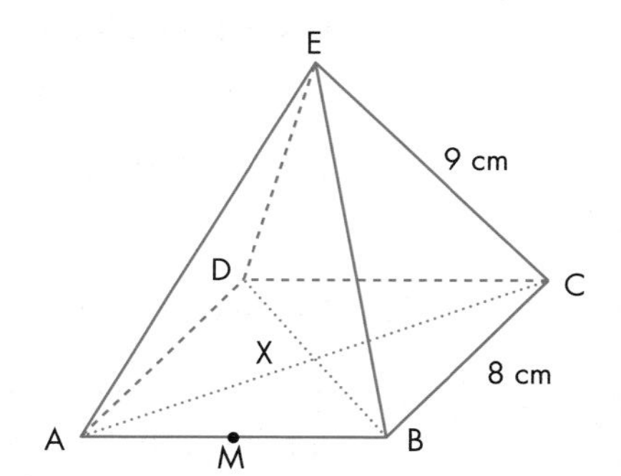

a Calculate the length of the diagonal AC.

b Calculate EX, the height of the pyramid.

c Using triangle ABE, calculate the length EM.

9 The diagram shows a cuboid with sides of 40 cm, 30 cm and 22.5 cm. M is the midpoint of the side FG. Calculate (or write down) these lengths, giving your answers to 3 significant figures if necessary.

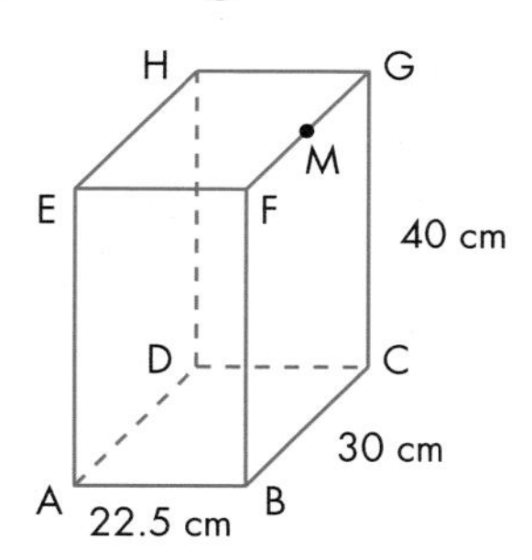

a AH **b** AG **c** AM **d** HM

5.5 Trigonometric ratios

This section will show you how to:

- use the three trigonometric ratios

Key words
adjacent side
cosine
hypotenuse
opposite side
sine
tangent
trigonometry

Trigonometry is concerned with the calculation of sides and angles in triangles, and involves the use of three important ratios: **sine**, **cosine** and **tangent**. These ratios are defined in terms of the sides of a right-angled triangle and an angle. The angle is often written as θ.

In a right-angled triangle:

- the side opposite the right angle is called the **hypotenuse** and is the longest side
- the side opposite the angle θ is called the **opposite side**
- the other side next to both the right angle and the angle θ is called the **adjacent side**.

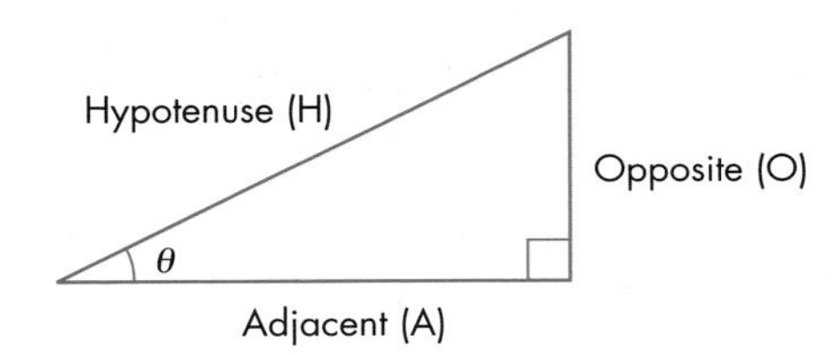

The sine, cosine and tangent ratios for θ are defined as:

$$\text{sine } \theta = \frac{\text{Opposite}}{\text{Hypotenuse}} \qquad \text{cosine } \theta = \frac{\text{Adjacent}}{\text{Hypotenuse}} \qquad \text{tangent } \theta = \frac{\text{Opposite}}{\text{Adjacent}}$$

These ratios are usually abbreviated as:

$$\sin \theta = \frac{O}{H} \qquad \cos \theta = \frac{A}{H} \qquad \tan \theta = \frac{O}{A}$$

These abbreviated forms are also used on calculator keys.

Memorising these formulae may be helped by a mnemonic such as,

Silly **O**ld **H**itler **C**ouldn't **A**dvance **H**is **T**roops **O**ver **A**frica

in which the first letter of each word is taken in order to give

$$S = \frac{O}{H} \qquad C = \frac{A}{H} \qquad T = \frac{O}{A}$$

Using your calculator

You will need to use a calculator to find trigonometric ratios.

Different calculators work in different ways, so make sure you know how to use your model.

Angles are not always measured in degrees. Sometimes radians or grads are used instead. You do not need to learn about those in your GCSE course. Calculators can be set to operate in any of these three units, so make sure your calculator is operating in degrees.

Use your calculator to find the sine of 60 degrees.

You will probably press the keys **sin** **6** **0** **=** in that order, but it might be different on your calculator.

The answer should be 0.8660… or $\frac{\sqrt{3}}{2}$. If it is the latter, make sure you can convert that to the decimal form.

3 cos 57° is a shorthand way of writing 3 × cos 57°.

On most calculators you do not need to use the × button and you can just press the keys in the way it is written: **3** **cos** **5** **7** **=**

Check to see whether your calculator works this way.

The answer should be 1.63.

EXAMPLE 6

Find 5.6 sin 30°.

This means 5.6 × sine of 30 degrees.

Remember that you may not need to press the × button.

5.6 sin 30° = 2.8

EXERCISE 5F

1 Find these values, rounding off your answers to 3 significant figures.

a sin 43° **b** sin 56° **c** sin 67.2° **d** sin 90°

e sin 45° **f** sin 20° **g** sin 22° **h** sin 0°

2 Find these values, rounding off your answers to 3 significant figures.

a cos 43° **b** cos 56° **c** cos 67.2° **d** cos 90°

e cos 45° **f** cos 20° **g** cos 22° **h** cos 0°

3 From your answers to questions **1** and **2**, what angle has the same value for sine and cosine?

B

B

4 **a** **i** What is sin 35°? **ii** What is cos 55°?

b **i** What is sin 12°? **ii** What is cos 78°?

c **i** What is cos 67°? **ii** What is sin 23°?

d What connects the values in parts **a**, **b** and **c**?

e Copy and complete these sentences.

i sin 15° is the same as cos …

ii cos 82° is the same as sin …

iii sin x is the same as cos …

5 Use your calculator to work out the values of the following.

a tan 43° **b** tan 56° **c** tan 67.2° **d** tan 90°

e tan 45° **f** tan 20° **g** tan 22° **h** tan 0°

6 Use your calculator to work out the values of the following.

a sin 73° **b** cos 26° **c** tan 65.2° **d** sin 88°

e cos 35° **f** tan 30° **g** sin 28° **h** cos 5°

7 What is so different about tan compared with both sin and cos?

8 Use your calculator to work out the values of the following.

a 5 sin 65° **b** 6 cos 42° **c** 6 sin 90° **d** 5 sin 0°

9 Use your calculator to work out the values of the following.

a 5 tan 65° **b** 6 tan 42° **c** 6 tan 90° **d** 5 tan 0°

10 Use your calculator to work out the values of the following.

a 4 sin 63° **b** 7 tan 52° **c** 5 tan 80° **d** 9 cos 8°

11 Use your calculator to work out the values of the following.

a $\frac{5}{\sin 63°}$ **b** $\frac{6}{\sin 32°}$ **c** $\frac{6}{\sin 90°}$ **d** $\frac{5}{\sin 30°}$

12 Use your calculator to work out the values of the following.

a $\frac{3}{\tan 64°}$ **b** $\frac{7}{\tan 42°}$ **c** $\frac{5}{\tan 89°}$ **d** $\frac{6}{\tan 40°}$

13 Use your calculator to work out the values of the following.

a 8 sin 75° **b** $\frac{19}{\sin 23°}$ **c** 7 cos 71° **d** $\frac{15}{\sin 81°}$

14 Use your calculator to work out the values of the following.

a 8 tan 75° **b** $\frac{19}{\tan 23°}$ **c** 7 tan 71° **d** $\frac{15}{\tan 81°}$

B

15 Using the following triangles calculate sin x, cos x, and tan x. Leave your answers as fractions.

a

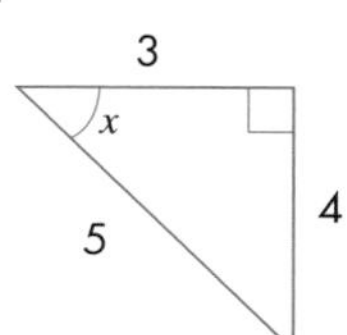

b

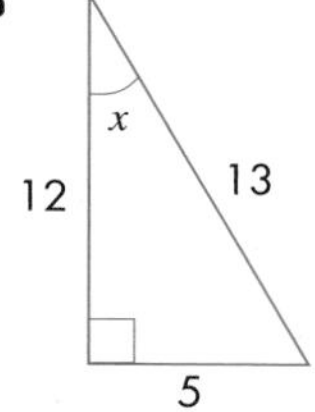

c

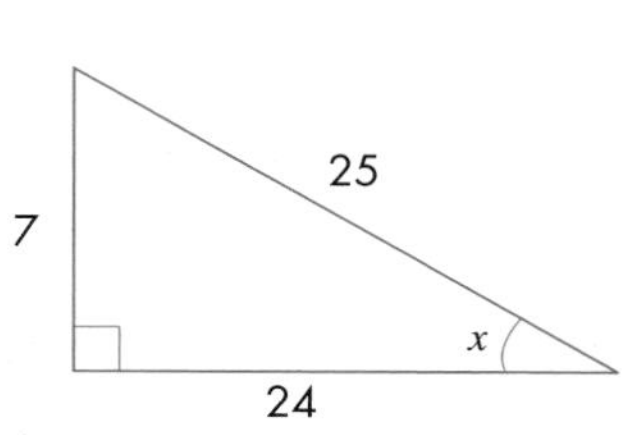

5.6 Calculating angles

This section will show you how to:
- use the trigonometric ratios to calculate an angle

Key words
inverse

What angle has a cosine of 0.6? We can use a calculator to find out.

'The angle with a cosine of 0.6' is written as $\cos^{-1} 0.6$ and is called the 'inverse cosine of 0.6'.

Find out where $\cos^{-1}$ is on your calculator.

You will probably find it on the same key as cos, but you will need to press SHIFT or INV or 2ndF first.

Look to see if $\cos^{-1}$ is written above the cos key.

Check that $\cos^{-1} 0.6 = 53.1301... = 53.1°$ (1 decimal place)

Check that $\cos 53.1° = 0.600$ (3 decimal places)

Check that you can find the inverse sine and the inverse tangent in the same way.

EXAMPLE 7

What angle has a sine of $\frac{3}{8}$?

You need to find $\sin^{-1} \frac{3}{8}$.

You could use the fraction button on your calculator or you could calculate $\sin^{-1} (3 \div 8)$.

If you use the fraction key you may not need a bracket, or your calculator may put one in automatically.

Try to do it in both of these ways and then use whichever you prefer.

The answer should be 22.0°.

EXAMPLE 8

Find the angle with a tangent of 0.75.

$\tan^{-1} 0.75 = 36.869\,897\,65 = 36.9°$ (1 decimal place)

EXERCISE 5G

Use your calculator to find the answers to the following. Give your answers to 1 decimal place.

B

1 What angles have the following sines?

a 0.5 **b** 0.785 **c** 0.64 **d** 0.877 **e** 0.999 **f** 0.707

2 What angles have the following cosines?

a 0.5 **b** 0.64 **c** 0.999 **d** 0.707 **e** 0.2 **f** 0.7

3 What angles have the following tangents?

a 0.6 **b** 0.38 **c** 0.895 **d** 1.05 **e** 2.67 **f** 4.38

4 What angles have the following sines?

a $4 \div 5$ **b** $2 \div 3$ **c** $7 \div 10$ **d** $5 \div 6$ **e** $1 \div 24$ **f** $5 \div 13$

5 What angles have the following cosines?

a $4 \div 5$ **b** $2 \div 3$ **c** $7 \div 10$ **d** $5 \div 6$ **e** $1 \div 24$ **f** $5 \div 13$

6 What angles have the following tangents?

a $3 \div 5$ **b** $7 \div 9$ **c** $2 \div 7$ **d** $9 \div 5$ **e** $11 \div 7$ **f** $6 \div 5$

7 What happens when you try to find the angle with a sine of 1.2? What is the largest value of sine you can put into your calculator without getting an error when you ask for the inverse sine? What is the smallest?

PS 8 **a** **i** What angle has a sine of 0.3? (Keep the answer in your calculator memory.)

ii What angle has a cosine of 0.3?

iii Add the two accurate answers of parts **i** and **ii** together.

b Will you always get the same answer to the above no matter what number you start with?

5.7 Using the sine and cosine functions

This section will show you how to:
- find lengths of sides and angles in right-angled triangles using the sine and cosine functions

Key words
cosine
sine

Sine function

Remember sine $\theta = \dfrac{\text{Opposite}}{\text{Hypotenuse}}$

We can use the **sine** ratio to calculate the lengths of sides and angles in right-angled triangles.

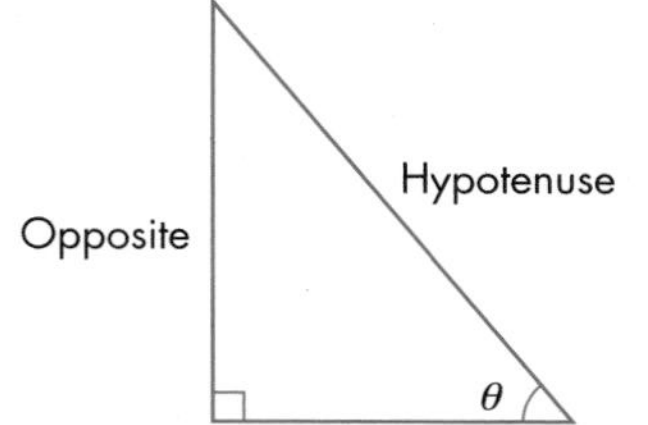

EXAMPLE 9

Find the angle θ, given that the opposite side is 7 cm and the hypotenuse is 10 cm.

Draw a diagram. (This is an essential step.)

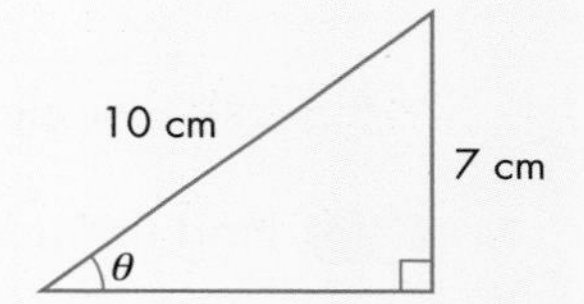

From the information given, use sine.

$$\sin \theta = \frac{O}{H} = \frac{7}{10} = 0.7$$

What angle has a sine of 0.7? To find out, use the inverse sine function on your calculator.

$$\sin^{-1} 0.7 = 44.4° \text{ (1 decimal place)}$$

EXAMPLE 10

Find the length of the side marked a in this triangle.

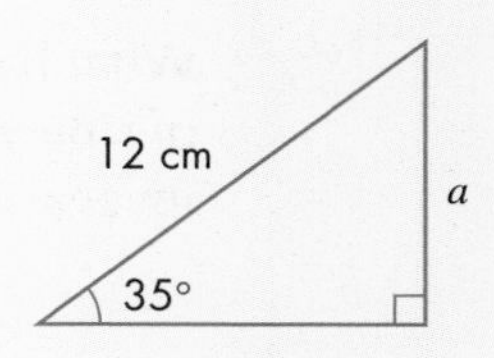

Side a is the opposite side, with 12 cm as the hypotenuse, so use sine.

$$\sin \theta = \frac{O}{H}$$

$$\sin 35° = \frac{a}{12}$$

So $a = 12 \sin 35° = 6.88$ cm (3 significant figures)

EXAMPLE 11

Find the length of the hypotenuse, h, in this triangle.

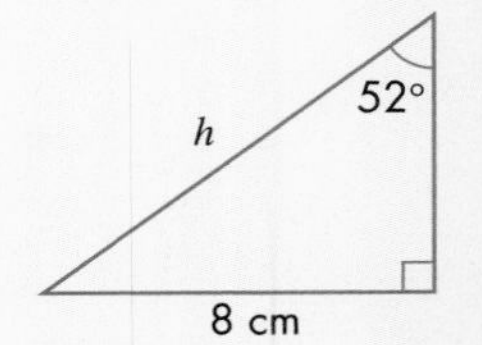

Note that although the angle is in the other corner, the opposite side is again given. So use sine.

$$\sin \theta = \frac{O}{H}$$

$$\sin 52° = \frac{8}{h}$$

So $h = \frac{8}{\sin 52°} = 10.2$ cm (3 significant figures)

EXERCISE 5H

B

1 Find the angle marked x in each of these triangles.

a

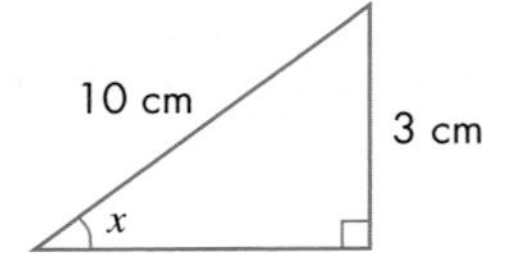

b

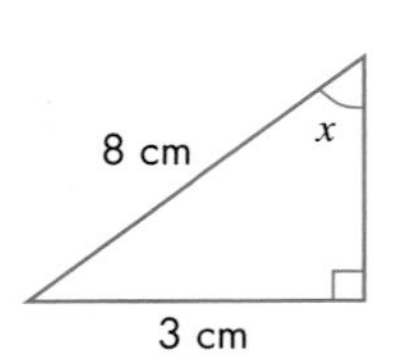

c

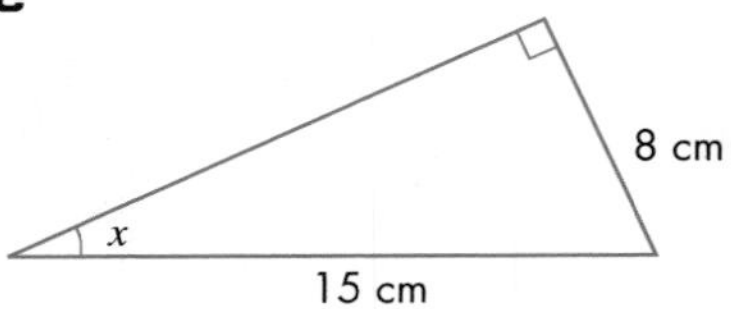

2 Find the side marked x in each of these triangles.

a

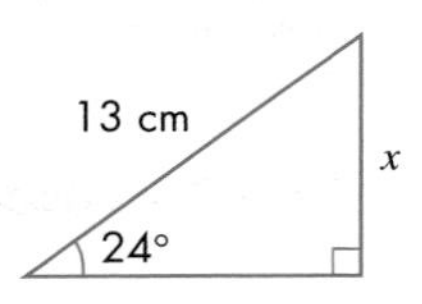

b

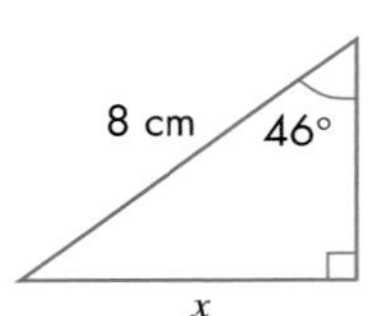

c

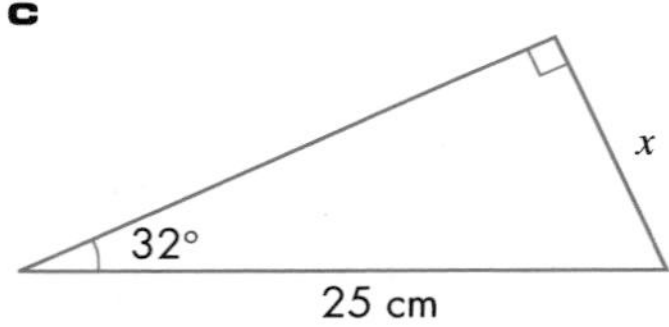

3 Find the side marked x in each of these triangles.

a

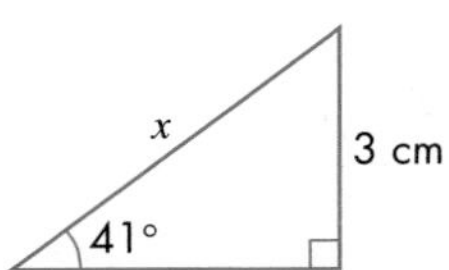

b

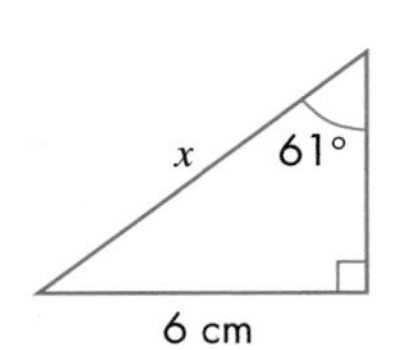

c

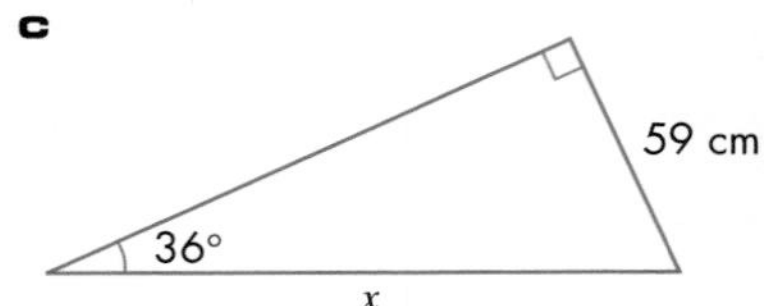

4 Find the side marked x in each of these triangles.

a

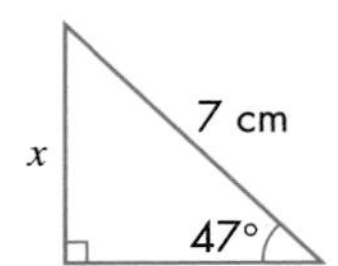

b

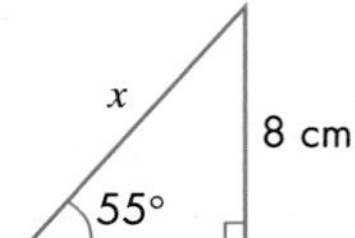

c

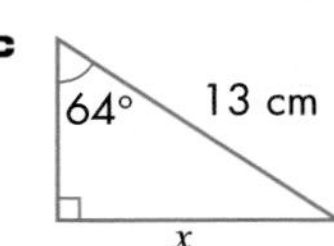

d

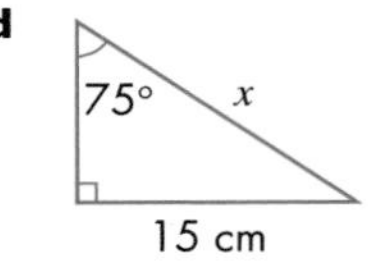

B

5 Find the value of x in each of these triangles.

a

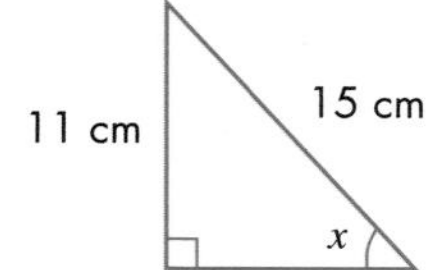

b

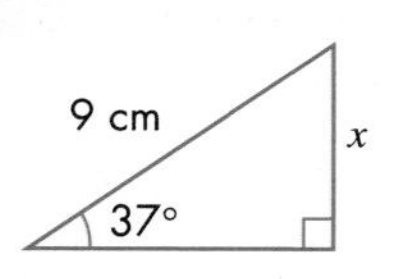

c

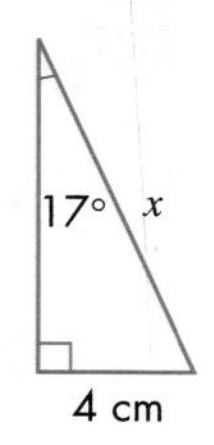

d

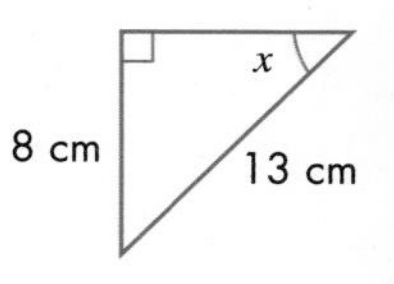

6 Angle θ has a sine of $\frac{3}{5}$. Calculate the missing lengths in these triangles.

a

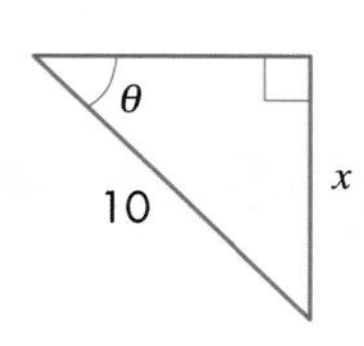

b

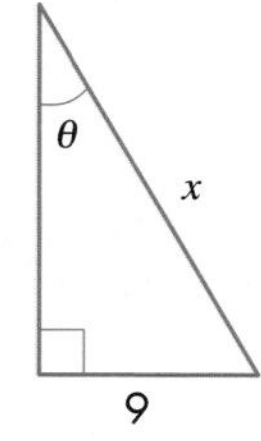

c

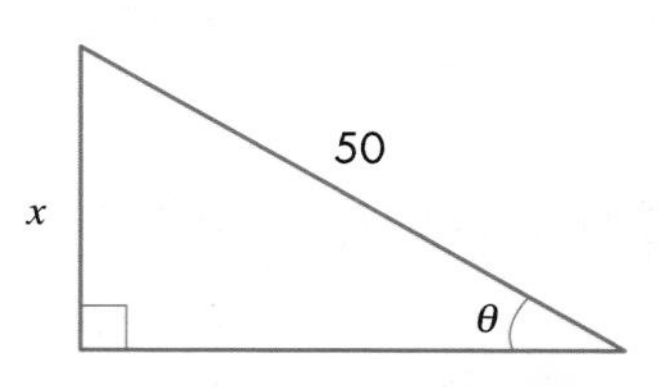

Cosine function

Remember cosine $\theta = \dfrac{\text{Adjacent}}{\text{Hypotenuse}}$

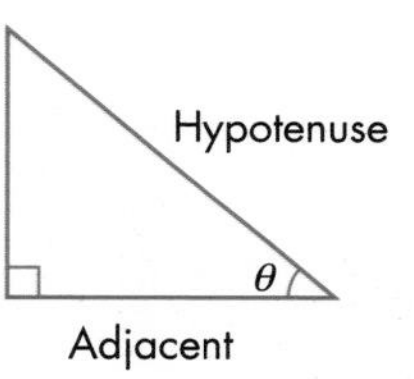

We can use the **cosine** ratio to calculate the lengths of sides and angles in right-angled triangles.

EXAMPLE 12

Find the angle θ, given that the adjacent side is 5 cm and the hypotenuse is 12 cm.

Draw a diagram. (This is an essential step.)

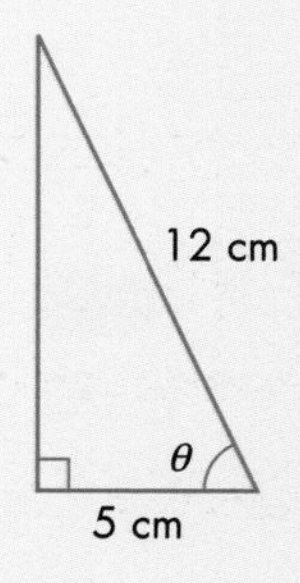

From the information given, use cosine.

$$\cos\,\theta = \frac{A}{H} = \frac{5}{12}$$

What angle has a cosine of $\frac{5}{12}$? To find out, use the inverse cosine function on your calculator.

$$\cos^{-1} = 65.4° \text{ (1 decimal place)}$$

EXAMPLE 13

Find the length of the side marked a in this triangle.

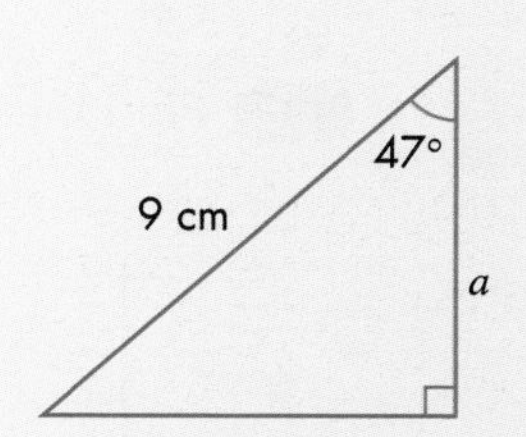

Side a is the adjacent side, with 9 cm as the hypotenuse, so use cosine.

$$\cos\,\theta = \frac{A}{H}$$

$$\cos 47° = \frac{a}{9}$$

So $a = 9 \cos 47° = 6.14$ cm (3 significant figures)

EXAMPLE 14

Find the length of the hypotenuse, h, in this triangle.

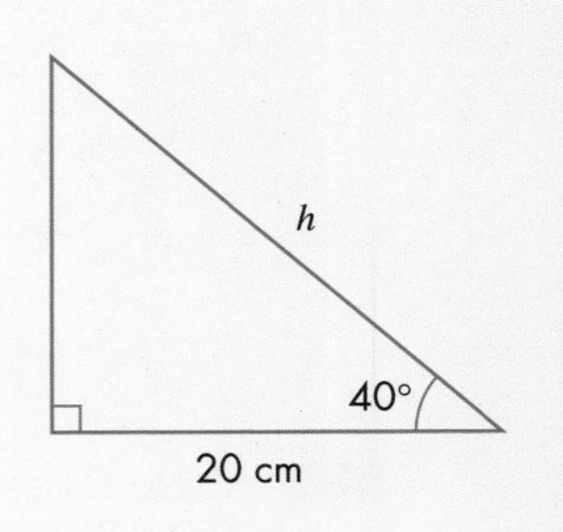

The adjacent side is given. So use cosine.

$$\cos\theta = \frac{A}{H}$$

$$\cos 40° = \frac{20}{h}$$

So $h = \frac{20}{\cos 40°} = 26.1$ cm (3 significant figures)

EXERCISE 5I

B

1 Find the angle marked x in each of these triangles.

a

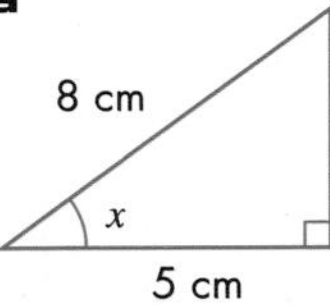

b

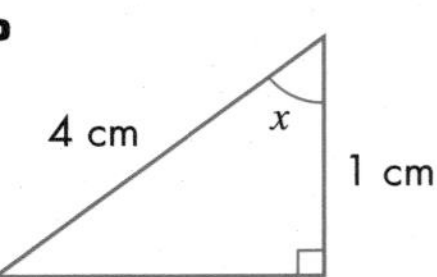

c

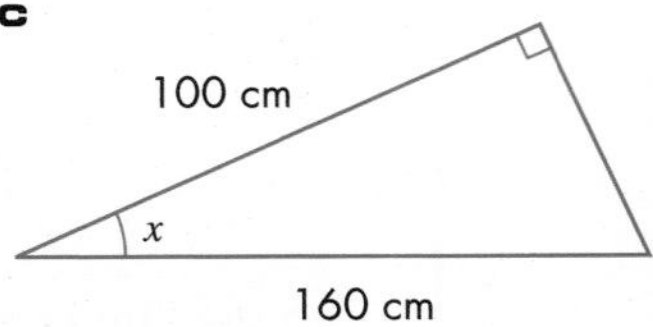

2 Find the side marked x in each of these triangles.

a

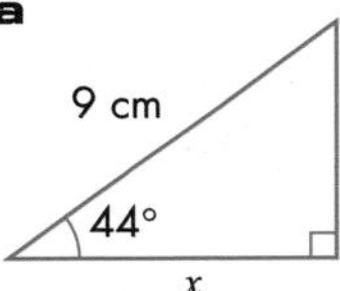

b

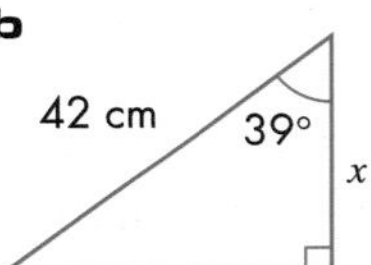

c

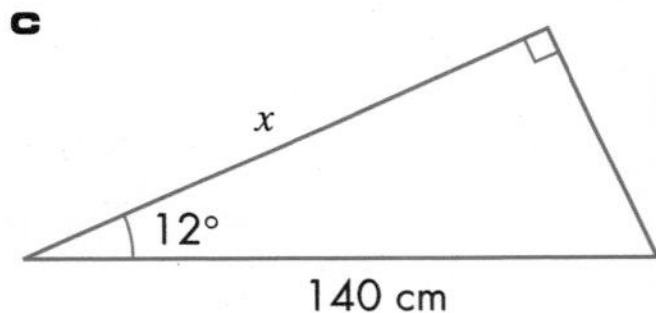

3 Find the side marked x in each of these triangles.

a

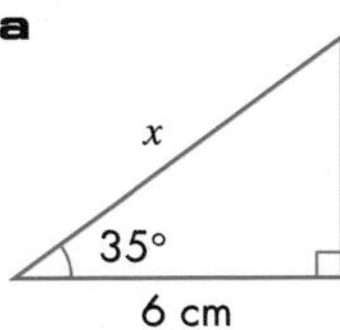

b

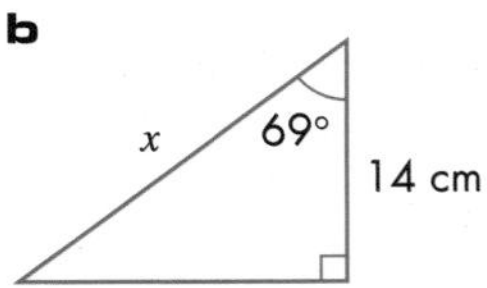

c

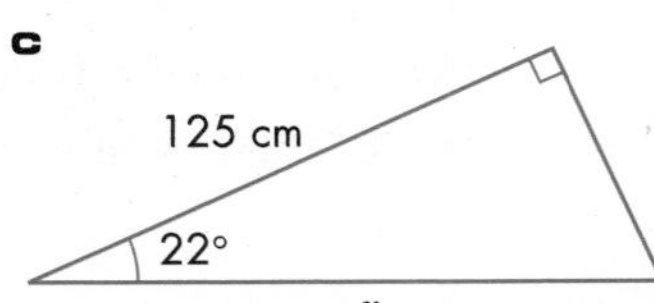

4 Find the side marked x in each of these triangles.

a

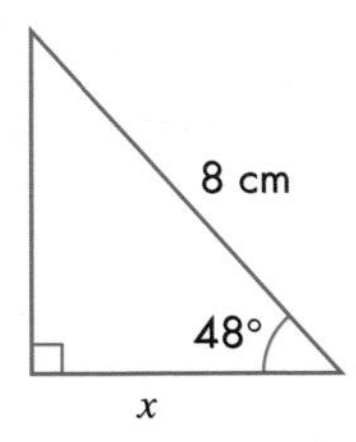

b

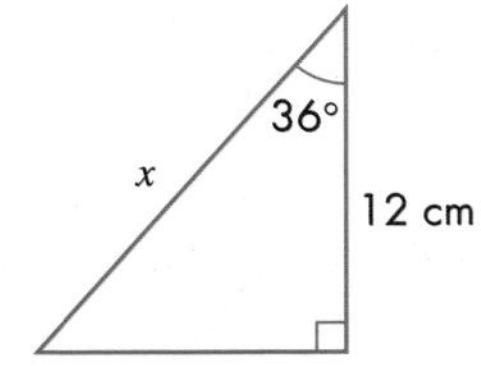

c

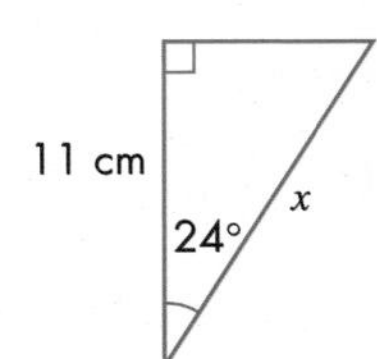

d

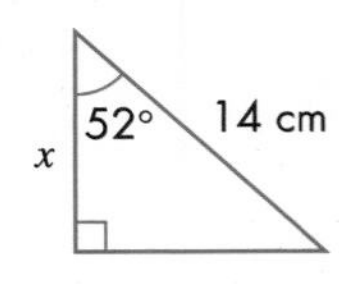

B

5 Find the value of x in each of these triangles.

a

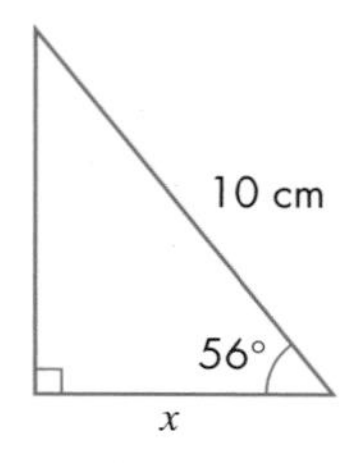

b

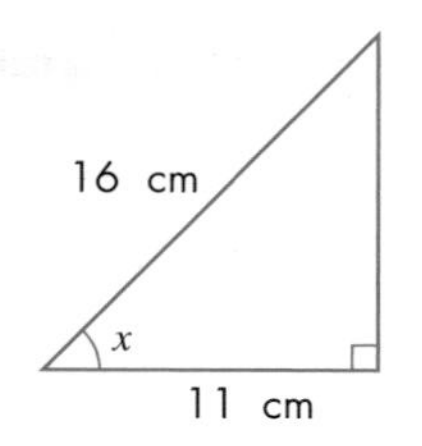

c

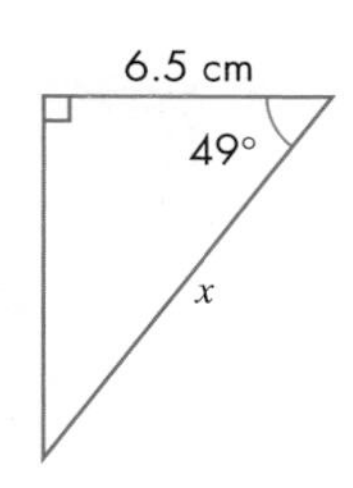

d

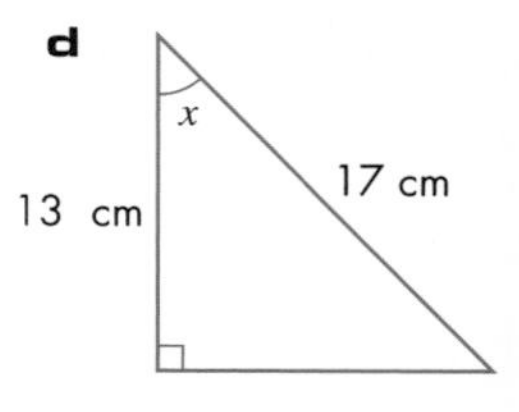

6 Angle θ has a cosine of $\frac{5}{13}$. Calculate the missing lengths in these triangles.

a

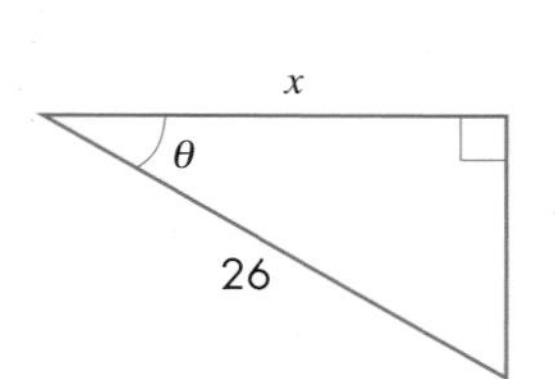

b

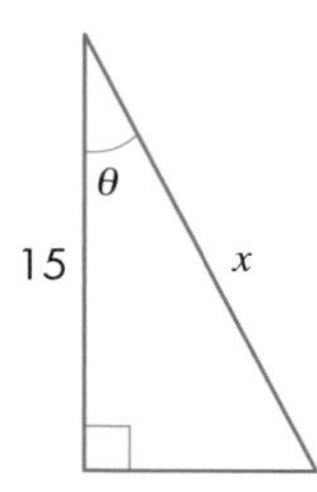

c

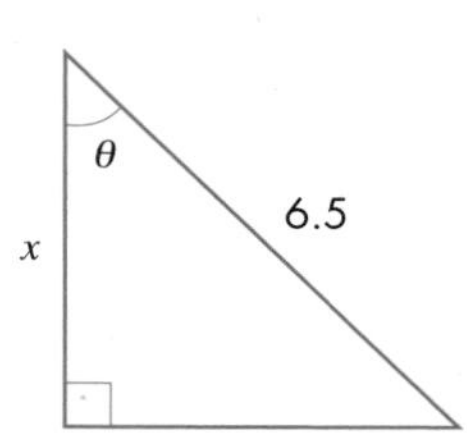

5.8 Using the tangent function

This section will show you how to:

- find lengths of sides and angles in right-angled triangles using the tangent function

Key words

tangent

Remember tangent $\theta = \dfrac{\text{Opposite}}{\text{Adjacent}}$

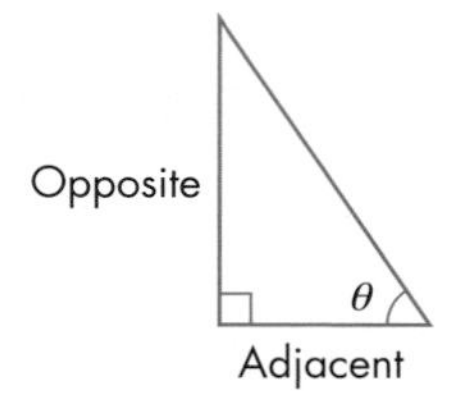

We can use the **tangent** ratio to calculate the lengths of sides and angles in right-angled triangles.

EXAMPLE 15

Find the angle θ, given that the opposite side is 3 cm and the adjacent side is 4 cm.

Draw a diagram. (This is an essential step.)

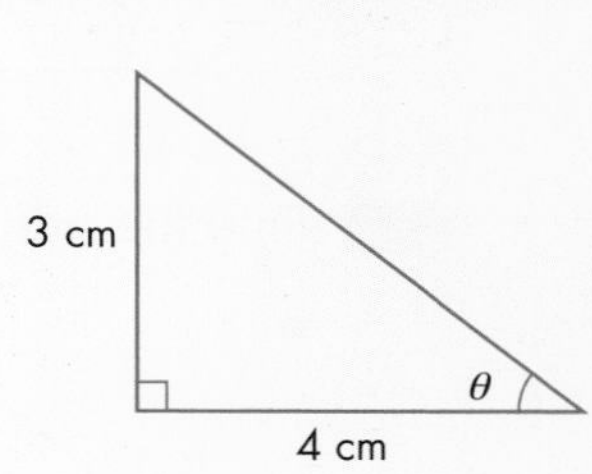

From the information given, use tangent.

$$\tan \theta = \frac{O}{A} = \frac{3}{4} = 0.75$$

What angle has a tangent of 0.75? To find out, use the inverse tangent function on your calculator.

$$\tan^{-1} 0.75 = 36.9° \text{ (1 decimal place)}$$

EXAMPLE 16

Find the length of the side marked x in this triangle.

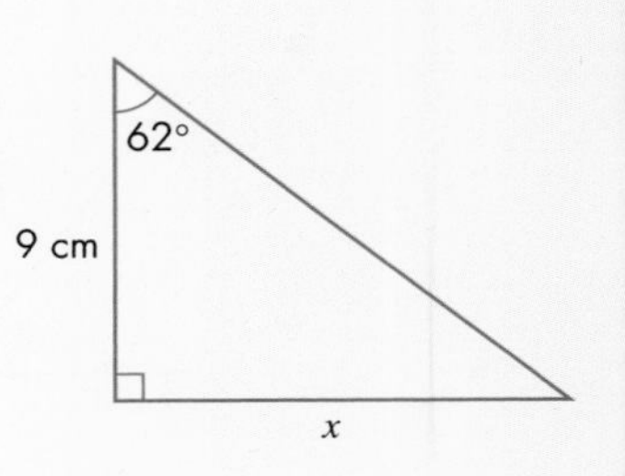

Side x is the opposite side, with 9 cm as the adjacent side, so use tangent.

$$\tan \theta = \frac{O}{A}$$

$$\tan 62° = \frac{x}{9}$$

So $x = 9 \tan 62° = 16.9$ cm (3 significant figures)

EXAMPLE 17

Find the length of the side marked a in this triangle.

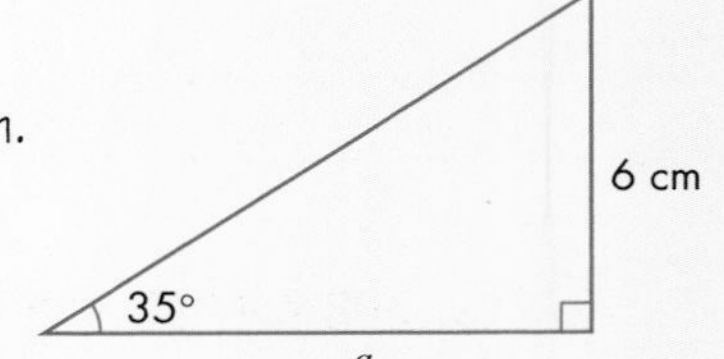

Side a is the adjacent side and the opposite side is given. So use tangent.

$$\tan \theta = \frac{O}{A}$$

$$\tan 35° = \frac{6}{a}$$

So $a = \dfrac{6}{\tan 35°} = 8.57$ cm (3 significant figures)

EXERCISE 5J

1 Find the angle marked x in each of these triangles.

a

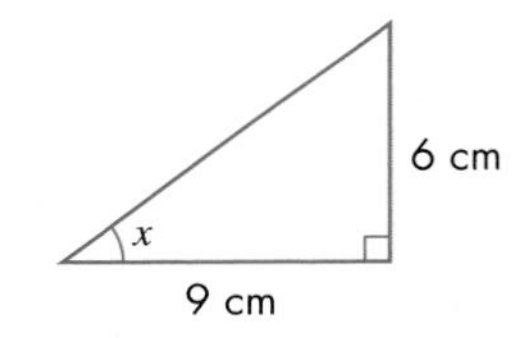

b

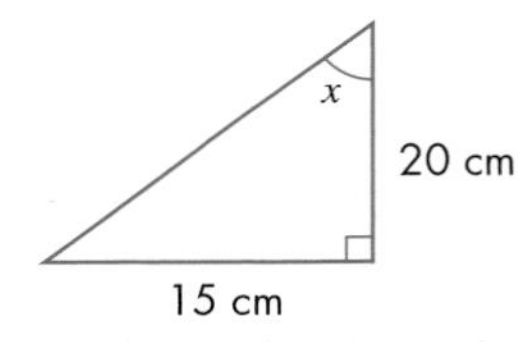

c

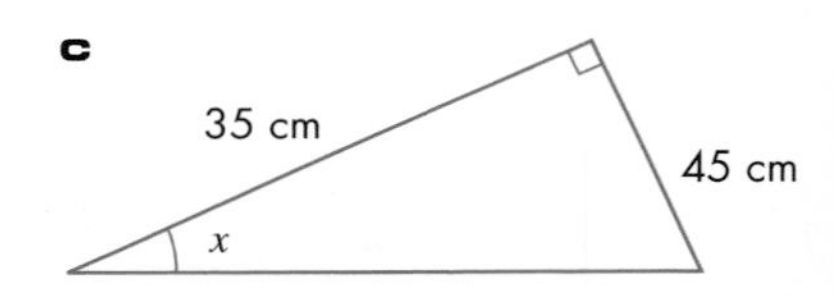

2 Find the side marked x in each of these triangles.

a

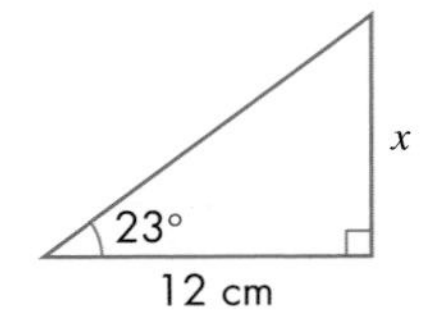

b

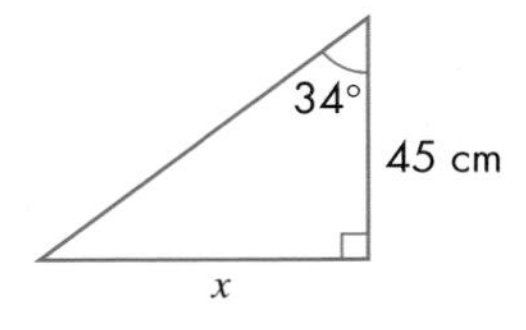

c

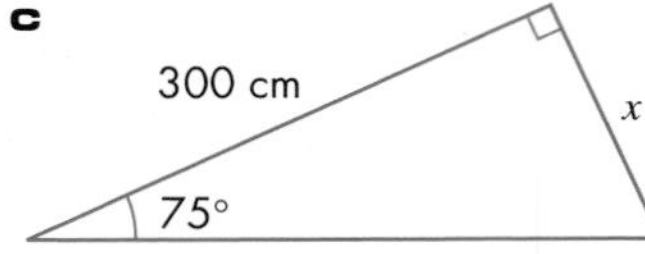

3 Find the side marked x in each of these triangles.

a

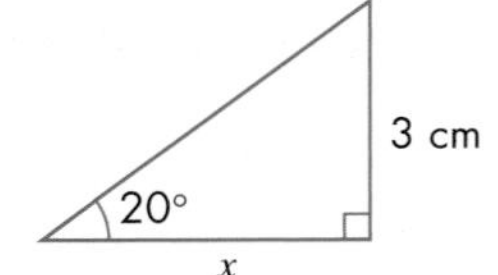

b

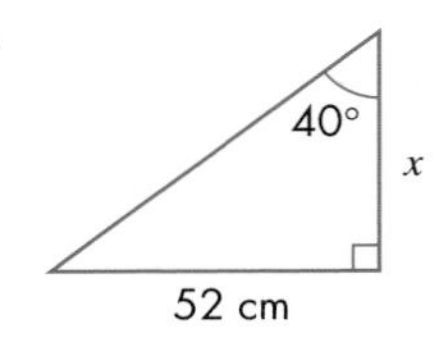

c

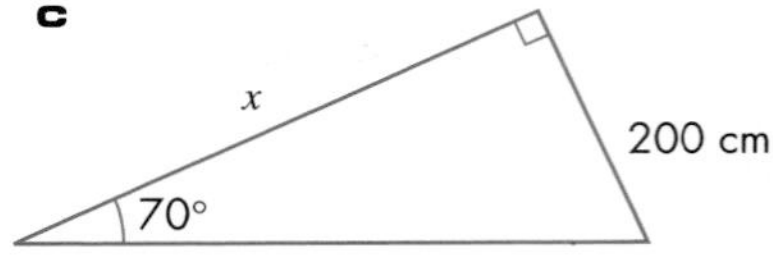

B

4 Find the side marked x in each of these triangles.

a
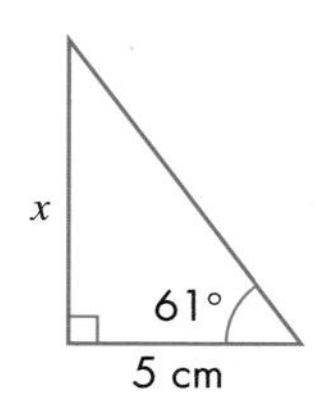

b
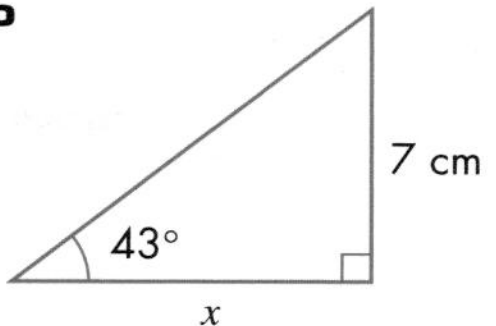

c
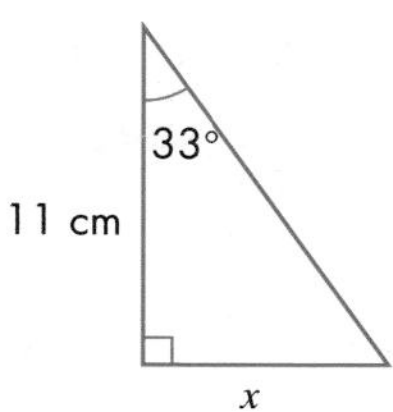

d
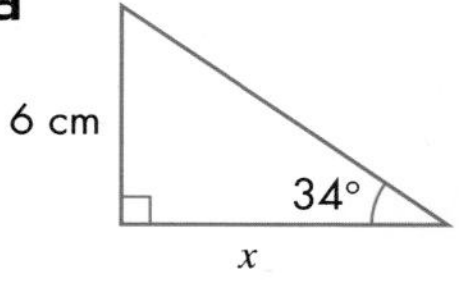

5 Find the value x in each of these triangles.

a
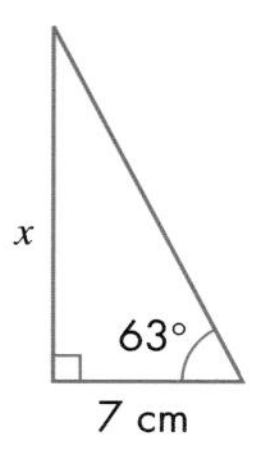

b
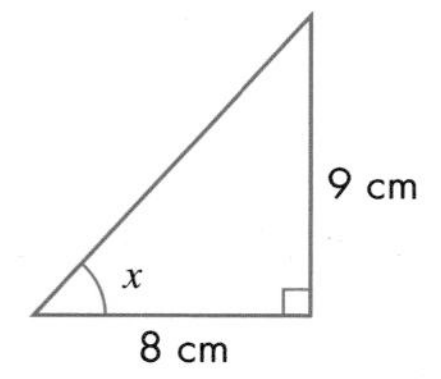

c
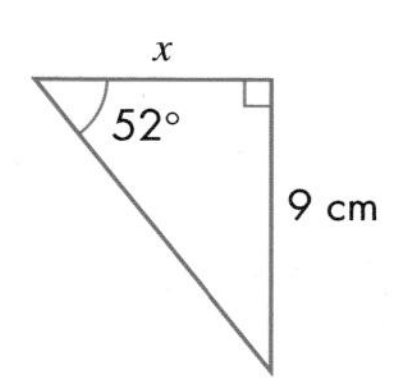

d
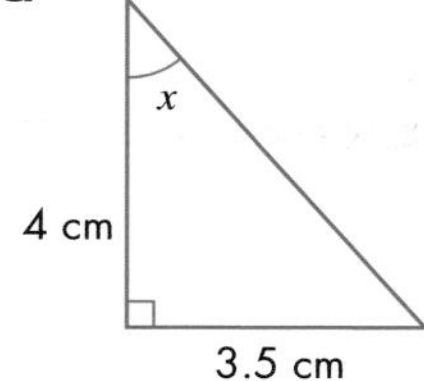

6 Angle θ has a tangent of $\frac{4}{3}$. Calculate the missing lengths in these triangles.

a
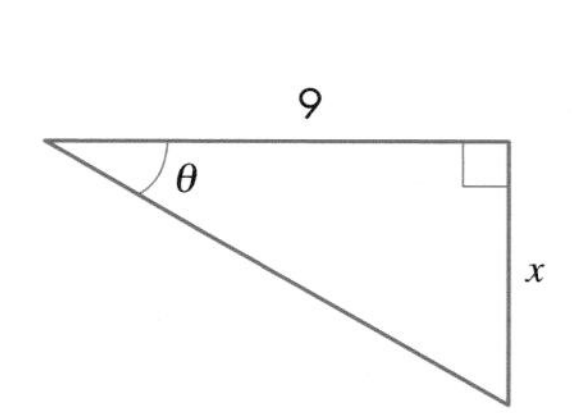

b
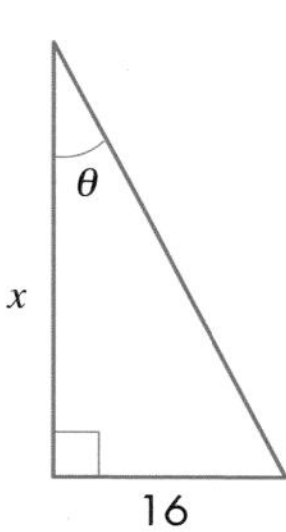

c
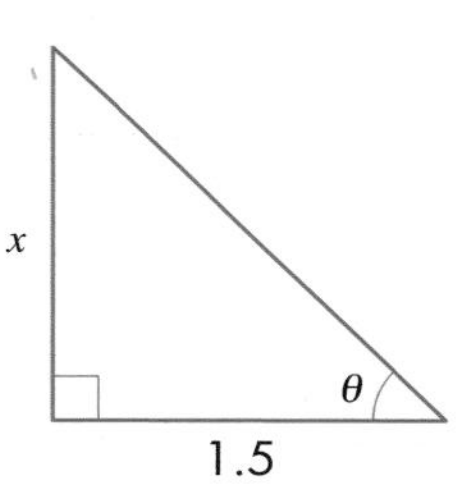

5.9 Which ratio to use

This section will show you how to:
- decide which trigonometric ratio to use in a right-angled triangle

Key words
cosine
sine
tangent

The difficulty with any trigonometric problem is knowing which ratio to use to solve it.

The following examples show you how to determine which ratio you need in any given situation.

EXAMPLE 18

Find the length of the side marked x in this triangle.

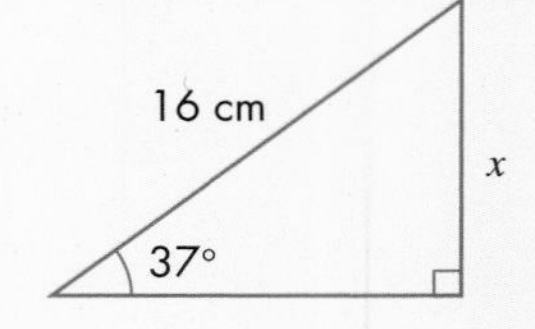

Step 1 Identify what information is given and what needs to be found. Namely, x is opposite the angle and 16 cm is the hypotenuse.

Step 2 Decide which ratio to use. Only one ratio uses opposite and hypotenuse: **sine**.

Step 3 Remember $\sin \theta = \frac{O}{H}$

Step 4 Put in the numbers and letters: $\sin 37° = \frac{x}{16}$

Step 5 Rearrange the equation and work out the answer: $x = 16 \sin 37° = 9.629\,040\,371$ cm

Step 6 Give the answer to an appropriate degree of accuracy: $x = 9.63$ cm (3 significant figures)

In reality, you do not write down every step as in Example 18. Step 1 can be done by marking the triangle. Steps 2 and 3 can be done in your head. Steps 4 to 6 are what you write down.

Remember that examiners will want to see evidence of working. Any reasonable attempt at identifying the sides and using a ratio will probably get you some method marks, but only if the fraction is the right way round.

The next examples are set out in a way that requires the *minimum* amount of working but gets *maximum* marks.

EXAMPLE 19

Find the length of the side marked x in this triangle.

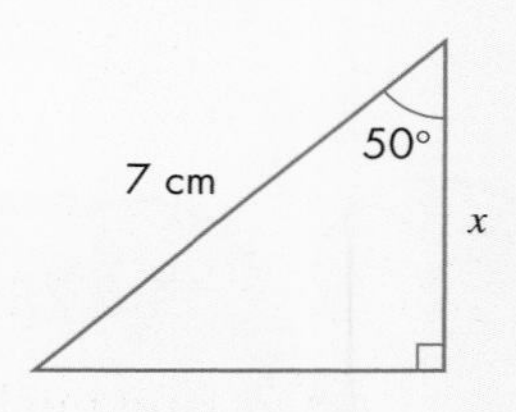

Mark on the triangle the side you know (H) and the side you want to find (A).

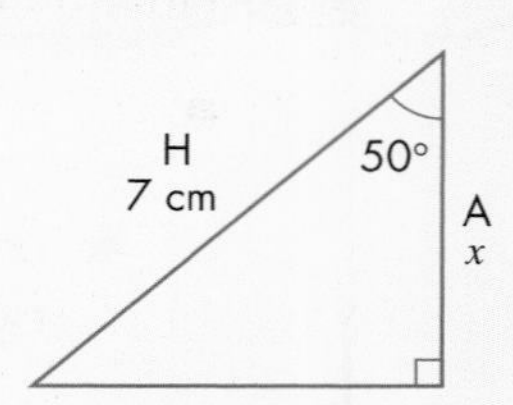

Recognise it is a **cosine** problem because you have A and H.

So $\cos 50° = \frac{x}{7}$

$x = 7 \cos 50° = 4.50$ cm (3 significant figures)

EXAMPLE 20

Find the angle marked x in this triangle.

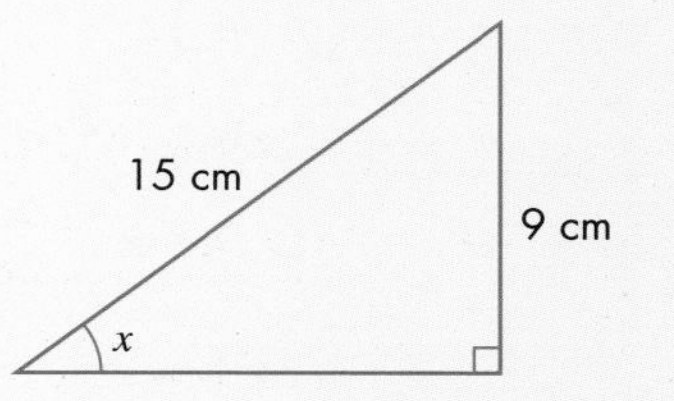

Mark on the triangle the sides you know.

Recognise it is a sine problem because you have O and H.

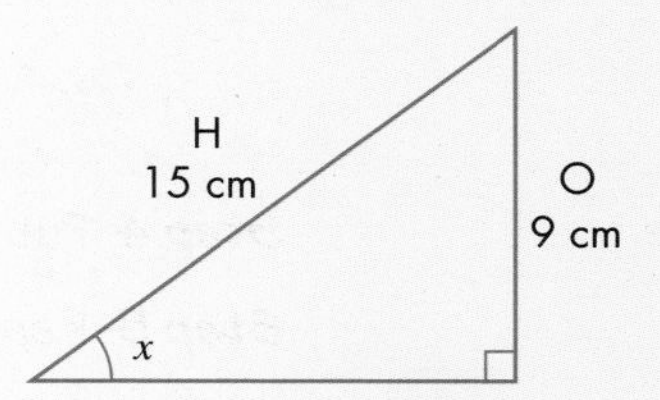

So $\sin x = \frac{9}{15} = 0.6$

$x = \sin^{-1} 0.6 = 36.9°$ (1 decimal place)

EXAMPLE 21

Find the angle marked x in this triangle.

Mark on the triangle the sides you know.

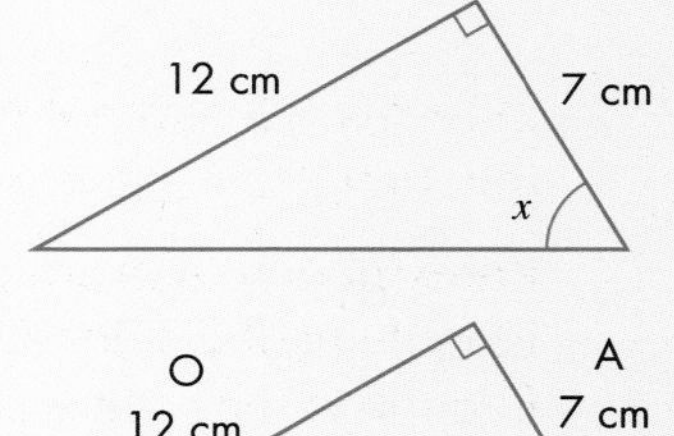

Recognise it is a **tangent** problem because you have O and A.

So $\tan x = \frac{12}{7}$

$x = \tan^{-1} \frac{12}{7} = 59.7°$ (1 decimal place)

EXERCISE 5K

1 Find the length marked x in each of these triangles.

a

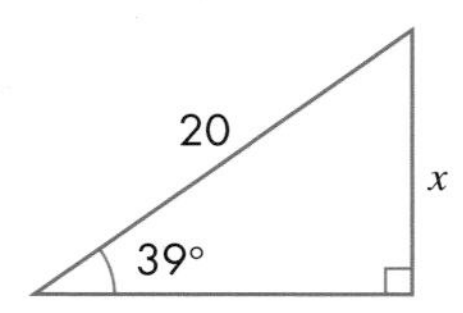

b

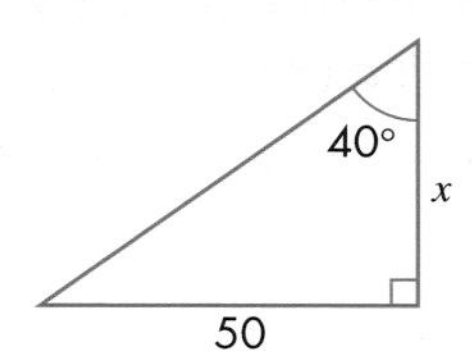

c

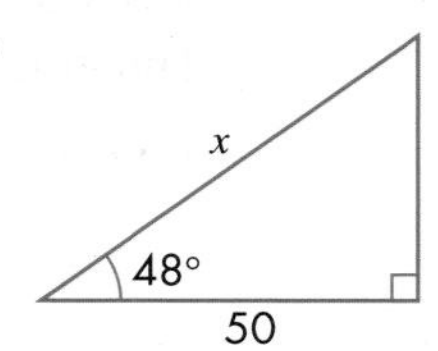

d

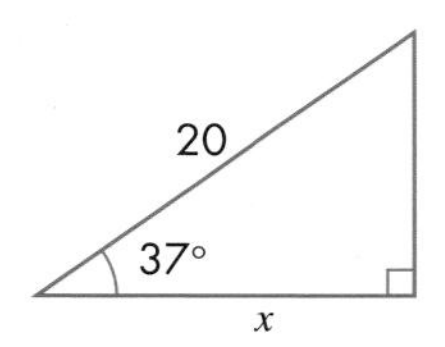

e

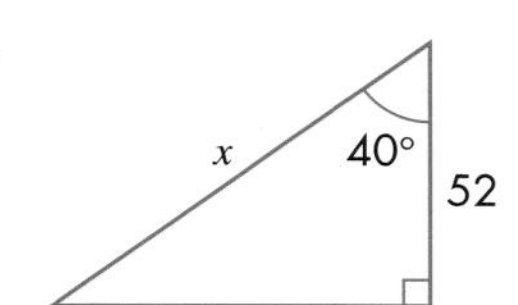

f

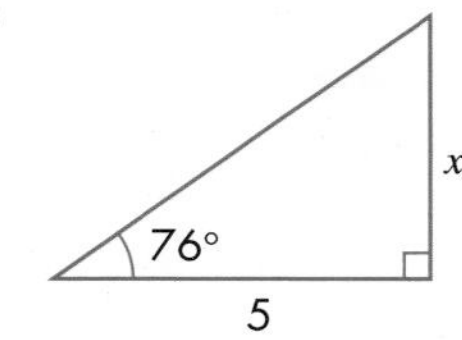

B

2 Find the angle marked x in each of these triangles.

a

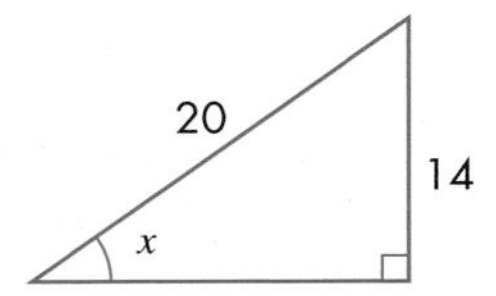

b

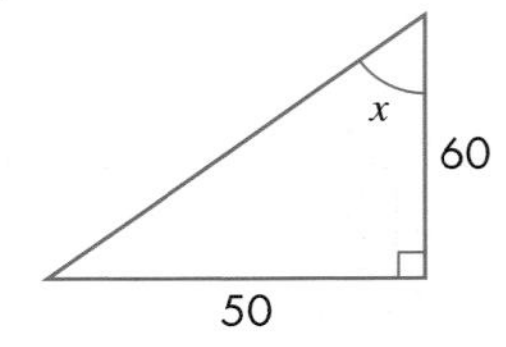

c

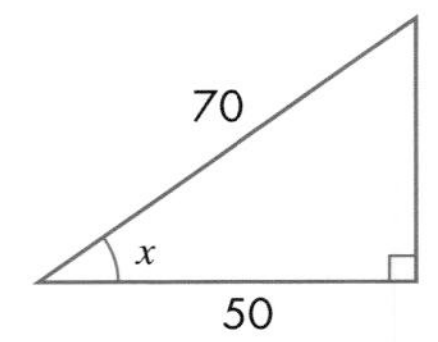

d

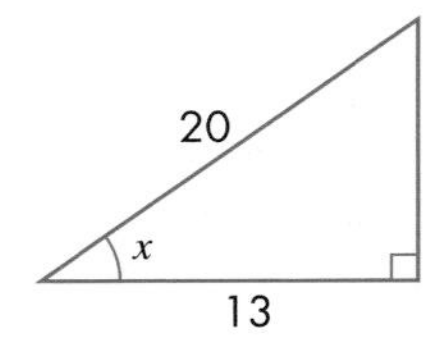

e

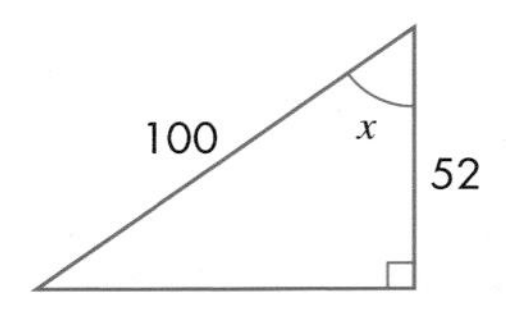

f

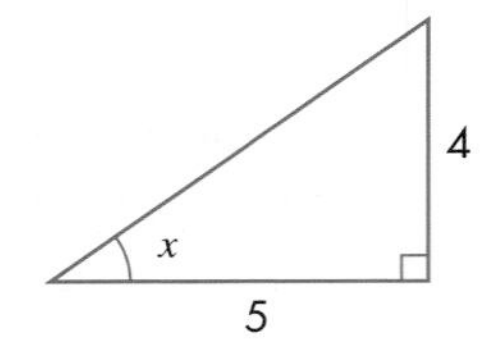

3 Find the angle or length marked x in each of these triangles.

a

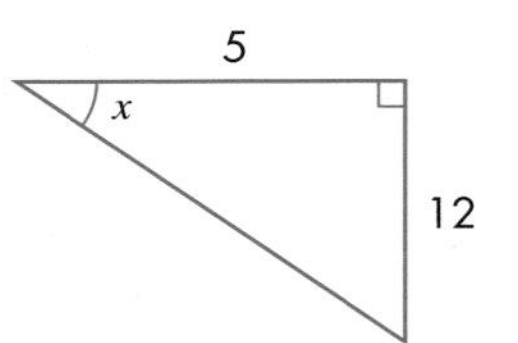

b

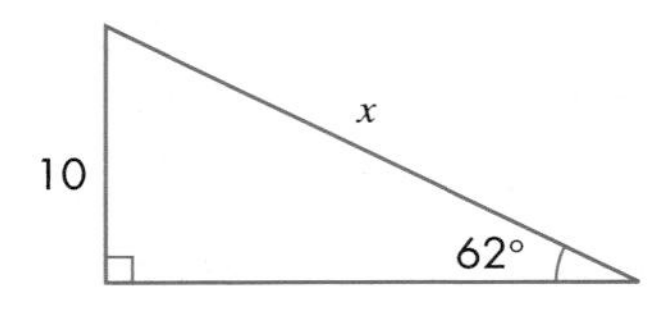

c

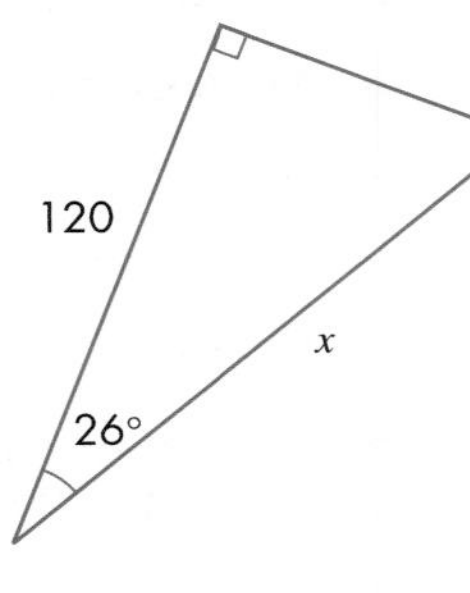
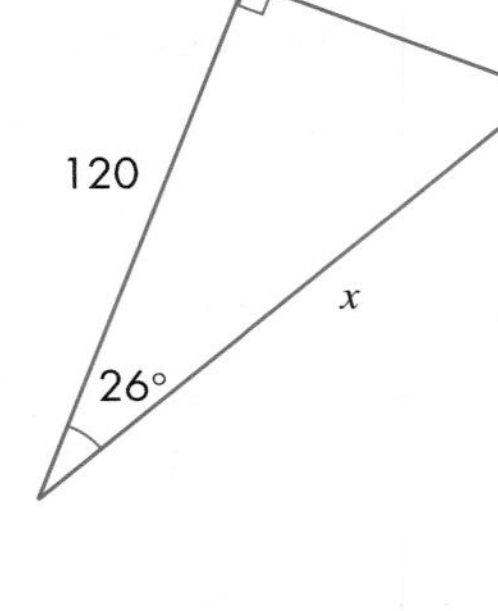

d

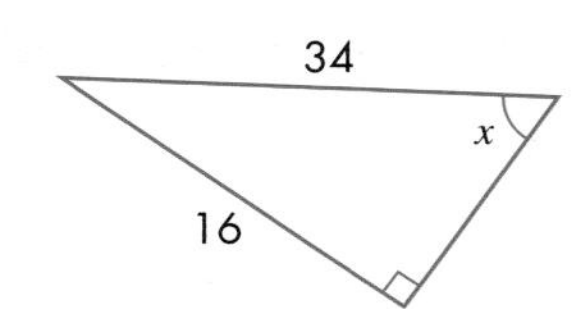

e

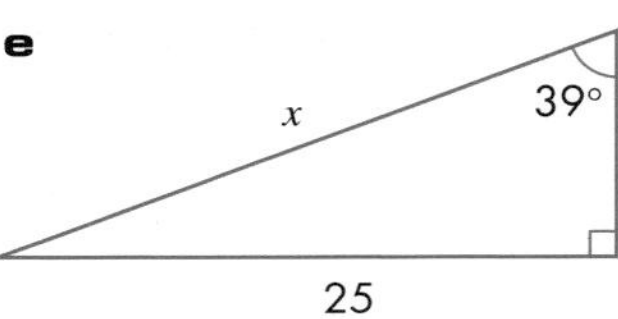

f

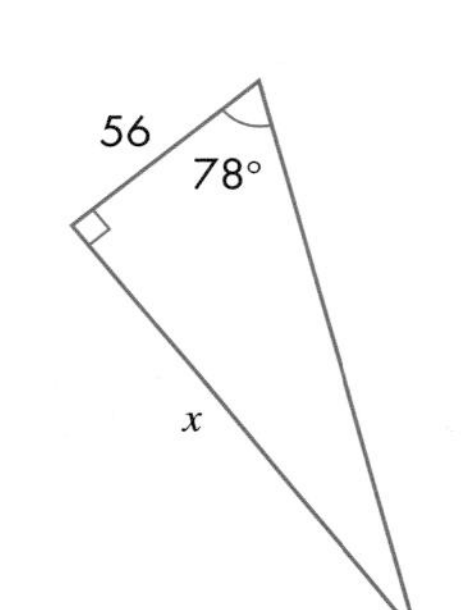

g

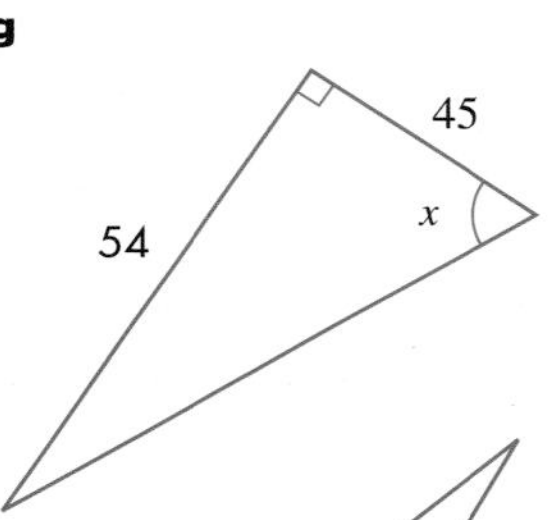

h

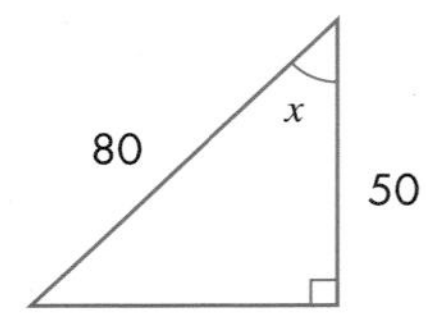

i

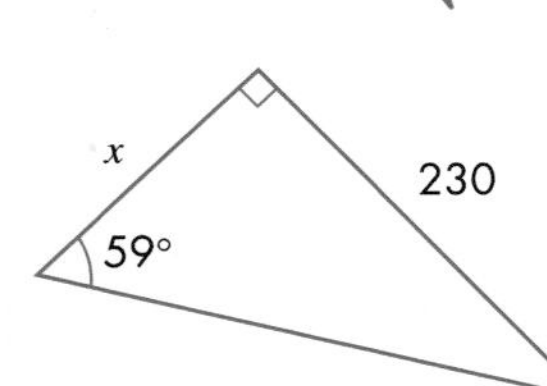

j

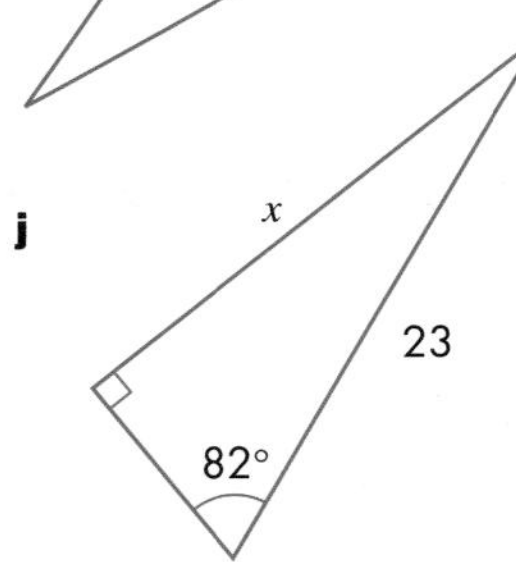

A

PS 4 **a** How does this diagram show that $\tan\theta = \dfrac{\sin\theta}{\cos\theta}$?

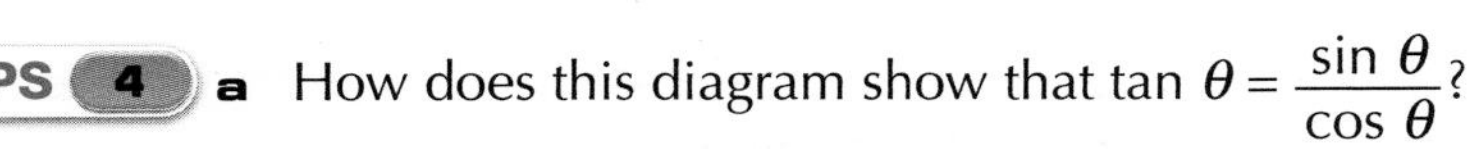

b How does the diagram show that $(\sin\theta)^2 + (\cos\theta)^2 = 1$?

c Choose a value for θ and check the two results in parts **a** and **b** are true.

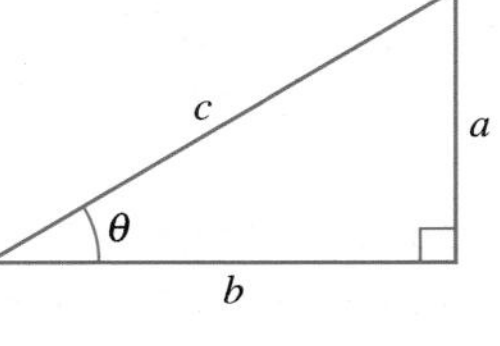

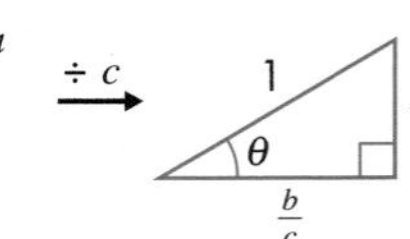

5.10 Solving problems using trigonometry 1

This section will show you how to:
- solve practical problems using trigonometry
- solve problems using an angle of elevation or an angle of depression

Key words
angle of depression
angle of elevation
trigonometry

Many **trigonometry** problems in GCSE examination papers do not come as straightforward triangles. Sometimes, solving a triangle is part of solving a practical problem. You should follow these steps when solving a practical problem using trigonometry.

- Draw the triangle required.
- Put on the information given (angles and sides).
- Put on x for the unknown angle or side.
- Mark on two of O, A or H as appropriate.
- Choose which ratio to use.
- Write out the equation with the numbers in.
- Rearrange the equation if necessary, then work out the answer.
- Give your answer to a sensible degree of accuracy. Answers given to 3 significant figures or to the nearest degree are acceptable in exams.

EXAMPLE 22

A window cleaner has a ladder which is 7 m long. The window cleaner leans it against a wall so that the foot of the ladder is 3 m from the wall. What angle does the ladder make with the wall?

Draw the situation as a right-angled triangle.

Then mark the sides and angle.

Recognise it is a sine problem because you have O and H.

So $\sin x = \frac{3}{7}$

$x = \sin^{-1} \frac{3}{7} = 25°$ (to the nearest degree)

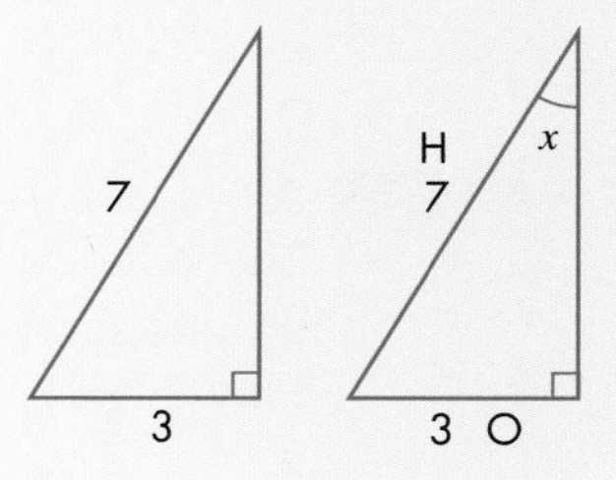

EXERCISE 5L

In these questions, give answers involving angles to the nearest degree.

1 A ladder, 6 m long, rests against a wall. The foot of the ladder is 2.5 m from the base of the wall. What angle does the ladder make with the ground?

FM 2 The ladder in question **1** has a 'safe angle' with the ground of between 70° and 80°. What are the safe limits for the distance of the foot of this ladder from the wall? How high up the wall does the ladder reach?

B

B

FM 3 A ladder, of length 10 m, is placed so that it reaches 7 m up the wall. What angle does it make with the ground?

FM 4 A ladder is placed so that it makes an angle of 76° with the ground. The foot of the ladder is 1.7 m from the foot of the wall. How high up the wall does the ladder reach?

PS 5 Calculate the angle that the diagonal makes with the long side of a rectangle which measures 10 cm by 6 cm.

FM 6 This diagram shows a frame for a bookcase.

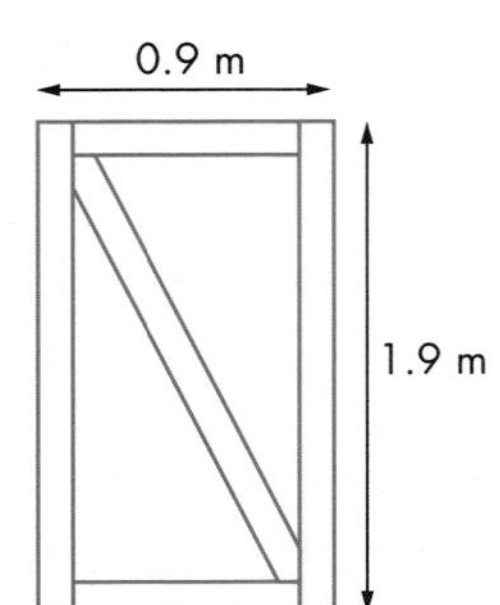

a What angle does the diagonal strut make with the long side?

b Use Pythagoras' theorem to calculate the length of the strut.

c Why might your answers be inaccurate in this case?

FM 7 This diagram shows a roof truss.

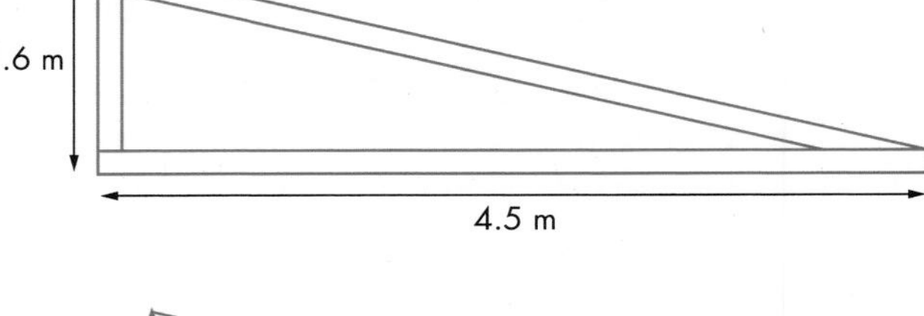

a What angle will the roof make with the horizontal?

b Calculate the length of the sloping strut.

FM 8 Alicia paces out 100 m from the base of a church. She then measures the angle to the top of the spire as 23°. How would Alicia find the height of the church spire?

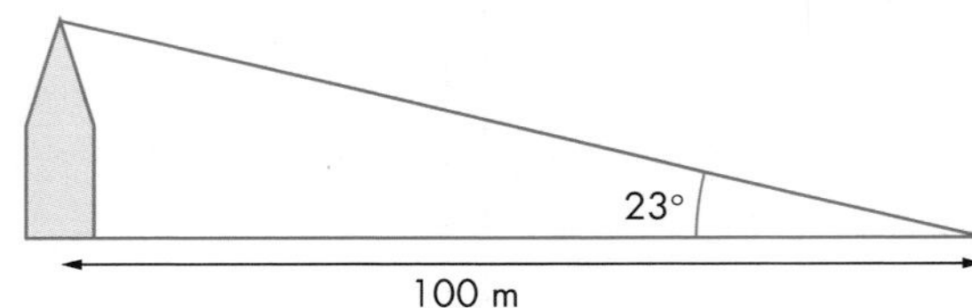

AU 9 A girl is flying a kite on a string 32 m long. The string, which is being held at 1 m above the ground, makes an angle of 39° with the horizontal. How high is the kite above the ground?

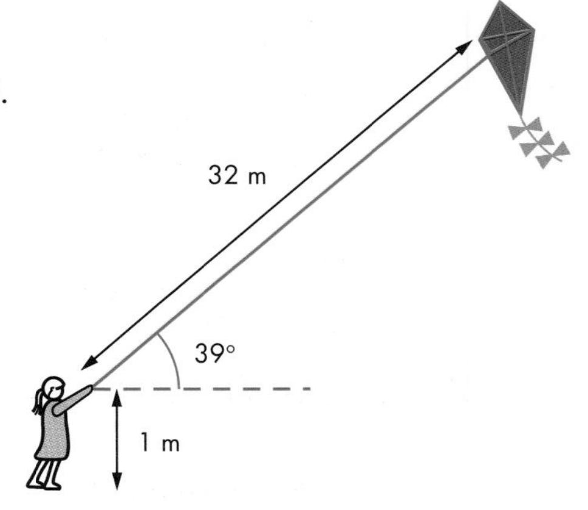

A

FM 10 Helena is standing on one bank of a wide river. She wants to find the width of the river. She cannot get to the other side.
She asks if you can use trigonometry to find the width of the river.

What can you suggest?

Angles of elevation and depression

When you look *up* at an aircraft in the sky, the angle through which your line of sight turns from looking straight ahead (the horizontal) is called the **angle of elevation**.

When you are standing on a high point and look *down* at a boat, the angle through which your line of sight turns from looking straight ahead (the horizontal) is called the **angle of depression**.

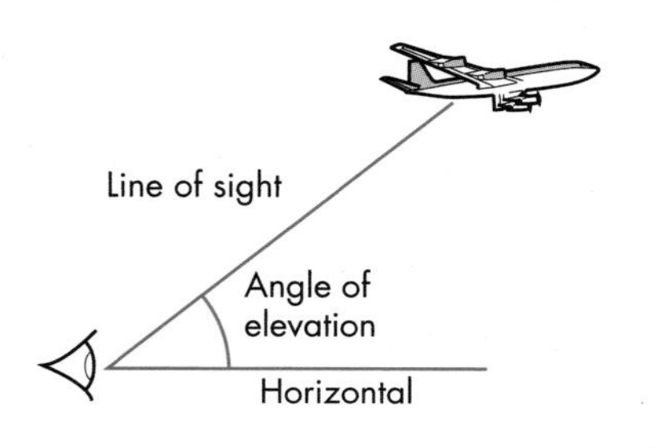

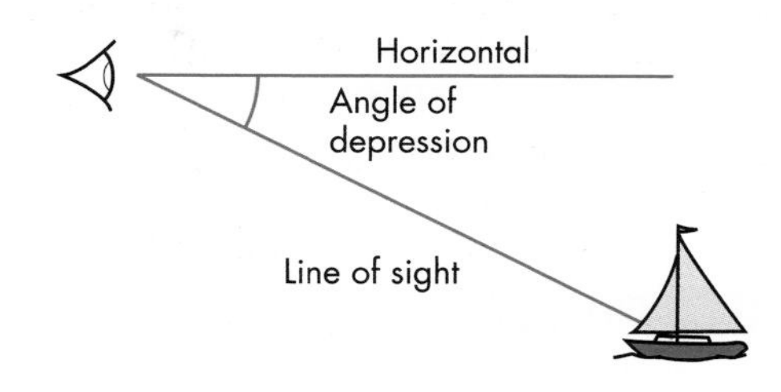

EXAMPLE 23

From the top of a vertical cliff, 100 m high, Andrew sees a boat out at sea. The angle of depression from Andrew to the boat is 42°. How far from the base of the cliff is the boat?

The diagram of the situation is shown in figure **i**.

From this, you get the triangle shown in figure **ii**.

i

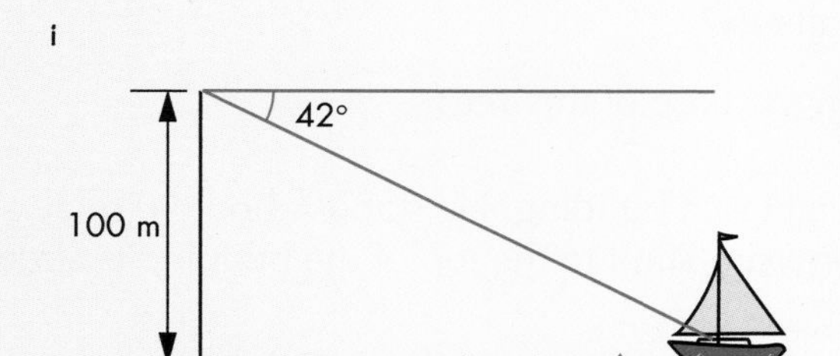

ii

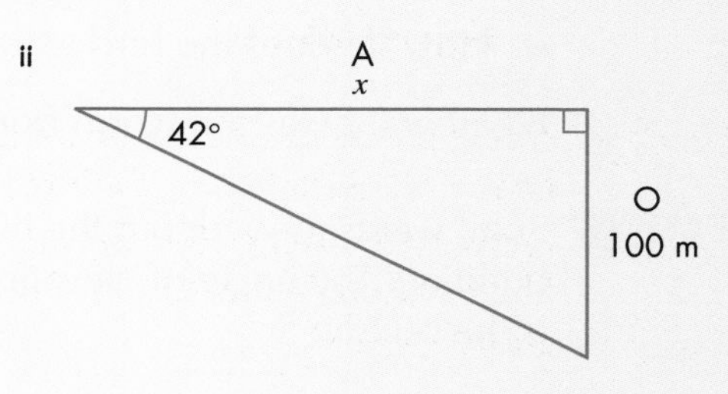

From figure **ii**, you see that this is a tangent problem.

So $\tan 42° = \frac{100}{x}$

$$x = \frac{100}{\tan 42°} = 111 \text{ m (3 significant figures)}$$

EXERCISE 5M

In these questions, give any answers involving angles to the nearest degree.

1 Eric sees an aircraft in the sky. The aircraft is at a horizontal distance of 25 km from Eric. The angle of elevation is 22°. How high is the aircraft?

2 An aircraft is flying at an altitude of 4000 m and is 10 km from the airport. If a passenger can see the airport, what is the angle of depression?

B

B

3 A man standing 200 m from the base of a television transmitter looks at the top of it and notices that the angle of elevation of the top is 65°. How high is the tower?

AU 4 **a** From the top of a vertical cliff, 200 m high, a boat has an angle of depression of 52°. How far from the base of the cliff is the boat?

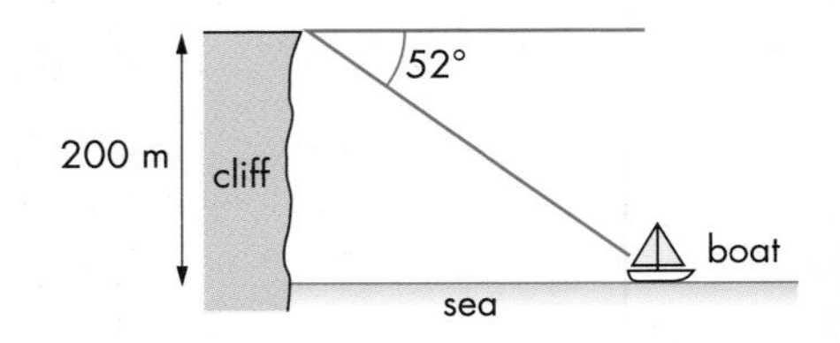

b A boat now sails away from the cliff so that the distance is doubled. Does that mean that the angle of depression is halved? Give a reason for your answer.

FM 5 From a boat, the angle of elevation of the foot of a lighthouse on the edge of a cliff is 34°.

a If the cliff is 150 m high, how far from the base of the cliff is the boat?

b If the lighthouse is 50 m high, what would be the angle of elevation of the top of the lighthouse from the boat?

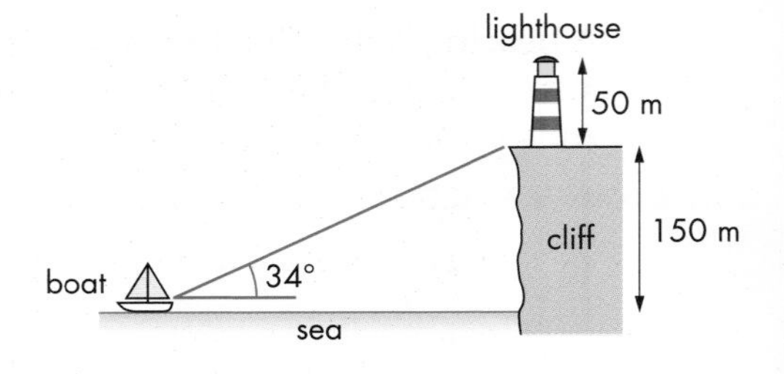

AU 6 A bird flies from the top of a 12 m tall tree, at an angle of depression of 34°, to catch a worm on the ground.

a How far does the bird actually fly?

b How far was the worm from the base of the tree?

FM 7 Sunil wants to work out the height of a building. He stands about 50 m away from a building. The angle of elevation from Sunil to the top of the building is about 15°. How tall is the building?

AU 8 The top of a ski run is 100 m above the finishing line. The run is 300 m long. What is the angle of depression of the ski run?

PS 9 Nessie and Cara are standing on opposite sides of a tree.

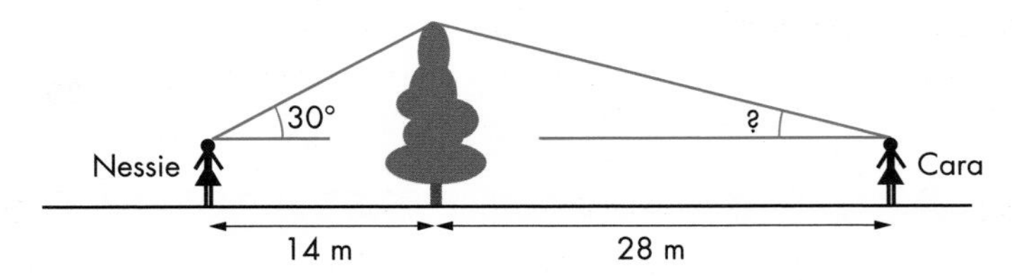

Nessie is 14 m away and the angle of elevation of the top of the tree is 30°.

Cara is 28 m away. She says the angle of elevation for her must be 15° because she is twice as far away.

Is she correct?

What do you think the angle of elevation is?

5.11 Solving problems using trigonometry 2

This section will show you how to:
- solve bearing problems using trigonometry
- use trigonometry to solve problems involving isosceles triangles

Key words
bearing
isosceles triangle
three-figure bearing
trigonometry

Trigonometry and bearings

A **bearing** is the direction to one place from another. The usual way of giving a bearing is as an angle measured from north in a clockwise direction. This is how a navigational compass and a surveyor's compass measure bearings.

A bearing is always written as a three-digit number, known as a **three-figure bearing**.

The diagram shows how this works, using the main compass points as examples.

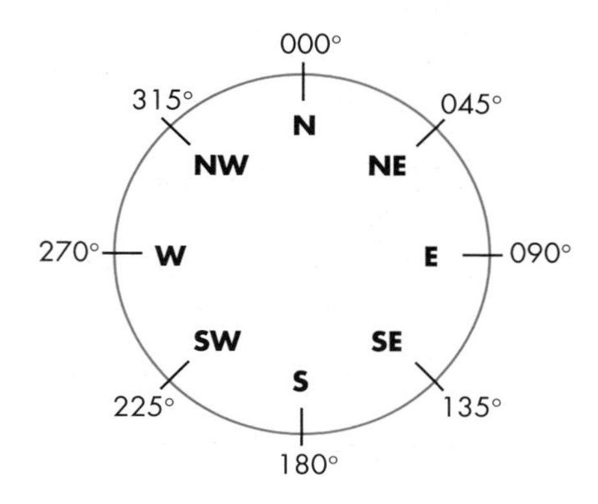

When working with bearings, follow these three rules.

- Always start from *north*.
- Always measure *clockwise*.
- Always give a bearing in degrees and as a *three-figure bearing*.

The difficulty with trigonometric problems involving bearings is dealing with those angles greater than 90° whose trigonometric ratios have negative values. To avoid this, we have to find a right-angled triangle that we can readily use. Example 24 shows you how to deal with such a situation.

EXAMPLE 24

A ship sails on a bearing of 120° for 50 km. How far east has it travelled?

The diagram of the situation is shown in figure **i**. From this, you can get the acute-angled triangle shown in figure **ii**.

i N, 120°, 50 km

ii A x cm, 30°, 50 cm H

From figure **ii**, you see that this is a cosine problem.

So $\cos 30° = \dfrac{x}{50}$

$x = 50 \cos 30° = 43.301 = 43.3$ km (3 significant figures)

So $x = \dfrac{50}{\cos 30°} = 43.301$

Distance east = 43.3 km (to 3 significant figures)

EXERCISE 5N

B

1 A ship sails for 75 km on a bearing of 078°.

a How far east has it travelled?

b How far north has it travelled?

2 Lopham is 17 miles from Wath on a bearing of 210°.

a How far south of Wath is Lopham?

b How far east of Lopham is Wath?

FM 3 A plane sets off from an airport and flies due east for 120 km, then turns to fly due south for 70 km before landing at Seddeth. Another pilot decides to fly the direct route from the airport to Seddeth. On what bearing should he fly?

PS 4 A helicopter leaves an army base and flies 60 km on a bearing of 278°.

a How far west has the helicopter flown?

b How far north has the helicopter flown?

5 A ship sails from a port on a bearing of 117° for 35 km before heading due north for 40 km and docking at Angle Bay.

a How far south had the ship sailed before turning?

b How far north had the ship sailed from the port to Angle Bay?

c How far east is Angle Bay from the port?

d What is the bearing from the port to Angle Bay?

AU 6 Mountain A is due west of a walker. Mountain B is due north of the walker. The guidebook says that mountain B is 4.3 km from mountain A, on a bearing of 058°. How far is the walker from mountain B?

PS 7 The shopping mall is 5.5 km east of my house and the supermarket is 3.8 km south. What is the bearing of the supermarket from the shopping mall?

8 The diagram shows the relative distances and bearings of three ships A, B and C.

a How far north of A is B? (Distance x on diagram.)

b How far north of B is C? (Distance y on diagram.)

c How far west of A is C? (Distance z on diagram.)

d What is the bearing of A from C? (Angle $w°$ on diagram.)

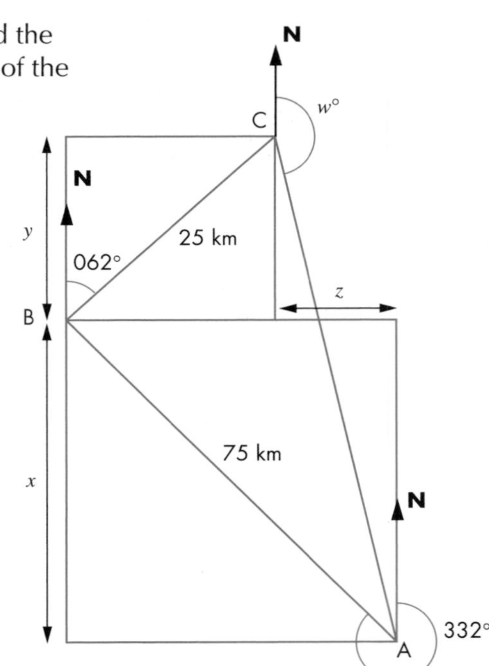

AU 9 A ship sails from port A for 42 km on a bearing of 130° to point B. It then changes course and sails for 24 km on a bearing of 040° to point C, where it breaks down and anchors. What distance and on what bearing will a helicopter have to fly from port A to go directly to the ship at C?

Trigonometry and isosceles triangles

Isosceles triangles often feature in **trigonometry** problems because such a triangle can be split into two right-angled triangles that are congruent.

EXAMPLE 25

a Find the length x in this isosceles triangle.

b Calculate the area of the triangle.

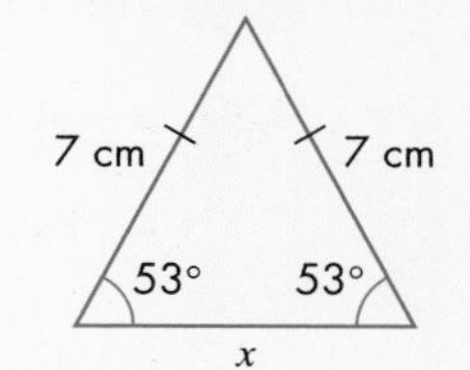

Draw a perpendicular from the apex of the triangle to its base, splitting the triangle into two congruent, right-angled triangles.

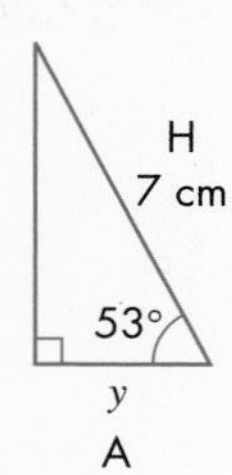

a To find the length y, which is $\frac{1}{2}$ of x, use cosine.

So, $\cos 53° = \frac{y}{7}$

$y = 7 \cos 53° = 4.212\,7051$ cm

So the length $x = 2y = 8.43$ cm (3 significant figures)

b To calculate the area of the original triangle, you first need to find its vertical height, h.

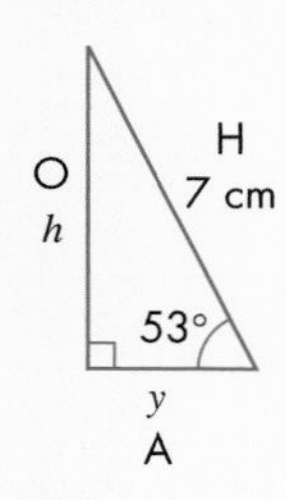

You have two choices, both of which involve the right-angled triangle of part **a**. We can use either Pythagoras' theorem ($h^2 + y^2 = 7^2$) or trigonometry. It is safer to use trigonometry again, since we are then still using known information.

This is a sine problem.

So, $\sin 53° = \frac{h}{7}$

$h = 7 \sin 53° = 5.590\,4486$ cm

(Keep the accurate figure in the calculator.)

The area of the triangle = $\frac{1}{2}$ × base × height. (We should use the most accurate figures we have for this calculation.)

$A = \frac{1}{2} \times 8.425\,4103 \times 5.590\,4486 = 23.6 \text{ cm}^2$ (3 significant figures)

You are not expected to write down these eight-figure numbers, just to use them.

EXERCISE 5O

B

1 Find the side or angle marked x.

a

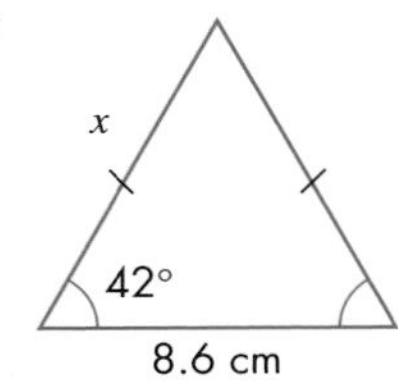

b

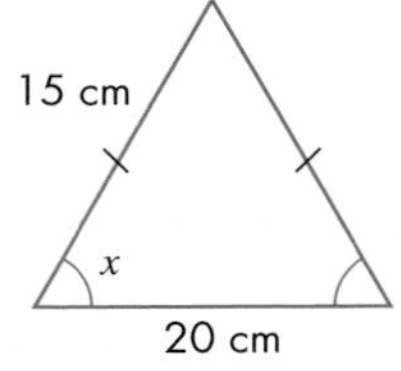

c

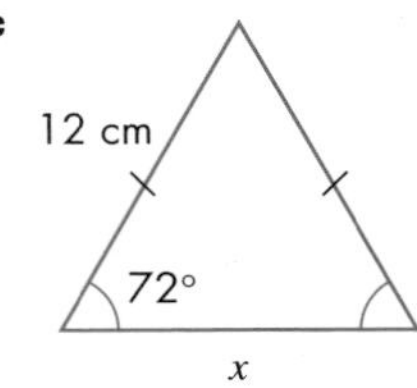

d

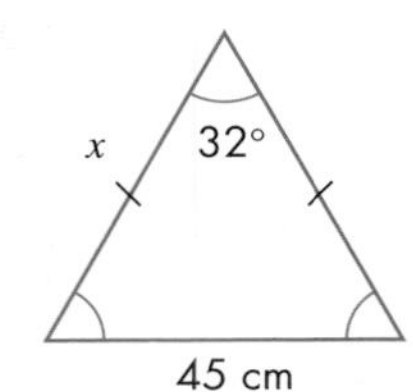

AU 2 This diagram below shows a roof truss. How wide is the roof?

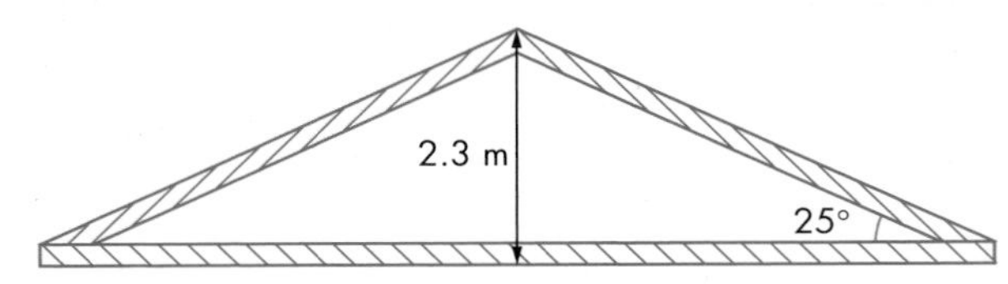

3 Calculate the area of each of these triangles.

a

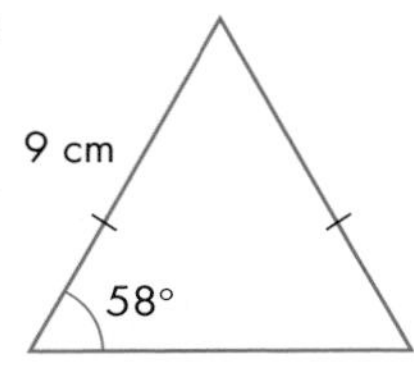

b

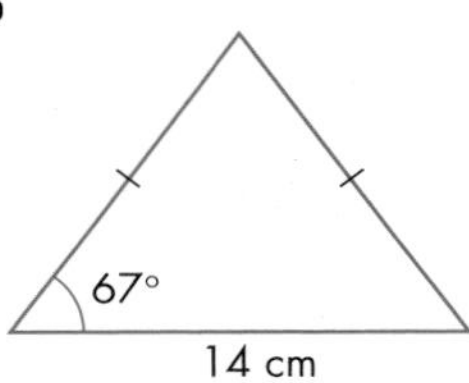

c

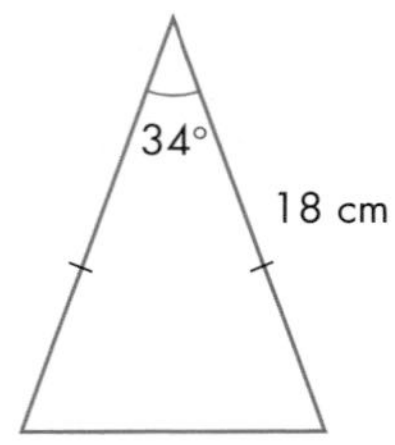

d

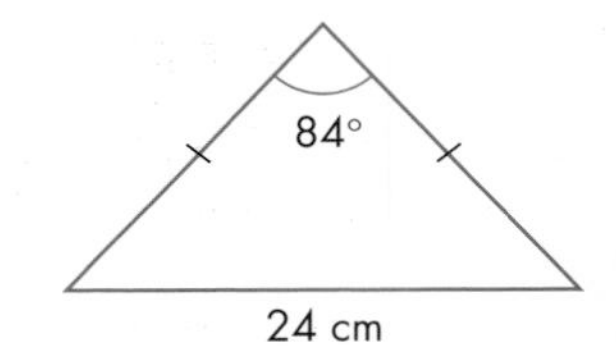

PS 4 An equilateral triangle has sides of length 10 cm.

A square is drawn on each side.

The corners of the squares are joined as shown.

What is the area of the resulting hexagon?

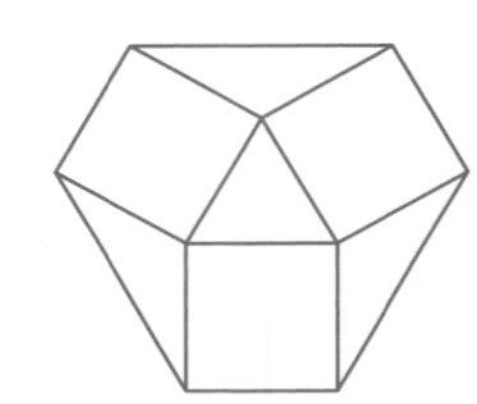

GRADE BOOSTER

- **C** You can use Pythagoras' theorem in right-angled triangles
- **C** You can solve problems in 2D using Pythagoras' theorem
- **B** You can solve problems in 3D using Pythagoras' theorem
- **B** You can use trigonometry to find lengths of sides and angles in right-angled triangles
- **A** You can use trigonometry to solve problems

What you should know now

- How to use Pythagoras' theorem
- How to solve problems using Pythagoras' theorem
- How to use the trigonometric ratios for sine, cosine and tangent in right-angled triangles
- How to solve problems using trigonometry
- How to solve problems using angles of elevation, angles of depression and bearings

EXAMINATION QUESTIONS

1 A football pitch ABCD is shown. The length of the pitch, AB = 120 m. The width of the pitch, BC = 90 m.

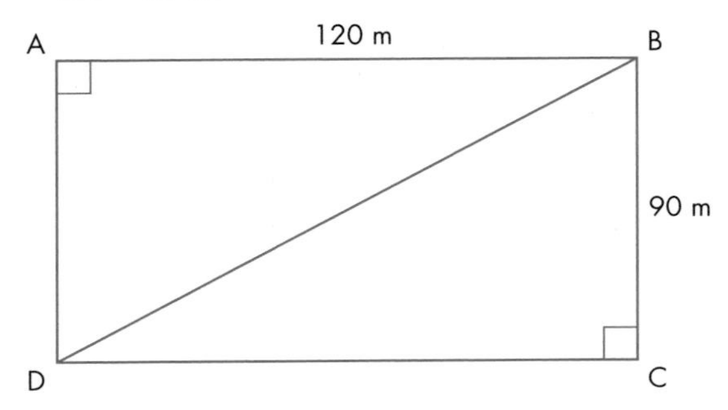

Calculate the length of the diagonal BD. Give your answer to 1 decimal place.

2 A ladder is leant against a wall. Its foot is 0.8 m from the wall and it reaches to a height of 4 m up the wall.

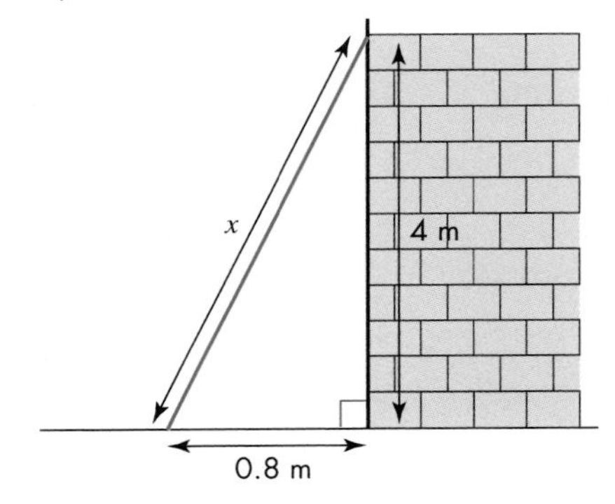

Calculate the length, in metres, of the ladder (marked x on the diagram). Give your answer to a suitable degree of accuracy.

3 In the diagram, ABC is a right-angled triangle. AC = 18 cm and AB = 12 cm.

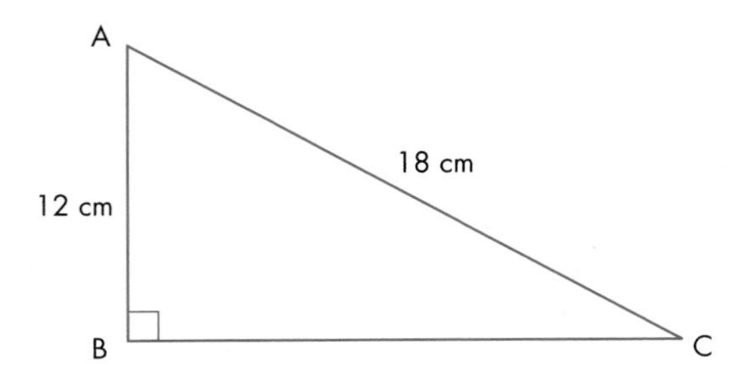

Calculate the length of BC.

4 **a** ABC is a right angled triangle. AB = 12 cm, BC = 8 cm. Find the size of angle CAB (marked x in the diagram). Give your answer to 1 decimal place.

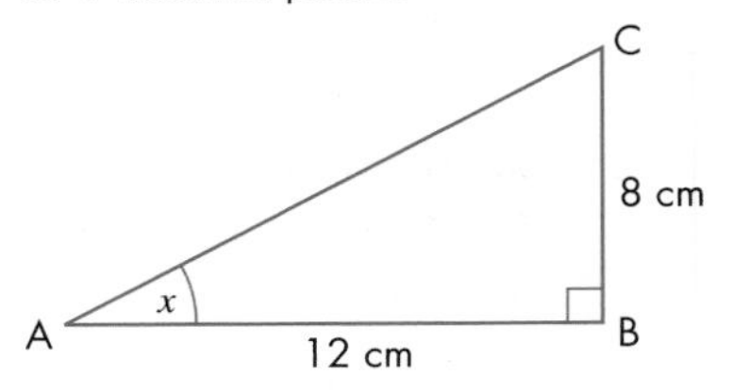

b PQR is a right-angled triangle. PQ = 15 cm, angle QPR = 32°. Find the length of PR (marked y in the diagram). Give your answer to 1 decimal place.

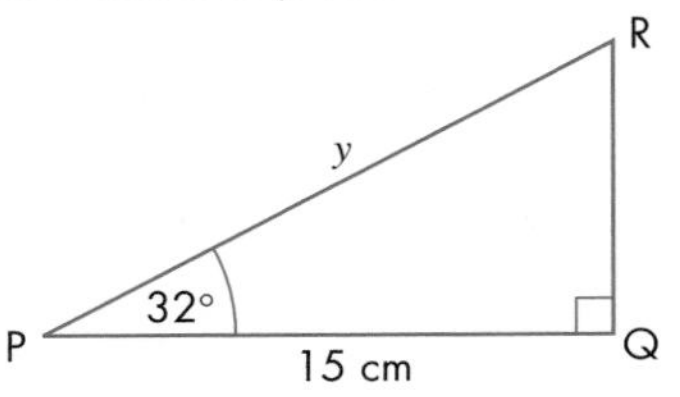

5 ABC is a right-angled triangle.

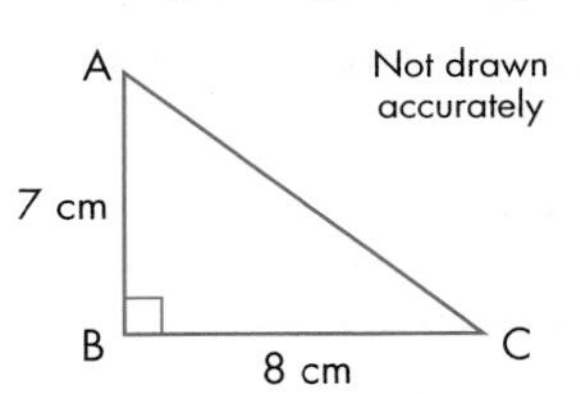

AB = 7 cm

BC = 8 cm

a Work out the area of the triangle. *(2 marks)*

b Work out the length of AC. *(3 marks)*

DEF is another right-angled triangle.

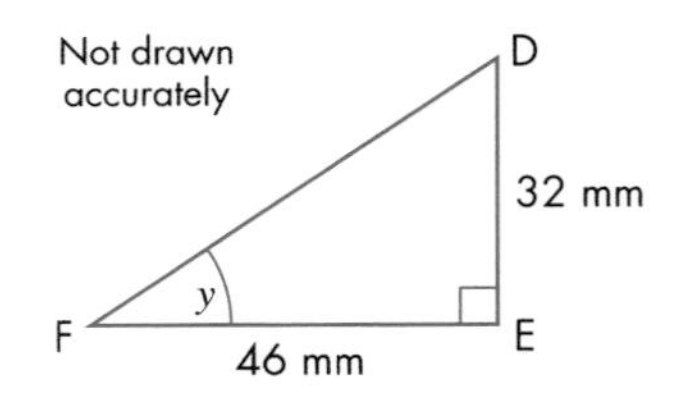

DE = 32 mm

FE = 46 mm

c Calculate the size of angle y.

Give your answer correct to 1 decimal place. *(3 marks)*

Edexcel, June 2008, Paper 4, Question 14

B C

6 

PQR is a right-angled triangle.

QR = 4 cm

PR = 10 cm

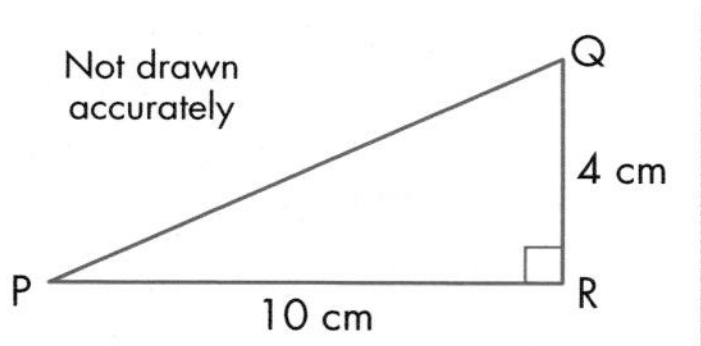

Work out the size of angle RPQ. Give your answer correct to 3 significant figures. *(3 marks)*

Edexcel, November 2008, Paper 15, Question 16

7 A lighthouse, L, is 3.2 km due West of a port, P. A ship, S, is 1.9 km due North of the lighthouse, L.

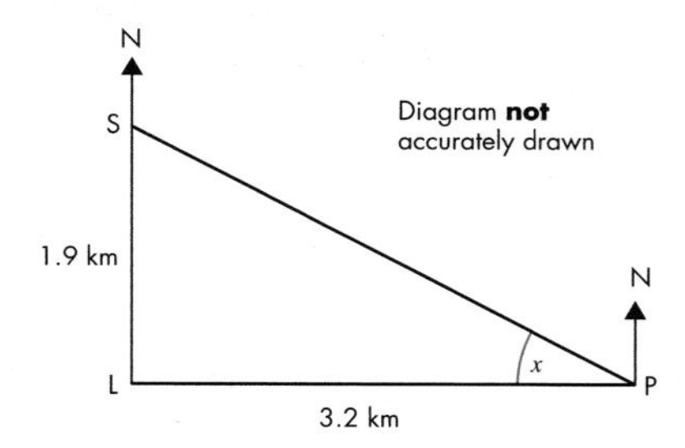

a Calculate the size of the angle marked x. Give your answer correct to 3 significant figures.

b Find the bearing of the port, P, from the ship, S. Give your answer correct to 3 significant figures.

Edexcel, June 2005, Paper 6 Higher, Question 10

Worked Examination Questions

1 a ABC is a right-angled triangle. AC = 19 cm and AB = 9 cm.

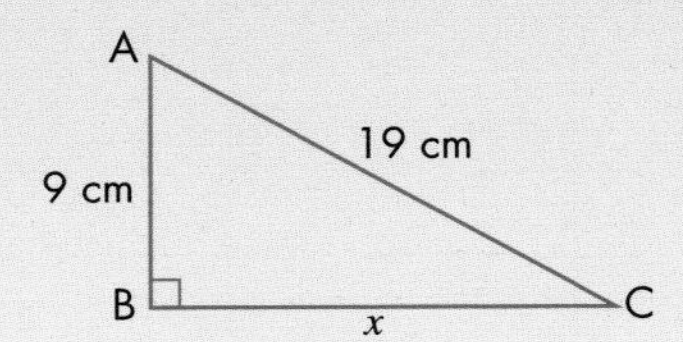

Calculate the length of BC.

b PQR is a right-angled triangle. PQ = 11 cm and QR = 24 cm.

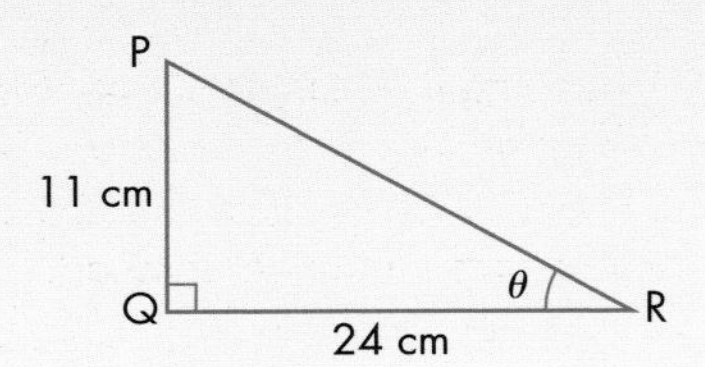

Calculate the size of angle PRQ.

c ABC and ACD are right-angled triangles. AD is parallel to BC.

AB = 12 cm, BC = 5 cm and AD = 33.8 cm.

Calculate the size of angle ADC.

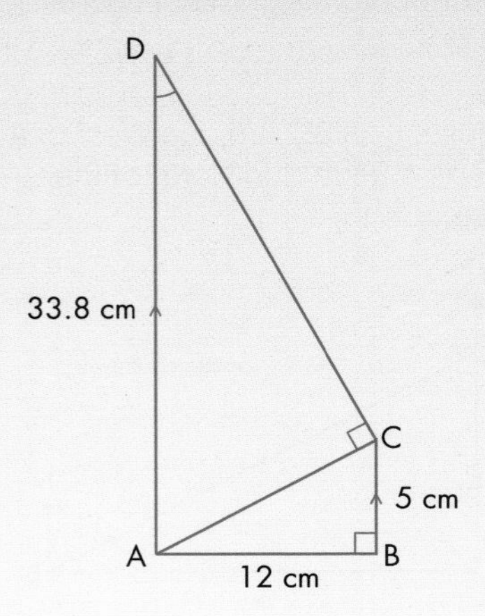

1 a Let BC = x

By Pythagoras' theorem

$x^2 = 19^2 - 9^2 \text{ cm}^2$ — This gets 1 mark for method.

$= 280 \text{ cm}^2$

So $x = \sqrt{280}$ — This gets 1 mark for method.

$= 16.7$ cm (3 sf) — This gets 1 mark for accuracy.

b Let $\angle PRQ = \theta$

So $\tan \theta = \frac{11}{24}$ — This gets 1 mark for method.

$\theta = \tan^{-1} \frac{11}{24} = 24.6°$ (1 dp) — This gets 1 mark for accuracy.

c In triangle ABC, let AC = y

By Pythagoras' theorem

$y^2 = 5^2 + 12^2 \text{ cm}^2$ — This gets 1 mark for method.

$= 169 \text{ cm}^2$

$y = \sqrt{169} = 13$ cm — This gets 1 mark for accuracy.

In triangle ACD, let $\angle ADC = z$

So, $\sin z = \frac{13}{33.8} = 0.3846$ — This gets 1 mark for method.

$z = \sin^{-1} 0.3846 = 22.6°$ (1 dp) — This gets 1 mark for accuracy.

Total: 9 marks

Worked Examination Questions

FM **2** A clock is designed to have circular face on a triangular surround.
The triangle is equilateral.

The face extends to the edge of the triangle.

The diameter of the clock face is 18.0 cm.

What is the perimeter of the triangle?

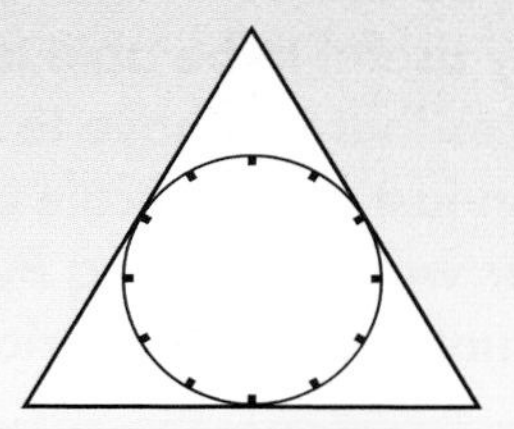

2 Find a right-angled triangle to use.

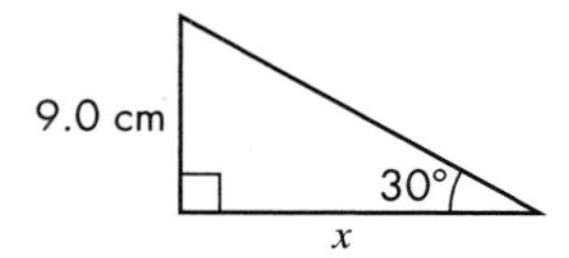

The radius is 9 cm. The angle is 30° because it is half the angle of an equilateral triangle. — Drawing the triangle scores 1 mark for method.

$\frac{9}{x} = \tan 30°$ — This gets 1 mark for method.

$x = \frac{9}{\tan 30°} = 15.58\ldots$ — This gets 1 mark for method and 1 mark for accuracy.

Perimeter of triangle = 15.88... × 6 = 93.5 cm or 94 cm — This gets 1 mark for accuracy.

Total: 5 marks

PS **3** Find the area of a regular hexagon of side 6 cm.

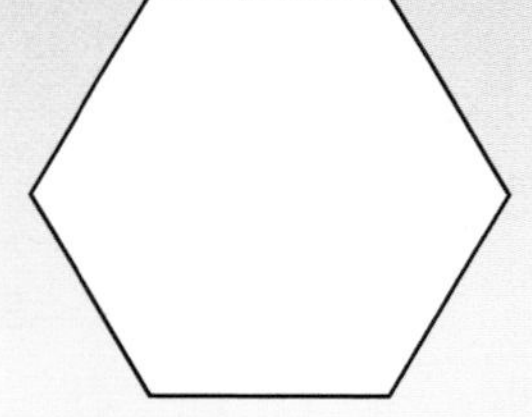

3 Divide the hexagon into smaller parts and add the separate areas together.
One way is to divide it into six equilateral triangles. — This gets 1 mark for method.
Use Pythagoras' theorem (or trigonometry) to find the height of the triangle.

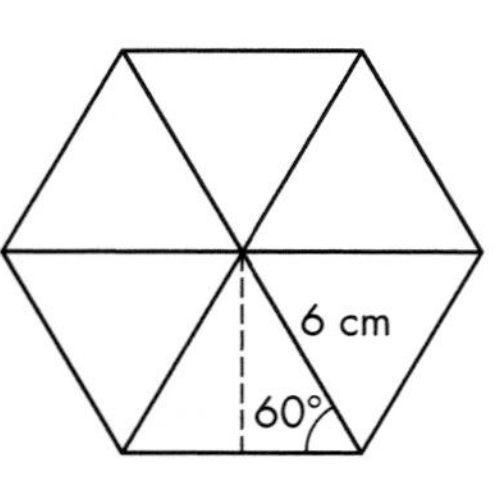

For example, height = $\sqrt{(62 - 32)} = 5.19\ldots$ — This gets 1 mark for method.

Area of one triangle = 3 × 5.19… = 15.58…

Area of hexagon = 6 × area of triangle = 94 cm^2 — This gets 1 mark for method and 1 mark for accuracy. Remember not to round your answer until the end.

Total: 5 marks

Problem Solving
Map work using Pythagoras

When you are out walking on the hills it can be very useful to be able to estimate various distances that you have to cover. Of course, you can just use the scale on the map, but another way is by using Pythagoras' theorem with small right-angled triangles.

Getting started

- You know that the **square** of 2 is $2^2 = 4$.
  Write down the **square** of each of these numbers.
  5 0.1 0.4 0.03
  Think of a number that has a square between 0.1 and 0.001.
- You know that the **square root** of 4 is $\sqrt{4} = 2$.
  Now write down the **square root** of each of these numbers.
  16 81 0.01 0.025
  Think of a number that has a square root between 50 and 60.
- On a set of coordinate axes, draw the points A(1, 2) and B(4, 6).
  What is the distance from point A to point B?

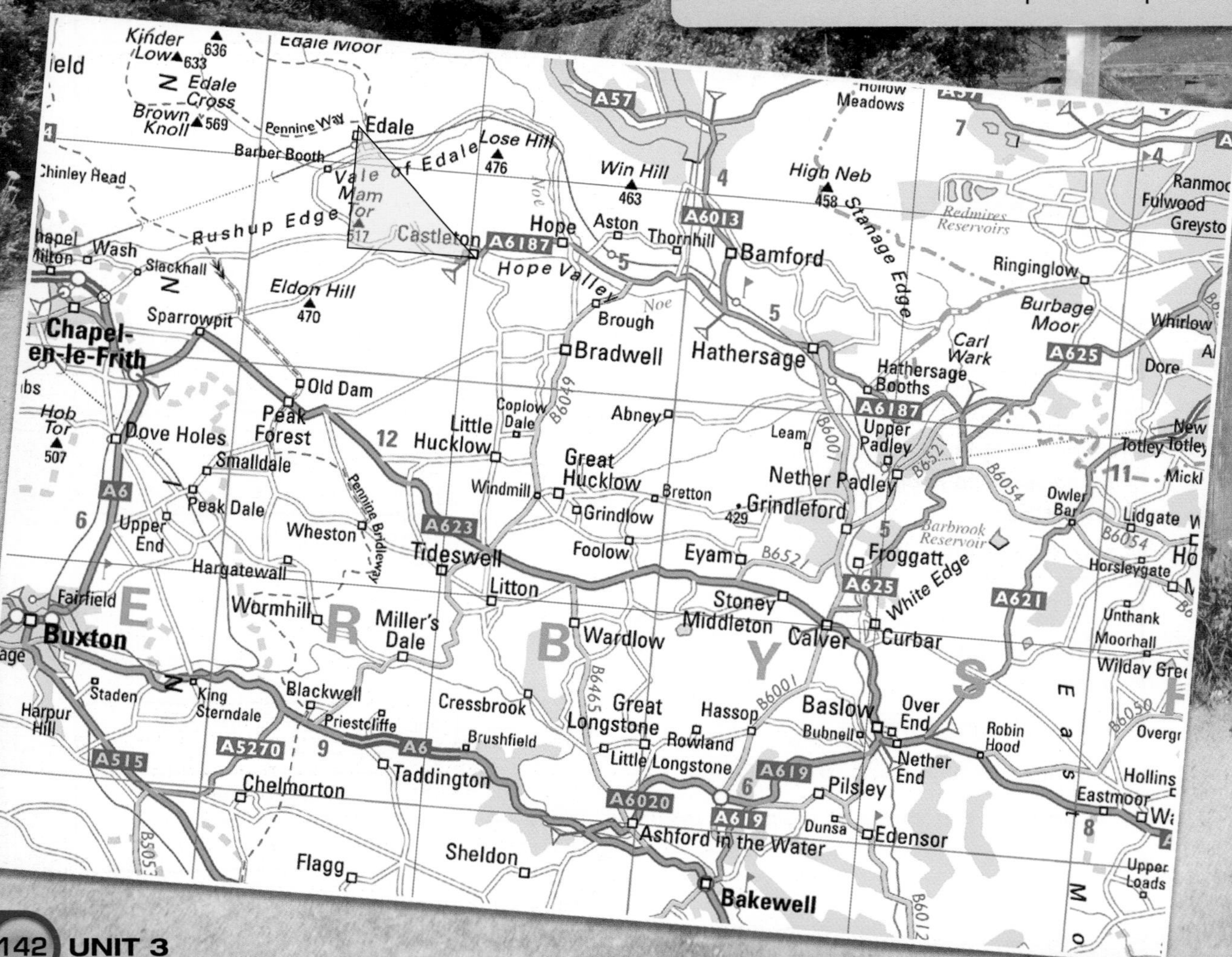

Your task

Freya and Chris often go out walking in the Peak District. On the left is a copy of the map they use. Use this map extract to complete these tasks.

1 Write five questions similar to the examples given on the right. Swap them with the person next to you.

Answer each other's questions, making sure you show your working clearly.

Now swap again and mark each other's answers. Give constructive feedback.

2 Plan a walk with a circular route that is between 20 and 35 km long.

If the average person walks at approximately 4.5 miles per hour, estimate the time it would take to complete your route.

Example

Freya and Chris were at Edale. They wanted to know the rough distance to Castleton. Freya decided to set herself a maths problem, using Pythagoras' theorem.

She looked at the map and imagined the yellow right-angled triangle.

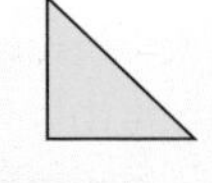

Using the fact that each square on the map represents an area 5 km by 5 km, she estimated each small side of the triangle to be 3 km.

Then she applied Pythagoras' theorem.

$3^2 + 3^2 = 9 + 9 = 18$

On the hillside, without a calculator, she estimated the square root of 18 to be just over 4, giving a distance of 4 km.

Another day, Freya and Chris were at Hucklow and wanted to know the distance to Hathersage.

Use Freya's method to estimate the distance from Hucklow to Hathersage.

Additional information

When working in distances, you need to work in either miles and other imperial units or kilometres and other metric units.

To change between these units there are some key conversion facts. Either use a textbook or the internet to find these.

Why this chapter matters

How many sides does a strip of paper have? Two or one?

Take a strip of paper about 20 cm by 2 cm.

A B

How many sides does it have? Easy! You can see that this has two sides, a topside and an underside. If you were to draw a line along one side of the strip, you would have one side with a line 20 cm long on it and one side blank.

Now mark the ends A and B, put a single twist in the strip of paper and tape (or glue) the two ends together, as shown.

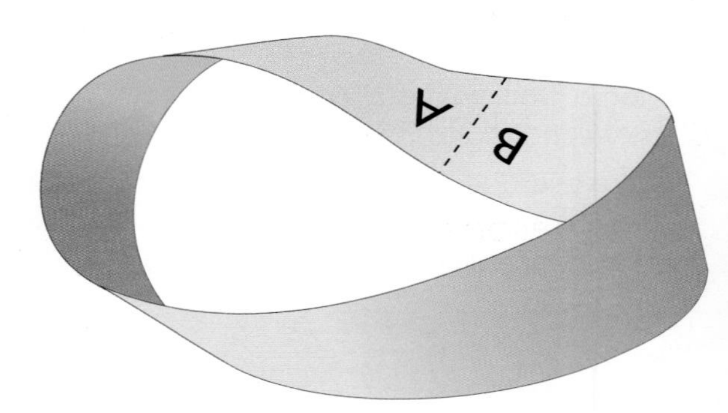

How many sides does this strip of paper have now?

Take a pen and draw a line on the paper, starting at any point you like. Continue the line along the length of the paper – you will eventually come back to your starting point. Your strip has only one side now! There is no blank side.

You have transformed a two-sided piece of paper into a one-sided piece of paper.

This curious shape is called a Möbius strip. It is named after August Ferdinand Möbius, a 19th-century German mathematician and astronomer. Möbius, along with others, caused a revolution in geometry.

Möbius strips have a number of surprising applications that exploit its remarkable property of one-sidedness, including conveyor belts in industry as well as in domestic vacuum cleaners. Have a look at the belt that turns the rotor in the vacuum cleaner at home.

The Möbius strip has become the universal symbol of recycling. The symbol was created in 1970 by Gary Anderson, who was a senior at the University of Southern California, as part of a contest sponsored by a paper company.

The Möbius strip is a form of transformation. In this chapter, you will look at some other transformations of shapes.

Chapter

Geometry: Transformation geometry

The grades given in this chapter are target grades.

1. Congruent triangles
2. Translations
3. Reflections
4. Rotations
5. Enlargement
6. Combined transformations

This chapter will show you ...

- D what is meant by a transformation
- D to B how to translate, reflect, rotate and enlarge 2D shapes
- B to A how to show that two triangles are congruent

Visual overview

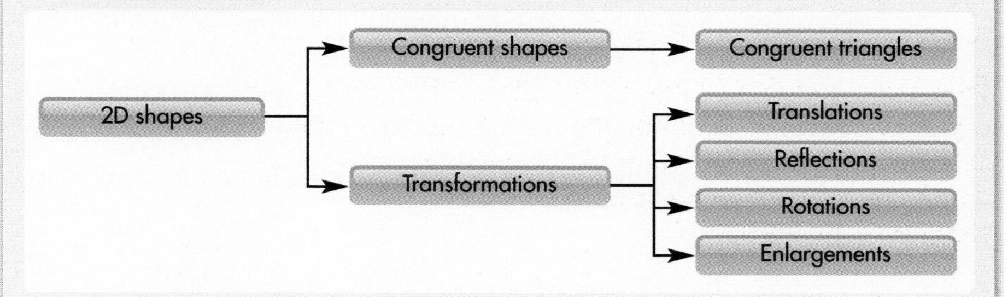

What you should already know

- How to recognise congruent shapes **(KS3 level 4, GCSE grade G)**
- How to find the lines of symmetry of a 2D shape **(KS3 level 4, GCSE grade F)**
- How to find the order of rotational symmetry of a 2D shape **(KS3 level 4, GCSE grade F)**
- How to draw the lines with equations $x = \pm a$, $y = \pm b$, $y = x$ and $y = -x$ **(KS3 level 5, GCSE grade E)**

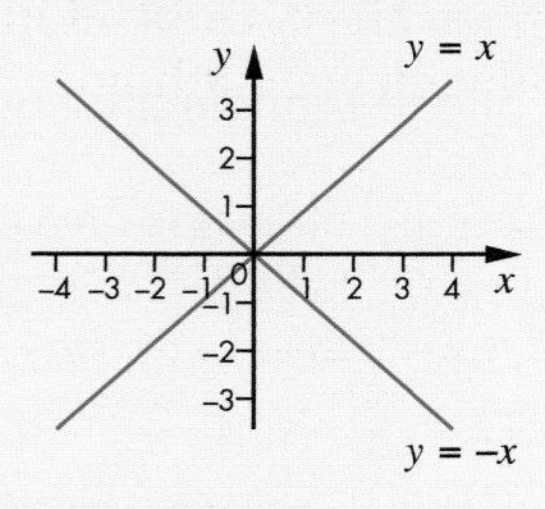

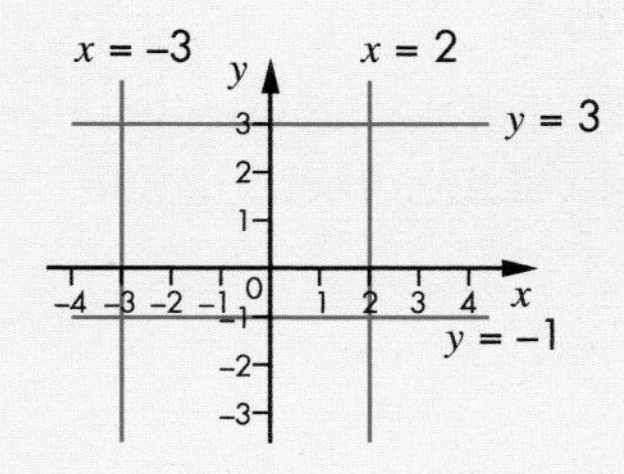

Quick check

Which of these shapes is not congruent to the others?

a 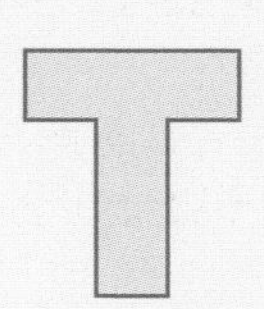

b 

c 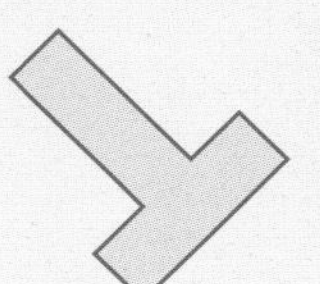

d 

6.1 Congruent triangles

This section will show you how to:
- show that two triangles are congruent

Key words
congruent

Two shapes are **congruent** if they are exactly the same size and shape.

For example, these triangles are all congruent.

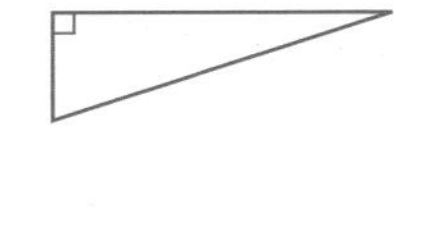

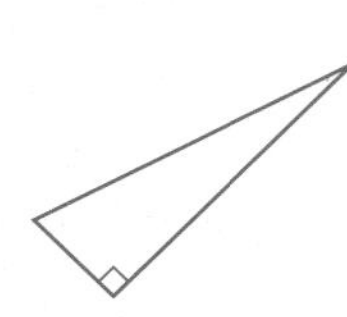
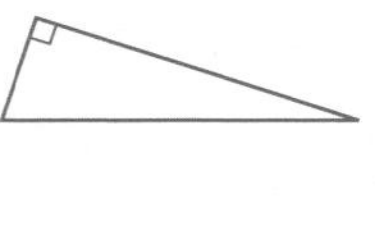

Notice that the triangles can be differently oriented (reflected or rotated).

Conditions for congruent triangles

Any one of the following four conditions is sufficient for two triangles to be congruent.

- **Condition 1**

  All three sides of one triangle are equal to the corresponding sides of the other triangle.

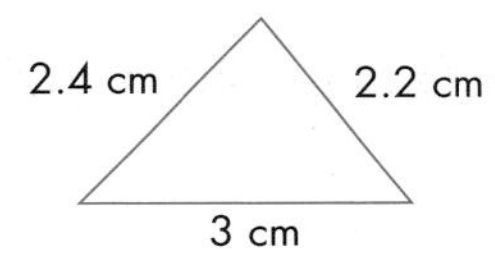

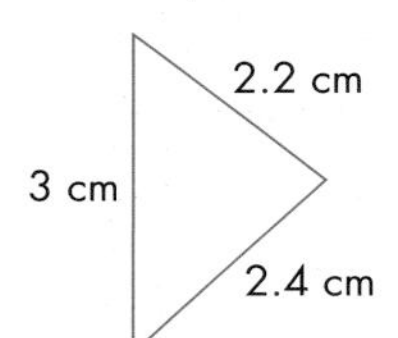

  This condition is known as SSS (side, side, side).

- **Condition 2**

  Two sides and the angle between them of one triangle are equal to the corresponding sides and angle of the other triangle.

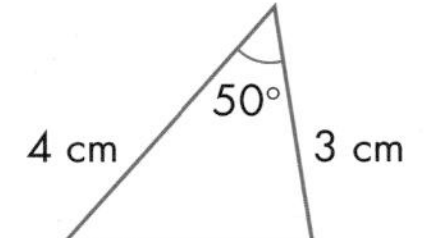

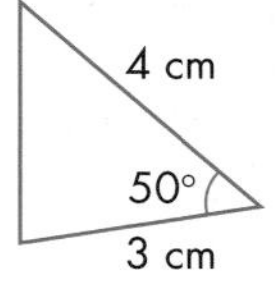

  This condition is known as SAS (side, angle, side).

- **Condition 3**

  Two angles and a side of one triangle are equal to the corresponding angles and side of the other triangle.

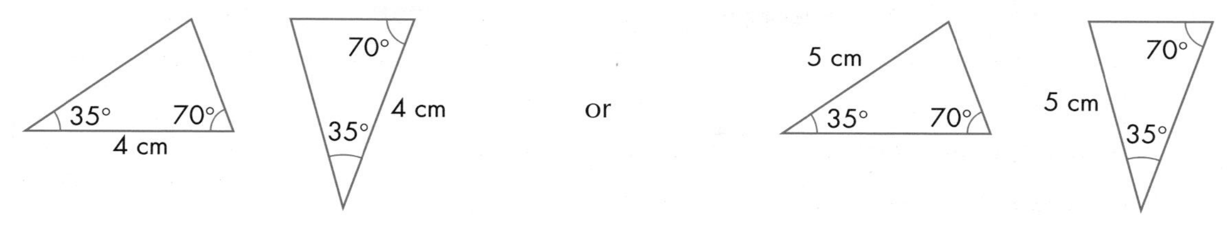

  This condition is known as ASA (angle, side, angle) or AAS (angle, angle, side).

- **Condition 4**

  Both triangles have a right angle, an equal hypotenuse and another equal side.

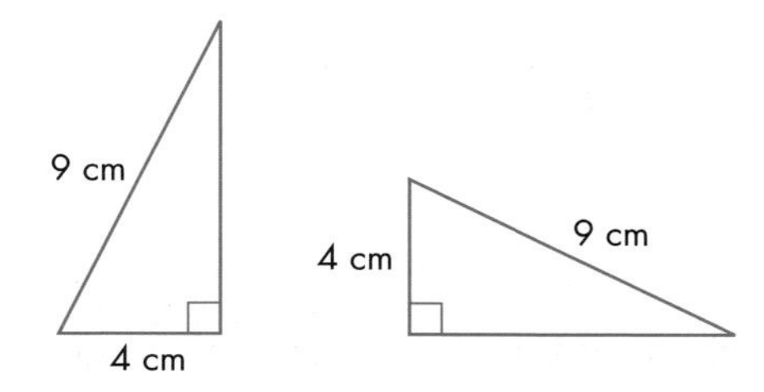

  This condition is known as RHS (right angle, hypotenuse, side).

Notation

Once you have shown that triangle ABC is congruent to triangle PQR by one of the above conditions, it means that:

$\angle A = \angle P$ AB = PQ

$\angle B = \angle Q$ BC = QR

$\angle C = \angle R$ AC = PR

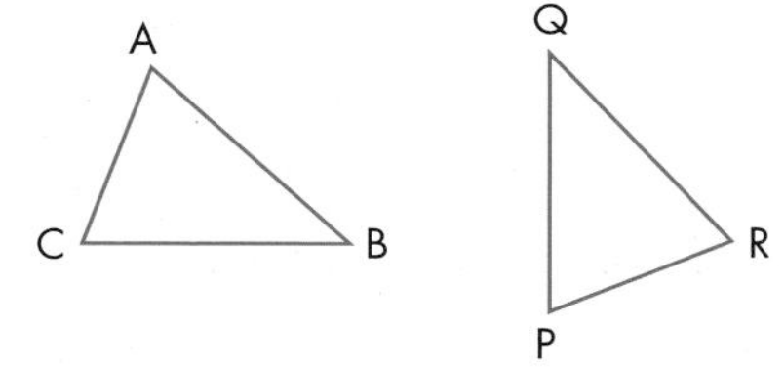

In other words, the points ABC correspond exactly to the points PQR in that order. Triangle ABC is congruent to triangle PQR can be written as $\Delta ABC \equiv \Delta PQR$.

EXAMPLE 1

ABCD is a kite. Show that triangle ABC is congruent to triangle ADC.

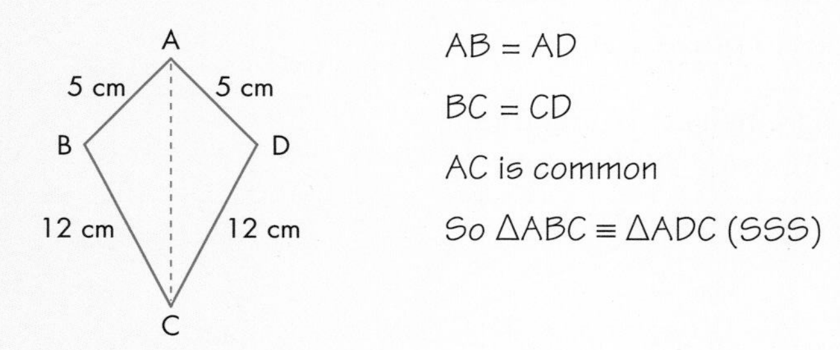

AB = AD

BC = CD

AC is common

So $\Delta ABC \equiv \Delta ADC$ (SSS)

EXERCISE 6A

B

1 The triangles in each pair are congruent. State the condition that shows that the triangles are congruent.

a

5 cm
100°
3 cm

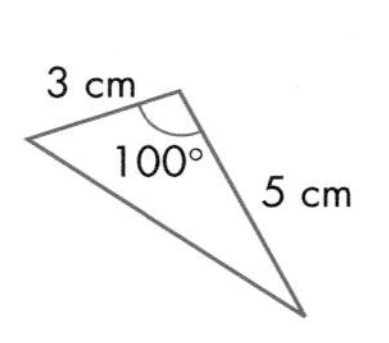

b

5 cm
4 cm
7 cm

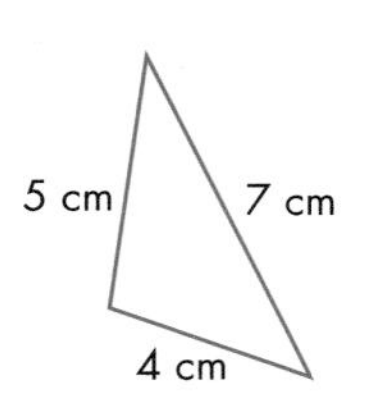

c

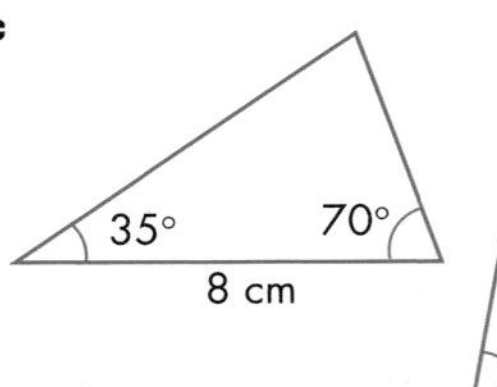

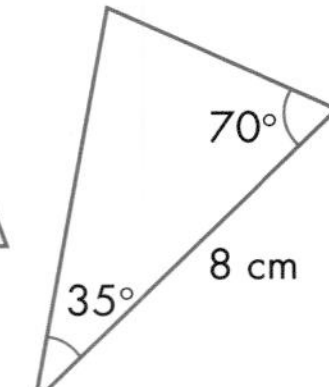

d

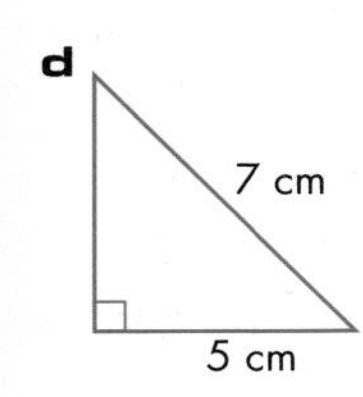

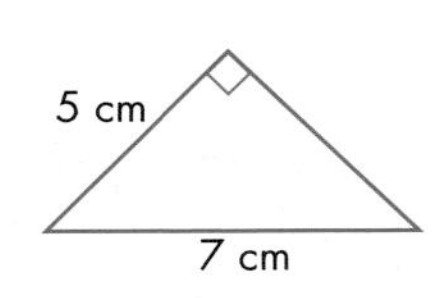

e

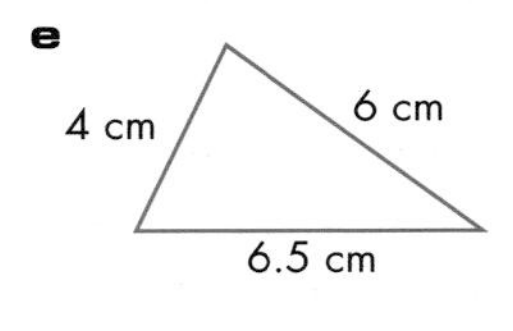

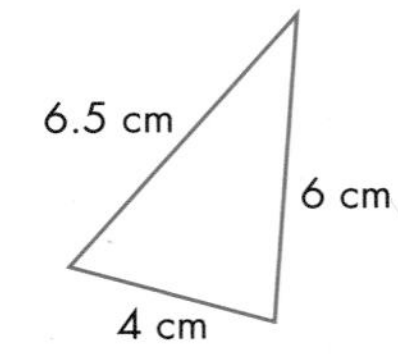

f

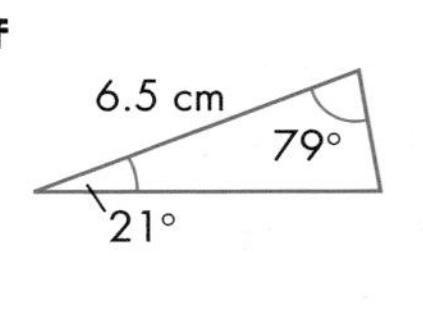

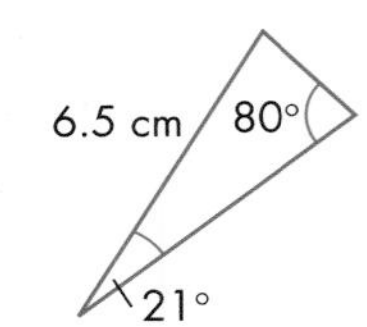

2 The triangles in each pair are congruent. State the condition that shows that the triangles are congruent and say which points correspond to which.

a ABC where AB = 8 cm, BC = 9 cm, AC = 7.4 cm
PQR where PQ = 9 cm, QR = 7.4 cm, PR = 8 cm

b ABC where AB = 5 cm, BC = 6 cm, angle B = 35°
PQR where PQ = 6 cm, QR = 50 mm, angle Q = 35°

3 Triangle ABC is congruent to triangle PQR, ∠A = 60°, ∠B = 80° and AB = 5 cm. Find these.

a ∠P **b** ∠Q **c** ∠R **d** PQ

4 ABCD is congruent to PQRS, ∠A= 110°, ∠B = 55°, ∠C = 85° and RS = 4 cm. Find these.

a ∠P **b** ∠Q **c** ∠R **d** ∠S **e** CD

A

5 Draw a rectangle EFGH. Draw in the diagonal EG. Prove that triangle EFG is congruent to triangle EHG.

6 Draw an isosceles triangle ABC where AB = AC. Draw the line from A to X, the midpoint of BC. Prove that triangle ABX is congruent to triangle ACX.

PS 7 In the diagram ABCD and DEFG are squares.
Use congruent triangles to prove that AE = CG.

AU 8 Jez says that these two triangles are congruent because two angles and a side are the same.

Explain why he is wrong.

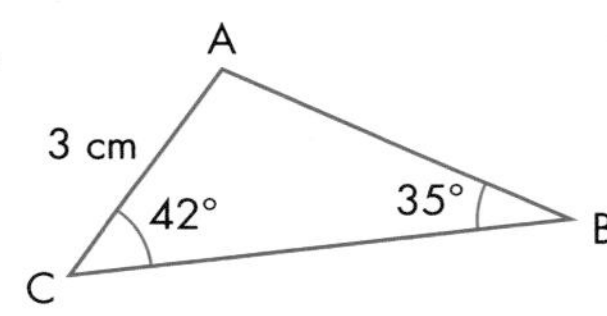

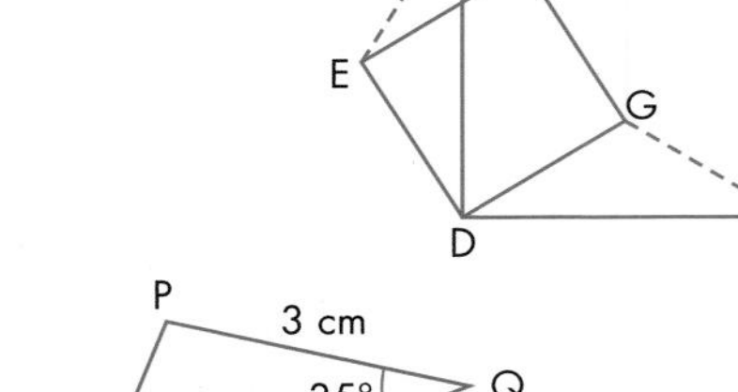

6.2 Translations

This section will show you how to:
- translate a 2D shape

Key words
transformation
translation
vector

A **transformation** changes the position or the size of a shape.

There are four basic ways of changing the position and size of 2D shapes: a **translation**, a reflection, a rotation or an enlargement. All of these transformations, except enlargement, keep shapes congruent.

A translation is the 'movement' of a shape from one place to another without reflecting it or rotating it. It is sometimes called a glide, since the shape appears to glide from one place to another. Every point in the shape moves in the same direction and through the same distance.

We describe translations by using **vectors**. A vector is represented by the combination of a horizontal shift and a vertical shift.

EXAMPLE 2

Use vectors to describe the translations of the following triangles.

a A to B

b B to C

c C to D

d D to A

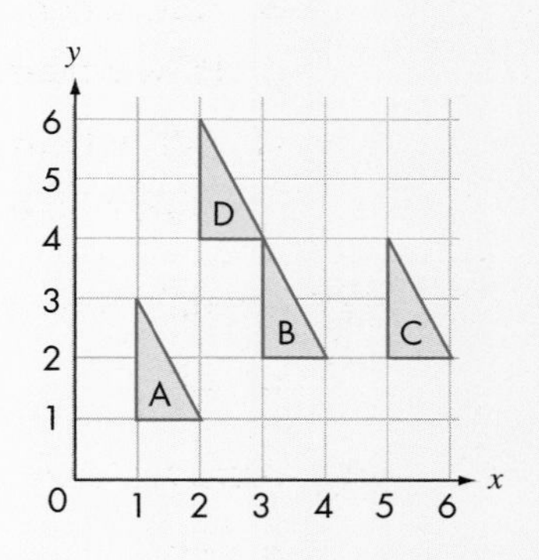

a The vector describing the translation from A to B is $\begin{pmatrix} 2 \\ 1 \end{pmatrix}$.

b The vector describing the translation from B to C is $\begin{pmatrix} 2 \\ 0 \end{pmatrix}$.

c The vector describing the translation from C to D is $\begin{pmatrix} -3 \\ 2 \end{pmatrix}$.

d The vector describing the translation from D to A is $\begin{pmatrix} -1 \\ 3 \end{pmatrix}$.

Note:
- The top number in the vector describes the horizontal movement. To the right +, to the left –.
- The bottom number in the vector describes the vertical movement. Upwards +, downwards –.
- These vectors are also called *direction vectors*.

EXERCISE 6B

1 Use vectors to describe the following translations of the shapes on the grid below.

a **i** A to B **ii** A to C **iii** A to D
iv A to E **v** A to F **vi** A to G

b **i** B to A **ii** B to C **iii** B to D
iv B to E **v** B to F **vi** B to G

c **i** C to A **ii** C to B **iii** C to D
iv C to E **v** C to F **vi** C to G

d **i** D to E **ii** E to B **iii** F to C
iv G to D **v** F to G **vi** G to E

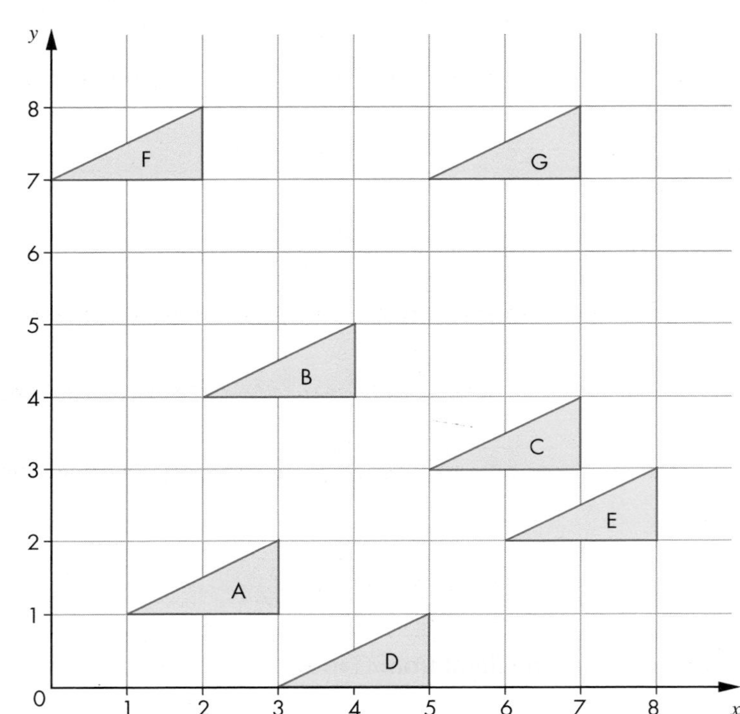

2 **a** Draw a set of coordinate axes and on it the triangle with coordinates A(1, 1), B(2, 1) and C(1, 3).

b Draw the image of ABC after a translation with vector $\begin{pmatrix} 2 \\ 3 \end{pmatrix}$. Label this triangle P.

c Draw the image of ABC after a translation with vector $\begin{pmatrix} -1 \\ 2 \end{pmatrix}$. Label this triangle Q.

d Draw the image of ABC after a translation with vector $\begin{pmatrix} 3 \\ -2 \end{pmatrix}$. Label this triangle R.

e Draw the image of ABC after a translation with vector $\begin{pmatrix} -2 \\ -4 \end{pmatrix}$. Label this triangle S.

3 Using your diagram from question **2**, use vectors to describe the translation that will move

a P to Q **b** Q to R **c** R to S **d** S to P

e R to P **f** S to Q **g** R to Q **h** P to S

PS 4 Draw a 10 × 10 coordinate grid and on it the triangle A(0, 0), B(1, 0) and C(0, 1). How many different translations are there that use integer values only and will move the triangle ABC to somewhere in the grid?

PS 5 In a game of *Snakes and ladders*, each of the snakes and ladders can be described by a translation.

Use the following vectors.

Ladders $\begin{pmatrix}1\\2\end{pmatrix}, \begin{pmatrix}2\\5\end{pmatrix}, \begin{pmatrix}-3\\4\end{pmatrix}, \begin{pmatrix}-2\\3\end{pmatrix}, \begin{pmatrix}3\\2\end{pmatrix}$

Snakes $\begin{pmatrix}1\\-3\end{pmatrix}, \begin{pmatrix}3\\-4\end{pmatrix}, \begin{pmatrix}-2\\-2\end{pmatrix}, \begin{pmatrix}-1\\-3\end{pmatrix}, \begin{pmatrix}2\\-5\end{pmatrix}$

Put all five ladders and all five snakes onto a 10 × 10 coordinate grid in order to design a *Snakes and ladders* game board.

AU 6 If a translation is given by:

$$\begin{pmatrix}x\\y\end{pmatrix}$$

describe the translation that would take the image back to the original position.

FM 7 A plane flies between three cities A, B and C. It uses direction vectors, with distances in kilometres.

The direction vector for the flight from A to B is $\begin{pmatrix}500\\200\end{pmatrix}$ and the direction vector for the flight from B to C is $\begin{pmatrix}-200\\300\end{pmatrix}$.

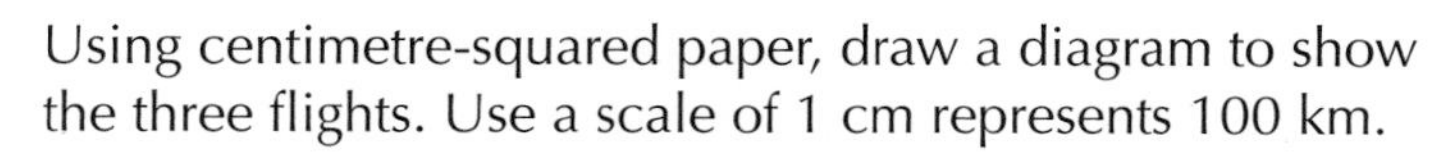

Using centimetre-squared paper, draw a diagram to show the three flights. Use a scale of 1 cm represents 100 km.

Work out the direction vector for the flight from C to A.

FM 8 A pleasure launch travels between three jetties X, Y and Z on a lake. It uses direction vectors, with distance in kilometres.

The direction vector from X to Y is $\begin{pmatrix}3\\-1\end{pmatrix}$ and the direction vector from Y to Z is $\begin{pmatrix}-2\\-3\end{pmatrix}$.

Using centimetre-squared paper, draw a diagram to show journeys between X, Y and Z. Use a scale of 1 cm represents 1 km. Work out the direction vector for the journey from Z to X.

6.3 Reflections

This section will show you how to:
- reflect a 2D shape in a mirror line

Key words
image
mirror line
object
reflection

A **reflection** transforms a shape so that it becomes a mirror image of itself.

EXAMPLE 3

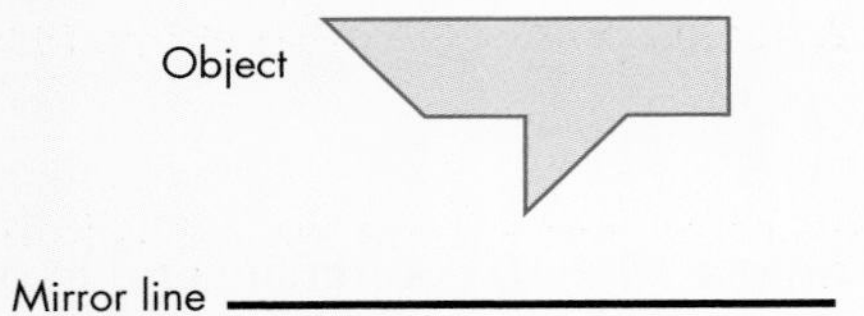

Notice the reflection of each point in the original shape, called the **object**, is perpendicular to the mirror line. So if you 'fold' the whole diagram along the **mirror line**, the object will coincide with its reflection, called its **image**.

EXERCISE 6C

1 Copy the diagram below and draw the reflection of the given triangle in the following lines.

a $x = 2$ **b** $x = -1$ **c** $x = 3$

d $y = 2$ **e** $y = -1$ **f** y-axis

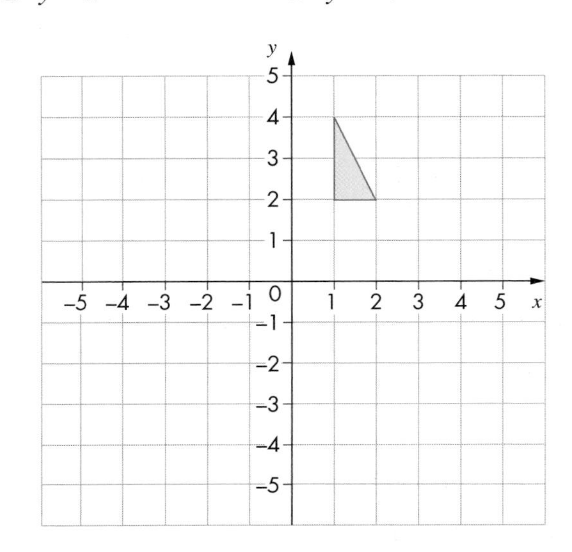

D

2 **a** Draw a pair of axes. Label the x-axis from –5 to 5 and the y-axis from –5 to 5.

b Draw the triangle with coordinates A(1, 1), B(3, 1), C(4, 5).

c Reflect the triangle ABC in the x-axis. Label the image P.

d Reflect triangle P in the y-axis. Label the image Q.

e Reflect triangle Q in the x-axis. Label the image R.

f Describe the reflection that will move triangle ABC to triangle R.

AU 3 **a** Draw a pair of axes. Label the x-axis from –5 to +5 and the y-axis from –5 to +5.

b Reflect the points A(2, 1), B(5, 0), C(–3, 3), D(3, –2) in the x-axis.

c What do you notice about the values of the coordinates of the reflected points?

d What would the coordinates of the reflected point be if the point (a, b) were reflected in the x-axis?

AU 4 **a** Draw a pair of axes. Label the x-axis from –5 to +5 and the y-axis from –5 to +5.

b Reflect the points A(2, 1), B(0, 5), C(3, –2), D(–4, –3) in the y-axis.

c What do you notice about the values of the coordinates of the reflected points?

d What would the coordinates of the reflected point be if the point (a, b) were reflected in the y-axis?

PS 5 By using the middle square as a starting square ABCD, describe how to keep reflecting the square to obtain the final shape in the diagram.

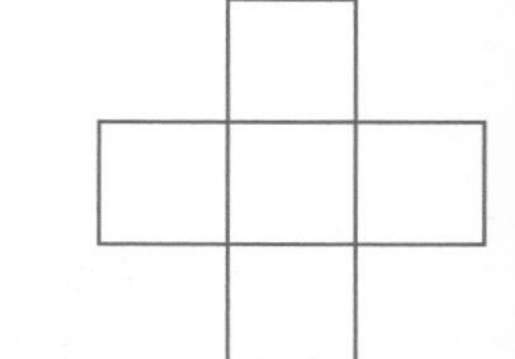

AU 6 Triangle A is drawn on a grid.

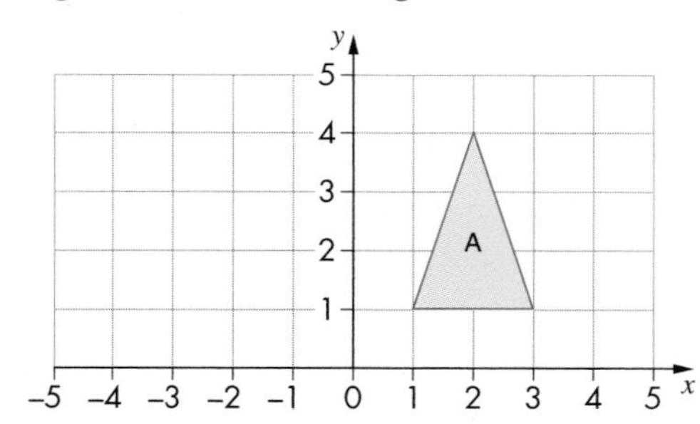

Triangle A is reflected to form a new triangle B.
The coordinates of B are (–4, 4), (–3, 1) and (–5, 1).

Work out the equation of the mirror line.

7 A designer used the following instructions to create a design.

- Start with any rectangle ABCD.
- Reflect the rectangle ABCD in the line AC.
- Reflect the rectangle ABCD in the line BD.

Draw a rectangle and use the above to create a design.

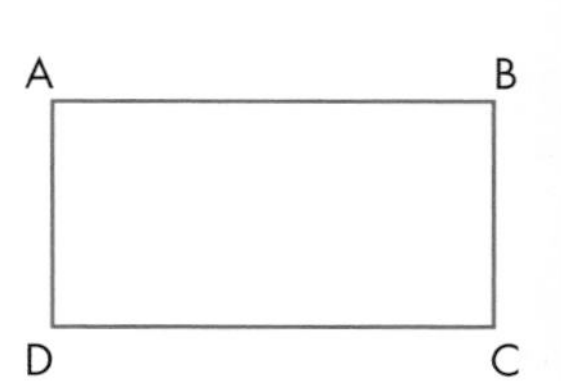

8 Draw each of these triangles on squared paper, leaving plenty of space on the opposite side of the given mirror line. Then draw the reflection of each triangle.

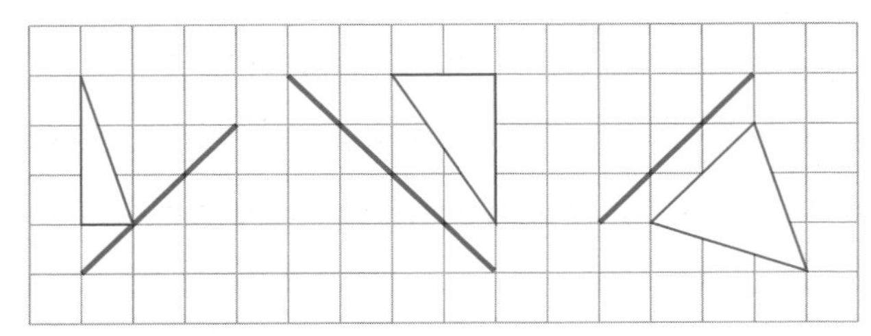

HINTS AND TIPS

Turn the page around so that the mirror lines are vertical or horizontal.

9 **a** Draw a pair of axes and the lines $y = x$ and $y = -x$, as shown.

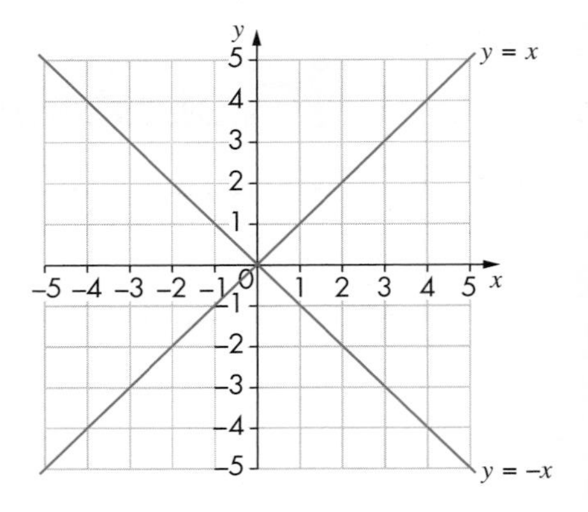

b Draw the triangle with coordinates A(2, 1), B(5, 1), C(5, 3).

c Draw the reflection of triangle ABC in the x-axis and label the image P.

d Draw the reflection of triangle P in the line $y = -x$ and label the image Q.

e Draw the reflection of triangle Q in the y-axis and label the image R.

f Draw the reflection of triangle R in the line $y = x$ and label the image S.

g Draw the reflection of triangle S in the x-axis and label the image T.

h Draw the reflection of triangle T in the line $y = -x$ and label the image U.

i Draw the reflection of triangle U in the y-axis and label the image W.

j What single reflection will move triangle W to triangle ABC?

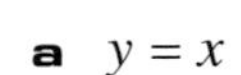

10 Copy the diagram and reflect the triangle in these lines.

a $y = x$ **b** $x = 1$

c $y = -x$ **d** $y = -1$

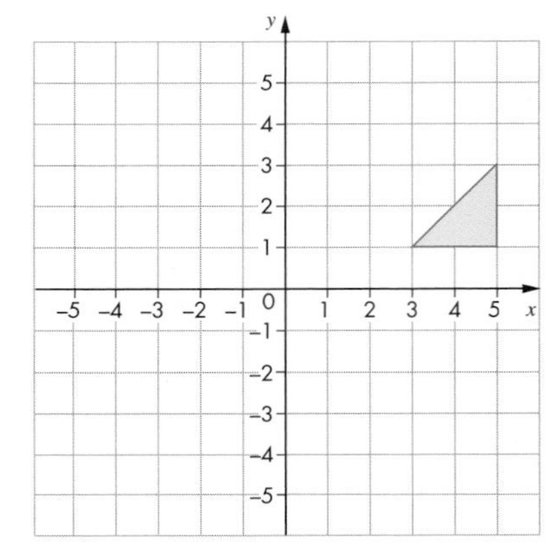

PS 11 **a** Draw a pair of axes. Label the x-axis from –5 to +5 and the y-axis from –5 to +5.

b Draw the line $y = x$.

c Reflect the points A(2, 1), B(5, 0), C(–3, 2), D(–2, –4) in the line $y = x$.

d What do you notice about the values of the coordinates of the reflected points?

e What would the coordinates of the reflected point be if the point (a, b) were reflected in the line $y = x$?

C

PS 12 **a** Draw a pair of axes. Label the x-axis from –5 to +5 and the y-axis from –5 to +5.

b Draw the line $y = -x$.

c Reflect the points A(2, 1), B(0, 5), C(3, –2), D(–4, –3) in the line $y = -x$.

d What do you notice about the values of the coordinates of the reflected points?

e What would the coordinates of the reflected point be if the point (a, b) were reflected in the line $y = -x$?

6.4 Rotations

This section will show you how to:

- rotate a 2D shape about a point

Key words

angle of rotation
anticlockwise
centre of rotation
clockwise
rotation

A **rotation** transforms a shape to a new position by turning it about a fixed point called the **centre of rotation**.

EXAMPLE 4

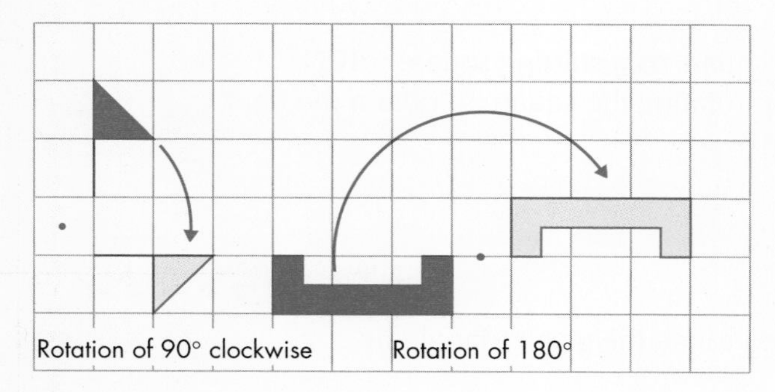

Note:

- The direction of turn or the **angle of rotation** is expressed as **clockwise** or **anticlockwise**.
- The position of the centre of rotation is always specified.
- The rotations 180° clockwise and 180° anticlockwise are the same.

The rotations that most often appear in examination questions are 90° and 180°.

EXERCISE 6D

D

1 On squared paper, draw each of these shapes and its centre of rotation, leaving plenty of space all round the shape.

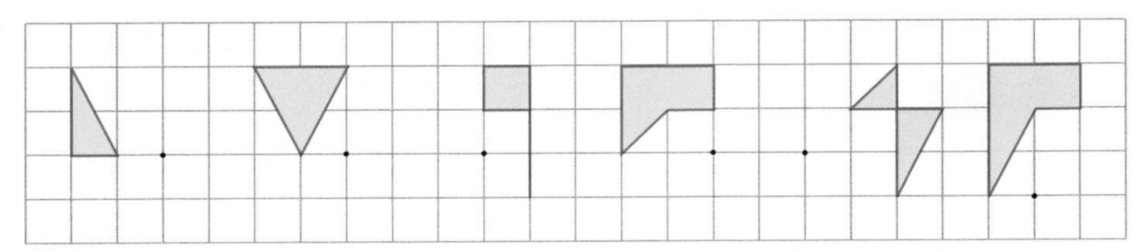

a Rotate each shape about its centre of rotation:

i first by 90° clockwise (call the image A)

ii then by 90° anticlockwise (call the image B).

b Describe, in each case, the rotation that would take:

i A back to its original position **ii** A to B.

2 A graphics designer came up with the following routine for creating a design.

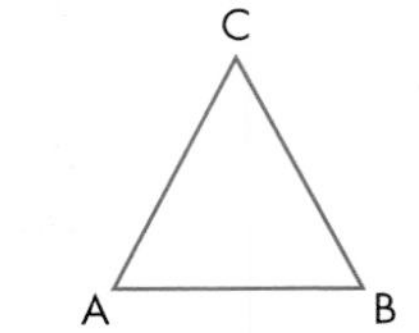

- Start with a triangle ABC.
- Reflect the triangle in the line AB.
- Rotate the whole shape about point C clockwise 90°, then a further clockwise 90°, then a further clockwise 90°.

From any triangle of your choice, create a design using the above routine.

PS 3 By using the middle square as a starting square ABCD, describe how to keep rotating the square to obtain the final shape in the diagram.

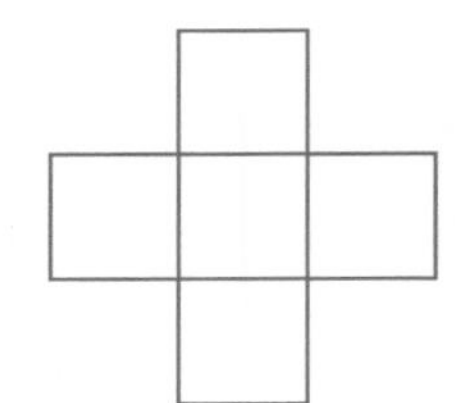

C

4 Copy the diagram and rotate the given triangle by the following.

a 90° clockwise about (0, 0)

b 180° about (3, 3)

c 90° anticlockwise about (0, 2)

d 180° about (–1, 0)

e 90° clockwise about (–1, –1)

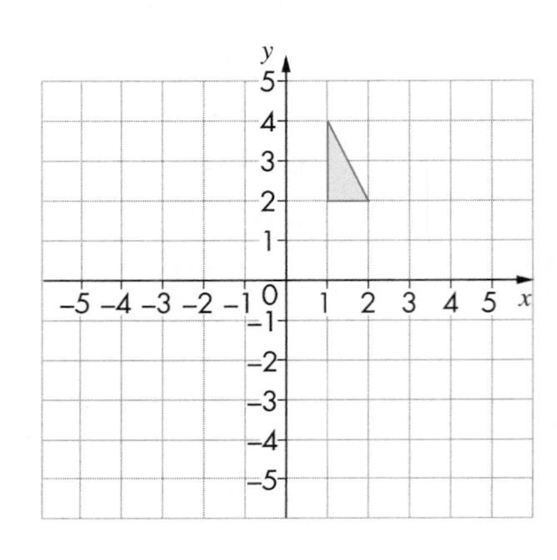

C

5 What other rotations are equivalent to these rotations?

a 270° clockwise **b** 90° clockwise

c 60° anticlockwise **d** 100° anticlockwise

6 **a** Draw a pair of axes where both the x-values and y-values are from –5 to 5.

b Draw the triangle ABC, where A = (1, 2), B = (2, 4) and C = (4, 1).

c **i** Rotate triangle ABC 90° clockwise about the origin (0, 0) and label the image A′, B′, C′, where A′ is the image of A, etc.

ii Write down the coordinates of A′, B′, C′.

iii What connection is there between A, B, C and A′, B′, C′?

iv Will this connection always be so for a 90° clockwise rotation about the origin?

7 Repeat question **6**, but rotate triangle ABC through 180°.

8 Repeat question **6**, but rotate triangle ABC 90° anticlockwise.

PS **9** Show that a reflection in the x-axis followed by a reflection in the y-axis is equivalent to a rotation of 180° about the origin.

PS **10** Show that a reflection in the line $y = x$ followed by a reflection in the line $y = -x$ is equivalent to a rotation of 180° about the origin.

11 **a** Draw a regular hexagon ABCDEF with centre O.

b Using O as the centre of rotation, describe a transformation that will result in the following movements.

i Triangle AOB to triangle BOC **ii** Triangle AOB to triangle COD

iii Triangle AOB to triangle DOE **iv** Triangle AOB to triangle EOF

c Describe the transformations that will move the rhombus ABCO to these positions.

i Rhombus BCDO **ii** Rhombus DEFO

AU **12** Triangle A, as shown on the grid, is rotated to form a new triangle B.

The coordinates of the vertices of B are (0, –2), (–3, –2) and (–3, –4).

Describe fully the rotation that maps triangle A onto triangle B.

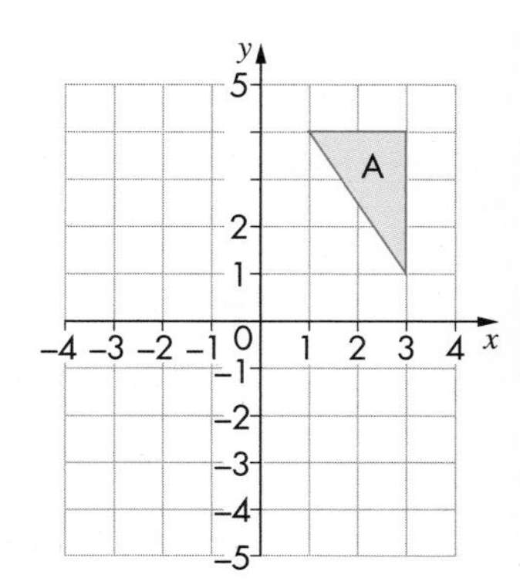

6.5 Enlargement

This section will show you how to:
- enlarge a 2D shape by a scale factor

Key words
centre of enlargement
enlargement
scale factor

An **enlargement** changes the size of a shape to give a similar image. It always has a **centre of enlargement** and a **scale factor**. Every length of the enlarged shape will be:

original length × scale factor

The distance of each image point on the enlargement from the centre of enlargement will be:

distance of original point from centre of enlargement × scale factor

EXAMPLE 5

The diagram shows the enlargement of triangle ABC by scale factor 3 about the centre of enlargement X.

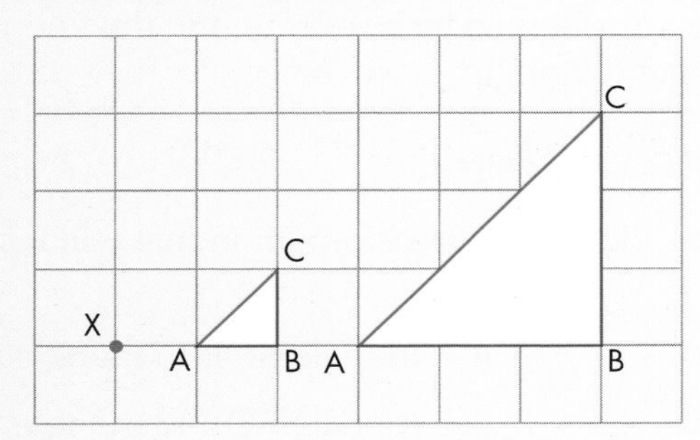

Note:

- Each length on the enlargement A′B′C′ is three times the corresponding length on the original shape.

  This means that the corresponding sides are in the same ratio:

  AB : A′B′ = AC : A′C′ = BC : B′C′ = 1 : 3

- The distance of any point on the enlargement from the centre of enlargement is three times the distance from the corresponding point on the original shape to the centre of enlargement.

There are two distinct ways to enlarge a shape: the ray method and the coordinate, or counting squares, method.

Ray method

This is the *only* way to construct an enlargement when the diagram is not on a grid.

EXAMPLE 6

Enlarge triangle ABC by scale factor 3 about the centre of enlargement X.

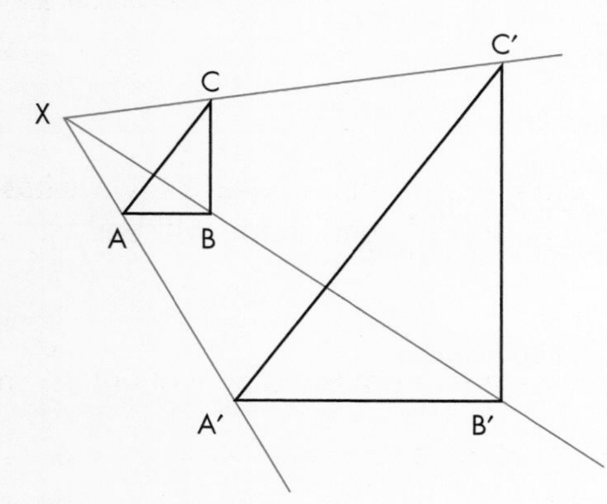

Notice that the rays have been drawn from the centre of enlargement to each vertex and beyond.

The distance from X to each vertex on triangle ABC is measured and multiplied by 3 to give the distance from X to each vertex A′, B′ and C′ for the enlarged triangle A′B′C′.

Once each image vertex has been found, the whole enlarged shape can then be drawn.

Check the measurements and see for yourself how the calculations have been done.

Notice again that the length of each side on the enlarged triangle is three times the length of the corresponding side on the original triangle.

Counting squares method

In this method, you use the coordinates of the vertices to 'count squares'.

EXAMPLE 7

Enlarge the triangle ABC by scale factor 3 from the centre of enlargement (1, 2).

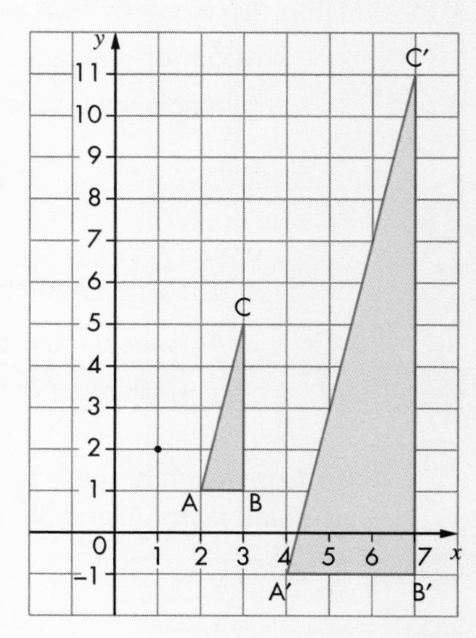

To find the coordinates of each image vertex, first work out the horizontal and vertical distances from each original vertex to the centre of enlargement.

Then multiply each of these distances by 3 to find the position of each image vertex.

For example, to find the coordinates of C′ work out the distance from the centre of enlargement (1, 2) to the point C(3, 5).

horizontal distance = 2
vertical distance = 3

Make these 3 times longer to give:

new horizontal distance = 6
new vertical distance = 9

So the coordinates of C′ are:

(1 + 6, 2 + 9) = (7, 11)

Notice again that the length of each side is three times as long in the enlargement.

Negative enlargement

A negative enlargement produces an image shape on the opposite side of the centre of enlargement to the original shape.

EXAMPLE 8

Triangle ABC has been enlarged by scale factor −2, with the centre of enlargement at (1, 0).

You can enlarge triangle ABC to give triangle A′B′C′ by either the ray method or the coordinate method. You calculate the new lengths on the opposite side of the centre of enlargement to the original shape.

Notice how a negative scale factor also inverts the original shape.

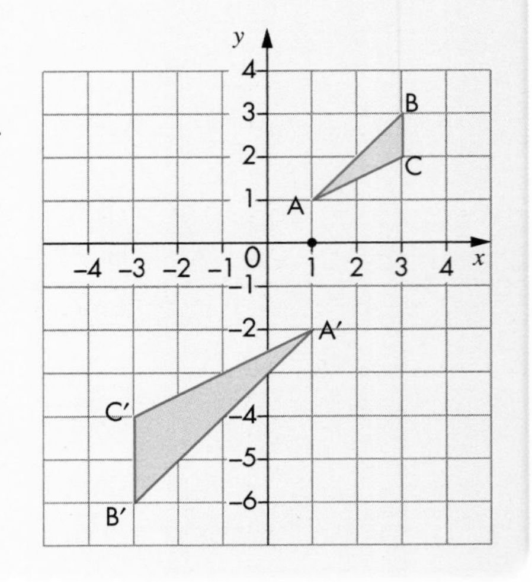

Fractional enlargement

Strange but true … you can have an enlargement in mathematics that is actually smaller than the original shape!

EXAMPLE 9

Triangle ABC has been enlarged by a scale factor of $\frac{1}{2}$ about the centre of enlargement O to give triangle A′B′C′.

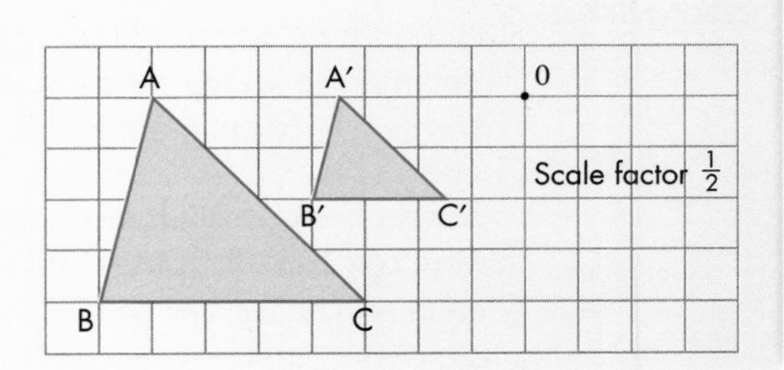

EXERCISE 6E

1 Copy each of these figures with its centre of enlargement. Then enlarge it by the given scale factor, using the ray method.

a

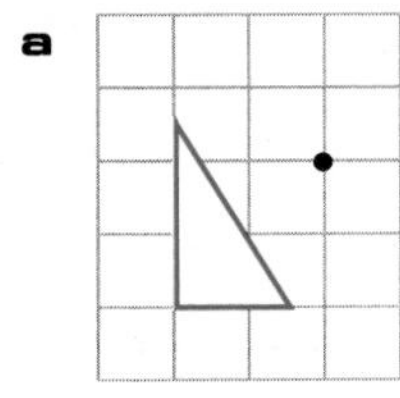

Scale factor 2

b

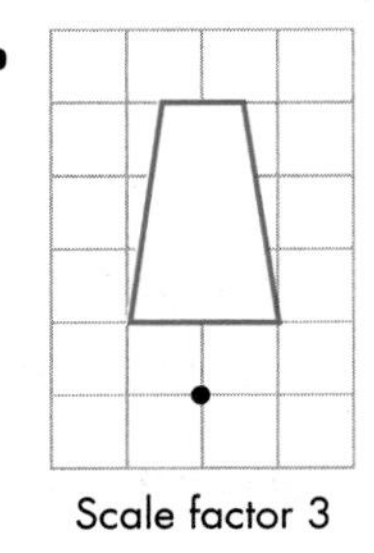

Scale factor 3

c

Scale factor 2

d

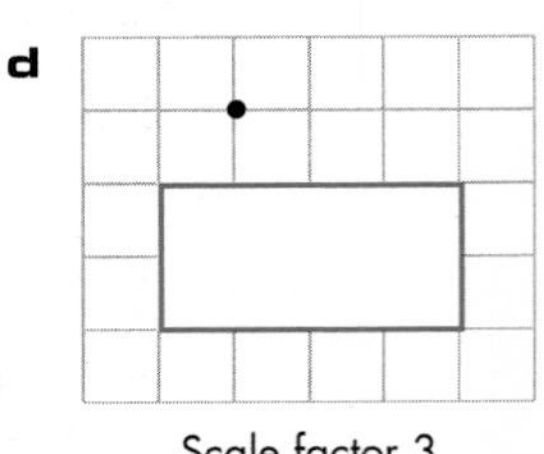

Scale factor 3

D

2 Copy each of these diagrams onto squared paper and enlarge it by scale factor 2, using the origin as the centre of enlargement.

a

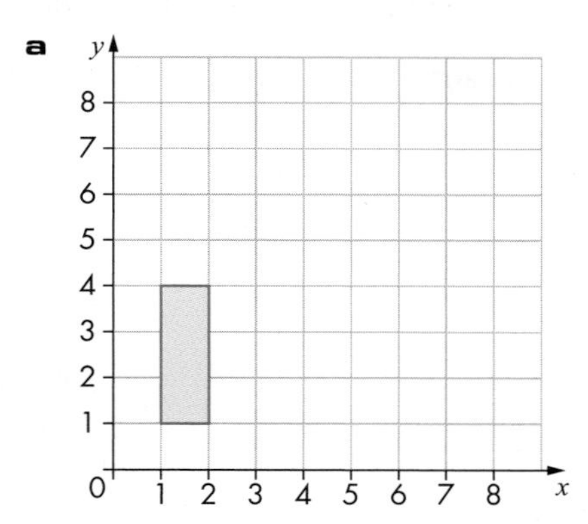

b

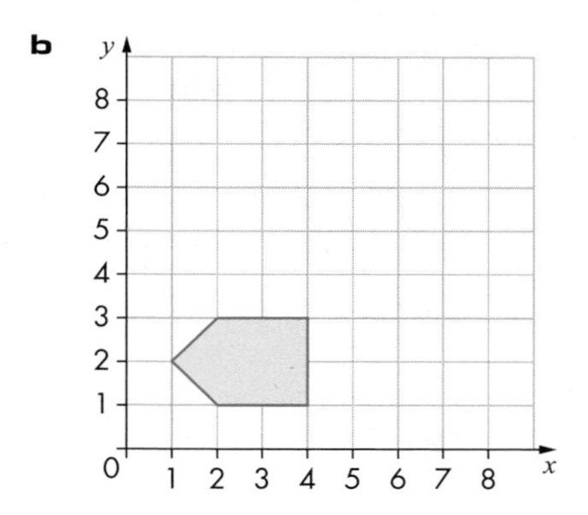

c

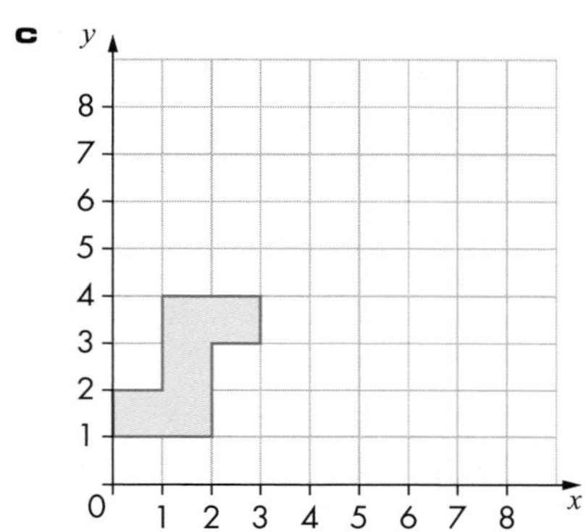

HINTS AND TIPS

Even if you are using a counting square method, you can always check by using the ray method.

3 Copy each of these diagrams onto squared paper and enlarge it by scale factor 2, using the given centre of enlargement.

a

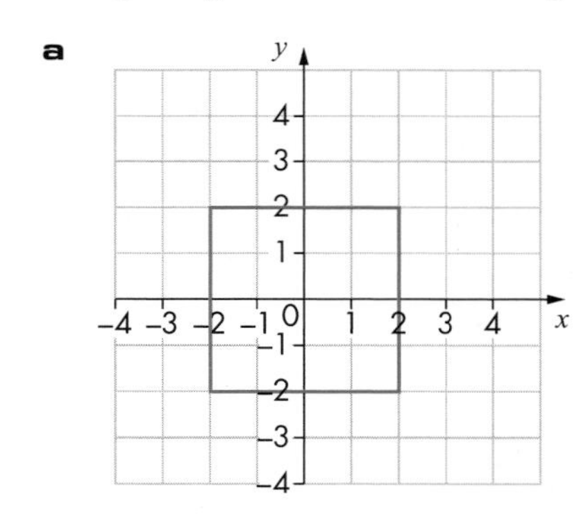

Centre of enlargement (–1,1)

b

Centre of enlargement (–2,–3)

4 A designer is told to use the following routine.

- Start with a rectangle ABCD.
- Reflect ABCD in the line AC.
- Rotate the whole new shape about C through 180°.
- Enlarge the whole shape scale factor 2, centre of enlargement point A.

A B D C

Start with any rectangle of your choice and create the design above.

C

5 Enlarge each of these shapes by a scale factor of $\frac{1}{2}$ about the given centre of enlargement.

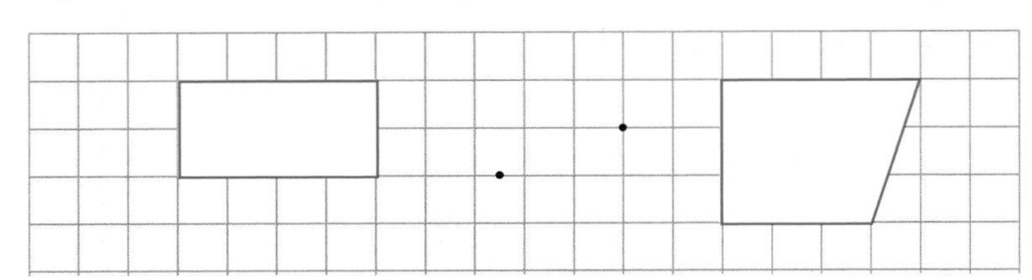

6 Copy this diagram onto squared paper.

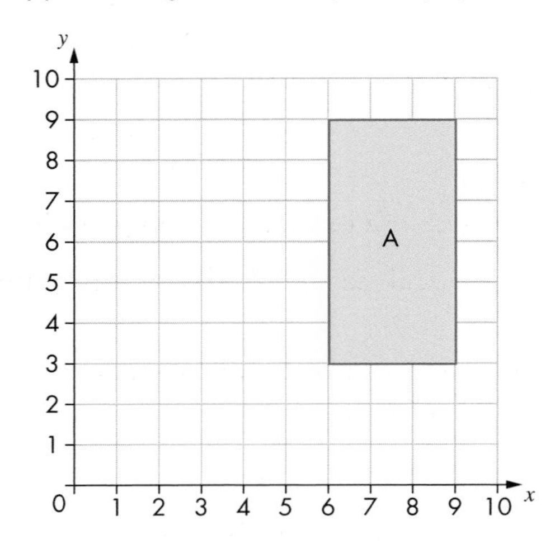

a Enlarge the rectangle A by scale factor $\frac{1}{3}$ about the origin. Label the image B.

b Write down the ratio of the lengths of the sides of rectangle A to the lengths of the sides of rectangle B.

c Work out the ratio of the perimeter of rectangle A to the perimeter of rectangle B.

d Work out the ratio of the area of rectangle A to the area of rectangle B.

B

AU 7 Copy this diagram onto squared paper.

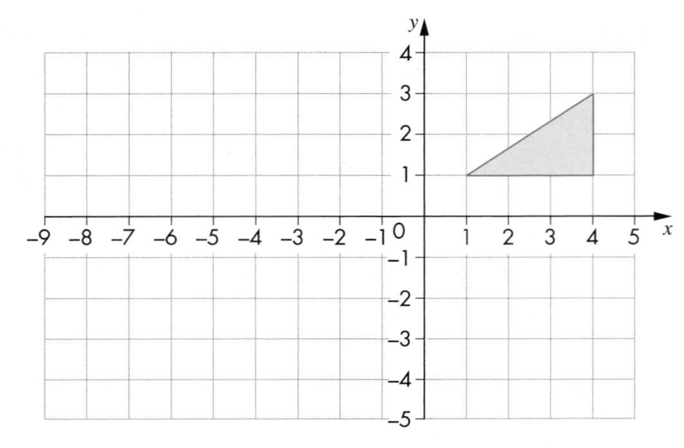

Enlarge the triangle by scale factor –2 about the origin.

B

8 Copy the diagram onto squared paper.

a Enlarge A by a scale factor of 3 about a centre (4, 5).

b Enlarge B by a scale factor $\frac{1}{2}$ about a centre (–1, –3).

c Enlarge B by scale factor $-\frac{1}{2}$ about a centre (–3, –1).

d What is the centre of enlargement and scale factor which maps B onto A?

e What is the centre of enlargement and scale factor which maps A onto B?

f What is the centre of enlargement and scale factor which maps the answer to part **b** to the answer to part **c**?

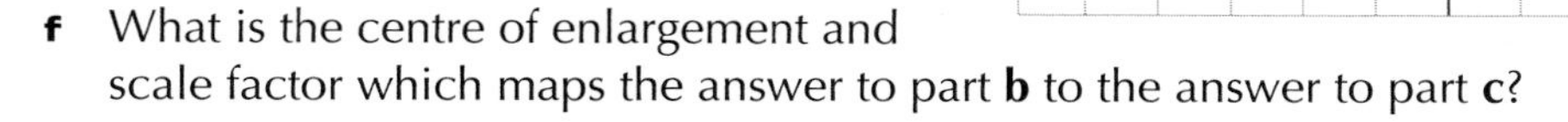

g What is the centre of enlargement and scale factor which maps the answer to part **c** to the answer to part **b**?

h What is the connection between the scale factors and the centres of enlargement in parts **d** and **e**, and in parts **f** and **g**?

PS **9** Triangle A has vertices with coordinates (2, 1), (4, 1) and (4, 4).
Triangle B has vertices with coordinates (–5, 1), (–5, 7) and (–1, 7).

Describe fully the single transformation that maps triangle A onto triangle B.

6.6 Combined transformations

This section will show you how to:

- combine transformations

Key words
enlargement
reflection
rotation
transformation
translation

Examination questions often require you to use more than one **transformation**. In this exercise, you will revise the transformations you have met so far.

Remember, to describe:

- a **translation** fully, you need to use a vector
- a **reflection** fully, you need to use a mirror line
- a **rotation** fully, you need a centre of rotation, an angle of rotation and the direction of turn
- an **enlargement** fully, you need a centre of enlargement and a scale factor

EXERCISE 6F

D

1 The point P(3, 4) is **reflected** in the x-axis, then rotated by 90° clockwise about the origin. What are the coordinates of the image of P?

2 A point Q(5, 2) is rotated by 180°, then reflected in the x-axis.

a What are the coordinates of the image point of Q?

b What single transformation would have taken point Q directly to the image point?

C

3 Describe fully the transformations that will map the shaded triangle onto each of the triangles A–F.

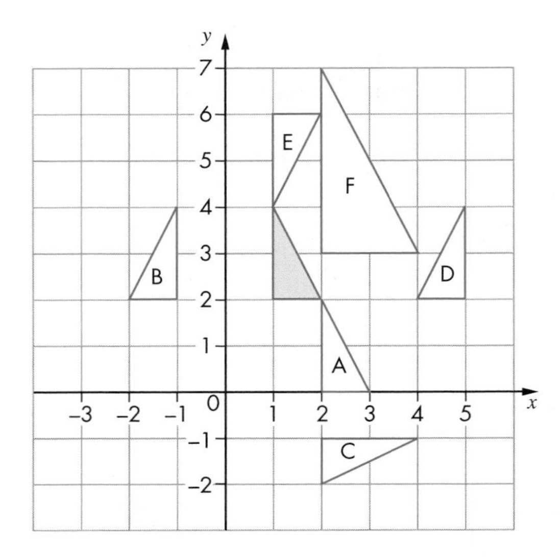

4 Describe fully the transformations that will result in the following movements.

a T_1 to T_2

b T_1 to T_6

c T_2 to T_3

d T_6 to T_2

e T_6 to T_5

f T_5 to T_4

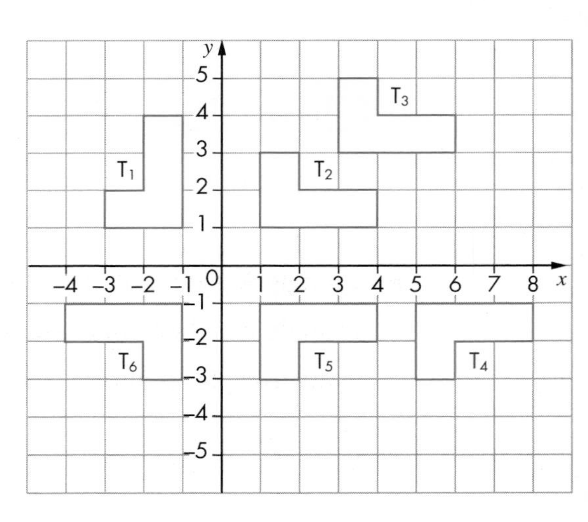

5 **a** Plot a triangle T with vertices (1, 1), (2, 1), (1, 3).

b Reflect triangle T in the y-axis and label the image T_b.

c Rotate triangle T_b 90° anticlockwise about the origin and label the image T_c.

d Reflect triangle T_c in the y-axis and label the image T_d.

e Describe fully the transformation that will move triangle T_d back to triangle T.

PS 6 Describe fully at least three different transformations that could move the square labelled S to the square labelled T.

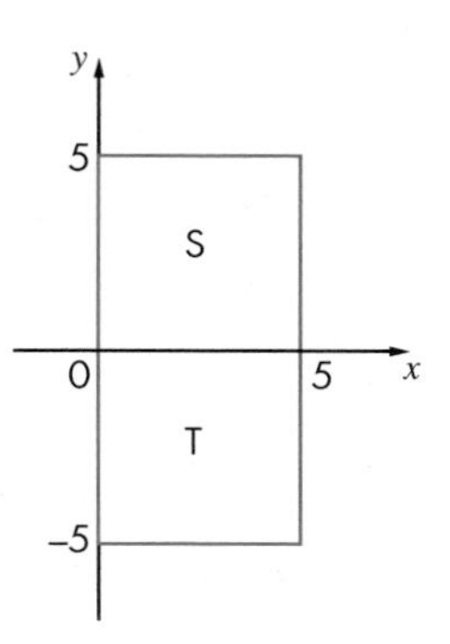

PS 7 The point A(4, 4) has been transformed to the point A′(4, –4). Describe as many different transformations as you can that could transform point A to point A′.

AU 8 Copy the diagram onto squared paper.

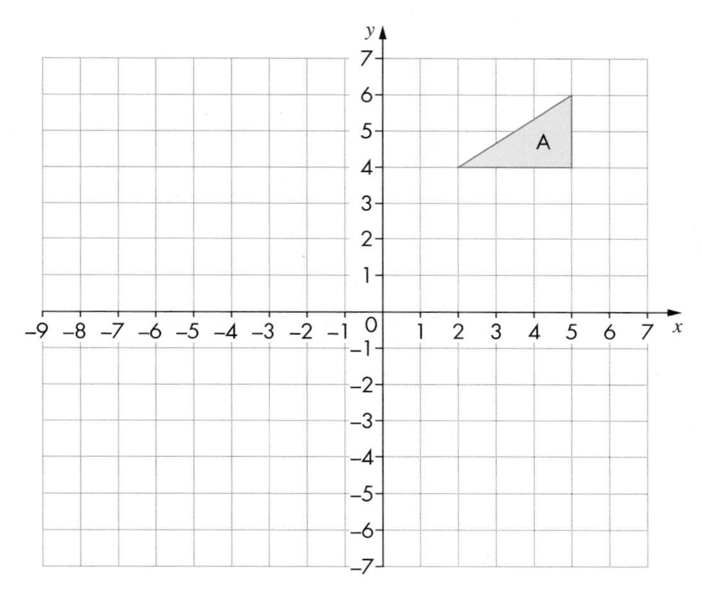

a Triangle A is translated by the vector $\begin{pmatrix} -1.5 \\ -3 \end{pmatrix}$ to give triangle B.

Triangle B is then enlarged by a scale factor –2 about the origin to give triangle C.

Draw triangles B and C on the diagram.

b Describe fully the single transformation that maps triangle C onto triangle A.

GRADE BOOSTER

- **D** You can reflect a 2D shape in a line $x = a$ or $y = b$
- **D** You can rotate a 2D shape about the origin
- **D** You can enlarge a 2D shape by a whole number scale factor
- **C** You can translate a 2D shape by a vector
- **C** You can reflect a 2D shape in the line $y = x$ or $y = -x$
- **C** You can rotate a 2D shape about any point
- **C** You can enlarge a 2D shape by a fractional scale factor
- **C** You can enlarge a 2D shape about any point
- **B** You know the conditions to show two triangles are congruent
- **B** You can enlarge a 2D shape by a negative scale factor
- **A** You can prove two triangles are congruent

What you should know now

- How to translate a 2D shape by a vector
- How to reflect a 2D shape in any line
- How to rotate a 2D shape about any point and through any angle
- How to enlarge a 2D shape about any point using a positive, fractional or negative scale factor
- How to show that two triangles are congruent

1

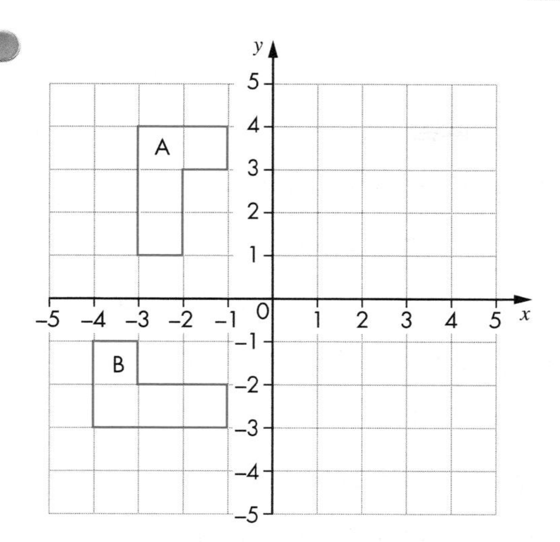

a On a copy of the diagram, reflect shape A in the y-axis. *(2 marks)*

b Describe fully the single transformation which takes shape A to shape B. *(3 marks)*

Edexcel, November 2008, Paper 3 Higher, Question 7

2

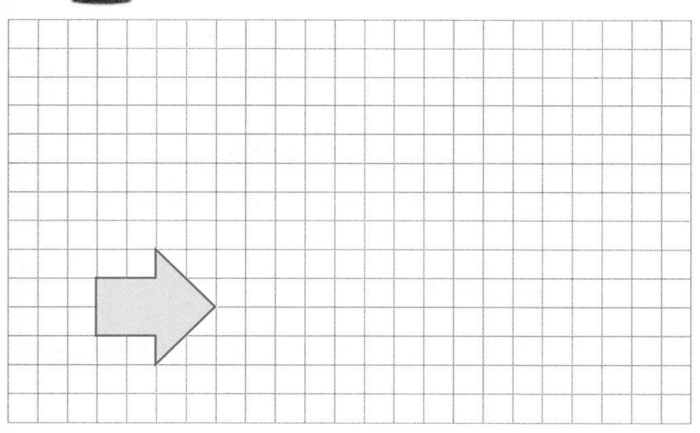

Copy the grid and enlarge the shaded shape with a scale factor of 2. *(2 marks)*

Edexcel, June 2008, Paper 15 Higher, Question 1(b)

3

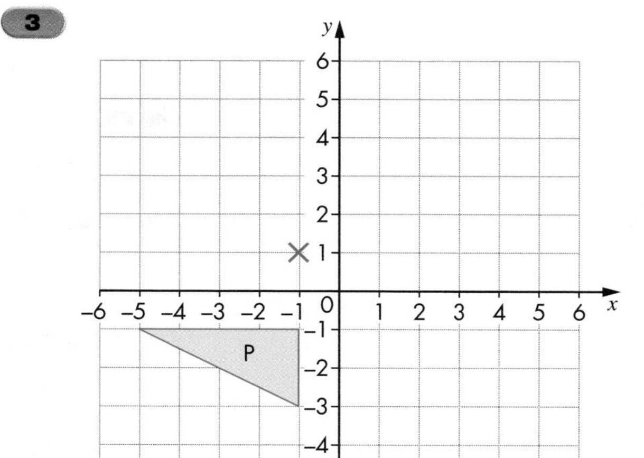

a Copy the diagram and rotate triangle P 180° about the point (–1, 1).

Label the new triangle A. *(2 marks)*

b Translate triangle P by the vector $\begin{pmatrix} 6 \\ -1 \end{pmatrix}$.

Label the new triangle B. *(1 mark)*

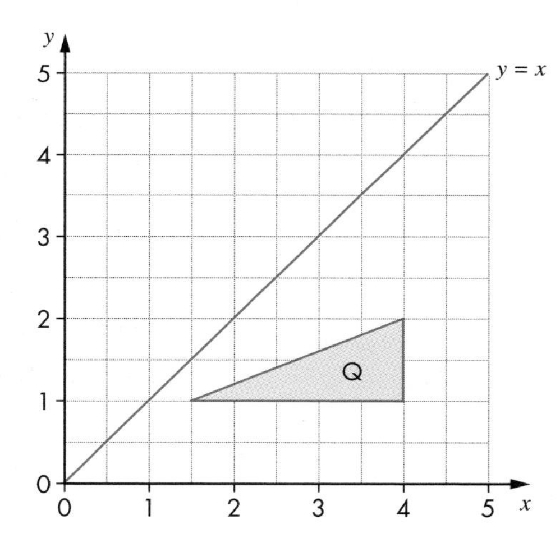

c Reflect triangle Q in the line $y = x$. Label the new triangle C.

Edexcel, May 2008, Paper 3 Higher, Question 14(a, b, c)

4

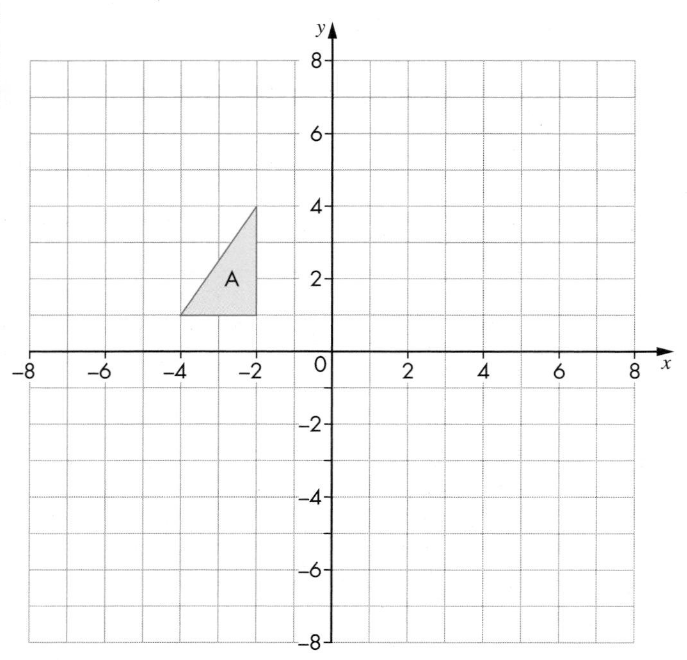

Triangle A is reflected in the x-axis to give triangle B.

Triangle B is reflected in the line $x = 1$ to give triangle C.

Describe the single transformation that takes triangle A to triangle C. *(3 marks)*

Edexcel A, June 2008, Paper 4 Higher, Question 15

5 In the diagram, AB = BC = CD = DA.

Prove that triangle ADB is congruent to triangle CDB.

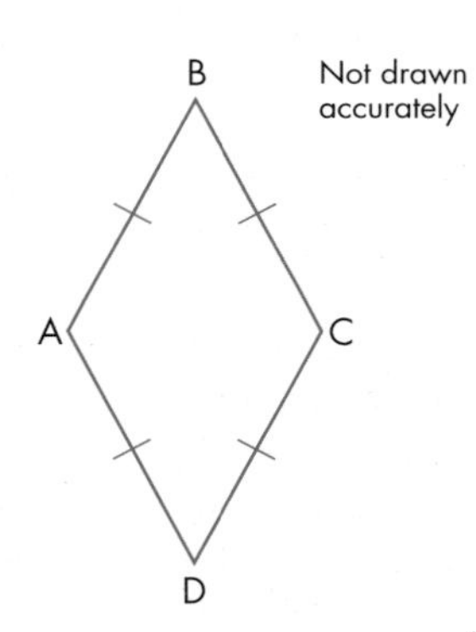

(3 marks)

Edexcel, November 2008, Paper 14 Higher, Question 12

Worked Examination Questions

AU **1** The grid shows several transformations of the shaded triangle.

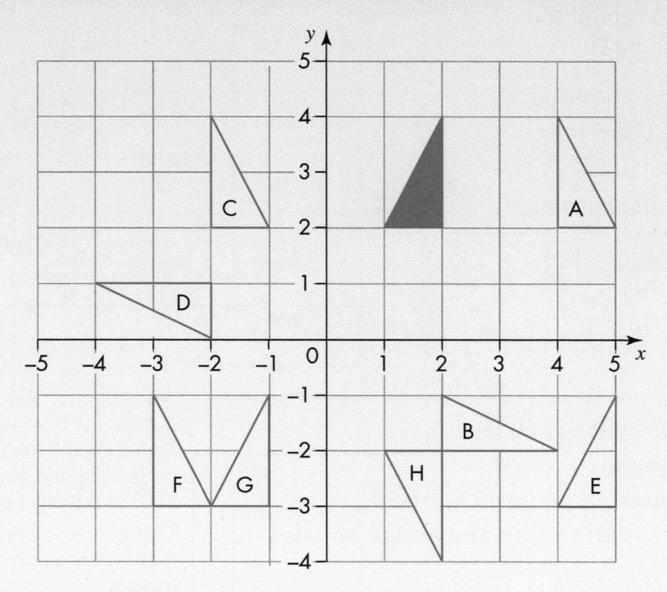

a Write down the letter of the shape:

i after the shaded triangle is reflected in the line $x = 3$

ii after the shaded triangle is translated by the vector $\begin{pmatrix} 3 \\ -5 \end{pmatrix}$

iii after the shaded triangle is rotated 90° clockwise about 0.

b Describe fully the single transformation that takes triangle F onto triangle G.

1 a i A

$x = 3$ is the vertical line passing through $x = 3$ on the x-axis. This scores 1 mark.

ii E

Move the triangle 3 squares to the right and 5 squares down. This scores 1 mark.

iii B

Use tracing paper to help you. Trace the shaded triangle, pivot the paper on 0 with your pencil point and rotate the paper through 90° clockwise.
This scores 1 mark.

b A reflection in the line $x = -2$.

The vertical mirror line passes through $x = -2$ on the x-axis. This scores 1 mark for method of identifying reflection and 1 mark for accuracy of mirror line.

Total: 5 marks

Worked Examination Questions

2 Triangle ABC has vertices A(6, 0), B(6, 9), C(9, 3).

a Rotate triangle ABC through 180° about the point (2, 4). Label the image triangle R.

b Enlarge triangle ABC by scale factor $\frac{1}{3}$ from the centre of enlargement (3, 0). Label the image triangle E.

c Describe fully the single transformation which maps triangle E to triangle R.

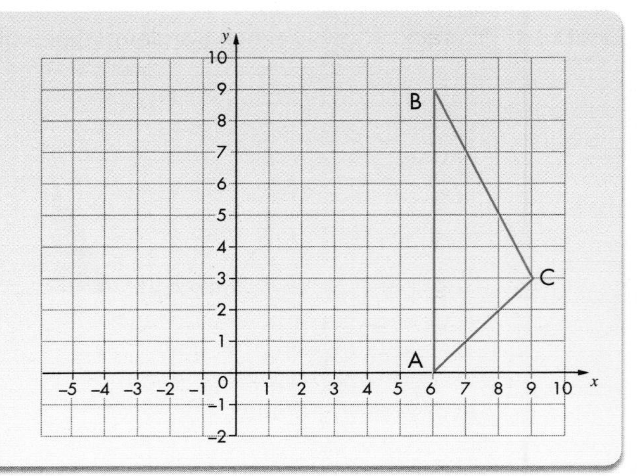

2 a Join each vertex to (2, 4) and rotate each line through 180° or use tracing paper.

This scores 1 method mark for rotating the triangle 180° about any point.

This scores 1 accuracy mark for the correct rotation.

b Use the ray method or the counting squares method. Remember, a fractional scale factor makes the image smaller.

This scores 1 method mark for enlarging the triangle by scale factor $\frac{1}{3}$ about any point.

This scores 1 accuracy mark for the correct enlargement.

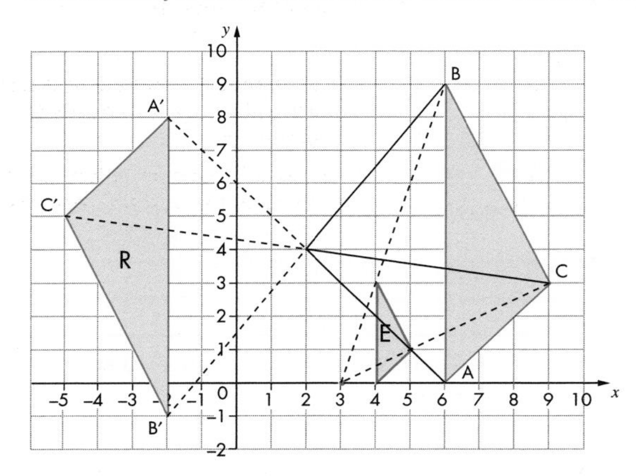

c

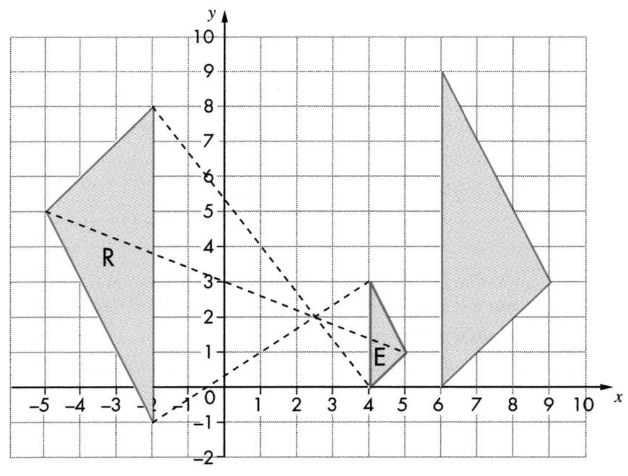

An enlargement of scale factor –3 about the point ($2\frac{1}{2}$, 2).
Draw in the ray lines to find the centre of enlargement.

This scores 1 mark for enlargement of scale factor –3.

This scores 1 mark for ($2\frac{1}{2}$, 2).

Total: 6 marks

Worked Examination Questions

3 In the diagram AB and CD are parallel.

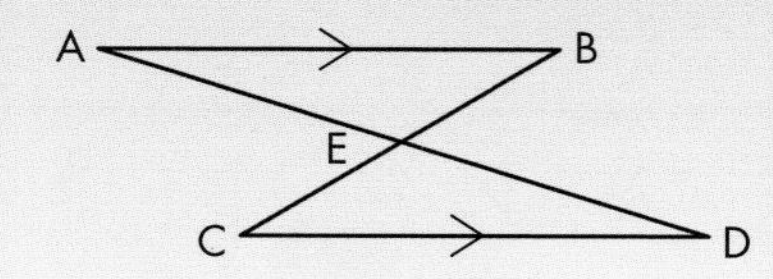

E is the midpoint of AD.

Prove triangle ABE is congruent to triangle CDE.

3 You will be expected to give reasons for each statement you make.

AE = DE (E is midpoint of AD)

∠BAE = ∠CDE (alternate angles)

∠AEB = ∠CED (opposite angles)

So ΔABE ≡ ΔCDE (ASA)

Total: 3 marks

This scores 1 mark for method.

This scores 1 mark for method.

Or you could use ∠ABE = ∠DCE (alternate angles).

This scores 1 accuracy mark for third statement with correct conclusion.

You would lose a mark if you missed out any of the reasons.

Problem Solving
Developing photographs

Enlargement is used in many aspects of life. How many can you think of?

Discuss these with the person sitting next to you or as a whole class.

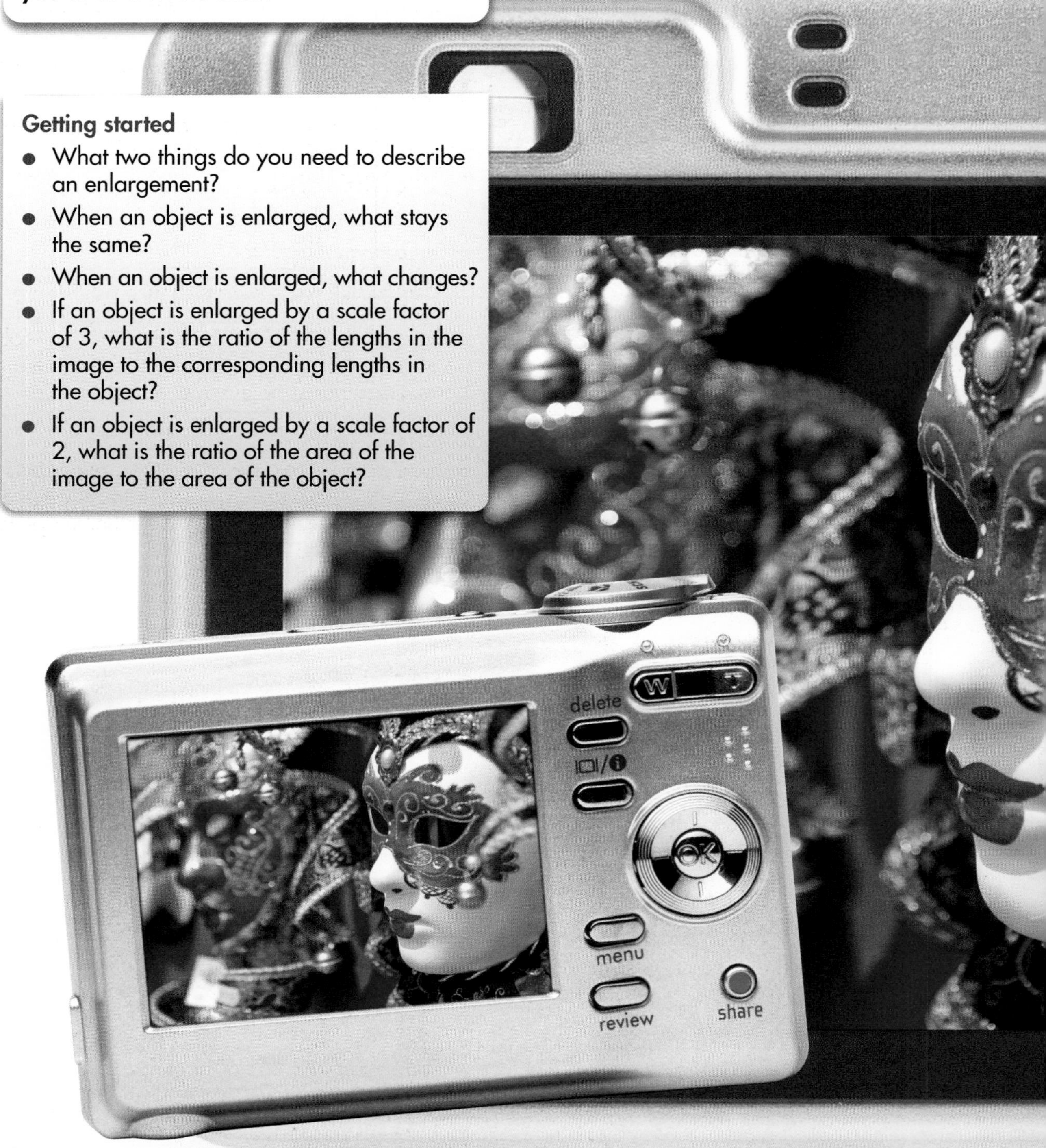

Getting started

- What two things do you need to describe an enlargement?
- When an object is enlarged, what stays the same?
- When an object is enlarged, what changes?
- If an object is enlarged by a scale factor of 3, what is the ratio of the lengths in the image to the corresponding lengths in the object?
- If an object is enlarged by a scale factor of 2, what is the ratio of the area of the image to the area of the object?

Getting started (continued)

Which set of photographs is the odd one out? Explain why.

Look at these images. How can you tell that they are all based on the same original photograph?

Which do you think is the original? Which are enlargements? Why?

Your task

Use a digital camera to take at least two photographs. Use them to make a display or poster to explain enlargement. To complete the task successfully, you should answer at least two of the following questions about enlargement.

1. What happens to the image when you move the centre of enlargement?
2. Compare what happens to the image when you have a scale factor that is:
   - a whole number
   - a number between 0 and 1
   - a negative number.
3. What is the relationship between the scale factor and the properties of the object and image? For example, consider how the scale factor affects the relationships between lengths or areas in the object and image.

Note

If you do not have a digital camera available, use a suitable computer program to produce and enlarge a simple image, logo or other shape for your display or poster.

Why this chapter matters

Mathematicians are interested in the way that numbers work, rather than just looking at specific calculations. Using letters to represent numbers, they can use formulae and solve the very complicated equations that occur in modern technology. Without the use of algebra, mankind would not have developed aircraft or walked on the Moon.

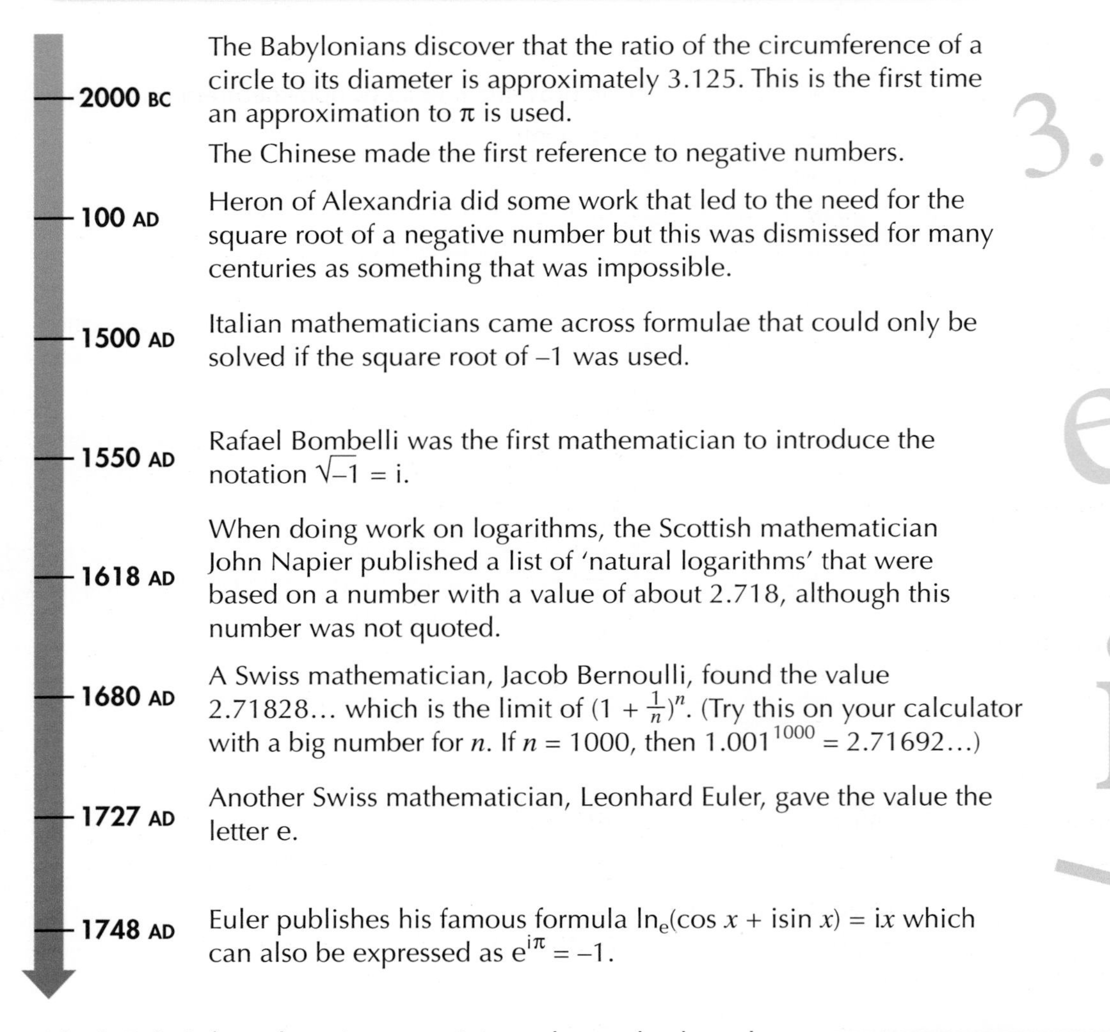

2000 BC The Babylonians discover that the ratio of the circumference of a circle to its diameter is approximately 3.125. This is the first time an approximation to π is used.

The Chinese made the first reference to negative numbers.

100 AD Heron of Alexandria did some work that led to the need for the square root of a negative number but this was dismissed for many centuries as something that was impossible.

1500 AD Italian mathematicians came across formulae that could only be solved if the square root of –1 was used.

1550 AD Rafael Bombelli was the first mathematician to introduce the notation $\sqrt{-1} = \text{i}$.

1618 AD When doing work on logarithms, the Scottish mathematician John Napier published a list of 'natural logarithms' that were based on a number with a value of about 2.718, although this number was not quoted.

1680 AD A Swiss mathematician, Jacob Bernoulli, found the value 2.71828… which is the limit of $(1 + \frac{1}{n})^n$. (Try this on your calculator with a big number for n. If $n = 1000$, then $1.001^{1000} = 2.71692\ldots$)

1727 AD Another Swiss mathematician, Leonhard Euler, gave the value the letter e.

1748 AD Euler publishes his famous formula $\ln_e(\cos x + \text{i}\sin x) = \text{i}x$ which can also be expressed as $e^{\text{i}\pi} = -1$.

Why is Euler's formula so important? Any advanced culture, be it human or alien, would need a counting system that included a unit of 1. It would also need a symbol to record zero. Circles occur all over the Universe, so the ratio π would be familiar. The same is true for would e, which is a constant that occurs naturally. So Euler's formula really is universal. Many people even think it proves the existence of God.

However, other people think that the fact that five of the most important numbers in mathematics, 0, 1, e, i and π, all occur in such a neat formula is pure coincidence.

Chapter

Algebra: Equations

1. Changing the subject of a formula
2. Solving linear equations
3. Setting up equations
4. Trial and improvement
5. Simultaneous linear equations
6. Solving problems using simultaneous equations
7. Linear and non-linear simultaneous equations

The grades given in this chapter are target grades.

This chapter will show you ...

- **C** how to set up linear equations
- **C** how to use trial and improvement to solve non-linear equations
- **B** how to solve linear equations
- **B** how to set up and solve simultaneous equations
- **A** how to rearrange formulae
- **A** how to solve simultaneous equations, one linear and one non-linear

Visual overview

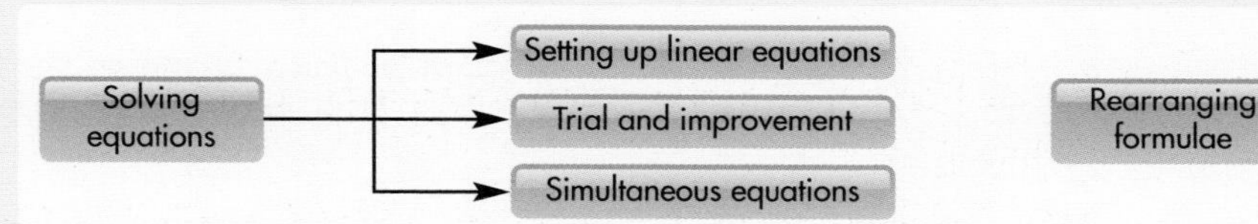

What you should already know

- The basic language of algebra **(KS3 level 5, GCSE grade E)**
- How to collect together like terms **(KS3 level 5, GCSE grade E)**
- How to multiply together terms such as $2m \times 3m$ **(KS3 level 5, GCSE grade E)**

Quick check

1 Make x the subject of each equation.

a $2y + x = 3$ **b** $x - 3y = 4$ **c** $4y - x = 3$

2 Expand the following.

a $2(x + 6)$ **b** $4(x - 3)$ **c** $6(2x - 1)$

3 Factorise each expression.

a $2x + 6$ **b** $x^2 - x$ **c** $10x^2 + 2x$

4 Simplify the following.

a $5y + 2y - y$ **b** $4x + 2 + 3x - 5$ **c** $3(x + 1) - 2(x - 1)$

d $3 \times 2x$ **e** $4y \times 2y$ **f** $c^2 \times 2c$

5 Solve the following equations.

a $x + 3 = 7$ **b** $5x = 30$ **c** $\frac{x}{8} = 8$

d $2x + 5 = 3$ **e** $4x - 3 = 13$ **f** $\frac{x}{3} - 2 = 9$

7.1 Changing the subject of a formula

This section will show you how to:
- change the subject of a formula where the subject occurs more than once

Key words
subject

You have already considered changing the **subject** of a formula in which the subject appears only once. This is like solving an equation but using letters. You have also solved equations in which the unknown appears on both sides of the equation. This requires you to collect the terms in the unknown (usually x) on one side and the numbers on the other.

You can do something similar, to **rearrange** formulae in which the subject appears more than once. The principle is the same. Collect all the subject terms on the same side and everything else on the other side. Most often, you then need to factorise the subject out of the resulting expression.

EXAMPLE 1

Make x the subject of this formula.

$$ax + b = cx + d$$

First, rearrange the formula to get all the x-terms on the left-hand side and all the other terms on the right-hand side. (The rule 'change sides – change signs' still applies.)

$$ax - cx = d - b$$

Factorise x out of the left-hand side to get:

$$x(a - c) = d - b$$

Divide by the expression in brackets, which gives:

$$x = \frac{d - b}{a - c}$$

EXAMPLE 2

Make p the subject of this formula.

$$5 = \frac{ap + b}{cp + d}$$

First, multiply both sides by the denominator of the algebraic fraction, which gives:

$$5(cp + d) = ap + b$$

Expand the brackets to get:

$$5cp + 5d = ap + b$$

Now continue as in Example 1:

$$5cp - ap = b - 5d$$

$$p(5c - a) = b - 5d$$

$$p = \frac{b + 5d}{5c - a}$$

EXERCISE 7A

In questions **1** to **10**, make the letter in brackets the subject of the formula.

1 $3(x + 2y) = 2(x - y)$ (x)

2 $3(x + 2y) = 2(x - y)$ (y)

3 $5 = \dfrac{a + b}{a - c}$ (a)

4 $p(a + b) = q(a - b)$ (a)

5 $p(a + b) = q(a - b)$ (b)

6 $A = 2\pi rh + \pi rk$ (r)

7 $v^2 = u^2 + av^2$ (v)

8 $s(t - r) = 2r - 3$ (r)

9 $s(t - r) = 2(r - 3)$ (r)

10 $R = \dfrac{x - 3}{x - 2}$ (x)

11 **a** The perimeter of the shape shown on the right is given by the formula $P = \pi r + 2kr$. Make r the subject of this formula.

b The area of the same shape is given by $A = \frac{1}{2}[\pi r^2 + r^2\sqrt{(k^2 - 1)}]$ Make r the subject of this formula.

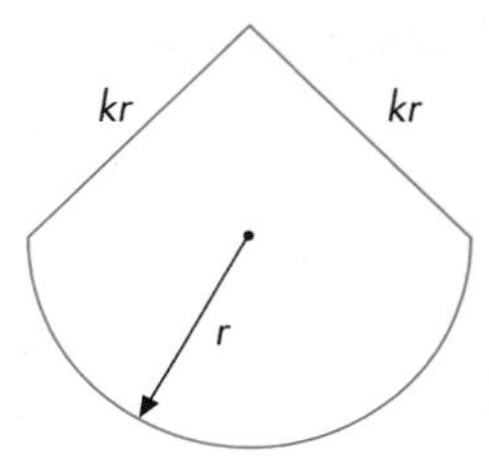

12 When £P is invested for Y years at a simple interest rate of R, the following formula gives the amount, A, at any time.

$$A = P + \frac{PRY}{100}$$

Make P the subject of this formula.

13 When two resistors with values a and b are connected in parallel, the total resistance is given by:

$$R = \frac{ab}{a + b}$$

a Make b the subject of the formula.

b Write the formula when a is the subject.

AU **14** **a** Make x the subject of this formula.

$$y = \frac{x + 2}{x - 2}$$

b Show that the formula $y = 1 + \dfrac{4}{x - 2}$ can be rearranged to give:

$$x = 2 + \frac{4}{y - 1}$$

c Combine the right-hand sides of each formula in part **b** into single fractions and simplify as much as possible.

d What do you notice?

15 The volume of the solid shown is given by:

$V = \frac{2}{3}\pi r^3 + \pi r^2 h$

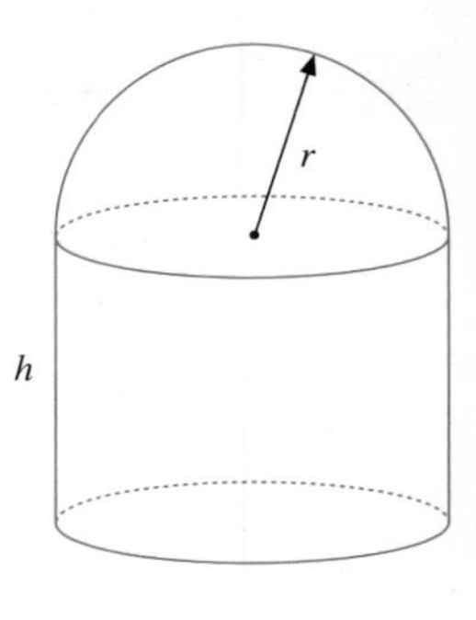

a Explain why it is not possible to make r the subject of this formula.

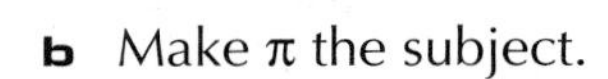

b Make π the subject.

c If $h = r$, can the formula be rearranged to make r the subject? If so, rearrange it to make r the subject.

16 Make x the subject of this formula.

$W = \frac{1}{2}z(x + y) + \frac{1}{2}y(x + z)$

PS 17 The following formulae in y can be rearranged to give the formulae in terms of x as shown.

$y = \dfrac{x + 1}{x + 2}$ gives $x = \dfrac{1 - 2y}{y - 1}$

$y = \dfrac{2x + 1}{x + 2}$ gives $x = \dfrac{1 - 2y}{y - 2}$

$y = \dfrac{3x + 2}{4x + 1}$ gives $x = \dfrac{2 - y}{4y - 3}$

$y = \dfrac{x + 5}{3x + 2}$ gives $x = \dfrac{5 - 2y}{3y - 1}$

Without rearranging the formula, write down $y = \dfrac{5x + 1}{2x + 3}$ as $x = \ldots$ and explain how you can do this without any algebra.

AU 18 *A formula used in GCSE mathematics is the cosine rule, which relates the three sides of any triangle with the angle between two of the sides.*

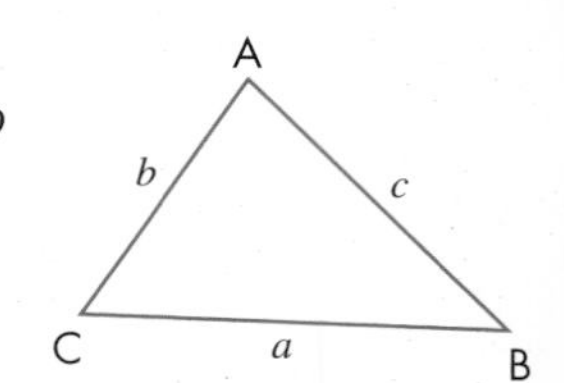

The formula is $a^2 = b^2 + c^2 - 2bc \cos A$.

This formula is known as an algebraic, cyclically symmetric formula.

That means that the various letters can be swapped with each other in a cycle, i.e. a becomes b, b becomes c and c becomes a.

a Write the formula so that it starts $b^2 = \ldots$

b When the formula is rearranged to make $\cos A$ the subject, then it becomes

$$\cos A = \frac{b^2 + c^2 - a^2}{2bc}$$

Write an equivalent formula that starts $\cos C = \ldots$

7.2 Solving linear equations

This section will show you how to:

- solve equations in which the variable (the letter) appears as part of the numerator of a fraction
- solve equations where you have to expand brackets first
- solve equations where the variable appears on both sides of the equals sign

Key words
brackets
do the same to both sides
equation
rearrange
solution
solve

Fractional equations

To **solve equations** with fractions you will need to multiply both sides of the equation by the denominator at some stage. It is important to do the inverse operations in the right order.

In Examples 3 and 5, you must eliminate the constant term first before multiplying by the denominator of the fraction. In Example 4, all of the left-hand side is part of the fraction, so multiply both sides by the denominator first. It is essential to check your answer in the original equation.

Work through the following examples noting how to **rearrange** the equations.

EXAMPLE 3

Solve this equation. $\frac{x}{3} + 1 = 5$

First subtract 1 from both sides: $\frac{x}{3} = 4$

Now multiply both sides by 3: $x = 12$

Check: $\frac{12}{3} + 1 = 4 + 1 = 5$

EXAMPLE 4

Solve this equation. $\frac{x-2}{5} = 3$

First multiply both sides by 5: $x - 2 = 15$

Now add 2 to both sides: $x = 17$

Check: $\frac{17-2}{5} = \frac{15}{5} = 3$

EXAMPLE 5

Solve this equation. $\frac{3x}{4} - 3 = 1$

First add 3 to both sides: $\frac{3x}{4} = 4$

Now multiply both sides by 4: $3x = 16$

Now divide both sides by 3: $x = \frac{16}{3} = 5\frac{1}{3}$

Check: $\frac{3 \times 5\frac{1}{3}}{4} - 3 = \frac{16}{4} - 3 = 4 - 3 = 1$

EXERCISE 7B

D

1 Solve these equations.

a $\frac{f}{5} + 2 = 8$ **b** $\frac{w}{3} - 5 = 2$ **c** $\frac{x}{8} + 3 = 12$

d $\frac{5t}{4} + 3 = 18$ **e** $\frac{3y}{2} - 1 = 8$ **f** $\frac{2x}{3} + 5 = 12$

g $\frac{t}{5} + 3 = 1$ **h** $\frac{x+3}{2} = 5$ **i** $\frac{t-5}{2} = 3$

j $\frac{3x+10}{2} = 8$ **k** $\frac{2x+1}{3} = 5$ **l** $\frac{5y-2}{4} = 3$

m $\frac{6y+3}{9} = 1$ **n** $\frac{2x-3}{5} = 4$ **o** $\frac{5t+3}{4} = 1$

C

AU **2** The solution to the equation $\frac{2x-3}{5} = 3$ is $x = 9$.

Make up *two* more *different* equations of the form $\frac{ax \pm b}{5} = d$

for which the answer is also 9, where a, b, c and d are positive whole numbers.

AU **3** A teacher asked her class to solve the equation $\frac{2x+4}{5} = 6$.

Amanda wrote:

$2x + 4 = 6 \times 5$

$2x + 4 - 4 = 30 - 4$

$2x = 26$

$2x \div 2 = 26 \div 2$

$x = 13$

Betsy wrote:

$\frac{2x}{5} = 6 + 4$

$2x = 6 + 4 + 5$

$2x = 15$

$2x - 2 = 15 - 2$

$x = 13$

When the teacher read out the correct answer of 13, both students ticked their work as correct.

a Which student used the correct method?

b Explain the mistakes the other student made.

Brackets

When you have an equation that contains **brackets**, you first must multiply out the brackets and then solve the resulting equation.

EXAMPLE 6

Solve $5(x + 3) = 25$

First multiply out the brackets to get:

$5x + 15 = 25$

Rearrange: $5x = 25 - 15 = 10$

Divide by 5: $\frac{5x}{5} = \frac{10}{5}$

$x = 2$

EXAMPLE 7

Solve $3(2x - 7) = 15$

Multiply out the brackets to get:

$6x - 21 = 15$

Add 21 to both sides: $6x = 36$

Divide both sides by 6: $x = 6$

EXERCISE 7C

1 Solve each of the following equations. Some of the answers may be decimals or negative numbers. Remember to check that each answer works for its original equation. Use your calculator if necessary.

a $2(x + 5) = 16$

b $5(x - 3) = 20$

c $3(t + 1) = 18$

d $4(2x + 5) = 44$

e $2(3y - 5) = 14$

f $5(4x + 3) = 135$

g $4(3t - 2) = 88$

h $6(2t + 5) = 42$

i $2(3x + 1) = 11$

j $4(5y - 2) = 42$

k $6(3k + 5) = 39$

l $5(2x + 3) = 27$

m $9(3x - 5) = 9$

n $2(x + 5) = 6$

o $5(x - 4) = -25$

p $3(t + 7) = 15$

q $2(3x + 11) = 10$

r $4(5t + 8) = 12$

HINTS AND TIPS

Once the brackets have been expanded the equations become straightforward. Remember to multiply *everything* inside the brackets with what is outside.

D

D

AU 2 Fill in values for a, b and c so that the answer to this equation is $x = 4$.

$a(bx + 3) = c$

PS 3 My son is x years old. In five years' time, I will be twice his age and both our ages will be multiples of 10. The sum of our ages will be between 50 and 100. How old am I now?

HINTS AND TIPS

Set up an equation and put it equal to 60, 70, 80, etc. Solve the equation and see if the answer fits the conditions.

Equations with the variable on both sides

When a letter (or variable) appears on both sides of an equation, it is best to use the '**do the same to both sides**' method of **solution**, and collect all the terms containing the letter on the left-hand side of the equation. But when there are more of the letters on the right-hand side, it is easier to turn the equation round. When an equation contains brackets, they must be multiplied out first.

EXAMPLE 8

Solve this equation. $5x + 4 = 3x + 10$

There are more xs on the left-hand side, so leave the equation as it is.

Subtract $3x$ from both sides: $2x + 4 = 10$

Subtract 4 from both sides: $2x = 6$

Divide both sides by 2: $x = 3$

EXAMPLE 9

Solve this equation. $2x + 3 = 6x - 5$

There are more xs on the right-hand side, so turn the equation round.

$6x - 5 = 2x + 3$

Subtract $2x$ from both sides: $4x - 5 = 3$

Add 5 to both sides: $4x = 8$

Divide both sides by 4: $x = 2$

EXAMPLE 10

Solve this equation. $3(2x + 5) + x = 2(2 - x) + 2$

Multiply out both brackets: $6x + 15 + x = 4 - 2x + 2$

Simplify both sides: $7x + 15 = 6 - 2x$

There are more xs on the left-hand side, so leave the equation as it is.

Add $2x$ to both sides: $9x + 15 = 6$

Subtract 15 from both sides: $9x = -9$

Divide both sides by 9: $x = -1$

EXERCISE 7D

D

1 Solve each of the following equations.

a $2x + 3 = x + 5$ **b** $5y + 4 = 3y + 6$

c $4a - 3 = 3a + 4$ **d** $5t + 3 = 2t + 15$

e $7p - 5 = 3p + 3$ **f** $6k + 5 = 2k + 1$

g $4m + 1 = m + 10$ **h** $8s - 1 = 6s - 5$

> **HINTS AND TIPS**
>
> **Remember:** 'Change sides, change signs'. Show all your working. Rearrange *before* you simplify. If you try to do these at the same time you could get it wrong.

PS 2 Terry says:

I am thinking of a number. I multiply it by 3 and subtract 2.

June says:

I am thinking of a number. I multiply it by 2 and add 5.

Terry and June find that they both thought of the same number and both got the same final answer.

What number did they think of?

> **HINTS AND TIPS**
>
> Set up equations; put them equal and solve.

C

3 Solve each of the following equations.

a $2(d + 3) = d + 12$ **b** $5(x - 2) = 3(x + 4)$

c $3(2y + 3) = 5(2y + 1)$ **d** $3(h - 6) = 2(5 - 2h)$

e $4(3b - 1) + 6 = 5(2b + 4)$ **f** $2(5c + 2) - 2c = 3(2c + 3) + 7$

AU 4 Explain why the equation $3(2x + 1) = 2(3x + 5)$ cannot be solved.

> **HINTS AND TIPS**
>
> Expand the brackets and collect terms on one side as usual. What happens?

PS 5 Wilson has eight coins of the same value and seven pennies.

Chloe has 11 coins of the same value as those that Wilson has and she also has five pennies.

Wilson says, "If you give me one of your coins and four pennies, we will have the same amount of money."

What is the value of the coins that Wilson and Chloe have?

> **HINTS AND TIPS**
>
> Call the coin x and set up the equations, e.g. Wilson has $8x + 7$, and then take one x and 4 from Chloe and add one x and 4 to Wilson. Then put the equations equal and solve.

AU 6 Explain why these are an infinite number of solutions to the equation:

$2(6x + 9) = 3(4x + 6)$

7.3 Setting up equations

This section will remind you how to:

- set up equations from given information, and then solve them

Key words
do the same to both sides
equation
rearrangement
solve

Equations are used to represent situations, so that you can **solve** real-life problems. Many real-life problems can be solved by setting them up as linear equations. You can **do the same to both sides** or use **rearrangement** to **solve** the problem.

EXAMPLE 11

A milkman sets off from the dairy with eight crates of milk, each containing b bottles.

He delivers 92 bottles to a large factory and finds that he has exactly 100 bottles left on his milk float. How many bottles were in each crate?

The equation is:

$8b - 92 = 100$

$8b = 192$ (Add 92 to both sides.)

$b = 24$ (Divide both sides by 8.)

EXAMPLE 12

The rectangle shown has a perimeter of 40 cm.

Find the value of x.

$3x + 1$

$x + 3$

The perimeter of the rectangle is:

$$3x + 1 + x + 3 + 3x + 1 + x + 3 = 40$$

This simplifies to: $8x + 8 = 40$

Subtract 8 from both sides: $8x = 32$

Divide both sides by 8: $x = 4$

EXERCISE 7E

Set up an equation to represent each situation described below. Then solve the equation. Remember to check each answer.

FM 1 A man buys a daily paper from Monday to Saturday for d pence. He buys a Sunday paper for £1.80. His weekly paper bill is £7.20.

What is the price of his daily paper?

2 The diagram shows a rectangle.

a What is the value of x?

b What is the value of y?

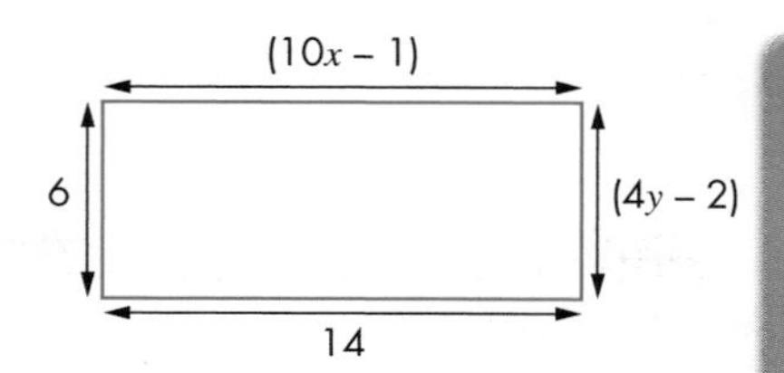

HINTS AND TIPS

Use the letter x for the variable unless you are given a letter to use. Once the equation is set up solve it by the methods shown above.

PS 3 In this rectangle, the length is 3 cm more than the width. The perimeter is 12 cm.

a What is the value of x?

b What is the area of the rectangle?

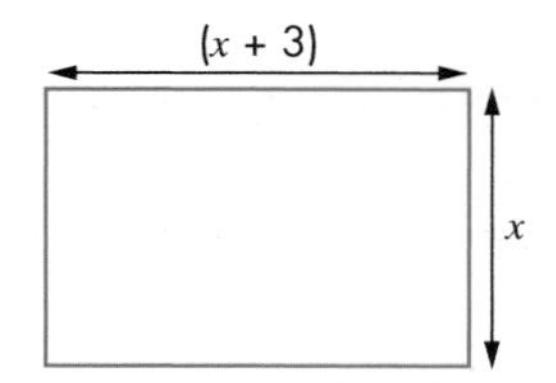

4 Mary has two bags, each of which contains the same number of sweets. She eats four sweets. She then finds that she has 30 sweets left. How many sweets were there in each bag to start with?

FM 5 A carpet costs £12.75 per square metre. The shop charges £35 for fitting. The final bill was £137.

How many square metres of carpet were fitted?

FM 6 Moshin bought eight garden chairs. When he got to the till he used a £10 voucher as part payment. His final bill was £56.

a Set this problem up as an equation, using c as the cost of one chair.

b Solve the equation to find the cost of one chair.

FM 7 This diagram shows the traffic flow through a one-way system in a town centre.

Cars enter at A and at each junction the fractions show the proportion of cars that take each route.

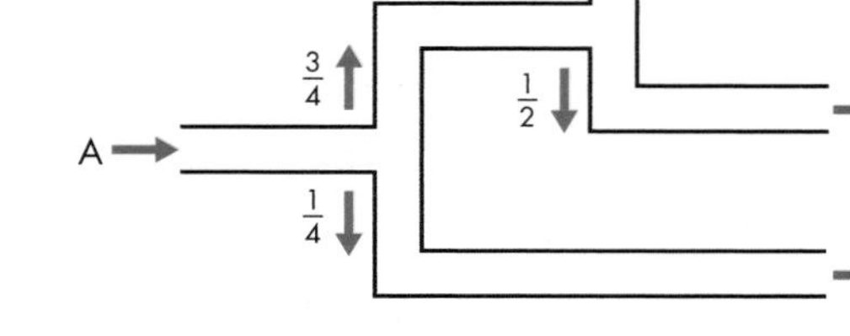

a 1200 cars enter at A. How many come out of each of the exits, B, C and D?

b If 300 cars exit at B, how many cars entered at A?

c If 500 cars exit at D, how many exit at B?

FM 8 A rectangular room is 3 m longer than it is wide. The perimeter is 16 m.

Carpet costs £9 per square metre. How much will it cost to carpet the room?

HINTS AND TIPS

Set up an equation to work out the length and width, then calculate the area.

PS 9 A boy is Y years old. His father is 25 years older than he is. The sum of their ages is 31. How old is the boy?

PS 10 Another boy is X years old. His sister is twice as old as he is. The sum of their ages is 27. How old is the boy?

11 The diagram shows a square.

Find x if the perimeter is 44 cm.

$(4x - 1)$

PS 12 Max thought of a number. He then multiplied his number by 3. He added 4 to the answer. He then doubled that answer to get a final value of 38. What number did he start with?

13 The angles of a triangle are $2x$, $x + 5°$ and $x + 35°$.

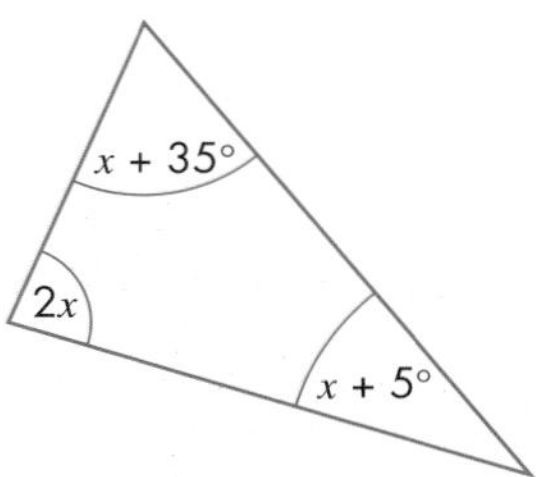

a Write down an equation to show this.

b Solve your equation to find the value of x.

FM 14 Five friends went for a meal in a restaurant. The bill was £x.
They decided to add a £10 tip and split the bill between them.

Each person paid £9.50.

a Set this problem up as an equation.

b Solve the equation to work out the bill before the tip was added.

AU 15 The diagram shows two number machines that perform the same operations.

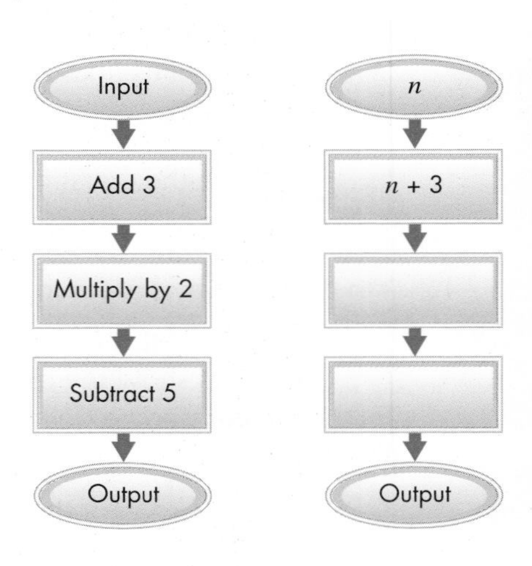

a Starting with an input value of 7, work through the left-hand machine to get the output.

b Find an input value that gives the same value for the output.

c Write down the algebraic expressions in the right-hand machine for an input of n. (The first operation has been filled in for you.)

d Set up an equation for the same input and output and show each step in solving the equation to get the answer in part **b**.

PS 16 A teacher asked her class to find three angles of a triangle that were consecutive even numbers.

Tammy wrote:

$$x + x + 2 + x + 4 = 180$$
$$3x + 6 = 180$$
$$3x = 174$$
$$x = 58$$

So the angles are 58°, 60° and 62°.

The teacher then asked the class to find four angles of a quadrilateral that are consecutive even numbers.

Can this be done? Explain your answer.

HINTS AND TIPS

Do the same type of working as Tammy did for a triangle. Work out the value of x. What happens?

FM **17** Mary has a large and a small bottle of cola. The large bottle holds 50 cl more than the small bottle.

From the large bottle she fills four cups and has 18 cl left over.

From the small bottle she fills three cups and has 1 cl left over.

How much cola does each bottle hold?

HINTS AND TIPS

Set up equations for both using x as the amount of pop in a cup. Put them equal but remember to add 50 to the small bottle equation to allow for the difference. Solve for x, then work out how much is in each bottle.

C

7.4 Trial and improvement

This section will show you how to:

- estimate the answers to some questions that do not have exact solutions, using the method of trial and improvement

Key words
comment
decimal place
guess
trial and improvement

Certain equations cannot be solved exactly. However, a close enough solution to such an equation can be found by the **trial-and-improvement method**. (Sometimes this is wrongly called the trial-and-error method.)

The idea is to keep trying different values in the equation to take it closer and closer to the 'true' solution. This step-by-step process is continued until a value is found that gives a solution that is close enough to the accuracy required.

The trial-and-improvement method is the way in which computers are programmed to solve equations.

EXAMPLE 13

Solve the equation $x^3 + x = 105$, giving the solution correct to 1 **decimal place**.

Step 1 You must find the two consecutive whole numbers between which x lies. You do this by intelligent guessing.

Try $x = 5$: $125 + 5 = 130$ — Too high – next trial needs to be much smaller.

Try $x = 4$: $64 + 4 = 68$ — Too low.

So now you know that the solution lies between $x = 4$ and $x = 5$.

EXAMPLE 13 continued

Step 2 You must find the two consecutive 1-decimal-place numbers between which x lies. Try 4.5, which is halfway between 4 and 5.

This gives 91.125 + 4.5 = 95.625 Too small.

Now attempt to improve this by trying 4.6.

This gives 97.336 + 4.6 = 101.936 Still too small.
Try 4.7 which gives 108.523. This is too high.

So the solution is between 4.6 and 4.7.

It looks as though 4.7 is closer but there is a very important final step.

Step 3 Now try the value that is halfway between the two 1-decimal-place values. In this case it is 4.65.

This gives 105.194 625.

This means that 4.6 is nearer the actual solution than 4.7.

Never assume that the one-decimal-place number that gives the closest value to the solution is the answer.

The diagram on the right shows why this is.

The approximate answer is $x = 4.6$ to 1 decimal place.

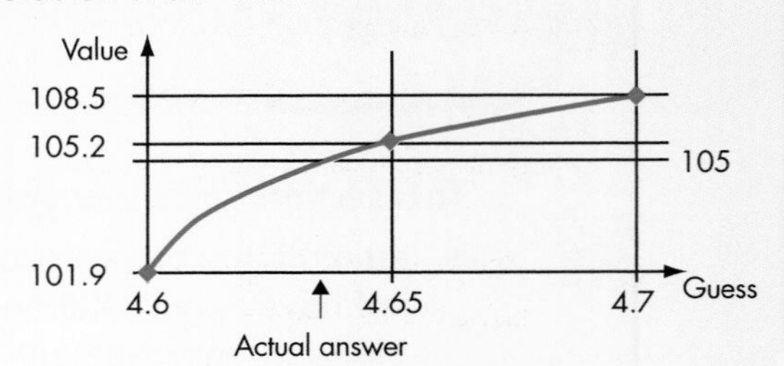

The best way to answer this type of question is to set up a table to show working. You will need three columns: **guess** (the trial), the equation to be solved and a **comment** – whether the value of the equation is too high or too low.

Guess	$x^3 + x$	Comment
4	68	Too low
5	130	Too high
4.5	95.625	Too low
4.6	101.936	Too low
4.7	108.523	Too high
4.65	105.194 625	Too high

EXERCISE 7F

C

1 Find the two consecutive *whole numbers* between which the solution to each of the following equations lies.

a $x^2 + x = 24$ **b** $x^3 + 2x = 80$ **c** $x^3 - x = 20$

2 Copy and complete the table by using trial and improvement to find an approximate solution to:

$x^3 + 2x = 50$

Give your answer correct to 1 decimal place.

Guess	$x^3 + 2x$	Comment
3	33	Too low
4	72	Too high

3 Copy and complete the table by using trial and improvement to find an approximate solution to:

$x^3 - 3x = 40$

Give your answer correct to 1 decimal place.

Guess	$x^3 + 3x$	Comment
4	52	Too high

C

4 Use trial and improvement to find an approximate solution to:

$2x^3 + x = 35$

Give your answer correct to 1 decimal place.

You are given that the solution lies between 2 and 3.

> **HINTS AND TIPS**
> Set up a table to show your working. This makes it easier for you to show method and the examiner to mark.

5 Use trial and improvement to find an exact solution to:

$4x^2 + 2x = 12$

Do not use a calculator.

6 Find a solution to each of the following equations, correct to 1 decimal place.

a $2x^3 + 3x = 35$ **b** $3x^3 - 4x = 52$ **c** $2x^3 + 5x = 79$

PS 7 A rectangle has an area of 100 cm^2. Its length is 5 cm longer than its width.

a Show that, if x is the width, then $x^2 + 5x = 100$.

b Find, correct to 1 decimal place, the dimensions of the rectangle.

8 Use trial and improvement to find a solution to the equation $x^2 + x = 40$.

> **HINTS AND TIPS**
> Call the length of the side with 'ratio 1' x, write down the other two sides in terms of x and then write down an equation for the volume = 500.

FM 9 Rob is designing a juice carton to hold $\frac{1}{2}$ litre (500 cm^3).

He wants the sides of the base in the ratio 1 : 2.

He wants the height to be 8 cm more than the shorter side of the base.

Use trial and improvement to find the dimensions of the carton.

AU 10 A cube of side x cm has a square hole of side $\frac{x}{2}$ and depth 8 cm cut from it.

The volume of the remaining solid is 1500 cm^3.

a Explain why $x^3 - 2x^2 = 1500$.

b Use trial and improvement to find the value of x to 1 decimal place.

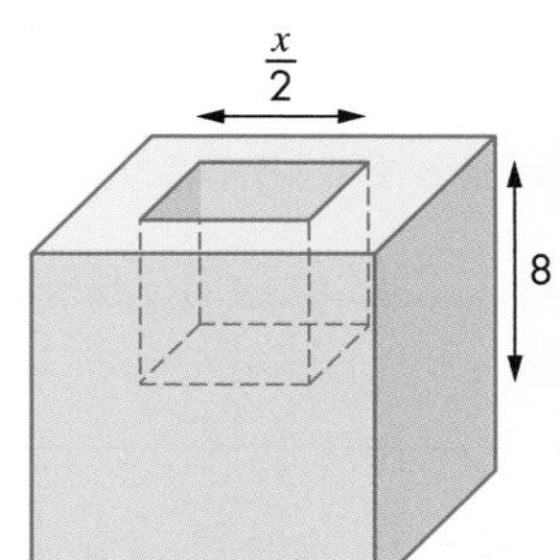

> **HINTS AND TIPS**
> Work out the volume of the cube and the hole and subtract them. The resulting expression is the volume of 1500.

PS 11 Two numbers a and b are such that $ab = 20$ and $a + b = 10$.

Use trial and improvement to find the two numbers to decimal places.

> **HINTS AND TIPS**
> Set up a table with three columns and headings a, $b = 10 - a$, ab.

7.5 Simultaneous linear equations

This section will show you how to:

- solve simultaneous linear equations in two variables

Key words
balance the coefficients
check
coefficient
eliminate
simultaneous equations
substitute
variable

A pair of **simultaneous equations** is exactly that — two equations (usually linear) for which you want the same solution, and which you therefore *solve together*. For example,

$x + y = 10$ has many solutions:

$x = 2, y = 8$ $\quad$ $x = 4, y = 6$ $\quad$ $x = 5, y = 5$ …

and $2x + y = 14$ has many solutions:

$x = 2, y = 10$ $\quad$ $x = 3, y = 8$ $\quad$ $x = 4, y = 6$ …

But only *one* solution, $x = 4$ and $y = 6$, satisfies both equations at the same time.

Elimination method

Here, you solve simultaneous equations by the *elimination method*. There are six steps in this method. **Step 1** is to **balance the coefficients** of one of the **variables**. **Step 2** is to **eliminate** this variable by adding or subtracting the equations. **Step 3** is to solve the resulting linear equation in the other variable. **Step 4** is to **substitute** the value found back into one of the previous equations. **Step 5** is to solve the resulting equation. **Step 6** is to **check** that the two values found satisfy the original equations.

EXAMPLE 14

Solve the equations: $6x + y = 15$ and $4x + y = 11$

Label the equations so that the method can be clearly explained.

$6x + y = 15$ $\quad$ (1)

$4x + y = 11$ $\quad$ (2)

Step 1: Since the y-term in both equations has the same **coefficient** there is no need to balance them.

Step 2: Subtract one equation from the other. (Equation (1) minus equation (2) will give positive values.)

(1) – (2) $\quad$ $2x = 4$

Step 3: Solve this equation: $\quad$ $x = 2$

Step 4: Substitute $x = 2$ into one of the original equations. (Usually the one with smallest numbers involved.)

So substitute into: $4x + y = 11$

which gives: $8 + y = 11$

Step 5: Solve this equation: $y = 3$

Step 6: Test the solution in the original equations. So substitute $x = 2$ and $y = 3$ into $6x + y$, which gives $12 + 3 = 15$ and into $4x + y$, which gives $8 + 3 = 11$. These are correct, so you can confidently say the solution is $x = 2$ and $y = 3$.

EXAMPLE 15

Solve these equations.

$5x + y = 22$ (1)

$2x - y = 6$ (2)

Step 1: Both equations have the same y-coefficient but with *different* signs so there is no need to balance them.

Step 2: As the signs are different, *add* the two equations, to eliminate the y-terms.

(1) + (2) $7x = 28$

Step 3: Solve this equation: $x = 4$

Step 4: Substitute $x = 4$ into one of the original equations, $5x + y = 22$,

which gives: $20 + y = 22$

Step 5: Solve this equation: $y = 2$

Step 6: Test the solution by putting $x = 4$ and $y = 2$ into the original equations, $2x - y$, which gives $8 - 2 = 6$ and $5x + y$ which gives $20 + 2 = 22$. These are correct, so the solution is $x = 4$ and $y = 2$.

Substitution method

This is an alternative method (which is covered again in Chapter 13). Which method you use depends very much on the coefficients of the variables and the way that the equations are written in the first place. There are five steps in the substitute method.

Step 1 is to rearrange one of the equations into the form $y = \dots$ or $x = \dots$.

Step 2 is to substitute the right-hand side of this equation into the other equation in place of the variable on the left-hand side.

Step 3 is to expand and solve this equation.

Step 4 is to substitute the value into the $y = \dots$ or $x = \dots$ equation.

Step 5 is to check that the values work in both original equations.

EXAMPLE 16

Solve the simultaneous equations: $y = 2x + 3$, $3x + 4y = 1$

Because the first equation is in the form $y = \dots$ it suggests that the substitution method should be used.

Again label the equations to help with explaining the method.

$y = 2x + 3$ (1)

$3x + 4y = 1$ (2)

Step 1: As equation (1) is in the form $y = \dots$ there is no need to rearrange an equation.

Step 2: Substitute the right-hand side of equation (1) into equation (2) for the variable y.

$$3x + 4(2x + 3) = 1$$

Step 3: Expand and solve the equation. $3x + 8x + 12 = 1$, $11x = -11$, $x = -1$

Step 4: Substitute $x = -1$ into $y = 2x + 3$: $y = -2 + 3 = 1$

Step 5: Test the values in $y = 2x + 3$ which gives $1 = -2 + 3$ and $3x + 4y = 1$, which gives $-3 + 4 = 1$. These are correct so the solution is $x = -1$ and $y = 1$.

EXERCISE 7G

B

1 Solve these simultaneous equations.

In question **1** parts **a** to **i** the coefficients of one of the variables are the same so there is no need to balance them. Subtract the equations when the identical terms have the same sign. Add the equations when the identical terms have opposite signs. In parts **j** to **l** use the substitution method.

a $4x + y = 17$
$2x + y = 9$

b $5x + 2y = 13$
$x + 2y = 9$

c $2x + y = 7$
$5x - y = 14$

d $3x + 2y = 11$
$2x - 2y = 14$

e $3x - 4y = 17$
$x - 4y = 3$

f $3x + 2y = 16$
$x - 2y = 4$

g $x + 3y = 9$
$x + y = 6$

h $2x + 5y = 16$
$2x + 3y = 8$

i $3x - y = 9$
$5x + y = 11$

j $2x + 5y = 37$
$y = 11 - 2x$

k $4x - 3y = 7$
$x = 13 - 3y$

l $4x - y = 17$
$x = 2 + y$

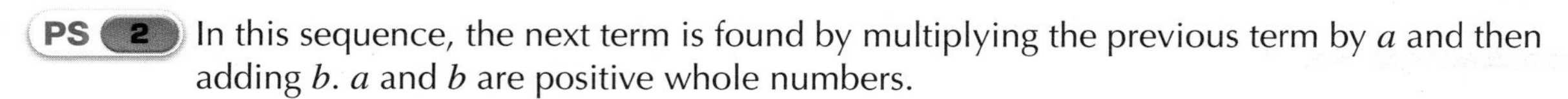

PS 2 In this sequence, the next term is found by multiplying the previous term by a and then adding b. a and b are positive whole numbers.

3 14 47 … …

a Explain why $3a + b = 14$

b Set up another equation in a and b.

c Solve the equations to solve for a and b.

d Work out the next two terms in the sequence.

B

Balancing coefficients in one equation only

You were able to solve all the pairs of equations in Exercise 7G, question **1** simply by adding or subtracting the equations in each pair, or just by substituting without rearranging. This does not always happen. The next examples show what to do when there are no identical terms to begin with, or when you need to rearrange.

EXAMPLE 17

Solve these equations. $3x + 2y = 18$ (1)

$2x - y = 5$ (2)

Step 1: Multiply equation (2) by 2. There are other ways to balance the coefficients but this is the easiest and leads to less work later. With practice, you will get used to which will be the best way to balance the coefficients.

$2 \times (2)$ $4x - 2y = 10$ (3)

Label this equation as number (3).

Be careful to multiply every term and not just the y-term, it sometimes helps to write:

$2 \times (2x - y = 5) \Rightarrow 4x - 2y = 10$ (3)

Step 2: As the signs of the y-terms are opposite, add the equations.

(1) + (3) $7x = 28$

Be careful to add the correct equations. This is why labelling them is useful.

Step 3: Solve this equation: $x = 4$

Step 4: Substitute $x = 4$ into any equation, say $2x - y = 5 \Rightarrow 8 - y = 5$

Step 5: Solve this equation: $y = 3$

Step 6: Check: (1), $3 \times 4 + 2 \times 3 = 18$ and (2), $2 \times 4 - 3 = 5$, which are correct so the solution is $x = 4$ and $y = 3$.

EXAMPLE 18

Solve the simultaneous equations: $3x + y = 5$ (1)

$5x - 2y = 10$ (2)

Step 1: Multiply the first equation by 2: $6x + 2y = 10$ (3)

Step 2: Add (1) + (3): $11x = 22$

Step 3: Solve: $x = 2$

Step 4: Substitute back: $3 \times 2 + y = 5$

Step 5: Solve: $y = -1$

Step 6: Check: (1) $3 \times 2 - 1 = 5$ and (2) $5 \times 2 - 2x -1 = 10 + 2 = 12$, which are correct.

EXERCISE 7H

B

1 Solve parts **a** to **c** by the substitution method and the rest by first changing one of the equations in each pair to obtain identical terms, and then adding or subtracting the equations to eliminate those terms.

a $5x + 2y = 4$
$4x - y = 11$

b $4x + 3y = 37$
$2x + y = 17$

c $x + 3y = 7$
$2x - y = 7$

d $2x + 3y = 19$
$6x + 2y = 22$

e $5x - 2y = 26$
$3x - y = 15$

f $10x - y = 3$
$3x + 2y = 17$

g $3x + 5y = 15$
$x + 3y = 7$

h $3x + 4y = 7$
$4x + 2y = 1$

i $5x - 2y = 24$
$3x + y = 21$

j $5x - 2y = 4$
$3x - 6y = 6$

k $2x + 3y = 13$
$4x + 7y = 31$

l $3x - 2y = 3$
$5x + 6y = 12$

AU **2** **a** Mary is solving the simultaneous equations $4x - 2y = 8$ and $2x - y = 4$.

She finds a solution of $x = 5$, $y = 6$ which works for both equations.

Explain why this is not a unique solution.

b Max is solving the simultaneous equations $6x + 2y = 9$ and $3x + y = 7$.

Why is it impossible to find a solution that works for both equations?

Balancing coefficients in both equations

There are also cases where *both* equations have to be changed to obtain identical terms. The next example shows you how this is done.

Note: The substitution method is not suitable for these types of equations as you end up with fractional terms.

EXAMPLE 19

Solve these equations. $4x + 3y = 27$ (1)

$5x - 2y = 5$ (2)

Both equations have to be changed to obtain identical terms in either x or y. However, you can see that if you make the y-coefficients the same, you will add the equations. This is always safer than subtraction, so this is obviously the better choice. We do this by multiplying the first equation by 2 (the y-coefficient of the other equation) and the second equation by 3 (the y-coefficient of the other equation).

Step 1: (1) × 2 or $2 \times (4x + 3y = 27) \Rightarrow 8x + 6y = 54$ (3)

(2) × 3 or $3 \times (5x - 2y = 5) \Rightarrow 15x - 6y = 15$ (4)

Label the new equations (3) and (4).

Step 2: Eliminate one of the variables: (3) + (4) $23x = 69$

Step 3: Solve the equation: $x = 3$

Step 4: Substitute into equation (1): $12 + 3y = 27$

Step 5: Solve the equation: $y = 5$

Step 6: Check: (1), $4 \times 3 + 3 \times 5 = 12 + 15 = 27$, and (2), $5 \times 3 - 2 \times 5 = 15 - 10 = 5$, which are correct so the solution is $x = 3$ and $y = 5$.

EXERCISE 7I

1 Solve the following simultaneous equations.

a $2x + 5y = 15$
$3x - 2y = 13$

b $2x + 3y = 30$
$5x + 7y = 71$

c $2x - 3y = 15$
$5x + 7y = 52$

d $3x - 2y = 15$
$2x - 3y = 5$

e $5x - 3y = 14$
$4x - 5y = 6$

f $3x + 2y = 28$
$2x + 7y = 47$

g $2x + y = 4$
$x - y = 5$

h $5x + 2y = 11$
$3x + 4y = 8$

i $x - 2y = 4$
$3x - y = -3$

j $3x + 2y = 2$
$2x + 6y = 13$

k $6x + 2y = 14$
$3x - 5y = 10$

l $2x + 4y = 15$
$x + 5y = 21$

m $3x - y = 5$
$x + 3y = -20$

n $3x - 4y = 4.5$
$2x + 2y = 10$

o $x - 5y = 15$
$3x - 7y = 17$

PS 2 Here are four equations.

A: $5x + 2y = 1$
B: $4x + y = 9$
C: $3x - y = 5$
D: $3x + 2y = 3$

Here are four sets of (x, y) values.

$(1, -2)$, $(-1, 3)$, $(2, 1)$, $(3, -3)$

Match each pair of (x, y) values to a pair of equations.

HINTS AND TIPS

You could solve each possible set of pairs but there are six to work out. Alternatively you can substitute values into the equations to see which work.

B

AU 3 Find the area of the triangle enclosed by these three equations.

$y - x = 2$ $\quad$ $x + y = 6$ $\quad$ $3x + y = 6$

AU 4 Find the area of the triangle enclosed by these three equations.

$x - 2y = 6$ $\quad$ $x + 2y = 6$ $\quad$ $x + y = 3$

HINTS AND TIPS

Find the point of intersection of each pair of equations, plot the points on a grid and use any method to work out the area of the resulting triangle.

7.6 Solving problems using simultaneous equations

This section will show you how to:

- solve problems, using simultaneous linear equations in two variables

Key words

balance, check, coefficient, eliminate, simultaneous equations, substitute, variable

You are now going to meet a type of problem that has to be expressed as a pair of simultaneous equations so that it can be solved. The next example shows you how to tackle such a problem.

EXAMPLE 20

On holiday last year, I was talking over breakfast to two families about how much it cost them to go to the theatre. They couldn't remember how much was charged for each adult or each child, but they could both remember what they had paid altogether.

The Advani family, consisting of Mr and Mrs Advani with their daughter Rupa, paid £23.

The Shaw family, consisting of Mrs Shaw with her two children, Len and Sue, paid £17.50.

How much would I have to pay for my wife, my four children and myself?

Make a pair of simultaneous equations from the situation, as follows.

Let x be the cost of an adult ticket, and y be the cost of a child's ticket. Then

$2x + y = 23$ for the Advani family

and $x + 2y = 17.5$ for the Shaw family

Now solve these equations just as you have done in the previous examples, to obtain:

$x = £9.50$ and $y = £4$.

You can now find the cost, which will be $(2 \times £9.50) + (4 \times £4) = £35$.

EXERCISE 7J

Read each situation carefully, then make a pair of simultaneous equations in order to solve the problem.

PS **1** Amul and Kim have £10.70 between them. Amul has £3.70 more than Kim. Let x be the amount Amul has and y be the amount Kim has. Set up a pair of simultaneous equations. How much does each have?

FM **2** The two people in front of me at the Post Office were both buying stamps. One person bought 10 second-class and five first-class stamps at a total cost of £4.20. The other bought eight second-class and 10 first-class stamps at a total cost of £5.40.

a Let x be the cost of a second-class stamp and y be the cost of a first-class stamp. Set up two simultaneous equations.

b How much did I pay for three second-class and four first-class stamps?

3 At a local tea room I couldn't help noticing that at one table, where the customers had eaten six buns and had three teas, the bill came to £4.35. At another table, the customers had eaten 11 buns and had seven teas at a total cost of £8.80.

a Let x be the cost of a bun and y be the cost of a cup of tea. Show the situation as a pair of simultaneous equations.

b My family and I had five buns and six teas. What did it cost us?

PS **4** The sum of my son's age and my age this year is 72.

Six years ago my age was double that of my son.

Let my age now be x and my son's age now be y.

a Explain why $x - 6 = 2(y - 6)$.

b Find the values of x and y.

5 In a tea shop, three teas and five buns cost £8.10

In the same tea shop three teas and three buns cost £6.30

a Using t to represent the cost of a tea and b to represent the cost of a bun, set up the above information as a pair of simultaneous equations.

b How much will I pay for four teas and six buns?

6 Three chews and four bubblies cost 72p. Five chews and two bubblies cost 64p. What would three chews and five bubblies cost?

FM **7** On a nut-and-bolt production line, all the nuts had the same mass and all the bolts had the same mass. An order of 50 nuts and 60 bolts had a mass of 10.6 kg. An order of 40 nuts and 30 bolts had a mass of 6.5 kg. What should the mass of an order of 60 nuts and 50 bolts be?

FM **8** My local taxi company charges a fixed amount plus so much per mile. When I took a six mile journey the cost was £3.70. When I took a 10 mile journey the cost was £5.10. My next journey is going to be eight miles. How much will this cost?

B

A

A

FM 9 Two members of the same church went to the same shop to buy material to make Christingles. One bought 200 oranges and 220 candles at a cost of £65.60. The other bought 210 oranges and 200 candles at a cost of £63.30. They only needed 200 of each. How much should it have cost them?

FM 10 When you book Bingham Hall for a conference you pay a fixed booking fee plus a charge for each delegate. AQA booked a conference for 65 delegates and was charged £192.50. OCR booked a conference for 40 delegates and was charged £180. EDEXCEL wants to book for 70 delegates. How much will they be charged?

FM 11 My mother-in-law uses this formula to cook a turkey:

$$T = a + bW$$

where T is the cooking time (minutes), W is the weight of the turkey (kg) and a and b are constants. She says it takes 4 hours 30 minutes to cook a 12 kg turkey, and 3 hours 10 minutes to cook an 8 kg turkey. How long will it take to cook a 5 kg turkey?

FM 12 Four sacks of potatoes and two sacks of carrots weigh 188 pounds.

Five sacks of potatoes and one sack of carrots weigh 202 pounds.

Baz buys seven sacks of potatoes and eight sacks of carrots.

Will he be able to carry them in his trailer, which has a safe working load of 450 pounds?

HINTS AND TIPS

Set up two simultaneous equations using p and c for the weight of a sack of potatoes and carrots respectively.

FM 13 Five bags of bark chipping and four trays of pansies cost £24.50.

Three bags of bark chippings and five trays of pansies cost £19.25.

Camilla wants six bags of bark chippings and eight trays of pansies.

She has £30. Will she have enough money?

HINTS AND TIPS

Set up a pair of simultaneous equations using b and p for the cost of bark chippings and pansies and solve them.

AU 14 A teacher asks her class to solve these two simultaneous equations.

$$y = x + 4 \quad (1)$$
$$2y - x = 10 \quad (2)$$

Carmen says to Jeff, "Let's save time, you work out the x-value and I'll work out the y-value." Jeff says, "Great idea."

This is Carmen's work.

$$y - x = 4 \quad (3)$$
$$2y - x = 10 \quad (2)$$
$$(2) - (3) \quad 3y = 6$$
$$y = 2$$

This is Jeff's work.

Substitute (1) into (2)

$$2(x + 4) - x = 10$$
$$2x + 8 - x = 10$$
$$3x = 18$$
$$x = 6$$

When the teacher reads out the answer as "two, six" the students mark their work correct.

Explain all the mistakes that Carmen and Jeff have made.

7.7 Linear and non-linear simultaneous equations

This section will show you how to:
- solve linear and non-linear simultaneous equations

Key words
linear
non-linear
substitute

You have already seen the method of substitution for solving **linear** simultaneous equations earlier in this chapter. Example 21 is a reminder.

EXAMPLE 21

Solve these simultaneous equations.

$$2x + 3y = 7 \quad (1)$$
$$x - 4y = 9 \quad (2)$$

First, rearrange equation (2) to obtain:

$$x = 9 + 4y$$

Substitute the expression for x into equation (1), which gives:

$$2(9 + 4y) + 3y = 7$$

Expand and solve this equation to obtain:

$$18 + 8y + 3y = 7$$
$$\Rightarrow 11y = -11$$
$$\Rightarrow y = -1$$

Now substitute for y into either equation (1) or (2) to find x. Using equation (1),

$$\Rightarrow 2x - 3 = 7$$
$$\Rightarrow x = 5$$

You can use a similar method when you need to solve a pair of equations, one of which is linear and the other of which is **non-linear**. But you must always **substitute** from the linear into the non-linear.

EXAMPLE 22

Solve these simultaneous equations.

$$x^2 + y^2 = 5$$
$$x + y = 3$$

Call the equations (1) and (2):

$$x^2 + y^2 = 5 \quad (1)$$
$$x + y = 3 \quad (2)$$

Rearrange equation (2) to obtain:

$$x = 3 - y$$

Substitute this into equation (1), which gives:

$$(3 - y)^2 + y^2 = 5$$

EXAMPLE 22 continued

Expand and rearrange into the general form of the quadratic equation:

$$9 - 6y + y^2 + y^2 = 5$$
$$2y^2 - 6y + 4 = 0$$

Cancel by 2:

$$y^2 - 3y + 2 = 0$$

Factorise:

$$(y - 1)(y - 2) = 0$$
$$\Rightarrow y = 1 \text{ or } 2$$

Substitute for y in equation (2):

When $y = 1$, $x = 2$ and when $y = 2$, $x = 1$

Note that you should always give answers as a pair of values in x and y.

EXAMPLE 23

Find the solutions of the pair of simultaneous equations: $y = x^2 + x - 2$ and $y = 2x + 4$

This example is slightly different, as both equations are given in terms of y, so substituting for y gives:

$$2x + 4 = x^2 + x - 2$$

Rearranging into the general quadratic:

$$x^2 - x - 6 = 0$$

Factorising and solving gives:

$$(x + 2)(x - 3) = 0$$
$$x = -2 \text{ or } 3$$

Substituting back to find y:

When $x = -2$, $y = 0$

When $x = 3$, $y = 10$

So the solutions are $(-2, 0)$ and $(3, 10)$.

EXERCISE 7K

A

1 Solve these pairs of linear simultaneous equations using the substitution method.

a $2x + y = 9$
$x - 2y = 7$

b $3x - 2y = 10$
$4x + y = 17$

c $x - 2y = 10$
$2x + 3y = 13$

A*

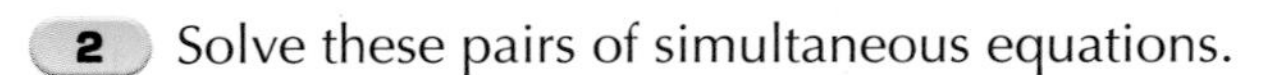

2 Solve these pairs of simultaneous equations.

a $xy = 2$
$y = x + 1$

b $xy = -4$
$2y = x + 6$

A*

3 Solve these pairs of simultaneous equations.

a $x^2 + y^2 = 25$
$x + y = 7$

b $x^2 + y^2 = 9$
$y = x + 3$

c $x^2 + y^2 = 13$
$5y + x = 13$

4 Solve these pairs of simultaneous equations.

a $y = x^2 + 2x - 3$
$y = 2x + 1$

b $y = x^2 - 2x - 5$
$y = x - 1$

c $y = x^2 - 2x$
$y = 2x - 3$

5 Solve these pairs of simultaneous equations.

a $y = x^2 + 3x - 3$ and $y = x$

b $x^2 + y^2 = 13$ and $x + y = 1$

c $x^2 + y^2 = 5$ and $y = x + 1$

d $y = x^2 - 3x + 1$ and $y = 2x - 5$

e $y = x^2 - 3$ and $y = x + 3$

f $y = x^2 - 3x - 2$ and $y = 2x - 6$

g $x^2 + y^2 = 41$ and $y = x + 1$

AU 6 **a** Solve the simultaneous equations: $y = x^2 + 3x - 4$ and $y = 5x - 5$

b Which of the sketches below represents the graphs of the equations in part **a**? Explain your choice.

i

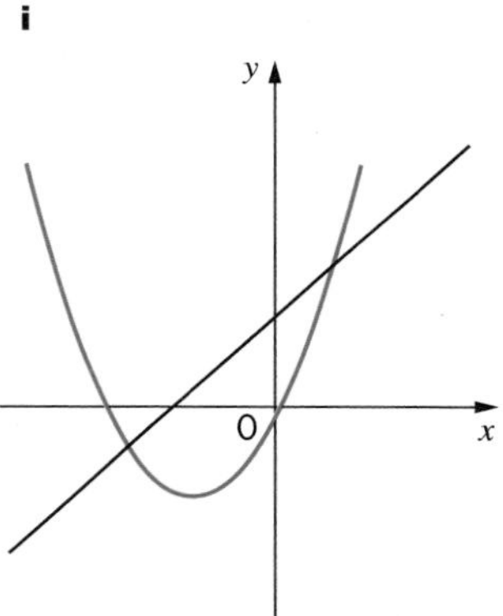

ii

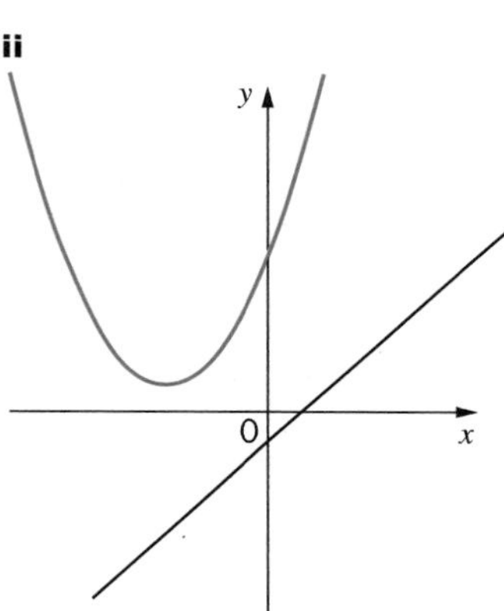

iii

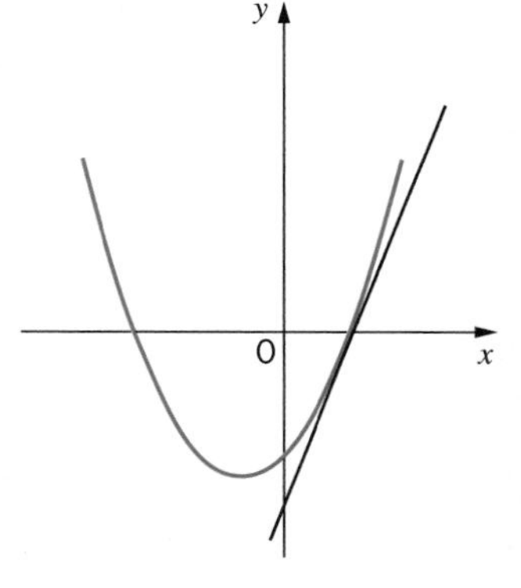

PS 7 The simultaneous equations $x^2 + y^2 = 5$ and $y = 2x + 5$ only has one solution.

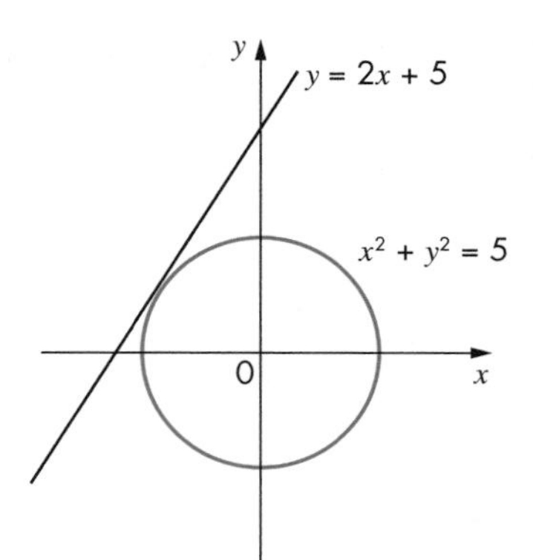

a Find the solution.

b Write down the intersection of each pair of graphs.

i $x^2 + y^2 = 5$ and $y = -2x + 5$

ii $x^2 + y^2 = 5$ and $y = -2x - 5$

iii $x^2 + y^2 = 5$ and $y = 2x - 5$

PS 8 Solve these pairs of simultaneous equations.

a $y = x^2 + x - 2$
$y = 5x - 6$

b $y = x^2 + 2x - 3$
$y = 4x - 4$

c What is the geometrical significance of the answers to parts **a** and **b**?

GRADE BOOSTER

- **D** You can solve linear equations
- **C** You can rearrange simple formulae
- **C** You can set up and solve linear equations from practical and real-life situations
- **C** You can use trial and improvement to solve non-linear equations
- **B** You can solve two simultaneous linear equations
- **A** You can solve a pair of simultaneous equations where one is linear and the other is non-linear
- **A** You can set up a real-life problem and solve it using simultaneous equations
- **A** You can rearrange a formula where the subject appears twice

What you should know now

- How to solve all types of linear equations
- How to rearrange a formula where the subject appears twice
- How to solve equations by trial and improvement
- How to set up and solve problems, using linear equations
- How to solve a pair of simultaneous equations where one is linear and one is non-linear

1 Make t the subject of the formula $2(t - 5) = y$ *(3 marks)*

Edexcel, May 2008, Paper 14 Higher, Question 12

2 Solve $2(x - 3) = 5$ *(3 marks)*

Edexcel, June 2008, Paper 15 Higher, Question 5

3 The equation $x^3 + 2x = 26$ has a solution between 2 and 3. Use a trial and improvement method to find this solution.

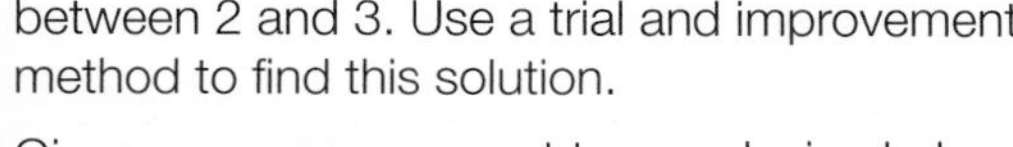

Give your answer correct to one decimal place.

You must show **all** your working. *(4 marks)*

Edexcel, June 2008, Paper 15 Higher, Question 6

4

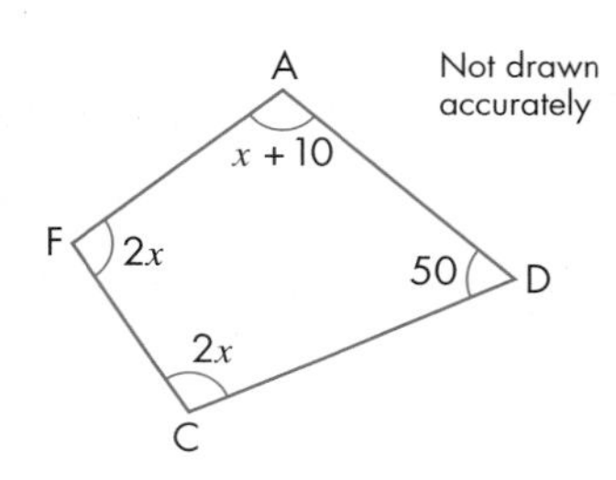

In this quadrilateral, the sizes of the angles, in degrees, are:

$x + 10$
$2x$
$2x$
50

a Use this information to write down an equation in terms of x. *(2 marks)*

b Work out the value of x. *(3 marks)*

Edexcel, June 2008, Paper 15 Higher, Question 8

5 The diagram shows a triangle.

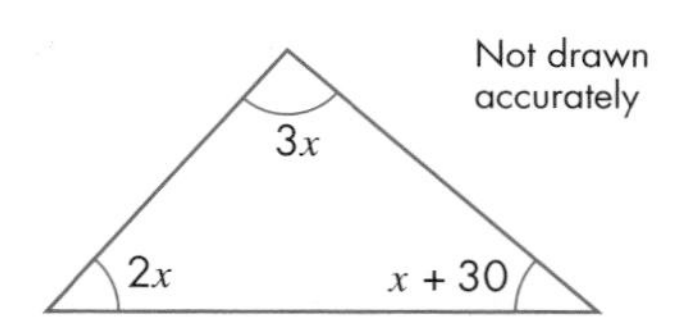

The sizes of the angles, in degrees, are:

$3x$
$2x$
$x + 30$

Work out the value of x. *(3 marks)*

Edexcel, November 2008, Paper 3 Higher, Question 12

6 The equation $x^3 + 4x = 26$ has a solution between 2 and 3.

Use a trial and improvement method to find this solution.

Give your answer correct to 1 decimal place.

You must show all your working. *(4 marks)*

Edexcel, November 2008, Paper 15 Higher, Question 5

7 Solve $\frac{5(2x + 1)}{3} = 4x + 7$ *(3 marks)*

Edexcel, June 2008, Paper 15 Higher, Question 16

8 Solve the simultaneous equations.

$2x + 3y = 0$
$x - 3y = 9$ *(3 marks)*

Edexcel, May 2008, Paper 3 Higher, Question 19

9 $v^2 = u^2 + 2as$
$u = 6$
$a = 2.5$
$s = 9$

a Work out a value of v *(3 marks)*

b Make s the subject of the formula
$v^2 = u^2 + 2as$ *(2 marks)*

Edexcel, November 2008, Paper 3 Higher, Question 16

10 Make b the subject of the formula

$a = \frac{2 - 7b}{b - 5}$ *(4 marks)*

Edexcel, May 2008, Paper 3 Higher, Question 22

11 Solve the simultaneous equations

$x^2 + y^2 = 5$
$y = 3x + 1$ *(6 marks)*

Edexcel, November 2008, Paper 3 Higher, Question 28

Worked Examination Questions

1 $4x + 3y = 6$

$3x - 2y = 13$

Solve these simultaneous equations algebraically. Show your method clearly.

1 $4x + 3y = 23$ (1)

$3x - 2y = 13$ (2)

Label the equations and decide on the best way to get the coefficients of one variable the same.

(1) × 2 $8x + 6y = 46$ (3)

(2) × 3 $9x - 6y = 39$ (4)

(3) + (4) $17x = 85$

Making the y-coefficients the same will be the most efficient way as the resulting equations will be added.

This gets 1 mark for method.

$x = 5$

Substitute into (1) $20 + 3y = 23$

$y = 1$

Solve the resulting equation. Substitute into one of the original equations. Work out the other value.

$4 \times 5 + 3 \times 1 = 23$ ✓

$3 \times 5 - 2 \times 1 = 13$ ✓

Check that these values work in the original equations.

Total: 6 marks

2 Temperatures can be measured in degrees Celsius (°C), degrees Fahrenheit (°F) or kelvin (K). The relationships between the scales of temperature are given by

$$C = \frac{5(F - 32)}{9}$$

$$K = C + 273$$

Express F

i in terms of C

ii in terms of K

2 i $9C = 5(F - 32)$ ✓

$9C = 5F - 160$

$5F = 9C + 160$

$F = \frac{9C + 160}{5}$

This gets 1 mark for multiplying both sides of the equation by 9; 1 mark for expanding the bracket and adding 160 to both sides; 1 mark for changing the equation round and dividing by 5.

ii $C = K - 273$

$F = \frac{9(K - 273) + 160}{5}$

$F = \frac{9K - 2457 + 160}{5}$

$F = \frac{9K - 2297}{5}$

Make C the subject of the second equation. Substitute for C in the answer to part **i**. Expand the bracket and tidy up the top line of the fraction. You get 1 mark each for the above steps for method and accuracy and 1 other mark.

Total: 6 marks

Worked Examination Questions

3 Make g the subject of the following formula.

$$\frac{t(3 + g)}{8 - g} = 2$$

3 $t(3 + g) = 2(8 - g)$ — Cross multiply to get rid of the fraction. This scores 1 mark for method.

$3t + gt = 16 - 2g$ — Expand the brackets. This scores 1 mark for accuracy.

$gt + 2g = 16 - 3t$

$g(t + 2) = 16 - 3t$ — Collect all the g terms on the left-hand side and other terms on the right-hand side. This scores 1 mark for method.

$g = \dfrac{16 - 3t}{t + 2}$ — Simplify, $gt + 2g = g(t + 2)$, and divide by $(t + 2)$. This scores 1 mark for accuracy.

Total: 4 marks

4 A rectangle with sides of 4 cm and $(3x + 1)$ cm has a smaller rectangle with sides of 3 cm and $(2x - 1)$ cm cut from it.

The area remaining is 20 cm^2.

Work out the value of x.

$4x - 1$

$2x - 1$

5

3

4 $5(4x - 1) - 3(2x + 1)$ — Write down the difference in the two areas to get 1 method mark.

$20x - 5 - 6x - 3 = 20$ — Expand the brackets and put the resulting expression equal to the area for 1 method mark.

$14x - 8 = 20$ — Simplify the expressions for 1 accuracy mark.

$x = 2$ — Solve the equation to get the 1 final mark for accuracy.

Total: 4 marks

7 Functional Maths
Choosing a mobile phone plan

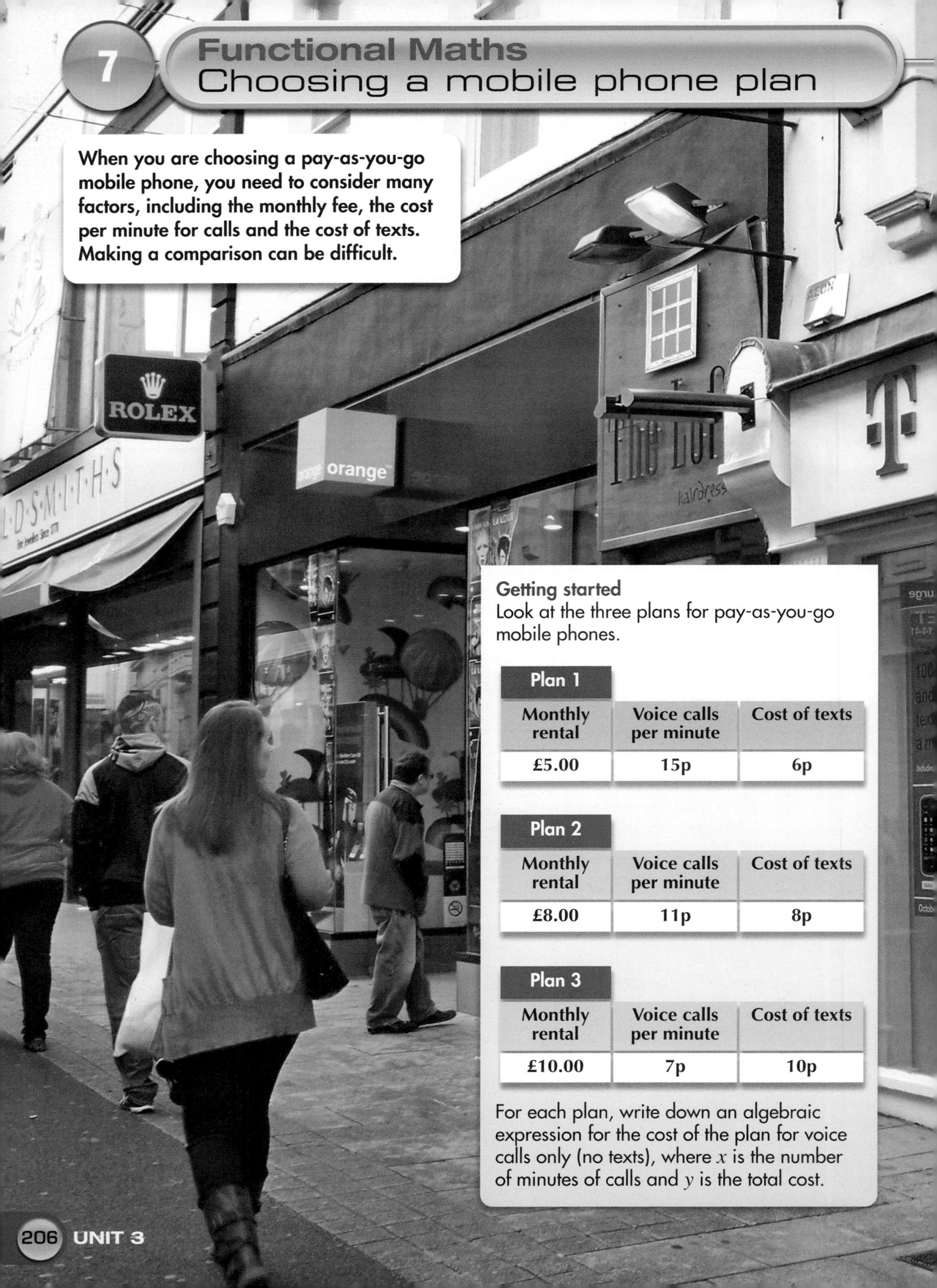

When you are choosing a pay-as-you-go mobile phone, you need to consider many factors, including the monthly fee, the cost per minute for calls and the cost of texts. Making a comparison can be difficult.

Getting started

Look at the three plans for pay-as-you-go mobile phones.

Plan 1

Monthly rental	Voice calls per minute	Cost of texts
£5.00	15p	6p

Plan 2

Monthly rental	Voice calls per minute	Cost of texts
£8.00	11p	8p

Plan 3

Monthly rental	Voice calls per minute	Cost of texts
£10.00	7p	10p

For each plan, write down an algebraic expression for the cost of the plan for voice calls only (no texts), where x is the number of minutes of calls and y is the total cost.

Your task

1 Compare the three mobile phone plans.

- Find the number of minutes of voice calls after which each plan becomes cheaper.

  Hint: Think about these questions.

  - After how many minutes does plan 2 become cheaper than plan 1?
  - After how many minutes does plan 3 become cheaper than plan 2?
  - After how many minutes does plan 3 become cheaper than plan 1?

- Decide which plan you would choose if you made 3 hours of voice calls per month and sent no texts.
- Write down an algebraic expression for the cost of the plan for 3 hours of voice calls and x texts, where y is the total cost.
- Work out the number of texts after which each plan becomes cheaper.

Using this information, decide which plan you would choose if you made 3 hours of voice calls per month and sent 250 texts.

Your task (continued)

2 Ask a few friends or relatives how many minutes of voice calls they make and how many texts they send per month.

Which of these plans would be best value for them? Write a short report for each of them, so that they can understand how you reached your conclusions.

Why this chapter matters

For anything, from a house to a landscape gardening project, the designer needs to construct plans accurately, to be sure that everything will fit together properly. This will also give the people putting it together a blueprint to work from.

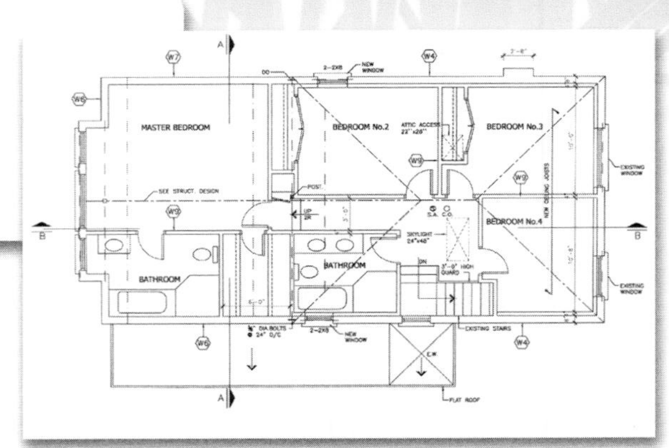

The need for accurate drawings is clear in bridge construction. Bridge engineers are responsible for producing practical bridge designs to meet the requirements of their employers. For example, a bridge intended to carry traffic over a newly constructed railway needs to be strong enough to bear the weight of the traffic and stable enough to counteract the effects of the moving traffic and strong winds. The designers produce a blueprint that has all the measurements, including heights, weights and angles, clearly marked on it. Construction workers then use this blueprint to build the bridge to the exact specifications set by the designers and engineers.

Generally, the construction workers work on both ends of the bridge at the same time, meeting in the middle. The blueprints are therefore essential for making sure that the bridge is safe and that the bridge meets in the middle.

Accurately-drawn blueprints were essential in the construction of the Golden Gate Bridge, which crosses the San Francisco Bay. When it was constructed in the 1930s, it was the longest suspension bridge in the world. The bridge engineers (who included Joseph Strauss and Charles Alton Ellis) had to draw precise blueprints to make sure that they had all the information necessary to build this innovative bridge and that it would be built correctly.

By contrast, a bridge built at a stadium for the Maccabiah Games in Israel was built without proper planning and without accurate blueprints. This led to the bridge collapsing soon after its construction in 1997, killing four athletes and injuring 64 people.

Just like a bridge engineer, you must be accurate in your construction, working with a freshly-sharpened pencil and a good pair of compasses, measuring and drawing angles carefully and drawing construction lines as faintly as possible.

In this chapter you will start with simple constructions of triangles, moving on to more complex bisectors and then to plotting loci, which are paths of whole sets of points obeying certain rules or criteria.

Chapter 8

Geometry: Constructions

The grades given in this chapter are target grades.

1 Constructing triangles

2 Bisectors

3 Defining a locus

4 Loci problems

This chapter will show you ...

- D how to construct triangles
- C how to bisect a line and an angle
- C how to construct perpendiculars
- C how to define a locus
- C how to solve locus problems

Visual overview

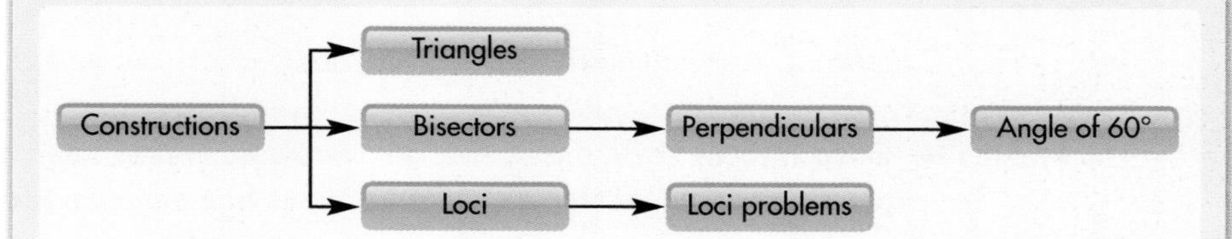

What you should already know

- How to measure lines and angles **(KS3 level 5, GCSE grade F)**
- How to use scale drawings **(KS3 level 5, GCSE grade E)**

Quick check

1 Measure the following lines.

a

b

c

2 Measure the following angles.

a

b

8.1 Constructing triangles

This section will show you how to:

- construct triangles, using compasses, a protractor and a straight edge

Key words
angle
compasses
construct
side

There are three ways of **constructing** a triangle. Which one you use depends on what information you are given about the triangle.

When carrying out geometric constructions, always use a sharp pencil (preferably grade 2H rather than HB) to give you thin, clear lines. These may be called faint or feint lines. The examiner will be marking your construction and will be looking for accuracy, which requires fine, clean lines and points as small as you can make them, while ensuring they are clearly visible.

All three sides known

EXAMPLE 1

Construct a triangle with **sides** that are 5 cm, 4 cm and 6 cm long.

- **Step 1:** Draw the longest side as the base. In this case, the base will be 6 cm, which you draw using a ruler. (The diagrams in this example are drawn at half-size.)

- **Step 2:** Deal with the second longest side, in this case the 5 cm side. Open the **compasses** to a radius of 5 cm (the length of the side), place the point on one end of the 6 cm line and draw a short faint arc, as shown here.

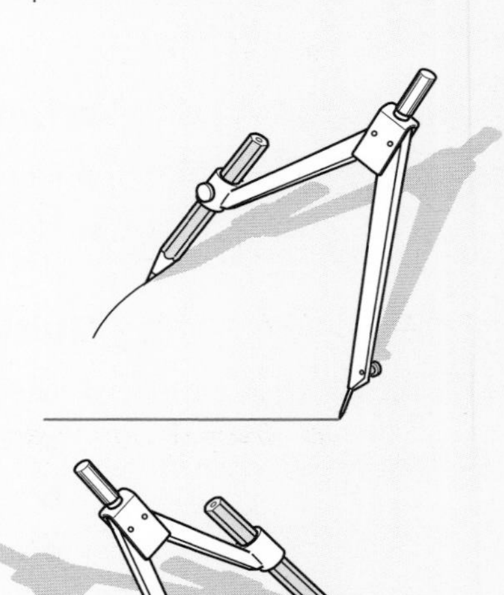

- **Step 3:** Deal with the shortest side, in this case the 4 cm side. Open the compasses to a radius of 4 cm, place the point on the other end of the 6 cm line and draw a second short faint arc to intersect the first arc, as shown here.

- **Step 4:** Complete the triangle by joining each end of the base line to the point where the two arcs intersect.

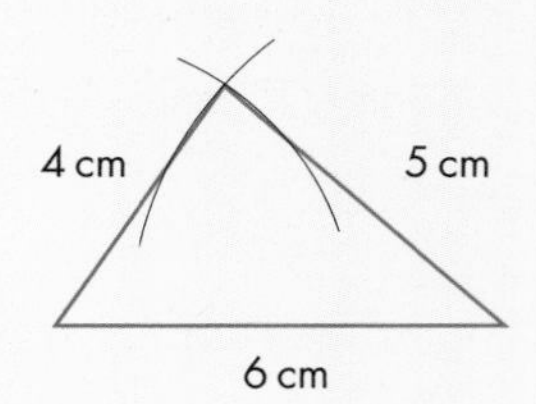

Note: The arcs are construction lines and so must be left in to show the examiner how you constructed the triangle.

Two sides and the included angle known

EXAMPLE 2

Draw a triangle ABC, in which AB is 6 cm, BC is 5 cm and the included **angle** ABC is 55°. (The diagrams in this example are drawn at half-size.)

- **Step 1:** Draw the longest side, AB, as the base. Label the ends of the base A and B.

A ———————— B

- **Step 2:** Place the protractor along AB with its centre on B and make a point on the diagram at the 55° mark.

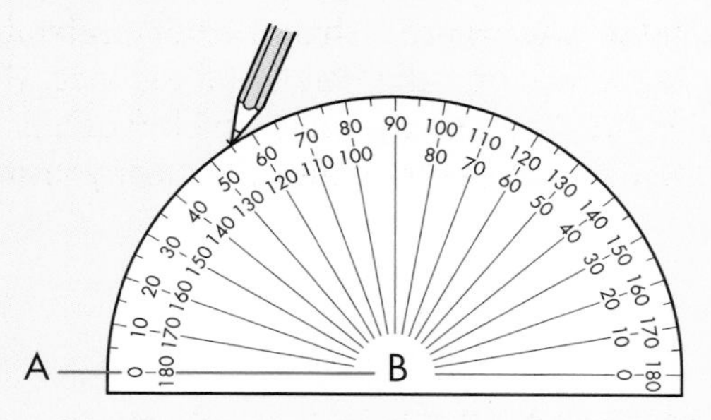

- **Step 3:** Draw a *faint* line from B through the 55° point. From B, using a pair of compasses, measure 5 cm along this line.
- Label the point where the arc cuts the line as C.

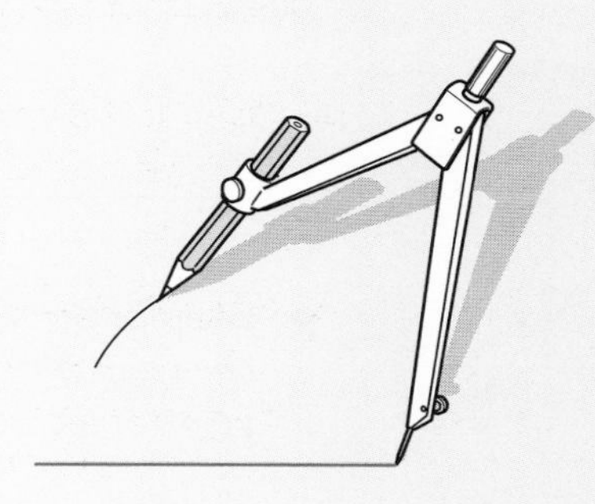

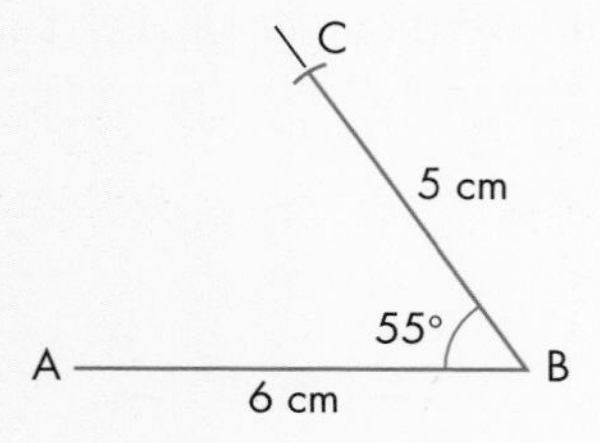

- **Step 4:** Join A and C to complete the triangle.

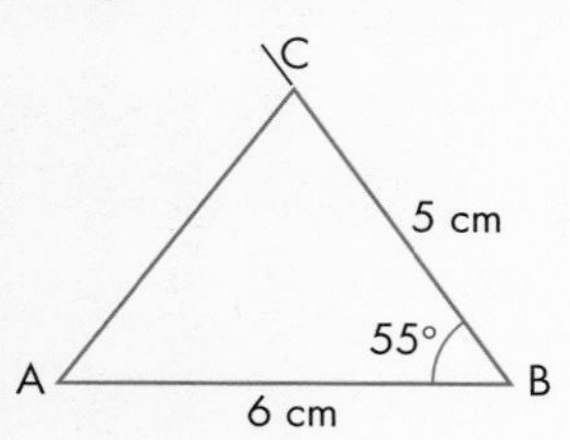

Note: Remember to use clean, sharp lines so that the examiner can see how the triangle has been constructed.

Two angles and a side known

When you know two angles of a triangle, you also know the third.

EXAMPLE 3

Draw a triangle ABC, in which AB is 7 cm, angle BAC is 40° and angle ABC is 65°.

- **Step 1:** As before, start by drawing the base, which here has to be 7 cm. Label the ends A and B.

- **Step 2:** Centre the protractor on A and mark the angle of 40°. Draw a clear, clean line from A through this point.

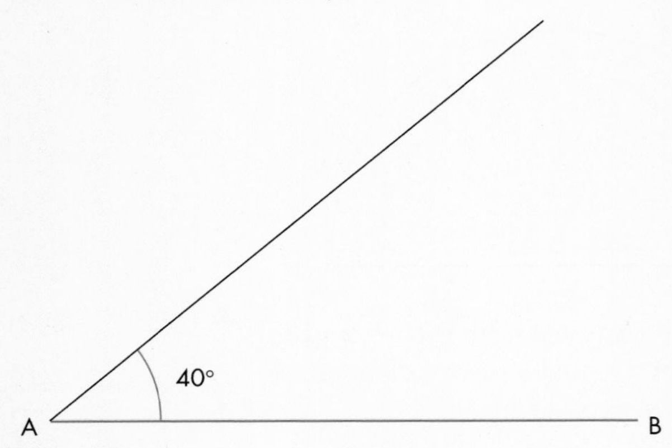

- **Step 3:** Centre the protractor on B and mark the angle of 65°. Draw a clear, clean line from B through this point, to intersect the 40° line drawn from A. Label the point of intersection as C.

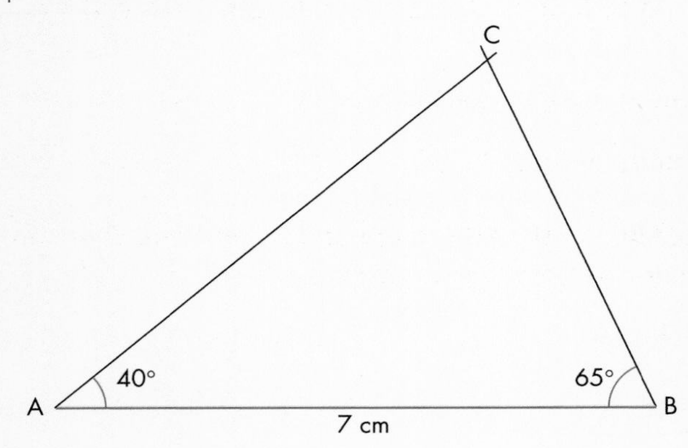

EXERCISE 8A

1 Draw the following triangles accurately and measure the sides and angles not given in the diagram.

> **HINTS AND TIPS**
>
> Always make a sketch if one is not given in the question.

a

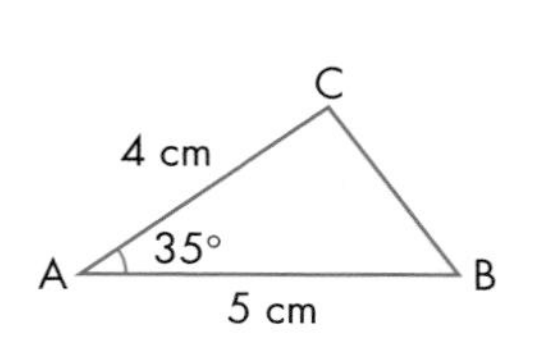

b

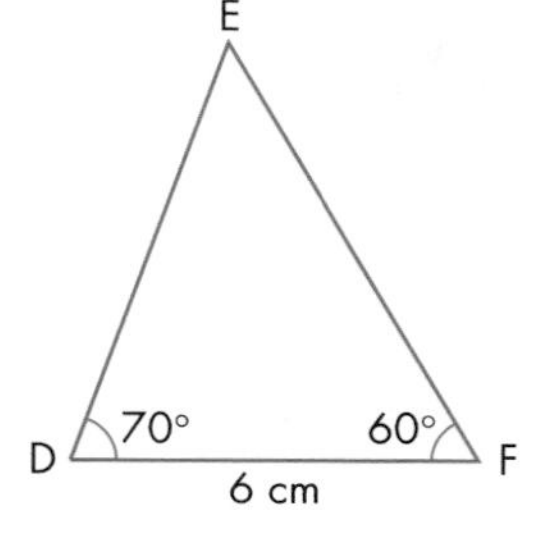

c

I
8 cm
4 cm
G
H
6 cm

d

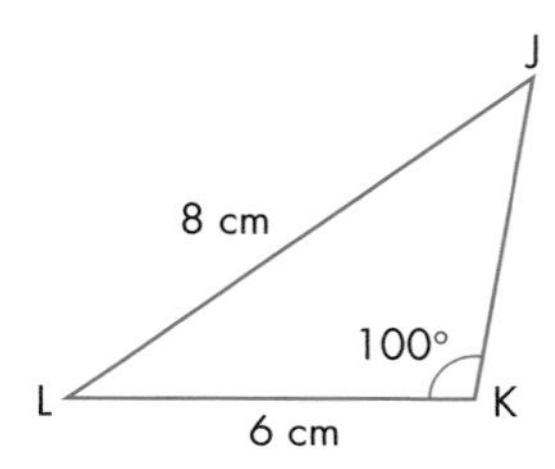

e

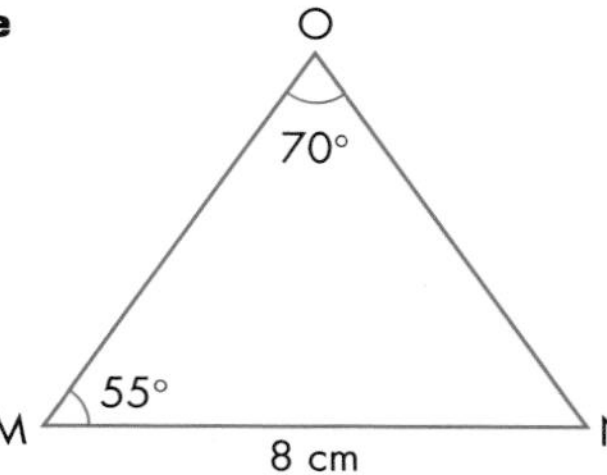

f

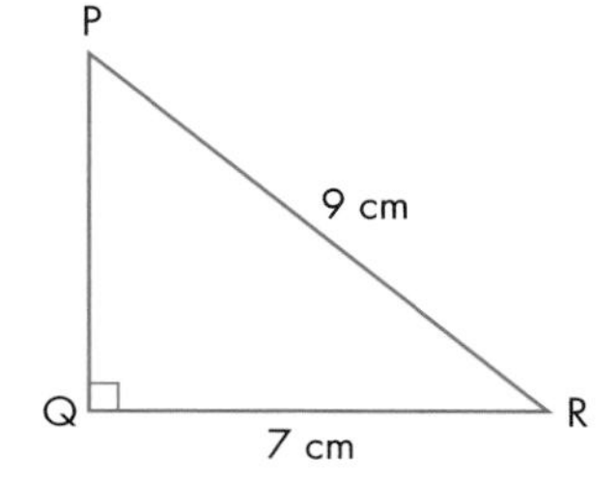

2 **a** Draw a triangle ABC, where AB = 7 cm, BC = 6 cm and AC = 5 cm.

b Measure the sizes of ∠ABC, ∠BCA and ∠CAB.

> **HINTS AND TIPS**
>
> Sketch the triangle first.

3 Draw an isosceles triangle that has two sides of length 7 cm and the included angle of 50°.

a Measure the length of the base of the triangle.

b What is the area of the triangle?

4 A triangle ABC has ∠ABC = 30°, AB = 6 cm and AC = 4 cm. There are two different triangles that can be drawn from this information.

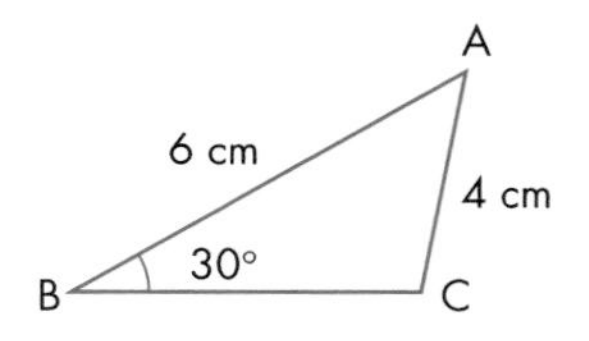

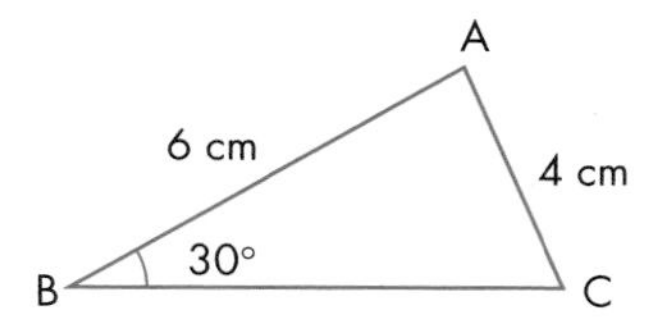

What are the two different lengths that BC can be?

5 Construct an equilateral triangle of side length 5 cm.

a Measure the height of the triangle.

b What is the area of this triangle?

D

D

6 Construct a parallelogram with sides of length 5 cm and 8 cm and with an angle of 120° between them.

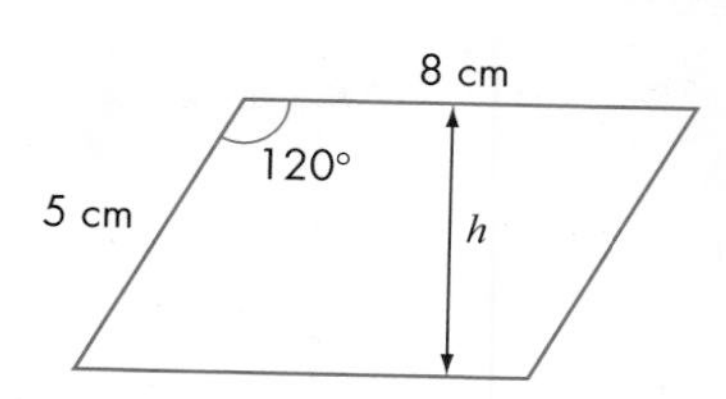

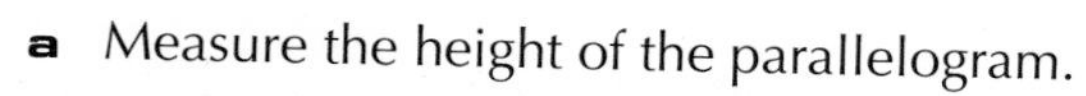

a Measure the height of the parallelogram.

b What is the area of the parallelogram?

7 Groundsmen painting white lines on a sports field may use a knotted rope, like the one shown below.

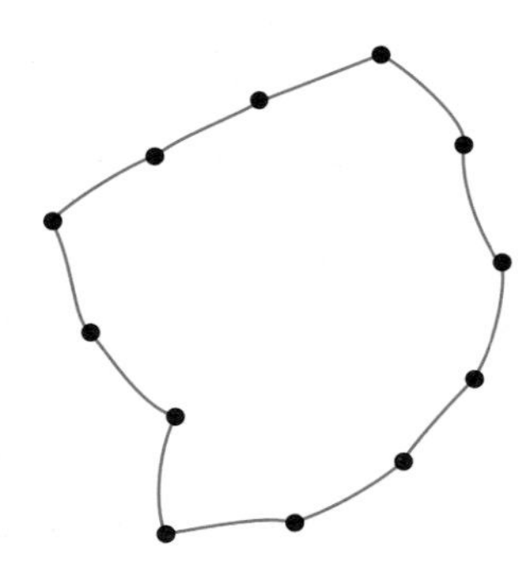

It has 12 equally-spaced knots.
It can be laid out to give a triangle, like this.

It will always be a right-angled triangle. This helps the groundsmen to draw lines perpendicular to each other.

Here are two more examples of such ropes.

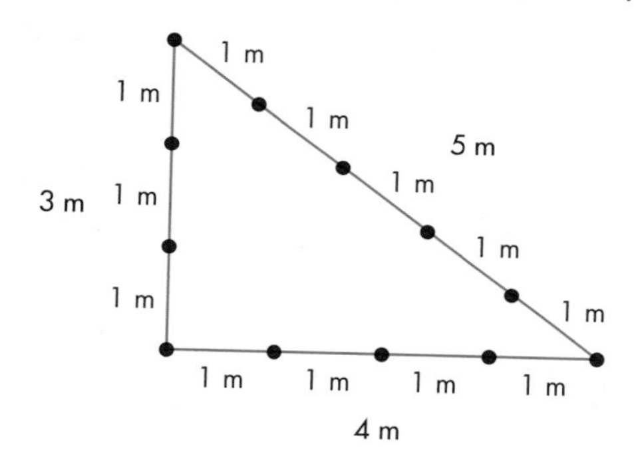

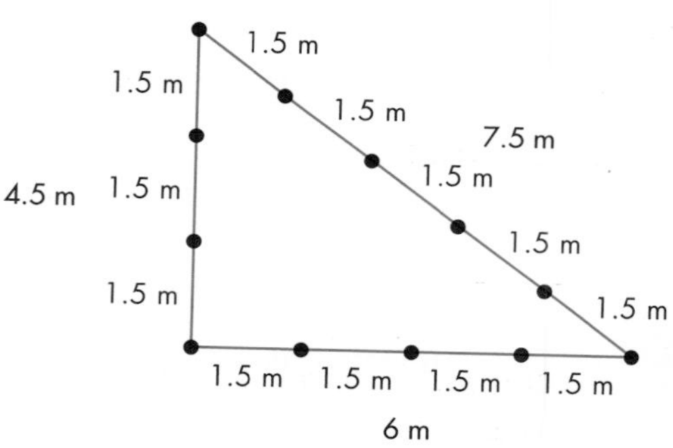

a Show, by constructing each of the above triangles (use a scale of 1 cm : 1 m), that each is a right-angled triangle.

b Choose a different triangle that you think might also be right-angled. Use the same knotted-rope idea to check.

PS 8 Construct the triangle with the largest area which has a total perimeter of 12 cm.

AU 9 Anil says that, as long as he knows all three angles of a triangle, he can draw it. Explain why Anil is wrong.

8.2 Bisectors

This section will show you how to:

- construct the bisectors of lines and angles
- construct angles of 60° and 90°

Key words
angle bisector
bisect
line bisector
perpendicular bisector

To **bisect** means to divide in half. So a bisector divides something into two equal parts.

- A **line bisector** divides a straight line into two equal lengths.
- An **angle bisector** is the straight line that divides an angle into two equal angles.

To construct a line bisector

It is usually more accurate to construct a line bisector than to measure its position (the midpoint of the line).

- **Step 1:** Here is a line to bisect.
- **Step 2:** Open your compasses to a radius of about three-quarters of the length of the line. Using each end of the line as a centre, and without changing the radius of your compasses, draw two intersecting arcs.

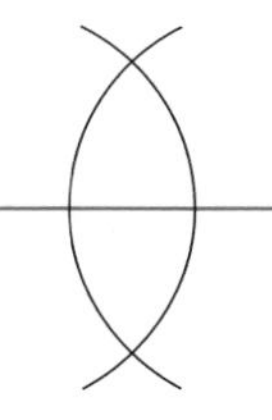

- **Step 3:** Join the two points at which the arcs intersect.
  This line is the **perpendicular bisector** of the original line.

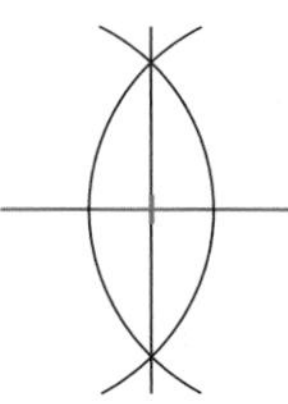

To construct an angle bisector

It is much more accurate to construct an angle bisector than to measure its position.

- **Step 1:** Here is an angle to bisect.

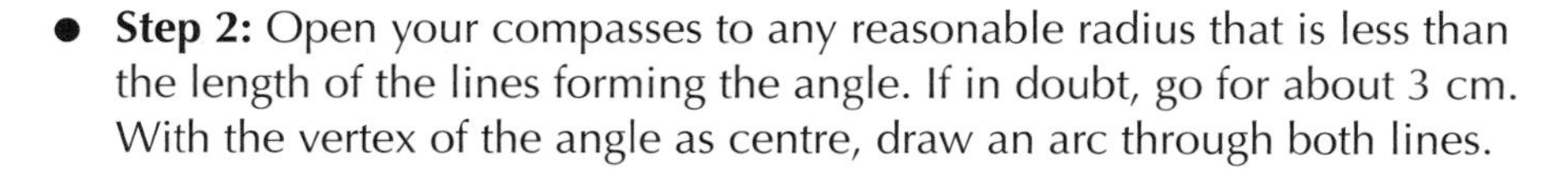

- **Step 2:** Open your compasses to any reasonable radius that is less than the length of the lines forming the angle. If in doubt, go for about 3 cm. With the vertex of the angle as centre, draw an arc through both lines.

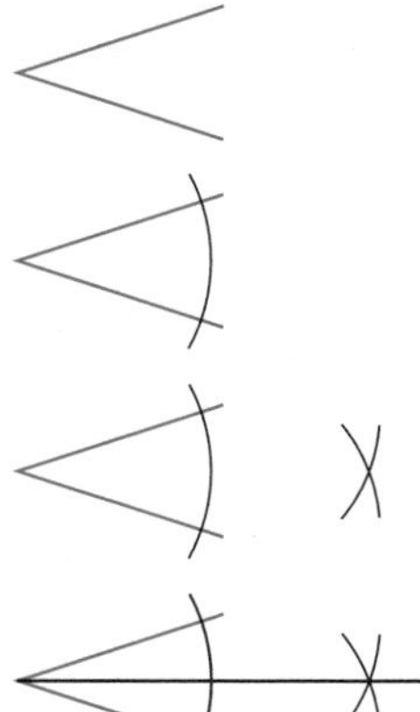

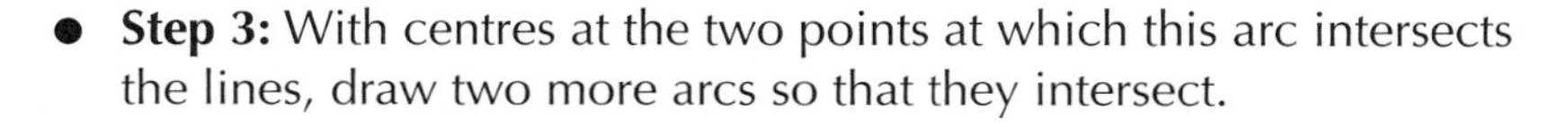

- **Step 3:** With centres at the two points at which this arc intersects the lines, draw two more arcs so that they intersect.

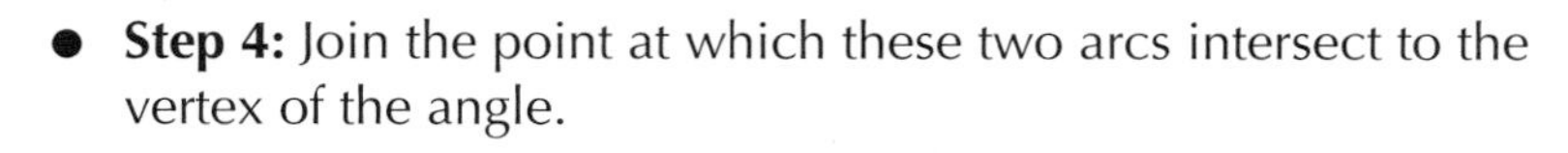

- **Step 4:** Join the point at which these two arcs intersect to the vertex of the angle.

This line is the angle bisector.

To construct an angle of 60°

It is more accurate to construct an angle of 60° than to measure and draw it with a protractor.

- **Step 1:** Draw a line and mark a point on it.

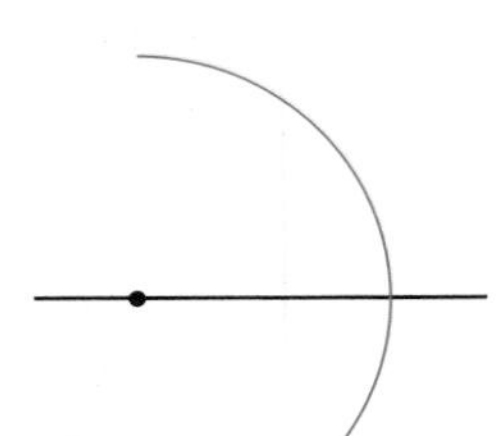

- **Step 2:** Open the compasses to a radius of about 4 cm.
  Using the point as the centre, draw an arc that crosses the line and extends almost above the point.

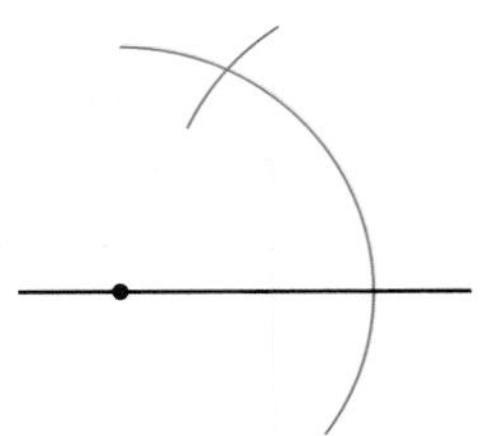

- **Step 3:** Keep the compasses set to the same radius.
  Using the point where the first arc crosses the line as a centre, draw another arc that intersects the first one.

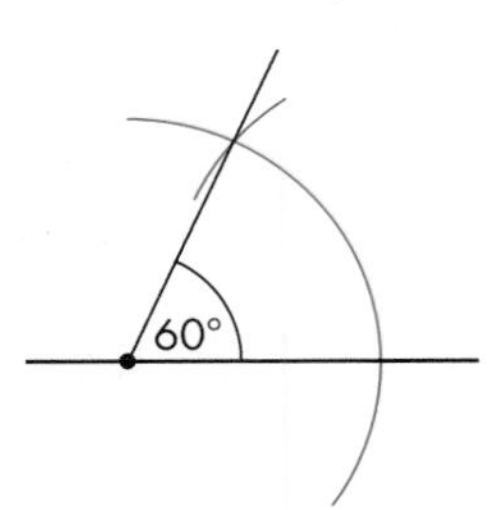

- **Step 4:** Join the original point to the point where the two arcs intersect.

- **Step 5:** Use a protractor to check that the angle is 60°.

To construct a perpendicular from a point on a line (an angle of 90°)

This construction will produce a perpendicular from a point A on a line.

- Open your compasses to about 2 or 3 cm.
  With point A as centre, draw two short arcs to intersect the line at each side of the point.

- Now extend the radius of your compasses to about 4 cm. With centres at the two points at which the arcs intersect the line, draw two arcs to intersect at X above the line.

- Join AX.

  AX is perpendicular to the line.

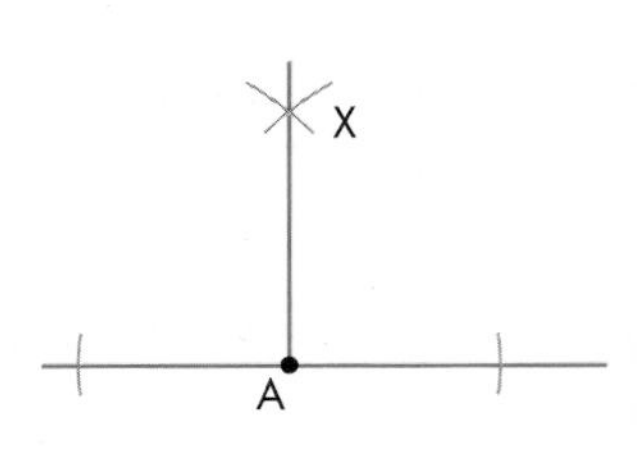

Note: If you needed to construct a 90° angle at the end of a line, you would first have to extend the line.

You could be even more accurate by also drawing two arcs underneath the line, which would give three points in line.

To construct a perpendicular from a point to a line

This construction will produce a perpendicular from a point A to a line.

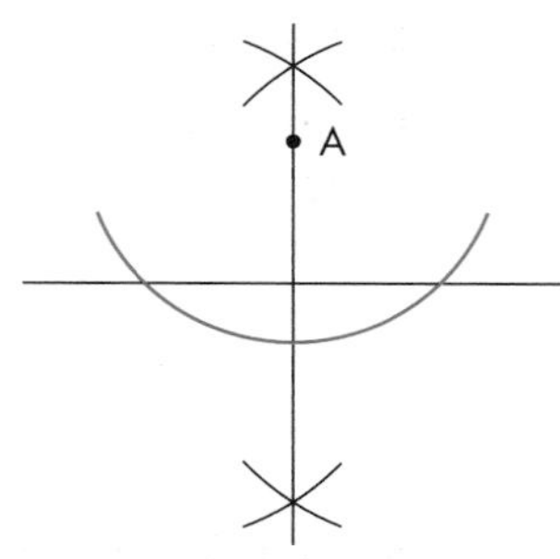

- With point A as centre, draw an arc which intersects the line at two points.
- With centres at these two points of intersection, draw two arcs to intersect each other both above and below the line.
- Join the two points at which the arcs intersect. The resulting line passes through point A and is perpendicular to the line.

Examination note: When a question says *construct*, you must *only* use compasses, not a protractor. When it says *draw*, you may use whatever you can to produce an accurate diagram. But also note, when constructing you may use your protractor to check your accuracy.

EXERCISE 8B

1 Draw a line 7 cm long and bisect it. Check your accuracy by seeing if each half is 3.5 cm.

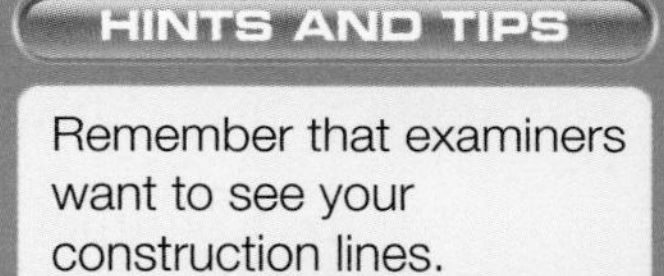

HINTS AND TIPS

Remember that examiners want to see your construction lines.

2 Draw a circle of about 4 cm radius.

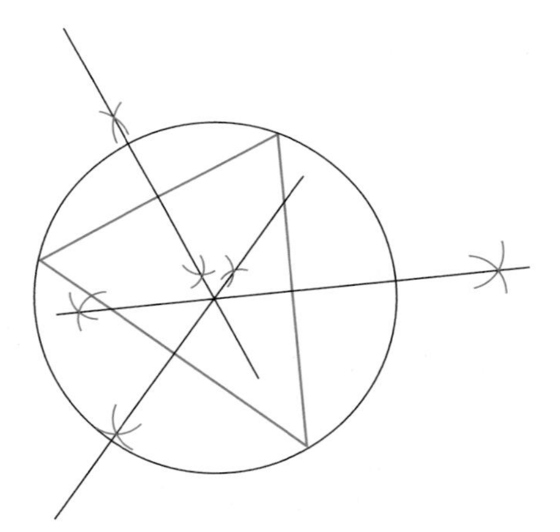

Draw a triangle inside the circle so that the corners of the triangle touch the circle.

Bisect each side of the triangle.

The bisectors should all meet at the same point, which should be the centre of the circle.

3 **a** Draw any triangle with sides that are between 5 cm and 10 cm.

b On each side construct the line bisector.

All your line bisectors should intersect at the same point.

c Using this point as the centre, draw a circle that goes through every vertex of the triangle.

C

C

4 Repeat question **3** with a different triangle and check that you get a similar result.

5 **a** Draw the following quadrilateral.

b Construct the line bisector of each side. These all should intersect at the same point.

c Use this point as the centre of a circle that goes through the quadrilateral at each vertex. Draw this circle.

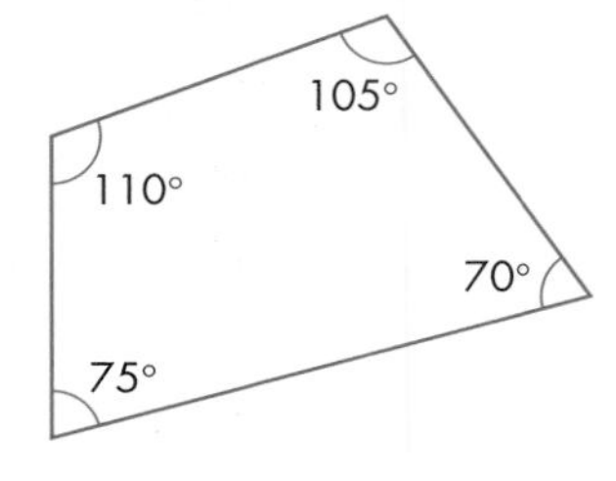

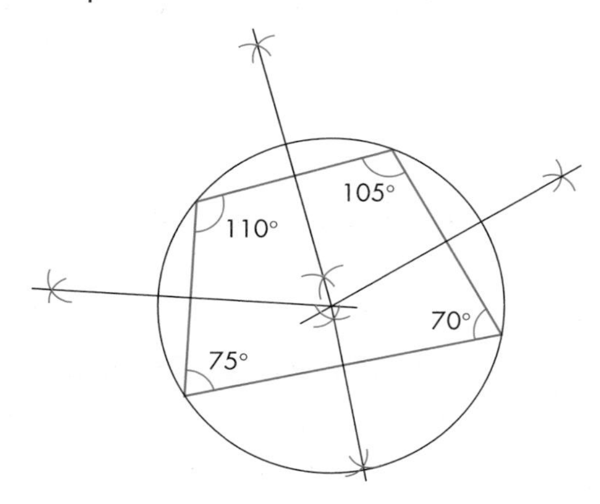

6 **a** Draw an angle of 50°.

b Construct the angle bisector.

c Check how accurate you have been by measuring each half. Both should be 25°.

7 Draw a circle with a radius of about 3 cm.

Draw a triangle so that the sides of the triangle are tangents to the circle.

Bisect each angle of the triangle.

The bisectors should all meet at the same point, which should be the centre of the circle.

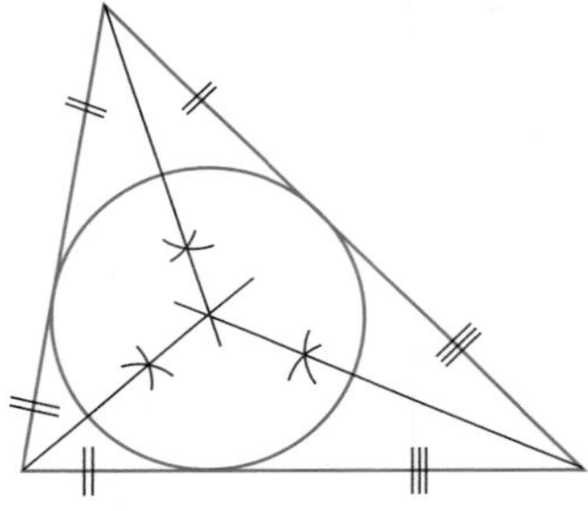

8 **a** Draw any triangle with sides that are between 5 cm and 10 cm.

b At each angle construct the angle bisector.
All three bisectors should intersect at the same point.

c Use this point as the centre of a circle that just touches the sides of the triangle.

9 Repeat question **8** with a different triangle.

FM 10 Gianni and Anna have children living in Bristol and Norwich. Gianni is about to start a new job in Birmingham. They are looking on a map of Britain for places they might move to.

Anna says, "I want to be the same distance from both children."

Gianni says, "I want to be as close to Birmingham as possible."

Find the largest city that would suit both Gianni and Anna. Use a map of the UK to help you.

C

PS **11** Draw a circle with radius about 4 cm.

Draw a quadrilateral, **not** a rectangle, inside the circle so that each vertex is on the circumference.

Construct the bisector of each side of the quadrilateral.

Where is the point where these bisectors all meet?

AU **12** Briefly outline how you would construct a triangle with angles 90°, 60° and 30°.

13 **a** Draw a line AB, 6 cm long, and construct an angle of 90° at A.

b Bisect this angle to construct an angle of 45°.

14 **a** Draw a line AB, 6 cm long, and construct an angle of 60° at A.

b Bisect this angle to construct an angle of 30°.

15 Draw a line AB, 6 cm long, and mark a point C, 4 cm above the middle of the line.

Construct the perpendicular from the point C to the line AB.

8.3 Defining a locus

This section will show you how to:
- draw a locus for a given rule

Key words
equidistant
loci
locus

A **locus** (plural **loci**) is the movement of a point according to a given rule.

EXAMPLE 4

A point P that moves so that it is always at a distance of 5 cm from a fixed point A will have a locus that is a circle of radius 5 cm.

You can express this mathematically by saying the locus of the point P is such that AP = 5 cm.

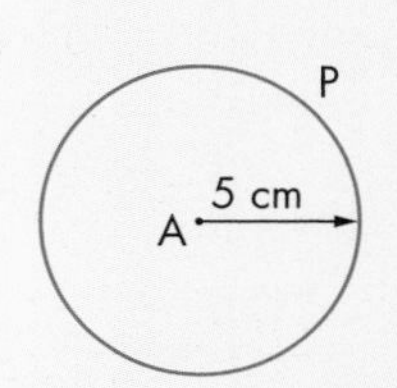

EXAMPLE 5

A point P that moves so that it is always the same distance from two fixed points A and B will have a locus that is the perpendicular bisector of the line joining A and B.

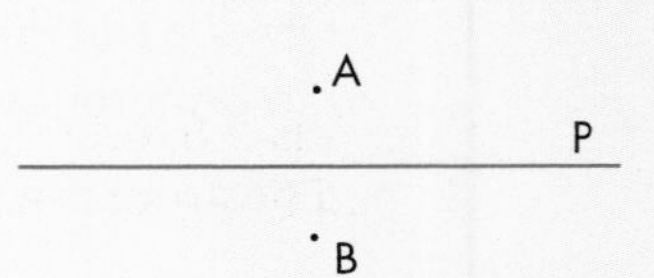

You can express this mathematically by saying

the locus of the point P is such that $AP = BP$.

A point that is always the same distance from two points is **equidistant** from the two points.

EXAMPLE 6

A point that is always 5 m from a long, straight wall will have a locus that is a line parallel to the wall and 5 m from it.

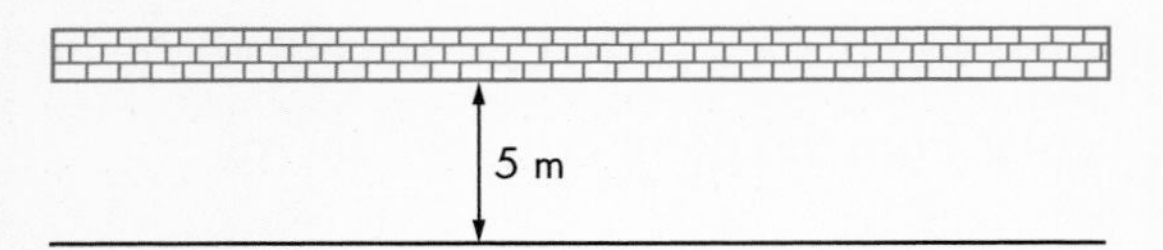

EXAMPLE 7

A point that moves so that it is always 5 cm from a line AB will have a locus that is a racetrack shape around the line.

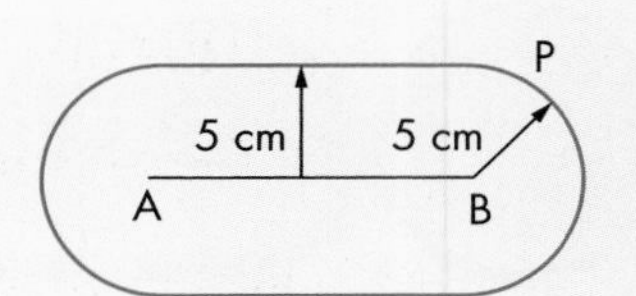

In your GCSE examination, you will usually get practical situations rather than abstract mathematical ones.

EXAMPLE 8

Imagine a grassy, flat field in which a horse is tethered to a stake by a rope that is 10 m long. What is the shape of the area that the horse can graze?

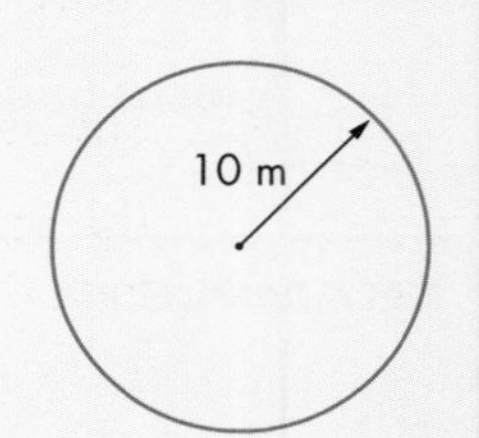

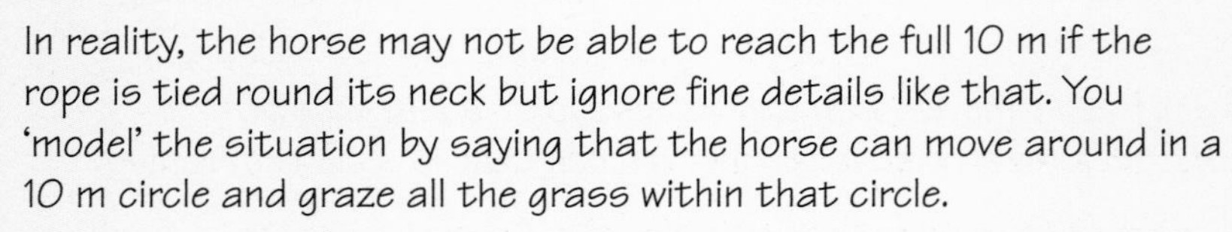

In reality, the horse may not be able to reach the full 10 m if the rope is tied round its neck but ignore fine details like that. You 'model' the situation by saying that the horse can move around in a 10 m circle and graze all the grass within that circle.

In this example, the locus is the whole of the area inside the circle.

You can express this mathematically as:

the locus of the point P is such that $AP \leqslant 10$ m.

EXERCISE 8C

1 A is a fixed point. Sketch the locus of the point P in each of these situations.

a AP = 2 cm **b** AP = 4 cm **c** AP = 5 cm

> **HINTS AND TIPS**
> Sketch the situation before doing an accurate drawing.

2 A and B are two fixed points 5 cm apart. Sketch the locus of the point P for each of these situations.

a AP = BP **b** AP = 4 cm and BP = 4 cm

c P is always within 2 cm of the line AB

FM 3 **a** A horse is tethered in a field on a rope 4 m long. Describe or sketch the area that the horse can graze.

b The horse is still tethered by the same rope but there is now a long, straight fence running 2 m from the stake. Sketch the area that the horse can now graze.

4 ABCD is a square of side 4 cm. In each of the following loci, the point P moves only inside the square. Sketch the locus in each case.

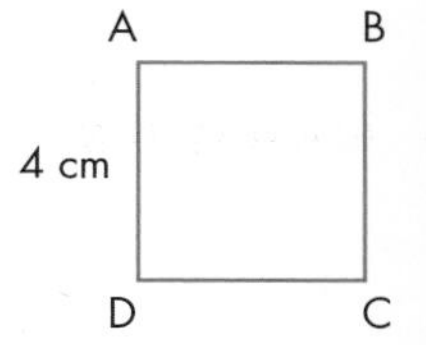

a AP = BP **b** $AP < BP$ **c** AP = CP

d $CP < 4$ cm **e** $CP > 2$ cm **f** $CP > 5$ cm

5 One of the following diagrams is the locus of a point on the rim of a bicycle wheel as it moves along a flat road. Which is it?

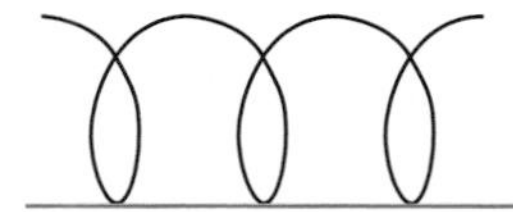

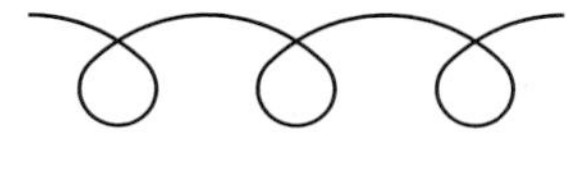

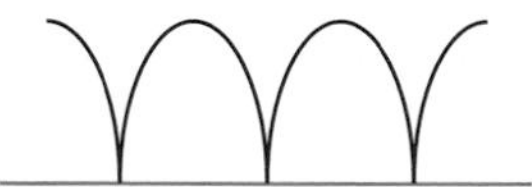

6 Draw the locus of the centre of the wheel for the bicycle in question **5**.

PS 7 ABC is a triangle.

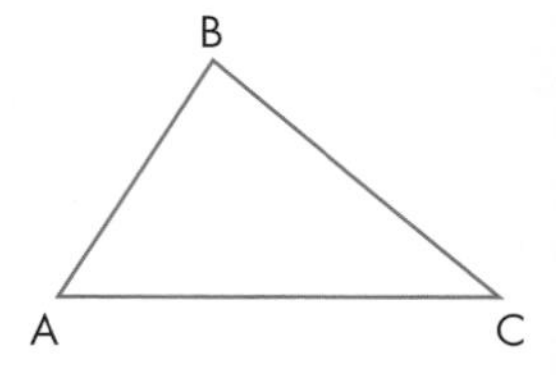

The region R is defined as the set of points inside the triangle such that:

- they are closer to the line AB than the line AC
- they are closer to the point A than the point C.

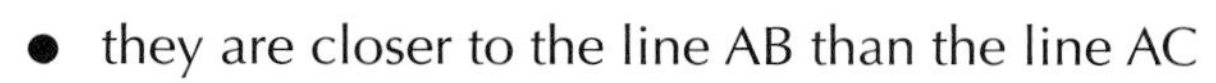

Using a ruler and compasses, construct the region R.

AU 8 ABCD is a rectangle.

Copy the diagram and draw the locus of all points that are 2 cm from the edges of the rectangle.

8.4 Loci problems

This section will show you how to:

- solve practical problems using loci

Key words
loci
scale

Most of the **loci** problems in your GCSE examination will be of a practical nature, as in the next example.

EXAMPLE 9

Imagine that a radio company wants to find a site for a transmitter. The transmitter must be the same distance from Doncaster and Leeds and within 20 miles of Sheffield.

In mathematical terms, this means they are concerned with the perpendicular bisector between Leeds and Doncaster and the area within a circle of radius 20 miles from Sheffield.

The diagram, drawn to a **scale** of 1 cm = 10 miles, illustrates the situation and shows that the transmitter can be built anywhere along the thick part of the blue line.

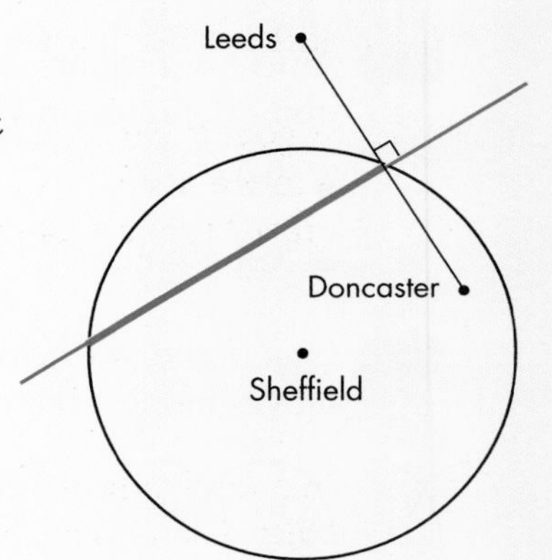

EXAMPLE 10

A radar station in Birmingham has a range of 150 km (that is, it can pick up any aircraft within a radius of 150 km). Another radar station in Norwich has a range of 100 km.

Can an aircraft be picked up by both radar stations at the same time?

The situation is represented by a circle of radius 150 km around Birmingham and another circle of radius 100 km around Norwich. The two circles overlap, so an aircraft could be picked up by both radar stations when it is in the overlap.

EXAMPLE 11

A dog is tethered by a rope, 3 m long, to the corner of a shed, 4 m by 2 m. What is the area that the dog can guard effectively?

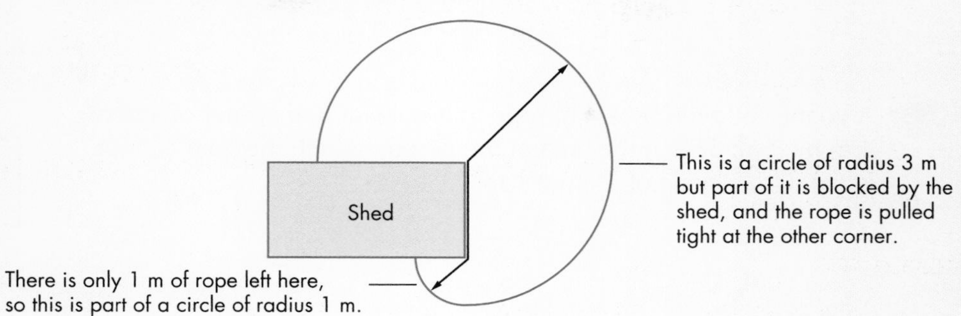

EXERCISE 8D

For questions **1** to **7**, you should start by sketching the picture given in each question on a 6×6 grid, each square of which is 1 cm by 1 cm. The scale for each question is given.

FM 1 A goat is tethered by a rope, 7 m long, in a corner of a field with a fence at each side. What is the locus of the area that the goat can graze? Use a scale of 1 cm ≡ 2 m.

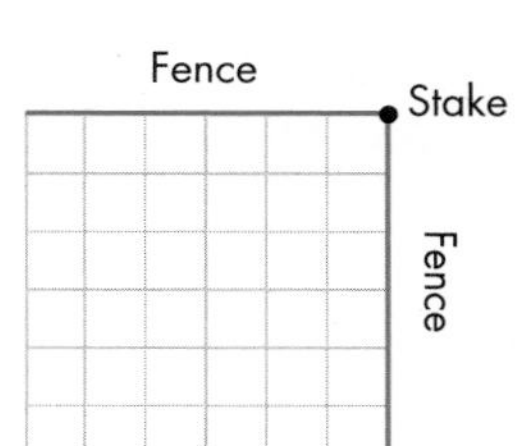

FM 2 In a field, a horse is tethered to a stake by a rope 6 m long. What is the locus of the area that the horse can graze? Use a scale of 1 cm ≡ 2 m.

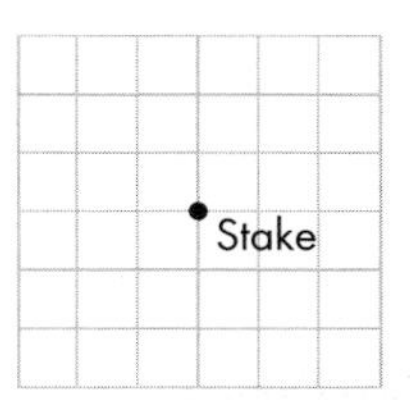

FM 3 A cow is tethered to a rail at the top of a fence 6 m long. The rope is 3 m long. Sketch the area that the cow can graze. Use a scale of 1 cm ≡ 2 m.

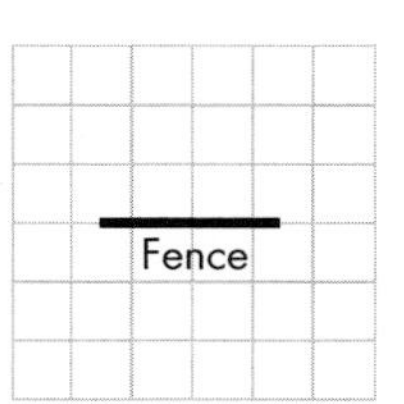

FM 4 A horse is tethered to a stake near a corner of a fenced field, at a point 4 m from each fence. The rope is 6 m long. Sketch the area that the horse can graze. Use a scale of 1 cm ≡ 2 m.

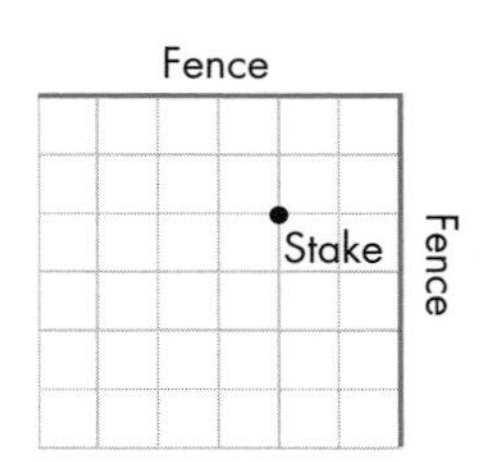

C

FM 5 A horse is tethered to a corner of a shed, 2 m by 1 m. The rope is 2 m long. Sketch the area that the horse can graze. Use a scale of 1 cm ≡ 1 m.

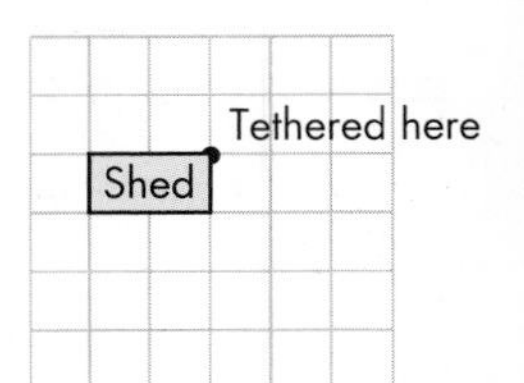

FM 6 A goat is tethered by a 4 m rope to a stake at one corner of a pen, 4 m by 3 m. Sketch the area of the pen on which the goat cannot graze. Use a scale of 1 cm ≡ 1 m.

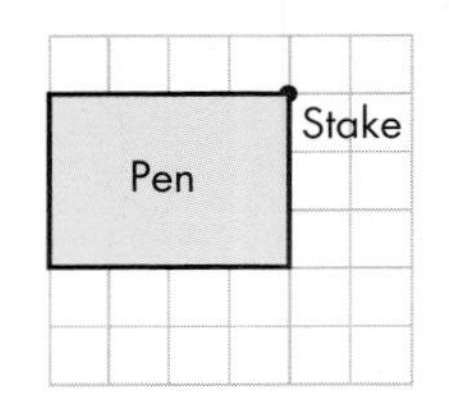

FM 7 A puppy is tethered to a stake by a rope, 1.5 m long, on a flat lawn on which are two raised brick flower beds. The stake is situated at one corner of a bed, as shown. Sketch the area that the puppy is free to roam in. Use a scale of 1 cm ≡ 1 m.

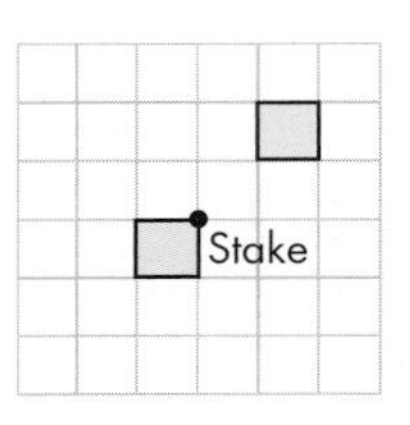

For questions **8** to **15**, you should use a copy of the map opposite. For each question, trace the map and mark on those points that are relevant to that question.

FM 8 A radio station broadcasts from London on a frequency of 1000 kHz with a range of 300 km. Another radio station broadcasts from Glasgow on the same frequency with a range of 200 km.

a Sketch the area to which each station can broadcast.

b Will they interfere with each other?

c If the Glasgow station increases its range to 400 km, will they then interfere with each other?

FM 9 The radar at Leeds airport has a range of 200 km. The radar at Exeter airport has a range of 200 km.

a Will a plane flying over Birmingham be detected by the Leeds radar?

b Sketch the area where a plane can be picked up by both radars at the same time.

FM 10 A radio transmitter is to be built according to these rules.

i It has to be the same distance from York and Birmingham.

ii It must be within 350 km of Glasgow.

iii It must be within 250 km of London.

a Sketch the line that is the same distance from York and Birmingham.

b Sketch the area that is within 350 km of Glasgow and 250 km of London.

c Show clearly the possible places at which the transmitter could be built.

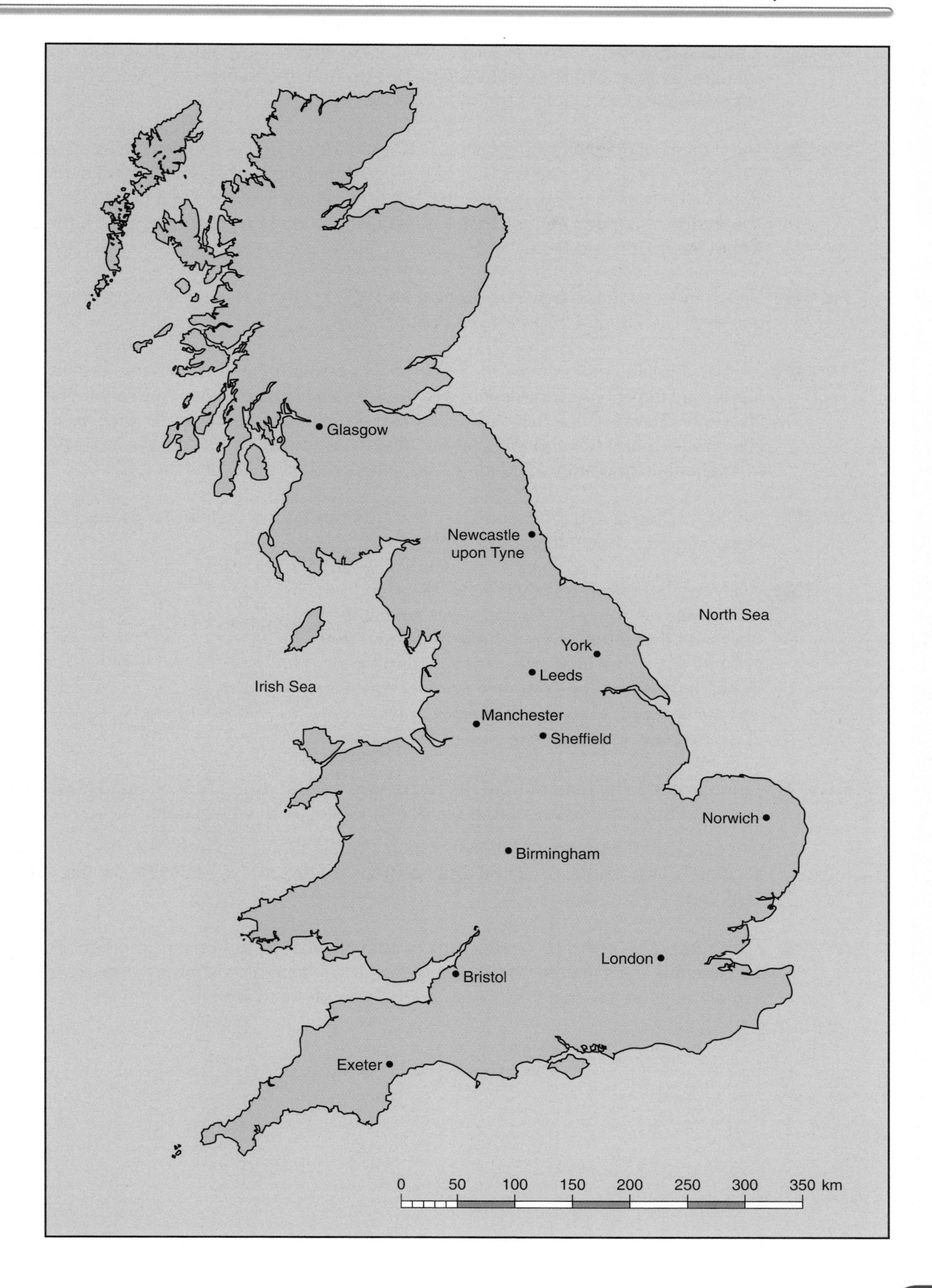
Glasgow
Newcastle upon Tyne
North Sea
York
Leeds
Irish Sea
Manchester
Sheffield
Norwich
Birmingham
London
Bristol
Exeter
0
50
100
150
200
250
300
350 km

C

FM 11 A radio transmitter centred at Birmingham is designed to give good reception in an area greater than 150 km and less than 250 km from the transmitter. Sketch the area of good reception.

FM 12 Three radio stations pick up a distress call from a boat in the Irish Sea. The station at Glasgow can tell from the strength of the signal that the boat is within 300 km of the station. The station at York can tell that the boat is between 200 km and 300 km from York. The station at London can tell that it is less than 400 km from London. Sketch the area where the boat could be.

FM 13 Sketch the area that is between 200 km and 300 km from Newcastle upon Tyne, and between 150 km and 250 km from Bristol.

FM 14 An oil rig is situated in the North Sea in such a position that it is the same distance from Newcastle upon Tyne and Manchester. It is also the same distance from Sheffield and Norwich. Draw the line that shows all the points that are the same distance from Newcastle upon Tyne and Manchester. Repeat for the points that are the same distance from Sheffield and Norwich and find out where the oil rig is located.

FM 15 Whilst looking at a map, Fred notices that his house is the same distance from Glasgow, Norwich and Exeter. Where is it?

16 Wathsea Harbour is as shown in the diagram. A boat sets off from point A and steers so that it stays the same distance from the sea wall and the West Pier. Another boat sets off from B and steers so that it stays the same distance from the East Pier and the sea wall. Copy the diagram. On your diagram show accurately the path of each boat.

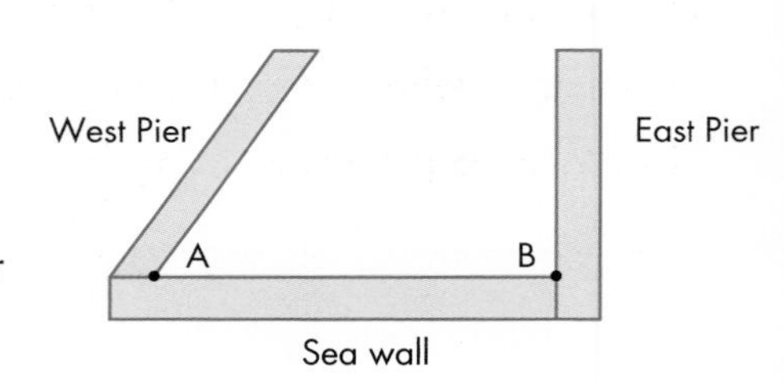

PS 17 Tariq wanted to fly himself from the Isle of Wight north, towards Scotland. He wanted to remain at the same distance from London as Bristol as far as he could.

Once he is past London and Bristol, which city should he aim toward to keep him, as accurately as possible, the same distance from London and Bristol? Use the map to help you.

AU 18 A distress call is heard by coastguards in both Newcastle and Bristol. The signal strength suggests that the call comes from a ship that is the same distance from both places. Explain how the coastguards could find the area of sea to search.

GRADE BOOSTER

- **D** You can construct a perpendicular from a point on a line
- **D** You can construct a perpendicular from a point to a line
- **D** You can construct angles of 60° and 90°
- **C** You can describe and draw the locus of a point from a given rule
- **C** You can use loci to solve problems
- **C** You can construct line and angle bisectors

What you should know now

- How to construct line and angle bisectors
- How to construct perpendiculars
- How to construct angles without using a protractor
- Understand what is meant by a locus
- How to solve problems, using loci

1 Use ruler and compasses to construct the bisector of angle ABC on a copy of this diagram.

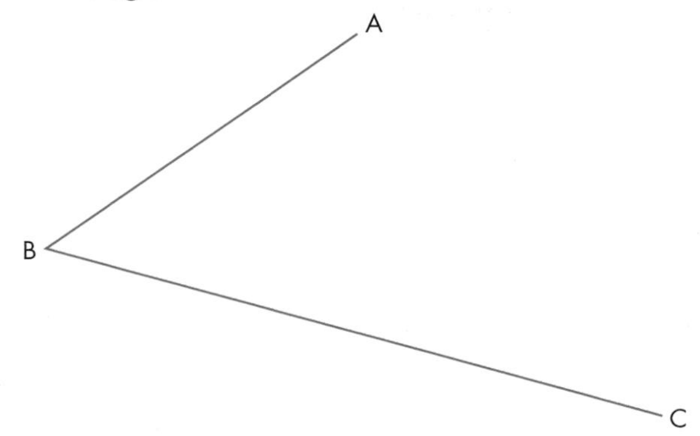

You must show all your construction lines.

(2 marks)

Edexcel, November 2008, Paper 14 Higher, Question 9

2 Use ruler and compasses to construct an equilateral triangle with sides of length 6 cm.

You must show all your construction lines.

(2 marks)

Edexcel, May 2008, Paper 3 Higher, Question 11

3 Copy the diagram and draw the locus of all points that are exactly 3 cm from the line PQ.

P Q

Edexcel, November 2008, Paper 15 Higher, Question 9

4

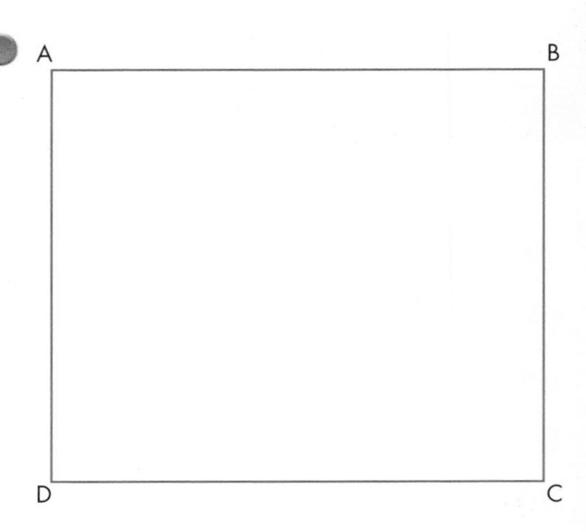

Copy the rectangle ABCD.

Shade the set of points, inside the rectangle, which are both:

more than 4 centimetres from the point A

and more than 1 centimetre from the line DC.

(4 marks)

Edexcel, June 2006, Paper 5 Higher, Question 4

Worked Examination Questions

1 Here is a sketch of a triangle. PR = 6.4 cm, QR = 7.7 cm and angle R = 35°.

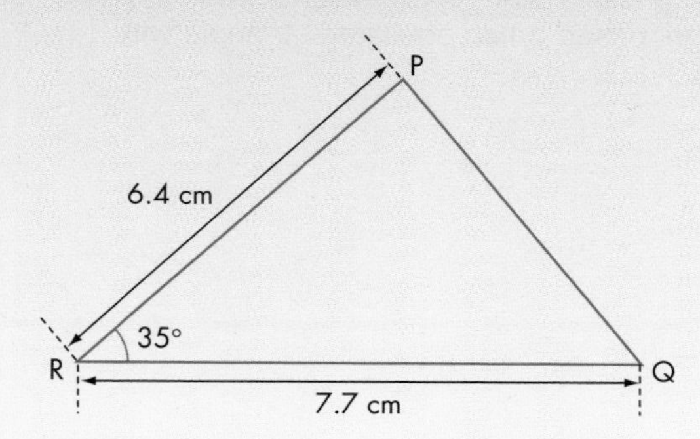

a Make an accurate drawing of the triangle.

b Measure the size of angle Q on your drawing.

1 **a** **Make an accurate drawing, using these steps.**

Step 1: Draw the base as a line 7.7 cm long.
You can draw this and measure it with a ruler, although using a pair of compasses is more accurate.

This scores 1 mark for the angle and line.

Step 2: Measure the angle at R as 35°.
Draw a faint line at this angle.

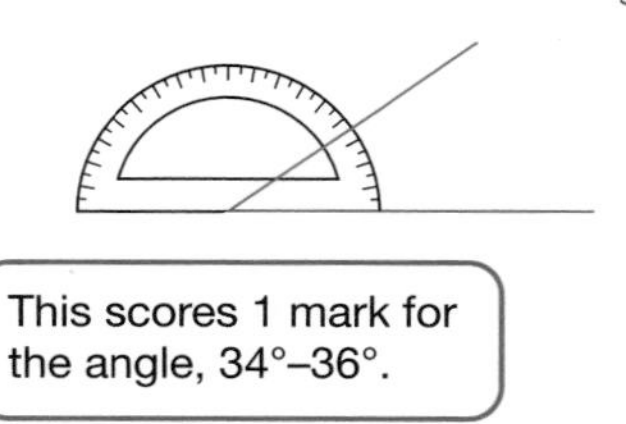

This scores 1 mark for the angle, 34°–36°.

P

Step 4: Join P to Q.

R Q

Step 3: Using a pair of compasses, draw an arc 6.4 cm long from R. Where this crosses the line from Step 2, make this P.

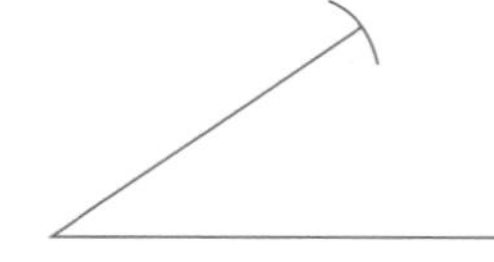

This scores 1 mark for the second side.

This scores 1 mark for completing the triangle

b **Measure the angle Q. It is 56°.**

This scores 1 mark for this angle 56° ± 2°. If your actual angle was not 56° but you measured it accurately, then you would still get the mark.

Total: 4 marks

Worked Examination Questions

FM **2** Some wind turbines follow a design based on **arcs** within an **equilateral triangle**.

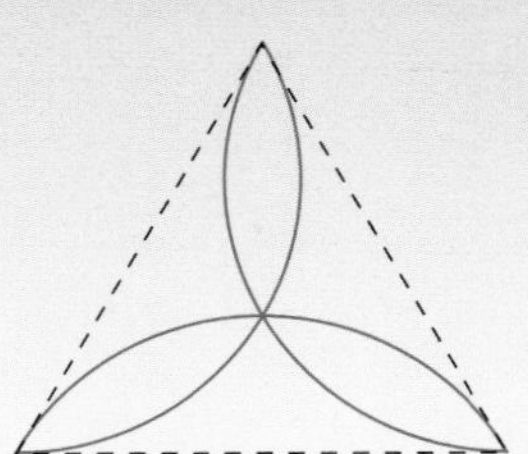

Construct this design, based on an equilateral triangle with sides of length 5 cm.

2 Start with the equilateral triangle.

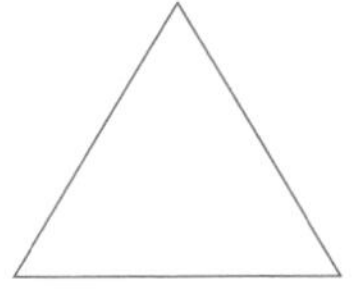

This scores 1 mark for method for correctly constructing this figure.

Construct the line bisectors of each side.

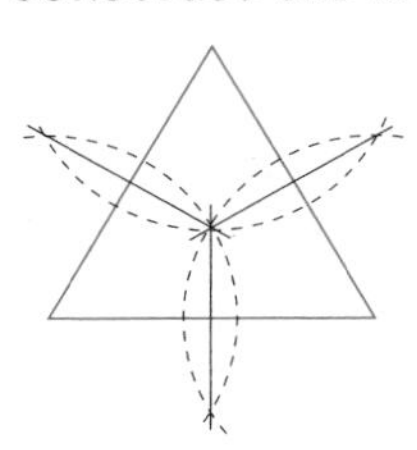

This scores 1 mark for method for showing correct construction for at least two arcs.

This scores 1 mark for accuracy for correctly constructing at least two of them.

Where these bisectors all meet gives the centre of the triangle.

This scores 1 mark for method for all bisectors meeting at same point.

Join each vertex to this centre point and bisect each of these lines, extending the new bisectors to intersect with the bisectors of the sides of the triangle.

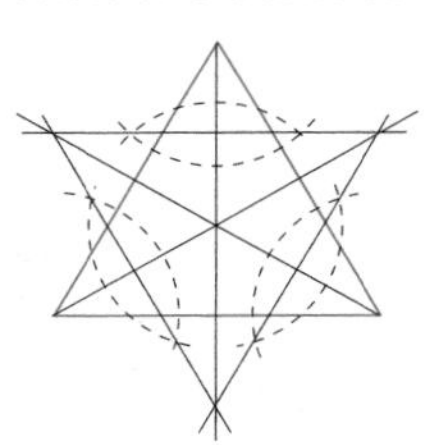

This scores 1 mark for method for attempting to construct bisectors on at least two of the correct lines.

This scores 1 mark for accuracy for correctly identifying at least two correct centres.

These points where the lines intersect are the centres of the arcs. Set the compasses to the distance between one of these points and the nearest vertex and draw the arcs from each point.

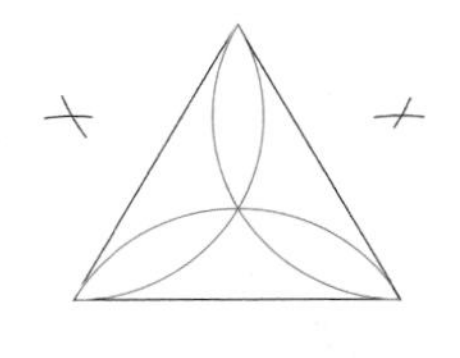

This scores 1 mark for accuracy for correctly constructing this final figure.

Total: 7 marks

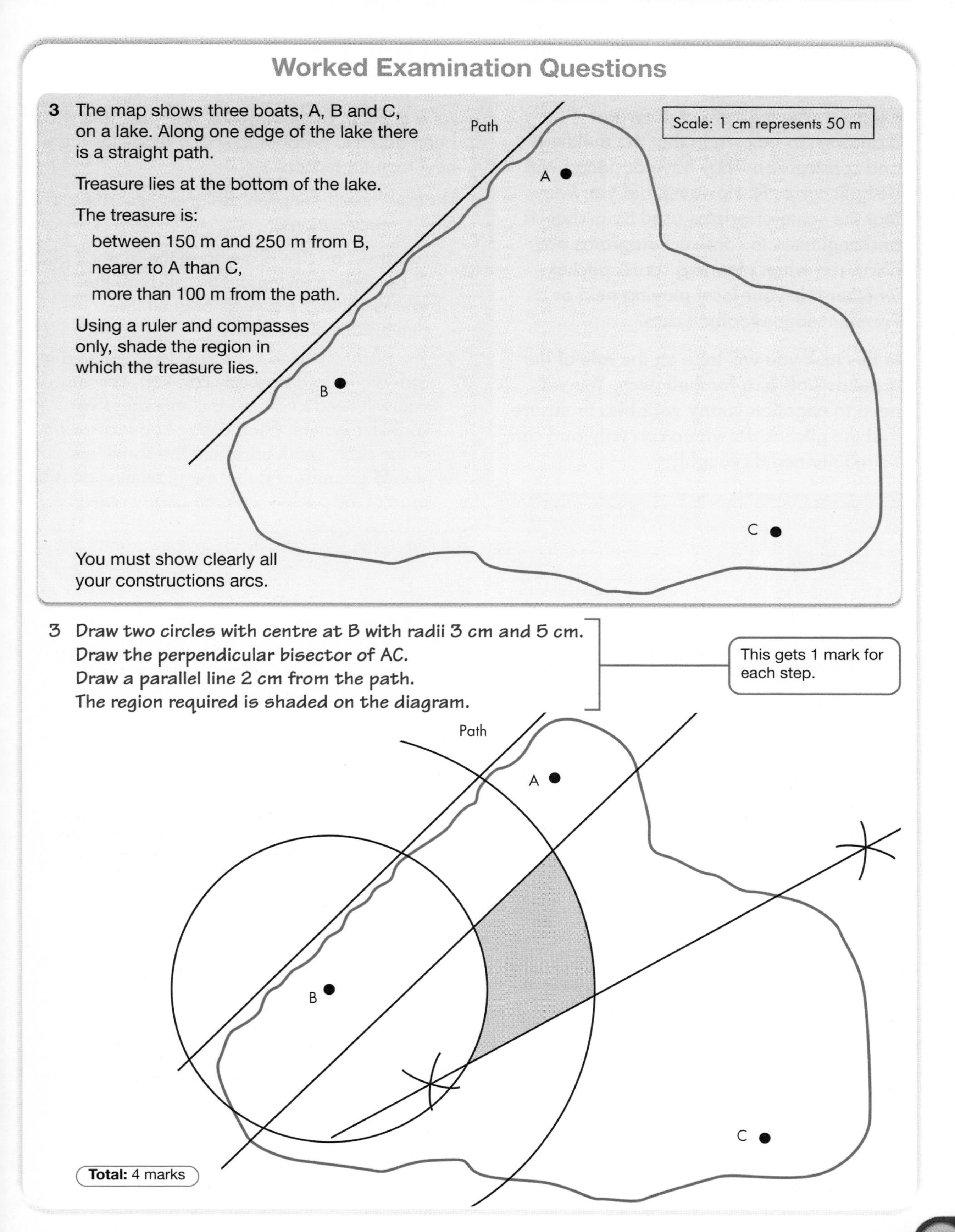

Worked Examination Questions

3 The map shows three boats, A, B and C, on a lake. Along one edge of the lake there is a straight path.

Treasure lies at the bottom of the lake.

The treasure is:

between 150 m and 250 m from B,

nearer to A than C,

more than 100 m from the path.

Using a ruler and compasses only, shade the region in which the treasure lies.

You must show clearly all your constructions arcs.

3 Draw two circles with centre at B with radii 3 cm and 5 cm.
Draw the perpendicular bisector of AC.
Draw a parallel line 2 cm from the path.
The region required is shaded on the diagram.

8 Problem Solving
Planning a football pitch

You already know that architects and engineers must construct accurate diagrams, to be certain that the buildings and constructions they have designed will be built correctly. However, did you know that the same principles used by architects and engineers to construct diagrams are also used when planning sports pitches, whether it is your local playing field or a Premier League football club?

In this task you will take on the role of the grounds staff of a football pitch. You will need to negotiate many variables to ensure that the pitch is drawn up correctly and can be maintained thoroughly.

Your task

As a member of the grounds staff you have been asked to prepare the pitch ready for the new football season.

The club needs the pitch designed according to FIFA's specifications.

1 Construct a scale drawing of the football pitch, to be used in laying out the pitch on the football field. Be sure to label all the dimensions of the pitch.

2 The pitch will need to be regularly watered in order to keep it in good condition. For this, you will need to design a comprehensive sprinkler system. On a copy of your drawing of the pitch, mark up where the sprinklers should go, ensuring that the maximum possible area of the pitch is watered at any one time.

Getting started

Discuss these questions with a partner.

- What shapes do you see on sports fields? Do these shapes vary, depending on the sports that take place on these pitches?
- What angles do you see on sports fields?
- How are shapes drawn on to sports fields?

FIFA specifications

The field

The field of play must be rectangular, divided into two halves by a halfway line. The centre mark is indicated at the midpoint of the halfway line and a circle with a radius of 9.15 m (10 yards) is marked around it.

The field dimensions should be as follows.

	Minimum	Maximum
Length	90 m (approx. 100 yards)	120 m (approx. 130 yards)
Width	45 m (approx. 50 yards)	90 m (approx. 100 yards)

The goal area

The goal mouth is 7.3 metres (8 yards) wide. The goal area is 5.5 m (6 yards) wide by 18.3 m (12 yards) long.

The penalty area

The penalty area is 16.5 m (18 yards) wide by 40.3 m (44 yards) long.

Within each penalty area there is a penalty mark 11 m (12 yards) from the midpoint between the goalposts and equidistant from them. An arc with a radius of 9.15 m (10 yards) from each penalty mark is drawn outside the penalty area.

The corner arc

A flagpost is placed in each corner. A quarter-circle with a radius of 1 m (approximately 1 yard) is drawn at each corner flag post, inside the field of play.

Hint: Use the internet to research football pitches and FIFA's specifications further.

Why this chapter matters

Thales of Miletus (624–547 BC) was a Greek philosopher and one of the Seven Sages of Greece. He is believed to have been the first person to use similar triangles to find the height of tall objects.

Thales discovered that, at a particular time of day, the height of an object and the length of its shadow were the same. He used this observation to calculate the height of the Egyptian pyramids. Later, he took this knowledge back to Greece. His observations are considered to be the forerunner of the technique of using similar triangles to solve such problems.

A clinometer is an instrument used to measure the height of objects from a distance. Using a clinometer, you can apply the geometry of triangles to determine the height visually, rather than by physically measuring it. Clinometers are commonly used to measure the heights of trees, buildings and towers, mountains and other objects for which taking physical measurements might be impractical.

Astronomers use the geometry of triangles to measure the distance to nearby stars. They take advantage of the Earth's journey in its orbit around the Sun to obtain the maximum distance between two measurements. They observe the star twice, from the same point on Earth and at the same time of day, but six months apart.

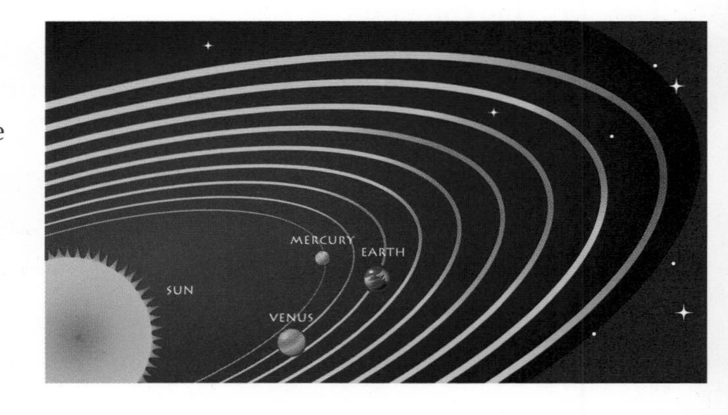

Telescopes and binoculars also use the geometry of triangles. The Hubble Space Telescope took this image of the Eagle Nebula. This star-forming region is located 6500 light years from Earth. It is only about 6 million years old and the dense clouds of interstellar gas are still collapsing to form new stars.

A light year is the distance that a ray of light travels in one year.

It is about 5 878 630 000 000 miles or just under 10^{13} km.

Chapter

Geometry: Similarity

The grades given in this chapter are target grades.

1 Similar triangles

2 Areas and volumes of similar shapes

This chapter will show you ...

C how to work out the scale factor for two similar shapes

B how to work out lengths of sides in similar figures

A to **A*** how to work out areas and volumes of similar shapes

Visual overview

What you should already know

- How to use and simplify ratios **KS3 level 6, GCSE grade D**
- How to enlarge a shape by a given scale factor **KS3 level 6, GCSE grade D–C**
- How to solve equations **KS3 level 6, GCSE grade D–C**

Quick check

1 Simplify the following ratios.

a 15 : 20 **b** 24 : 30

c $6 : 1\frac{1}{2}$ **d** 7.5 : 5

2 Solve the following equations.

a $\frac{x}{4} = \frac{7}{2}$ **b** $\frac{x}{2} = \frac{5}{4}$

c $\frac{x}{4} = \frac{x-2}{3}$ **d** $\frac{x+4}{x} = \frac{5}{3}$

9.1 Similar triangles

This section will show you how to:
- show two triangles are similar
- work out the scale factor between similar triangles

Key words
ratio
scale factor
similar
similar triangles

Triangles are **similar** if their corresponding angles are equal. Their corresponding sides are then in the same **ratio**.

These two right-angled triangles are **similar triangles**.

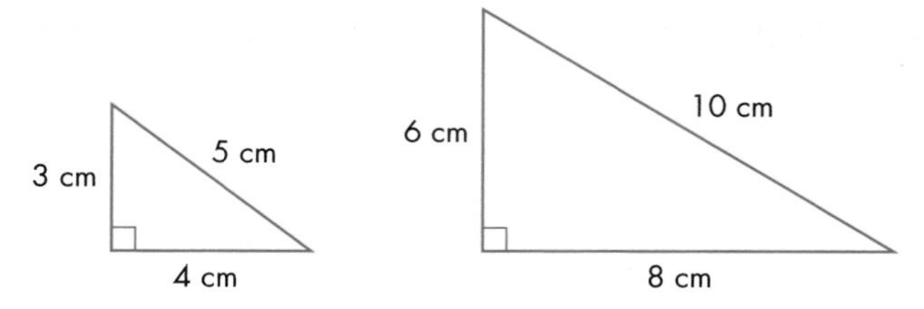

The scale factor of the enlargement = 2

The ratios of the lengths of corresponding sides all cancel to the same ratio.

$3 : 6 = 4 : 8 = 5 : 10 = 1 : 2$

All corresponding angles are equal.

EXAMPLE 1

The triangles ABC and PQR are similar. Find the length of the side PR.

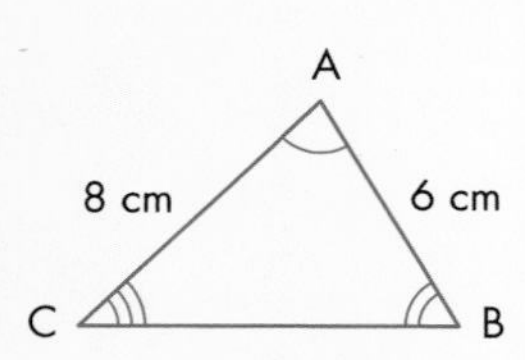

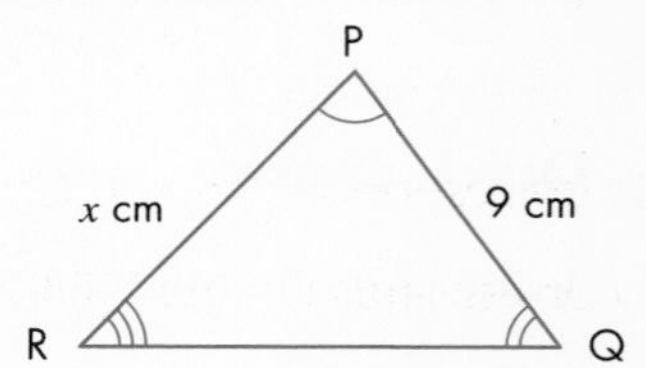

Take two pairs of corresponding sides, one pair of which must contain the unknown side. Form each pair into a fraction, so that x is on top. Since these fractions must be equal,

$$\frac{PR}{AC} = \frac{PQ}{AB}$$

$$\frac{x}{8} = \frac{9}{6}$$

To find x:

$$x = \frac{9 \times 8}{6} \text{ cm} \quad \Rightarrow \quad x = \frac{72}{6} = 12 \text{ cm}$$

EXERCISE 9A 

1 These diagrams are drawn to scale. What is the **scale factor** of the enlargement in each case?

a

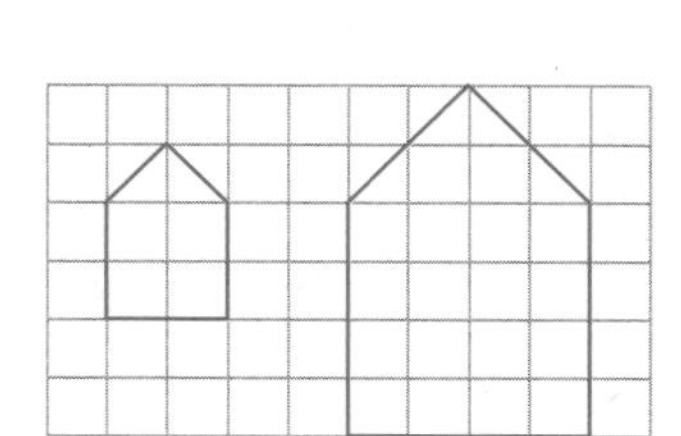

b

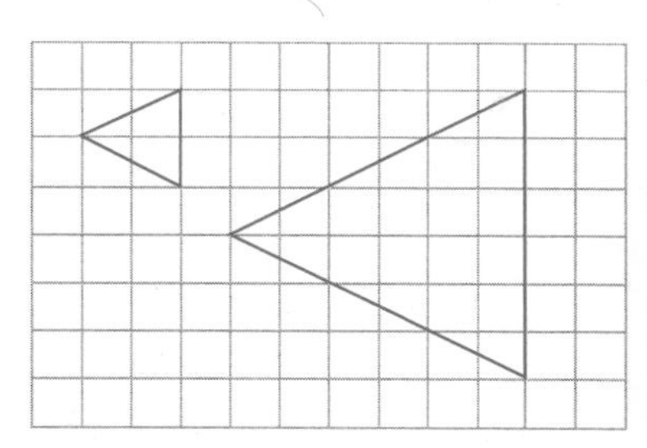

> **HINTS AND TIPS**
>
> If you need to revise enlargements, look back at Section 6.5.

C

AU 2 Are these pairs of shapes similar? If so, give the scale factor. If not, give a reason.

a

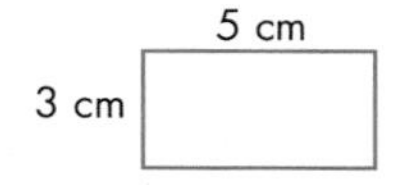

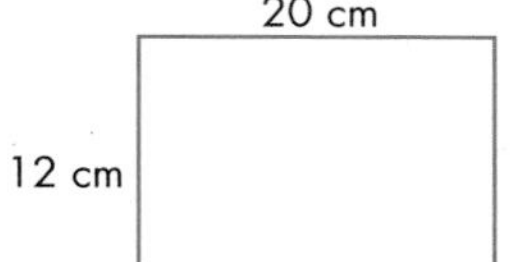

b

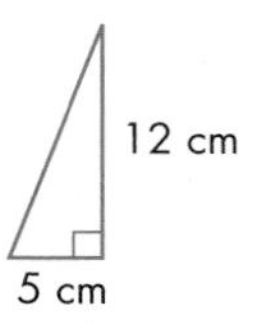

22 cm

15 cm

3 **a** Explain why these triangles are similar.

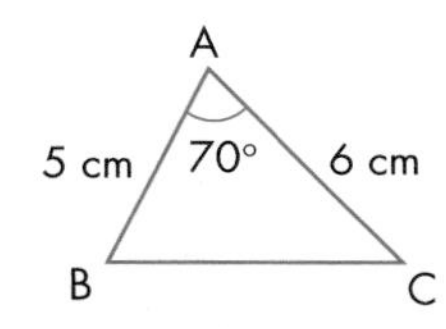

P

15 cm

70°

18 cm

Q

R

b Give the ratio of the sides.

c Which angle corresponds to angle C?

d Which side corresponds to side QP?

4 **a** Explain why these triangles are similar.

b Which angle corresponds to angle A?

c Which side corresponds to side AC?

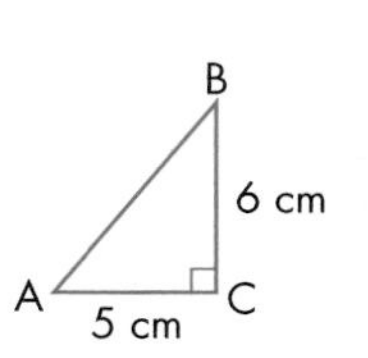

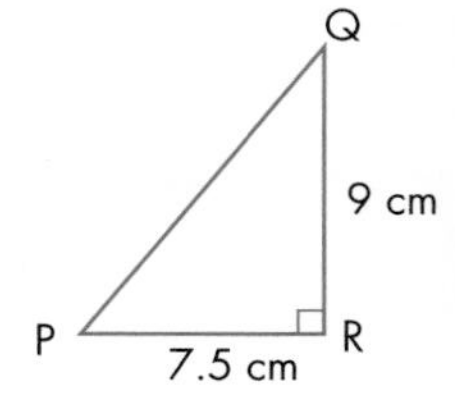

5 **a** Explain why triangle ABC is similar to triangle AQR.

b Which angle corresponds to the angle at B?

c Which side of triangle AQR corresponds to side AC of triangle ABC? Your answers to question **4** may help you.

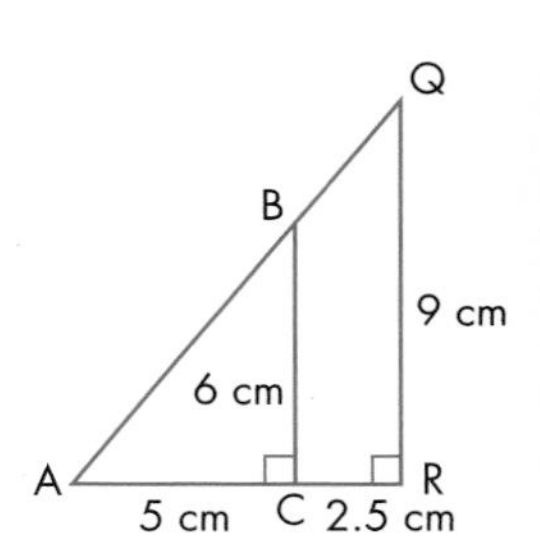

B

6 In the diagrams **a** to **f**, each pair of shapes are similar but not drawn to scale.

a Find x.

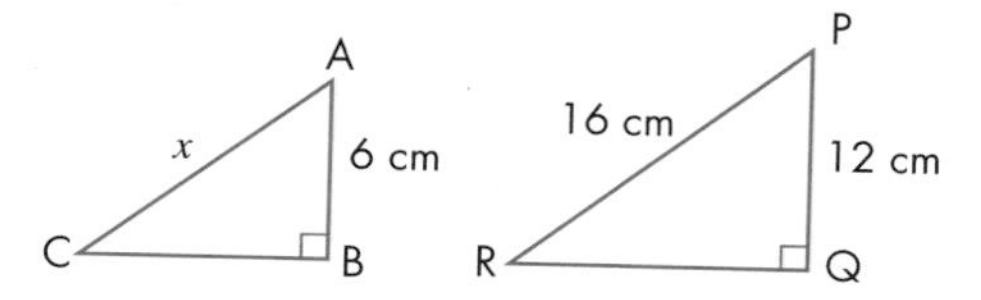

b Find PQ.

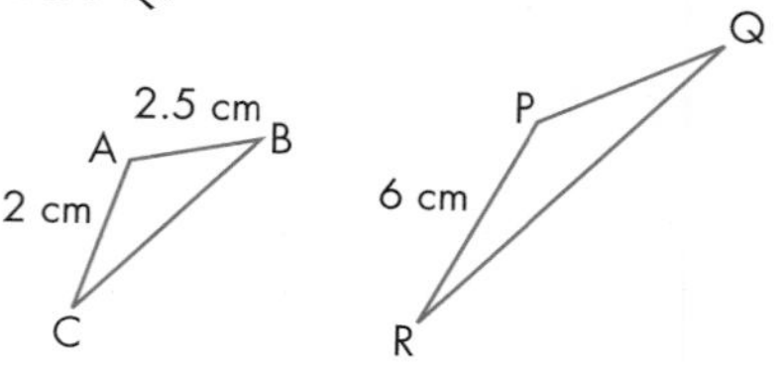

c Find x and y.

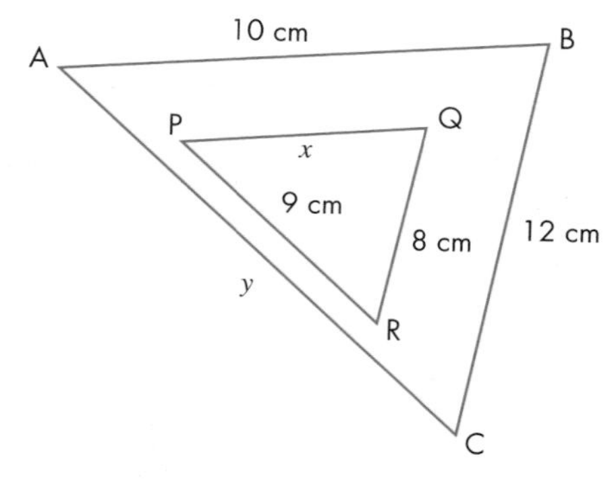

d Find x and y.

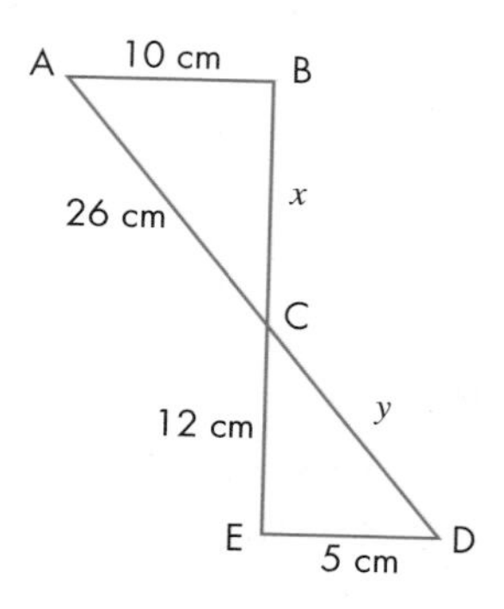

e Find the lengths of AB and PQ.

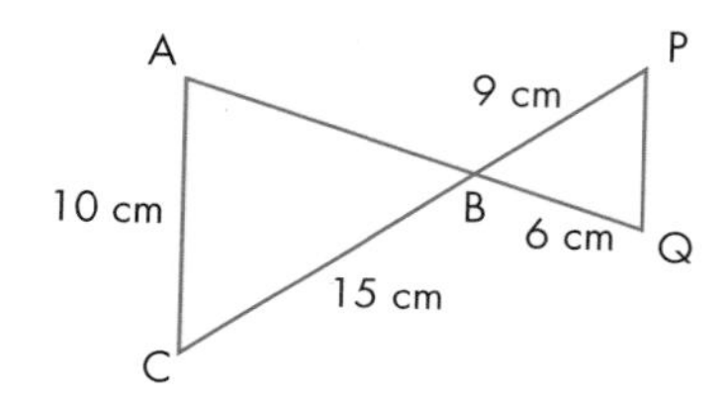

f Find the length of QR.

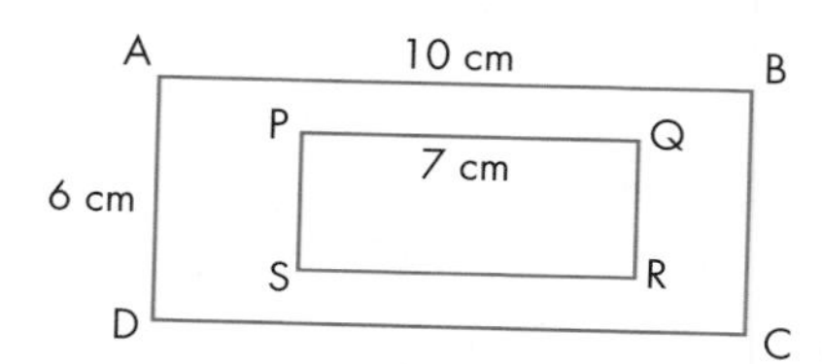

7 **a** Explain why these two triangles are similar.

b What is the ratio of their sides?

c Use Pythagoras' theorem to calculate the length of side AC of triangle ABC.

d Write down the length of the side PR of triangle PQR.

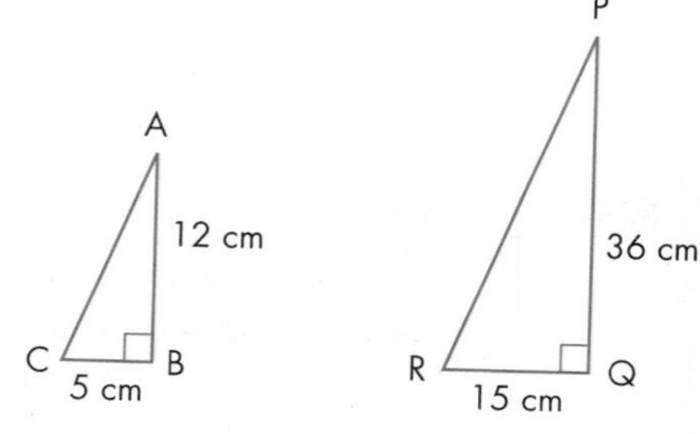

FM 8 Sean is standing next to a tree.

His height is 1.6 m and he casts a shadow that has a length of 2.4 m.

The tree casts a shadow that has a length of 7.8 m.

Use what you know about similar triangles to work out the height of the tree, h.

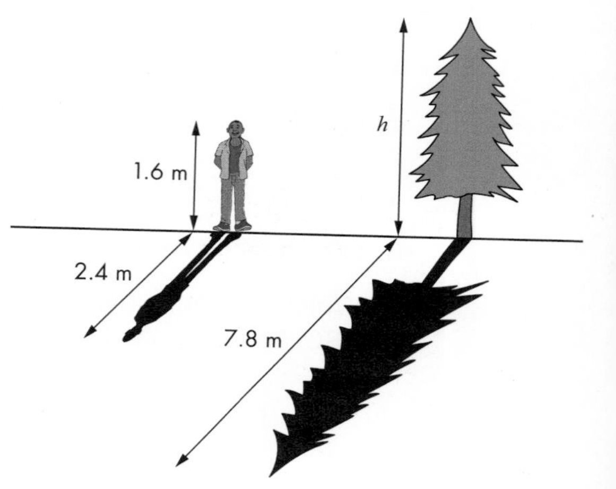

B

PS 9 Here are two rectangles.

Explain why the two rectangles are not similar.

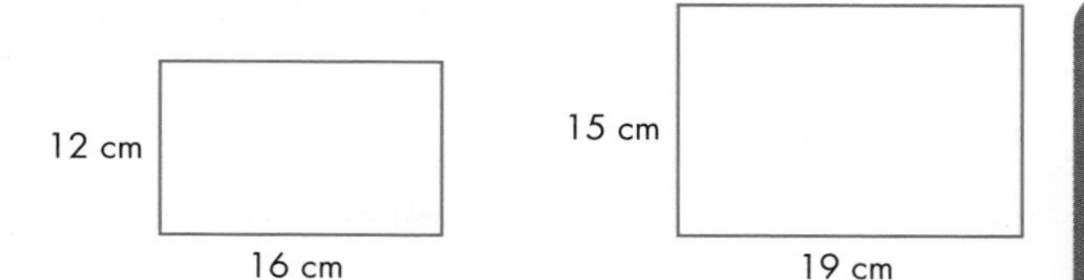

AU 10 Triangle ABC is similar to triangle CDE.

Jay says that the length of DE is 14 cm.

Explain why Jay is wrong.

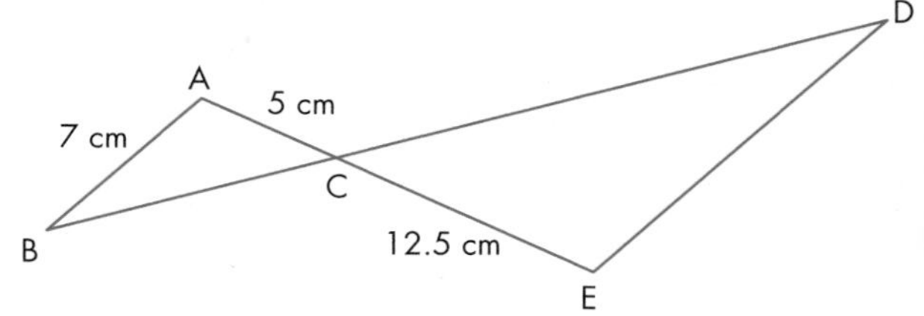

Further examples of similar triangles

EXAMPLE 2

Find the lengths marked x and y in the diagram (not drawn to scale).

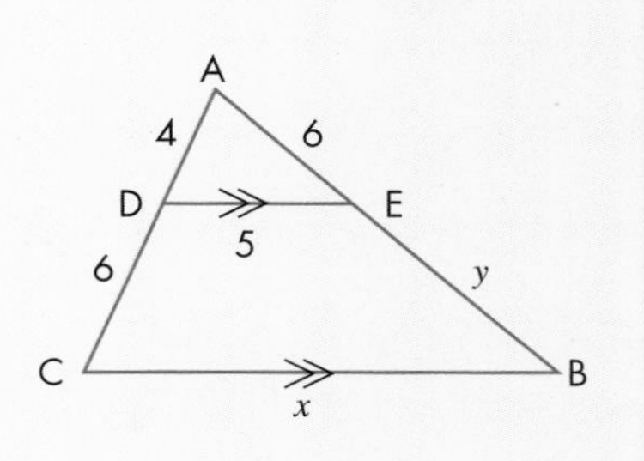

Triangles AED and ABC are similar. So using the corresponding sides CB, DE with AC, AD gives,

$$\frac{x}{5} = \frac{10}{4}$$

$$\Rightarrow x = \frac{10 \times 5}{4} = 12.5$$

Using the corresponding sides AE, AB with AD, AC gives,

$$\frac{y+6}{6} = \frac{10}{4} \Rightarrow y + 6 = \frac{10 \times 6}{4} = 15$$

$$\Rightarrow \quad y = 15 - 6 = 9$$

EXAMPLE 3

Ahmed wants to work out the height of a tall building. He walks 100 paces from the building and sticks a pole, 2 m long, vertically into the ground. He then walks another 10 paces on the same line and notices that when he looks from ground level, the top of the pole and the top of the building are in line. How tall is the building?

First, draw a diagram of the situation and label it.

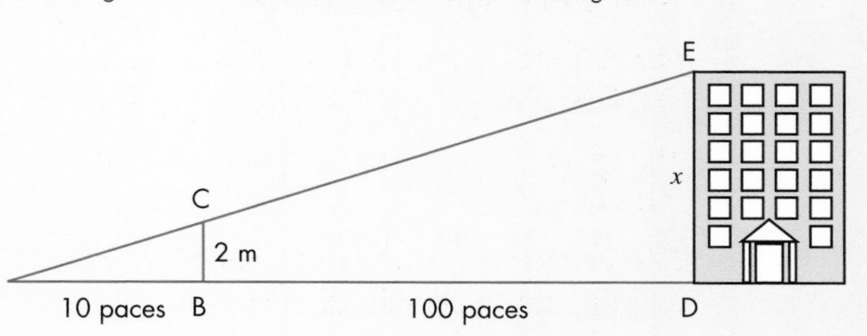

Using corresponding sides ED, CB with AD, AB gives,

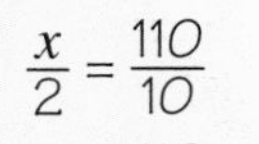

$$\frac{x}{2} = \frac{110}{10}$$

$$\Rightarrow \quad x = \frac{110 \times 2}{10} = 22 \text{ m}$$

Hence the building is 22 m high.

EXERCISE 9B 

1 In each of the cases below, state a pair of similar triangles and find the length marked x. Separate the similar triangles if it makes it easier for you.

a

A, B, C, D, E
4 cm
3 cm
8 cm
x

b

A, B, C, D, E
5 cm
4 cm
10 cm
x

2 **a** Find the value of x.

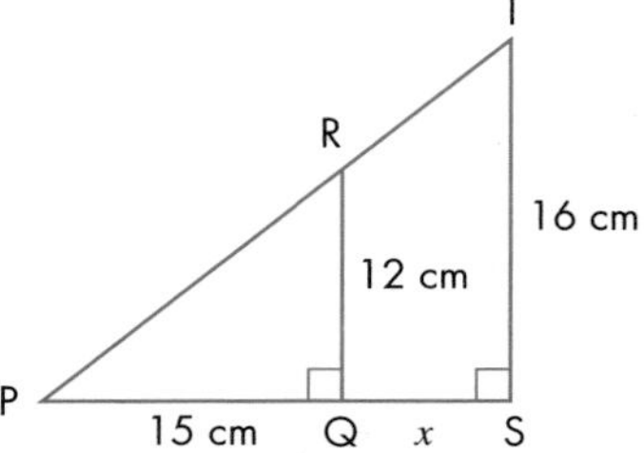

b Find the length of CE.

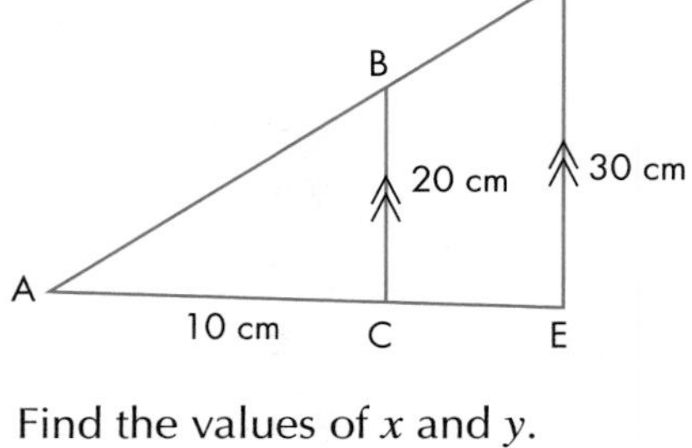

c Find the values of x and y.

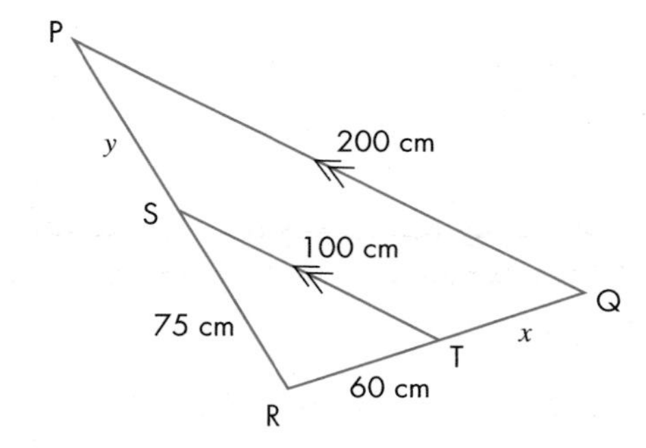

d Find the values of x and y.

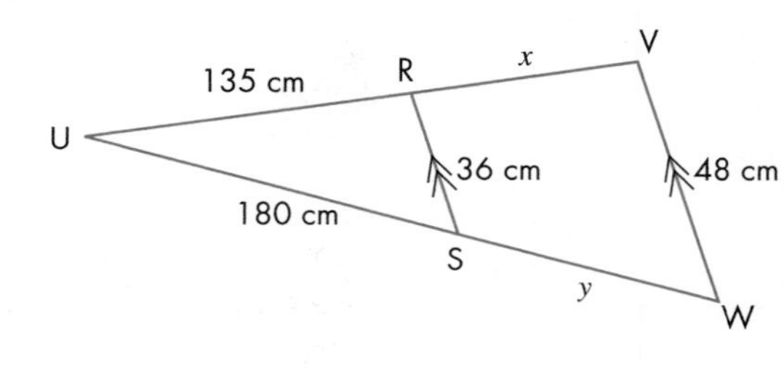

e Find the lengths of DC and EB.

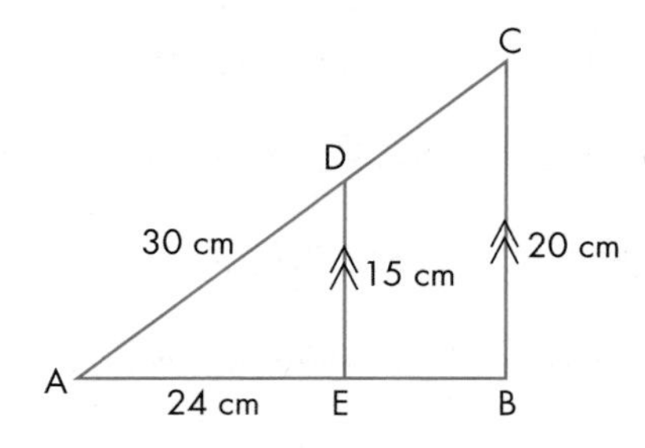

FM 3 This diagram shows a method of working out the height of a tower.

A stick, 2 m long, is placed vertically 120 m from the base of a tower so that the top of the tower and the top of the stick are in line with a point on the ground 3 m from the base of the stick. How high is the tower?

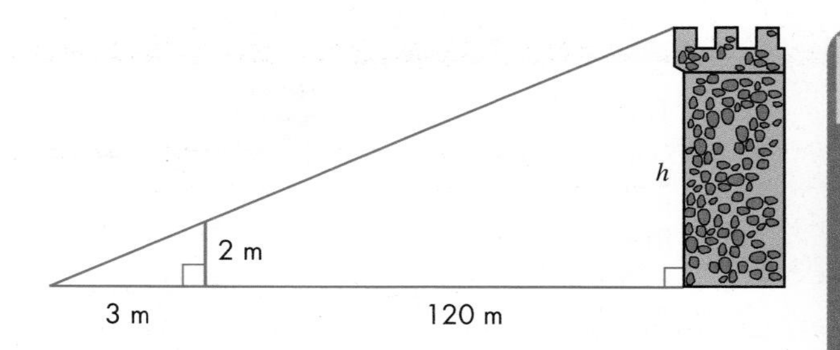

FM 4 It is known that a factory chimney is 330 feet high. Patrick paces out distances as shown in the diagram, so that the top of the chimney and the top of the flag pole are in line with each other. How high is the flag pole?

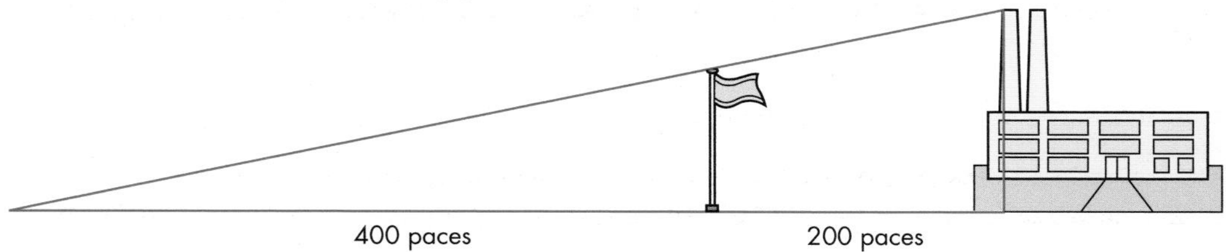

FM 5 The height of a golf flag is 1.5 m. Use the diagram to find the height of the tree.

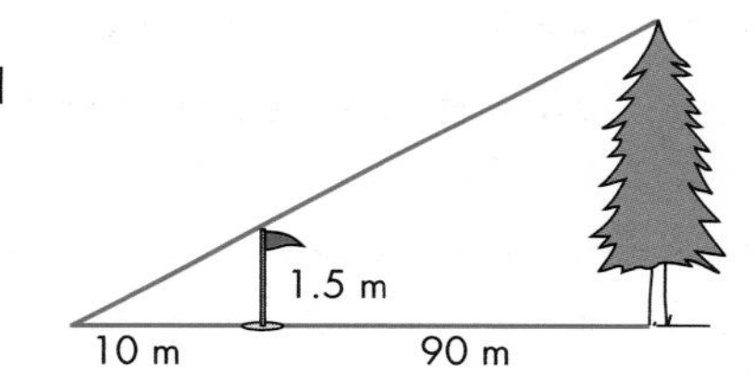

FM 6 Find the height of a pole that casts a shadow of 1.5 m when at the same time a man of height 165 cm casts a shadow of 75 cm.

7 Bob, a builder, is making this wooden frame for a roof.

In the diagram, triangle ABC is similar to triangle AXY.

AB = 1.5 m, BX = 3.5 m and XY = 6 m

Work out the length of wood that Bob needs to make BC.

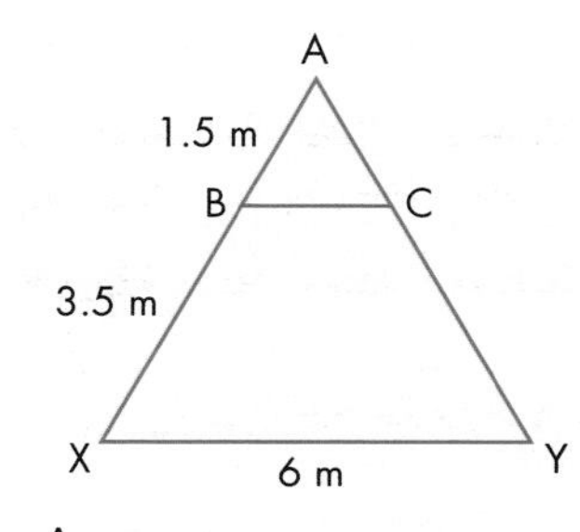

PS 8 Triangle ABC is similar to triangle ACD.

AC = 9 cm and CD = 6 cm

Work out the length of BC.

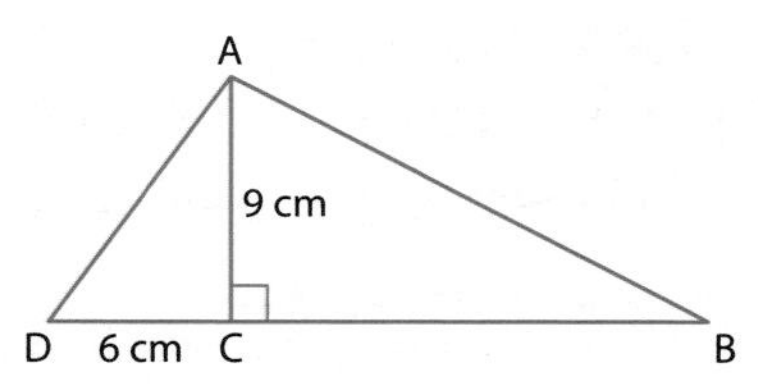

AU 9 In the diagram triangle ABC is similar to triangle AXY.

Which of the following is the correct length of BX?

Explain how you decide.

a 2 cm **b** 3 cm

c 4 cm **d** 5 cm

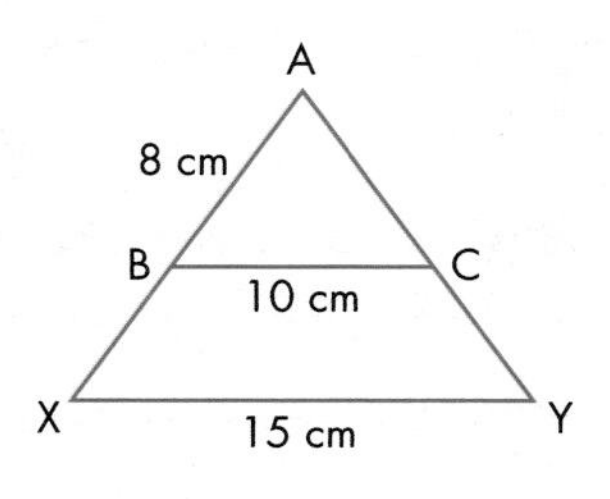

More complicated problems

The information given in a similar triangle situation can be more complicated than anything you have met so far, and you will need to have good algebraic skills to deal with it. Example 4 is typical of the more complicated problem you may be asked to solve, so follow it through carefully.

EXAMPLE 4

Find the value of x in this triangle.

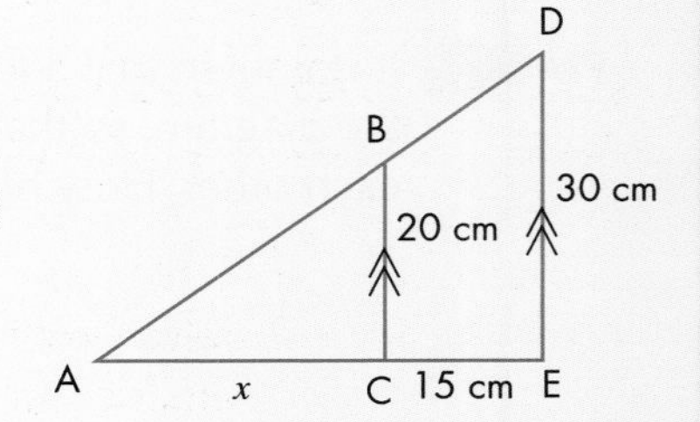

You know that triangle ABC is similar to triangle ADE.

Splitting up the triangles may help you to see what will be needed.

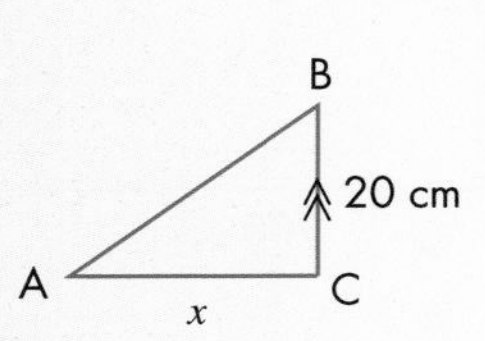

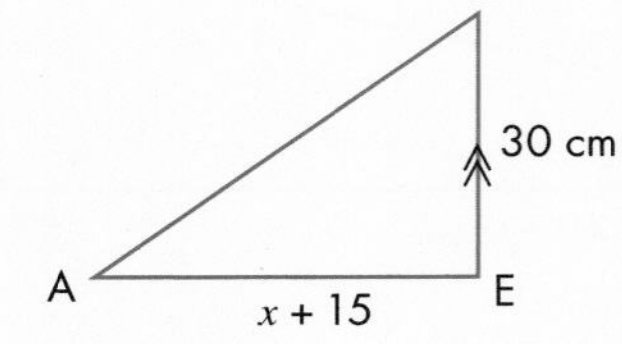

So your equation will be:

$$\frac{x+15}{x} = \frac{30}{20}$$

Cross multiplying (moving each of the two bottom terms to the opposite side and multiplying) gives:

$$20x + 300 = 30x$$

$$\Rightarrow \quad 300 = 10x \Rightarrow x = 30 \text{ cm}$$

EXERCISE 9C

Find the lengths x or x and y in the diagrams **1** to **6**.

1

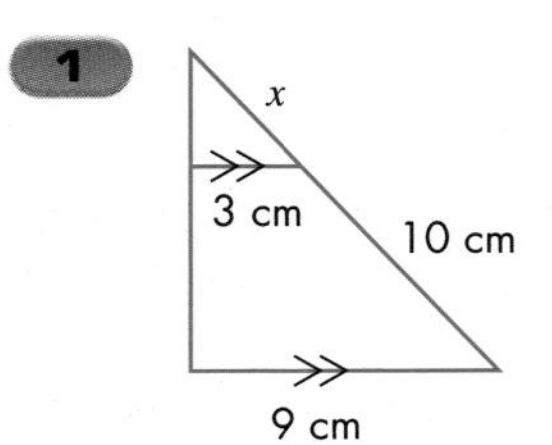

2

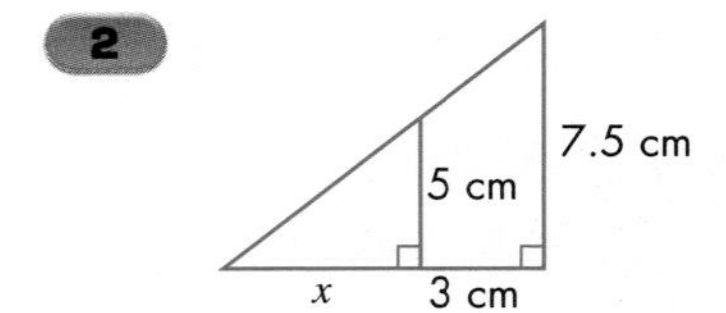

3

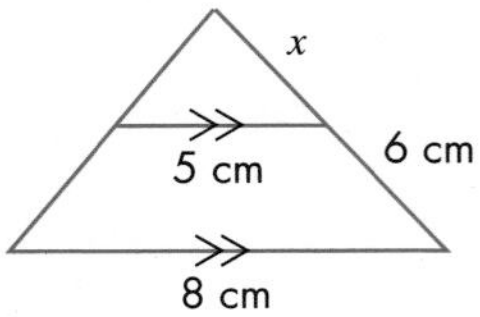

4

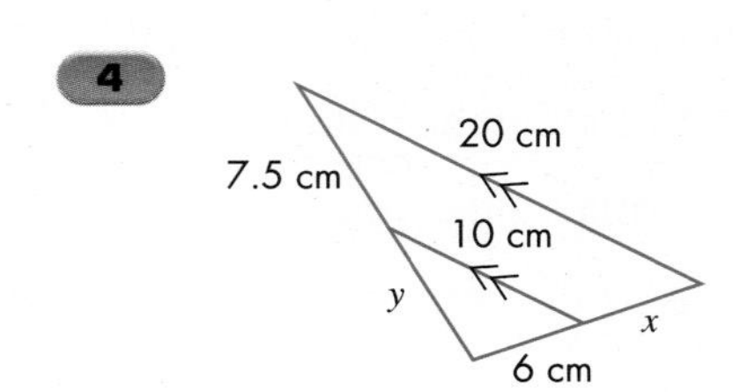

B

5

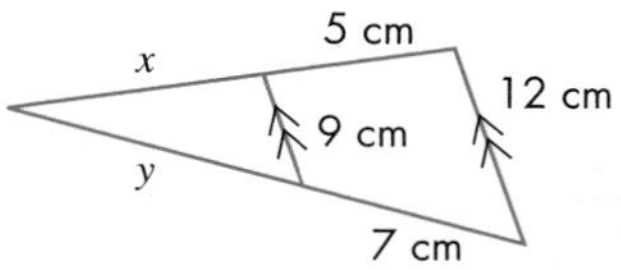

5 cm, 12 cm, 9 cm, 7 cm, x, y

6

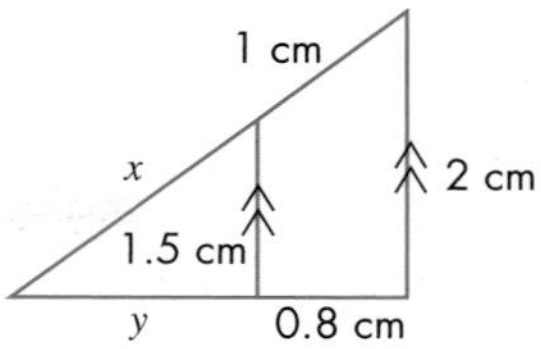

1 cm, 2 cm, 1.5 cm, 0.8 cm, x, y

9.2 Areas and volumes of similar shapes

This section will show you how to:

- solve problems involving the area and volume of similar shapes

Key words

area ratio
area scale factor
length ratio
linear scale factor
volume ratio
volume scale factor

There are relationships between the lengths, areas and volumes of similar shapes.

You saw in section 6.5 that when a 2D shape is enlarged by a given scale factor to form a new, similar shape, the corresponding lengths of the original shape and the new shape are all in the same ratio, which is equal to the scale factor. This scale factor of the lengths is called the **length ratio** or **linear scale factor**.

Two similar shapes also have an **area ratio**, which is equal to the ratio of the squares of their corresponding lengths. The area ratio, or **area scale factor**, is the square of the length ratio.

Likewise, two 3D shapes are similar if their corresponding lengths are in the same ratio. Their **volume ratio** is equal to the ratio of the cubes of their corresponding lengths. The volume ratio, or **volume scale factor**, is the cube of the length ratio.

Generally, the relationship between similar shapes can be expressed as:

Length ratio $x : y$ Area ratio $x^2 : y^2$ Volume ratio $x^3 : y^3$

EXAMPLE 5

A model yacht is made to a scale of $\frac{1}{20}$ of the size of the real yacht. The area of the sail of the model is 150 cm^2. What is the area of the sail of the real yacht?

At first sight, it may appear that you do not have enough information to solve this problem, but it can be done as follows.

Linear scale factor = 1 : 20
Area scale factor = 1 : 400 (square of the linear scale factor)
Area of real sail = 400 × area of model sail
= 400 × 150 cm^2
= 60 000 cm^2 = 6 m^2

EXAMPLE 6

A bottle has a base radius of 4 cm, a height of 15 cm and a capacity of 650 cm^3. A similar bottle has a base radius of 3 cm.

a What is the length ratio?

b What is the volume ratio?

c What is the volume of the smaller bottle?

a The length ratio is given by the ratio of the two radii, that is 4 : 3.

b The volume ratio is therefore $4^3 : 3^3 = 64 : 27$.

c Let v be the volume of the smaller bottle. Then the volume ratio is:

$$\frac{\text{volume of smaller bottle}}{\text{volume of larger bottle}} = \frac{v}{650} = \frac{27}{64}$$

$$\Rightarrow v = \frac{27 \times 650}{64} = 274 \text{ cm}^3 \text{ (3 significant figures)}$$

EXAMPLE 7

The cost of a tin of paint , height 12 cm, is £3.20 and its label has an area of 24 cm^2.

a If the cost is based on the amount of paint in the tin, what is the cost of a similar tin, 18 cm high?

b Assuming the labels are similar, what will be the area of the label on the larger tin?

a The cost of the paint is proportional to the volume of the tin.

Length ratio = 12 : 18 = 2 : 3

Volume ratio = $2^3 : 3^3 = 8 : 27$

Let P be the cost of the larger tin. Then the cost ratio is:

$$\frac{\text{cost of larger tin}}{\text{cost of smaller tin}} = \frac{P}{3.2}$$

Therefore,

$$\frac{P}{3.2} = \frac{27}{8}$$

$$\Rightarrow P = \frac{27 \times 3.2}{8} = £10.80$$

b Area ratio = $2^2 : 3^2 = 4 : 9$

Let A be the area of the larger label. Then the area ratio is:

$$\frac{\text{larger label area}}{\text{smaller label area}} = \frac{A}{24}$$

Therefore,

$$\frac{A}{24} = \frac{9}{4}$$

$$\Rightarrow A = \frac{9 \times 24}{4} = 54 \text{ cm}^2$$

EXERCISE 9D

A

1 The length ratio between two similar solids is 2 : 5.

a What is the area ratio between the solids?

b What is the volume ratio between the solids?

2 The length ratio between two similar solids is 4 : 7.

a What is the area ratio between the solids?

b What is the volume ratio between the solids?

3 Copy and complete this table.

Linear scale factor	Linear ratio	Linear fraction	Area scale factor	Volume scale factor
2	1 : 2	$\frac{2}{1}$		
3				
$\frac{1}{4}$	4 : 1	$\frac{1}{4}$		$\frac{1}{64}$
			25	
				$\frac{1}{1000}$

4 A shape has an area of 15 cm^2. What is the area of a similar shape with lengths that are three times the corresponding lengths of the first shape?

FM 5 A toy brick has a surface area of 14 cm^2. What would be the surface area of a similar toy brick with lengths that are:

a twice the corresponding lengths of the first brick

b three times the corresponding lengths of the first brick?

6 A rug has an area of 12 m^2. What area would be covered by rugs with lengths that are:

a twice the corresponding lengths of the first rug

b half the corresponding lengths of the first rug?

7 A brick has a volume of 300 cm^3. What would be the volume of a similar brick whose lengths are:

a twice the corresponding lengths of the first brick

b three times the corresponding lengths of the first brick?

FM 8 A tin of paint, 6 cm high, holds a half a litre of paint. How much paint would go into a similar tin which is 12 cm high?

FM 9 A model statue is 10 cm high and has a volume of 100 cm^3. The real statue is 2.4 m high. What is the volume of the real statue? Give your answer in m^3.

A

10 A small tin of paint costs 75p. What is the cost of a larger similar tin with height twice that of the smaller tin? Assume that the cost is based only on the volume of paint in the tin.

11 A small trinket box of width 2 cm has a volume of 10 cm³. What is the width of a similar trinket box with a volume of 80 cm³?

FM 12 A cinema sells popcorn in two different-sized tubs that are similar in shape.

Show that it is true that the big tub is better value.

PS 13 The diameters of two ball bearings are given below.

Work out:

a the ratio of their radii

b the ratio of their surface areas

c the ratio of their volumes.

AU 14 Cuboid A is similar to cuboid B.

The length of cuboid A is 10 cm and the length of cuboid B is 5 cm.

The volume of cuboid A is 720 cm³.

Shona says that the volume of cuboid B must be 360 cm³.

Explain why she is wrong.

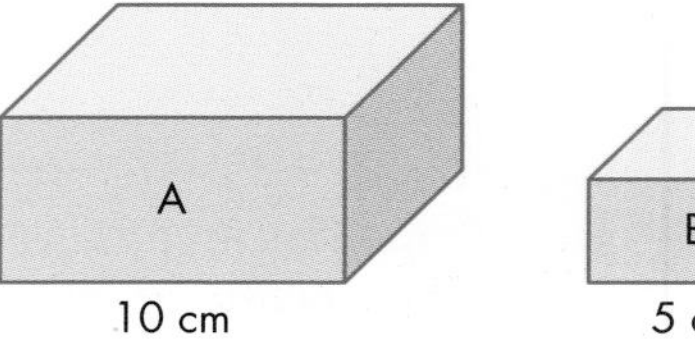

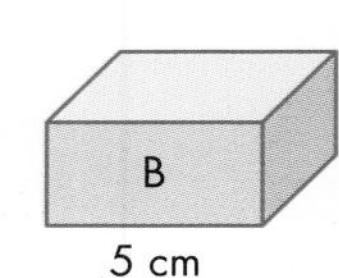

More complex problems using area and volume ratios

In some problems involving similar shapes, the length ratio is not given, so we have to start with the area ratio or the volume ratio. We usually then need to find the length ratio in order to proceed with the solution.

EXAMPLE 8

A manufacturer makes a range of clown hats that are all similar in shape. The smallest hat is 8 cm tall and uses 180 cm² of card. What will be the height of a hat made from 300 cm² of card?

The area ratio is 180 : 300

Therefore, the length ratio is $\sqrt{180} : \sqrt{300}$ (do not calculate these yet)

Let the height of the larger hat be H, then

$$\frac{H}{8} = \frac{\sqrt{300}}{\sqrt{180}} = \sqrt{\frac{300}{180}}$$

$$\Rightarrow H = 8 \times \sqrt{\frac{300}{180}} = 10.3 \text{ cm (1 decimal place)}$$

EXAMPLE 9

A supermarket stocks similar small and large tins of soup. The areas of their labels are 110 cm^2 and 190 cm^2 respectively. The weight of a small tin is 450 g. What is the weight of a large tin?

The area ratio is 110 : 190

Therefore, the length ratio is $\sqrt{110} : \sqrt{190}$ (do not calculate these yet)

So the volume (weight) ratio is $(\sqrt{110})^3 : (\sqrt{190})^3$.

Let the weight of a large tin be W, then

$$\frac{W}{450} = \frac{(\sqrt{190})^3}{(\sqrt{110})^3} = \left(\sqrt{\frac{190}{110}}\right)^3$$

$$\Rightarrow \quad W = 450 \times \left(\sqrt{\frac{190}{110}}\right)^3 = 1020 \text{ g} \qquad \text{(3 significant figures)}$$

EXAMPLE 10

Two similar tins hold respectively 1.5 litres and 2.5 litres of paint. The area of the label on the smaller tin is 85 cm^2. What is the area of the label on the larger tin?

The volume ratio is 1.5 : 2.5

Therefore, the length ratio is $\sqrt[3]{1.5} : \sqrt[3]{2.5}$ (do not calculate these yet)

So the area ratio is $(\sqrt[3]{1.5})^2 : (\sqrt[3]{2.5})^2$

Let the area of the label on the larger tin be A, then

$$\frac{A}{85} = \frac{(\sqrt[3]{2.5})^2}{(\sqrt[3]{1.5})^2} = \left(\sqrt[3]{\frac{2.5}{1.5}}\right)^2$$

$$\Rightarrow \quad A = 85 \times \left(\sqrt[3]{\frac{2.5}{1.5}}\right)^2 = 119 \text{ cm}^2 \qquad \text{(3 significant figures)}$$

EXERCISE 9E

FM 1 A firm produces three sizes of similar-shaped labels for its products. Their areas are 150 cm^2, 250 cm^2 and 400 cm^2. The 250 cm^2 label just fits around a can of height 8 cm. Find the heights of similar cans around which the other two labels would just fit.

A*

2 A firm makes similar gift boxes in three different sizes: small, medium and large. The areas of their lids are as follows.

Small: 30 cm^2 Medium: 50 cm^2 Large: 75 cm^2

The medium box is 5.5 cm high. Find the heights of the other two sizes.

3 A cone of height 8 cm can be made from a piece of card with an area of 140 cm^2. What is the height of a similar cone made from a similar piece of card with an area of 200 cm^2?

4 It takes 5.6 litres of paint to paint a chimney which is 3 m high. What is the tallest similar chimney that can be painted with 8 litres of paint?

5 A piece of card, 1200 cm^2 in area, will make a tube 13 cm long. What is the length of a similar tube made from a similar piece of card with an area of 500 cm^2?

6 All television screens (of the same style) are similar. If a screen of area 220 cm^2 has a diagonal length of 21 cm, what will be the diagonal length of a screen of area 350 cm^2?

7 Two similar statues, made from the same bronze, are placed in a school. One weighs 300 g, the other weighs 2 kg. The height of the smaller statue is 9 cm. What is the height of the larger statue?

FM 8 A supermarket sells similar cans of pasta rings in three different sizes: small, medium and large. The sizes of the labels around the cans are as follows.

Small can: 24 cm^2 Medium can: 46 cm^2 Large can: 78 cm^2

The medium size can is 6 cm tall with a weight of 380 g. Calculate these quantities.

a The heights of the other two sizes

b The weights of the other two sizes

9 A statue weighs 840 kg. A similar statue was made out of the same material but two-fifths the height of the first one. What was the weight of the smaller statue?

10 A model stands on a base of area 12 cm^2. A smaller but similar model, made of the same material, stands on a base of area 7.5 cm^2. Calculate the weight of the smaller model if the larger one is 3.5 kg.

FM 11 Steve fills two similar jugs with orange juice.

The first jug holds 1.5 litres of juice and has a base diameter of 8 cm.

The second jug holds 2 litres of juice. Work out the base diameter of the second jug.

PS 12 The total surface areas of two similar cuboids are 500 cm^2 and 800 cm^2.

If the width of one of the cuboids is 10 cm, calculate the two possible widths for the other cuboid.

AU 13 The volumes of two similar cylinders are 256 cm^3 and 864 cm^3.

Which of the following gives the ratio of their surface areas?

a 2 : 3 **b** 4 : 9 **c** 8 : 27

GRADE BOOSTER

C You can work out unknown lengths in 2D shapes, using scale factors

B You can use ratios and equations to find unknown lengths in similar triangles

A You can solve problems, using area and volume scale factors

A* You can solve more complex problems, using area and volume scale factors

---

What you should know now

- How to find the ratios between two similar shapes
- How to work out unknown lengths, areas and volumes of similar 3D shapes
- How to solve practical problems, using similar shapes
- How to solve problems, using area and volume ratios

1 Shapes ABCD and EFGH are mathematically similar.

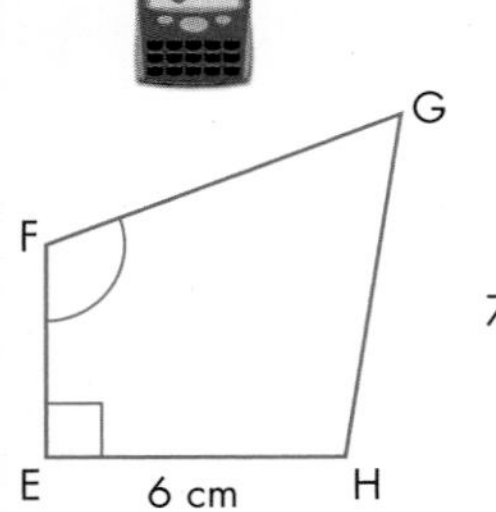

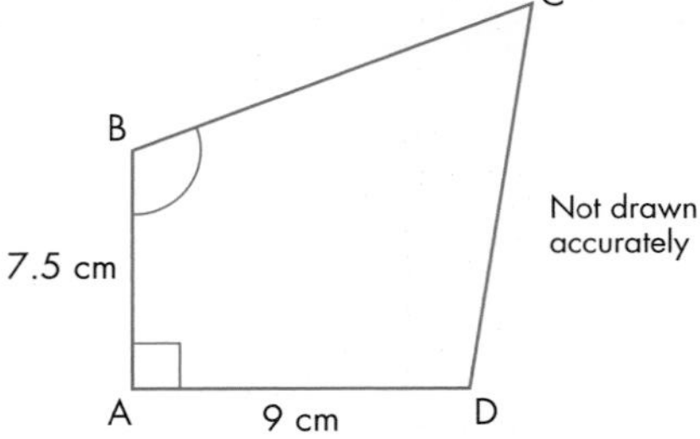

a Calculate the length of BC. *(2 marks)*

b Calculate the length of EF. *(2 marks)*

Edexcel, November 2008, Paper 15, Question 13(a, b)

2

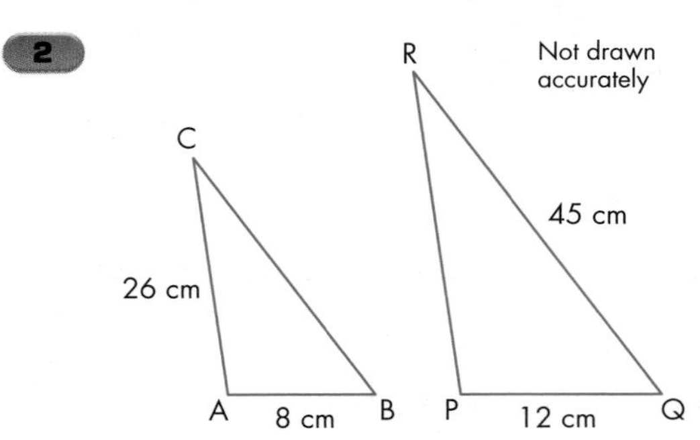

The two triangles ABC and PQR are mathematically similar.

Angle A = angle P

Angle B = angle Q

AB = 8 cm, AC = 26 cm, PQ = 12 cm, QR = 45 cm.

a Work out the length of PR. *(2 marks)*

b Work out the length of BC. *(2 marks)*

Edexcel, June 2006, Paper 5, Question 11

3

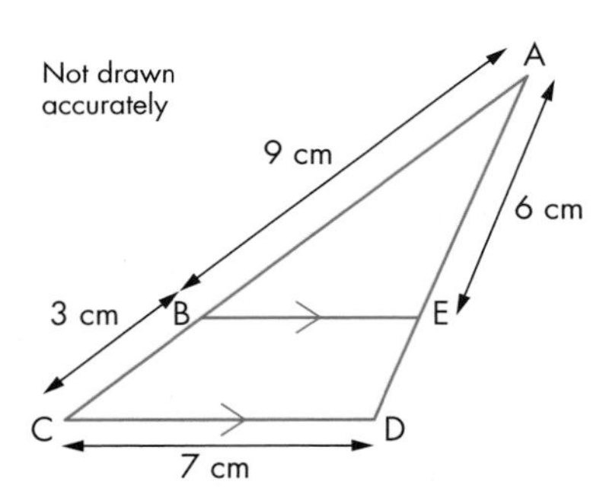

BE is parallel to CD.

AB = 9 cm, BC = 3 cm, CD = 7 cm, AE = 6 cm.

a Calculate the length of ED. *(2 marks)*

b Calculate the length of BE. *(2 marks)*

Edexcel, June 2005, Paper 6, Question 11

4

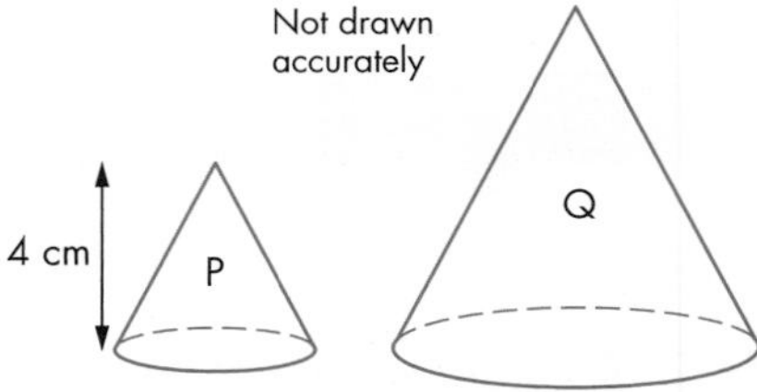

Two cones, P and Q, are mathematically similar.

The total surface area of cone P is 24 cm^2.

The total surface area of cone Q is 96 cm^2.

The height of cone P is 4 cm.

a Work out the height of cone Q. *(2 marks)*

The volume of cone P is 12 cm^3.

b Work out the volume of cone Q. *(2 marks)*

Edexcel, June 2007, Paper 5, Question 20

5

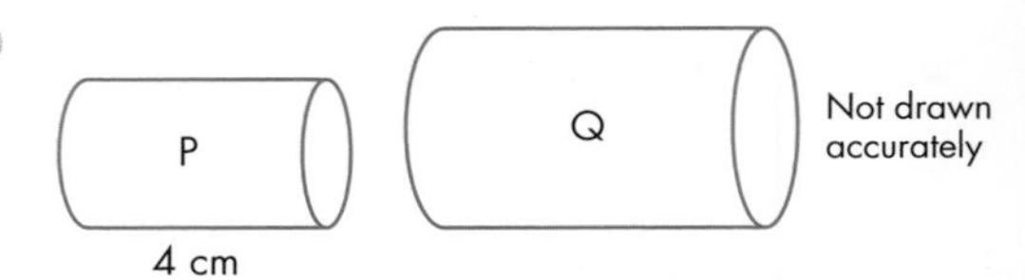

The cylinders P and Q are mathematically similar.

The total surface area of cylinder P is 90π cm^2.

The total surface area of cylinder Q is 810π cm^2.

The length of cylinder P is 4 cm.

a Work out the length of cylinder Q. *(3 marks)*

The volume of cylinder P is 100π cm^3.

b Work out the volume of cylinder Q.
Give your answer as a multiple of π. *(2 marks)*

Edexcel, June 2005, Paper 5, Question 18

Worked Examination Questions

PS **1** Triangles ABC and CDE are similar.

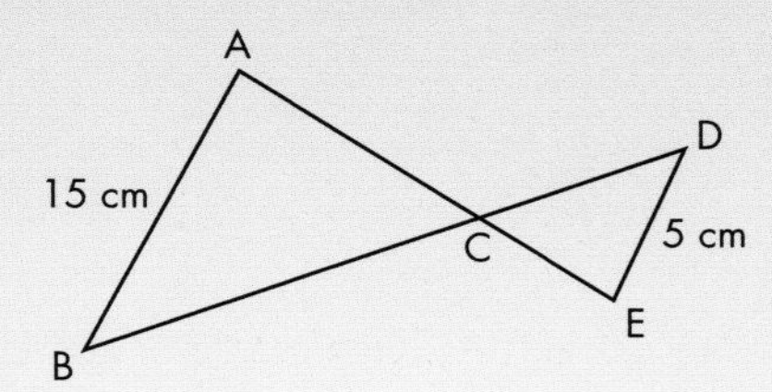

AB = 15 cm and DE = 5 cm

The length of BD is 24 cm.

Work out the lengths of BC and CD.

1 AB : DE = 15 : 5 = 3 : 1 — AB and DE are corresponding sides in similar triangles. This gets 1 mark for method.

So the ratio BC : CD is also 3 : 1 — BC and CD are another pair of corresponding sides.

BD = 24 cm, divide this in the ratio 3 : 1 — The ratio has 4 parts, so 24 ÷ 4 = 6.

So BC = 18 cm — This gets 1 mark for accuracy.

and CD = 6 cm — This gets 1 mark for accuracy.

Total: 3 marks

Worked Examination Questions

AU **2** A manufacturer makes cylindrical boxes in two different sizes.

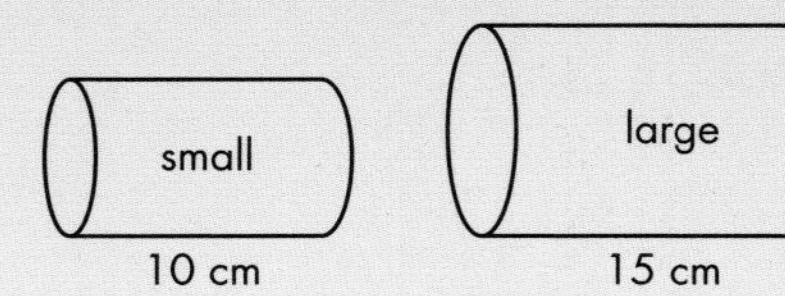

The two boxes are similar.

The length of the small box is 10 cm.

The length of the large box is 15 cm.

Some information about the boxes is given in the table.

	length	surface area	volume
small	10 cm	224 cm^2	270 cm^3
large	15 cm		

Complete the table.

2 Ratio of lengths = 10 : 15 = 2 : 3

> Or scale factor = 1.5

So ratio of areas = 4 : 9

> Square the length ratios or area scale factor = $1.5^2 = 2.25$
> This gets 1 mark for method.

Surface area of large box = $\frac{224 \times 9}{4} = 504 \text{ cm}^2$

> Or 224 × 2.25
> This gets 1 mark for accuracy.

Ratio of volumes = 8 : 27

> Cube the length ratios or volume scale factor = $1.5^3 = 3.375$
> This gets 1 mark for method.

Volume of large box = $\frac{270 \times 27}{8} = 911 \text{ cm}^3$ (3 sf)

> Or 270 × 3.375
> This gets 1 mark for accuracy.

	length	surface area	volume
small	10 cm	224 cm^2	270 cm^3
large	15 cm	504 cm^2	911 cm^3 (3 sf)

Total: 4 marks

Worked Examination Questions

FM **3** A camping gas container is in the shape of a cylinder with a hemispherical top.
The dimensions of the container are shown in the diagram.

It is decided to increase the volume of the container by **20%**.
The new container is mathematically **similar** to the old one.

Calculate the base **diameter** of the new container.

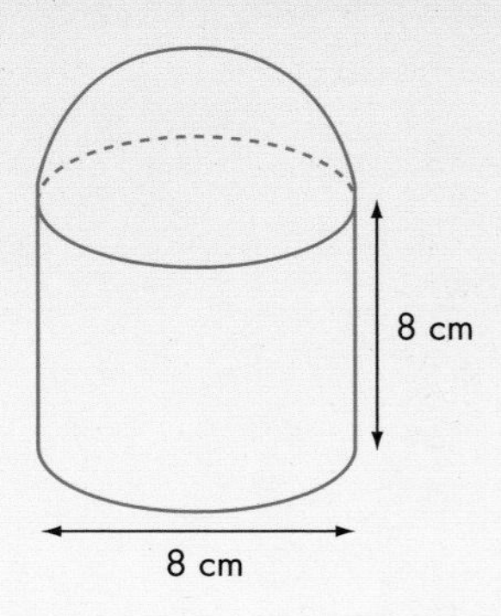

3 Old volume : New volume = 100% : 120% = 1 : 1.2 — First find the volume scale factor. This gets 1 mark for method.

$\sqrt[3]{1} : \sqrt[3]{1.2} = 1 : 1.06265$ — Take the cube root to get the linear scale factor. This gets 1 mark for method.

New diameter = Old diameter × 1.06265
= 8 × 1.06265 = 8.5 cm — Multiply the old diameter by the linear scale factor to get the new diameter. This gets 1 mark for accuracy.

Total: 3 marks

Functional Maths
Making a scale model

Professional model-makers make scale models for all kinds of purposes. These models help people to visualise what things will look like in reality. An architect may need a scale model of a shopping centre, to show local people what it would be like. A film-maker may want a scale model of a dinosaur, to use for special effects.

Getting started

A transport museum displays scale models. Sometimes it takes these models into schools to show children.

Museum department	Model	Scale
Aeroplanes	*Concorde*	1 : 50
Trains	*Orient Express* single cabin	1 : 5
Automobiles	Rolls Royce *Phantom*	1 : 8

- Why didn't the model-maker use the same scale for *Concorde* as for the Rolls Royce *Phantom*?
- The dimensions of the bed in the model of the *Orient Express* cabin are 38 cm × 17 cm. Work out the dimensions of a bed on the real *Orient Express*.
- The wing area of the model *Concorde* is 1432 cm^2. Work out the wing area for the real *Concorde*. Give your answer in square metres (m^2).

Your task

The transport museum decides to have two further models made: one for its spacecraft department and one for its bicycle department.

It is your task to commission the two models. You must decide what type of spacecraft and bicycle you wish to exhibit. You must also choose appropriate scales for the models, so that they may sometimes be taken into schools.

Then write the descriptions that will appear on display boards next to each of the models in the museum. Include some facts about the modes of transport and also compare some of the dimensions (for example, lengths, heights, areas and volumes) of each model and the real thing.

Extension

Try using the internet to search for 'spacecraft dimensions' or 'bicycle dimensions'.

Why this chapter matters

Trigonometry has a wide variety of applications. It is used in practical fields such as navigation, land surveying, building, engineering and astronomy.

In surveying, trigonometry is used extensively in triangulation. This is a process for establishing the location of a point by measuring angles or bearings to this point, from two other known points at either end of a fixed baseline.

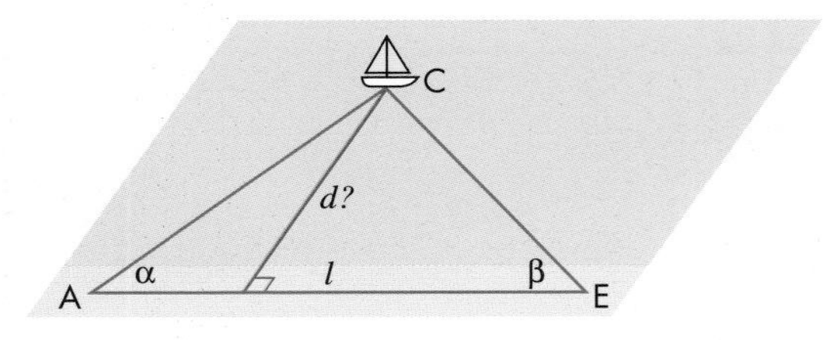

Triangulation can be used to calculate the position and distance from the shore to a ship. The observer at A measures the angle α between the shore and the ship, and the observer at B does likewise for angle β. With the length, l, or the position of A and B known, then the sine rule can be applied to find the position of the ship at C and the distance, d.

Marine sextants like this are used to measure the angle of the Sun or stars with respect to the horizon. Using trigonometry and a marine chronometer, the ship's navigator can determine the position of the ship on the sea.

The Canadarm2 robotic manipulator in the International Space Station is operated by controlling the angles of its joints. Calculating the final position of the astronaut at the end of the arm requires repeated use of trigonometry in three dimensions.

Chapter

Geometry: Trigonometry

1. Some 2D problems
2. Some 3D problems
3. Trigonometric ratios of angles between 90° and 360°
4. Solving any triangle
5. Trigonometric ratios in surd form
6. Using sine to find the area of a triangle

The grades given in this chapter are target grades.

This chapter will show you ...

- **A to A*** how to use trigonometric ratios to solve more complex 2D problems and 3D problems
- **A to A*** how to use the sine and cosine rules to solve problems involving non right-angled triangles
- **A to A*** how to find the area of a triangle using formula $A = \frac{1}{2}ab \sin C$
- **A*** how to find the trigonometric ratios for any angle from 0° to 360°

Visual overview

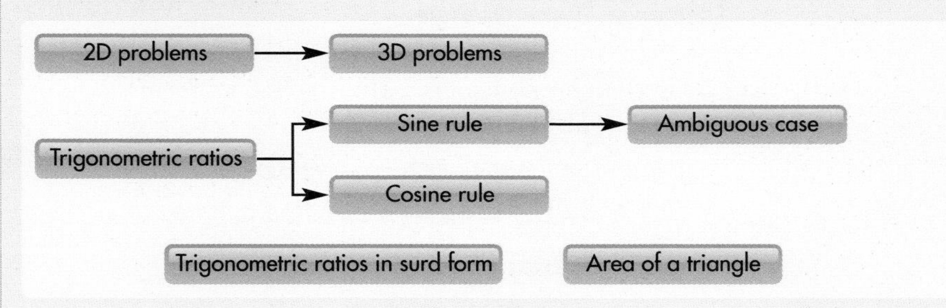

What you should already know

- How to find the sides of right-angled triangles using Pythagoras' theorem **(KS3 level 7, GCSE grade C)**
- How to find angles and sides of right-angled triangles using sine, cosine and tangent **(KS3 level 8, GCSE grade B)**

Quick check

Calculate the value of x in each of these right-angled triangles.

Give your answers to 3 significant figures.

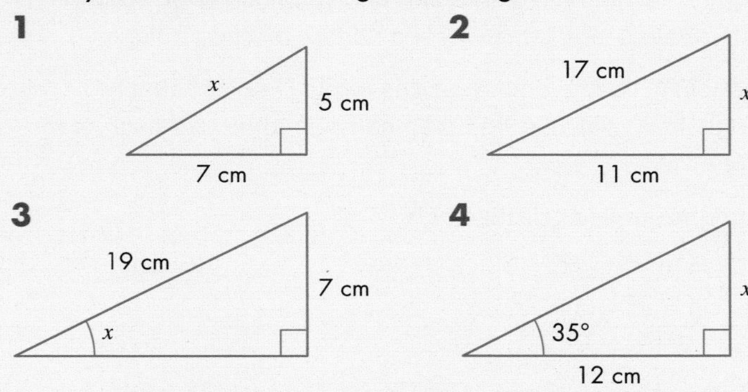

10.1 Some 2D problems

This section will show you how to:

- use trigonometric ratios and Pythagoras' theorem to solve more complex two-dimensional problems

Key words
cosine
Pythagoras' theorem
sine
tangent

This lesson brings together previous work on **Pythagoras' theorem**, circle theorems and trigonometric ratios – **sine** (sin), **cosine** (cos) and **tangent** (tan).

EXAMPLE 1

In triangle ABC , AB = 6 cm, BC = 9 cm and angle ABC = 52°. Calculate:

a the length of the perpendicular from A to BC

b the area of the triangle.

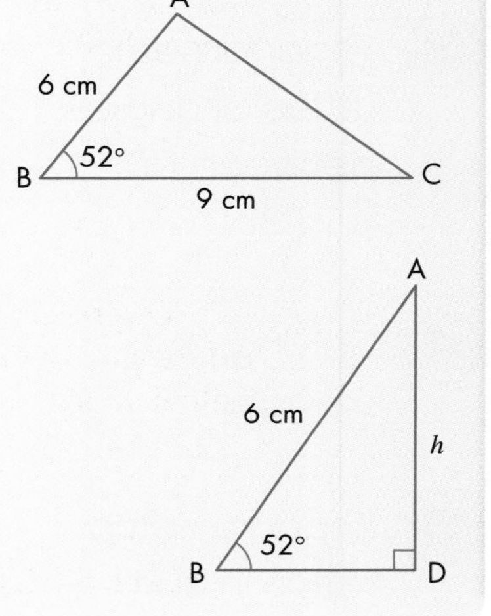

a Drop the perpendicular from A to BC to form the right-angled triangle ADB.

Let h be the length of the perpendicular AD. Then,

$h = 6 \sin 52° = 4.73$ (3 significant figures)

b The area of triangle ABC is given by,

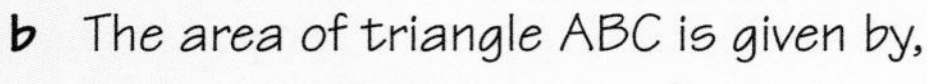

Area = $\frac{1}{2} \times$ base $\times$ height

$= \frac{1}{2} \times 9 \times h = 21.3 \text{ cm}^2$ (3 significant figures)

EXAMPLE 2

SR is a diameter of a circle of radius 25 cm.
PQ is a chord at right angles to SR. X is the midpoint of PQ. The length of XR is 1 cm. Calculate the length of the arc PQ.

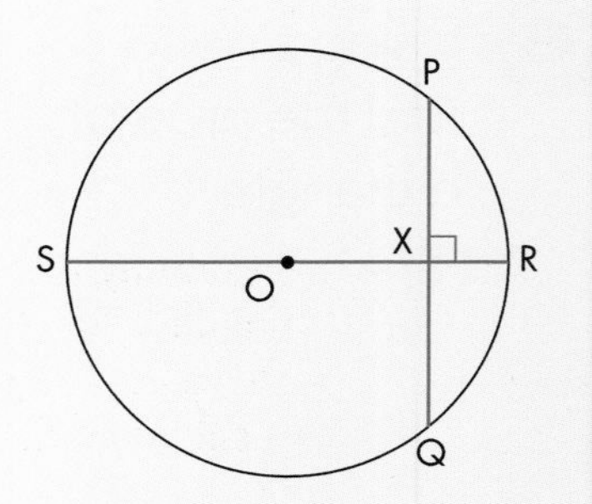

To find the length of the arc PQ, you need first to find the angle it subtends at the centre of the circle.

So join P to the centre of the circle O to obtain the angle POX, which is equal to half the angle subtended by PQ at O.

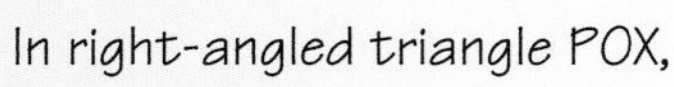

In right-angled triangle POX,

OX = OR – XR

OX = 25 – 1 = 24 cm

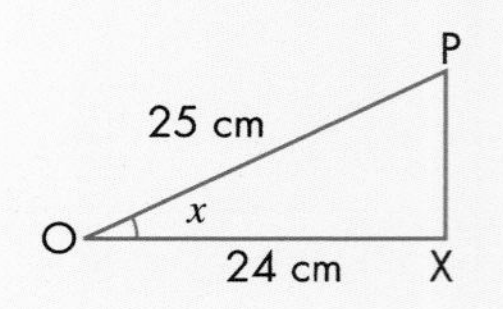

Therefore,

$$\cos x = \frac{24}{25}$$

$$\Rightarrow x = \cos^{-1} 0.96 = 16.26°$$

So, the angle subtended at the centre by the arc PQ is $2 \times 16.26° = 32.52°$, giving the length of the arc PQ as:

$$\frac{32.52}{360} \times 2 \times \pi \times 25 = 14.2 \text{ cm} \qquad \text{(3 significant figures)}$$

EXERCISE 10A

1 AC and BC are tangents to a circle of radius 7 cm. Calculate the length of AB.

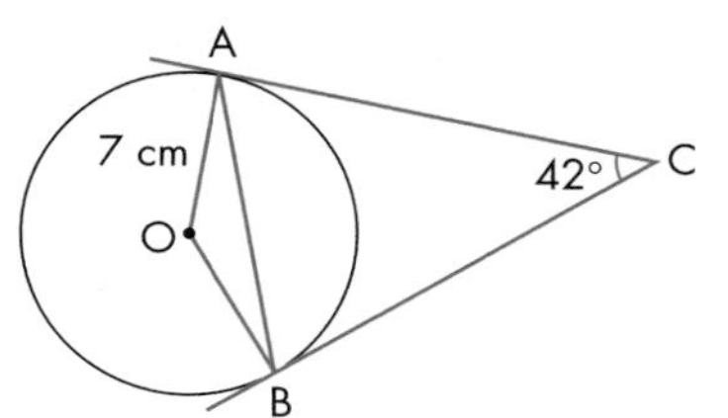

2 CD, length 20 cm, is a diameter of a circle. AB, length 12 cm, is a chord at right angles to DC. Calculate the angle AOB.

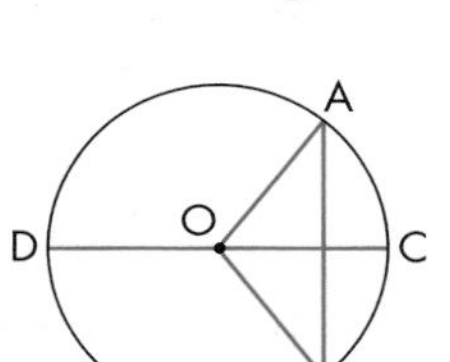

3 Calculate the length of AB in the diagram.

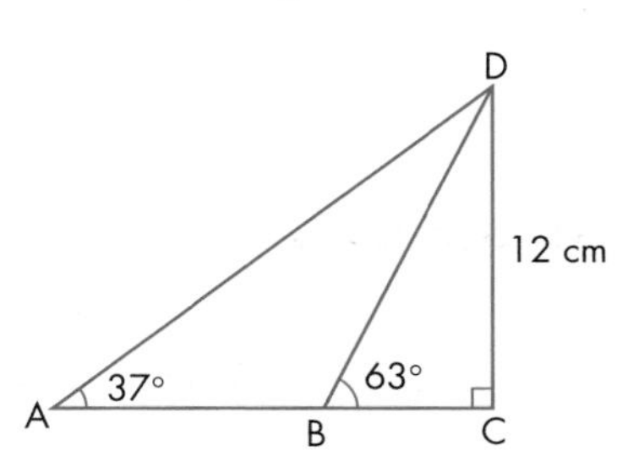

FM 4 A building has a ledge halfway up, as shown in the diagram. Alf measures the length AB as 100 m, the angle CAB as 31° and the angle EAB as 42°. Use this information to calculate the width of the ledge CD.

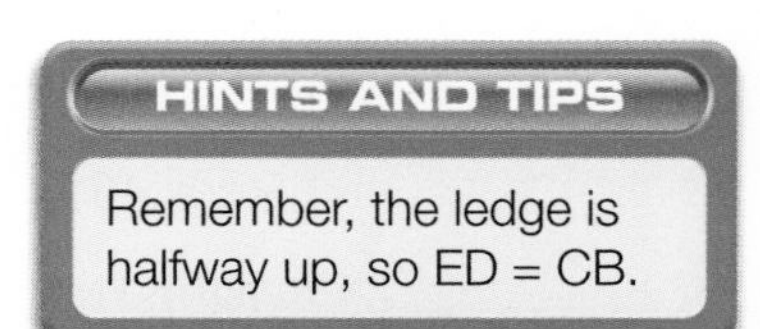

E
C
D
42°
A
31°
B
100 m

A*

5 AB and CD are two equal, perpendicular chords of a circle that intersect at X. The circle is of radius 6 cm and the angle COA is 113°. Calculate:

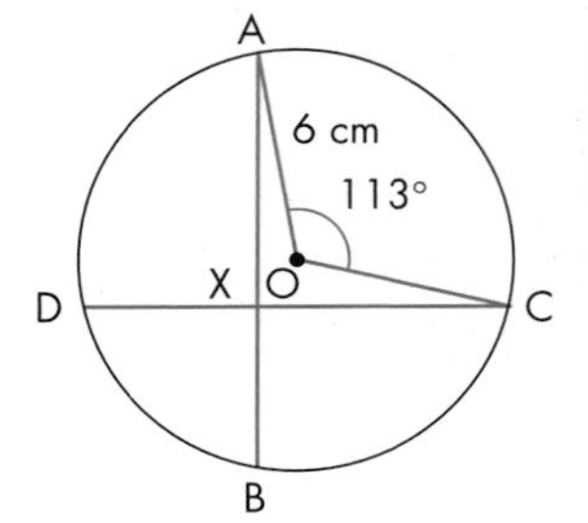

HINTS AND TIPS

AX = XC

a the length AC

b the angle XAO

c the length XB.

6 A vertical flagpole PQ is held by a wooden framework, as shown in the diagram. The framework is in the same vertical plane. Angle SRP = 25°, SQ = 6 m and PR = 4 m. Calculate the size of the angle QRP.

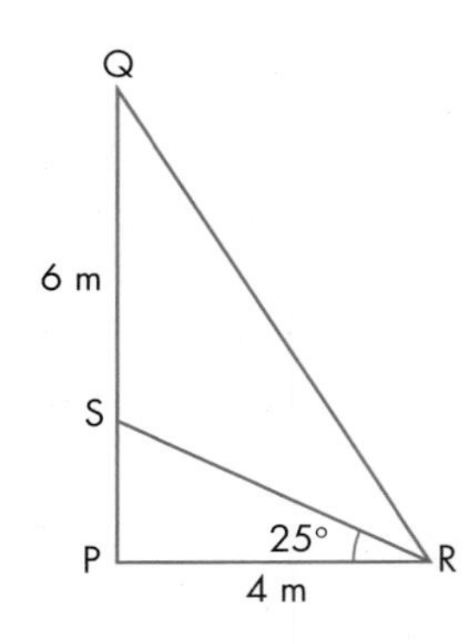

FM 7 A mine descends from ground level for 500 m at an angle of 13° to the horizontal and then continues for another 300 m at an angle of 17° to the horizontal, as shown in the diagram. A mining company decides to drill a vertical shaft to join up with the bottom of the mine as shown. How far along the surface from the opening, marked x on the diagram, do they need to drill down?

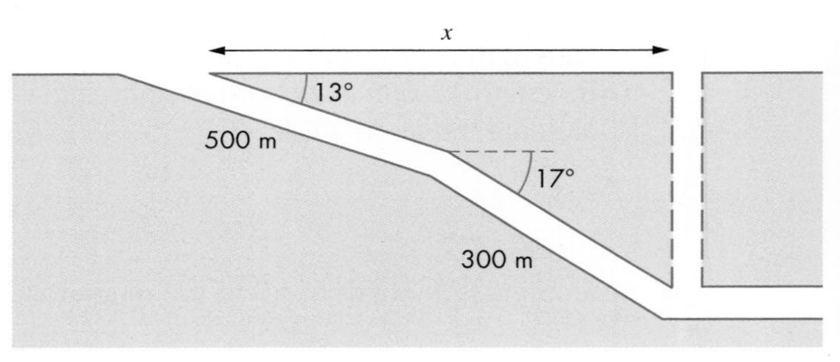

PS 8 **a** Use Pythagoras' theorem to work out the length of AC.

Leave your answer in surd form.

b Write down the values of:

i cos 45° **ii** sin 45° **iii** tan 45°

leaving your answers in surd form.

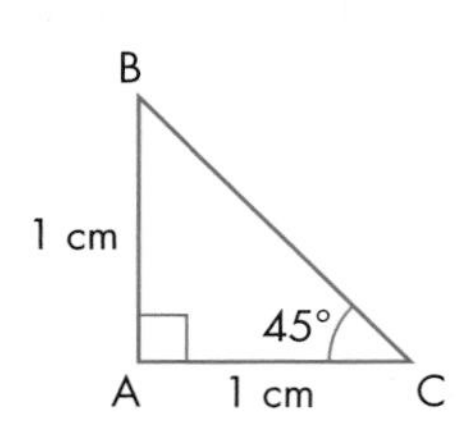

AU 9 In the diagram, AD = 5 cm, AC = 8 cm and AB = 12 cm.

Calculate angle CAB.

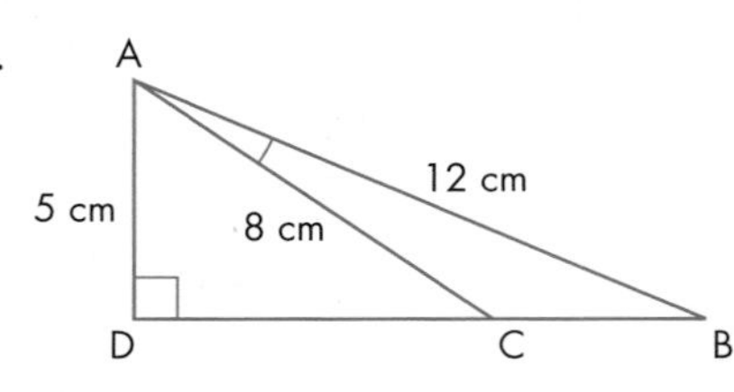

10.2 Some 3D problems

This section will show you how to:

- use trigonometric ratios and Pythagoras' theorem to solve more complex three-dimensional problems

Key words
cosine
Pythagoras' theorem
sine
tangent

Solving a problem set in three dimensions nearly always involves identifying a right-angled triangle that contains the length or angle required. This triangle will have to contain (apart from the right angle) two known measures from which the required calculation can be made.

It is essential to extract the triangle you are going to use from its 3D situation and redraw it as a separate, plain, right-angled triangle. (It is rarely the case that the required triangle appears as a true right-angled triangle in its 3D representation. Even if it does, you should still redraw it as a separate figure.)

Annotate the redrawn triangle with the known quantities and the unknown quantity that is to be found. Then use the trigonometric ratios **sine** (sin), **cosine** (cos) and **tangent** (tan), **Pythagoras' theorem** and the circle theorems to solve the triangle.

EXAMPLE 3

A, B and C are three points at ground level. They are in the same horizontal plane. C is 50 km east of B. B is north of A. C is on a bearing of 050° from A.

An aircraft, flying in an easterly direction, passes over B and over C at the same height. When it passes over B, the angle of elevation from A is 12°. Find the angle of elevation of the aircraft from A when it is over C.

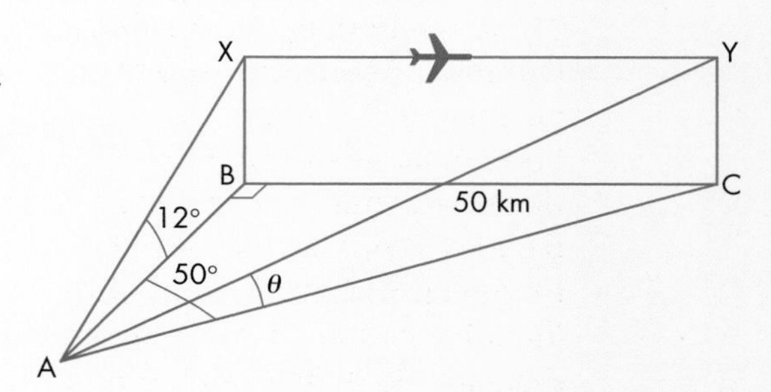

First, draw a diagram containing all the known information.

Next, use the right-angled triangle ABC to calculate AB and AC.

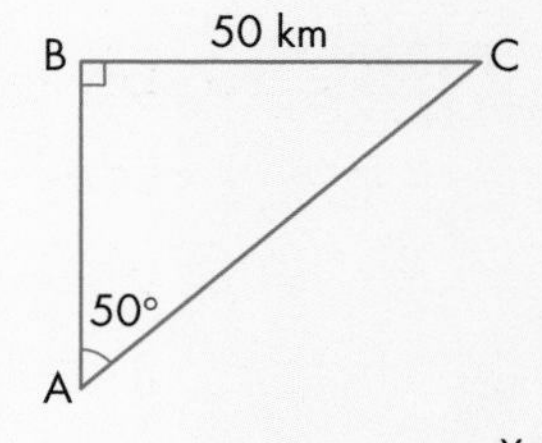

$$AB = \frac{50}{\tan 50°} = 41.95 \text{ km} \qquad \text{(4 significant figures)}$$

$$AC = \frac{50}{\sin 50°} = 65.27 \text{ km} \qquad \text{(4 significant figures)}$$

Then use the right-angled triangle ABX to calculate BX, and hence CY.

$BX = 41.95 \tan 12° = 8.917$ km (4 significant figures)

A 12° 41.95 km B X

Finally, use the right-angled triangle ACY to calculate the required angle of elevation, θ.

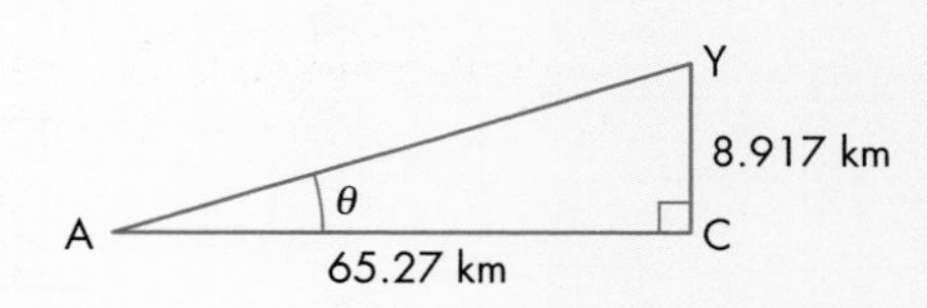

$$\tan \theta = \frac{8.917}{65.27} = 0.1366$$

$$\Rightarrow \theta = \tan^{-1} 0.1366 = 7.8° \qquad \text{(1 decimal place)}$$

Always write down intermediate working values to at least 4 significant figures, or use the answer on your calculator display to avoid inaccuracy in the final answer.

EXAMPLE 4

The diagram shows a cuboid 22.5 cm by 40 cm by 30 cm.
M is the midpoint of FG.

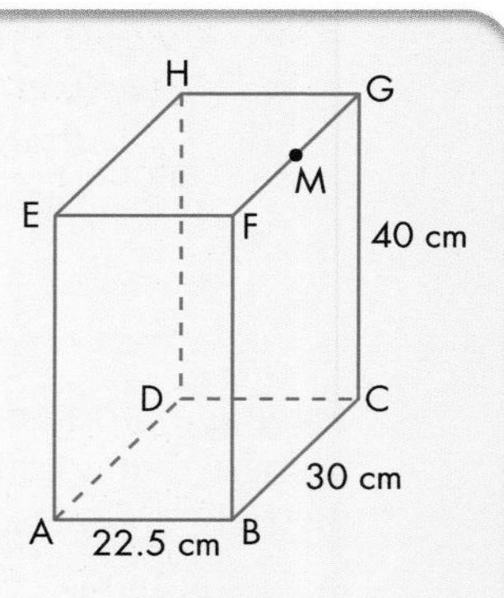

Calculate these angles.

a ABE

b ECA

c EMH

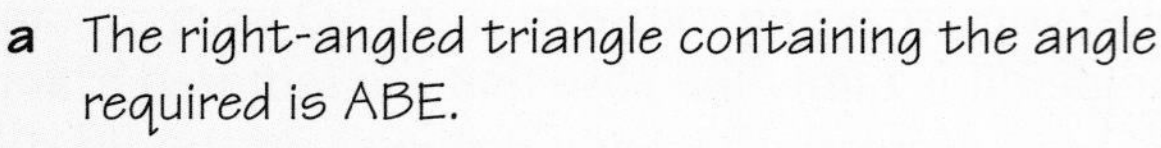

a The right-angled triangle containing the angle required is ABE.

Solving for α gives,

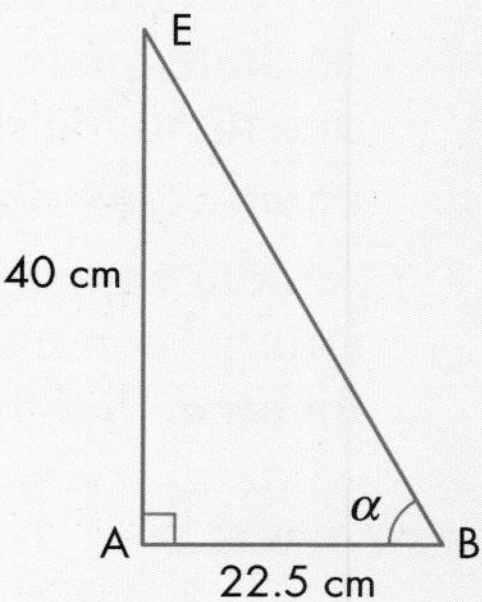

$$\tan \alpha = \frac{40}{22.5} = 1.7777$$

$$\Rightarrow \alpha = \tan^{-1} 1.7777 = 60.6° \quad \text{(3 significant figures)}$$

b The right-angled triangle containing the angle required is ACE, but for which only AE is known. Therefore, you need to find AC by applying Pythagoras' theorem to the right-angled triangle ABC.

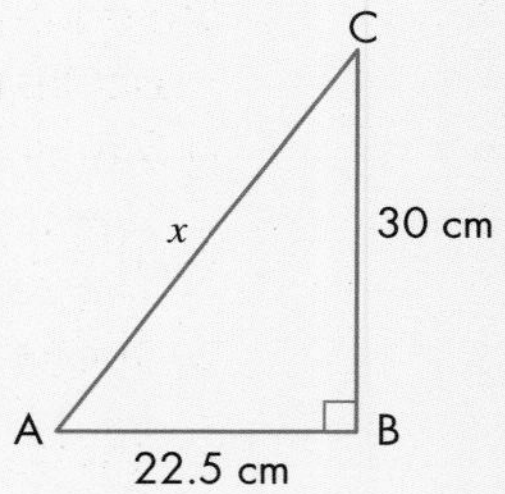

$$x^2 = (22.5)^2 + (30)^2 \text{ cm}^2$$

$$\Rightarrow x = 37.5 \text{ cm}$$

Returning to triangle ACE,

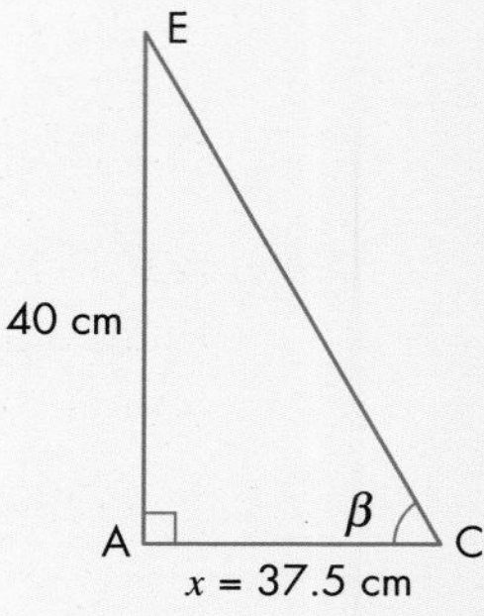

$$\tan \beta = \frac{40}{37.5} = 1.0666$$

$$\Rightarrow \beta = 46.8° \quad \text{(3 significant figures)}$$

c EMH is an isosceles triangle.

Drop the perpendicular from M to N, the midpoint of HE, to form two right-angled triangles. Angle HMN equals angle EMN, and HN = NE = 15 cm.

Taking triangle MEN,

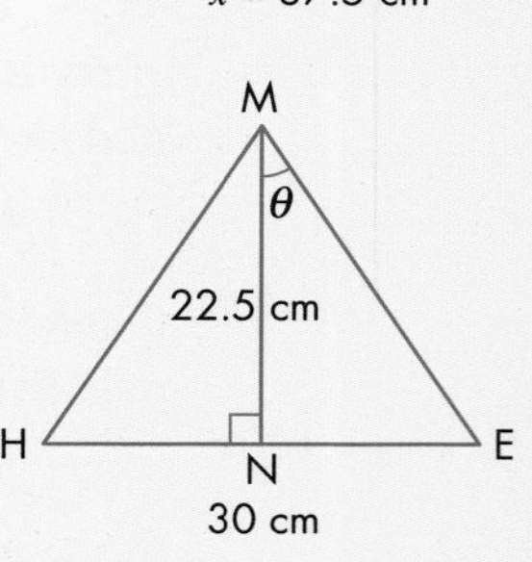

$$\tan \theta = \frac{15}{22.5} = 0.66666$$

$$\Rightarrow \theta = \tan^{-1} 0.66666 = 33.7°$$

Therefore, angle HME is 2 × 33.7° = 67.4° (3 significant figures)

EXERCISE 10B 

1 A vertical flagpole AP stands at the corner of a rectangular courtyard ABCD.

Calculate the angle of elevation of P from C.

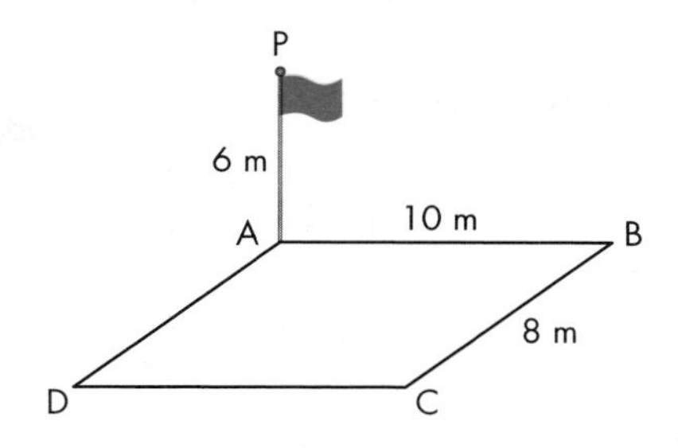

2 The diagram shows a pyramid. The base is a horizontal rectangle ABCD, 20 cm by 15 cm. The length of each sloping edge is 24 cm. The apex, V, is over the centre of the rectangular base. Calculate:

a the size of the angle VAC

b the height of the pyramid

c the volume of the pyramid

d the size of the angle between the face VAD and the base ABCD.

FM 3 The diagram shows the roof of a building. The base ABCD is a horizontal rectangle 7 m by 4 m. The triangular ends are equilateral triangles. Each side of the roof is an isosceles trapezium. The length of the top of the roof, EF, is 5 m. Calculate:

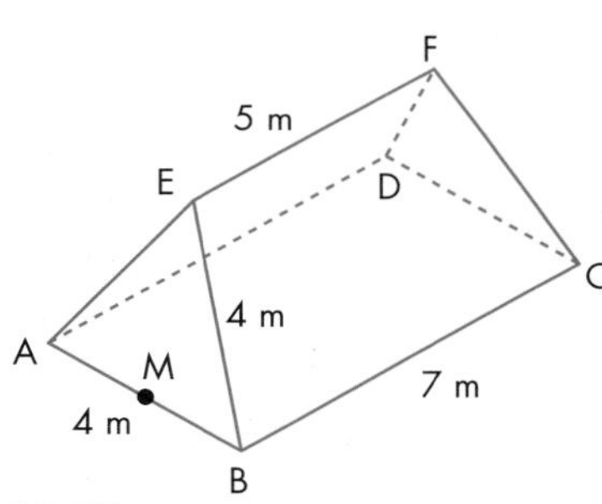

a the length EM, where M is the midpoint of AB

b the size of angle EBC

c the size of the angle between the face EAB and the base ABCD

d the surface area of the roof (excluding the base).

e Tiles cost £25 per square metre. How much would it cost to tile the roof?

4 ABCD is a vertical rectangular plane. EDC is a horizontal triangular plane. Angle CDE = 90°, AB = 10 cm, BC = 4 cm and ED = 9 cm. Calculate:

a angle AED

b angle DEC

c EC

d angle BEC.

5 The diagram shows a tetrahedron, each face of which is an equilateral triangle of side 6 m. The lines AN and BM meet the sides CB and AC at a right angle. The lines AN and BM intersect at X, which is directly below the vertex, D. Calculate:

a the distance AX

b the angle between the side DBC and the base ABC.

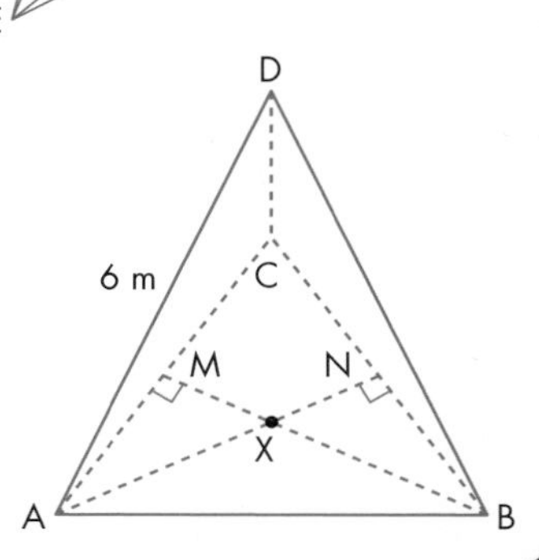

A*

PS 6 The lengths of the sides of a cuboid are a, b and c.

Show that the length of the diagonal XY is:

$\sqrt{a^2 + b^2 + c^2}$

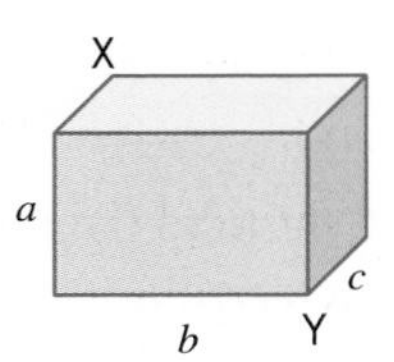

AU 7 In the diagram, XABCD is a right pyramid with a rectangular base.

Ellie says that the angle between the edge XD and the base ABCD is 56.3°.

Work out the correct answer to show that Ellie is wrong.

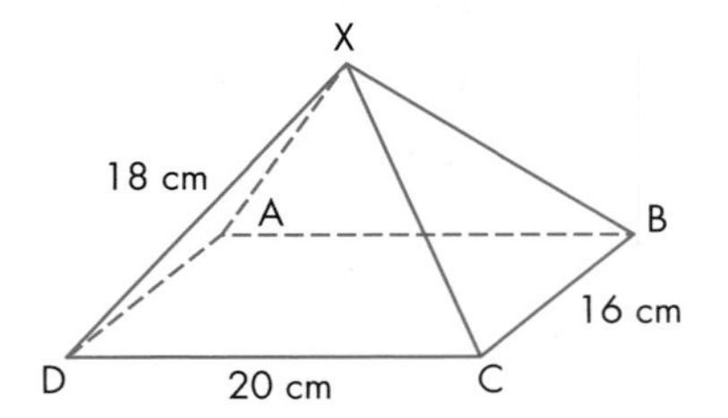

10.3 Trigonometric ratios of angles between 90° and 360°

This section will show you how to:

- find the sine, cosine and tangent of any angle from 0° to 360°

Key words
circular function
cosine
sine
tangent

ACTIVITY

a Copy and complete this table, using your calculator and rounding to three decimal places.

x	sin x	x	sin x	x	sin x	x	sin x
0°		180°		180°		360°	
15°		165°		195°		335°	
30°		150°		210°		320°	
45°		135°		225°		315°	
60°		120°		240°		300°	
75°		105°		255°		285°	
90°		90°		270°		270°	

b Comment on what you notice about the **sine** of each acute angle, and the sines of its corresponding non-acute angles.

c Draw a graph of sin x against x. Take x from 0° to 360° and sin x from –1 to 1.

d Comment on any symmetries your graph has.

You should have discovered these three facts.

- When $90° < x < 180°$, $\sin x = \sin (180° - x)$
  For example, $\sin 153° = \sin (180° - 153°) = \sin 27° = 0.454$
- When $180° < x < 270°$, $\sin x = -\sin (x - 180°)$
  For example, $\sin 214° = -\sin (214° - 180°) = -\sin 34° = -0.559$
- When $270° < x < 360°$, $\sin x = -\sin (360° - x)$
  For example, $\sin 287° = -\sin (360° - 287°) = -\sin 73° = -0.956$

Note:

- Each and every value of sine between –1 and 1 gives *two* angles between 0° and 360°.
- When the value of sine is positive, both angles are between 0° and 180°.
- When the value of sine is negative, both angles are between 180° and 360°.
- You can use the sine graph from 0° to 360° to check values approximately.

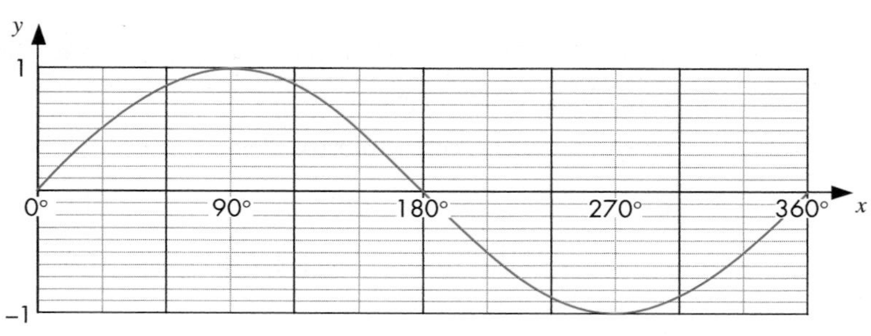

Sine x

EXAMPLE 5

Find the angles with a sine of 0.56.

You know that both angles are between 0° and 180°.

Using your calculator to find $\sin^{-1} 0.56$, you obtain 34.1°.

The other angle is, therefore,

$180° - 34.1° = 145.9°$

So, the angles are 34.1° and 145.9°.

EXAMPLE 6

Find the angles with a sine of –0.197.

You know that both angles are between 180° and 360°.

Using your calculator to find $\sin^{-1} 0.197$, you obtain 11.4°.

So the angles are

$180° + 11.4°$ and $360° - 11.4°$

which give 191.4° and 348.6°.

You can always use your calculator to check your answer to this type of problem by first keying in the angle and the appropriate trigonometric function (which would be sine in the above examples).

EXERCISE 10C

A*

1 State the two angles between 0° and 360° for each of these sine values.

a 0.6 **b** 0.8 **c** 0.75 **d** –0.7

e –0.25 **f** –0.32 **g** –0.175 **h** –0.814

i 0.471 **j** –0.097 **k** 0.553 **l** –0.5

AU 2 Which of these values is the odd one out and why?

sin 36° sin 144° sin 234° sin 324°

PS 3 The graph of sine x is cyclic, which means that it repeats forever in each direction.

a Write down one value of x greater than 360° for which the sine value is 0.978 147 600 73.

b Write down one value of x less than 0° for which the sine value is 0.978 147 600 73.

c Describe any symmetries of the graph of $y = \sin x$.

ACTIVITY

a Copy and complete this table, using your calculator and rounding to 3 decimal places.

x	$\cos x$	x	$\cos x$	x	$\cos x$	x	$\cos x$
0°		180°		180°		360°	
15°		165°		195°		335°	
30°		150°		210°		320°	
45°		135°		225°		315°	
60°		120°		240°		300°	
75°		105°		255°		285°	
90°		90°		270°		270°	

b Comment on what you notice about the cosines of the angles.

c Draw a graph of $\cos x$ against x. Take x from 0° to 360° and $\cos x$ from –1 to 1.

d Comment on the symmetry of the graph.

You should have discovered these three facts.

- When $90° < x < 180°$, $\cos x = -\cos(180 - x)°$
  For example, $\cos 161° = -\cos(180° - 161°) = -\cos 19° = -0.946$ (3 significant figures)
- When $180° < x < 270°$, $\cos x = -\cos(x - 180°)$
  For example, $\cos 245° = -\cos(245° - 180°) = -\cos 65° = -0.423$ (3 significant figures)
- When $270° < x < 360°$, $\cos x = \cos(360° - x)$
  For example, $\cos 310° = \cos(360° - 310°) = \cos 50° = 0.643$ (3 significant figures)

Note:

- Each and every value of cosine between –1 and 1 gives *two* angles between 0° and 360°.
- When the value of cosine is positive, one angle is between 0° and 90°, and the other is between 270° and 360°.
- When the value of cosine is negative, both angles are between 90° and 270°.
- You can use the cosine graph from 0° to 360° to check values approximately.

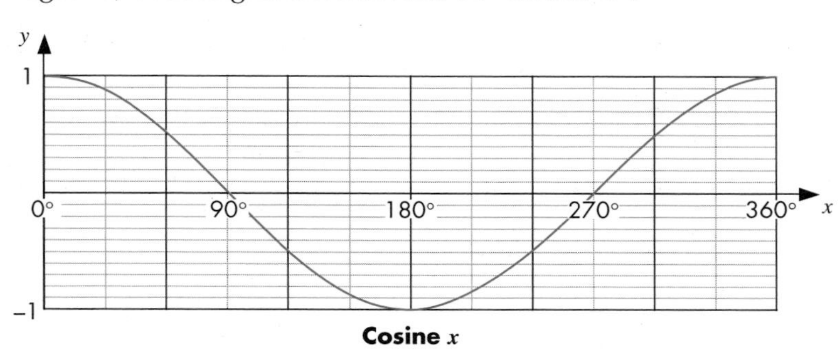

EXAMPLE 7

Find the angles with a cosine of 0.75.

One angle is between 0° and 90°, and the other is between 270° and 360°.

Using your calculator to find $\cos^{-1} 0.75$, you obtain 41.4°.

The other angle is, therefore,

$360° - 41.4° = 318.6°$

So, the angles are 41.4° and 318.6°.

EXAMPLE 8

Find the angles with a cosine of –0.285.

You know that both angles are between 90° and 270°.

Using your calculator to find $\cos^{-1} 0.285$, you obtain 73.4°.

The two angles are, therefore,

$180° - 73.4°$ and $180° + 73.4°$

which give 106.6° and 253.4°.

Here again, you can use your calculator to check your answer, by keying in cosine.

EXERCISE 10D

A*

1 State the two angles between 0° and 360° for each of these cosine values.

a 0.6 **b** 0.58 **c** 0.458 **d** 0.575

e 0.185 **f** –0.8 **g** –0.25 **h** –0.175

i –0.361 **j** –0.974 **k** 0.196 **l** 0.714

AU **2** Which of these values is the odd one out and why?

cos 58° cos 118° cos 238° cos 262°

PS **3** The graph of cosine x is cyclic, which means that it repeats forever in each direction.

a Write down one value of x greater than 360° for which the cosine value is –0.669 130 606 36.

b Write down one value of x less than 0° for which the cosine value is –0.669 130 606 36.

c Describe any symmetries of the graph of $y = \cos x$.

EXERCISE 10E

A*

1 Write down the sine of each of these angles.

a 135° **b** 269° **c** 305° **d** 133°

2 Write down the cosine of each of these angles.

a 129° **b** 209° **c** 95° **d** 357°

3 Write down the two possible values of x ($0° < x < 360°$) for each equation. Give your answers to 1 decimal place.

a $\sin x = 0.361$ **b** $\sin x = -0.486$ **c** $\cos x = 0.641$

d $\cos x = -0.866$ **e** $\sin x = 0.874$ **f** $\cos x = 0.874$

4 Find two angles such that the sine of each is 0.5.

5 cos 41° = 0.755. What is cos 139°?

6 Write down the value of each of the following, correct to 3 significant figures.

a sin 50° + cos 50° **b** cos 120° – sin 120° **c** sin 136° + cos 223°

d sin 175° + cos 257° **e** sin 114° – sin 210° **f** cos 123° + sin 177°

AU **7** It is suggested that $(\sin x)^2 + (\cos x)^2 = 1$ is true for all values of x. Test out this suggestion to see if you agree.

PS 8 Suppose the sine key on your calculator is broken, but not the cosine key. Show how you could calculate these.

a sin 25° **b** sin 130°

PS 9 Find a solution to each of these equations.

a $\sin(x + 20°) = 0.5$ **b** $\cos(5x) = 0.45$

PS 10 Use any suitable method to find the solution to the equation $\sin x = (\cos x)^2$.

ACTIVITY

a Try to find tan 90°. What do you notice?

Which is the closest angle to 90° for which you can find the **tangent** on your calculator?

What is the largest value for tangent that you can get on your calculator?

b Find values of tan x where $0° < x < 360°$. Draw a graph of your results.

State some rules for finding both angles between 0° and 360° that have any given tangent.

EXAMPLE 9

Find the angles between 0° and 360° with a tangent of 0.875.

One angle is between 0° and 90°, and the other is between 180° and 270°.

Using your calculator to find $\tan^{-1} 0.875$, you obtain 41.2°.

The other angle is, therefore,

$180° + 41.2° = 221.2°$

So, the angles are 41.2° and 221.2°.

EXAMPLE 10

Find the angles between 0° and 360° with a tangent of –1.5.

You know that one angle is between 90° and 180°, and that the other is between 270° and 360°.

Using your calculator to find $\tan^{-1} 1.5$, you obtain 56.3°.

The angles are, therefore,

$180° - 56.3°$ and $360° - 56.3°$

which give 123.7° and 303.7°.

EXERCISE 10F

A*

1 State the angles between 0° and 360° which have each of these tangent values.

a 0.258	**b** 0.785	**c** 1.19	**d** 1.875	**e** 2.55
f –0.358	**g** –0.634	**h** –0.987	**i** –1.67	**j** –3.68
k 1.397	**l** 0.907	**m** –0.355	**n** –1.153	**o** 4.15
p –2.05	**q** –0.098	**r** 0.998	**s** 1.208	**t** –2.5

AU 2 Which of these values is the odd one out and why?

tan 45° tan 135° tan 235° tan 315°

PS 3 The graph of tan x is cyclic, which means that it repeats forever in each direction.

a Write down one value of x greater than 360° for which the tangent value is 2.144 506 920 51.

b Write down one value of x less than 0° for which the tangent value is 2.144 506 920 51.

c Describe any symmetries of the graph of $y = \tan x$.

You may see the trigonometric functions sine, cosine and tangent referred to as **circular functions**. You could use the internet to find out more.

10.4 Solving any triangle

This section will show you how to:

- use the sine rule and the cosine rule to find sides and angles in any triangle

Key words

cosine rule
included angle
sine rule

We have already established that any triangle has six measurements: three sides and three angles. To solve a triangle (that is, to find any unknown angles or sides), we need to know at least three of the measurements. Any combination of three measurements – except that of all three angles – is enough to work out the rest. In a right-angled triangle, one of the known measurements is, of course, the right angle.

When we need to solve a triangle which contains no right angle, we can use one or the other of two rules, depending on what is known about the triangle. These are the **sine rule** and the **cosine rule**.

The sine rule

Take a triangle ABC and draw the perpendicular from A to the opposite side BC.

From right-angled triangle ADB, $h = c \sin B$

From right-angled triangle ADC, $h = b \sin C$

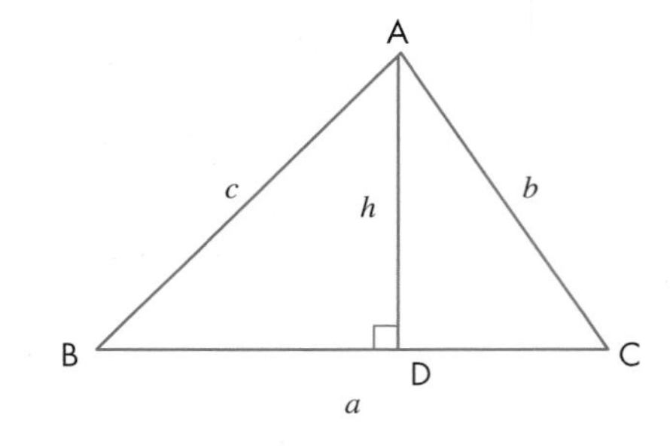

Therefore,

$$c \sin B = b \sin C$$

which can be rearranged to give:

$$\frac{c}{\sin C} = \frac{b}{\sin B}$$

By drawing a perpendicular from each of the other two vertices to the opposite side (or by algebraic symmetry), we see that

$$\frac{a}{\sin A} = \frac{c}{\sin C} \quad \text{and that} \quad \frac{a}{\sin A} = \frac{b}{\sin B}$$

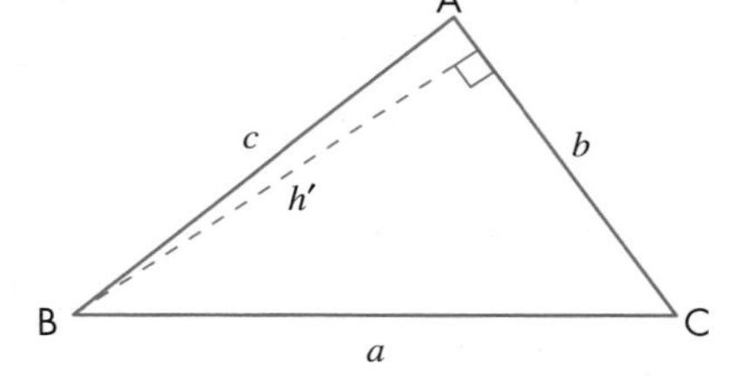

These are usually combined in the form

$$\frac{a}{\sin A} = \frac{b}{\sin B} = \frac{c}{\sin C}$$

which can be inverted to give:

$$\frac{\sin A}{a} = \frac{\sin B}{b} = \frac{\sin C}{c}$$

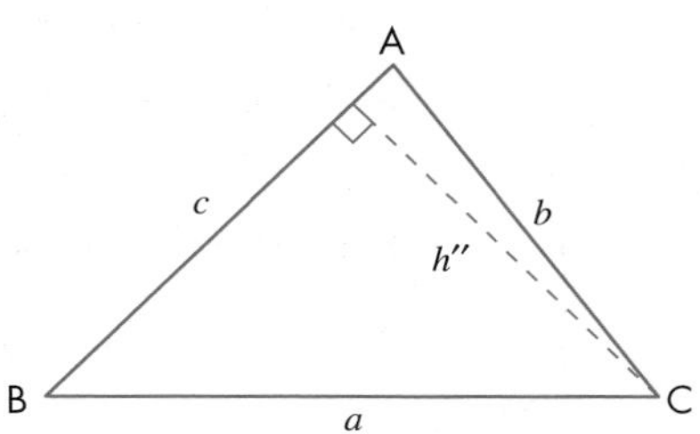

Usually, a triangle is not conveniently labelled, as in these diagrams. So, when using the sine rule, it is easier to remember to proceed as follows: take each side in turn, divide it by the sine of the angle opposite and then equate the resulting quotients.

Note:

- When you are calculating a *side*, use the rule with the *sides on top*.
- When you are calculating an *angle*, use the rule with the *sines on top*.

EXAMPLE 11

In triangle ABC, find the value of x.

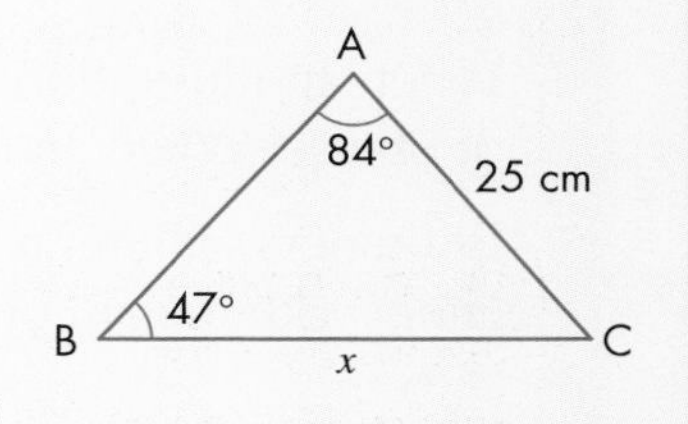

Use the sine rule with sides on top, which gives:

$$\frac{x}{\sin 84°} = \frac{25}{\sin 47°}$$

$$\Rightarrow x = \frac{25 \sin 84°}{\sin 47°} = 34.0 \text{ cm} \quad \text{(3 significant figures)}$$

EXAMPLE 12

In the triangle ABC, find the value of the acute angle x.

Use the sine rule with sines on top, which gives:

$$\frac{\sin x}{7} = \frac{7 \sin 40°}{6}$$

$$\Rightarrow \sin x = \frac{7 \sin 40°}{6} = 0.7499$$

$$\Rightarrow x = \sin^{-1} 0.7499 = 48.6° \quad \text{(3 significant figures)}$$

The ambiguous case

It is possible to find the sine of an angle that is greater than 90° (see section 10.3).

For example, $\sin 30° = \sin 150° = 0.5$. (Notice that the two angles add up to 180°.)

So $\sin 25° = \sin 155°$ and $\sin 100° = \sin 80°$

EXAMPLE 13

In triangle ABC, AB = 9 cm, AC = 7 cm and angle ABC = 40°. Find the angle ACB.

As you sketch triangle ABC, note that C can have two positions, giving two different configurations.

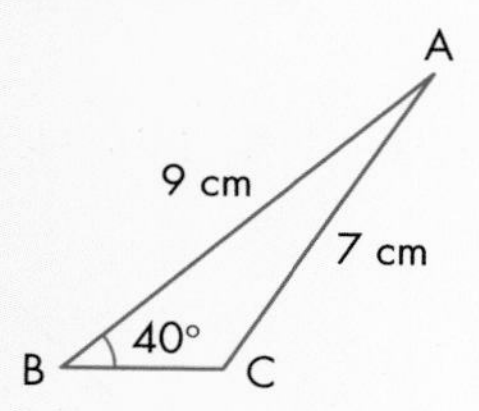

or

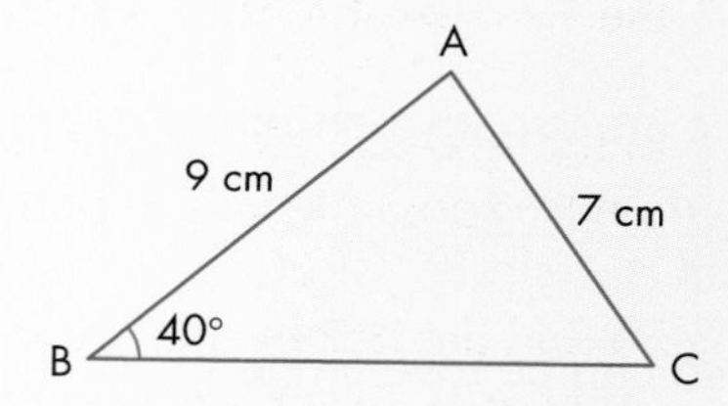

But you still proceed as in the normal sine rule situation, obtaining:

$$\frac{\sin C}{9} = \frac{\sin 40°}{7}$$

$$\Rightarrow \sin C = \frac{9 \sin 40°}{7}$$

$$= 0.8264$$

Keying inverse sine on the calculator gives C = 55.7°. But there is another angle with a sine of 0.8264, given by (180° – 55.7°) = 124.3°.

These two values for C give the two different situations shown above.

When an illustration of the triangle is given, it will be clear whether the required angle is acute or obtuse. When an illustration is not given, the more likely answer is an acute angle.

Examiners will not try to catch you out with the ambiguous case. They will indicate clearly, either with the aid of a diagram or by stating it, what is required.

EXERCISE 10G

1 Find the length x in each of these triangles.

a

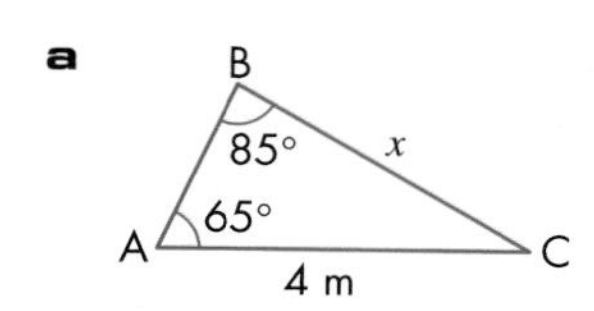

b

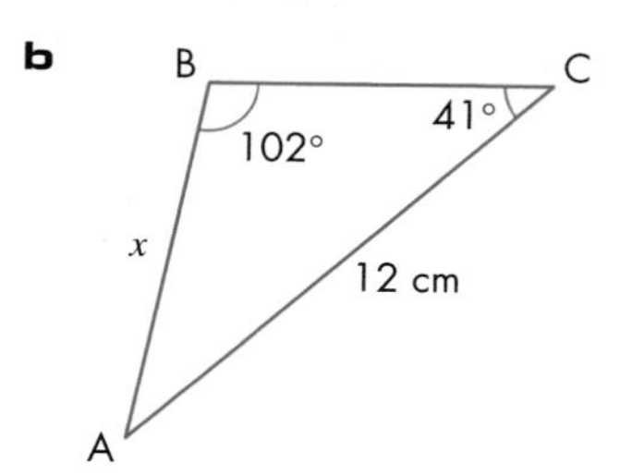

c

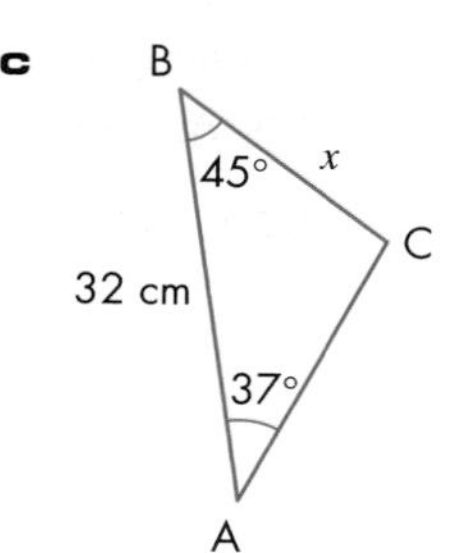

A

2 Find the angle x in each of these triangles.

a

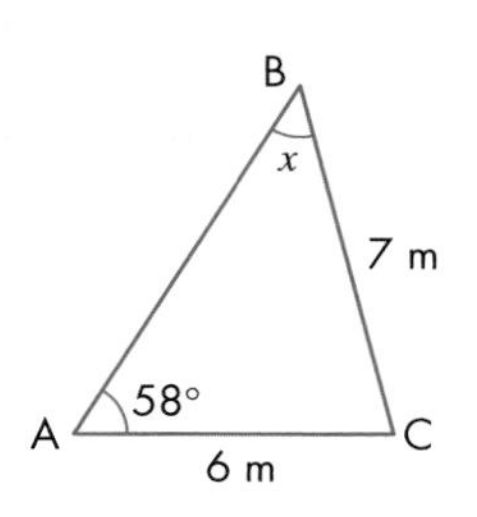

b

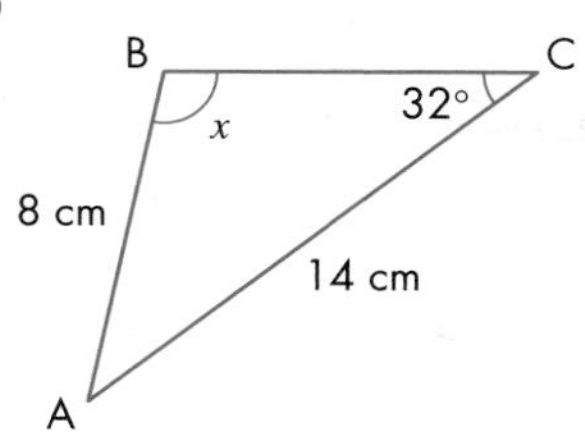

c

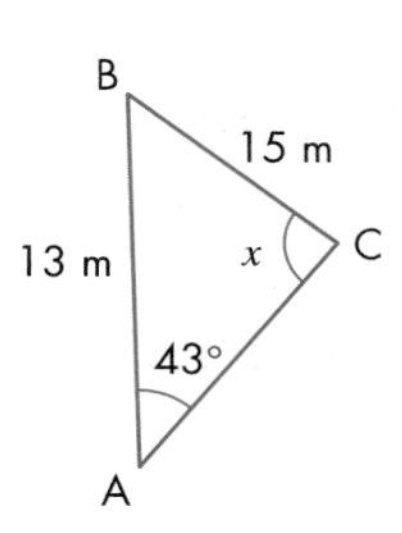

A*

3 In triangle ABC, the angle at A is 38°, the side AB is 10 cm and the side BC is 8 cm. Find the two possible values of the angle at C.

4 In triangle ABC, the angle at A is 42°, the side AB is 16 cm and the side BC is 14 cm. Find the two possible values of the side AC.

FM 5 To find the height of a tower standing on a small hill, Mary made some measurements (see diagram).

From a point B, the angle of elevation of C is 20°, the angle of elevation of A is 50°, and the distance BC is 25 m.

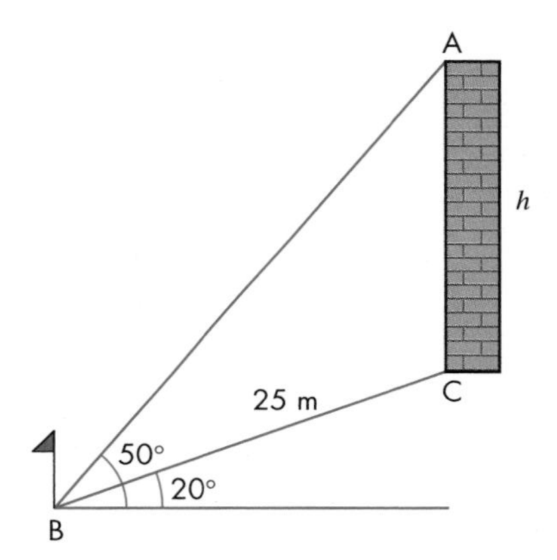

a Calculate these angles.

i ABC

ii BAC

b Using the sine rule and triangle ABC, calculate the height h of the tower.

PS 6 Use the information on this sketch to calculate the width, w, of the river.

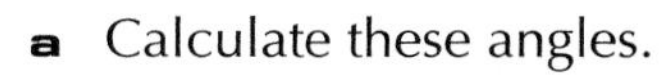

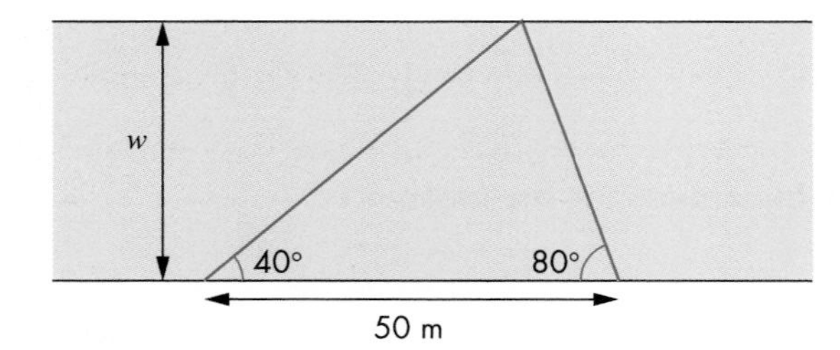

FM 7 An old building is unsafe and is protected by a fence. A demolition company is employed to demolish the building and has to work out the height BD, marked h on the diagram.

Calculate the value of h, using the given information.

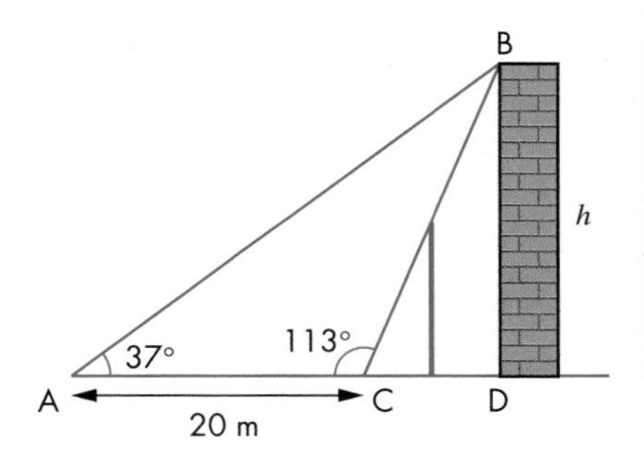

8 A weight is hung from a horizontal beam using two strings. The shorter string is 2.5 m long and makes an angle of 71° with the horizontal. The longer string makes an angle of 43° with the horizontal. What is the length of the longer string?

FM 9 An aircraft is flying over an army base. Suddenly, two searchlights, 3 km apart, are switched on. The two beams of light meet on the aircraft at an angle of 125° vertically above the line joining the searchlights. One of the beams of light makes an angle of 31° with the horizontal. Calculate the height of the aircraft.

FM 10 Two ships leave a port in directions that are 41° from each other. After half an hour, the ships are 11 km apart. If the speed of the slower ship is 7 km/h, what is the speed of the faster ship?

FM 11 A rescue helicopter is based at an airfield at A.

The helicopter is sent out to rescue a man who has had an accident on a mountain at M, due north of A.

The helicopter then flies on a bearing of 145° to a hospital at H as shown on the diagram.

Calculate the direct distance from the mountain to the hospital.

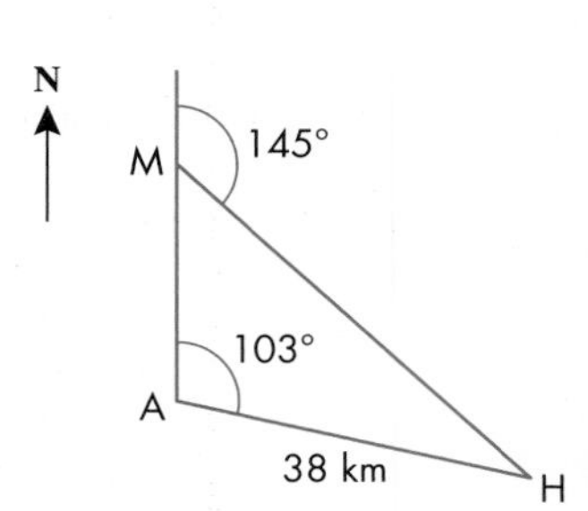

AU 12 Choose four values of θ, $0° < \theta < 90°$, to show that $\sin \theta = \sin (180° - \theta)$.

13 Triangle ABC has an obtuse angle at B.

Calculate the size of angle ABC.

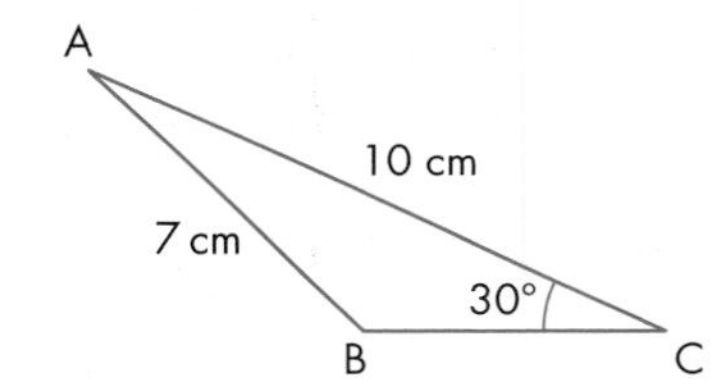

PS 14 For any triangle ABC, prove the sine rule:

$$\frac{a}{\sin A} = \frac{b}{\sin B} = \frac{c}{\sin C}$$

The cosine rule

Take the triangle, shown on the right, where D is the foot of the perpendicular to BC from A.

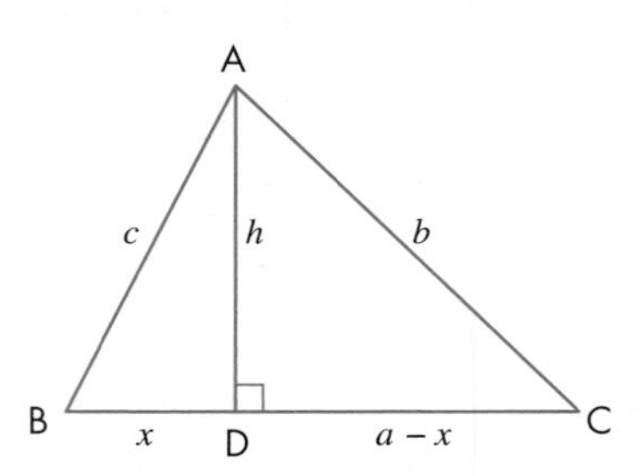

Using Pythagoras' theorem on triangle BDA:

$$h^2 = c^2 - x^2$$

Using Pythagoras' theorem on triangle ADC:

$$h^2 = b^2 - (a - x)^2$$

Therefore,

$$c^2 - x^2 = b^2 - (a - x)^2$$

$$c^2 - x^2 = b^2 - a^2 + 2ax - x^2$$

$$\Rightarrow c^2 = b^2 - a^2 + 2ax$$

From triangle BDA, $x = c \cos B$.

Hence,

$$c^2 = b^2 - a^2 + 2ac \cos B$$

Rearranging gives:

$$b^2 = a^2 + c^2 - 2ac \cos B$$

By algebraic symmetry:

$$a^2 = b^2 + c^2 - 2bc \cos A \quad \text{and} \quad c^2 = a^2 + b^2 - 2ab \cos C$$

This is the cosine rule, which can be best remembered by the diagram on the right, where:

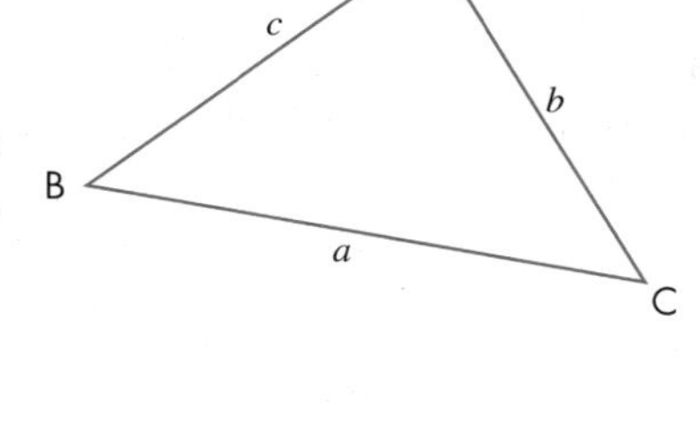

$$a^2 = b^2 + c^2 - 2bc \cos A$$

Note the symmetry of the rule and how the rule works using two adjacent sides and the angle between them (the **included angle**).

The formula can be rearranged to find any of the three angles.

$$\cos A = \frac{b^2 + c^2 - a^2}{2bc}$$

$$\cos B = \frac{a^2 + c^2 - b^2}{2ac}$$

$$\cos C = \frac{a^2 + b^2 - c^2}{2ab}$$

Note that the cosine rule $a^2 = b^2 + c^2 - 2bc \cos A$ is given in the formula sheets in the GCSE examination but the rearranged formula for the angle is not given. You are advised to learn this as trying to rearrange usually ends up with an incorrect formula.

EXAMPLE 14

Find x in this triangle.

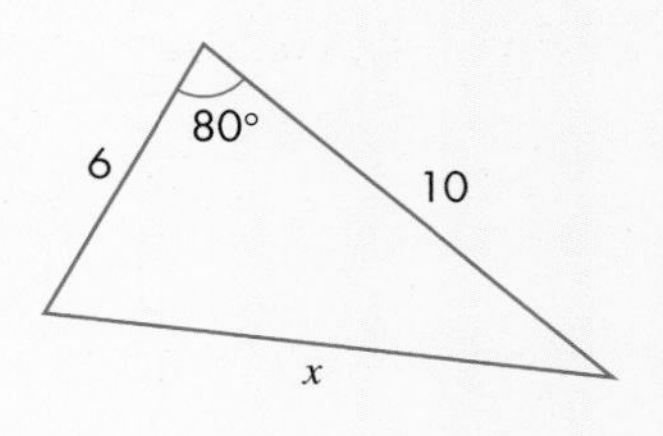

By the cosine rule:

$$x^2 = 6^2 + 10^2 - 2 \times 6 \times 10 \times \cos 80°$$

$$x^2 = 115.16$$

$$\Rightarrow x = 10.7 \quad \text{(3 significant figures)}$$

EXAMPLE 15

Find x in this triangle.

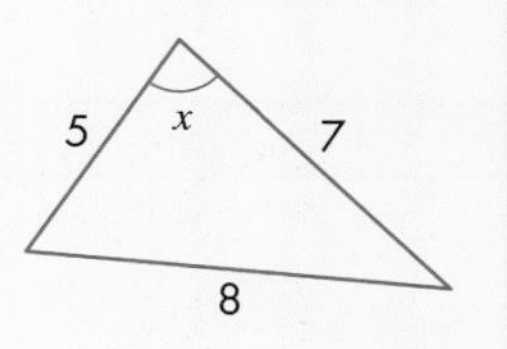

By the cosine rule:

$$\cos x = \frac{5^2 + 7^2 - 8^2}{2 \times 5 \times 7} = 0.1428$$

$$\Rightarrow x = 81.8° \quad \text{(3 significant figures)}$$

It is possible to find the cosine of an angle that is greater than 90° (see section 10.3). For example, $\cos 120° = -\cos 60° = -0.5$. (Notice the minus sign; the two angles add up to 180°.)

So $\cos 150° = -\cos 30° = -0.866$

EXAMPLE 16

A ship sails from a port on a bearing of 055° for 40 km. It then changes course to 123° for another 50 km. On what course should the ship be steered to get it straight back to the port?

Previously, you have solved this type of problem using right-angled triangles. This method could be applied here but it would involve at least six separate calculations.

With the aid of the cosine and sine rules, however, you can reduce the solution to two separate calculations, as follows.

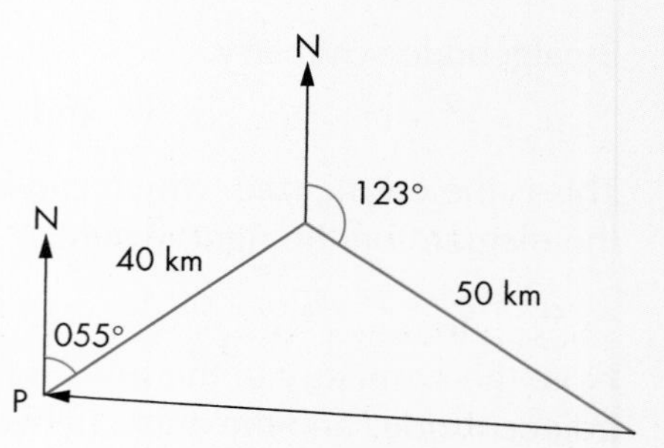

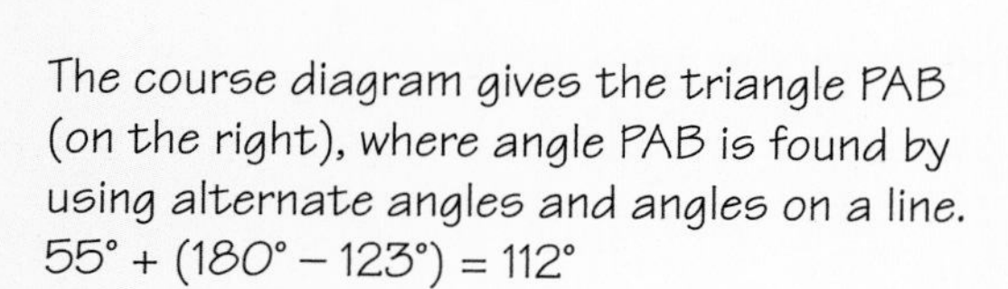

The course diagram gives the triangle PAB (on the right), where angle PAB is found by using alternate angles and angles on a line.
55° + (180° – 123°) = 112°

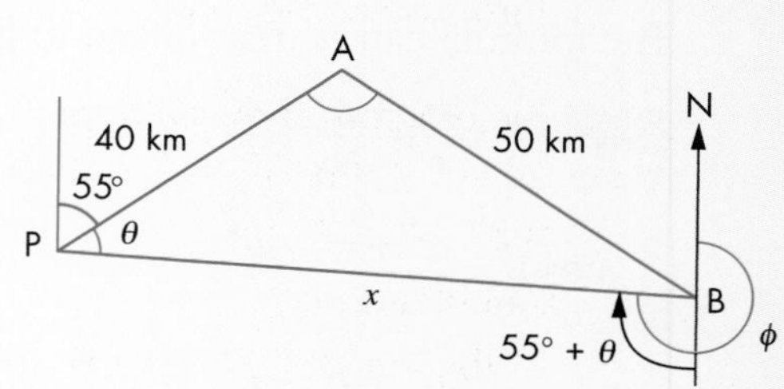

Let ϕ be the bearing to be steered, then

$$\phi = \theta + 55° + 180°$$

To find θ, you first have to obtain PB(= x), using the cosine rule.

$$x^2 = 40^2 + 50^2 - 2 \times 40 \times 50 \times \cos 112° \text{ km}^2$$

(Remember: the cosine of 112° is negative.)

$$\Rightarrow x^2 = 5598.43 \text{ km}^2$$

$$\Rightarrow x = 74.82 \text{ km}$$

You can now find θ from the sine rule.

$$\frac{\sin \theta}{50} = \frac{\sin 112°}{74.82}$$

$$\Rightarrow \sin \theta = \frac{50 \times \sin 112°}{74.82} = 0.6196$$

$$\Rightarrow \theta = 38.3°$$

So the ship should be steered on a bearing of:

38.3° + 55° + 180° = 273.3°

EXERCISE 10H

1 Find the length x in each of these triangles.

a

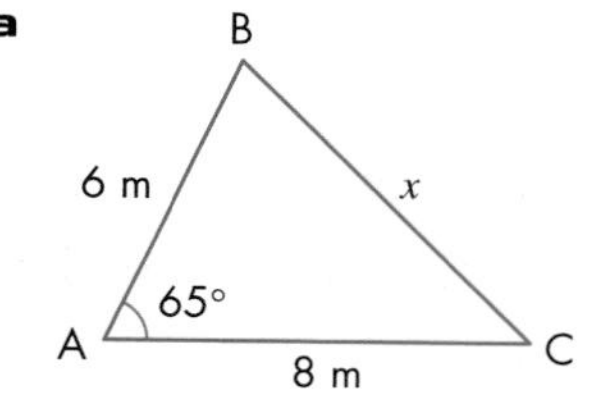

b

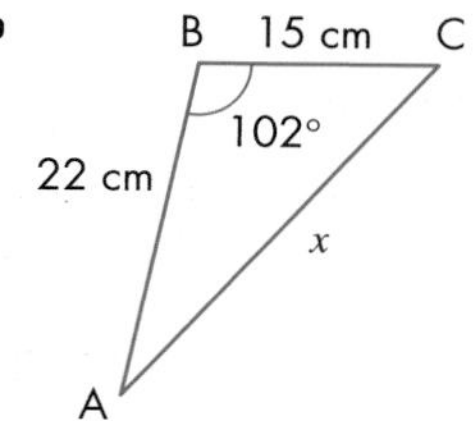

c

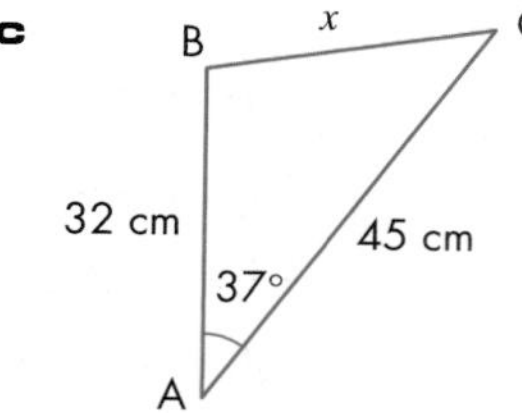

2 Find the angle x in each of these triangles.

a

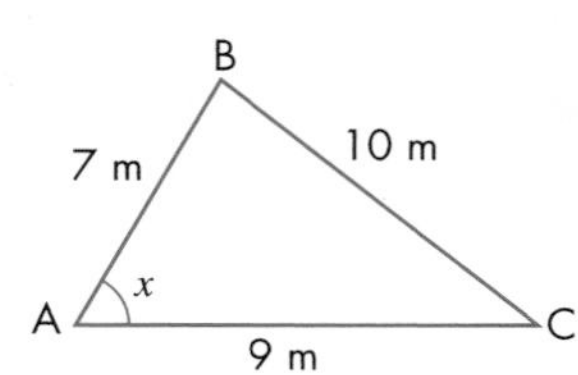

b

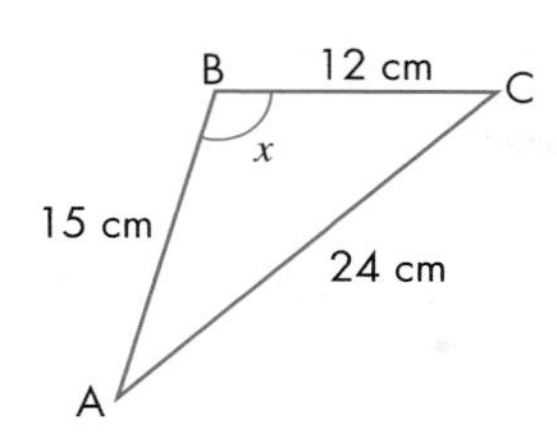

c

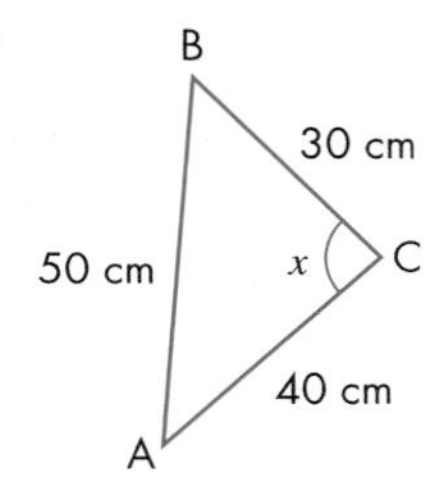

d Explain the significance of the answer to part **c**.

3 In triangle ABC, AB = 5 cm, BC = 6 cm and angle ABC = 55°. Find AC.

4 A triangle has two sides of length 40 cm and an angle of 110°. Work out the length of the third side of the triangle.

5 The diagram shows a trapezium ABCD.
AB = 6.7 cm, AD = 7.2 cm, CB = 9.3 cm and angle DAB = 100°.

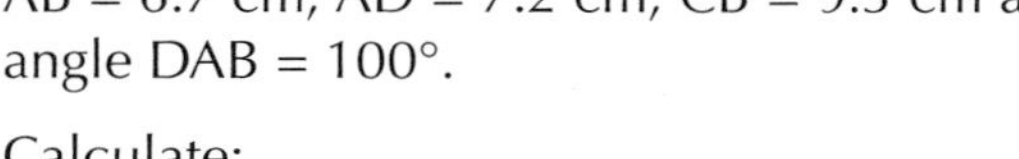

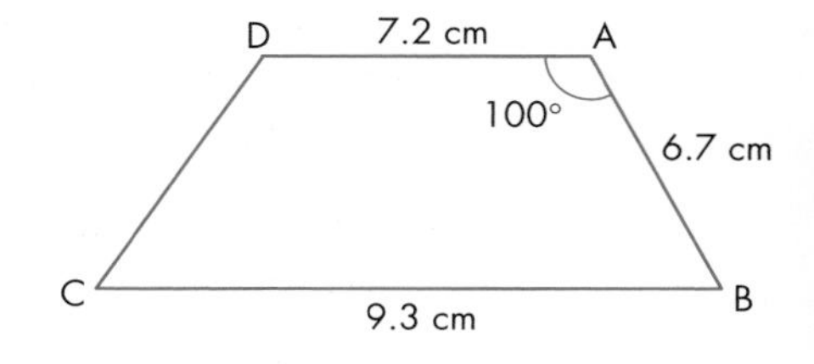

Calculate:

a the length DB **b** angle DBA

c angle DBC **d** the length DC

e the area of the trapezium.

6 A quadrilateral ABCD has AD = 6 cm, DC = 9 cm, AB = 10 cm and BC = 12 cm. Angle ADC = 120°. Calculate angle ABC.

7 A triangle has two sides of length 30 cm and an angle of 50°. Unfortunately, the position of the angle is not known. Sketch the two possible triangles and use them to work out the two possible lengths of the third side of the triangle.

FM 8 A ship sails from a port on a bearing of 050° for 50 km then turns on a bearing of 150° for 40 km. A crewman is taken ill, so the ship drops anchor. What course and distance should a rescue helicopter from the port fly to reach the ship in the shortest possible time?

HINTS AND TIPS

Bearings can be revised on page 133.

9 The three sides of a triangle are given as $3a$, $5a$ and $7a$. Calculate the smallest angle in the triangle.

10 ABCD is a trapezium where AB is parallel to CD. AB = 4 cm, BC = 5 cm, CD = 8 cm, DA = 6 cm. A line BX is parallel to AD and cuts DC at X. Calculate:

a angle BCD **b** the length BD.

PS 11 Two ships, X and Y, leave a port at 9 am.

Ship X travels at an average speed of 20 km/h on a bearing of 075° from the port.

Ship Y travels at an average speed of 25 km/h on a bearing of 130° from the port.

Calculate the distance between the two ships at 11 am.

A*

AU 12 Choose four values of θ, $0° < \theta < 90°$, to show that $\cos \theta = -\cos (180° - \theta)$.

AU 13 Calculate the size of the largest angle in the triangle ABC.

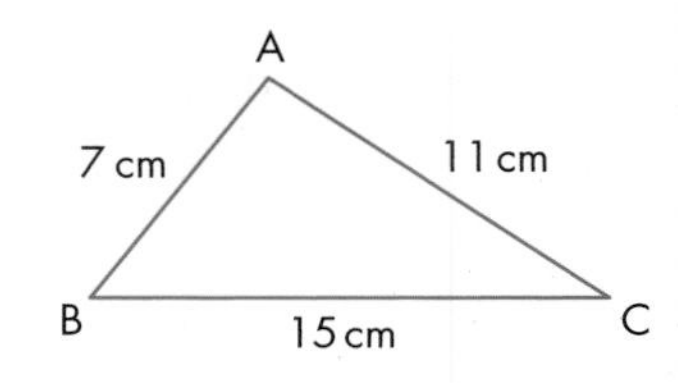

Choosing the correct rule

When solving triangles, there are only four situations that can occur, each of which can be solved completely in three stages.

Two sides and the included angle

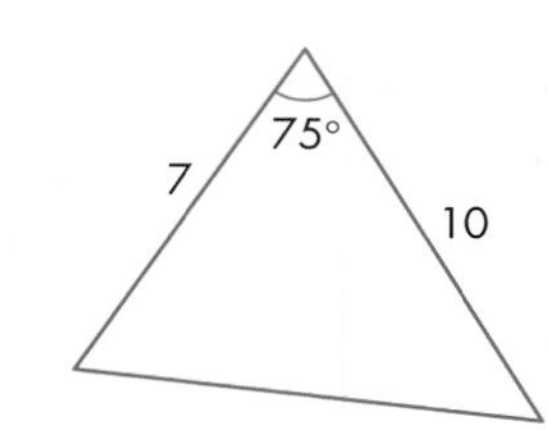

1 Use the cosine rule to find the third side.

2 Use the sine rule to find either of the other angles.

3 Use the sum of the angles in a triangle to find the third angle.

Two angles and a side

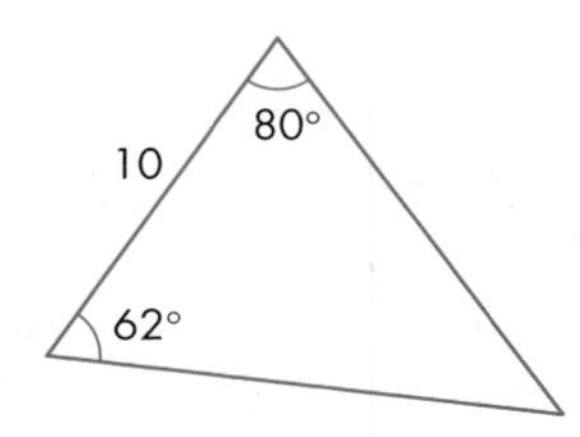

1 Use the sum of the angles in a triangle to find the third angle.

2, 3 Use the sine rule to find the other two sides.

Three sides

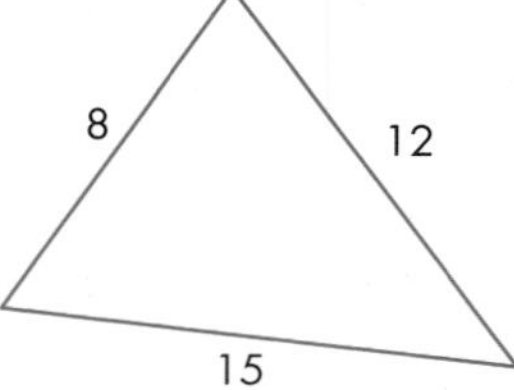

1 Use the cosine rule to find one angle.

2 Use the sine rule to find another angle.

3 Use the sum of the angles in a triangle to find the third angle.

Two sides and a non-included angle

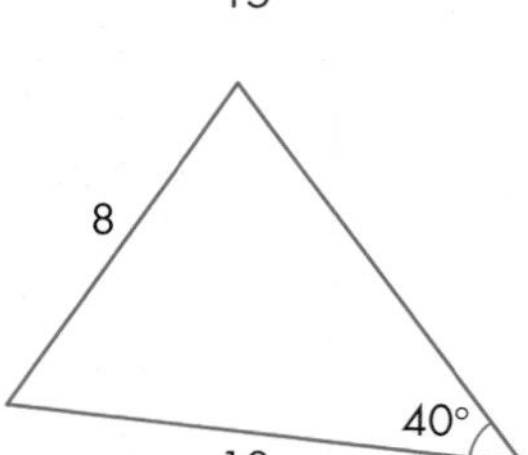

This is the ambiguous case already covered (page 272).

1 Use the sine rule to find the two possible values of the appropriate angle.

2 Use the sum of the angles in a triangle to find the two possible values of the third angle.

3 Use the sine rule to find the two possible values for the length of the third side.

Note: Apply the sine rule wherever you can – it is always easier to use than the cosine rule. You should never need to use the cosine rule more than once.

EXERCISE 10I

1 Find the length or angle x in each of these triangles.

a

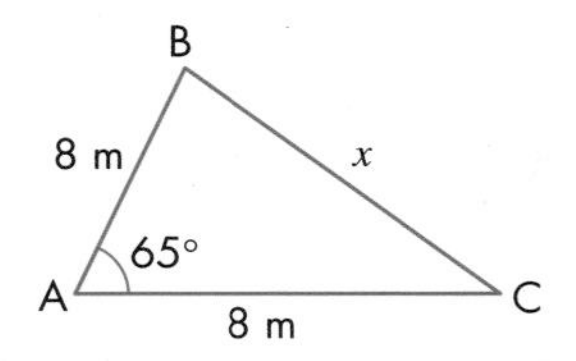

b

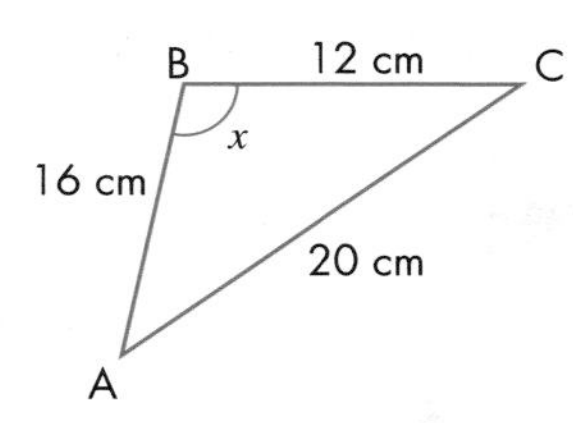

c

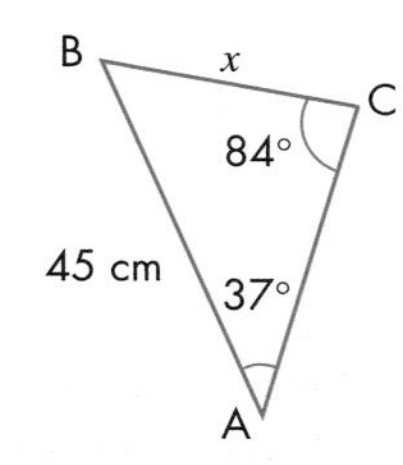

d

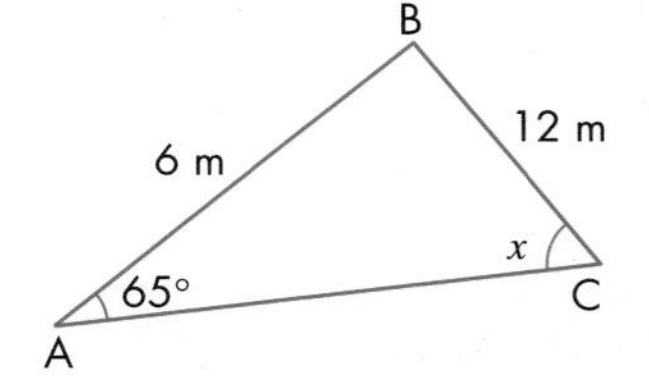

e

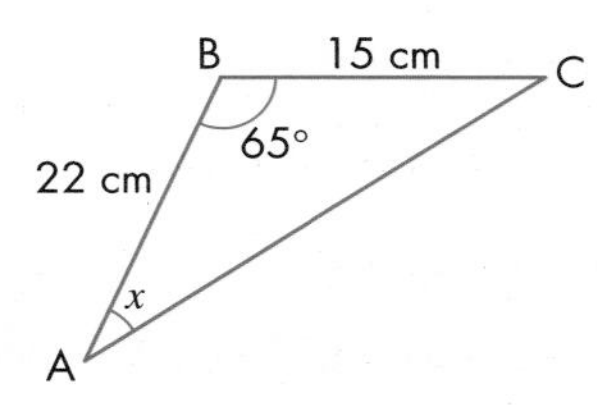

f

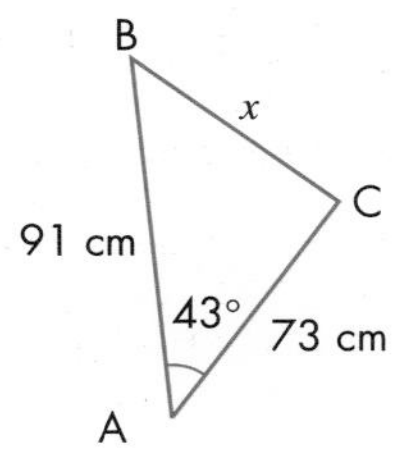

g

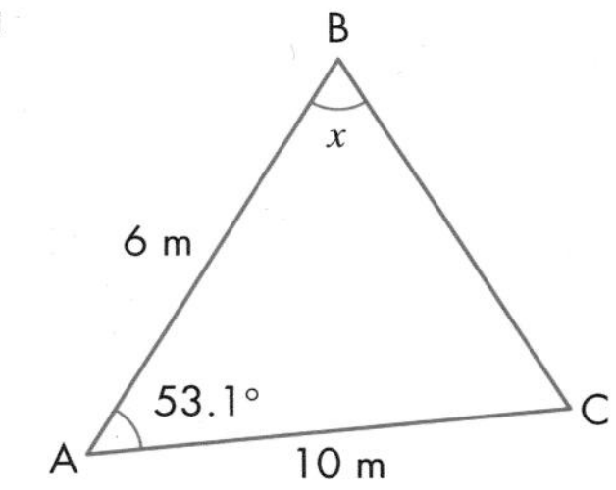

h

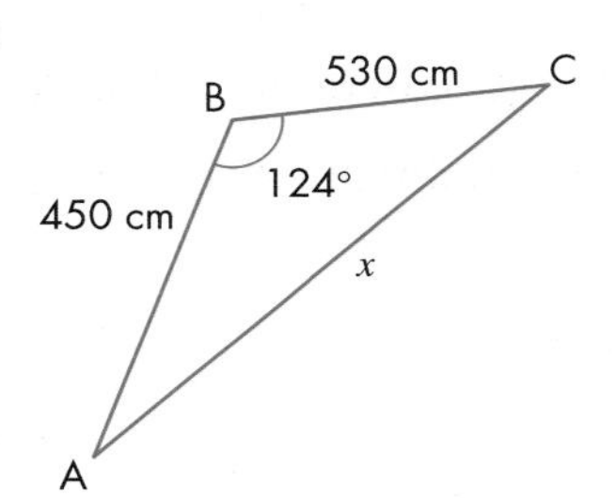

i

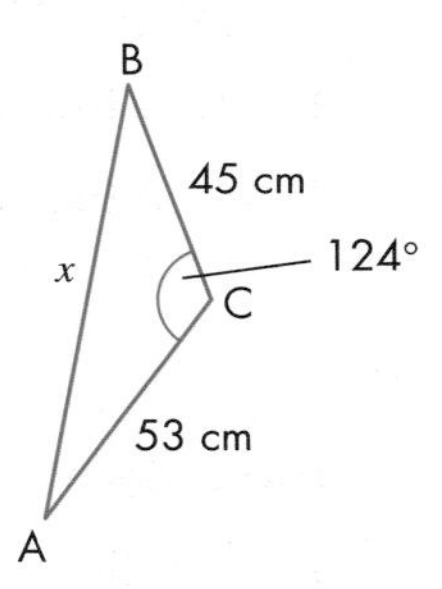

2 The hands of a clock have lengths 3 cm and 5 cm. Find the distance between the tips of the hands at 4 o'clock.

3 A spacecraft is seen hovering at a point which is in the same vertical plane as two towns, X and F, which are on the same level. Its distances from X and F are 8.5 km and 12 km respectively. The angle of elevation of the spacecraft when observed from F is 43°. Calculate the distance between the two towns.

FM 4 Two boats, Mary Jo and Suzie, leave port at the same time. Mary Jo sails at 10 knots on a bearing of 065°. Suzie sails on a bearing of 120° and after 1 hour Mary Jo is on a bearing of 330° from Suzie. What is Suzie's speed? (A knot is a nautical mile per hour.)

> **HINTS AND TIPS**
>
> Bearings can be revised on page 133.

5 Two ships leave port at the same time, Darling Dave sailing at 12 knots on a bearing of 055°, and Merry Mary at 18 knots on a bearing of 280°.

a How far apart are the two ships after 1 hour?

b What is the bearing of Merry Mary from Darling Dave?

A*

PS 6 Triangle ABC has sides with lengths a, b and c, as shown in the diagram.

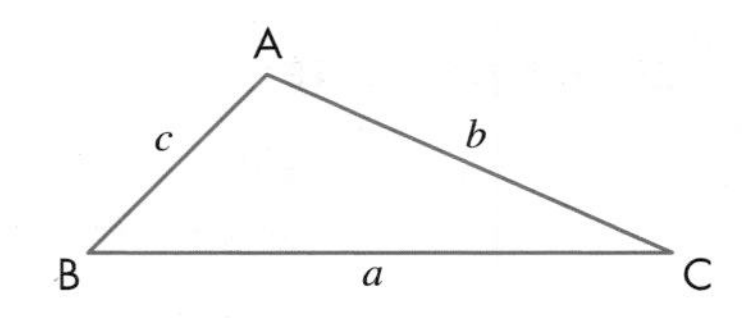

a What can you say about the angle BAC, if $b^2 + c^2 - a^2 = 0$?

b What can you say about the angle BAC, if $b^2 + c^2 - a^2 > 0$?

c What can you say about the angle BAC, if $b^2 + c^2 - a^2 < 0$?

AU 7 The diagram shows a sketch of a field ABCD.
A farmer wants to put a new fence round the perimeter of the field.

Calculate the perimeter of the field.

Give your answer to an appropriate degree of accuracy.

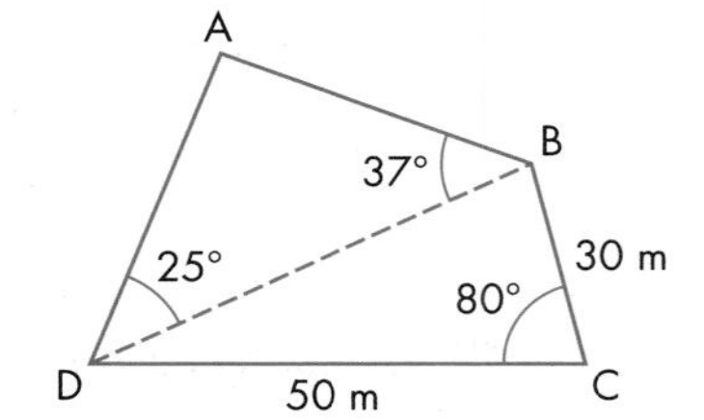

10.5 Trigonometric ratios in surd form

This section will show you how to:
- work out trigonometric ratios in surd form

Key words
cosine
Pythagoras' theorem
sine
surd
surd form
tangent

Solving triangles often involves finding square roots. Unless a number has an exact square root, the value on a calculator can only be an approximation. The exact value of the square root of 2, for example, is often written as $\sqrt{2}$. This expression, using the square root symbol, is a **surd**, and answers given in this way are in **surd form**.

EXAMPLE 17

Using an equilateral triangle with sides of 2 units, write down expressions for the sine, cosine and tangent of 60° and 30°. Give answers in surd form.

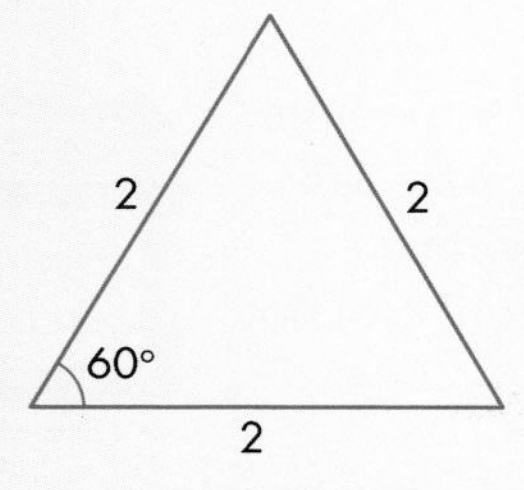

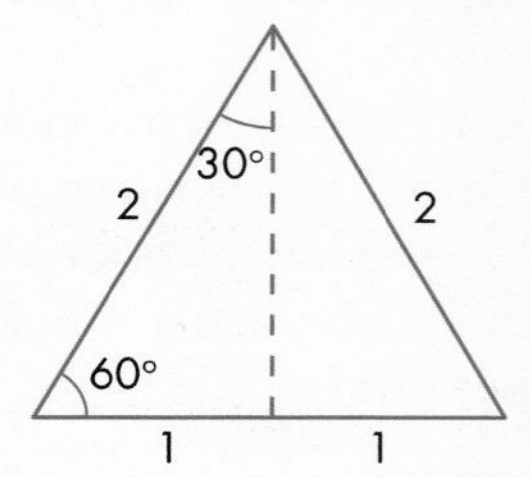

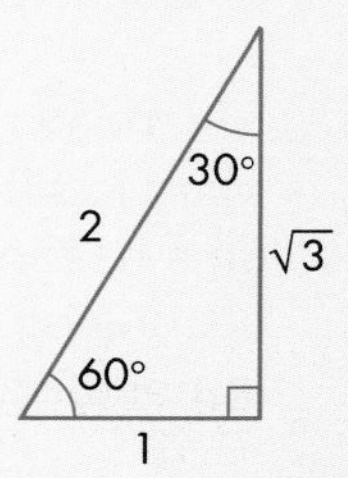

Divide the equilateral triangle into two equal right-angled triangles. Taking one of them, use **Pythagoras' theorem** and the definition of **sine, cosine** and **tangent** to obtain:

$\sin 60° = \frac{\sqrt{3}}{2}$ $\cos 60° = \frac{1}{2}$ $\tan 60° = \sqrt{3}$

and $\sin 30° = \frac{1}{2}$ $\cos 30° = \frac{\sqrt{3}}{2}$ $\tan 30° = \frac{1}{\sqrt{3}} = \frac{\sqrt{3}}{3}$

EXAMPLE 18

Using a right-angled isosceles triangle in which the equal sides are 1 unit, find the sine, cosine and tangent of 45°. Give answers in surd form.

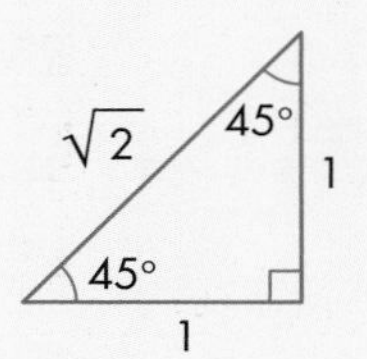

By Pythagoras' theorem, the hypotenuse of the triangle is $\sqrt{2}$ units.

From the definition of sine, cosine and tangent,

$$\sin 45° = \frac{1}{\sqrt{2}} = \frac{\sqrt{2}}{2} \qquad \cos 45° = \frac{1}{\sqrt{2}} = \frac{\sqrt{2}}{2} \qquad \tan 45° = 1$$

These results can be summarised in a table.

You do not need to learn these results, but it is useful to know how to work them out.

θ	$\cos\theta$	$\sin\theta$	$\tan\theta$
30°	$\frac{\sqrt{3}}{2}$	$\frac{1}{2}$	$\frac{\sqrt{3}}{3}$
45°	$\frac{\sqrt{2}}{2}$	$\frac{\sqrt{2}}{2}$	1
60°	$\frac{1}{2}$	$\frac{\sqrt{3}}{2}$	$\sqrt{3}$

When solving problems, you can write trigonometric ratios as numerical values or in surd form, in which case you do not need a calculator.

EXAMPLE 19

ABC is a right-angled triangle.

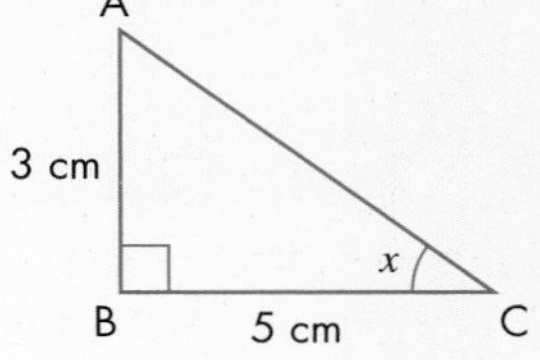

a Write down the value of $\tan x$.

b Work out the values of $\cos x$ and $\sin x$, giving your answers in simplified surd form.

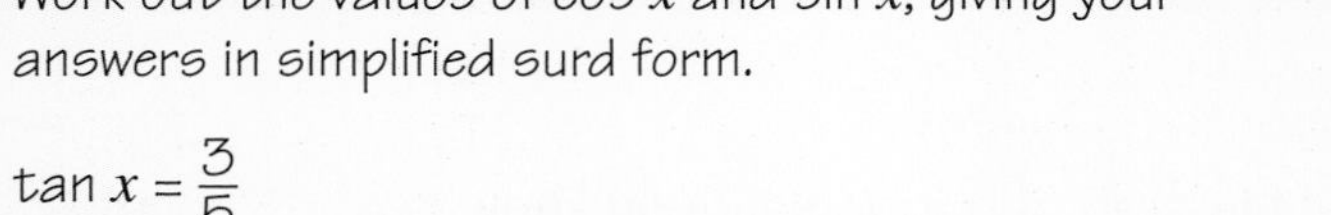

a $\tan x = \frac{3}{5}$

b Using Pythagoras' theorem, AC = $\sqrt{34}$.

So $\cos x = \frac{5}{\sqrt{34}} = \frac{5\sqrt{34}}{34}$

and $\sin x = \frac{3}{\sqrt{34}} = \frac{3\sqrt{34}}{34}$

EXERCISE 10J

AU 1 The sine of angle x is $\frac{4}{5}$. Work out the cosine of angle x.

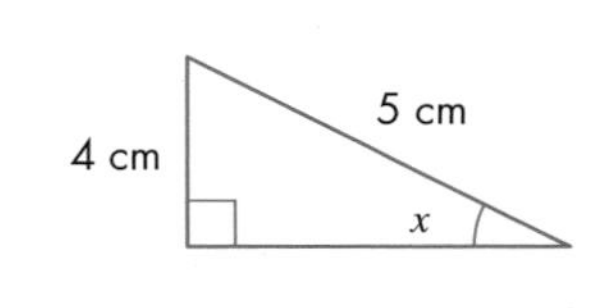

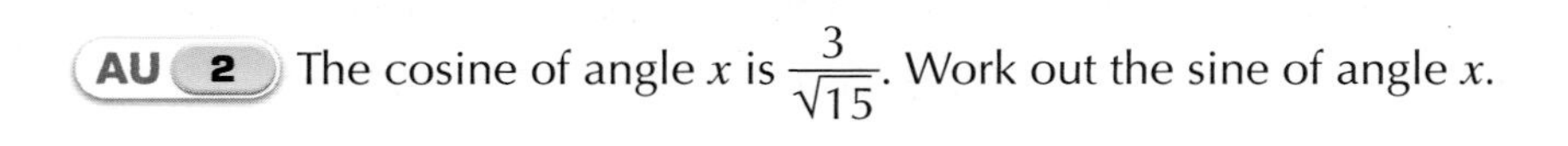

AU 2 The cosine of angle x is $\frac{3}{\sqrt{15}}$. Work out the sine of angle x.

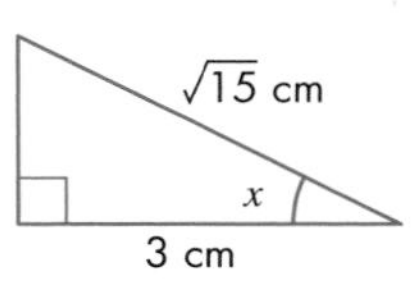

A*

A*

PS **3** The lengths of the two short sides of a right-angled triangle are $\sqrt{6}$ and $\sqrt{13}$. Write down the exact value of the hypotenuse of this triangle, and the exact value of the sine, cosine and tangent of the smallest angle in the triangle.

PS **4** The tangent of angle A is $\frac{6}{11}$. Use this to write down possible lengths of two sides of the triangle.

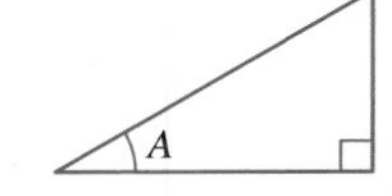

a Calculate the length of the third side of the triangle.

b Write down the exact values of sin A and cos A.

PS **5** Calculate the exact value of the area of an equilateral triangle of side 6 cm.

6 Work out the exact value of the area of a right-angled isosceles triangle with hypotenuse 40 cm.

7 Work out the length of AB in the triangle ABC.

Give your answer in simplified surd form.

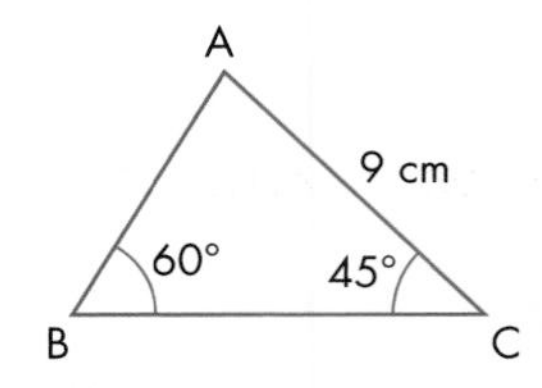

10.6 Using sine to find the area of a triangle

This section will show you how to:

- work out the area of a triangle if you know two sides and the included angle

Key words

area
area sine rule
cosine rule
included angle
sine rule

In triangle ABC, the vertical height is BD and the base is AC.

Let BD = h and AC = b, then the **area** of the triangle is given by:

$\frac{1}{2} \times \text{AC} \times \text{BD} = \frac{1}{2}bh$

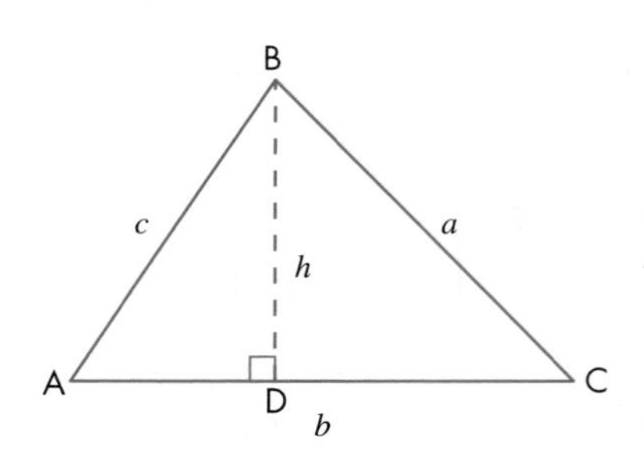

However, in triangle BCD,

$h = \text{BC} \sin C = a \sin C$

where BC = a.

Substituting into $\frac{1}{2}bh$ gives:

$\frac{1}{2}b \times (a \sin C) = \frac{1}{2}ab \sin C$

as the area of the triangle.

By taking the perpendicular from A to its opposite side BC, and the perpendicular from C to its opposite side AB, we can show that the area of the triangle is also given by:

$\frac{1}{2}ac \sin B$ and $\frac{1}{2}bc \sin A$

Note the pattern: the area is given by the product of two sides multiplied by the sine of the **included angle**. This is the **area sine rule**. Starting from any of the three forms, it is also possible to use the **sine rule** to establish the other two.

EXAMPLE 20

Find the area of triangle ABC.

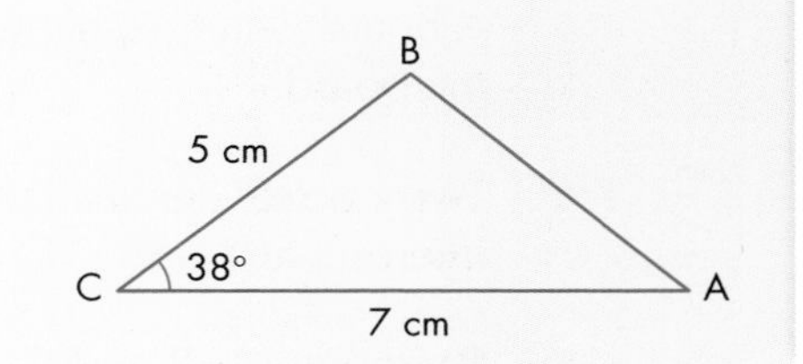

Area = $\frac{1}{2}ab \sin C$

Area = $\frac{1}{2} \times 5 \times 7 \times \sin 38° = 10.8 \text{ cm}^2$

(3 significant figures)

EXAMPLE 21

Find the area of triangle ABC.

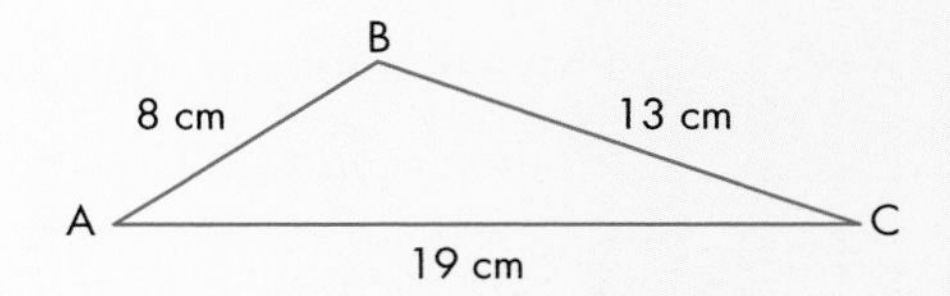

You have all three sides but no angle. So first you must find an angle in order to apply the area sine rule.

Find angle C, using the **cosine rule**.

$$\cos C = \frac{a^2 + b^2 - c^2}{2ab}$$

$$= \frac{13^2 + 19^2 - 8^2}{2 \times 13 \times 19} = 0.9433$$

$$\Rightarrow C = \cos^{-1} 0.9433 = 19.4°$$

(Keep the exact value in your calculator memory.)

Now apply the area sine rule.

$$\frac{1}{2}ab \sin C = \frac{1}{2} \times 13 \times 19 \times \sin 19.4°$$

$$= 41.0 \text{ cm}^2 \quad \text{(3 significant figures)}$$

EXERCISE 10K

1 Find the area of each of the following triangles.

a Triangle ABC where BC = 7 cm, AC = 8 cm and angle ACB = 59°

b Triangle ABC where angle BAC = 86°, AC = 6.7 cm and AB = 8 cm

c Triangle PQR where QR = 27 cm, PR = 19 cm and angle QRP = 109°

d Triangle XYZ where XY = 231 cm, XZ = 191 cm and angle YXZ = 73°

e Triangle LMN where LN = 63 cm, LM = 39 cm and angle NLM = 85°

2 The area of triangle ABC is 27 cm^2. If BC = 14 cm and angle BCA = 115°, find AC.

3 The area of triangle LMN is 113 cm^2, LM = 16 cm and MN = 21 cm. Angle LMN is acute. Calculate these angles.

a LMN **b** MNL

4 In a quadrilateral ABCD, DC = 4 cm, BD = 11 cm, angle BAD = 32°, angle ABD = 48° and angle BDC = 61°. Calculate the area of the quadrilateral.

5 A board is in the shape of a triangle with sides 60 cm, 70 cm and 80 cm. Find the area of the board.

6 Two circles, centres P and Q, have radii of 6 cm and 7 cm respectively. The circles intersect at X and Y. Given that PQ = 9 cm, find the area of triangle PXQ.

7 The points A, B and C are on the circumference of a circle, centre O and radius 7 cm. AB = 4 cm and BC = 3.5 cm. Calculate:

a angle AOB **b** the area of quadrilateral OABC.

PS 8 Prove that for any triangle ABC,

area = $\frac{1}{2}ab \sin C$

9 **a** ABC is a right-angled isosceles triangle with short sides of 1 cm. Write down the value of sin 45°.

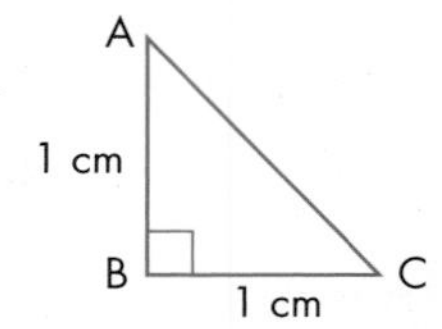

b Calculate the area of triangle PQR. Give your answer in surd form.

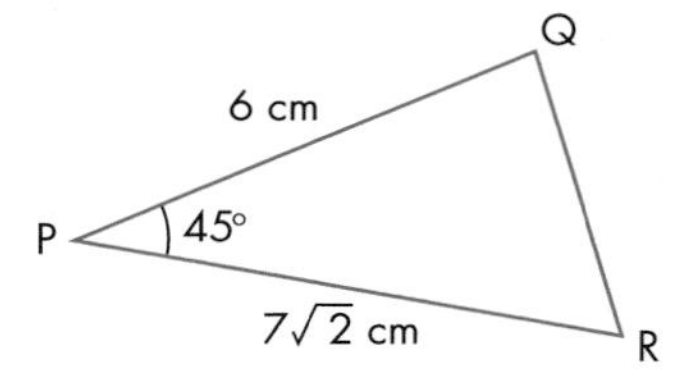

10 Sanjay is making a kite.
The diagram shows a sketch of his kite.

Calculate the area of the material required to make the kite.

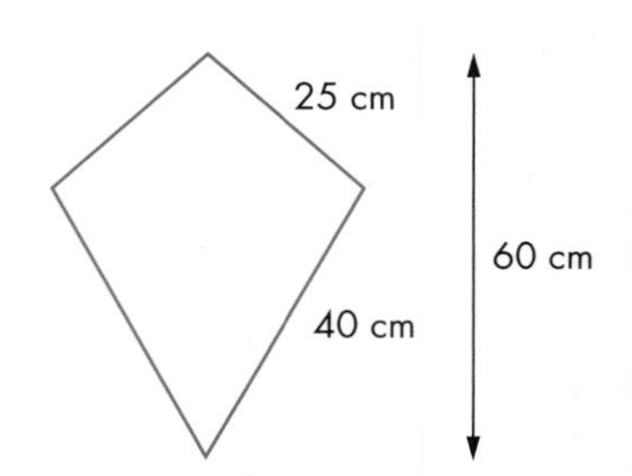

PS 11 An equilateral triangle has sides of length a.

Work out the area of the triangle, giving your answer in surd form.

AU 12 The lengths of all the sides of a square-based pyramid are 10 cm.

Which of these possible answers correctly gives the total surface area of the pyramid?

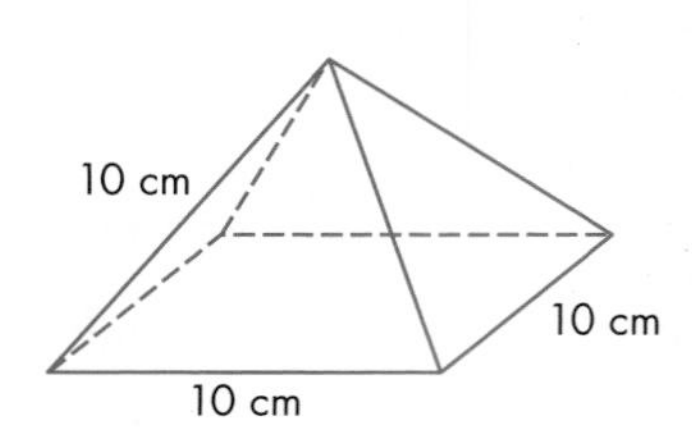

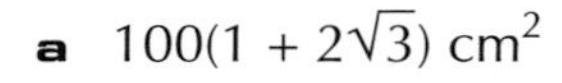

a $100(1 + 2\sqrt{3})$ cm^2 **b** $100(2 + \sqrt{3})$ cm^2

c $100(1 + \sqrt{3})$ cm^2 **d** $100(1 + 2\sqrt{2})$cm^2

GRADE BOOSTER

A You can solve more complex 2D problems, using Pythagoras' theorem and trigonometry

A You can use the sine and cosine rules to calculate missing angles or sides in non-right-angled triangles

A You can find the area of a triangle using the formula area = $\frac{1}{2}ab \sin C$

A* You can use the sine and cosine rules to solve more complex problems involving non-right-angled triangles

A* You can solve 3D problems, using Pythagoras' theorem and trigonometric ratios

A* You can find the trigonometric ratios for any angle from 0° to 360°

What you should know now

- How to use the sine and cosine rules
- How to find the area of a triangle, using area = $\frac{1}{2}ab \sin C$
- How to find the trigonometric ratio for any angle from 0° to 360°

1

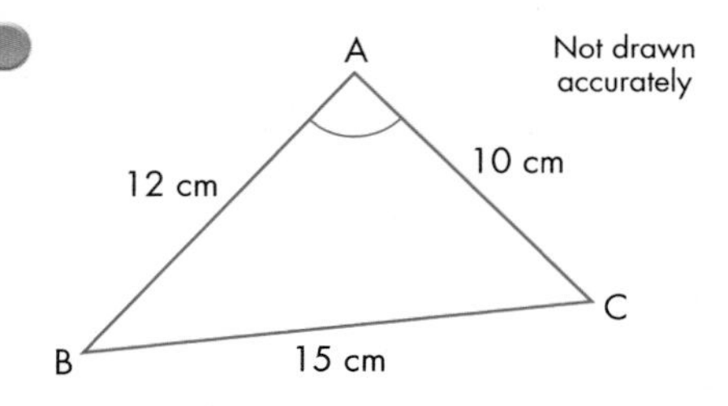

ABC is a triangle.

AB = 12 m, AC = 10 m, BC = 15 m.

Calculate the size of angle BAC.

Give your answer correct to one decimal place. *(3 marks)*

Edexcel, June 2006, Paper 4 Higher, Question 24

2

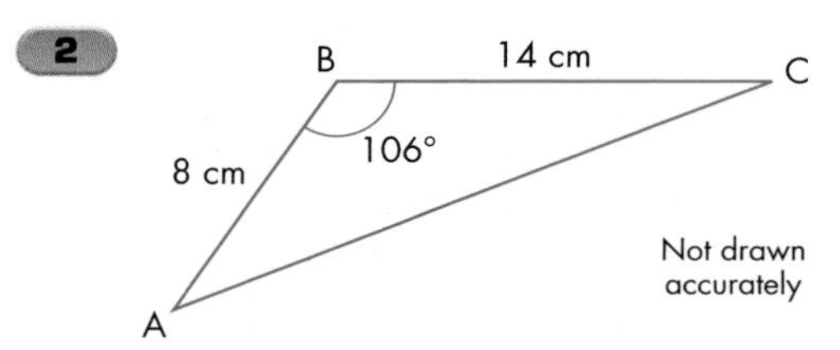

ABC is a triangle.

AB = 8 cm

BC = 14 cm

Angle ABC = 106°

Calculate the area of the triangle.

Give your answer correct to 3 significant figures. *(3 marks)*

Edexcel, June 2005, Paper 6, Question 13

3

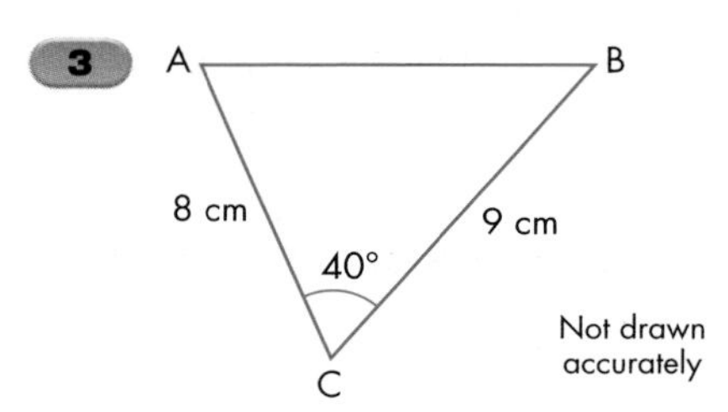

ABC is a triangle.

AC = 8 cm

BC = 9 cm

Angle ACB = 40°

Calculate the length of AB.

Give your answer correct to 3 significant figures. *(3 marks)*

Edexcel, June 2007, Paper 6, Question 21

4 The diagram shows an equilateral triangle of side 2 m.

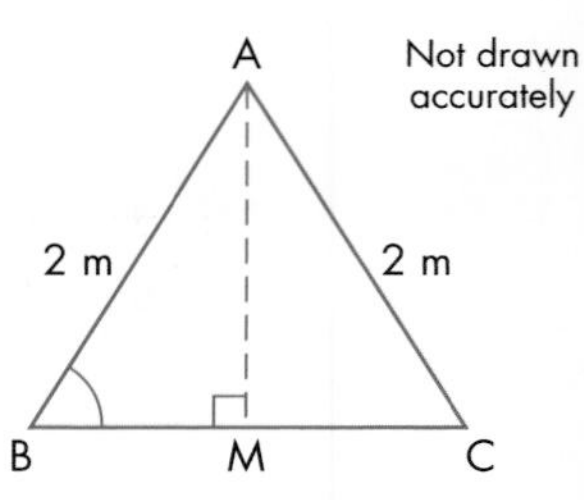

a **i** Use the diagram to show that $\cos 60° = \frac{1}{2}$.

ii Use the diagram to find the exact value of sin 60°.

Give your answer as a surd. *(4 marks)*

b Use the exact values of cos 60° and sin 60° to show that $(\cos 60°)^2 + (\sin 60°)^2 = 1$. *(2 marks)*

Edexcel, June 2009, Paper 4H, Question 20

5

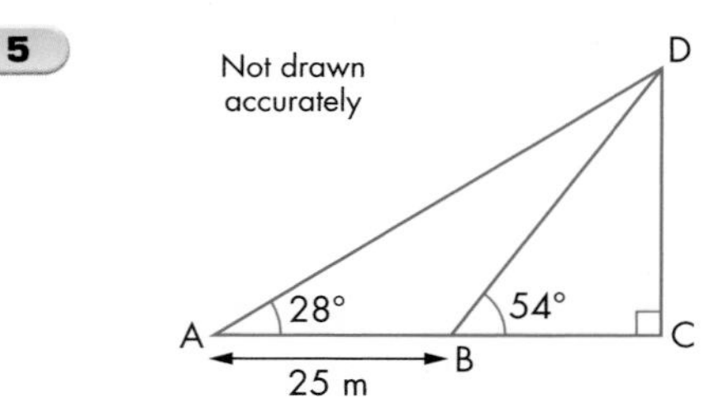

The diagram shows a vertical tower DC on horizontal ground ABC.

ABC is a straight line.

The angle of elevation of D from A is 28°.

The angle of elevation of D from B is 54°.

AB = 25 m.

Calculate the height of the tower.

Give your answer correct to 3 significant figures. *(5 marks)*

Edexcel, June 2006, Paper 6, Question 23

6 The diagram shows a sketch of the curve $y = \sin x°$ for $0 \leqslant x \leqslant 360$

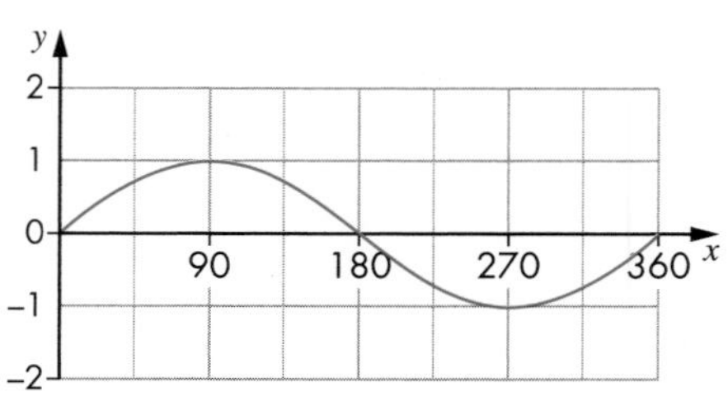

The exact value of $\sin 60° = \frac{\sqrt{3}}{2}$

Write down the exact value of

a sin 120° **b** sin 240°.

A* A

Worked Examination Questions

AU **1** The diagram shows a cuboid ABCDEFGH.
Calculate the size of angle AGE.

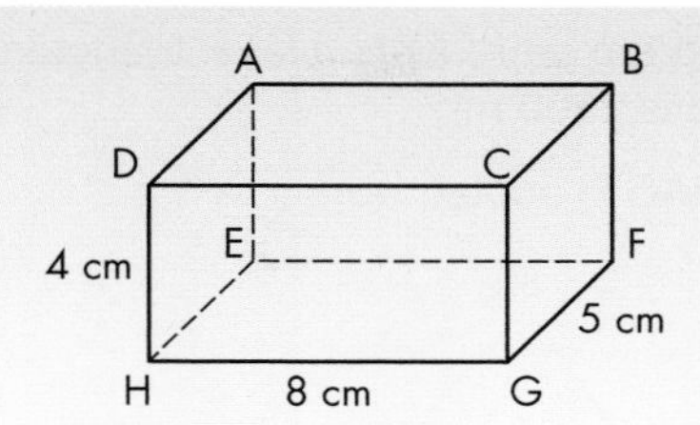

1 First draw the right-angled triangle EFG to find the length of EG.
Mark it x on the diagram.

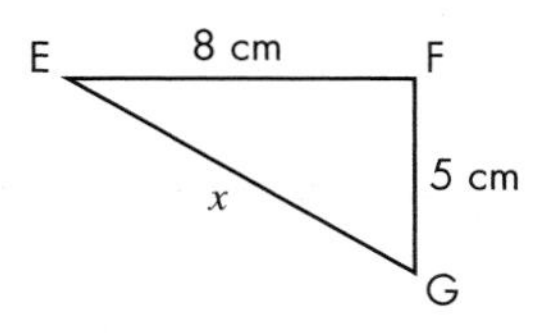

Find the length of EG by Pythagoras' theorem.

$x^2 = 8^2 + 5^2 = 89$ — Store $\sqrt{89}$ in your calculator or write down the answer to 4 significant figures. This line gets 1 mark for method.

So $x = \sqrt{89} = 9.434$ cm — Answer gets 1 mark for accuracy ($\sqrt{89}$ is acceptable).

Now draw the right-angled triangle AGE and mark the required angle y.

$\tan y = \frac{O}{A}$

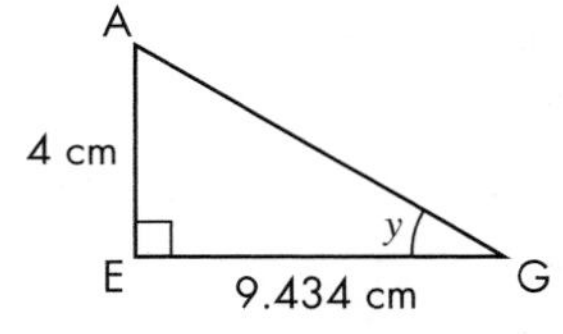

Use tangent to find angle y.

$\tan y = \dfrac{4}{9.434} = 0.4240$ — This gets 1 mark for method for this line.

So $y = \tan^{-1} 0.420$

$y = 23.0°$ (3 significant figures) — This gets 1 mark for accuracy for the answer.

Total: 4 marks

Worked Examination Questions

FM **2** The diagram represents a level triangular piece of land. AB = 61 m, AC = 76 m and the area of the land is 2300 m^2.

Angle BAC is acute.

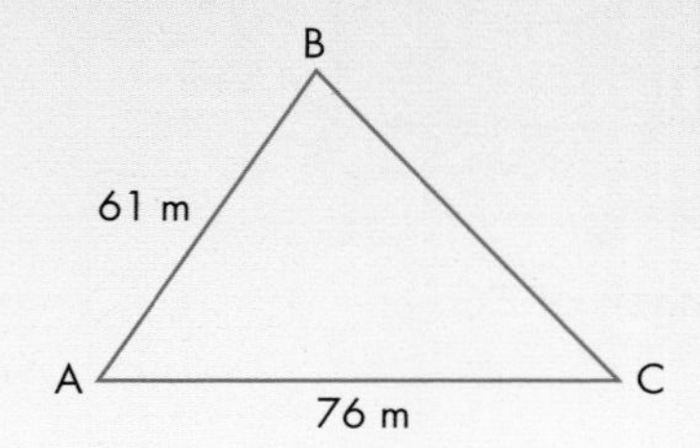

Calculate the length of BC. Give your answer to an appropriate degree of accuracy.

2 $\frac{1}{2} \times 61 \times 76 \times \sin A = 2300$

> Since the third side is unknown, we need to find an angle.
> Find angle BAC (= A) so that we can use the cosine rule.
> Use area = $\frac{1}{2}bc \sin A$.

$\therefore \sin \text{BAC} = \frac{4600}{4636} = 0.9922 \ldots$

> Use area = $\frac{1}{2}bc \sin A$ to set up an equation and solve it to get angle A. You are given that A is acute so there is no problem with any ambiguity.
> This gets 1 mark for method for this line.

$\therefore A = 82.86°$

> This gets 1 mark for accuracy for the answer.
> Now use the cosine rule to find BC.

$\text{BC}^2 = 61^2 + 76^2 - 2 \times 61 \times 76 \times \cos 82.9$
$= 8343.75$

> Use the cosine rule to work out the side BC.
> If possible keep values in your calculator display but if you have to write down values then use at least 4 significant figures for trigonometric ratios and at least 1 decimal place for angles. This will avoid any inaccuracy in the final answer.
> This gets 1 mark for method for this line.

BC = 91.3 m (3 significant figures)

> The answer gets 1 mark for accuracy.

Total: 4 marks

Worked Examination Questions

PS **3** A **tetrahedron** has one face which is an equilateral triangle of side 6 cm and three faces which are isosceles triangles with sides 6 cm, 9 cm and 9 cm.

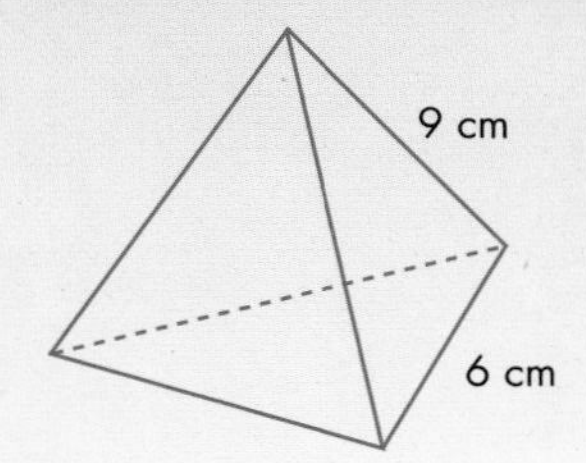

Calculate the surface area of the tetrahedron.

HINTS AND TIPS

A tetrahedron is a triangle-based pyramid.

3 First work out the area of the base, which has angles of 60°.
Use area = $\frac{1}{2}ac \sin B$.

Area base = $\frac{1}{2} \times 6 \times 6 \times \sin 60° = 15.59 \text{ cm}^2$ (4 significant figures) — This gets 1 mark for accuracy of the answer.

Next, work out the top angle in one of the isosceles triangles, using the cosine rule.

$$\cos x = \frac{9^2 + 9^2 - 6^2}{2 \times 9 \times 9} = 0.7778$$

— This gets 1 mark for method.

So $x = 38.9°$ — Keep the exact value in your calculator. This gets 1 mark for accuracy.

Work out the area of one side face and then add all faces together.

Area side face = $\frac{1}{2} \times 9 \times 9 \times \sin 38.9°$
= 25.46 cm^2 (4 significant figures) — This gets 1 mark for method.

Total area = $3 \times 25.46 + 15.59 = 92.0 \text{ cm}^2$ (3 significant figures) — This answer gets 1 mark for accuracy. Remember to include the units (cm^2) in your final answer.

Total: 4 marks

10

Functional Maths
Building tree-houses

Forest researchers collect data about trees. Sometimes they measure heights. To do this, they use a piece of equipment that extends as far as the tree top. However, the equipment can be bulky and may require more than one person to operate it. Instead, many forest researchers simply use a clinometer, which is a small instrument used for measuring angles of elevation and depression, and trigonometry.

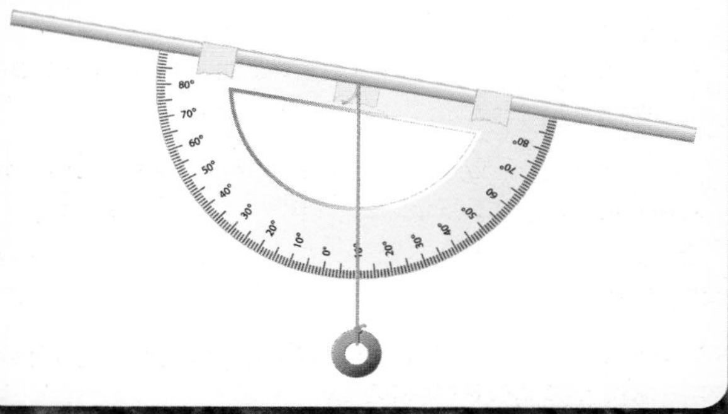

Getting started

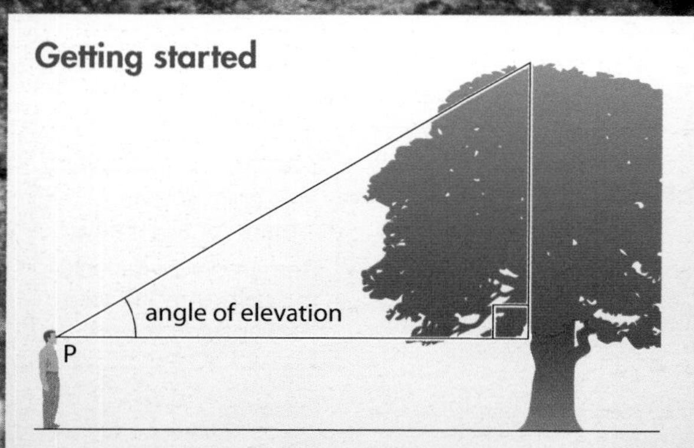

A forest researcher measures the angle of elevation of the top of the tree.

- What other measurement should the researcher find, to help him calculate the height of the tree?
- Which trigonometric ratio would the researcher use?

English oaks			
Tree number	Distance from researcher to base of trunk (m)	Angle of elevation of top of tree	Angle of elevation of lowest branch
001	16.3	45°	15°
002	10	58°	20°
003	24.5	49°	14°
004	15	55°	17°
005	12.4	52°	21°

Your task

The table shows information a researcher collected for the English oaks in one particular forest.

The forest owner would like to install camping tree-houses in three of the English oak trees. She would like each tree-house to be at a different height, with a different incline for each ladder, to appeal to a range of holiday-makers. For safety reasons:

- a tree-house must be no higher than 5.5 m from the ground
- the angle between the tree-house ladder and the ground must be 75° or less.

Look at the table and choose three trees for the tree-houses. Decide on the height above the ground and the length of the ladder required for each tree-house.

Then, write a leaflet advertising your three tree-houses to holiday-makers.

Remember to include information about heights and ladders. What other mathematical information might you include?

Why this chapter matters

Like most mathematics, quadratic equations have their origins in ancient Egypt.

The Egyptians did not have a formal system of algebra but could solve problems that involved quadratics.This problem was written in hieroglyphics on the Berlin Papyrus which was written some time around 2160–1700BC:

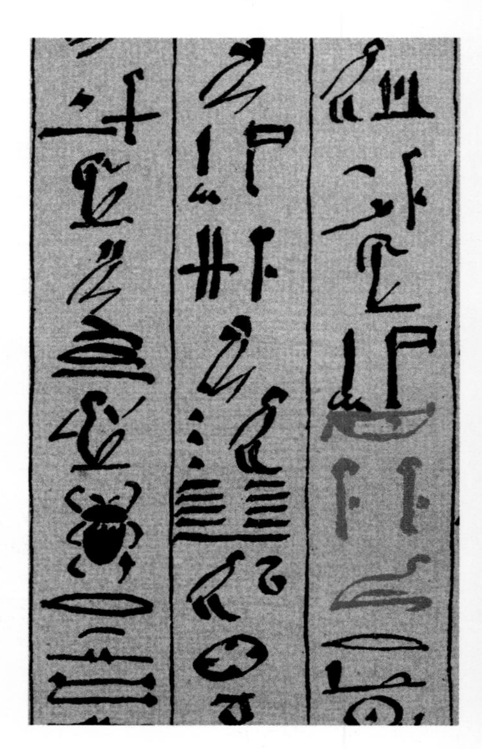

The area of a square of 100 is equal to that of two smaller squares. The side of one is $\frac{1}{2} + \frac{1}{4}$ the side of the other.

Today we would express this as: $x^2 + y^2 = 100$

$y = \frac{3}{4}x$

Euclid

In about 300BC, Euclid developed a geometrical method for solving quadratics. This work was developed by Hindu mathematicians, but it was not until much later, in 1145AD that the Arabic mathematician Abraham bar Hiyya Ha-Nasi, published the book *Liber embadorum*, which gave a complete solution of the quadratic equation.

On 26 June 2003, the quadratic equation was the subject of a debate in Parliament. The National Union of Teachers had suggested that students should be allowed to give up mathematics at the age of 14, and making them do 'abstract and irrelevant' things such as learning about quadratic equations did not serve any purpose.

Mr Tony McWalter, the MP for Hemel Hempstead, defending the teaching of quadratic equations, said:

> "Someone who thinks that the quadratic equation is an empty manipulation, devoid of any other significance, is someone who is content with leaving the many in ignorance. I believe that he or she is also pleading for the lowering of standards. A quadratic equation is not like a bleak room, devoid of furniture, in which one is asked to squat. It is a door to a room full of the unparalleled riches of human intellectual achievement. If you do not go through that door, or if it is said that it is an uninteresting thing to do, much that passes for human wisdom will be forever denied you."

Chapter

Algebra: Quadratics

1. Expanding brackets
2. Quadratic factorisation
3. Solving quadratic equations by factorisation
4. Solving a quadratic equation by the quadratic formula
5. Solving a quadratic equation by completing the square
6. Problems involving quadratic equations

The grades given in this chapter are target grades.

This chapter will show you ...

- **C** how to expand two linear brackets to obtain a quadratic expression
- **B** how to factorise a quadratic expression
- **A** to **A*** how to solve quadratic equations by factorisation and completing the square
- **A** how to solve quadratic equations by the quadratic formula
- **A*** how to solve problems with quadratic equations

Visual overview

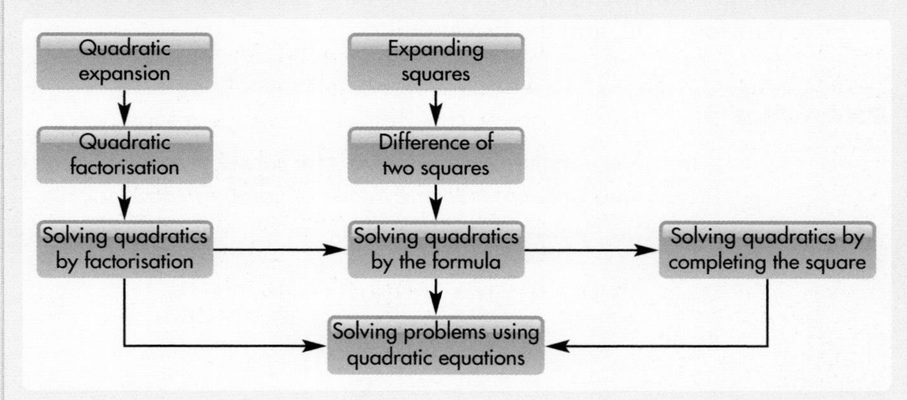

What you should already know

- The basic language of algebra **(KS3 level 5, GCSE grade E)**
- How to collect together like terms **(KS3 level 5, GCSE grade E)**
- How to multiply together two algebraic expressions **(KS3 level 5, GCSE grade E)**
- How to solve simple linear equations **(KS3 level 5 to 6, GCSE grade E to C)**

Quick check

1 Simplify the following.

a $-2x - x$ **b** $3x - x$ **c** $-5x + 2x$

d $2m \times 3m$ **e** $3x \times -2x$ **f** $-4p \times 3p$

2 Solve these equations.

a $x + 6 = 0$ **b** $2x + 1 = 0$ **c** $3x - 2 = 0$

11.1 Expanding brackets

This section will show you how to:

- expand two linear brackets to obtain a quadratic expression

Key words
coefficient
linear
quadratic expression

Quadratic expansion

A **quadratic expression** is one in which the highest power of the variables is 2. For example,

y^2 $\quad 3t^2 + 5t$ $\quad 5m^2 + 3m + 8$

An expression such as $(3y + 2)(4y - 5)$ can be expanded to give a quadratic expression.

Multiplying out such pairs of brackets is usually called *quadratic expansion*.

The rule for expanding expressions such as $(t + 5)(3t - 4)$ is similar to that for expanding single brackets: multiply everything in one set of brackets by everything in the other set of brackets.

There are several methods for doing this. Examples 1 to 3 show the three main methods: expansion, FOIL and the box method.

EXAMPLE 1

In the expansion method, split the terms in the first set of brackets, make each of them multiply both terms in the second set of brackets, then simplify the outcome.

Expand $(x + 3)(x + 4)$

$$(x + 3)(x + 4) = x(x + 4) + 3(x + 4)$$
$$= x^2 + 4x + 3x + 12$$
$$= x^2 + 7x + 12$$

EXAMPLE 2

FOIL stands for First, Outer, Inner and Last. This is the order of multiplying the terms from each set of brackets.

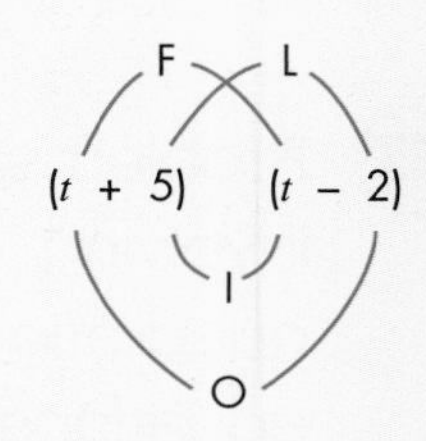

Expand $(t + 5)(t - 2)$

First terms give: $t \times t = t^2$

Outer terms give: $t \times -2 = -2t$.

Inner terms give: $5 \times t = 5t$

Last terms give: $+5 \times -2 = -10$

$$(t + 5)(t - 2) = t^2 - 2t + 5t - 10$$
$$= t^2 + 3t - 10$$

EXAMPLE 3

The box method is similar to that used to do long multiplication.

Expand $(k - 3)(k - 2)$

$$(k - 3)(k - 2) = k^2 - 2k - 3k + 6$$
$$= k^2 - 5k + 6$$

$\times$	k	-3
k	k^2	$-3k$
-2	$-2k$	$+6$

Warning: Be careful with the signs. This is the main place where mistakes are made in questions involving the expansion of brackets.

EXERCISE 11A

Expand the expressions in questions **1–17**.

HINTS AND TIPS

Use whichever method you prefer. There is no fixed method in GCSE examinations. Examiners give credit for all methods. Whatever method you use, it is important to show the examiner that you know there are four terms in the expansion before it is simplified.

1 $(x + 3)(x + 2)$

2 $(t + 4)(t + 3)$

3 $(w + 1)(w + 3)$

4 $(m + 5)(m + 1)$

5 $(k + 3)(k + 5)$

6 $(a + 4)(a + 1)$

7 $(x + 4)(x - 2)$

8 $(t + 5)(t - 3)$

9 $(w + 3)(w - 1)$

10 $(f + 2)(f - 3)$

11 $(g + 1)(g - 4)$

12 $(y + 4)(y - 3)$

13 $(x - 3)(x + 4)$

14 $(p - 2)(p + 1)$

15 $(k - 4)(k + 2)$

16 $(y - 2)(y + 5)$

17 $(a - 1)(a + 3)$

HINTS AND TIPS

A common error is to get minus signs wrong.
$-2x - 3x = -5x$ and
$-2 \times -3 = +6$

The expansions of the expressions in questions **18–26** follow a pattern. Work out the first few and try to spot the pattern that will allow you immediately to write down the answers to the rest.

18 $(x + 3)(x - 3)$

19 $(t + 5)(t - 5)$

20 $(m + 4)(m - 4)$

21 $(t + 2)(t - 2)$

22 $(y + 8)(y - 8)$

23 $(p + 1)(p - 1)$

24 $(5 + x)(5 - x)$

25 $(7 + g)(7 - g)$

26 $(x - 6)(x + 6)$

C

PS 27 This rectangle is made up of four parts with areas of x^2, $2x$, $3x$ and 6 square units.

Work out expressions for the sides of the rectangle, in terms of x.

x^2	$2x$
$3x$	6

PS 28 This square has an area of x^2 square units.
It is split into four rectangles.

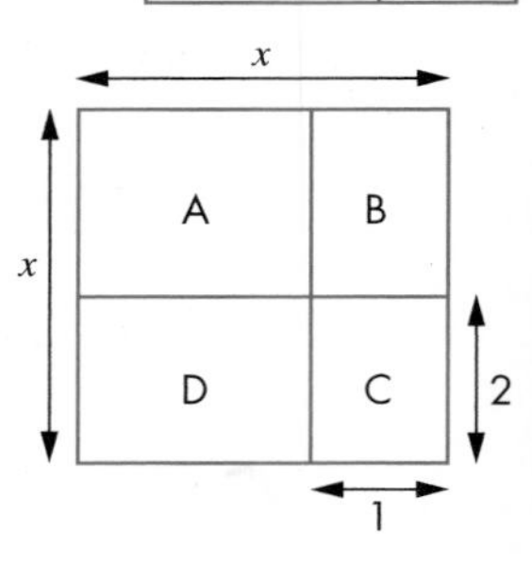

a Fill in the table below to show the dimensions and area of each rectangle.

Rectangle	Length	Width	Area
A	$x - 1$	$x - 2$	$(x - 1)(x - 2)$
B			
C			
D			

b Add together the areas of rectangles B, C and D.
Expand any brackets and collect terms together.

c Use the results to explain why $(x - 1)(x - 2) = x^2 - 3x + 2$.

AU 29 **a** Expand $(x - 3)(x + 3)$

b Use the result in **a** to write down the answers to these. (Do not use a calculator or do a long multiplication.)

i 97×103 **ii** 197×203

Quadratic expansion with non-unit coefficients

All the algebraic terms in x^2 in Exercise 10A have a **coefficient** of 1 or –1. The next two examples show what to do if you have to expand brackets containing terms in x^2 with coefficients that are not 1 or –1.

EXAMPLE 4

Expand $(2t + 3)(3t + 1)$

$$(2t + 3)(3t + 1) = 6t^2 + 2t + 9t + 3$$
$$= 6t^2 + 11t + 3$$

$\times$	$2t$	$+3$
$3t$	$6t^2$	$+9t$
$+1$	$+2t$	$+3$

EXAMPLE 5

Expand $(4x - 1)(3x - 5)$

$$(4x - 1)(3x - 5) = 4x(3x - 5) - (3x - 5)$$ [**Note:** $(3x - 5)$ is the same as $1(3x - 5)$.]
$$= 12x^2 - 20x - 3x + 5$$
$$= 12x^2 - 23x + 5$$

EXERCISE 11B

Expand the expressions in questions **1–21**.

1 $(2x + 3)(3x + 1)$

2 $(3y + 2)(4y + 3)$

3 $(3t + 1)(2t + 5)$

4 $(4t + 3)(2t - 1)$

5 $(5m + 2)(2m - 3)$

6 $(4k + 3)(3k - 5)$

7 $(3p - 2)(2p + 5)$

8 $(5w + 2)(2w + 3)$

9 $(2a - 3)(3a + 1)$

10 $(4r - 3)(2r - 1)$

11 $(3g - 2)(5g - 2)$

12 $(4d - 1)(3d + 2)$

13 $(5 + 2p)(3 + 4p)$

14 $(2 + 3t)(1 + 2t)$

15 $(4 + 3p)(2p + 1)$

16 $(6 + 5t)(1 - 2t)$

17 $(4 + 3n)(3 - 2n)$

18 $(2 + 3f)(2f - 3)$

19 $(3 - 2q)(4 + 5q)$

20 $(1 - 3p)(3 + 2p)$

21 $(4 - 2t)(3t + 1)$

HINTS AND TIPS

Always give answers in the form $\pm ax^2 \pm bx \pm c$ even if the quadratic coefficient is negative.

PS 22 Expand:

a $(x + 1)(x + 1)$

b $(x - 1)(x - 1)$

c $(x + 1)(x - 1)$

d Use the results in parts **a**, **b** and **c** to show that $(p + q)^2 \equiv p^2 + 2pq + q^2$ is an identity.

HINTS AND TIPS

Take $p = x + 1$ and $q = x - 1$.

AU 23 **a** Without expanding the brackets, match each expression on the left with an expression on the right. One is done for you.

$(3x - 2)(2x + 1)$	$4x^2 - 4x + 1$
$(2x - 1)(2x - 1)$	$6x^2 - x - 2$
$(6x - 3)(x + 1)$	$6x^2 + 7x + 2$
$(4x + 1)(x - 1)$	$6x^2 + 3x - 3$
$(3x + 2)(2x + 1)$	$4x^2 - 3x - 1$

b Taking any expression on the left, explain how you can match it with an expression on the right without expanding the brackets.

EXERCISE 11C

Try to spot the pattern in each of the expressions in questions **1–15** so that you can immediately write down the expansion.

B

1 $(2x + 1)(2x - 1)$

2 $(3t + 2)(3t - 2)$

3 $(5y + 3)(5y - 3)$

4 $(4m + 3)(4m - 3)$

5 $(2k - 3)(2k + 3)$

6 $(4h - 1)(4h + 1)$

7 $(2 + 3x)(2 - 3x)$

8 $(5 + 2t)(5 - 2t)$

9 $(6 - 5y)(6 + 5y)$

10 $(a + b)(a - b)$

11 $(3t + k)(3t - k)$

12 $(2m - 3p)(2m + 3p)$

13 $(5k + g)(5k - g)$

14 $(ab + cd)(ab - cd)$

15 $(a^2 + b^2)(a^2 - b^2)$

PS 16 Imagine a square of side a units with a square of side b units cut from one corner.

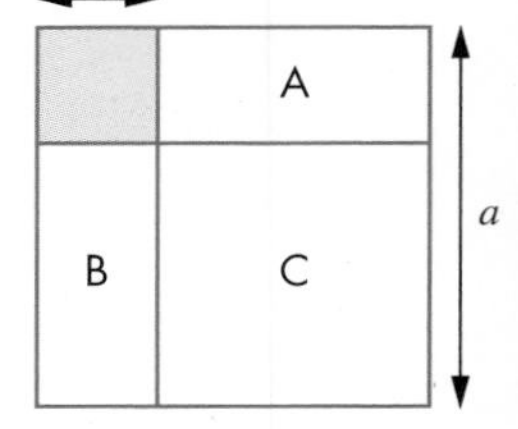

a What is the area remaining after the small square is cut away?

b The remaining area is cut into rectangles, A, B and C, and rearranged as shown.

Write down the dimensions and area of the rectangle formed by A, B and C.

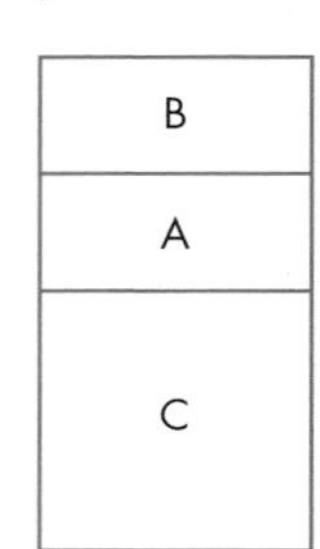

c Explain why $a^2 - b^2 = (a + b)(a - b)$.

AU 17 Explain why the areas of the shaded regions are the same.

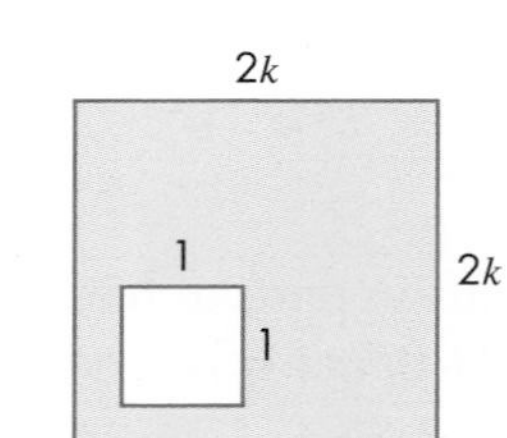

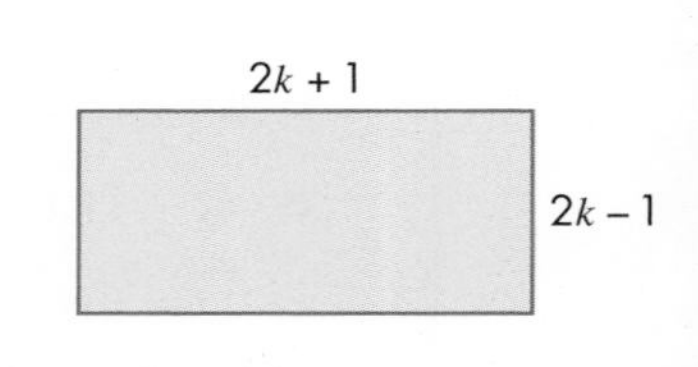

Expanding squares

Whenever you see a **linear** bracketed term squared you must write the brackets down twice and then use whichever method you prefer to expand.

EXAMPLE 6

Expand $(x + 3)^2$

$$\begin{aligned}(x + 3)^2 &= (x + 3)(x + 3)\\ &= x(x + 3) + 3(x + 3)\\ &= x^2 + 3x + 3x + 9\\ &= x^2 + 6x + 9\end{aligned}$$

EXAMPLE 7

Expand $(3x - 2)^2$

$(3x - 2)^2 = (3x - 2)(3x - 2)$

$= 9x^2 - 6x - 6x + 4$

$= 9x^2 - 12x + 4$

F L

(3x – 2) (3x – 2)

I

O

EXERCISE 11D

Expand the squares in questions **1–24** and simplify.

HINTS AND TIPS

Remember *always* write down the brackets twice. Do not try to take any short cuts.

1 $(x + 5)^2$ **2** $(m + 4)^2$ **3** $(6 + t)^2$

4 $(3 + p)^2$ **5** $(m - 3)^2$ **6** $(t - 5)^2$

7 $(4 - m)^2$ **8** $(7 - k)^2$

9 $(3x + 1)^2$ **10** $(4t + 3)^2$ **11** $(2 + 5y)^2$ **12** $(3 + 2m)^2$

13 $(4t - 3)^2$ **14** $(3x - 2)^2$ **15** $(2 - 5t)^2$ **16** $(6 - 5r)^2$

17 $(x + y)^2$ **18** $(m - n)^2$ **19** $(2t + y)^2$ **20** $(m - 3n)^2$

21 $(x + 2)^2 - 4$ **22** $(x - 5)^2 - 25$ **23** $(x + 6)^2 - 36$ **24** $(x - 2)^2 - 4$

PS 25 A teacher asks her class to expand $(3x + 1)^2$.

Bernice's answer is $9x^2 + 1$.

Pete's answer is $3x^2 + 6x + 1$.

a Explain the mistakes that Bernice has made.

b Explain the mistakes that Pete has made.

c Work out the correct answer.

AU 26 Use the diagram to show algebraically and diagrammatically that:

$(2x - 1)^2 = 4x^2 - 4x + 1$

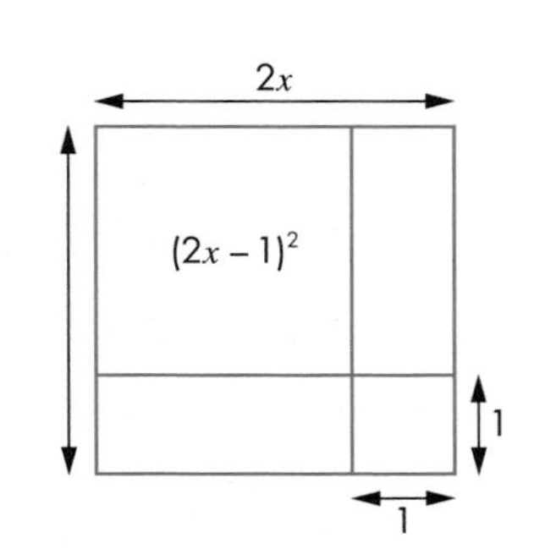

C

B

11.2 Quadratic factorisation

This topic will also be assessed in Unit 2.

This section will show you how to:

- factorise a quadratic expression into two linear brackets

Key words
brackets
coefficient
difference of two squares
factorisation
quadratic expression

Factorisation involves putting a **quadratic expression** back into its **brackets** (if possible). We start with the factorisation of quadratic expressions of the type:

$$x^2 + ax + b$$

where a and b are integers.

Sometimes it is easy to put a quadratic expression back into its brackets, other times it seems hard. However, there are some simple rules that will help you to factorise.

- The expression inside each set of brackets will start with an x, and the signs in the quadratic expression show which signs to put after the xs.
- When the second sign in the expression is a plus, the signs in both sets of brackets are the same as the first sign.

  $x^2 + ax + b = (x + ?)(x + ?)$ Since everything is positive.

  $x^2 - ax + b = (x - ?)(x - ?)$ Since –ve × –ve = +ve
- When the *second* sign is a *minus*, the signs in the brackets are *different*.

  $x^2 + ax - b = (x + ?)(x - ?)$ Since +ve × –ve = –ve

  $x^2 - ax - b = (x + ?)(x - ?)$
- Next, look at the *last* number, b, in the expression. When multiplied together, the two numbers in the brackets must give b.
- Finally, look at the **coefficient** *of* x, a. The *sum* of the two *numbers* in the brackets will give a.

EXAMPLE 8

Factorise $x^2 - x - 6$

Because of the signs we know the brackets must be $(x + ?)(x - ?)$.

Two numbers that have a product of -6 and a sum of -1 are -3 and $+2$.

So, $x^2 - x - 6 = (x + 2)(x - 3)$

EXAMPLE 9

Factorise $x^2 - 9x + 20$

Because of the signs we know the brackets must be $(x - ?)(x - ?)$.

Two numbers that have a product of $+20$ and a sum of -9 are -4 and -5.

So, $x^2 - 9x + 20 = (x - 4)(x - 5)$

EXERCISE 11E

Factorise the expressions in questions **1–40**.

1 $x^2 + 5x + 6$

2 $t^2 + 5t + 4$

3 $m^2 + 7m + 10$

4 $k^2 + 10k + 24$

5 $p^2 + 14p + 24$

6 $r^2 + 9r + 18$

7 $w^2 + 11w + 18$

8 $x^2 + 7x + 12$

9 $a^2 + 8a + 12$

10 $k^2 + 10k + 21$

11 $f^2 + 22f + 21$

12 $b^2 + 20b + 96$

13 $t^2 - 5t + 6$

14 $d^2 - 5d + 4$

15 $g^2 - 7g + 10$

16 $x^2 - 15x + 36$

17 $c^2 - 18c + 32$

18 $t^2 - 13t + 36$

19 $y^2 - 16y + 48$

20 $j^2 - 14j + 48$

21 $p^2 - 8p + 15$

22 $y^2 + 5y - 6$

23 $t^2 + 2t - 8$

24 $x^2 + 3x - 10$

25 $m^2 - 4m - 12$

26 $r^2 - 6r - 7$

27 $n^2 - 3n - 18$

28 $m^2 - 7m - 44$

29 $w^2 - 2w - 24$

30 $t^2 - t - 90$

31 $h^2 - h - 72$

32 $t^2 - 2t - 63$

33 $d^2 + 2d + 1$

34 $y^2 + 20y + 100$

35 $t^2 - 8t + 16$

36 $m^2 - 18m + 81$

37 $x^2 - 24x + 144$

38 $d^2 - d - 12$

39 $t^2 - t - 20$

40 $q^2 - q - 56$

> **HINTS AND TIPS**
>
> First decide on the signs in the brackets, then look at the numbers.

PS 41 This rectangle is made up of four parts. Two of the parts have areas of x^2 and 6 square units.

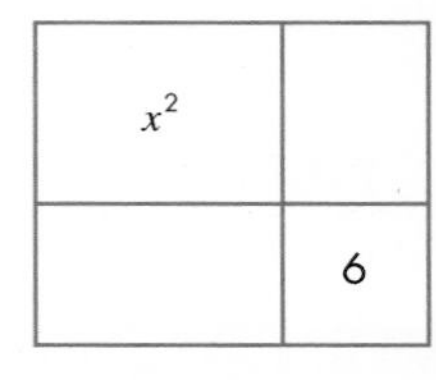

The sides of the rectangle are of the form $x + a$ and $x + b$.

There are two possible answers for a and b.

Work out both answers and copy and complete the areas in the other parts of the rectangle.

AU 42 **a** Expand $(x + a)(x + b)$

b If $x^2 + 7x + 12 = (x + p)(x + q)$, use your answer to part **a** to write down the values of:

i $p + q$ **ii** pq

c Explain how you can tell that $x^2 + 12x + 7$ will not factorise.

Difference of two squares

In Exercise 10C, you multiplied out, for example, $(a + b)(a - b)$ and obtained $a^2 - b^2$. This type of quadratic expression, with only two terms, both of which are perfect squares separated by a minus sign, is called the **difference of two squares**. You should have found that all the expansions in Exercise 10C involved the differences of two squares.

The exercise illustrates a system of factorisation that will *always* work for the difference of two squares such as these.

$x^2 - 9$ $\quad$ $x^2 - 25$ $\quad$ $x^2 - 4$ $\quad$ $x^2 - 100$

There are three conditions that must be met if the difference of two squares works.

- There must be two terms.
- They must separated by a negative sign.
- Each term must be a perfect square, say x^2 and n^2.

When these three conditions are met, the factorisation is:

$$x^2 - n^2 = (x + n)(x - n)$$

EXAMPLE 10

Factorise $x^2 - 36$

- Recognise the difference of two squares x^2 and 6^2.
- So it factorises to $(x + 6)(x - 6)$.

Expanding the brackets shows that they do come from the original expression.

EXAMPLE 11

Factorise $9x^2 - 169$

- Recognise the difference of two squares $(3x)^2$ and 13^2.
- So it factorises to $(3x + 13)(3x - 13)$.

EXERCISE 11F

Each of the expressions in questions **1–9** is the difference of two squares. Factorise them.

HINTS AND TIPS

Learn how to spot the difference of two squares as it occurs a lot in GCSE examinations.

1 $x^2 - 9$ $\quad$ **2** $t^2 - 25$

3 $m^2 - 16$ $\quad$ **4** $9 - x^2$

5 $49 - t^2$ $\quad$ **6** $k^2 - 100$

7 $4 - y^2$ $\quad$ **8** $x^2 - 64$ $\quad$ **9** $t^2 - 81$

B

B

PS 10 **a** A square has a side of x units.

What is the area of the square?

b A rectangle, A, 2 units wide, is cut from the square and placed at the side of the remaining rectangle, B.

A square, C, is then cut from the bottom of rectangle A to leave a final rectangle, D.

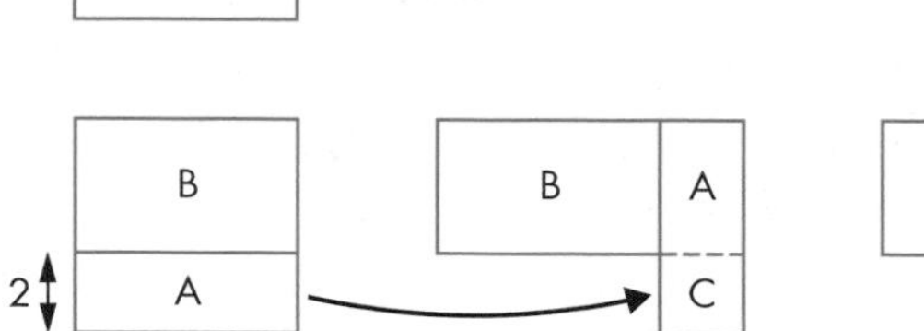

i What is the height of rectangle B?

ii What is the width of rectangle D?

iii What is the area of rectangle B plus rectangle A?

iv What is the area of square C?

c By working out the area of rectangle D, explain why $x^2 - 4 = (x + 2)(x - 2)$.

AU 11 **a** Expand and simplify: $(x + 2)^2 - (x + 1)^2$

b Factorise: $a^2 - b^2$

c In your answer for part **b**, replace a with $(x + 2)$ and b with $(x + 1)$. Expand and simplify the answer.

d What can you say about the answers to parts **a** and **c**?

e Simplify: $(x + 1)^2 - (x - 1)^2$

Each of the expressions in questions **12–20** is the difference of two squares. Factorise them.

A

12 $x^2 - y^2$ **13** $x^2 - 4y^2$ **14** $x^2 - 9y^2$ **15** $9x^2 - 1$ **16** $16x^2 - 9$

17 $25x^2 - 64$ **18** $4x^2 - 9y^2$ **19** $9t^2 - 4w^2$ **20** $16y^2 - 25x^2$

Factorising $ax^2 + bx + c$

We can adapt the method for factorising $x^2 + ax + b$ to take into account the factors of the coefficient of x^2.

EXAMPLE 12

Factorise $3x^2 + 8x + 4$

- First, note that both signs are positive. So the signs in the brackets must be $(?x + ?)(?x + ?)$.
- As 3 has only 3×1 as factors, the brackets must be $(3x + ?)(x + ?)$.
- Next, note that the factors of 4 are 4×1 and 2×2.
- Now find which pair of factors of 4 combine with 3 and 1 to give 8.

(3)	|	4	(2)
(1)	|	1	(2)

You can see that the combination 3×2 and 1×2 adds up to 8.

- So, the complete factorisation becomes $(3x + 2)(x + 2)$.

EXAMPLE 13

Factorise $6x^2 - 7x - 10$

- First, note that both signs are negative. So the signs in the brackets must be $(?x + ?)(?x - ?)$.
- As 6 has 6×1 and 3×2 as factors, the brackets could be $(6x \pm ?)(x \pm ?)$ or $(3x \pm ?)(2x \pm ?)$.
- Next, note that the factors of 10 are 5×2 and 1×10.
- Now find which pair of factors of 10 combine with the factors of 6 to give -7.

3	(6)	|	±1	(±2)
2	(1)	|	±10	(±5)

You can see that the combination 6×-2 and 1×5 adds up to -7.

- So, the complete factorisation becomes $(6x + 5)(x - 2)$.

Although this seems to be very complicated, it becomes quite easy with practice and experience.

EXERCISE 11G

Factorise the expressions in questions **1–12**.

1 $2x^2 + 5x + 2$

2 $7x^2 + 8x + 1$

3 $4x^2 + 3x - 7$

4 $24t^2 + 19t + 2$

5 $15t^2 + 2t - 1$

6 $16x^2 - 8x + 1$

7 $6y^2 + 33y - 63$

8 $4y^2 + 8y - 96$

9 $8x^2 + 10x - 3$

10 $6t^2 + 13t + 5$

11 $3x^2 - 16x - 12$

12 $7x^2 - 37x + 10$

PS 13 This rectangle is made up of four parts, with areas of $12x^2$, $3x$, $8x$ and 2 square units.

Work out expressions for the sides of the rectangle, in terms of x.

$12x^2$	$3x$
$8x$	2

AU 14 Three students are asked to factorise the expression $6x^2 + 30x + 36$. These are their answers.

Adam	**Bertie**	**Cara**
$(6x + 12)(x + 3)$	$(3x + 6)(2x + 6)$	$(2x + 4)(3x + 9)$

All the answers are correctly factorised.

a Explain why one quadratic expression can have three different factorisations.

b Which of the following is the most complete factorisation?

$2(3x + 6)(x + 3)$ $\quad$ $6(x + 2)(x + 3)$ $\quad$ $3(x + 2)(2x + 6)$

Explain your choice.

11.3 Solving quadratic equations by factorisation

This section will show you how to:
- solve a quadratic equation by factorisation

Key words
factors
solve

Solving the quadratic equation $x^2 + ax + b = 0$

To **solve** a quadratic equation such as $x^2 - 2x - 3 = 0$, you first have to be able to factorise it. Work through Examples 14 to 16 below to see how this is done.

EXAMPLE 14

Solve $x^2 + 6x + 5 = 0$

This factorises into $(x + 5)(x + 1) = 0$.

The only way this expression can ever equal 0 is if the value of one of the brackets is 0.
Hence either $(x + 5) = 0$ or $(x + 1) = 0$

$\Rightarrow x + 5 = 0$ or $x + 1 = 0$

$\Rightarrow x = -5$ or $x = -1$

So the solution is $x = -5$ or $x = -1$.

EXAMPLE 15

Solve $x^2 + 3x - 10 = 0$

This factorises into $(x + 5)(x - 2) = 0$.

Hence either $(x + 5) = 0$ or $(x - 2) = 0$

$\Rightarrow x + 5 = 0$ or $x - 2 = 0$

$\Rightarrow x = -5$ or $x = 2$.

So the solution is $x = -5$ or $x = 2$.

EXAMPLE 16

Solve $x^2 - 6x + 9 = 0$

This factorises into $(x - 3)(x - 3) = 0$.

The equation has repeated roots.

That is: $(x - 3)^2 = 0$

Hence, there is only one solution, $x = 3$.

EXERCISE 11H

Solve the equations in questions **1–12**.

1 $(x + 2)(x + 5) = 0$

2 $(t + 3)(t + 1) = 0$

3 $(a + 6)(a + 4) = 0$

4 $(x + 3)(x - 2) = 0$

5 $(x + 1)(x - 3) = 0$

6 $(t + 4)(t - 5) = 0$

7 $(x - 1)(x + 2) = 0$

8 $(x - 2)(x + 5) = 0$

9 $(a - 7)(a + 4) = 0$

10 $(x - 3)(x - 2) = 0$

11 $(x - 1)(x - 5) = 0$

12 $(a - 4)(a - 3) = 0$

First factorise, then solve the equations in questions **13–26**.

13 $x^2 + 5x + 4 = 0$

14 $x^2 + 11x + 18 = 0$

15 $x^2 - 6x + 8 = 0$

16 $x^2 - 8x + 15 = 0$

17 $x^2 - 3x - 10 = 0$

18 $x^2 - 2x - 15 = 0$

19 $t^2 + 4t - 12 = 0$

20 $t^2 + 3t - 18 = 0$

21 $x^2 - x - 2 = 0$

22 $x^2 + 4x + 4 = 0$

23 $m^2 + 10m + 25 = 0$

24 $t^2 - 8t + 16 = 0$

25 $t^2 + 8t + 12 = 0$

26 $a^2 - 14a + 49 = 0$

PS **27** A woman is x years old. Her husband is three years younger. The product of their ages is 550.

a Set up a quadratic equation to represent this situation.

b How old is the woman?

HINTS AND TIPS

If one solution to a real-life problem is negative, reject it and only give the positive answer.

PS **28** A rectangular field is 40 m longer than it is wide.
The area is 48 000 square metres.

The farmer wants to place a fence all around the field.

How long will the fence be?

HINTS AND TIPS

Let the width be x, set up a quadratic equation and solve it to get x.

First rearrange the equations in questions **29–37**, then solve them.

HINTS AND TIPS

You cannot solve a quadratic equation by factorisation unless it is in the form
$x^2 + ax + b = 0$

29 $x^2 + 10x = -24$

30 $x^2 - 18x = -32$

31 $x^2 + 2x = 24$

32 $x^2 + 3x = 54$

33 $t^2 + 7t = 30$

34 $x^2 - 7x = 44$

35 $t^2 - t = 72$

36 $x^2 = 17x - 72$

37 $x^2 + 1 = 2x$

AU **38** A teacher asks her class to solve $x^2 - 3x = 4$.

This is Mario's answer.

$x^2 - 3x - 4 = 0$

$(x - 4)(x + 1) = 0$

Hence $x - 4 = 0$ or $x + 1 = 0$

$x = 4$ or -1

This is Sylvan's answer.

$x(x - 3) = 4$

Hence $x = 4$ or $x - 3 = 4 \Rightarrow x = -3 + 4 = -1$

When the teacher reads out the answer of $x = 4$ or -1, both students mark their work as correct.

Who used the correct method and what mistakes did the other student make?

A

Solving the general quadratic equation by factorisation

The general quadratic equation is of the form $ax^2 + bx + c = 0$ where a, b and c are positive or negative whole numbers. (It is easier to make sure that a is always positive.) Before any quadratic equation can be solved by factorisation, it must be rearranged to this form.

The method is similar to that used to solve equations of the form $x^2 + ax + b = 0$. That is, you have to find two **factors** of $ax^2 + bx + c$ with a product of 0.

EXAMPLE 17

Solve these quadratic equations. **a** $12x^2 - 28x = -15$ **b** $30x^2 - 5x - 5 = 0$

a First, rearrange the equation to the general form.

$12x^2 - 28x + 15 = 0$

This factorises into $(2x - 3)(6x - 5) = 0$.

The only way this product can equal 0 is if the value of one of the brackets is 0. Hence:

either $2x - 3 = 0$ or $6x - 5 = 0$

$\Rightarrow 2x = 3$ or $6x = 5$

$\Rightarrow x = \frac{3}{2}$ or $x = \frac{5}{6}$

So the solution is $x = 1\frac{1}{2}$ or $x = \frac{5}{6}$

Note: It is almost always the case that if a solution is a fraction which is then changed into a rounded-off decimal number, the original equation cannot be evaluated exactly, using that decimal number. So it is preferable to leave the solution in its fraction form. This is called the *rational form.*

b This equation is already in the general form and it will factorise to $(15x + 5)(2x - 1) = 0$ or $(3x + 1)(10x - 5) = 0$.

Look again at the equation. There is a common factor of 5 which can be taken out to give:

$5(6x^2 - x - 1 = 0)$

This is much easier to factorise to $5(3x + 1)(2x - 1) = 0$, which can be solved to give $x = -\frac{1}{3}$ or $x = \frac{1}{2}$

Special cases

Sometimes the values of b and c are zero. (Note that if a is zero the equation is no longer a quadratic equation but a linear equation. These were covered in Chapter 7.)

EXAMPLE 18

Solve these quadratic equations. **a** $3x^2 - 4 = 0$ **b** $4x^2 - 25 = 0$ **c** $6x^2 - x = 0$

a Rearrange to get $3x^2 = 4$.

Divide both sides by 3: $x^2 = \frac{4}{3}$

Take the square root on both sides: $x = \pm\sqrt{\frac{4}{3}} = \pm\frac{2}{\sqrt{3}} = \pm\frac{2\sqrt{3}}{3}$

Note: A square root can be positive or negative. The answer is in *surd form* (see Chapter 10).

b You can use the method of part **a** or you should recognise this as the difference of two squares (page 302). This can be factorised to $(2x - 5)(2x + 5) = 0$.

Each set of brackets can be put equal to zero.

$2x - 5 = 0 \Rightarrow x = +\frac{5}{2}$

$2x + 5 = 0 \Rightarrow x = -\frac{5}{2}$ So the solution is $x = \pm\frac{5}{2}$

c There is a common factor of x, so factorise as $x(6x - 1) = 0$.

There is only one set of brackets this time but each factor can be equal to zero, so $x = 0$ or $6x - 1 = 0$.

Hence, $x = 0$ or $\frac{1}{6}$

EXERCISE 11I

Give your answers either in rational form or as mixed numbers.

1 Solve these equations.

a $3x^2 + 8x - 3 = 0$
b $6x^2 - 5x - 4 = 0$
c $5x^2 - 9x - 2 = 0$
d $4t^2 - 4t - 35 = 0$
e $18t^2 + 9t + 1 = 0$
f $3t^2 - 14t + 8 = 0$
g $6x^2 + 15x - 9 = 0$
h $12x^2 - 16x - 35 = 0$
i $15t^2 + 4t - 35 = 0$
j $28x^2 - 85x + 63 = 0$
k $24x^2 - 19x + 2 = 0$
l $16t^2 - 1 = 0$
m $4x^2 + 9x = 0$
n $25t^2 - 49 = 0$
o $9m^2 - 24m - 9 = 0$

HINTS AND TIPS

Look out for the special cases where b or c is zero.

2 Rearrange these equations into the general form and then solve them.

a $x^2 - x = 42$
b $8x(x + 1) = 30$
c $(x + 1)(x - 2) = 40$
d $13x^2 = 11 - 2x$

A

e $(x + 1)(x - 2) = 4$

f $10x^2 - x = 2$

g $8x^2 + 6x + 3 = 2x^2 + x + 2$

h $25x^2 = 10 - 45x$

i $8x - 16 - x^2 = 0$

j $(2x + 1)(5x + 2) = (2x - 2)(x - 2)$

k $5x + 5 = 30x^2 + 15x + 5$

l $2m^2 = 50$

m $6x^2 + 30 = 5 - 3x^2 - 30x$

n $4x^2 + 4x - 49 = 4x$

o $2t^2 - t = 15$

AU 3 Here are three equations.

A: $(x - 1)^2 = 0$ B: $3x + 2 = 5$ C: $x^2 - 4x = 5$

a Give some mathematical fact that equations A and B have in common.

b Give a mathematical reason why equation B is different from equations A and C.

PS 4 Pythagoras' theorem states that the sum of the squares of the two short sides of a right-angled triangle equals the square of the long side (hypotenuse).

A right-angled triangle has sides $5x - 1$, $2x + 3$ and $x + 1$ cm.

a Show that: $20x^2 - 24x - 9 = 0$

b Find the area of the triangle.

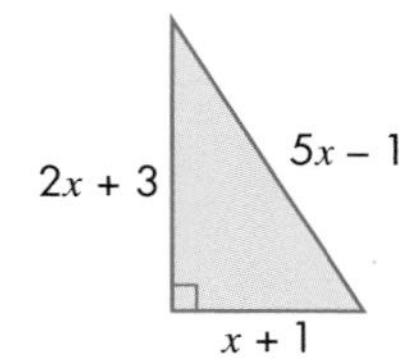

11.4 Solving a quadratic equation by the quadratic formula

This section will show you how to:

- solve a quadratic equation by using the quadratic formula

Key words
coefficient
constant term
quadratic formula
soluble
solve

Many quadratic equations cannot be solved by factorisation because they do not have simple factors. Try to factorise, for example, $x^2 - 4x - 3 = 0$ or $3x^2 - 6x + 2 = 0$. You will find it is impossible.

One way to **solve** this type of equation is to use the **quadratic formula**. This formula can be used to solve *any* quadratic equation that is **soluble**. (Some are not, which the quadratic formula would immediately show. See Chapter 7.)

The solution of the equation $ax^2 + bx + c = 0$ is given by:

$$x = \frac{-b \pm \sqrt{b^2 - 4ac}}{2a}$$

where a and b are the **coefficients** of x^2 and x respectively and c is the **constant** term.

This is the quadratic formula. It is given on the formula sheet of GCSE examinations but it is best to learn it.

The symbol ± states that the square root has a positive and a negative value, *both* of which must be used in solving for x.

EXAMPLE 19

Solve $5x^2 - 11x - 4 = 0$, giving solutions correct to 2 decimal places.

Take the quadratic formula:

$$x = \frac{-b \pm \sqrt{b^2 - 4ac}}{2a}$$

and put $a = 5$, $b = -11$ and $c = -4$, which gives:

$$x = \frac{(-11) \pm \sqrt{(-11)^2 - 4(5)(-4)}}{2(5)}$$

Note that the values for a, b and c have been put into the formula in brackets. This is to avoid mistakes in calculation. It is a very common mistake to get the sign of b wrong or to think that -11^2 is -121. Using brackets will help you do the calculation correctly.

$$x = \frac{11 \pm \sqrt{121 + 80}}{10} = \frac{11 \pm \sqrt{201}}{10}$$

$\Rightarrow x = 2.52$ or -0.32

Note: The calculation has been done in stages. With a calculator it is possible just to work out the answer, but make sure you can use your calculator properly. If not, break the calculation down. Remember the rule 'if you try to do two things at once, you will probably get one of them wrong'.

Examination tip: If you are asked to solve a quadratic equation to one or two decimal places, you can be sure that it can be solved *only* by the quadratic formula.

EXERCISE 11J

Use the quadratic formula to solve the equations in questions **1** to **15**. Give your answers to 2 decimal places.

HINTS AND TIPS

Use brackets when substituting and do not try to work two things out at the same time.

1 $2x^2 + x - 8 = 0$

2 $3x^2 + 5x + 1 = 0$

3 $x^2 - x - 10 = 0$

4 $5x^2 + 2x - 1 = 0$

5 $7x^2 + 12x + 2 = 0$

6 $3x^2 + 11x + 9 = 0$

7 $4x^2 + 9x + 3 = 0$

8 $6x^2 + 22x + 19 = 0$

9 $x^2 + 3x - 6 = 0$

10 $3x^2 - 7x + 1 = 0$

11 $2x^2 + 11x + 4 = 0$

12 $4x^2 + 5x - 3 = 0$

13 $4x^2 - 9x + 4 = 0$

14 $7x^2 + 3x - 2 = 0$

15 $5x^2 - 10x + 1 = 0$

A

A

FM 16 A rectangular lawn is 2 m longer than it is wide.

The area of the lawn is 21 m^2. The gardener wants to edge the lawn with edging strips, which are sold in lengths of $1\frac{1}{2}$ m. How many will she need to buy?

AU 17 Shaun is solving a quadratic equation, using the formula.

He correctly substitutes values for a, b and c to get:

$$x = \frac{3 \pm \sqrt{37}}{2}$$

What is the equation Shaun is trying to solve?

PS 18 Terry uses the quadratic formula to solve $4x^2 - 4x + 1 = 0$.

June uses factorisation to solve $4x^2 - 4x + 1 = 0$.

They both find something unusual in their solutions.

Explain what this is, and why.

11.5 Solving a quadratic equation by completing the square

This section will show you how to:
- solve a quadratic equation by completing the square

Key words
completing the square
square root
surd

Another method for solving quadratic equations is **completing the square**. This method can be used to give answers to a specified number of decimal places or to leave answers in **surd** form.

You will remember that:

$$(x + a)^2 = x^2 + 2ax + a^2$$

which can be rearranged to give:

$$x^2 + 2ax = (x + a)^2 - a^2$$

This is the basic principle behind completing the square.

There are three basic steps in rewriting $x^2 + px + q$ in the form $(x + a)^2 + b$.

Step 1: Ignore q and just look at the first two terms, $x^2 + px$.

Step 2: Rewrite $x^2 + px$ as $\left(x + \frac{p}{2}\right)^2 - \left(\frac{p}{2}\right)^2$.

Step 3: Bring q back to get $x^2 + px + q = \left(x + \frac{p}{2}\right)^2 - \left(\frac{p}{2}\right)^2 + q$.

Note: p is always even so the numbers involved are whole numbers.

EXAMPLE 20

Rewrite the following in the form $(x \pm a) \pm b$.

a $x^2 + 6x - 7$

b $x^2 - 8x + 3$

a Ignore −7 for the moment.

Rewrite $x^2 + 6x$ as $(x + 3)^2 - 9$

(Expand $(x + 3)^2 - 9 = x^2 + 6x + 9 - 9 = x^2 + 6x$. The 9 is subtracted to get rid of the constant term when the brackets are expanded.)

Now bring the −7 back, so $x^2 + 6x - 7 = (x + 3)^2 - 9 - 7$

Combine the constant terms to get the final answer: $x^2 + 6x - 7 = (x + 3)^2 - 16$

b Ignore +3 for the moment.

Rewrite $x^2 - 8x$ as $(x - 4)^2 - 16$

(Note that you still subtract $(-4)^2$, as $(-4)^2 = +16$)

Now bring the +3 back, so $x^2 - 8x + 3 = (x - 4)^2 - 16 + 3$

Combine the constant terms to get the final answer: $x^2 - 8x + 3 = (x - 4)^2 - 13$

EXAMPLE 21

Rewrite $x^2 + 4x - 7$ in the form $(x + a)^2 - b$. Hence solve the equation $x^2 + 4x - 7 = 0$, giving your answers to 2 decimal places.

Note that:

$$x^2 + 4x = (x + 2)^2 - 4$$

So:

$$x^2 + 4x - 7 = (x + 2)^2 - 4 - 7 = (x + 2)^2 - 11$$

When $x^2 + 4x - 7 = 0$, you can rewrite the equations completing the square as: $(x + 2)^2 - 11 = 0$

Rearranging gives $(x + 2)^2 = 11$

Taking the **square root** of both sides gives:

$x + 2 = \pm\sqrt{11}$ This answer is in surd form and could be left like this, but you

$\Rightarrow x = -2 \pm \sqrt{11}$ are asked to evaluate it to 2 decimal places.

$\Rightarrow x = 1.32$ or -5.32 (to 2 decimal places)

EXAMPLE 22

Solve $x^2 - 6x - 1 = 0$ by completing the square. Leave your answer in the form $a \pm \sqrt{b}$.

$x^2 - 6x = (x - 3)^2 - 9$

So $x^2 - 6x - 1 = (x - 3)^2 - 9 - 1 = (x - 3)^2 - 10$

When $x^2 - 6x - 1 = 0$, then $(x - 3)^2 - 10 = 0$

$\Rightarrow (x - 3)^2 = 10$

Taking the square root of both sides gives:

$x - 3 = \pm\sqrt{10}$

$\Rightarrow x = 3 \pm\sqrt{10}$

EXERCISE 11K

1 Write an equivalent expression in the form $(x \pm a)^2 - b$.

a $x^2 + 4x$ **b** $x^2 + 14x$ **c** $x^2 - 6x$ **d** $x^2 + 6x$

e $x^2 - 4x$ **f** $x^2 + 6x$ **g** $x^2 - 10x$ **h** $x^2 + 20x$

i $x^2 + 10x$ **j** $x^2 + 8x$ **k** $x^2 - 2x$ **l** $x^2 + 2x$

2 Write an equivalent expression in the form $(x \pm a)^2 - b$.

Question **1** will help with **a** to **h**.

a $x^2 + 4x - 1$ **b** $x^2 + 14x - 5$ **c** $x^2 - 6x + 3$ **d** $x^2 + 6x + 7$

e $x^2 - 4x - 1$ **f** $x^2 + 6x + 3$ **g** $x^2 - 10x - 5$ **h** $x^2 + 20x - 1$

i $x^2 + 8x - 6$ **j** $x^2 + 2x - 1$ **k** $x^2 - 2x - 7$ **l** $x^2 + 2x - 9$

3 Solve the following equations by completing the square. Leave your answers in surd form where appropriate. The answers to question **2** will help.

a $x^2 + 4x - 1 = 0$ **b** $x^2 + 14x - 5 = 0$ **c** $x^2 - 6x + 3 = 0$

d $x^2 + 6x + 7 = 0$ **e** $x^2 - 4x - 1 = 0$ **f** $x^2 + 6x + 3 = 0$

g $x^2 - 10x - 5 = 0$ **h** $x^2 + 20x - 1 = 0$ **i** $x^2 + 8x - 6 = 0$

j $x^2 + 2x - 1 = 0$ **k** $x^2 - 2x - 7 = 0$ **l** $x^2 + 2x - 9 = 0$

4 Solve by completing the square. Give your answers to 2 decimal places.

a $x^2 + 2x - 5 = 0$ **b** $x^2 - 4x - 7 = 0$ **c** $x^2 + 2x - 9 = 0$

5 Prove that the solutions to the equation $x^2 + bx + c = 0$ are:

$$-\frac{b}{2} \pm \sqrt{\left(\frac{b^2}{4} - c\right)}$$

A

A*

A*

AU 6 Dave rewrites the expression $x^2 + px + q$ by completing the square.
He correctly does this and gets $(x - 7)^2 - 52$.

What are the values of p and q?

PS 7 **a** Frankie writes the steps to solve $x^2 + 6x + 7 = 0$ by completing the square on sticky notes. Unfortunately he drops them and they get out of order. Can you put the notes in the correct order?

Take –2 over the equals sign	Take +3 over the equals sign	Write $x^2 + 6x + 7 = 0$ as $(x + 3)^2 - 2 = 0$	Take the square root of both sides

b Write down the stages as in part **a** needed to solve the equation $x^2 - 4x - 3 = 0$

c Solve the equations below, giving the answers in surd form.

i $x^2 + 6x + 7 = 0$ **ii** $x^2 - 4x - 3 = 0$

PS 8 Rearrange the following statements to give the complete solution, using the method of completing the square to the equation: $ax^2 + bx + c = 0$

A: $x = -\frac{b}{2a} \pm \sqrt{\frac{b^2}{4a^2} - \frac{c}{a}}$

B: $\left(\left(x + \frac{b}{2a}\right)^2 - \frac{b^2}{4a^2}\right) + \frac{c}{a} = 0$

C: $a\left(\left(x + \frac{b}{2a}\right)^2 - \frac{b^2}{4a^2}\right) + c = 0$

D: $\left(x + \frac{b}{2a}\right)^2 - \frac{b^2}{4a^2} + \frac{c}{a}$

E: $\left(x + \frac{b}{2a}\right)^2 - \frac{b^2}{4a^2} + \frac{c}{a} = 0$

F: $x = -\frac{b}{2a} \pm \sqrt{\frac{b^2}{4a^2} - \frac{4ac}{4a^2}}$

G: $x = -\frac{b}{2a} \pm \frac{1}{2a}\sqrt{b^2 - 4ac}$

H: $a\left(x^2 + \frac{b}{a}x\right) + c = 0$

I: $x = \frac{-b \pm \sqrt{b^2 - 4ac}}{2a}$

J: $x + \frac{b}{2a} = \pm\sqrt{\frac{b^2}{4a^2} - \frac{c}{a}}$

11.6 Problems involving quadratic equations

This section will show you how to:

- recognise why some quadratic equations cannot be factorised
- solve practical problems, using quadratic equations

Key words
discriminant

Quadratic equations with no solution

The quantity $(b^2 - 4ac)$ in the quadratic formula is known as the **discriminant**.

When $b^2 > 4ac$, $(b^2 - 4ac)$ is positive. This has been the case in almost all of the quadratics you have solved so far and it means there are two solutions.

When $b^2 = 4ac$, $(b^2 - 4ac)$ is zero. This has been the case in some of the quadratics you have solved so far. It means there is only one solution (the repeated root).

When $b^2 < 4ac$, $(b^2 - 4ac)$ is negative. So you need to find the square root of a negative number.

Such a square root cannot be found (at GCSE level) and therefore there are no solutions. You will not be asked about this in examinations but if it happens then you will have made a mistake and should check your working.

EXAMPLE 23

Find the discriminant $b^2 - 4ac$ of the equation $x^2 + 3x + 5 = 0$ and explain what the result tells you.

$b^2 - 4ac = (3)^2 - 4(1)(5) = 9 - 20 = -11.$

This means there are no solutions for x.

You will meet quadratic graphs in Chapter 12. All quadratic equations can be shown as graphs that have a characteristic shape known as a parabola.

Here are the graphs of the three types of quadratic equations: one with two solutions $(b^2 - 4ac > 0)$, one with one solution $(b^2 - 4ac = 0)$ and one with no solutions $(b^2 - 4ac < 0)$.

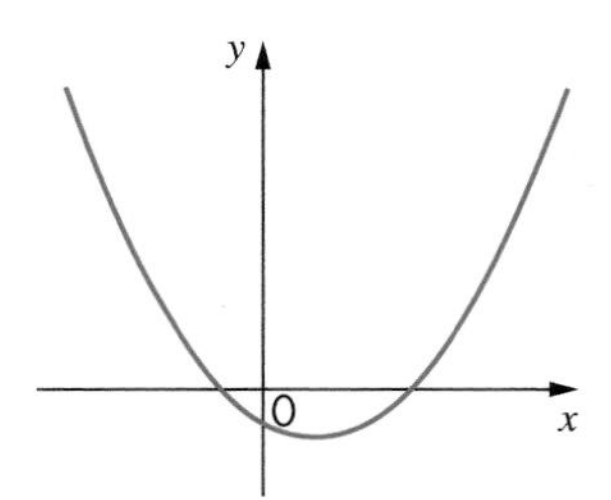

$b^2 - 4ac > 0$ two solutions, crosses the x-axis twice

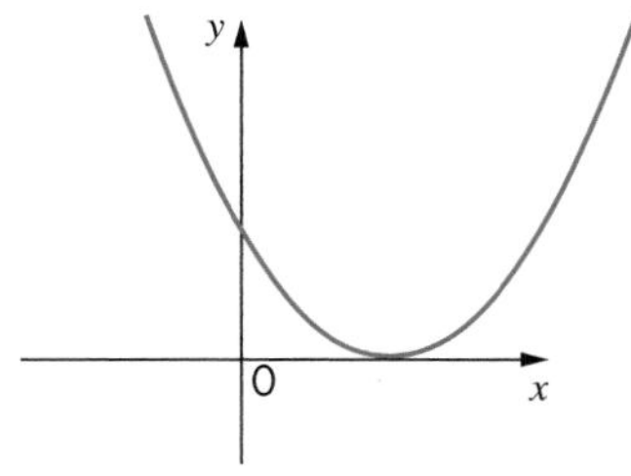

$b^2 - 4ac = 0$ one solution, just touches the x-axis

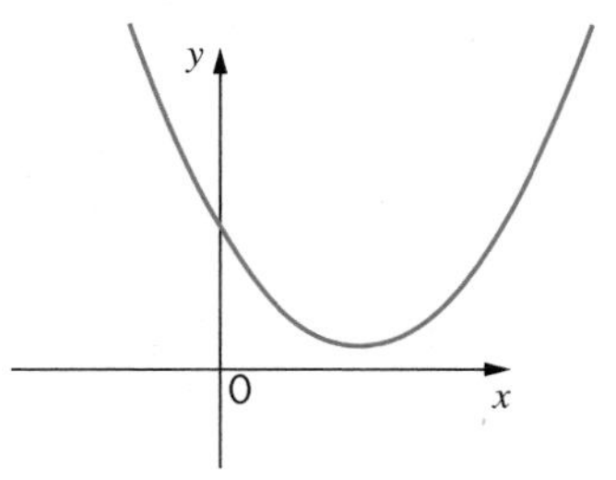

$b^2 - 4ac < 0$ no solution, does not cross the x-axis

EXERCISE 11L

A*

Work out the discriminant $b^2 - 4ac$ of the equations in questions **1** to **12**. In each case say how many solutions the equation has.

1 $3x^2 + 2x - 4 = 0$

2 $2x^2 - 7x - 2 = 0$

3 $5x^2 - 8x + 2 = 0$

4 $3x^2 + x - 7 = 0$

5 $16x^2 - 23x + 6 = 0$

6 $x^2 - 2x - 16 = 0$

7 $5x^2 + 5x + 3 = 0$

8 $4x^2 + 3x + 2 = 0$

9 $5x^2 - x - 2 = 0$

10 $x^2 + 6x - 1 = 0$

11 $17x^2 - x + 2 = 0$

12 $x^2 + 5x - 3 = 0$

PS 13 Bill works out the discriminant of the quadratic equation $x^2 + bx - c = 0$ as $b^2 - 4ac = 13$. There are four possible equations that could lead to this discriminant. What are they?

Problems solved by quadratic equations

You are likely to have to solve a problem which involves generating a quadratic equation and finding its solution.

EXAMPLE 24

Find the sides of the right-angled triangle shown in the diagram.

x + 5

13

x − 2

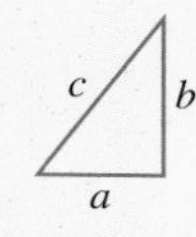

The sides of a right-angled triangle are connected by Pythagoras' theorem, which says that $c^2 = a^2 + b^2$.

$(x + 5)^2 + (x - 2)^2 = 13^2$

$(x^2 + 10x + 25) + (x^2 - 4x + 4) = 169$

$2x^2 + 6x + 29 = 169$

$2x^2 + 6x - 140 = 0$

Divide by a factor of 2: $x^2 + 3x - 70 = 0$

This factorises to: $(x + 10)(x - 7) = 0$

This gives $x = -10$ or 7.

Reject the negative value as it would give negative lengths.

Hence the sides of the triangle are 5, 12 and 13.

Note: You may know the Pythagorean triple 5, 12, 13 and guessed the answer but you would be expected to show working. Most 'real-life' problems will end up with a quadratic that factorises, as the questions are complicated enough without expecting you to use the quadratic formula.

EXAMPLE 25

Solve this equation. $2x - \frac{3}{x} = 5$

Multiply through by x to give: $2x^2 - 3 = 5x$

Rearrange into the general form: $2x^2 - 5x - 3 = 0$

This factorises to: $(2x + 1)(x - 3) = 0$

So $x = -\frac{1}{2}$ or $x = 3$.

EXAMPLE 26

A coach driver undertook a journey of 300 km. Her actual average speed turned out to be 10 km/h less than expected. Therefore, she took 1 hour longer over the journey than expected. Find her actual average speed.

Let the driver's actual average speed be x km/h.

So the estimated speed would have been $(x + 10)$ km/h.

Time taken = $\frac{\text{distance travelled}}{\text{speed}}$

At x km/h, she did the journey in $\frac{300}{x}$ hours.

At $(x + 10)$ km/h, she would have done the journey in $\frac{300}{x + 10}$ hours.

Since the journey took 1 hour longer than expected, then:

time taken = $\frac{300}{x + 10} + 1 = \frac{300 + x + 10}{x + 10} = \frac{300 + x}{x + 10}$

So = $\frac{300}{x} = \frac{310 + x}{x + 10} \Rightarrow 300(x + 10) = x(310 + x) \Rightarrow 300x + 3000 = 310x + x^2$

Rearranging into the general form gives: $x^2 + 10x - 3000 = 0$

This factorises into: $(x + 60)(x - 50) = 0 \Rightarrow x = -60$ or 50

The coach's average speed could not be –60 km/h, so it has to be 50 km/h.

EXERCISE 11M

PS 1 The length of a rectangle is 5 m more than its width. Its area is 300 m^2. Find the actual dimensions of the rectangle.

FM 2 The average mass of a group of people is 45.2 kg. A newcomer to the group weighs 51 kg, which increases the average mass by 0.2 kg. How many people are now in the group?

3 Solve the equation $x + \frac{3}{x} = 7$. Give your answers correct to 2 decimal places.

A*

4 Solve the equation $2x + \frac{5}{x} = 11$.

PS 5 A tennis court has an area of 224 m^2. If the length were decreased by 1 m and the width increased by 1 m, the area would be increased by 1 m^2. Find the dimensions of the court.

PS 6 On a journey of 400 km, the driver of a train calculates that if he were to increase his average speed by 2 km/h, he would take 20 minutes less. Find his average speed.

PS 7 The difference of the squares of two positive numbers, whose difference is 2, is 184. Find these two numbers.

PS 8 The length of a carpet is 1 m more than its width. Its area is 9 m^2. Find the dimensions of the carpet to 2 decimal places.

FM 9 Helen worked out that she could save 30 minutes on a 45 km journey if she travelled at an average speed which was 15 km/h faster than that at which she had planned to travel. Find the speed at which Helen had originally planned to travel.

FM 10 Claire intended to spend £3.20 on balloons for her party. But each balloon cost her 2p more than she expected, so she had to buy eight fewer balloons. Find the cost of each balloon.

PS 11 The sum of a number and its reciprocal is 2.05. What are the two numbers?

FM 12 A woman buys goods for £$60x$ and sells them for £$(600 - 6x)$ at a loss of x%. Find x.

FM 13 A train has a scheduled time for its journey. If the train averages 50 km/h, it arrives 12 minutes early. If the train averages 45 km/h, it arrives 20 minutes late. Find how long the train should take for the journey.

14 A rectangular garden measures 15 m by 11 m and is surrounded by a path of uniform width of area 41.25 m^2. Find the width of the path.

15 A rectangular room is 3 m longer than it is wide.

It cost £364 to carpet the room. Carpet costs £16 per square metre.

How wide is the room?

HINTS AND TIPS

Calculate the area and set up a quadratic equation to work out the width.

GRADE BOOSTER

- **C** You can expand a pair of linear brackets to get a quadratic expression
- **B** You can factorise a quadratic expression of the form $x^2 + ax + b$
- **B** You can solve a quadratic equation of the form $x^2 + ax + b = 0$
- **A** You can factorise a quadratic expression of the form $ax^2 + bx + c$
- **A** You can solve a quadratic equation of the form $ax^2 + bx + c = 0$ by factorisation
- **A** You can solve a quadratic equation of the form $ax^2 + bx + c = 0$ by using the quadratic formula
- **A*** You can solve a quadratic equation by completing the square

What you should know now

- How to expand linear brackets
- How to solve quadratic equations by factorisation, the quadratic formula and completing the square

EXAMINATION QUESTIONS

1 Expand and simplify $(e + 3)(e + 4)$ *(2 marks)*

Edexcel, June 2007, Paper 11 Higher, Question 6(c)

2 Expand and simplify $(x + 3)(x - 5)$ *(2 marks)*

Edexcel, March 2008, Paper 11 Higher, Question 5(b)

3 Expand and simplify $(x - 2)(x + 1)$ *(2 marks)*

Edexcel, March 2008, Paper 10 Higher, Question 7

4 Expand and simplify $(2x + 5)(x - 2)$ *(2 marks)*

Edexcel, November 2008, Paper 3 Higher, Question 22(b)

5 Expand and simplify $(y + 2)(y + 3)$ *(2 marks)*

Edexcel, November 2008, Paper 10 Higher, Question 6(a)

6 Factorise $t^2 - 16$ *(1 mark)*

Edexcel, November 2008, Paper 11 Higher, Question 6(c)

7 Expand and simplify $(x + 4)(x - 3)$ *(2 marks)*

Edexcel, November 2008, Paper 11 Higher, Question 6

8 Expand and simplify $(x + 5)(x + 8)$ *(2 marks)*

Edexcel, March 2009, Paper 10 Higher, Question 4(b)

9 Expand and simplify $(3x - 5)(x + 1)$ *(2 marks)*

Edexcel, March 2007, Paper 11 Higher, Question 7

10 **a** Expand and simplify $(x + 3)(x + 4)$ *(2 marks)*

b Factorise $y^2 + 8y + 15$ *(2 marks)*

Edexcel, June 2008, Paper 4 Higher, Question 11(d & e)

11 Factorise $a^2 - 16a + 64$ *(2 marks)*

Edexcel, June 2007, Paper 11 Higher, Question 6(c)

12 Factorise $x^2 + 2x - 15$ *(2 marks)*

Edexcel, November 2007, Paper 11 Higher, Question 6

13 Expand and simplify $(3x + 4)(5x - 1)$ *(2 marks)*

Edexcel, November 2007, Paper 11 Higher, Question 9

14 Factorise $x^2 - 6x + 5$ *(2 marks)*

Edexcel, March 2008, Paper 11 Higher (5543H/11A), Question 8

15 Factorise $x^2 - 36$ *(1 mark)*

Edexcel, May 2008, Paper 3 Higher, Question 15(b)

16 Expand and simplify $(2x - 1)(5x + 3)$ *(3 marks)*

Edexcel, June 2008, Paper 11 Higher, Question 7

17 Expand and simplify $(2x - 3)(3x + 2)$ *(2 marks)*

Edexcel, March 2008, Paper 11 Higher, Question 9(a)

18 **a** Factorise $x^2 - y^2$ *(1 mark)*

Hence, or otherwise,

b Factorise $(x + 1)^2 - (y + 1)^2$ *(2 marks)*

Edexcel, March 2009, Paper 10 Higher, Question 8

19 **a** Show that the equation

$$\frac{5}{x + 2} = \frac{4 - 3x}{x - 1}$$

Can be rearranged to give
$3x^2 + 7x - 13 = 0$ *(3 marks)*

b Solve $3x^2 + 7x - 13 = 0$

Give your solutions correct to 2 decimal places. *(3 marks)*

Edexcel, June 2008, Paper 4 Higher, Question 23

20 The diagram below shows a six-sided shape.

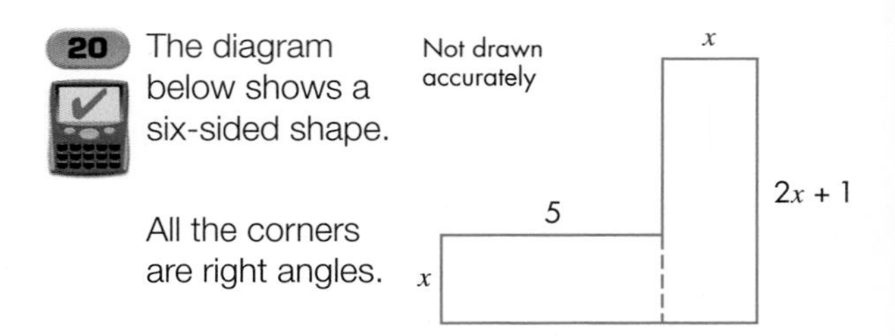

All the corners are right angles.

All the measurements are given in centimetres.

The area of the shape is 95 cm^2.

a Show that $2x^2 + 6x - 95 = 0$ *(3 marks)*

b Solve the equation $2x^2 + 6x - 95 = 0$

Give your solutions correct to 3 significant figures. *(3 marks)*

Edexcel, November 2008, Paper 4 Higher, Question 19

A B C

Worked Examination Questions

1 You are given that:

$(2x + b)^2 + c = ax^2 - 4x - 5$

Calculate the values a, b and c

1 $4x^2 + 4bx + b^2 + c = ax^2 - 4x - 5$

Expand and simplify the left-hand side. This scores 1 mark for method and 1 mark for accuracy.

$4x^2 = ax^2 \Rightarrow a = 4$

$4bx = -4x \Rightarrow b = -1$

$b^2 + c = -5 \Rightarrow c = -6$

Equate the terms in x^2, x and the constant term. Solve the resulting equations. This scores 1 mark for method and 1 mark for accuracy.

Total: 4 marks

2 **a** Factorise: $2n^2 + 9n + 9$

b Hence, or otherwise, write 299 as the product of two prime factors.

2 a $(2n + 3)(n + 3)$

Factorise in the normal way. This scores 1 mark for method and 1 mark for accuracy.

b Let $n = 10$

$2 \times 10^2 + 9 \times 10 + 9 = 299$

$(2 \times 10 + 3)(10 + 3) = 23 \times 13$

Substitute $n = 10$ to find the prime factors. This scores 1 mark for method and 1 mark for accuracy.

Total: 4 marks

Worked Examination Questions

FM **3** The area of a rectangular room is 24 m^2.
The height of the room is 2.4 m
The length of the room is x metres.
The width of the room is 2 m shorter than the length.

a Show that $x^2 - 2x - 24 = 0$

b The two long walls and one of the short walls are to be wallpapered.
A roll of wallpaper is 12 m long and 80 cm wide.
How many rolls of paper will be needed?

3 a $x(x - 2) = 24$
$x^2 - 2x - 24 = 0$

Set up an equation using the area and rearrange into a standard quadratic. This gets 1 mark.

b $(x - 6)(x + 4) = 0$
$x = 6$ (ignore -4)

Solve the equation to find x. This gets 1 mark for method and 1 mark for accuracy.

Total length that needs to be wallpapered = 16 m
So number of strips = $16 \div 0.8 = 20$

Work out how much wall needs to be wallpapered. This gets 1 mark for method. Work out how many strips are needed.

Number of strips in a roll $12 \div 2.4 = 5$
So number of rolls = $20 \div 5 = 4$ rolls.

Work out how many strips there are per roll and how many rolls will be needed. This gets 1 mark for method and 1 mark for accuracy.

Total: 6 marks

AU **4** **a** Show that $(a + b)(a - b) \equiv a^2 - b^2$

b Hence simplify $(3x + 1)^2 - (2x - 1)^2$

4 a $(a + b)(a - b) = a^2 + \cancel{ab} - \cancel{ab} - b^2 = a^2 - b^2$

This is just a lead in to part **b**. Make sure you show all four terms and show that two of them cancel out. This gets 1 mark.

b Using the identity from part **a**:

$(3x + 1)^2 - (2x - 1)^2 = (3x + 1 + 2x - 1) \times (3x + 1 - (2x - 1))$

As this is an identity it is true for all values whether numerical or algebraic so substitute the brackets for **a** and **b**. This gets 1 mark for method.

$= 5x(x + 2)$

$= 5x^2 + 10x$

Expand and simplify the brackets. This gets 1 mark for method and 1 mark for accuracy. This is easier than expanding both squared brackets and collecting terms.

Total: 4 marks

Worked Examination Questions

PS **5** The sketch shows four quadratic graphs.

Here are four quadratic equations.

P: $x^2 + 2x + 1 = 0$

Q: $-x^2 + x - 12 = 0$

R: $x^2 + 2x + 6 = 0$

S: $x^2 + 7x + 12 = 0$

Match the graph to the equations.

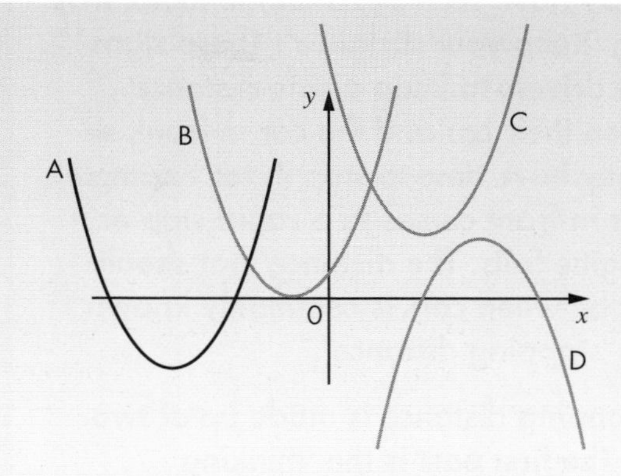

5 *Graph D matches equation Q.* — Start with the obvious one. This gets 1 mark.

Graph B matches equation P. — P factorises to $(x + 1)^2 = 0$ so only has one solution. This gets 1 mark.

Graph A matches equation S. — S factorises to $(x + 3)(x + 4) = 0$ which has solutions –3 or –4. This gets 1 mark.

Graph C matches equation R. — This is the only pair left but the equation does not factorise, and if you try to solve by using the quadratic formula, you end up with the root of a negative number. So there are no solutions.

Total: 3 marks

11 Functional Maths
Stopping distances

You may have seen signs on the motorway saying 'Keep your distance'. These signs advise drivers to keep a safe distance between their car and the car in front, so that they have time to stop if, for example, the car in front comes to a rapid stop or if the engine fails. The distance that should be left between cars is commonly known as the 'stopping distance'.

The stopping distance is made up of two parts. The first part is the 'thinking distance', which is the time it takes for the brain to react and for you to apply the brakes. The second part is the braking distance, which is the distance it takes for the car to come to a complete stop, once the brakes have been applied.

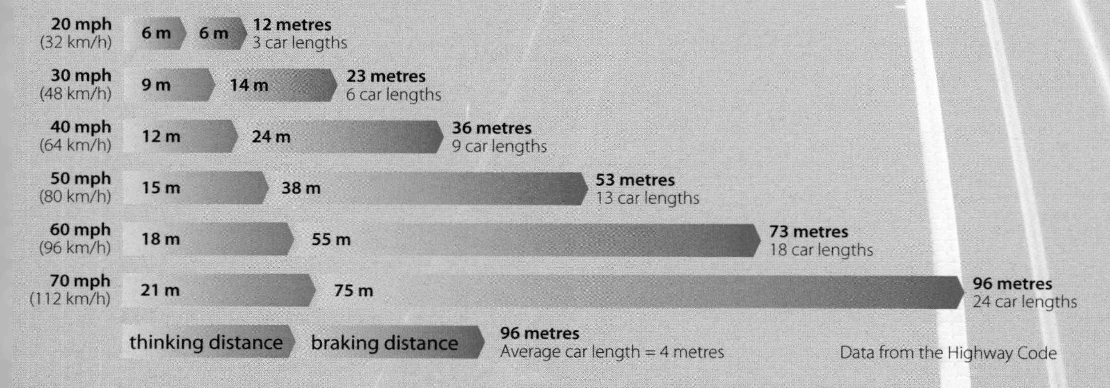

Notice that this diagram is based on 'typical' stopping distances. There are many factors that affect the stopping distance, such as road conditions, how good the brakes are and how heavy the car is, and so the stopping distance will never be the same in every situation.

Your task

This table shows the probability of a crash being fatal at certain speeds.

Speed at which crash happens (mph)	Probability of crash being fatal
70	0.60
60	0.50
50	0.42
40	0.34
30	0.26
20	0.17
10	0.09
5	0.05

James is learning to drive. His driving instructor has been teaching him about typical stopping distances, but James is not convinced that the distances given by his instructor, which are taken from the *Highway Code*, are right.

His instructor sets him a challenge: show whether the distances given in the *Highway Code* are correct.

James finds this formula for calculating stopping distances:

$$d = \frac{s^2}{20} + s$$

where d is the stopping distance (in feet) and s is the speed (in miles per hour).

Remember: 1 foot = 30 cm.

Does he prove his instructor to be right or wrong?

Getting started

Consider the following points to help you complete this task.

- How could you represent thinking, braking and stopping distances as a graph?
- Can you apply James's formula to real-life scenarios?
  - For example:
    If you are driving down a straight road and suddenly, 175 feet in front of you, a pallet falls off a lorry, what is the maximum speed from which you would be able to stop safely?
- How might the weather affect James's formula?
- What are the risks of fatality?

Why this chapter matters

There are many curves that can be seen in everyday life. Did you know that all these curves can be represented mathematically?

Below are a few examples of simple curves that you may have come across. Can you think of others?

Many road signs are circular.

A chain hanging freely between two supports forms a curve called a catenary.

The examples above all show circular-based curves. However, in mathematics, curves can take many shapes. These can be demonstrated using a cone, as shown on the right and below. If you can make a cone out of plasticine or modelling clay, then you can prove this principle yourself. As you look at these curves, try to think of where you have seen them in your own life.

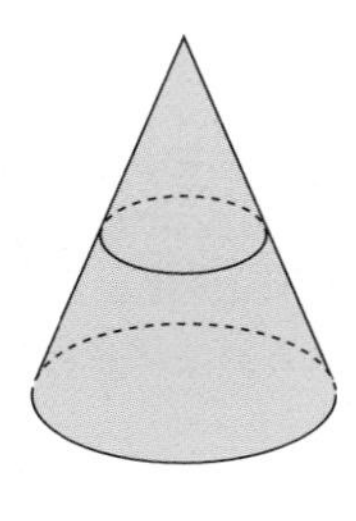

If you slice the cone parallel to the base, the shape you are left with is a circle.

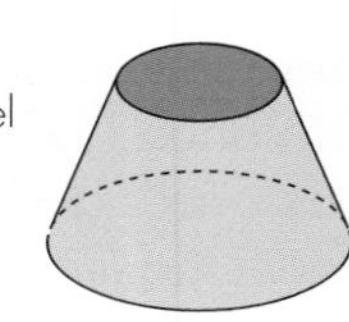

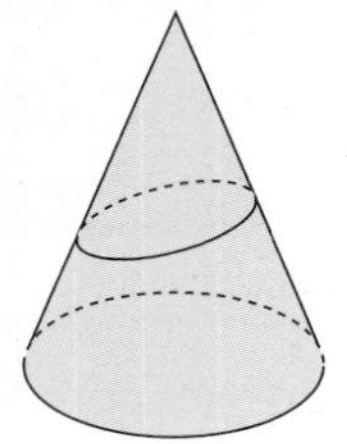

If you slice the cone at an angle to the base, the shape you are left with is an ellipse

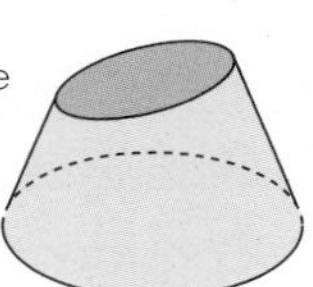

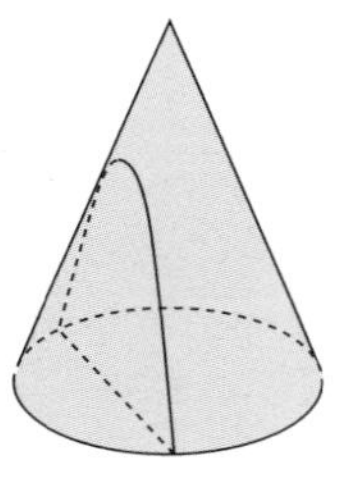

If you slice the cone vertically, the shape you are left with is a hyperbola.

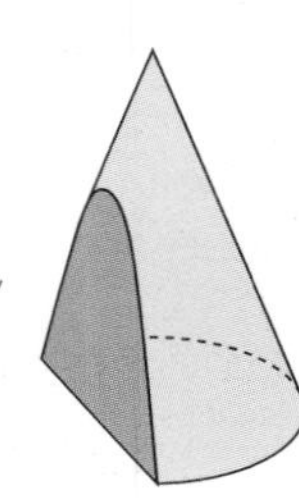

The curve that will be particularly important in this chapter is the parabola. Car headlights are shaped like parabolas.

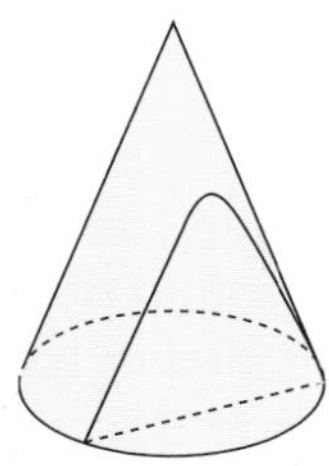

If you slice the cone parallel to its side, the shape you are left with is a parabola.

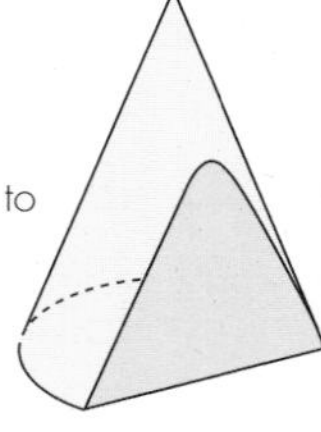

All parabolas are quadratic graphs. During the course of this chapter you will be looking at how to use quadratic equations to draw graphs that have this kind of curve.

The suspension cables on the Humber Bridge are also parabolas.

Chapter

12 Algebra: Graphs and their equations

1. Drawing graphs by the gradient-intercept method
2. Finding the equation of a line from its graph
3. Quadratic graphs
4. The significant points of a quadratic graph
5. The circular function graphs
6. Solving one linear and one non-linear equation by the method of intersection
7. Solving equations by the method of intersection

The grades given in this chapter are target grades.

This chapter will show you ...

- **C** how to draw graphs of linear equations
- **C** how to use graphs to find the solution to linear equations
- **C** how to draw quadratic graphs
- **B** how to find the equation of a linear graph
- **A** how to recognise and find the significant points of a quadratic graph
- **A** how to use graphs to find the solutions to one linear and one non-linear pair of simultaneous equations
- **A*** how to use the method of intersection to solve one quadratic equation, using the graph of another quadratic equation and an appropriate straight line
- **A*** how to use the sine and cosine graphs to find angles with the same sine and cosine between 0° and 360°

Visual overview

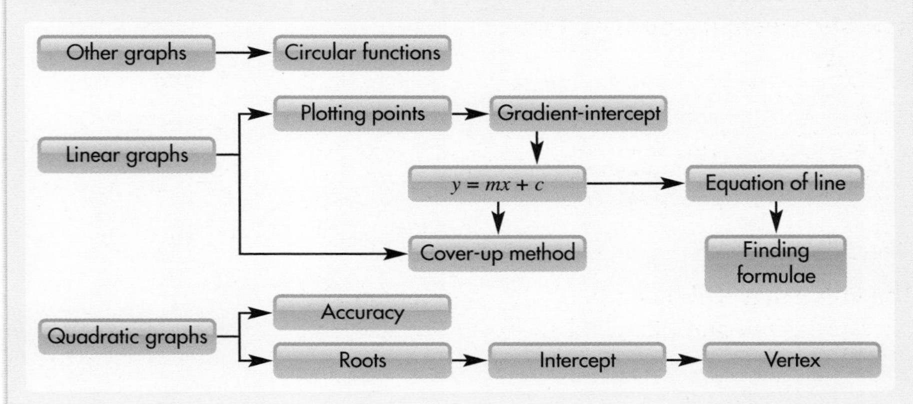

What you should already know

- How to read and plot coordinates **(KS3 level 4, GCSE grade F)**
- How to substitute into simple algebraic functions **(KS3 level 4, GCSE grade F)**
- How to plot a graph from a given table of values **(KS3 level 5, GCSE grade E)**

continued

Quick check

1 This table shows values of $y = 2x + 3$ for $-2 \leqslant x \leqslant 5$.

x	–2	–1	0	1	2	3	4	5
y	–1	1	3	5	7	9	11	

a Complete the table for $x = 5$.

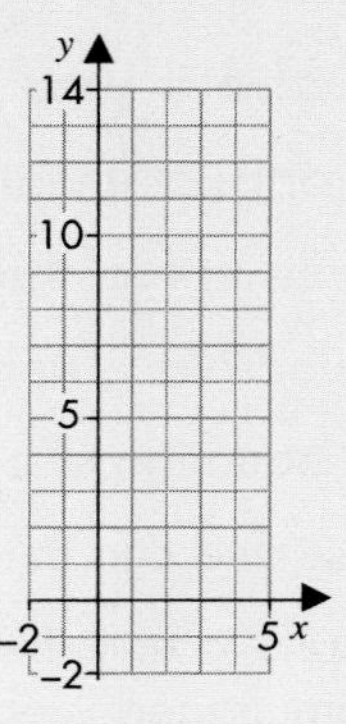

b Copy these axes and plot the points to draw the graph of $y = 2x + 3$.

12.1 Drawing graphs by the gradient-intercept method

This section will show you how to:

- draw graphs using the gradient-intercept method

Key words
coefficient
constant term
cover-up method
gradient-intercept
$y = mx + c$

The ideas that you have discovered in the last activity lead to another way of plotting lines, known as the **gradient-intercept** method.

EXAMPLE 1

Draw the graph of $y = 3x - 1$, using the gradient-intercept method.

- Because the **constant term** is −1, you know that the graph goes through the y-axis at −1. Mark this point with a dot or a cross (**A** on diagram **i**).
- The number in front of x (called the **coefficient** of x) gives the relationship between y and x. 3 is the coefficient of x and this tells you that the y-value is 3 times the x-value, so the gradient of the line is 3. For an x-step of one unit, there is a y-step of three. Starting at −1 on the y-axis, move one square across and three squares up and mark this point with a dot or a cross (**B** on diagram **i**).

Repeat this from every new point. You can also move one square back and three squares down. When enough points have been marked, join the dots (or crosses) to make the graph (diagram **ii**). Note that if the points are not in a straight line, you have made a mistake.

i

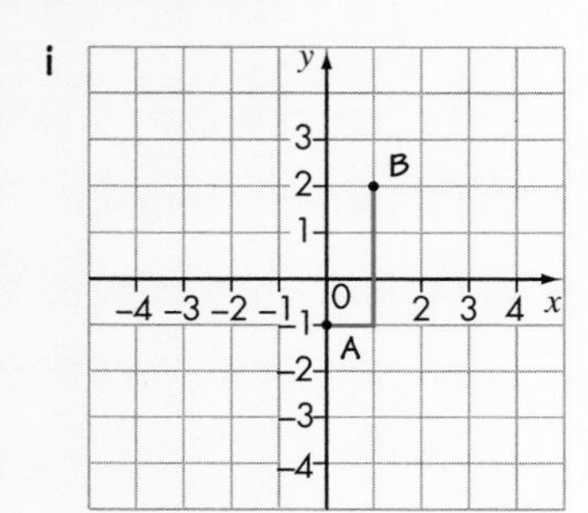

ii

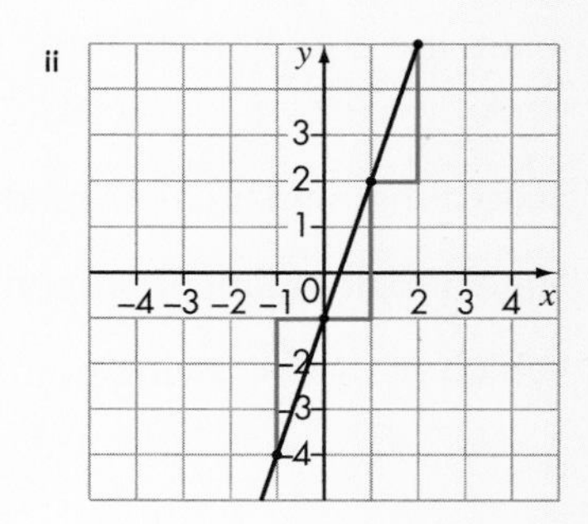

In any equation of the focus $y = mx + c$, the constant term, c, is the intercept on the y-axis and the coefficient of x, m, is the gradient of the line.

EXERCISE 12A

C

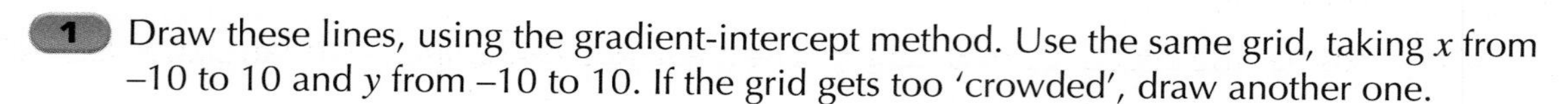

1 Draw these lines, using the gradient-intercept method. Use the same grid, taking x from –10 to 10 and y from –10 to 10. If the grid gets too 'crowded', draw another one.

a $y = 2x + 6$ **b** $y = 3x - 4$ **c** $y = \frac{1}{2}x + 5$

d $y = x + 7$ **e** $y = 4x - 3$ **f** $y = 2x - 7$

g $y = \frac{1}{4}x - 3$ **h** $y = \frac{2}{3}x + 4$ **i** $y = 6x - 5$

j $y = x + 8$ **k** $y = \frac{4}{5}x - 2$ **l** $y = 3x - 9$

2 **a** Using the gradient-intercept method, draw the following lines on the same grid. Use axes with ranges $-6 \leqslant x \leqslant 6$ and $-8 \leqslant y \leqslant 8$.

i $y = 3x + 1$ **ii** $y = 2x + 3$

b Where do the lines cross?

3 **a** Using the gradient-intercept method, draw the following lines on the same grid. Use axes with ranges $-14 \leqslant x \leqslant 4$ and $-2 \leqslant y \leqslant 6$.

i $y = \dfrac{x}{3} + 3$ **ii** $y = \dfrac{x}{4} + 2$

b Where do the lines cross?

4 **a** Using the gradient-intercept method, draw the following lines on the same grid. Use axes with ranges $-4 \leqslant x \leqslant 6$ and $-6 \leqslant y \leqslant 8$.

i $y = x + 3$ **ii** $y = 2x$

b Where do the lines cross?

AU 5 Here are the equations of three lines.

A: $y = 3x - 1$ B: $2y = 6x - 4$ C: $y = 2x - 2$

a State a mathematical property that lines A and B have in common.

b State a mathematical property that lines B and C have in common.

c Which of the following points is the intersection of lines A and C?

(1, –4) (–1, –4) (1, 4)

B

PS 6 **a** What is the gradient of line A?

b What is the gradient of line B?

c What angle is there between lines A and B?

d What relationship do the gradients of A and B have with each other?

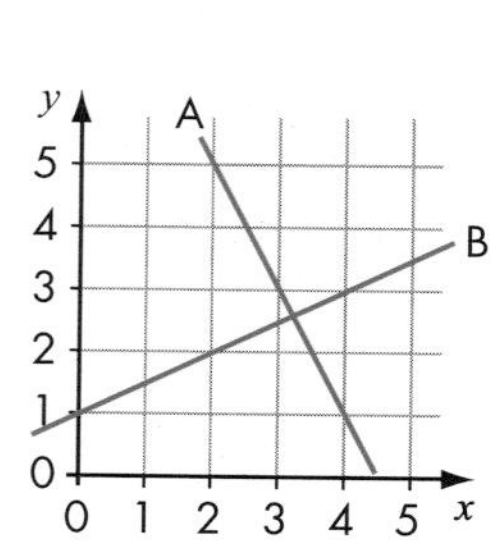

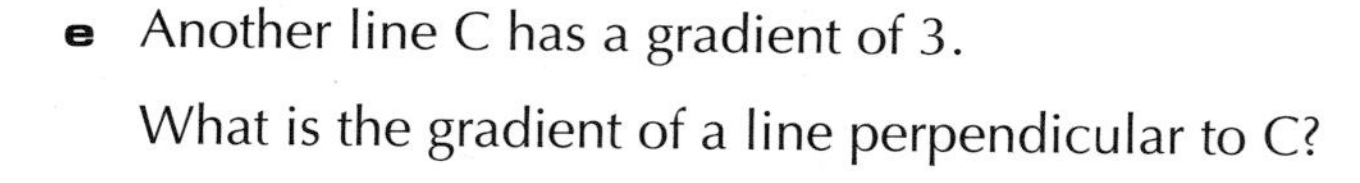

e Another line C has a gradient of 3.

What is the gradient of a line perpendicular to C?

Cover-up method for drawing graphs

The x-axis has the equation $y = 0$. This means that all points on the x-axis have a y-value of 0.

The y-axis has the equation $x = 0$. This means that all points on the y-axis have an x-value of 0.

You can use these facts to draw any line that has an equation of the form:

$$ax + by = c$$

EXAMPLE 2

Draw the graph of $4x + 5y = 20$.

Because the value of x is 0 on the y-axis, you can solve the equation for y:

$$4(0) + 5y = 20$$
$$5y = 20$$
$$\Rightarrow y = 4$$

Hence, the line passes through the point (0, 4) on the y-axis (diagram **A**).

Because the value of y is 0 on the x-axis, you can also solve the equation for x:

$$4x + 5(0) = 20$$
$$4x = 20$$
$$\Rightarrow x = 5$$

Hence, the line passes through the point (5, 0) on the x-axis (diagram **B**). You need only two points to draw a line. (Normally, you would like a third point but in this case you can accept two.) Draw the graph by joining the points (0, 4) and (5, 0) (diagram **C**).

A

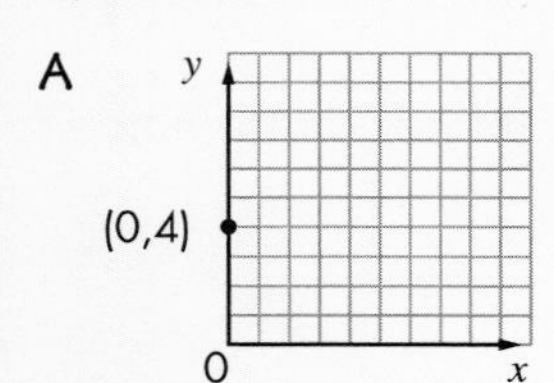

B

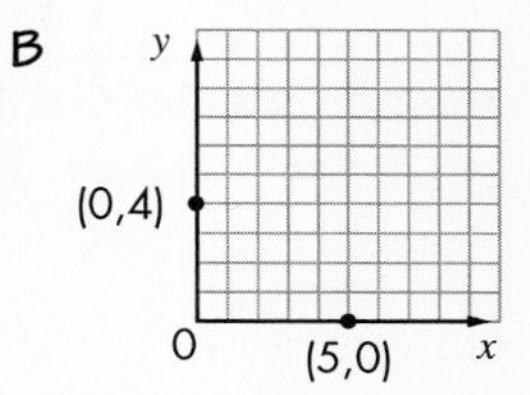

C

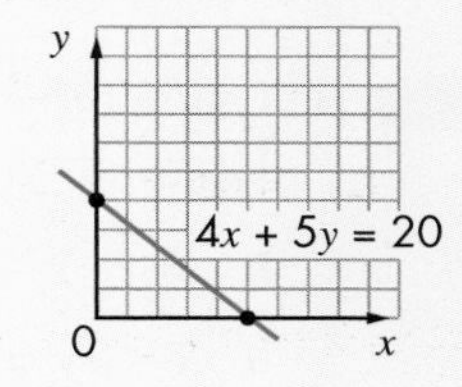

This type of equation can be drawn very easily, without much working at all, using the **cover-up method**.

Start with the equation:	$4x + 5y = 20$
Cover up the x-term:	$+ 5y = 20$
Solve the equation (when $x = 0$):	$y = 4$
Now cover up the y-term:	$4x + \quad = 20$
Solve the equation (when $y = 0$):	$x = 5$

This gives the points (0, 4) on the y-axis and (5, 0) on the x-axis.

EXAMPLE 3

Draw the graph of $2x - 3y = 12$.

Start with the equation: $2x - 3y = 12$

Cover up the x-term: $\square - 3y = 12$

Solve the equation (when $x = 0$): $y = -4$

Now cover up the y-term: $2x + \square = 12$

Solve the equation (when $y = 0$): $x = 6$

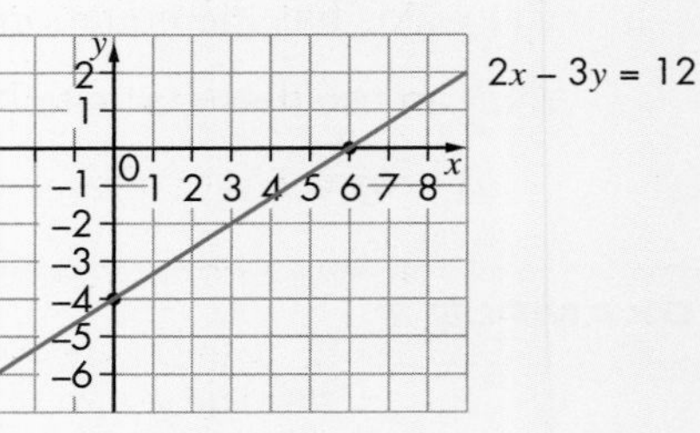

This gives the points $(0, -4)$ on the y-axis and $(6, 0)$ on the x-axis.

EXERCISE 12B

B

1 Draw these lines, using the cover-up method. Use the same grid, taking x from –10 to 10 and y from –10 to 10. If the grid gets too 'crowded', draw another.

a $3x + 2y = 6$ **b** $4x + 3y = 12$ **c** $4x - 5y = 20$

d $x + y = 10$ **e** $3x - 2y = 18$ **f** $x - y = 4$

g $5x - 2y = 15$ **h** $2x - 3y = 15$ **i** $6x + 5y = 30$

j $x + y = -5$ **k** $x + y = 3$ **l** $x - y = -4$

2 **a** Using the cover-up method, draw the following lines on the same grid. Use axes with ranges $-2 \leqslant x \leqslant 6$ and $-2 \leqslant y \leqslant 6$.

i $2x + y = 4$

ii $x - 2y = 2$

b Where do the lines cross?

3 **a** Using the cover-up method, draw the following lines on the same grid.

Use axes with ranges $-2 < x < 6$ and $-3 < y < 6$.

i $x + 2y = 6$

ii $2x - y = 2$

b Where do the lines cross?

4 **a** Using the cover-up method, draw the following lines on the same grid.

Use axes with ranges $-6 \leqslant x \leqslant 8$ and $-2 \leqslant y \leqslant 8$.

i $x + y = 6$

ii $x - y = 2$

b Where do the lines cross?

B

AU 5 Here are the equations of three lines.

A: $2x + 6y = 12$ B: $x - 2y = 6$ C: $x + 3y = -9$

a State a mathematical property that lines A and B have in common.

b State a mathematical property that lines B and C have in common.

c State a mathematical property that lines A and C have in common.

d The line A crosses the y-axis at (0, 2).

The line C crosses the x-axis at (–9, 0).

Find values of a and b so that this line passes through these two points.

$ax + by = 18$

PS 6 The diagram shows an octagon ABCDEFGH.

The equation of the line through A and B is $y = 3$.

The equation of the line through B and C is $x + y = 4$.

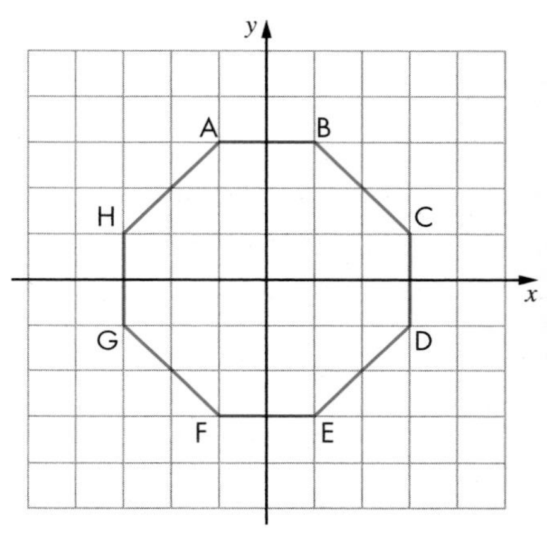

a Write down the equation of the lines through:

i C and D ii D and E iii E and F

iv F and G v G and H vi H and A

b The gradient of the line through F and B is 3.

Write down the gradient of the lines through:

i A and E ii G and C iii H and D

12.2 Finding the equation of a line from its graph

This section will show you how to:

- find the equation of a line, using its gradient and intercept

Key words

coefficient

gradient

intercept

The equation $y = mx + c$

When a graph can be expressed in the form $y = mx + c$, the **coefficient** of x, m, is the **gradient**, and the constant term, c, is the **intercept** on the y-axis.

This means that if you know the gradient, m, of a line and its intercept, c, on the y-axis, you can write down the equation of the line immediately.

For example, if $m = 3$ and $c = -5$, the equation of the line is $y = 3x - 5$.

All linear graphs can be expressed in the form $y = mx + c$.

This gives a method of finding the equation of any line drawn on a pair of coordinate axes.

EXAMPLE 4

Find the equation of the line shown in diagram **A**.

A

B

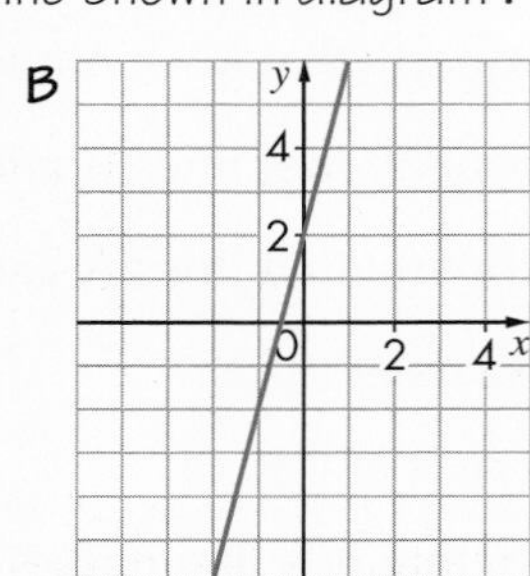

C

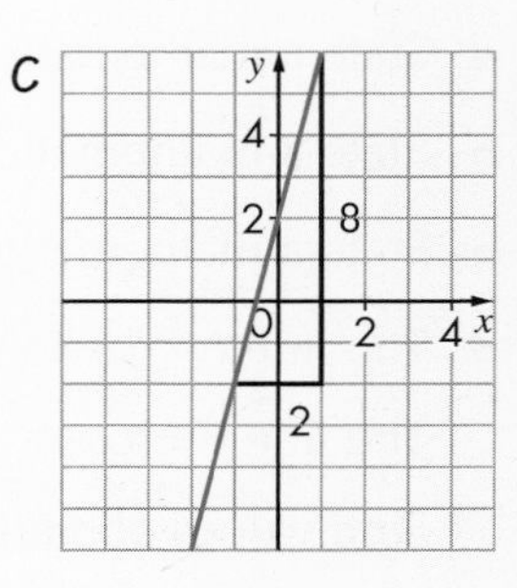

First, find where the graph crosses the y-axis (diagram **B**).

So $c = 2$

Next, measure the gradient of the line (diagram **C**).

y-step = 8

x-step = 2

gradient = 8 ÷ 2 = 4

So $m = 4$

Finally, write down the equation of the line: $y = 4x + 2$

EXERCISE 12C

1 Give the equation of each of these lines, all of which have positive gradients. (Each square represents one unit.)

a

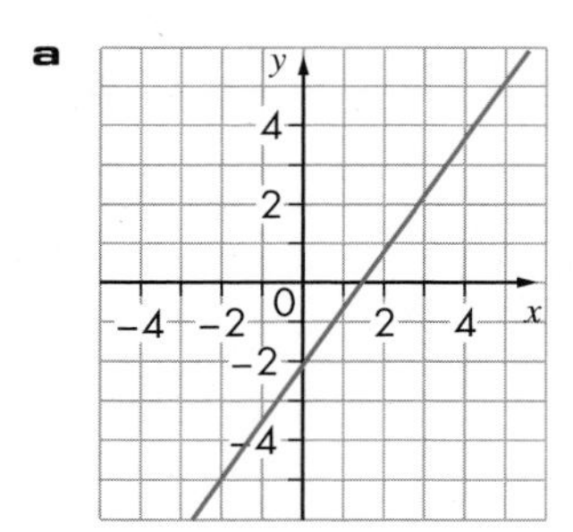

b

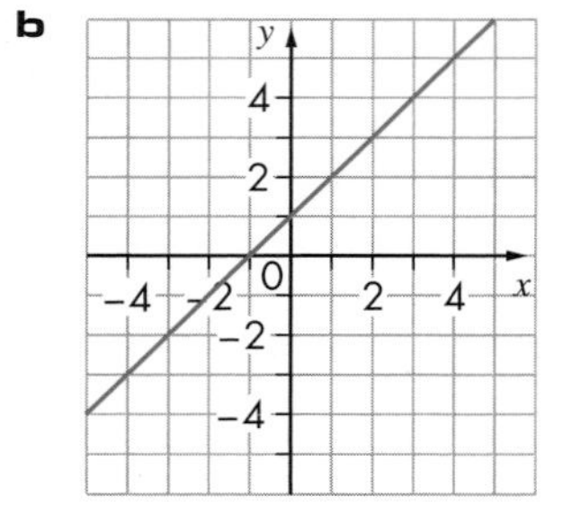

c

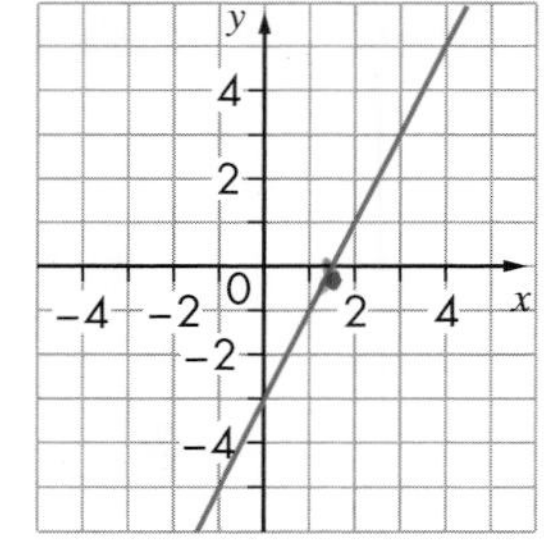

d

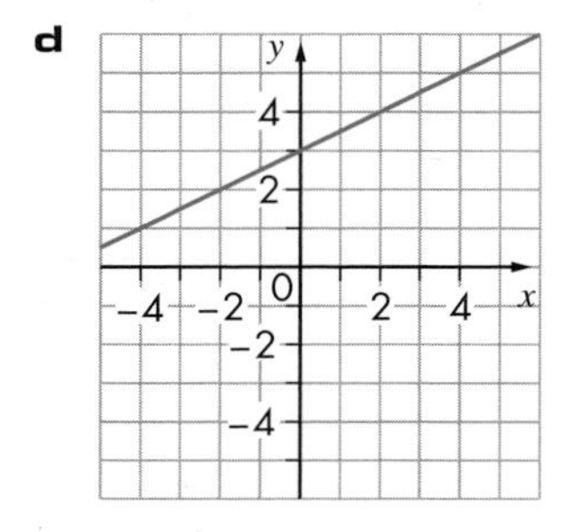

e

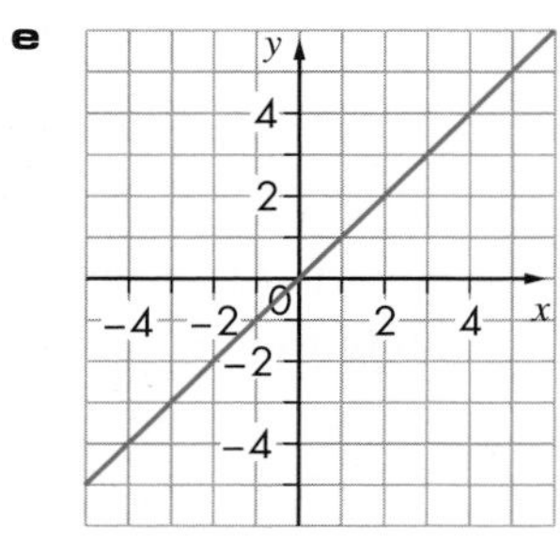

f

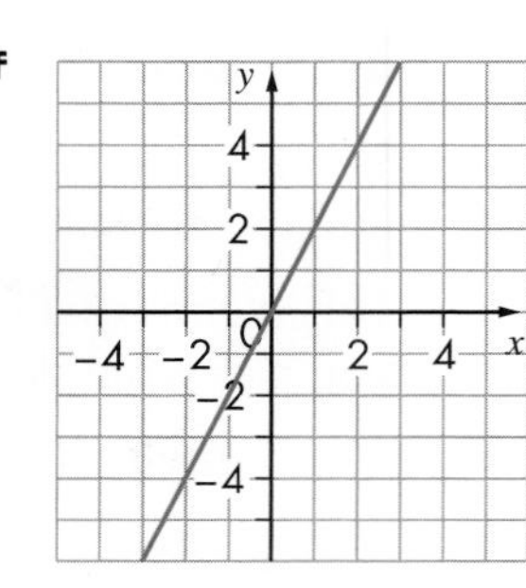

B

PS 2 In each of these grids, there are two lines. (Each square represents one unit.)

a

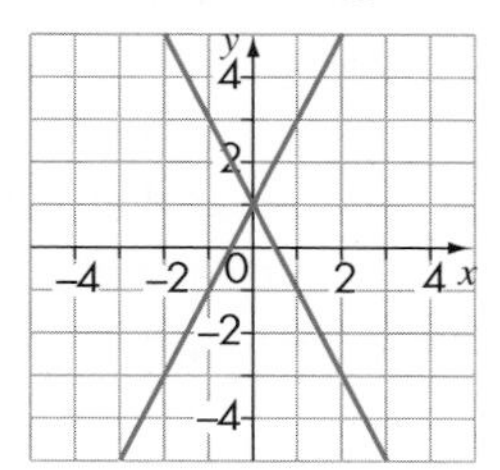

b

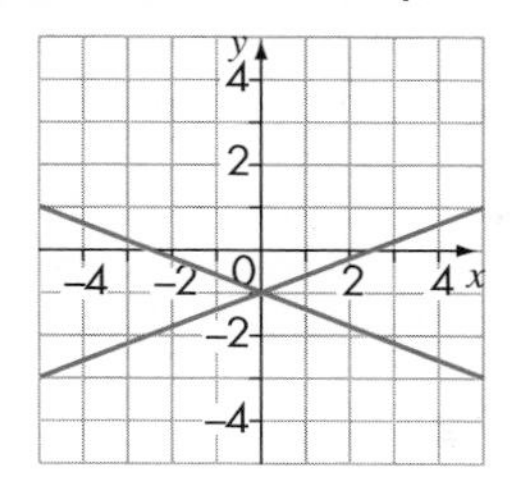

c

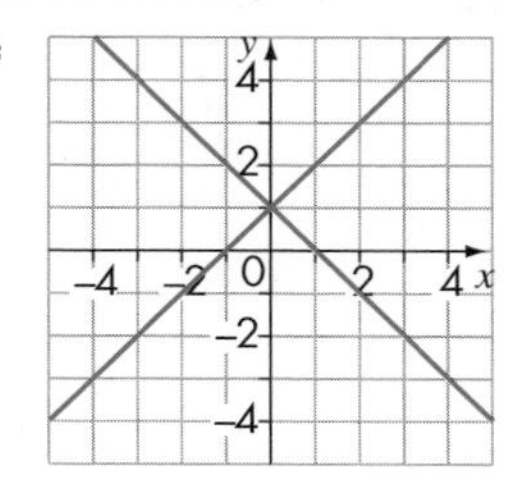

For each grid:

i find the equation of each of the lines

ii describe any symmetries that you can see

iii describe any connection between the gradients of each pair of lines.

AU 3 A straight line passes through the points (1, 3) and (2, 5).

a Explain how you can tell that the line also passes through (0, 1).

b Explain how you can tell that the line has a gradient of 2.

c Work out the equation of the line that passes through (1, 5) and (2, 8).

4 Give the equation of each of these lines, all of which have negative gradients. (Each square represents one unit.)

a

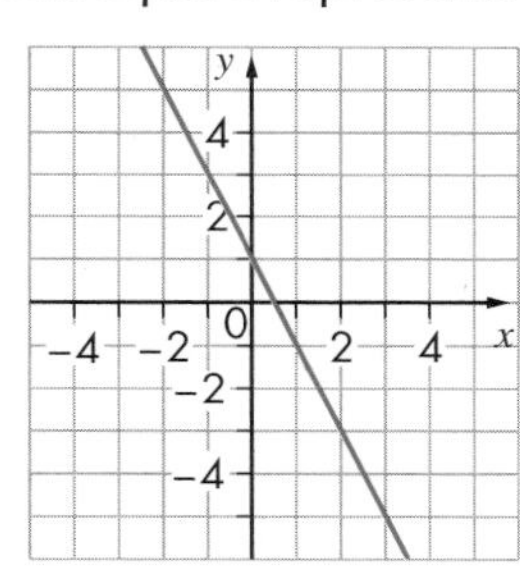

b

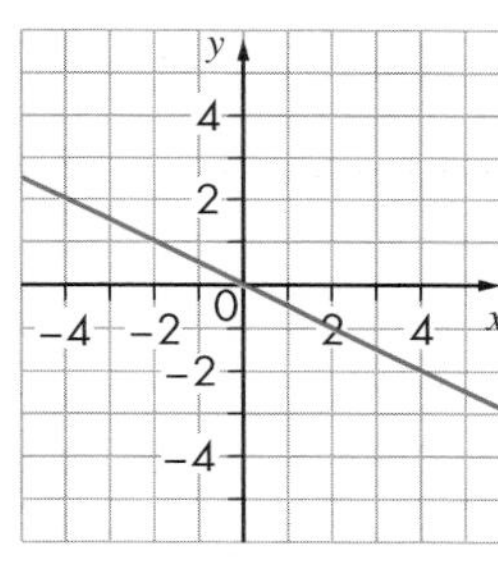

c

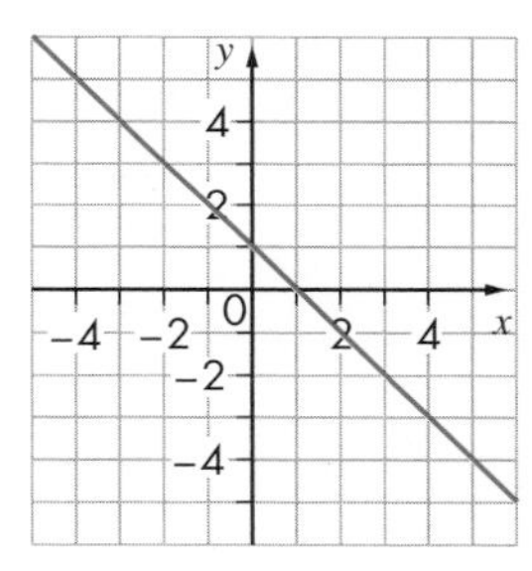

d

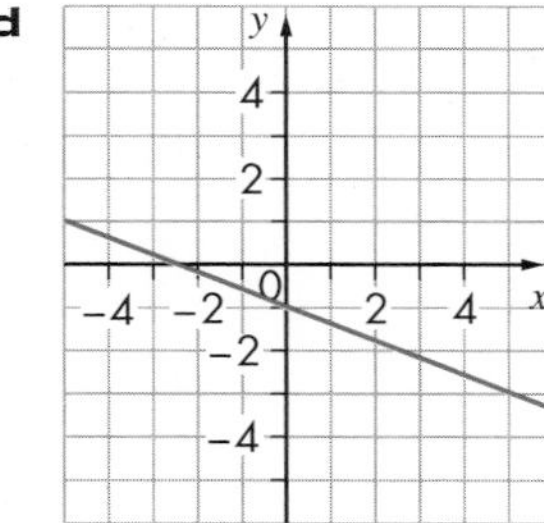

e

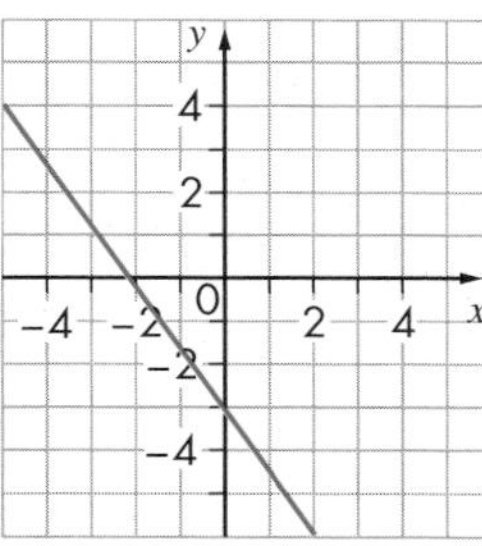

PS 5 In each of these grids, there are three lines. One of them is $y = x$. (Each square represents one unit.)

a

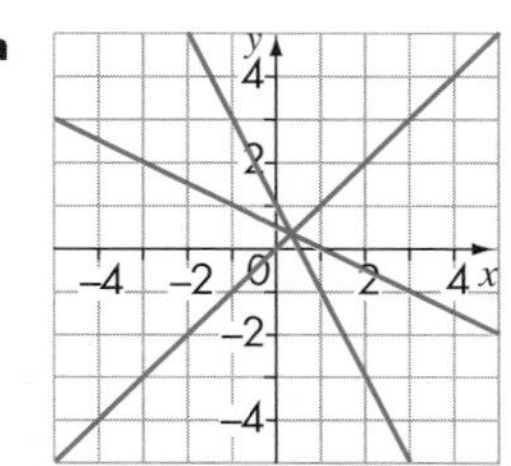

b

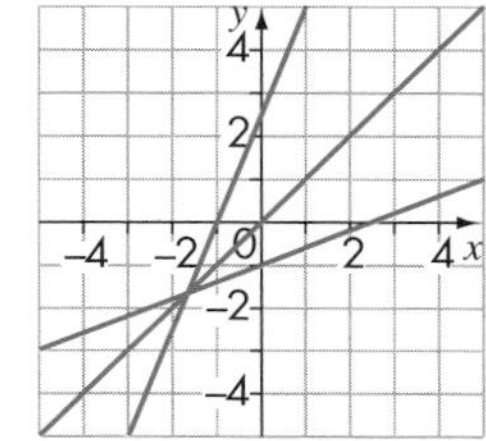

c

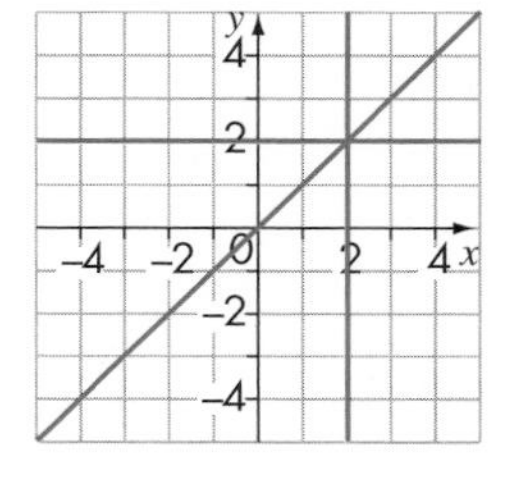

For each grid:

i find the equation of each of the other two lines

ii describe any symmetries that you can see

iii describe any connection between the gradients of each group of lines.

12.3 Quadratic graphs

This section will show you how to:
- draw and read values from quadratic graphs

Key words
parabola
quadratic

A **quadratic** graph has a term in x^2 in its equation. All of the following are quadratic equations and each would produce a quadratic graph.

$y = x^2$ $\quad y = x^2 + 5$ $\quad y = x^2 - 3x$

$y = x^2 + 5x + 6$ $\quad y = 3x^2 - 5x + 4$

EXAMPLE 5

Draw the graph of $y = x^2 + 5x + 6$ for $-5 \leqslant x \leqslant 3$.

Make a table, as shown below. Work out the values in each row (x^2, $5x$, 6) separately, adding them together to obtain the values of y. Then plot the points from the table.

x	–5	–4	–3	–2	–1	0	1	2	3
y^2	25	16	9	4	1	0	1	4	9
$+5x$	–25	–20	–15	–10	–5	0	5	10	15
+6	6	6	6	6	6	6	6	6	6
y	6	2	0	0	2	6	12	20	30

Note that in an examination paper you may be given only the first and last rows, with some values filled in. For example,

x	–5	–4	–3	–2	–1	0	1	2	3
y	6		0		2				30

In this case, you would either construct your own table, or work out the remaining y-values with a calculator.

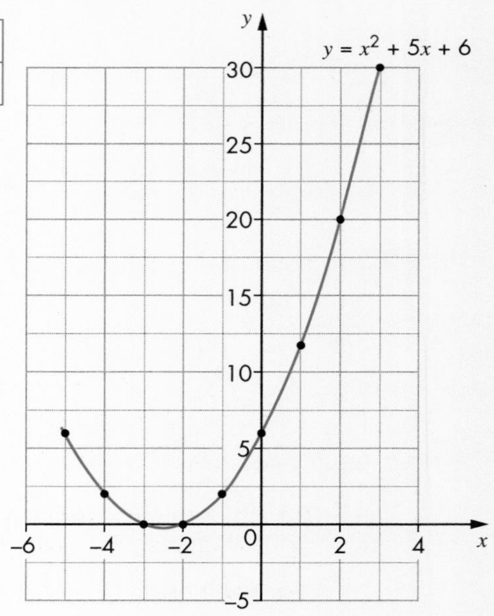

EXAMPLE 6

a Complete the table for $y = 3x^2 - 5x + 4$ for $-1 \leqslant x \leqslant 3$, then draw the graph.

x	–5	–0.5	0	0.5	1	1.5	2	2.5	3
y	12		0	2.25	2			10.25	16

b Use your graph to find the value of y when $x = 2.2$.

c Use your graph to find the values of x that give a y-value of 9.

a The table gives only some values. So you either set up your own table with $3x^2$, $-5x$ and +4, or calculate each y-value. For example, on the majority of scientific calculators, the value for –0.5 will be worked out as:

[3] [×] [(] [(–)] [0] [•] [5] [)] [x²] [–] [5] [×] [(–)] [0] [•] [5] [+] [4] [=]

Check that you get an answer of 7.25

If you want to make sure that you are doing the correct arithmetic with your calculator, try some values for x for which you know the answer. For example, try $x = 0.5$, and see whether your answer is 2.25

The complete table should be:

x	–1	–0.5	0	0.5	1	1.5	2	2.5	3
y	12	7.25	4	2.25	2	3.25	6	10.25	16

The graph is shown on the right.

b To find the corresponding y-value for any value of x, you start on the x-axis at that x-value, go up to the curve, across to the y-axis and read off the y-value. This procedure is marked on the graph with arrows.

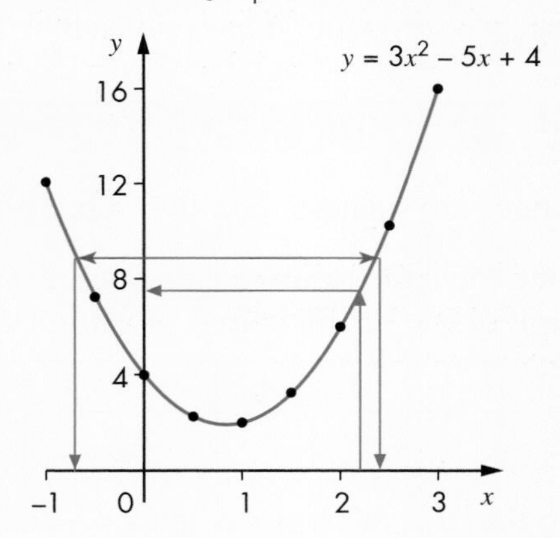

Always show these arrows because even if you make a mistake and misread the scales, you may still get a mark.

When $x = 2.2$, $y = 7.5$.

c This time start at 9 on the y-axis and read off the two x-values that correspond to a y-value of 9. Again, this procedure is marked on the graph with arrows.

When $y = 9$, $x = -0.7$ or $x = 2.4$.

A quadratic curve drawn correctly will always give a smooth curve, called a **parabola**.

Drawing accurate graphs

Although it is difficult to draw accurate curves, examiners work to a tolerance of only 1 mm. Here are some of the more common ways in which marks are lost in an examination.

- When the points are too far apart, a curve tends to 'wobble'.
- Drawing curves in small sections leads to 'feathering'.
- The place where a curve should turn smoothly is drawn 'flat'.
- A line is drawn through a point that, clearly, has been incorrectly plotted.

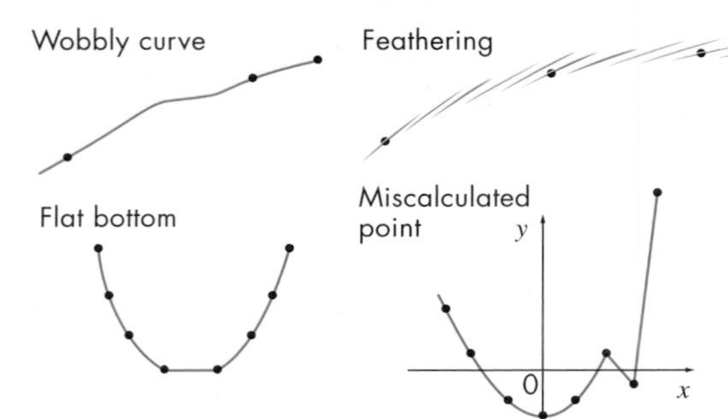

Here are some tips which will make it easier for you to draw smooth, curved lines.

- If you are *right-handed*, turn your paper or exercise book round so that you draw from left to right. Your hand is steadier this way than when you are trying to draw from right to left or away from your body. If you are *left-handed*, you should find drawing from right to left the more accurate way.
- Move your pencil over the points as a practice run without drawing the curve.
- Do one continuous curve and only stop at a plotted point.
- Use a *sharp* pencil and do not press too heavily, so that you may easily rub out mistakes.

Normally, in an examination, grids are provided with the axes clearly marked, so the examiner can place a transparent master over a graph and see immediately whether any lines are badly drawn or points are misplotted. Remember: a tolerance of 1 mm is all that you are allowed.

You do not need to work out all values in a table. You need only to work out the y-value. The other rows in the table are just working lines to break down the calculation. Learn how to calculate y-values with a calculator as there is no credit given for setting up tables in examinations.

EXERCISE 12D

In this exercise, suitable ranges are suggested for the axes. You can use any type of graph paper.

C

1 **a** Copy and complete the table or use a calculator to work out values for the graph of $y = 3x^2$ for values of x from –3 to 3.

x	–3	–2	–1	0	1	2	3
y	27		3			12	

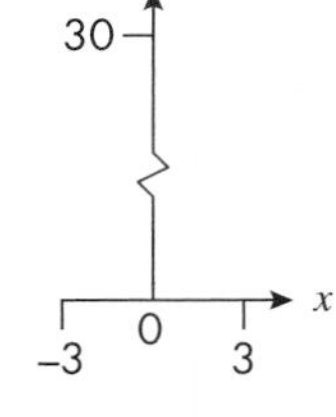

b Use your graph to find the value of y when $x = –1.5$.

c Use your graph to find the values of x that give a y-value of 10.

2 **a** Copy and complete the table or use a calculator to work out values for the graph of $y = x^2 + 2$ for values of x from –5 to 5.

x	–5	–4	–3	–2	–1	0	1	2	3	4	5
$y = x^2 + 2$	27		11					6			

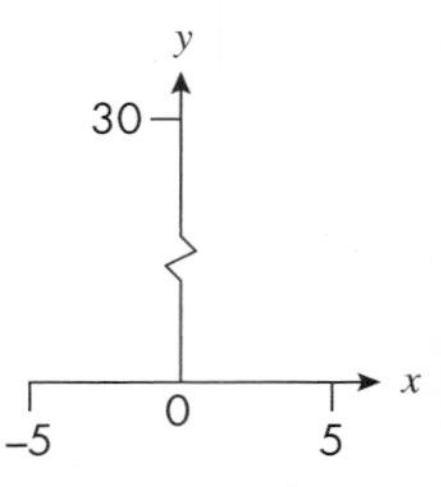

b Use your graph to find the value of y when $x = –2.5$.

c Use your graph to find the values of x that give a y-value of 14.

3 **a** Copy and complete the table or use a calculator to work out values for the graph of $y = x^2 - 2x - 8$ for values of x from –5 to 5.

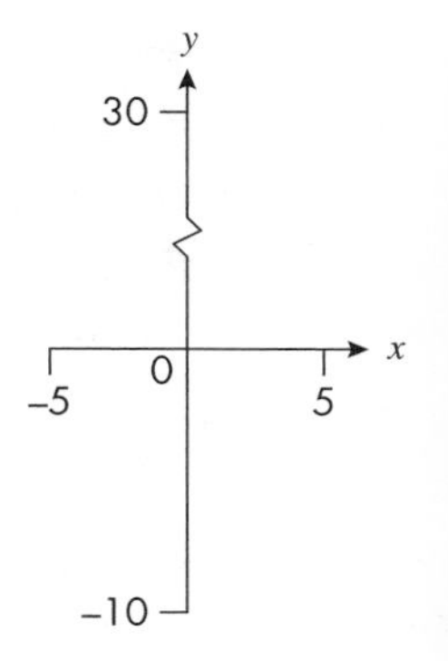

x	–5	–4	–3	–2	–1	0	1	2	3	4	5
x^2	25		9					4			
$-2x$	10							–4			
–8	–8							–8			
y	27							–8			

b Use your graph to find the value of y when $x = 0.5$.

c Use your graph to find the values of x that give a y-value of –3.

4 **a** Copy and complete the table or use a calculator to work out the values for the graph of $y = x^2 + 2x - 1$ for values of x from –3 to 3.

x	–3	–2	–1	0	1	2	3
x^2	9				1	4	
$+2x$	–6		–2			4	
–1	–1	–1				–1	
y	2					7	

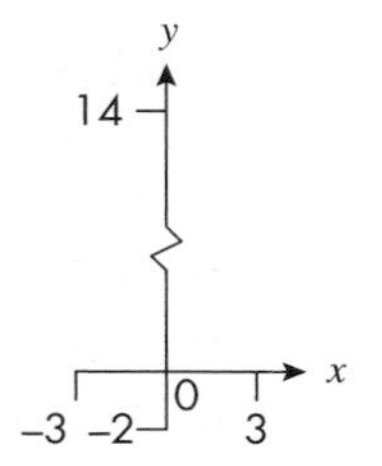

b Use your graph to find the y-value when $x = -2.5$.

c Use your graph to find the values of x that give a y-value of 1.

d On the same axes, draw the graph of $y = \frac{x}{2} + 2$.

e Where do the graphs $y = x^2 + 2x - 1$ and $y = \frac{x}{2} + 2$ cross?

5 **a** Copy and complete the table or use a calculator to work out values for the graph of $y = x^2 - x + 6$ for values of x from –3 to 3.

x	–3	–2	–1	0	1	2	3
x^2	9				1	4	
$-x$	3					–2	
+6	6					6	
y	18					8	

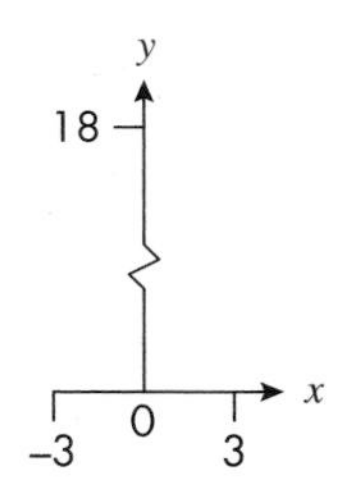

b Use your graph to find the y-value when $x = 2.5$.

c Use your graph to find the values of x that give a y-value of 8.

d Copy and complete the table or use a calculator to draw the graph of $y = x^2 + 5$ on the same axes.

x	–3	–2	–1	0	1	2	3
y	14		6				14

e Where do the graphs $y = x^2 - x + 6$ and $y = x^2 + 5$ cross?

C

6 **a** Copy and complete the table or use a calculator to work out values for the graph of $y = x^2 + 2x + 1$ for values of x from –3 to 3.

x	–3	–2	–1	0	1	2	3
x^2	9				1	4	
$+2x$	–6					4	
$+1$	1					1	
y	4						

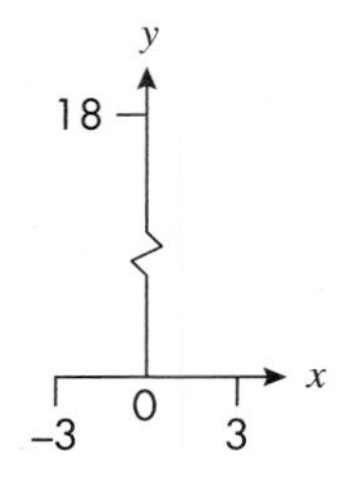

b Use your graph to find the y-value when $x = 1.7$.

c Use your graph to find the values of x that give a y-value of 2.

d On the same axes, draw the graph of $y = 2x + 2$.

e Where do the graphs $y = x^2 + 2x + 1$ and $y = 2x + 2$ cross?

7 **a** Copy and complete the table or use a calculator to work out values for the graph of $y = 2x^2 - 5x - 3$ for values of x from –2 to 4.

x	–2	–1.5	–1	–0.5	0	0.5	1	1.5	2	2.5	3	3.5	4
y	15	9			–3	–5				–3			9

b Where does the graph cross the x-axis?

PS 8 The diagram shows a side elevation of a cone with a cut parallel to one side.

The cone is divided into horizontal sections.

A plan view of the cone is shown.

Construction lines have been drawn to link the elevation and the plan.

Two of the intersecting points have been drawn on the plan.

Two points have also been drawn where the construction lines from the side elevation intersect with the construction lines from the plan.

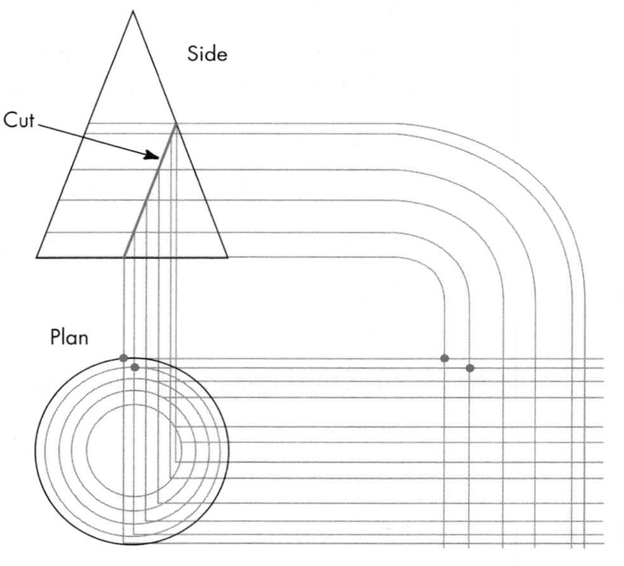

a Plot the rest of the points on the plan and join them with a smooth curve to see the plan view of the parabola.

b Plot the rest of the points on the intersecting lines and join them with a smooth curve to see the parabola.

PS 9 Copy the grid onto centimetre-squared paper.

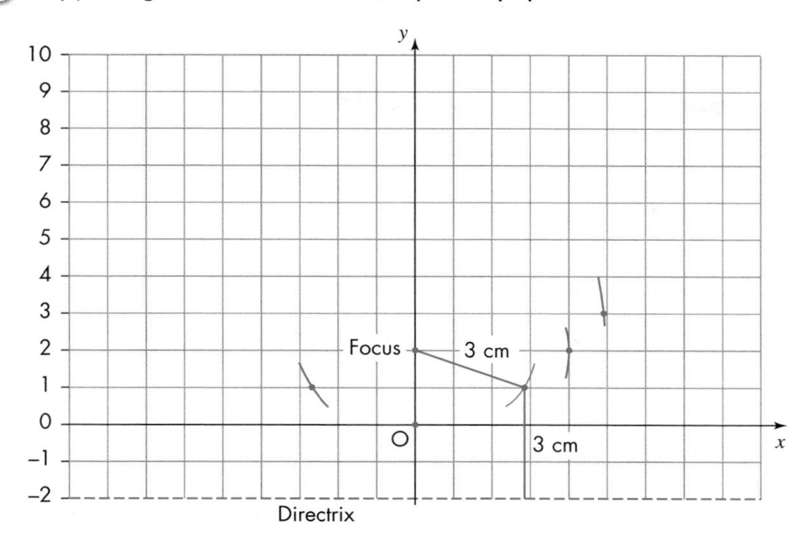

Mark a point at (0, 2). This is the focus.

Draw the line $y = -2$. This is the directrix.

A property of all parabolas is that all the points on a parabola are the same distance from the focus and the directrix.

The origin is 2 units away from both and this will be the lowest point of the parabola.

Set a pair of compasses to a radius of 3 cm. Using the focus as the centre, draw arcs on both sides to intersect with the line $y = 1$, which is 3 cm from the directrix.

Now set the compasses at 4 cm and draw arcs from the focus to intersect with $y = 2$.

Repeat with the compasses set to 5 cm, 6 cm, etc.

Once you have drawn all the points, join them with a smooth curve to show a parabola.

The parabola drawn has the equation $y = \frac{1}{8}x^2$.

AU 10 Here are the equations of three quadratic equations.

Parabola A: $y = 2x^2$

Parabola B: $y = -x^2$

Parabola C: $y = x^2 + 2$

Give a reason why each line may be the odd one out.

C

B

12.4 The significant points of a quadratic graph

This section will show you how to:
- recognise and calculate the significant points of a quadratic graph

Key words
intercept
maximum
minimum
roots
vertex

A quadratic graph has four points that are of interest to a mathematician. These are the points A, B, C and D on the diagram. The x-values at A and B are called the **roots**, and are where the graph crosses the x-axis. C is the point where the graph crosses the y-axis (the **intercept**) and D is the **vertex**, which is the lowest or highest point of the graph.

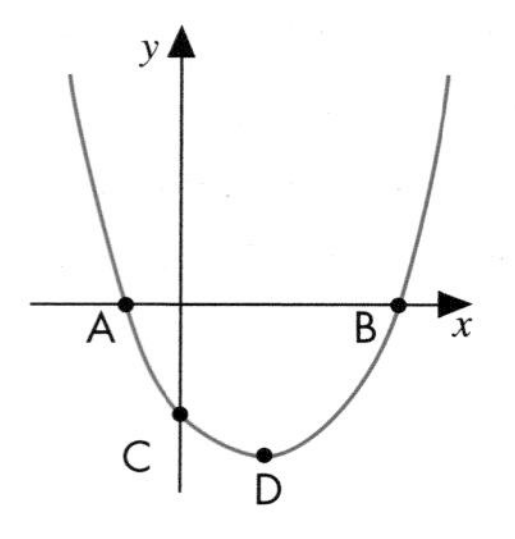

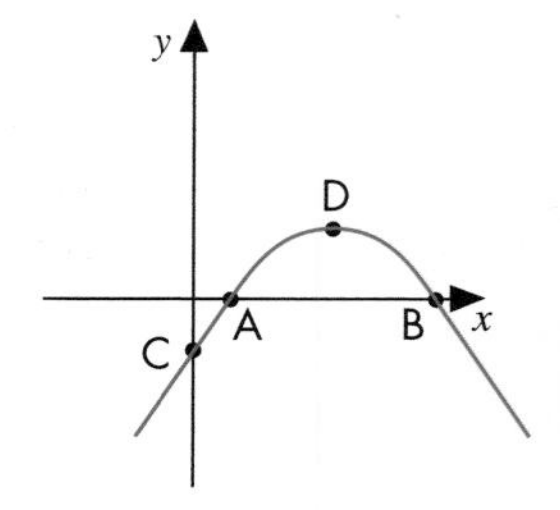

The roots

If you look at your answer to question **7** in Exercise 12D, you will see that the graph crosses the x-axis at $x = -0.5$ and $x = 3$. Since the x-axis is the line $y = 0$, the y-value at any point on the x-axis is zero. So, you have found the solution to the equation:

$0 = 2x^2 - 5x - 3$ that is $2x^2 - 5x - 3 = 0$

Equations of this type are known as *quadratic equations*.

You can solve quadratic equations by finding the values of x that make them true. Such values are called the roots of the equation. On the graph, these occur where the curve cuts the x-axis. So the roots of the quadratic equation $2x^2 - 5x - 3 = 0$ are -0.5 and 3.

Let's check these values.

For $x = 3.0$ $\quad 2(3)^2 - 5(3) - 3 = 18 - 15 - 3 = 0$

For $x = 0.5$ $\quad 2(-0.5)^2 - 5(-0.5) - 3 = 0.5 + 2.5 - 3 = 0$

You can find the roots of a quadratic equation by drawing its graph and finding where the graph crosses the x-axis.

EXAMPLE 7

a Draw the graph of $y = x^2 - 3x - 4$ for $-2 \leqslant x \leqslant 5$.

b Use your graph to find the roots of the equation $x^2 - 3x - 4 = 0$.

a Set up a table.

x	−2	−1	0	1	2	3	4	5
y^2	4	1	0	1	4	9	16	25
$-3x$	6	3	0	−3	−6	−9	−12	−15
-4	−4	−4	−4	−4	−4	−4	−4	−4
y	6	0	−4	−6	−6	−4	0	6

Draw the graph.

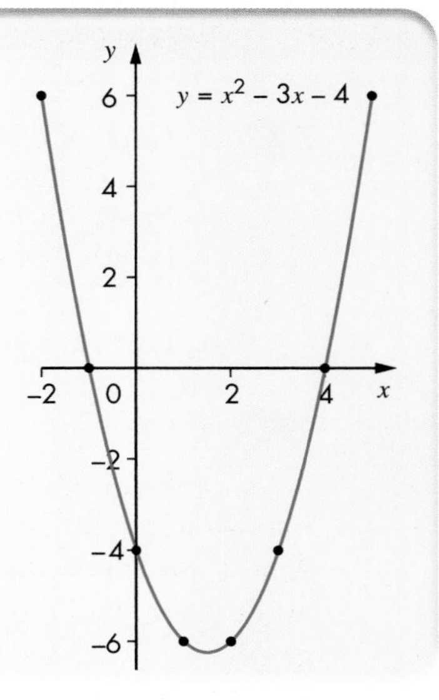

b The points where the graph crosses the x-axis are −1 and 4.

So, the roots of $x^2 - 3x - 4 = 0$ are $x = -1$ and $x = 4$.

Note that sometimes the quadratic graph may not cross the x-axis. In this case there are no roots. You will meet equations of this type later in the course.

The y-intercept

If you look at all the quadratic graphs you have drawn so far you will see a connection between the equation and the point where the graph crosses the y-axis. Very simply, the constant term of the equation $y = ax^2 + bx + c$ (that is, the value c) is where the graph crosses the y-axis. The intercept is at $(0, c)$.

The vertex

The lowest (or highest) point of a quadratic graph is called the *vertex*.

If it is the highest point, it is called the **maximum**.

If it is the lowest point, it is called the **minimum**.

It is difficult to find a general rule for this point, but the x-coordinate is always half-way between the roots. The easiest way to find the y-value is to substitute the x-value into the original equation.

EXAMPLE 8

a Write the equation $x^2 - 3x - 4 = 0$ in the form $(x - p)^2 - q = 0$.

b What is the least value of the graph $y = x^2 - 3x - 4$?

a $(x - p)^2 - q = x^2 - 2px + p^2 - q = x^2 - 3x - 4$

So $p = 1\frac{1}{2}$ and $p^2 - q = -4$, $q = p^2 + 4 = (1\frac{1}{2})^2 + 4 = 6\frac{1}{4}$

b Looking at the graph drawn in Example 7 you can see that the minimum point is at $(1\frac{1}{2}, -6\frac{1}{4})$, so the least value is $-6\frac{1}{4}$.

You should be able to see the connection between the vertex point and the equation written in the 'completing the square' form.

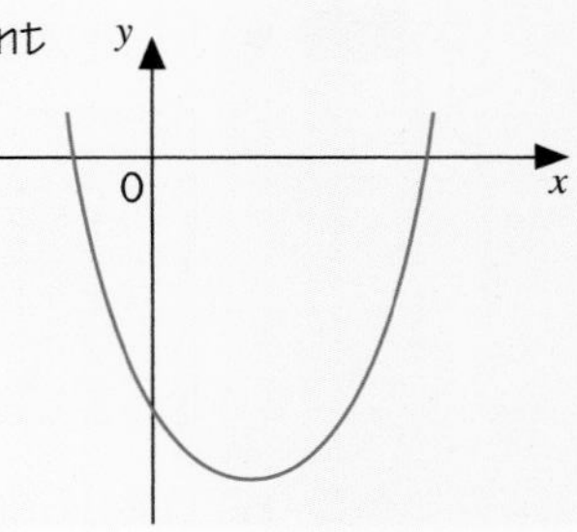

As a general rule when a quadratic is written in the form $(x - p)^2 + q$ then the minimum point is (p, q). Note the sign change of p.

Note: If the x^2 term is negative then the graph will be inverted and the vertex will be a maximum.

EXERCISE 12E

B

1 **a** Copy and complete the table to draw the graph of $y = x^2 - 4$ for $-4 \leqslant x \leqslant 4$.

x	–4	–3	–2	–1	0	1	2	3	4
y	12			–3				5	

b Use your graph to find the roots of $x^2 - 4 = 0$.

2 **a** Copy and complete the table and draw the graph of $y = x^2 - 9$ for $-4 \leqslant x \leqslant 4$.

x	–4	–3	–2	–1	0	1	2	3	4
y	7				–9			0	

b Use your graph to find the roots of $x^2 - 9 = 0$.

PS **3** **a** Look at the equations of the graphs you drew in questions **1** and **2**. Is there a connection between the numbers in each equation and its roots?

b Before you draw the graphs in parts **c** and **d**, try to predict what their roots will be.

c Copy and complete the table and draw the graph of $y = x^2 - 1$ for $-4 \leqslant x \leqslant 4$.

x	–4	–3	–2	–1	0	1	2	3	4
y	15				–1			8	

d Copy and complete the table and draw the graph of $y = x^2 - 5$ for $-4 \leqslant x \leqslant 4$.

x	–4	–3	–2	–1	0	1	2	3	4
y	11		–1					4	

e Were your predictions correct?

4 **a** Copy and complete the table and draw the graph of $y = x^2 + 4x$ for $-5 \leqslant x \leqslant 2$.

x	–5	–4	–3	–2	–1	0	1	2
x^2	25			4			1	
$+4x$	–20			–8			4	
y	5			–4			5	

b Use your graph to find the roots of the equation $x^2 + 4x = 0$.

5 **a** Copy and complete the table and draw the graph of $y = x^2 - 6x$ for $-2 \leqslant x \leqslant 8$.

x	–2	–1	0	1	2	3	4	5	6	7	8
x^2	4			1			16				
$-6x$	12			–6			–24				
y	16			–5			–8				

b Use your graph to find the roots of the equation $x^2 - 6x = 0$.

B

6 **a** Copy and complete the table and draw the graph of $y = x^2 + 3x$ for $-5 \leqslant x \leqslant 3$.

x	–5	–4	–3	–2	–1	0	1	2	3
y	10			–2				10	

b Use your graph to find the roots of the equation $x^2 + 3x = 0$.

PS **7** **a** Look at the equations of the graphs you drew in questions **4**, **5** and **6**. Is there a connection between the numbers in each equation and the roots?

b Before you draw the graphs in parts **c** and **d**, try to predict what their roots will be.

c Copy and complete the table and draw the graph of $y = x^2 - 3x$ for $-2 \leqslant x \leqslant 5$.

x	–2	–1	0	1	2	3	4	5
y	10			–2				10

d Copy and complete the table and draw the graph of $y = x^2 + 5x$ for $-6 \leqslant x \leqslant 2$.

x	–6	–5	–4	–3	–2	–1	0	1	2
y	6			–6				6	

e Were your predictions correct?

8 **a** Copy and complete the table and draw the graph of $y = x^2 - 4x + 4$ for $-1 \leqslant x \leqslant 5$.

x	–1	0	1	2	3	4	5
y	9				1		

b Use your graph to find the roots of the equation $x^2 - 4x + 4 = 0$.

c What happens with the roots?

9 **a** Copy and complete the table and draw the graph of $y = x^2 - 6x + 3$ for $-1 \leqslant x \leqslant 7$.

x	–1	0	1	2	3	4	5	6	7
y	10			–5			–2		

b Use your graph to find the roots of the equation $x^2 - 6x + 3 = 0$.

A

10 **a** Copy and complete the table and draw the graph of $y = 2x^2 + 5x - 6$ for $-5 \leqslant x \leqslant 2$.

x	–5	–4	–3	–2	–1	0	1	2
y								

b Use your graph to find the roots of the equation $2x^2 + 5x - 6 = 0$.

PS 11 Look back at questions **1** to **7**.

a Write down the point of intersection of the graph with the y-axis for each one.

b Write down the coordinates of the minimum point (vertex) of each graph for each one.

c Explain the connection between these points and the original equation.

12 **a** Write the equation $y = x^2 - 4x + 4$ in the form $y = (x - p)^2 + q$.

b Write down the minimum value of the equation $y = x^2 - 4x + 4$.

13 **a** Write the equation $y = x^2 - 6x + 3$ in the form $y = (x - p)^2 + q$.

b Write down the minimum value of the equation $y = x^2 - 6x + 3$.

14 **a** Write the equation $y = x^2 - 8x + 2$ in the form $y = (x - p)^2 + q$.

b Write down the minimum value of the equation $y = x^2 - 8x + 2$.

15 **a** Write the equation $y = -x^2 + 2x - 6$ in the form $y = -(x - p)^2 + q$.

b Write down the maximum value of the equation $y = -x^2 + 2x - 6$.

PS 16 Look at your answers to questions **12** to **13**.

a What is the connection between the maximum or minimum point and the values in the equation when written as $(x - a)^2 + b$?

b Without drawing the curve, predict the minimum point of the graph:

$y = x^2 + 10x - 3$.

AU 17 Masood draws a quadratic graph which has a minimum point at $(3, -7)$.

He forgets to label it and later cannot remember what the quadratic function was.

He knows it is of the form $y = x^2 + px + q$.

Can you help him?

AU 18 **a** The graph $y = x^2 + 4x + 2$ has a minimum point at $(-2, 2)$.

Write down the minimum point of the graph $y = x^2 + 4x - 3$.

b The graph $y = x^2 - 2ax + b$ has a minimum point at $(a, b - a^2)$.

Write down the minimum points of:

i $y = x^2 - 2ax + 2b$

ii $y = x^2 - 4ax + b$

12.5 The circular function graphs

This section will show you how to:

- use the symmetry of the graphs $y = \sin x$, and $y = \cos x$ in answering questions
- understand that for every value of sine and cosine between 1 and –1 there are two angles between 0° and 360°

Key words
cosine
cyclic
inverse cosine
inverse sine
line symmetry
rotational symmetry
sine

You have just met **sine** and **cosine** graphs.

These graphs have some special properties.

- They are **cyclic**. This means that they repeat indefinitely in both directions.
- For every value of sine or cosine between –1 and 1 there are two angles between 0° and 360°, and an infinite number of angles altogether.
- The sine graph has **rotational symmetry** about (180°, 0) and has **line symmetry** between 0° and 180° about $x = 90°$, and between 180° and 360° about $x = 270°$.
- The cosine graph has line symmetry about $x = 180°$, and has rotational symmetry between 0° and 180° about (90°, 0) and between 180° and 360° about (270°, 0).

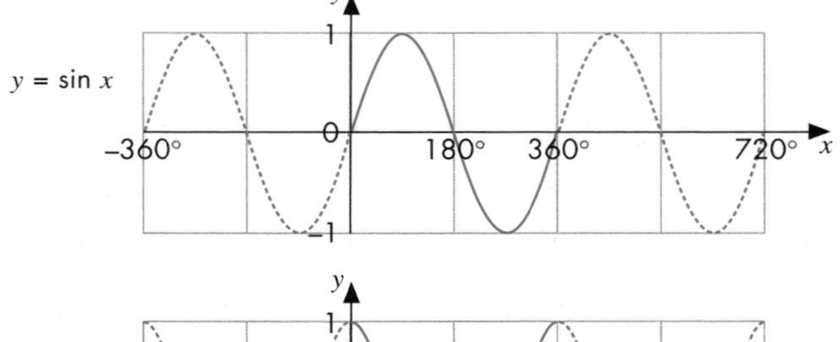

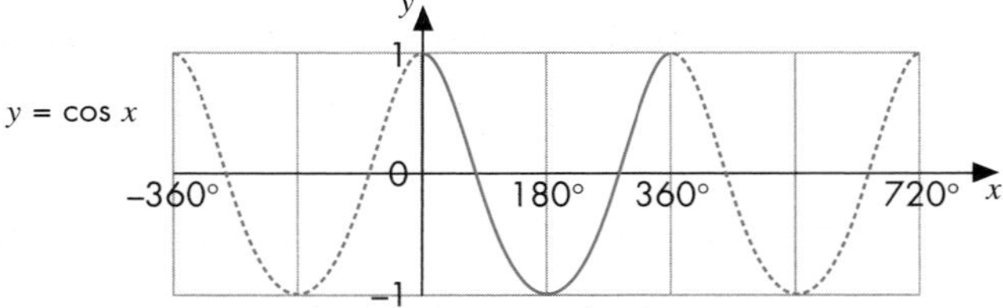

The graphs can be used to find angles with certain values of sine and cosine.

EXAMPLE 9

Given that sin 42° = 0.669, find another angle between 0° and 360° that also has a sine of 0.669.

Plot the approximate value 0.669 on the sine graph and use the symmetry to work out the other value.

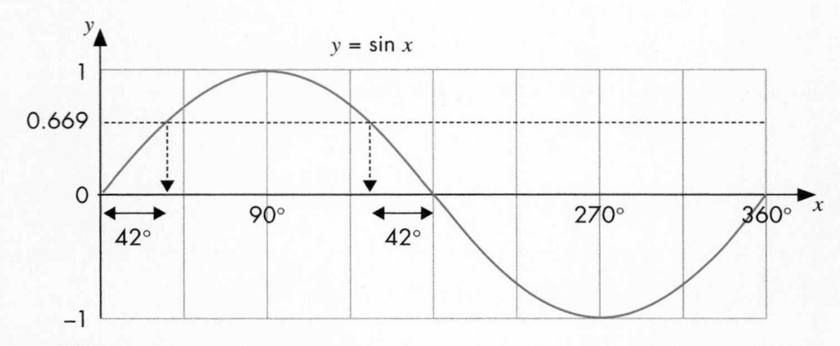

The other value is 180° – 42° = 138°

EXAMPLE 10

Given that cos 110° = –0.342, find two angles between 0° and 360° that have a cosine of +0.342.

Plot the approximate values –0.342 and 0.342 on the cosine graph and use the symmetry to work out the values.

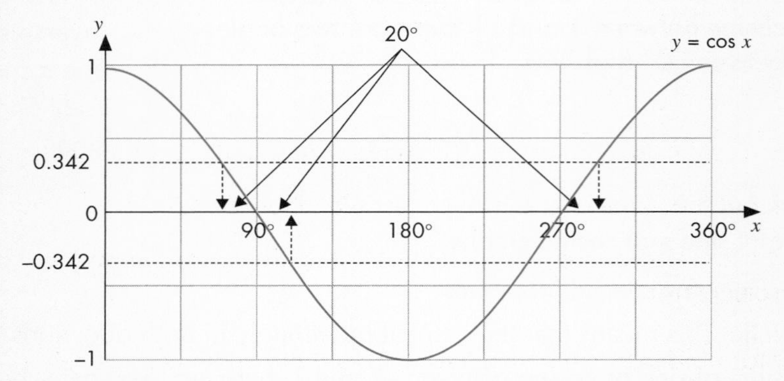

The required values are 90° – 20° = 70° and 270° + 20° = 290°

EXERCISE 12F

A*

1 Given that sin 65° = 0.906, find another angle between 0° and 360° that also has a sine of 0.906.

2 Given that sin 213° = –0.545, find another angle between 0° and 360° that also has a sine of –0.545.

3 Given that cos 36° = 0.809, find another angle between 0° and 360° that also has a cosine of 0.809.

4 Given that cos 165° = –0.966, find another angle between 0° and 360° that also has a cosine of –0.966.

5 Given that sin 30° = 0.5, find two angles between 0° and 360° that have a sine of –0.5.

6 Given that cos 45° = 0.707, find two angles between 0° and 360° that have a cosine of –0.707.

PS **7** **a** Choose an acute angle a. Write down the values of:

i sin a **ii** cos (90° – a).

b Repeat with another acute angle b.

c Write down a rule connecting the sine of an acute angle x and the cosine of the complementary angle (i.e. the difference with 90°).

d Find a similar rule for the cosine of x and the sine of its complementary angle.

A*

8 Given that $\sin 26° = 0.438$:

a write down an angle between 0° and 90° that has a cosine of 0.438

b find two angles between 0° and 360° that have a sine of –0.438

c find two angles between 0° and 360° that have a cosine of –0.438.

AU 9 A formula used to work out the angle of a triangle is

$$\cos A = \frac{b^2 + c^2 - a^2}{2bc}$$

where a, b and c are the sides of the triangle and angle A is the angle opposite side a.

a Work out the value of $\cos A$ for a triangle where $a = 20$, $b = 11$ and $c = 13$.

b What is the size of angle A to the nearest degree? Use the **inverse cosine function** on your calculator.

PS 10 Another formula that can be used to work out the angle of a triangle is

$$\sin A = \frac{a \sin B}{b}$$

where a and b are sides of the triangle and A and B are angles opposite the sides a and b respectively.

a Work out the value of $\sin A$ for this triangle, where $a = 16$, $b = 14$ and $B = 46°$.

b Use your calculator and the **inverse sine** function to find the value of B.

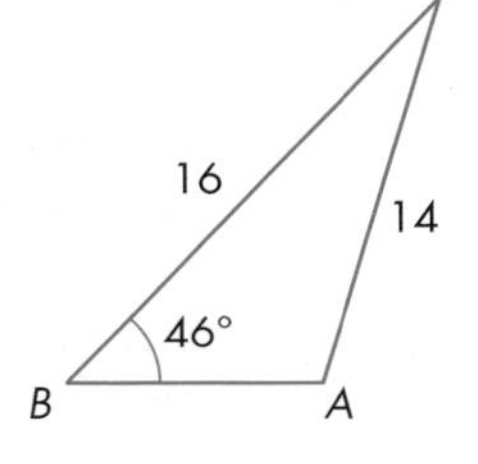

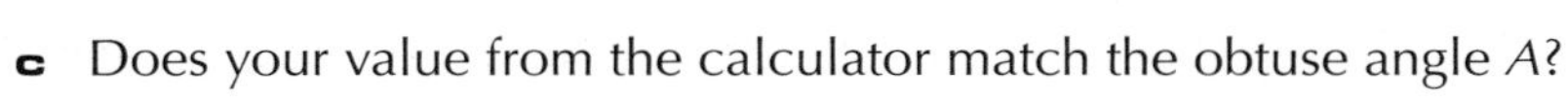

c Does your value from the calculator match the obtuse angle A?

d If not, explain why not.

PS 11 Mike used the same rule as in question 10 to work out the size of angle A in this triangle.

a Work out the value of $\sin A$ for the triangle shown, where $a = 16$, $b = 12$ and $B = 58°$.

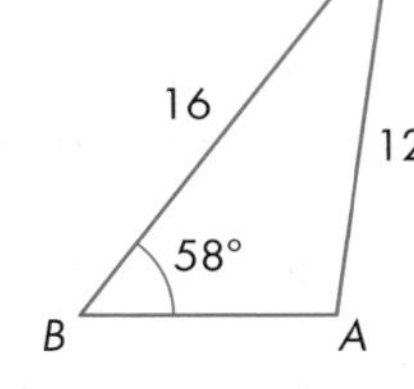

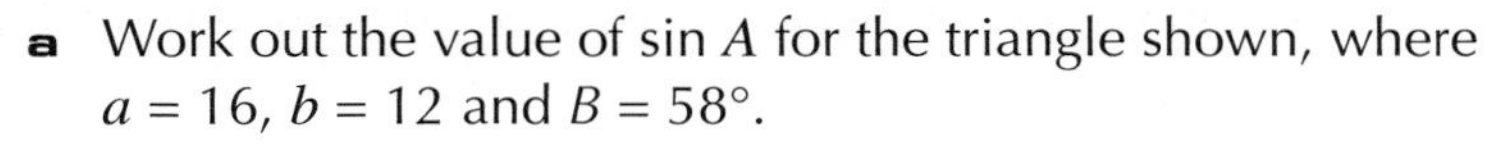

b Use your calculator and the inverse sine function to find the value of B.

c What happens? Can you explain why?

AU 12 State if the following rules are true or false.

a $\sin x = \sin (180° - x)$

b $\sin x = -\sin (360° - x)$

c $\cos x = \cos (360° - x)$

d $\sin x = -\sin (180° + x)$

e $\cos(180° - x) = \cos (180° + x)$

12.6 Solving one linear and one non-linear equation by the method of intersection

This section will show you how to:

- solve a pair of simultaneous equations where one is linear and one is non-linear, using graphs

Key words
linear
non-linear
simultaneous equations

In Chapter 7, section 7, you learned how to use an algebraic method for solving a pair of **simultaneous equations** where one is **linear** (a straight line) and one is **non-linear** (a curve). In this section, you will learn how to do this graphically. You have seen in Chapter 7, section 5 how to find the solution to a pair of linear simultaneous equations. The same principle applies here. The point where the graphs cross gives the solution. However, in most cases, there are two solutions, because the straight line will cross the curve twice.

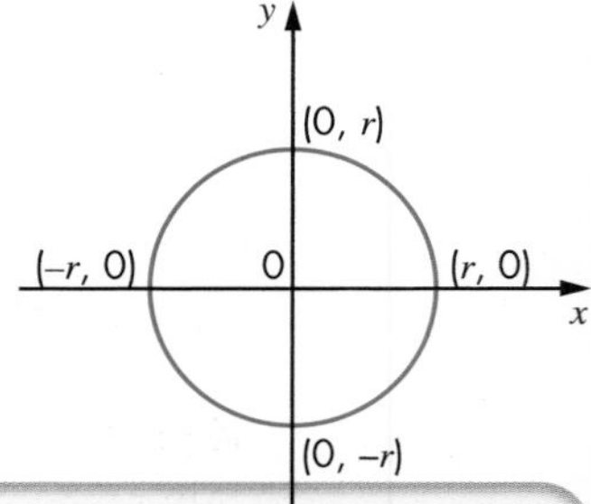

Most of the non-linear graphs will be quadratic graphs, but there is one other type you can meet. This is an equation of the form $x^2 + y^2 = r^2$, which is a circle, with the centre as the origin and a radius of r.

EXAMPLE 11

Find the approximate solutions of the pair of equations $y = x^2 + x - 2$ and $y = 2x + 3$ by graphical means.

Set up a table for the quadratic.

x	–4	–3	–2	–1	0	1	2	3	4
y	10	4	0	–2	–2	0	4	10	18

Draw both graphs on the same set of axes.

From the graph, the approximate solutions can be seen to be (–1.8, –0.6) and (2.8, 8.6).

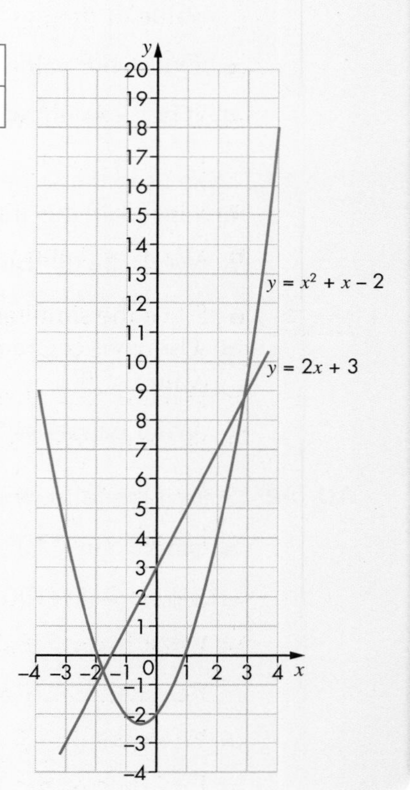

EXAMPLE 12

Find the approximate solutions of the pair of equations $x^2 + y^2 = 25$ and $y = x + 2$ by graphical means.

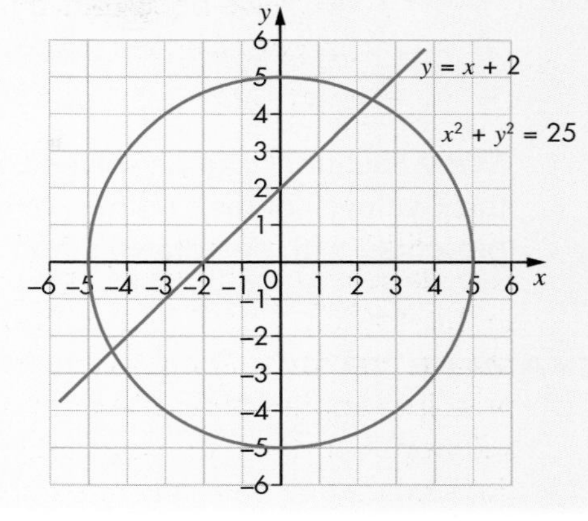

The curve is a circle of radius 5 centred on the origin.

From the graph, the approximate solutions can be seen to be (–4.4, –2.4), (2.4, 4.4).

EXERCISE 12G

1 Use graphical methods to find the approximate or exact solutions to the following pairs of simultaneous equations. In this question, suitable ranges for the axes are given. In an examination a grid will be supplied.

- **a** $y = x^2 + 3x - 2$ and $y = x$ $(-5 \leqslant x \leqslant 5, -5 \leqslant y \leqslant 5)$
- **b** $y = x^2 - 3x - 6$ and $y = 2x$ $(-4 \leqslant x \leqslant 8, -10 \leqslant y \leqslant 20)$
- **c** $x^2 + y^2 = 25$ and $x + y = 1$ $(-6 \leqslant x \leqslant 6, -6 \leqslant y \leqslant 6)$
- **d** $x^2 + y^2 = 4$ and $y = x + 1$ $(-5 \leqslant x \leqslant 5, -5 \leqslant y \leqslant 5)$
- **e** $y = x^2 - 3x + 1$ and $y = 2x - 1$ $(0 \leqslant x \leqslant 6, -4 \leqslant y \leqslant 12)$
- **f** $y = x^2 - 3$ and $y = x + 3$ $(-5 \leqslant x \leqslant 5, -4 \leqslant y \leqslant 8)$
- **g** $y = x^2 - 3x - 2$ and $y = 2x - 3$ $(-5 \leqslant x \leqslant 5, -5 \leqslant y \leqslant 10)$
- **h** $x^2 + y^2 = 9$ and $y = x - 1$ $[-5 \leqslant x \leqslant 5, -5 \leqslant y \leqslant 5)$

A

PS 2
- **a** Solve the simultaneous equations $y = x^2 + 3x - 4$ and $y = 5x - 5$ $(-5 \leqslant x \leqslant 5, -8 \leqslant y \leqslant 8)$.
- **b** What is special about the intersection of these two graphs?
- **c** Show that $5x - 5 = x^2 + 3x - 4$ can be rearranged to $x^2 - 2x + 1 = 0$.
- **d** Factorise and solve $x^2 - 2x + 1 = 0$.
- **e** Explain how the solution in part **d** relates to the intersection of the graphs.

AU 3
- **a** Solve the simultaneous equations $y = x^2 + 2x + 3$ and $y = x - 1$ $(-5 \leqslant x \leqslant 5, -5 \leqslant y \leqslant 8)$.
- **b** What is special about the intersection of these two graphs?
- **c** Rearrange $x - 1 = x^2 + 2x + 3$ into the general quadratic form $ax^2 + bx + c = 0$.
- **d** Work out the discriminant $b^2 - 4ac$ for the quadratic in part **c**.
- **e** Explain how the value of the discriminant relates to the intersection of the graphs.

12.7 Solving equations by the method of intersection

This section will show you how to:

- solve equations by the method of intersecting graphs

Many equations can be solved by drawing two intersecting graphs on the same axes and using the x-value(s) of their point(s) of intersection. (In the GCSE examination, you are likely to be presented with one drawn graph and asked to draw a straight line to solve a new equation.)

EXAMPLE 13

Show how each equation given below can be solved using the graph of $y = x^3 - 2x - 2$ and its intersection with another graph. In each case, give the equation of the other graph and the solution(s).

a $x^3 - 2x - 4 = 0$ **b** $x^3 - 3x - 1 = 0$

a This method will give the required graph.

Step 1: Write down the original (given) equation. $y = x^3 - 2x - 2$

Step 2: Write down the (new) equation to be solved in reverse. $0 = x^3 - 2x - 4$

Step 3: Subtract these equations. $y = \quad + 2$

Step 4: Draw this line on the original graph to solve the new equation.

The graphs of $y = x^3 - 2x - 2$ and $y = 2$ are drawn on the same axes.

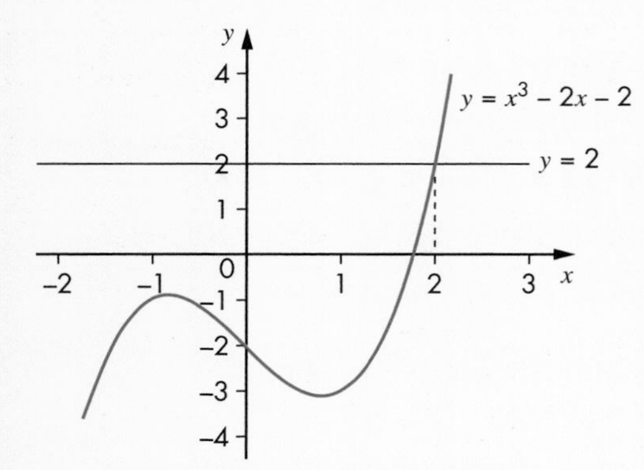

The intersection of these two graphs is the solution of

$$x^3 - 2x - 4 = 0.$$

The solution is $x = 2$.

This works because you are drawing a straight line on the same axes as the original graph, and solving for x and y where they intersect.

At the points of intersection the y-values will be the same and so will the x-values.

So you can say: original equation = straight line

Rearranging this gives: (original equation) – (straight line) = 0

You have been asked to solve: (new equation) = 0

So (original equation) – (straight line) = (new equation)

Rearranging this again gives: (original equation) – (new equation) = straight line

Note: In GCSE exams the curve is always drawn already and you will only have to draw the straight line.

EXAMPLE 13 continued

b Write down given graph: $y = x^3 - 2x - 2$

Write down new equation: $0 = x^3 - 3x - 1$

Subtract: $y = \quad + x - 1$

The graphs of $y = x^3 - 2x - 2$ and $y = x - 1$ are then drawn on the same axes.

The intersection of the two graphs is the solution of $x^3 - 3x - 1 = 0$.

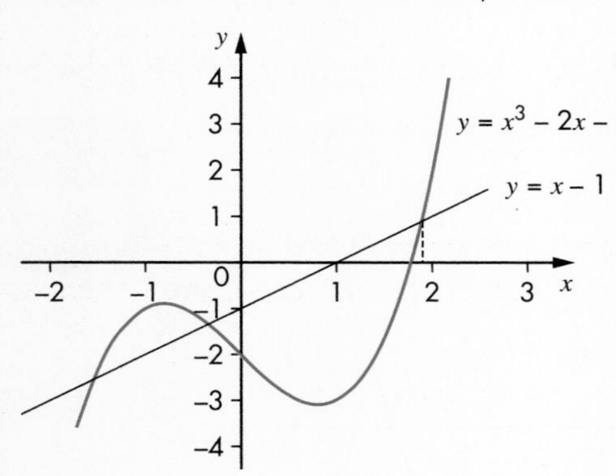

The solutions are $x = -1.5, -0.3$ and 1.9.

EXAMPLE 14

The graph shows the curve $y = x^2 + 3x - 2$.

By drawing a suitable straight line, solve these equations.

a $x^2 + 3x - 1 = 0$ **b** $x^2 + 2x - 3 = 0$

a Given graph: $y = x^2 + 3x - 2$

New equation: $0 = x^2 + 3x - 1$

Subtract: $y = \quad -1$

Draw: $y = -1$

Solutions: $x = 0.3, -3.3$

b Given graph: $y = x^2 + 3x - 2$

New equation: $0 = x^2 + 2x - 3$

Subtract: $y = \quad +x \quad +1$

Draw: $y = x + 1$

Solutions: $x = 1, -3$

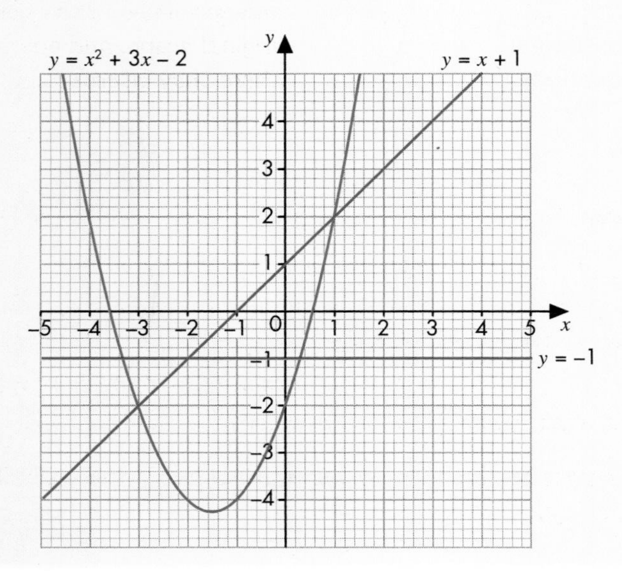

EXERCISE 12H

In questions **1** to **5**, use the graphs given here. Trace the graphs or place a ruler over them in the position of the line. Solution values only need to be given to 1 decimal place. In questions **6** to **10**, either draw the graphs yourself or use a graphics calculator to draw them.

A*

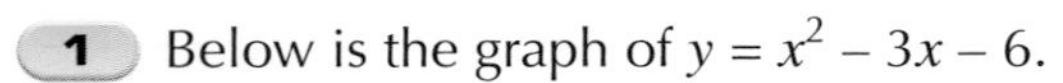

1 Below is the graph of $y = x^2 - 3x - 6$.

a Solve these equations.

i $x^2 - 3x - 6 = 0$ **ii** $x^2 - 3x - 6 = 4$ **iii** $x^2 - 3x - 2 = 0$

b By drawing a suitable straight line solve $2x^2 - 6x + 2 = 0$.

HINTS AND TIPS

Cancel by 2 first.

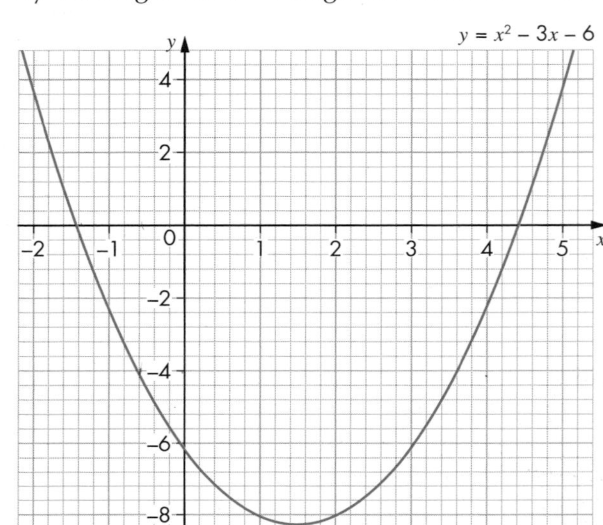

2 Below is the graph of

$y = x^2 + 4x - 5$.

a Solve $x^2 + 4x - 5 = 0$.

b By drawing suitable straight lines solve these equations.

i $x^2 + 4x - 5 = 2$

ii $x^2 + 4x - 4 = 0$

iii $3x^2 + 12x + 6 = 0$

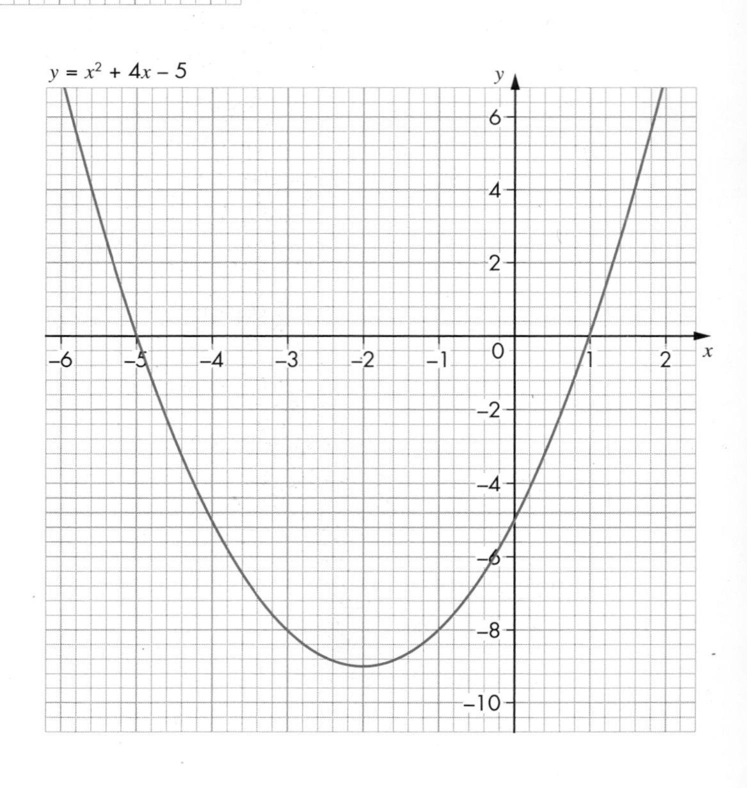

3 Below are the graphs of $y = x^2 - 5x + 3$ and $y = x + 3$.

a Solve these equations. **i** $x^2 - 6x = 0$ **ii** $x^2 - 5x + 3 = 0$

b By drawing suitable straight lines solve these equations.

i $x^2 - 5x + 3 = 2$ **ii** $x^2 - 5x - 2 = 0$

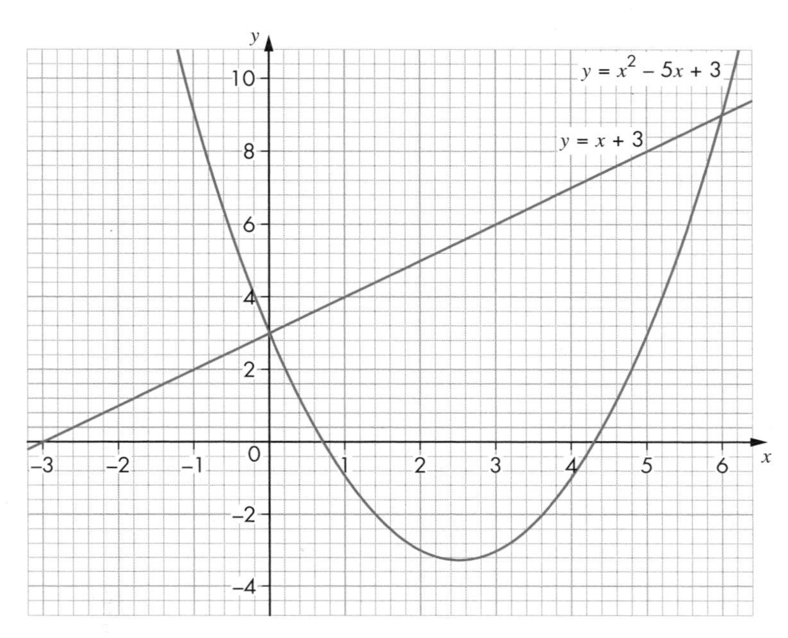

4 Below are the graphs of $y = x^2 - 2$ and $y = x + 2$.

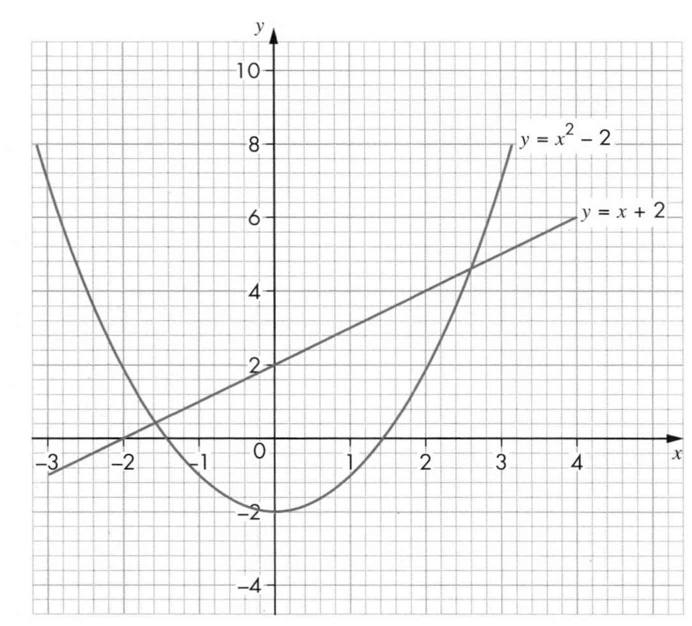

a Solve these equations. **i** $x^2 - x - 4 = 0$ **ii** $x^2 - 2 = 0$

b By drawing suitable straight lines solve these equations.

i $x^2 - 2 = 3$ **ii** $x^2 - 4 = 0$

A*

A*

PS 5 Below are the graphs of $y = x^3 - 2x^2$, $y = 2x + 1$ and $y = x - 1$.

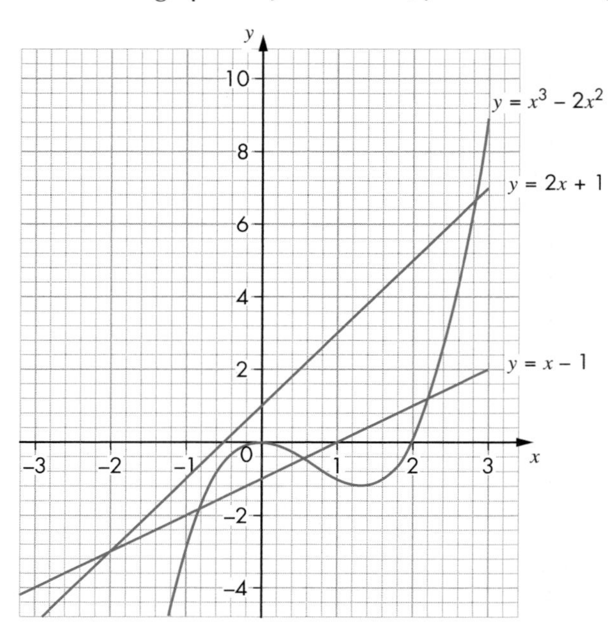

Solve these equations.

a $x^3 - 2x^2 = 0$

b $x^3 - 2x^2 = 3$

c $x^3 - 2x^2 + 1 = 0$

d $x^3 - 2x^2 - 2x - 1 = 0$

e $x^3 - 2x^2 - x + 1 = 0$

6 Draw the graph of $y = x^2 - 4x - 2$.

a Solve $x^2 - 4x - 2 = 0$.

b By drawing a suitable straight line solve $x^2 - 4x - 5 = 0$.

7 Draw the graph of $y = 2x^2 - 5$.

a Solve $2x^2 - 5 = 0$.

b By drawing a suitable straight line solve $2x^2 - 3 = 0$.

8 Draw the graphs of $y = x^2 - 3$ and $y = x + 2$ on the same axes. Use the graphs to solve these equations.

a $x^2 - 5 = 0$

b $x^2 - x - 5 = 0$

9 Draw the graphs of $y = x^2 - 3x - 2$ and $y = 2x - 3$ on the same axes. Use the graphs to solve these equations.

a $x^2 - 3x - 1 = 0$

b $x^2 - 5x + 1 = 0$

10 Draw the graphs of $y = x^3 - 2x^2 + 3x - 4$ and $y = 3x - 1$ on the same axes. Use the graphs to solve these equations.

a $x^3 - 2x^2 + 3x - 6 = 0$

b $x^3 - 2x^2 - 3 = 0$

A*

AU 11 The graph shows the lines A: $y = x^2 + 3x - 2$; B: $y = x$; C: $y = x + 2$; D: $y + x = 3$ and E: $y + x + 1 = 0$.

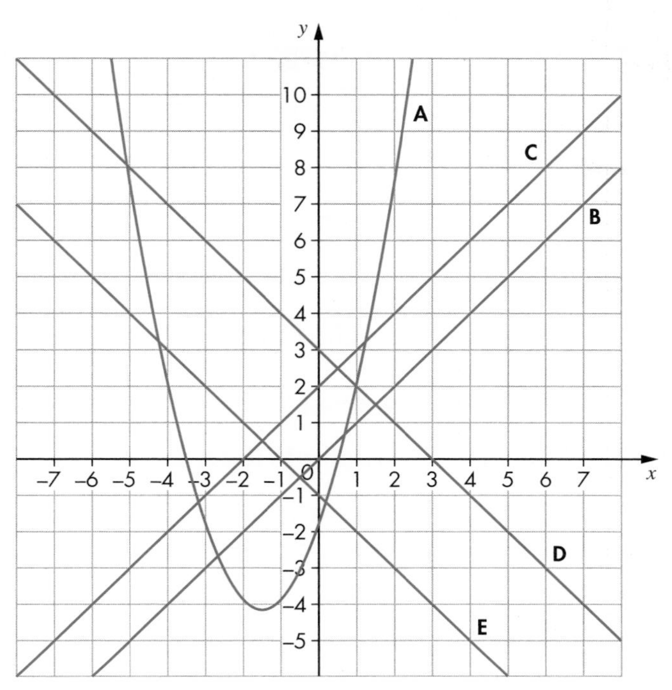

a Which pair of lines has a common solution of (0.5, 2.5)?

b Which pair of lines has the solutions of (1, 2) and (–5, 8)?

c What quadratic equation has an approximate solution of (–4.2, 3.2) and (0.2, –1.2)?

d The minimum point of the graph $y = x^2 + 3x - 2$ is at (–1.5, –4.25).

What is the minimum point of the graph $y = x^2 + 3x - 8$?

PS 12 Jamil was given a sketch of the graph $y = x^2 + 3x + 5$ and asked to draw an appropriate straight line to solve $x^2 + x - 2 = 0$.

This is Jamil's working:

$$\begin{array}{ll} \text{Original} & y = x^2 + 3x + 5 \\ \text{New} & \underline{0 = x^2 + x - 2} \\ & y = 2x - 7 \end{array}$$

When Jamil drew the line $y = 2x - 7$, it did not intersect with the parabola $y = x^2 + 3x + 5$.

He concluded that the equation $x^2 + x - 2 = 0$ did not have any solutions.

a Show by factorisation that the equation $x^2 + x - 2 = 0$ has solutions –2 and 1.

b Explain the error that Jamil made.

c What line should Jamil have drawn?

GRADE BOOSTER

- **D** You can draw straight lines by plotting points
- **C** You can draw straight lines using the gradient-intercept method
- **C** You can draw quadratic graphs from their tables of values
- **B** You can solve quadratic equations from their graphs
- **B to A*** You can recognise the significant points of a quadratic graph
- **A*** You can solve equations, using the intersection of two graphs
- **A*** You can use trigonometric graphs to solve sine and cosine problems
- **A*** You can find two angles between 0° and 360° for any given value of a trigonometric ratio (positive or negative)

What you should know now

- How to draw linear graphs
- How to draw quadratic graphs

1 On a copy of the grid below, draw the graph of $y = 2x + 1$.

Use values of x from –2 to +2.

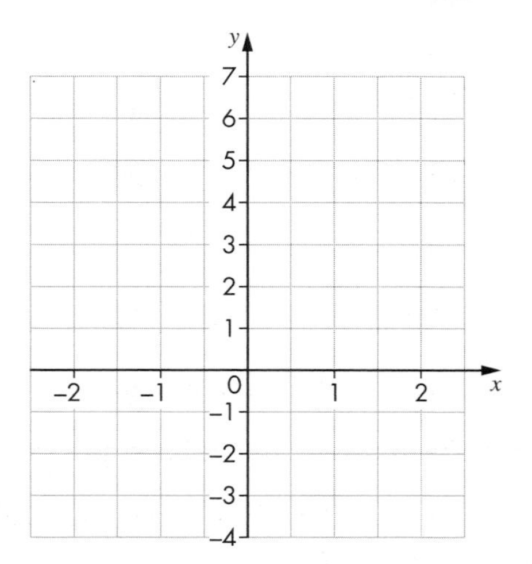

(2 marks)

Edexcel, March 2007, Paper 11 Higher, Question 2

2 On a copy of the coordinate grid below, draw the graph of $y = 2x - 3$.

Use values of x from –2 to +2.

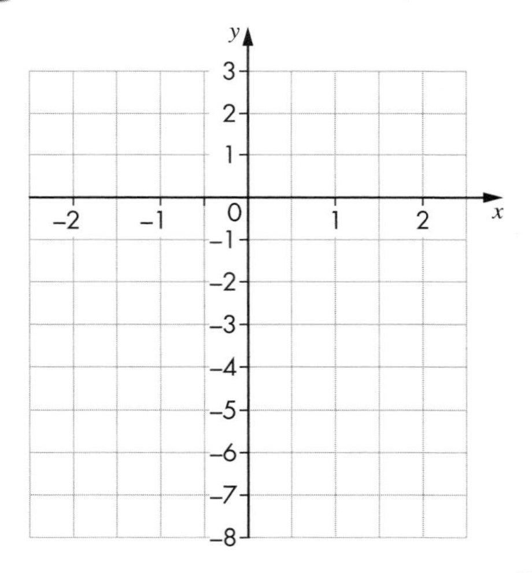

(3 marks)

Edexcel, June 2007, Paper 11 Higher, Question 3

3 On a copy of the grid, draw the graph of $y = 4x - 3$.

Use values of x from –1 to +3.

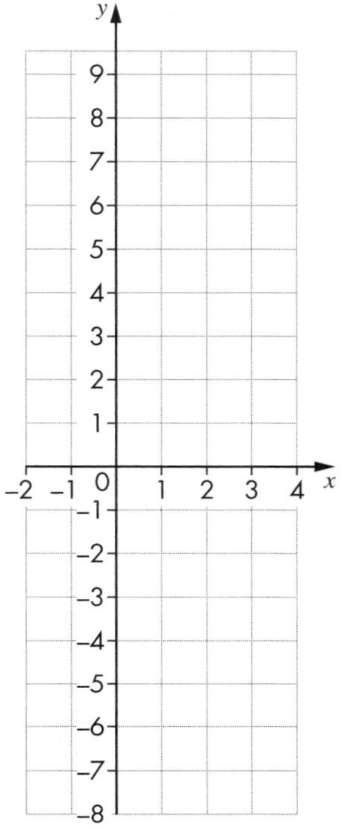

(3 marks)

Edexcel, June 2008, Paper 11 Higher, Question 5

4

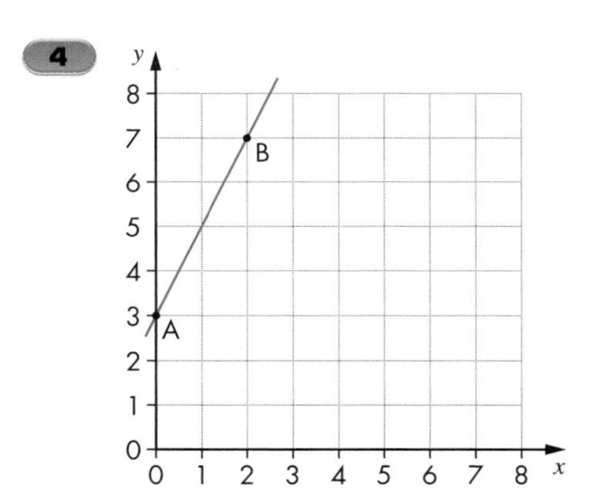

A has coordinates (0, 3).

B has coordinates (2, 7).

Work out the gradient of the line that passes through A and B. (2 marks)

Edexcel, November 2008, Paper 15 Higher, Question 14

5 On a copy of the grid, draw the graph of $x + y = 6$.

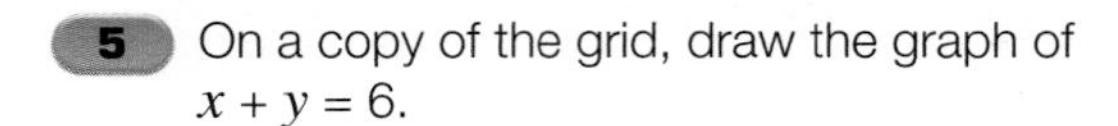

Use values of x from $x = 0$ to $x = 6$.

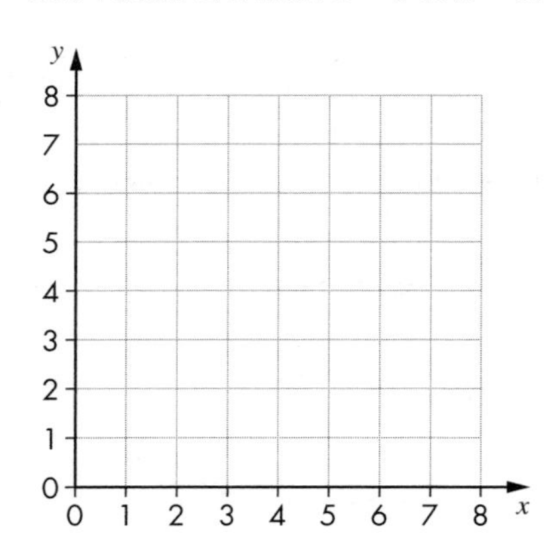

(3 marks)

Edexcel, November 2008, Paper 11 Higher, Question 5

6 **a** Complete the table of values for $y = x^2 - 4x + 2$. *(2 marks)*

x	–1	0	1	2	3	4	5
y		2	–1		–1		7

b On a copy of the grid below, draw the graph of $y = x^2 - 4x + 2$.

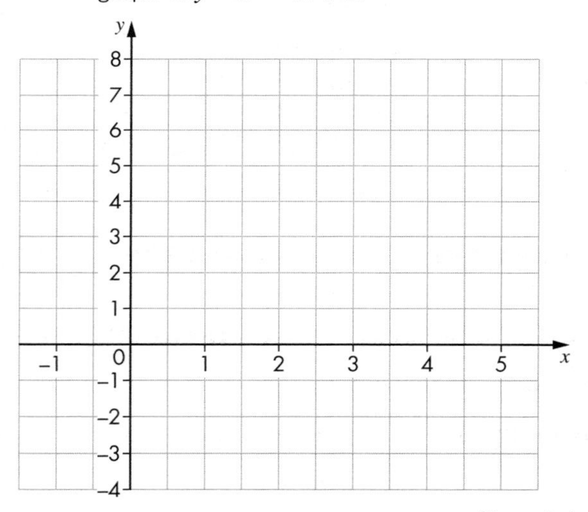

(2 marks)

Edexcel, May 2008, Paper 3 Higher, Question 20

7 **a** Complete the table of values for $y = x^2 - 4x - 2$. *(2 marks)*

x	–1	0	1	2	3	4	5
y		–2	–5			–2	3

b On a copy of the grid below, draw the graph of $y = x^2 - 4x - 2$.

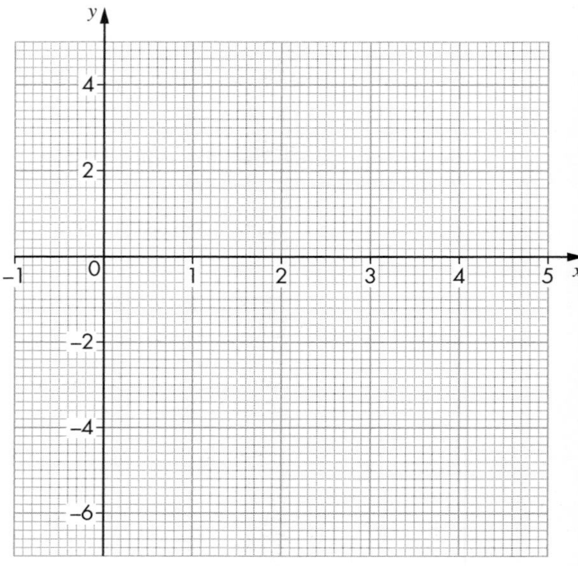

(2 marks)

c Use your graph to estimate the values of x when $y = -3$. *(2 marks)*

Edexcel, November 2008, Paper 4 Higher, Question 11

8 The diagram shows a sketch of the graph $y = ab^x$.

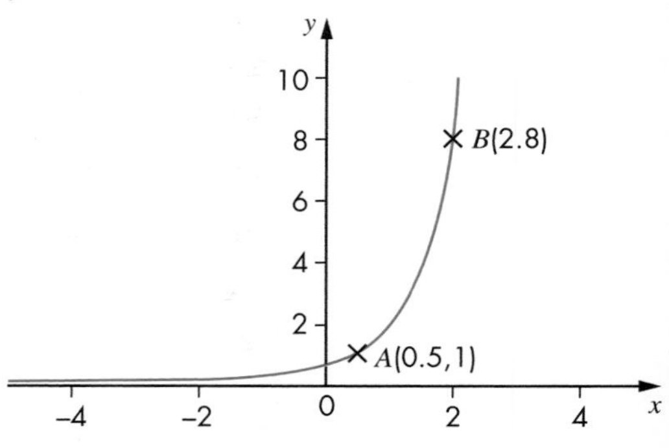

The curve passes through the points A (0.5, 1) and B (2, 8).

The point C (–0.5, k) lies on the curve.

Find the value of k. *(4 marks)*

Edexcel, November 2008, Paper 4 Higher, Question 2

Worked Examination Questions

1 Three corners of a square A(0, 4), B(4, 5) and D(1, 0) are shown on the grid.

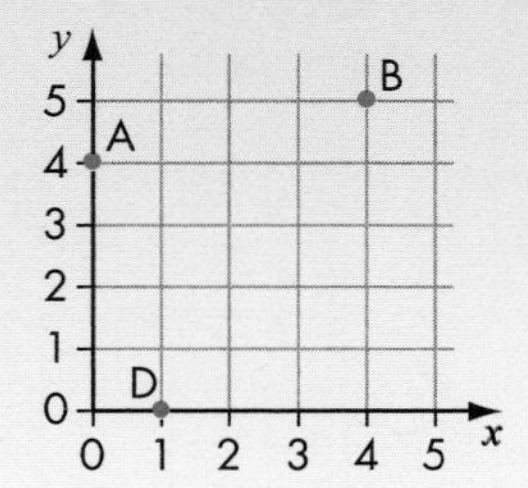

Show algebraically that the point C is at (5, 1).

1 Line AB has gradient and intercept (0, 4).

So has equation $y = \frac{1}{4}x + 4$

Line AD is $y = -4x + 4$

> Find the equations of lines AB and AD. This scores 1 mark for method and 1 mark for accuracy.

Line BC is of the form $y = -4x + c$ and passes through (4, 5).

Substituting, $5 = -4 \times 4 + c \Rightarrow c = 21$

Line DC is of the form $y = \frac{1}{4}x + c$ and passes through (1, 0).

Substituting, $0 = 1 \times \frac{1}{4} + c \Rightarrow c = -\frac{1}{4}$

> Find the equations of lines through B and D parallel to AD and AB. This scores 1 mark for method and 1 mark for accuracy.

Point C is the intersection of $y = -4x + 21$ and $y = \frac{1}{4}x - \frac{1}{4}$

> Find the intersection point of the lines. This scores 1 mark for method and 1 mark for accuracy.

So $\frac{1}{4}x - \frac{1}{4} = -4x + 21, \Rightarrow 4\frac{1}{4}x = 21\frac{1}{4} \Rightarrow x = 5, y = 1.$

Total: 6 marks

Worked Examination Questions

AU **2** **a** Find the equation of the line shown.

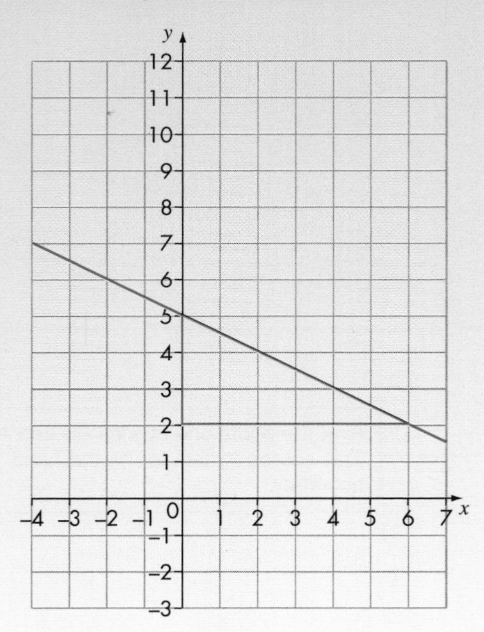

b Find the equation of the line shown.

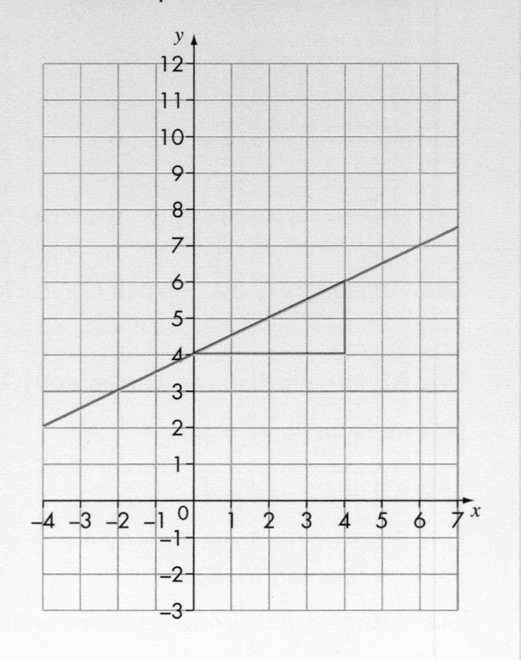

2 a Intercept is at (0, 5)

First identify the point where the line crosses the y-axis. This is the intercept, c. This gets 1 mark.

Gradient $= -\frac{3}{6} = -\frac{1}{2}$

Draw a right-angled triangle using grid lines as two sides of the triangle and part of the line as the hypotenuse. (Shown in red on diagram.) This gets 1 mark for method. Measure the y-step and the x-step of the triangle and divide the y-step by the x-step to get the gradient, m. As the line slopes down from left to right the gradient is negative.

Equation of the line is $y = -\frac{1}{2}x + 5$

Put the two numbers into the equation $y = mx + c$ to get the equation of the line. This gets 1 mark for accuracy.

b Intercept is at (0, 4)

First identify the point where the line crosses the y-axis. This is the intercept, c. This gets 1 mark.

Gradient $= \frac{2}{4} = \frac{1}{2}$

Draw a right-angled triangle using grid lines as two sides of the triangle and part of the line as the hypotenuse. (Shown in red on diagram.) This gets 1 mark for method. Measure the y-step and the x-step of the triangle and divide the y-step by the x-step to get the gradient, m. As the line slopes down from left to right the gradient is negative.

Equation of the line is $y = \frac{1}{2}x + 4$

Put the two numbers into the equation $y = mx + c$ to get the equation of the line. This gets 1 mark for accuracy.

Total: 6 marks

Worked Examination Questions

3 The grid below shows the graph of $y = x^2 - x - 6$.

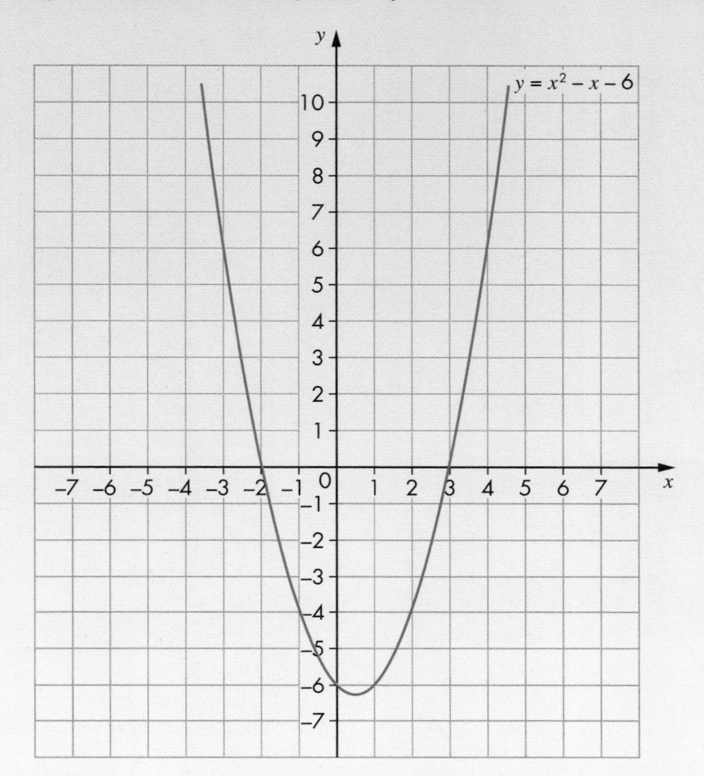

a Write down the solutions of the quadratic equation $x^2 - x - 6 = 0$.

b Find the least value of the graph of $y = x^2 - x - 6$.

3 a The graph cuts the x-axis at $x = -2$ and $x = 3$, so these are the solutions.

The roots or solutions of a quadratic equation are always the points where the graph crosses the x-axis. This gets 2 marks, one for each solution.

b The minimum value occurs halfway between (–2, 0) and (3, 0), so it is at $x = 0.5$.

The x-coordinate of the minimum value is halfway between the points where the graph crosses the axis. Identifying the point gets 1 mark.

Then $y = (0.5)2 - 0.5 - 6 = -6.25$

Substituting into the equation gets 1 mark for method.

So the minimum value is –6.25

Getting the minimum value scores 1 mark for accuracy.

Total: 5 marks

12

Problem Solving
Quadratics in bridges

The forms of many suspension bridges are based on quadratic functions. Their shape is a quadratic curve.

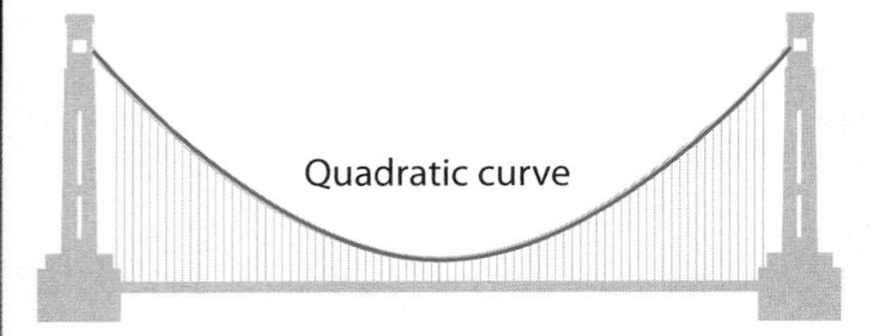

In this task you will investigate the quadratic functions that can be used to describe stable suspension bridges.

Getting started

Use these questions to familiarise yourself with how quadratic expressions can be used to represent bridges.

- What would you need to know, to be able to describe the shape of a bridge, in terms of a quadratic formula?
- Think about bridges and other landmarks in your local area. How many of these are based on quadratic equations and form parabolas? Use images to illustrate your findings.

Your task

Below you can see the dimensions of the Clifton Suspension Bridge. Use these dimensions to construct a diagram of the bridge.

Then, using your diagram, estimate the quadratic equation for the curve of the bridge. Represent this equation appropriately.

Write a report explaining the mathematical process that you used to solve this problem. State any assumptions that you made and explain whether you could have found different answers if you had changed your assumptions.

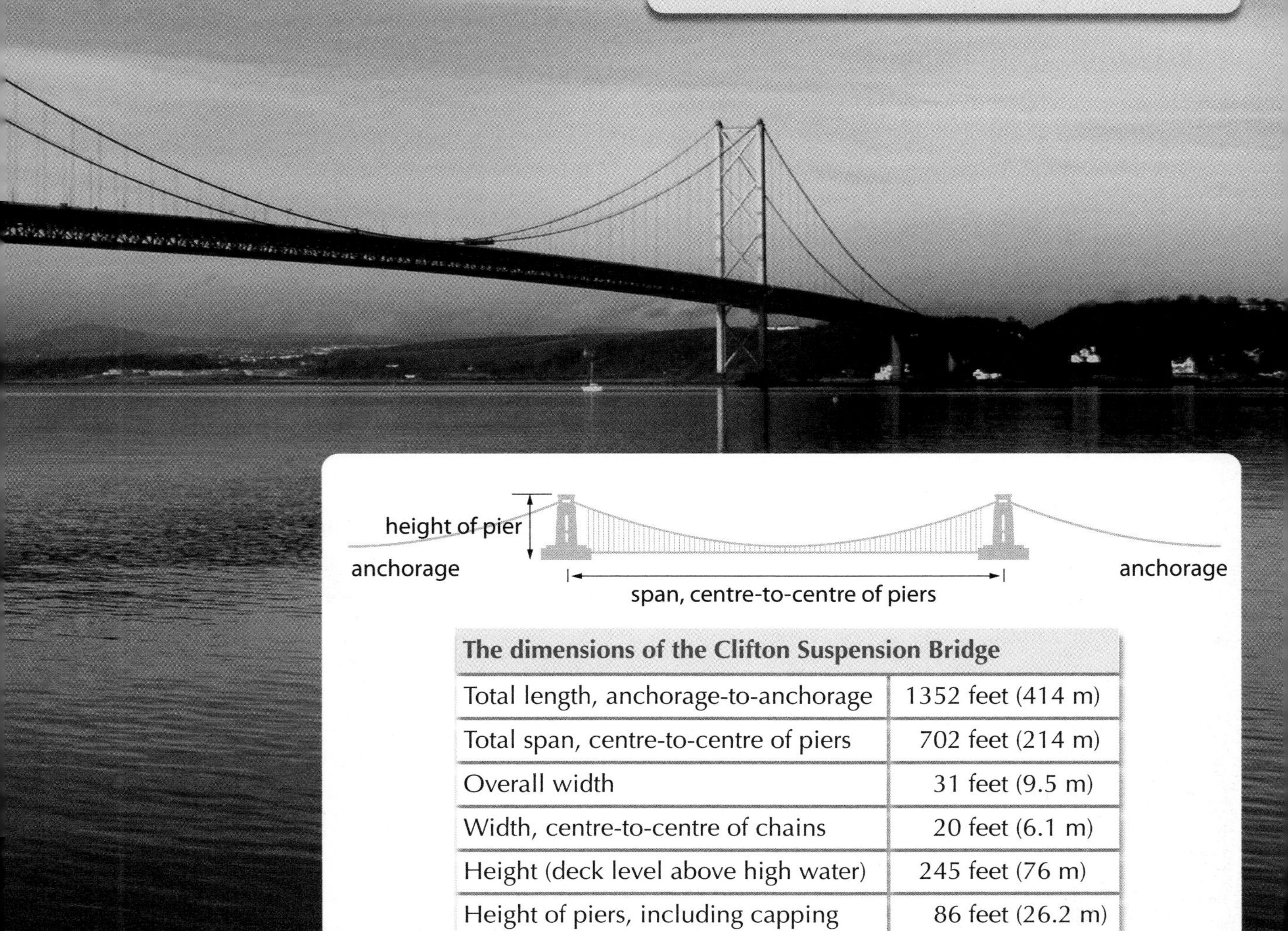

The dimensions of the Clifton Suspension Bridge	
Total length, anchorage-to-anchorage	1352 feet (414 m)
Total span, centre-to-centre of piers	702 feet (214 m)
Overall width	31 feet (9.5 m)
Width, centre-to-centre of chains	20 feet (6.1 m)
Height (deck level above high water)	245 feet (76 m)
Height of piers, including capping	86 feet (26.2 m)
Dip of chains	70 feet (21.3 m)

Why this chapter matters

The word comes from the Latin, *fractus*, meaning broken. You are already familiar with the idea of a fraction, comprising one or more of the equal parts into which a whole may be broken. But how does this relate to algebra?

The Babylonians – the first fractions

Fractions can be traced back to the Babylonians, in around 1800BC.

Their fractions were based on the number 60, as was their whole number system. However, the symbols could not accurately represent fractions, or show that they were fractions rather than standard numbers.

Egyptian fractions

The Egyptians (around 1000BC) were known to use fractions but generally these were unit fractions (fractions with a numerator of 1), for example $\frac{1}{2}$, $\frac{1}{3}$, $\frac{1}{4}$.

The Egyptians used this symbol ⬭ to represent the "one".

Then using the number symbols shown in Chapter 2 they made their fractions: ⬭ III $= \frac{1}{3}$

They also had special symbols for $\frac{1}{2}$, $\frac{2}{3}$, and $\frac{3}{4}$.

$= \frac{1}{2}$ | $= \frac{2}{3}$ | $= \frac{3}{4}$

Fractions in India

In India, around 500AD a system called *brahmi* was devised using symbols for the numbers. Fractions were written as a symbol above a symbol but without a line as used now.

Brahmi symbols

1	2	3	4	5	6	7	8	9
—	=	≡	+	h	ϥ	ʔ	ϭ	ʔ

So $\frac{5}{9}$ was written as: h over ʔ

Arabic fractions

The Arabs, probably around 1200AD, built on the number system – including the fractions – developed by the Indians. The line was first introduced into fractions in Arabia – sometimes drawn horizontally and sometimes slanting – leading to the clear fractions that we use today.

$$\frac{3}{4} \quad \frac{1}{8} \quad \frac{1}{2} \quad \frac{2}{3} \quad \frac{1}{4} \quad \frac{5}{10}$$

Algebraic fractions

Algebra is just mathematics expressed in general terms, and this is how mathematicians like to work at advanced stages of study.

Here is an equation used in the study of dynamics.

$$v^2 = u^2 + {}^2as$$

You can rearrange this to: $s = \dfrac{v^2 - u^2}{2a}$

and end up with something that looks just like an algebraic fraction!

Daniel Bernoulli (1700–1782) was a Dutch-Swiss mathematician famous for his studies of fluid mechanics, probability and statistics. Bernoulli's equation is well known in the study of the flow of liquids or gases.

$$\frac{v^2}{2} + gz + \frac{p}{\rho} = \text{constant}$$

where v is the speed of flow of the fluid, g is the acceleration due to gravity, z is the height of the point above a reference plane, p is the pressure on the flow and ρ is the density of the fluid.

Bernoulli established that if all the variables are combined, according to his formula, the end value is always the same, a constant. He used algebraic proof to reach a conclusion that has been accepted in the study of the movement of fluids and which has applications in many fields, including aerodynamics.

There are many other famous equations covering many different topics, but they all use algebraic variables to represent the different components to establish an algebraic proof in general terms. Another common feature is the use of algebraic fractions.

Algebra: Fractions and proof

The grades given in this chapter are target grades.

1. Algebraic fractions
2. Algebraic proof

This chapter will show you ...

- **B** Some common sequences of numbers
- **B** to **A*** How to combine fractions algebraically and solve equations with algebraic fractions
- **B** to **A*** How to prove results, using rigorous and logical mathematical arguments

Visual overview

Algebraic fractions → Algebraic proof ← Transposing formulae

What you should already know

- How to substitute numbers into an algebraic expression **(KS3 level 5, GCSE grade E)**
- How to state a rule for a simple linear sequence in words **(KS3 level 6, GCSE grade D)**
- How to factorise simple linear expressions **(KS3 level 6, GCSE grade D)**
- How to expand a pair of linear brackets to get a quadratic equation **(KS3 level 7, GCSE grade C)**

Quick check

1 Write down the next three terms of these sequences.

a 2, 5, 8, 11, 14, …

b 1, 4, 9, 16, 25, 36, ….

2 Work out the value of the expression $3n - 2$ for:

a $n = 1$ **b** $n = 2$ **c** $n = 3$

3 Factorise:

a $2x + 6$ **b** $x^2 - x$ **c** $10x^2 + 2x$

4 Expand:

a $(x + 6)(x + 2)$ **b** $(2x + 1)(x - 3)$ **c** $(x - 2)^2$

5 Make x the subject of:

a $2y + x = 3$ **b** $x - 3y = 4$ **c** $4y - x = 3$

13.1 Algebraic fractions

This topic will also be assessed in Unit 2.

This section will show you how to:

- simplify algebraic fractions
- solve equations containing algebraic fractions

Key words
brackets
cancel
cross-multiply
expression
factorise

The following four rules are used to work out the value of fractions.

Addition: $\frac{a}{b} + \frac{c}{d} = \frac{ad + bc}{bd}$

Subtraction: $\frac{a}{b} - \frac{c}{d} = \frac{ad - bc}{bd}$

Multiplication: $\frac{a}{b} \times \frac{c}{d} = \frac{ac}{bd}$

Division: $\frac{a}{b} \div \frac{c}{d} = \frac{ad}{bc}$

Note that a, b, c and d can be numbers, other letters or algebraic **expressions**. Remember:

- use **brackets**, if necessary
- **factorise** if you can
- **cancel** if you can.

EXAMPLE 1

Simplify **a** $\frac{1}{x} + \frac{x}{2y}$ **b** $\frac{2}{b} - \frac{a}{2b}$

a Using the addition rule: $\frac{1}{x} + \frac{x}{2x} = \frac{(1)(2y) + (x)(x)}{(x)(2y)} = \frac{2y + x^2}{2xy}$

b Using the subtraction rule: $\frac{2}{b} - \frac{a}{2b} = \frac{(2)(2b) - (a)(b)}{(b)(2b)} = \frac{4b - ab}{2b^2}$

$$= \frac{\cancel{b}(4 - a)}{2b^{\cancel{2}}} = \frac{4 - a}{2b}$$

Note: There are different ways of working out fraction calculations. Part **b** could have been done by making the denominator of each fraction the same.

$$\frac{(2)2}{(2)b} - \frac{a}{2b} = \frac{4 - a}{2b}$$

EXAMPLE 2

Simplify **a** $\frac{x}{3} \times \frac{x+2}{x-2}$ **b** $\frac{x}{3} \div \frac{2x}{7}$

a Using the multiplication rule: $\frac{x}{3} \times \frac{x+2}{x-2} = \frac{(x)(x+2)}{(2)(x-2)} = \frac{x^2+2x}{3x-6)}$

Remember that the line that separates the top from the bottom of an algebraic fraction acts as brackets as well as a division sign. Note that it is sometimes preferable to leave an algebraic fraction in a factorised form.

b Using the division rule: $\frac{x}{3} \div \frac{2x}{7} = \frac{(x)(7)}{(3)(2x)} = \frac{7}{6}$

EXAMPLE 3

Solve this equation. $\frac{x+1}{3} - \frac{x-3}{2} = 1$

Use the rule for combining fractions, and then **cross-multiply** to take the denominator of the left-hand side to the right-hand side.

$$\frac{(2)(x+1) - (3)(x-3)}{(2)(3)} = 1$$

$$2(x+1) - 3(x-3) = 6 \ (= 1 \times 2 \times 3)$$

Note the brackets. These will avoid problems with signs and help you to expand to get a linear equation.

$$2x + 2 - 3x + 9 = 6 \Rightarrow -x = -5 \Rightarrow x = 5$$

EXAMPLE 4

Solve this equation. $\frac{3}{x-1} - \frac{2}{x+1} = 1$

Use the rule for combining fractions, and cross-multiply to take the denominator of the left-hand side to the right-hand side, as in Example 3. Use brackets to help with expanding and to avoid problems with minus signs.

$$3(x+1) - 2(x-1) = (x-1)(x+1)$$

$$3x + 3 - 2x + 2 = x^2 - 1$$ (Right-hand side is the difference of two squares.)

Rearrange into the general quadratic form (see Chapter 11).

$$x^2 - x - 6 = 0$$

Factorise and solve $(x-3)(x+2) = 0 \Rightarrow x = 3$ or -2

Note that when your equation is rearranged into the quadratic form it should factorise. If it doesn't, then you have almost certainly made a mistake. If the question required an answer as a decimal or a surd it would say so.

EXAMPLE 5

Simplify this expression. $\dfrac{2x^2 + x - 3}{4x^2 - 9}$

Factorise the numerator and denominator: $\dfrac{(2x + 3)(x - 1)}{(2x + 3)(2x - 3)}$

Denominator is the difference of two squares.

Cancel any common factors: $\dfrac{\cancel{(2x + 3)}(x - 1)}{\cancel{(2x + 3)}(3x - 3)}$

If at this stage there isn't a common factor on top and bottom, you should check your factorisations.

The remaining fraction is the answer: $\dfrac{(x - 1)}{(2x - 3)}$

EXERCISE 13A

1 Simplify each of these.

a $\dfrac{x}{2} + \dfrac{x}{3}$ **b** $\dfrac{3x}{4} + \dfrac{x}{5}$ **c** $\dfrac{3x}{4} + \dfrac{2x}{5}$

d $\dfrac{x}{2} + \dfrac{y}{3}$ **e** $\dfrac{xy}{4} + \dfrac{2}{x}$ **f** $\dfrac{x + 1}{2} + \dfrac{x + 2}{3}$

g $\dfrac{2x + 1}{2} + \dfrac{3x + 1}{4}$ **h** $\dfrac{x}{5} + \dfrac{2x + 1}{3}$ **i** $\dfrac{x - 2}{2} + \dfrac{x - 3}{4}$

j $\dfrac{x - 4}{4} + \dfrac{2x - 3}{2}$

2 Simplify each of these.

a $\dfrac{x}{2} - \dfrac{x}{3}$ **b** $\dfrac{3x}{4} - \dfrac{x}{5}$ **c** $\dfrac{3x}{4} - \dfrac{2x}{5}$

d $\dfrac{x}{2} - \dfrac{y}{3}$ **e** $\dfrac{xy}{4} - \dfrac{2}{y}$ **f** $\dfrac{x + 1}{2} - \dfrac{x + 2}{3}$

g $\dfrac{2x + 1}{2} - \dfrac{3x + 1}{4}$ **h** $\dfrac{x}{5} - \dfrac{2x + 1}{3}$ **i** $\dfrac{x - 2}{2} - \dfrac{x - 3}{4}$

j $\dfrac{x - 4}{4} - \dfrac{2x - 3}{2}$

3 Solve the following equations.

a $\dfrac{x + 1}{2} + \dfrac{x + 2}{5} = 3$ **b** $\dfrac{x + 2}{4} + \dfrac{x + 1}{7} = 3$ **c** $\dfrac{4x + 1}{3} - \dfrac{x + 2}{4} = 2$

d $\dfrac{2x - 1}{3} + \dfrac{3x + 1}{4} = 7$ **e** $\dfrac{2x + 1}{2} - \dfrac{x + 1}{7} = 1$ **f** $\dfrac{3x + 1}{5} - \dfrac{5x - 1}{7} = 0$

B

4 Simplify each of these.

a $\frac{x}{2} \times \frac{x}{3}$ **b** $\frac{2x}{7} \times \frac{3y}{4}$ **c** $\frac{4x}{3y} \times \frac{2y}{x}$

d $\frac{4y^2}{9x} \times \frac{3x^2}{2y}$ **e** $\frac{x}{2} \times \frac{x-2}{5}$ **f** $\frac{x-3}{15} \times \frac{5}{2x-6}$

g $\frac{2x+1}{2} \times \frac{3x+1}{4}$ **h** $\frac{x}{5} \times \frac{2x+1}{3}$ **i** $\frac{x-2}{2} \times \frac{4}{x-3}$

j $\frac{x-5}{10} \times \frac{5}{x^2-5x}$

A

5 Simplify each of these.

a $\frac{x}{2} \div \frac{x}{3}$ **b** $\frac{2x}{7} \div \frac{4y}{14}$ **c** $\frac{4x}{3y} \div \frac{x}{2y}$

d $\frac{4y^2}{9x} \div \frac{2y}{3x^2}$ **e** $\frac{x}{2} \div \frac{x-2}{5}$ **f** $\frac{x-3}{15} \div \frac{5}{2x-6}$

g $\frac{2x+1}{2} \div \frac{4x+2}{4}$ **h** $\frac{x}{6} \div \frac{2x^2+x}{3}$ **i** $\frac{x-2}{12} \div \frac{4}{x-3}$

j $\frac{x-5}{10} \div \frac{x^2-5x}{5}$

6 Simplify each of these. Factorise and cancel where appropriate.

a $\frac{3x}{4} + \frac{x}{4}$ **b** $\frac{3x}{4} - \frac{x}{4}$ **c** $\frac{3x}{4} \times \frac{x}{4}$

d $\frac{3x}{4} \div \frac{x}{4}$ **e** $\frac{3x+1}{2} + \frac{x-2}{5}$ **f** $\frac{3x+1}{2} - \frac{x-2}{5}$

g $\frac{3x+1}{2} \times \frac{x-2}{5}$ **h** $\frac{x^2-9}{10} \times \frac{5}{x-3}$ **i** $\frac{2x+3}{5} \div \frac{6x+9}{10}$

j $\frac{2x^2}{9} - \frac{2y^2}{3}$

7 Show that each algebraic fraction simplifies to the given expression.

a $\frac{2}{x+1} + \frac{5}{x+2} = 3$ simplifies to $3x^2 + 2x - 3 = 0$

b $\frac{4}{x-2} + \frac{7}{x+1} = 3$ simplifies to $3x^2 - 14x + 4 = 0$

c $\frac{3}{4x+1} - \frac{4}{x+2} = 2$ simplifies to $8x^2 + 31x + 2 = 0$

d $\frac{2}{2x-1} - \frac{6}{x+1} = 11$ simplifies to $22x^2 + 21x - 19 = 0$

e $\frac{3}{2x-1} - \frac{4}{3x-1} = 1$ simplifies to $x^2 - x = 0$

A*

PS 8 For homework a teacher asks her class to simplify the expression $\frac{x^2 - x - 2}{x^2 + x - 6}$.

This is Tom's answer:

$$\frac{\cancel{x^2} - x - \cancel{2}^{\,-1}}{\cancel{x^2} + x - \cancel{6}_{\,+3}}$$

$$= \frac{-x - 1}{x + 3} = \frac{x + 1}{x + 3}$$

When she marked the homework, the teacher was in a hurry and only checked the answer, which was correct.

Tom made several mistakes. What are they?

AU 9 An expression of the form $\frac{ax^2 + bx - c}{dx^2 - e}$ simplifies to $\frac{x - 1}{2x - 3}$.

What was the original expression?

10 Solve the following equations.

a $\frac{4}{x + 1} + \frac{5}{x + 2} = 2$

b $\frac{18}{4x - 1} - \frac{1}{x + 1} = 1$

c $\frac{2x - 1}{2} - \frac{6}{x + 1} = 1$

d $\frac{3}{2x - 1} - \frac{4}{3x - 1} = 1$

11 Simplify the following expressions.

a $\frac{x^2 + 2x - 3}{2x^2 + 7x + 3}$

b $\frac{4x^2 - 1}{2x^2 + 5x - 3}$

c $\frac{6x^2 + x - 2}{9x^2 - 4}$

d $\frac{4x^2 + x - 3}{4x^2 - 7x + 3}$

e $\frac{4x^2 - 25}{8x^2 - 22x + 5}$

13.2 Algebraic proof

This section will show you how to:
- recognise and continue some special number sequences

Key words
proof
show
verify

You will have met the fact that the sum of any two odd numbers is always an even number before, but can you prove it?

You can take any two odd numbers, add them together and get a number that divides exactly by 2. This does not prove the result, even if everyone in your class, or your school, or the whole of Britain, did this for a different pair of starting odd numbers. Unless you tried every pair of odd numbers (and there is an infinite number of them) you cannot be 100% certain this result is always true.

This is how to prove the result.

Let n be any whole number.

Whatever whole number is represented by n, $2n$ has to be even. So, $2n + 1$ represents any odd number.

Let one odd number be $2n + 1$, and let the other odd number be $2m + 1$.

The sum of these is:

$$(2n + 1) + (2m + 1) = 2n + 2m + 1 + 1$$
$$= 2n + 2m + 2$$
$$= 2(n + m + 1), \text{ which must be even.}$$

This proves the result, as n and m can be any numbers.

In an algebraic **proof**, every step must be shown clearly and the algebra must be done properly.

There are three levels of 'proof': **Verify** that …, **Show** that …, and Prove that …

- At the lowest level (verification), all you have to do is substitute numbers into the result to show that it works.
- At the middle level, you have to show that both sides of the result are the same algebraically.
- At the highest level (proof), you have to manipulate the left-hand side of the result to become its right-hand side.

The following example demonstrates these three different procedures.

EXAMPLE 6

You are given that $n^2 + (n + 1)^2 - (n + 2)^2 = (n - 3)(n + 1)$.

a Verify that this result is true.

b Show that this result is true.

c Prove that this result is true.

a Choose a number for n, say $n = 5$. Put this value into both sides of the expression, which gives:

$$5^2 + (5 + 1)^2 - (5 + 2)^2 = (5 + 3)(5 + 1)$$
$$25 + 36 - 49 = 2 \times 6$$
$$12 = 12$$

Hence, the result is true.

b Expand the LHS and RHS of the expression to get:

$$n^2 + n^2 + 2n + 1 - (n^2 + 4n + 4) = n^2 - 2n - 3$$
$$n^2 - 2n - 3 = n^2 - 2n - 3$$

That is, both sides are algebraically the same.

c Expand the LHS of the expression to get: $n^2 + n^2 + 2n + 1 - (n^2 + 4n + 4)$

Collect like terms, which gives $n^2 + n^2 - n^2 + 2n - 4n + 1 - 4 = n^2 - 2n - 3$

Factorise the collected result: $n^2 - 2n - 3 = (n - 3)(n + 1)$, which is the RHS of the original expression.

EXERCISE 13B

AU 1 **a** Choose any odd number and any even number. Add these together. Is the result odd or even? Does this always work for any odd number and even number you choose?

b Let any odd number be represented by $2n + 1$. Let any even number be represented by $2m$, where m and n are integers. Prove that the sum of an odd number and an even number always gives an odd number.

AU 2 Prove the following results.

a The sum of two even numbers is even.

b The product of two even numbers is even.

c The product of an odd number and an even number is even.

d The product of two odd numbers is odd.

e The sum of four consecutive numbers is always even.

f Half the sum of four consecutive numbers is always odd.

A*

AU 3 A Fibonacci sequence is formed by adding the previous two terms to get the next term. For example, starting with 3 and 4, the series is:

3, 4, 7, 11, 18, 29, 47, 76, 123, 199, …

a Continue the Fibonacci sequence 1, 1, 2, … up to 10 terms.

b Continue the Fibonacci sequence $a, b, a + b, a + 2b, 2a + 3b, \ldots$ up to 10 terms.

c Prove that the difference between the 8th term and the 5th term of any Fibonacci sequence is twice the 6th term.

AU 4 The nth term in the sequence of triangular numbers 1, 3, 6, 10, 15, 21, 28, … is given by $\frac{1}{2}n(n + 1)$.

a Show that the sum of the 11th and 12th terms is a perfect square.

b Explain why the $(n + 1)$th term of the triangular number sequence is given by $\frac{1}{2}(n + 1)(n + 2)$.

c Prove that the sum of any two consecutive triangular numbers is always a square number.

AU 5 The diagram shows part of a 10 × 10 'hundred square'.

12	13	14	15
22	23	24	25
32	33	34	35
42	43	44	45

a One 2 × 2 square is marked.

i Work out the difference between the product of the bottom-left and top-right values and the product of the top-left and bottom-right values:

$22 \times 13 - 12 \times 23$

ii Repeat this for any other 2 × 2 square of your choosing.

b Prove that this will always give an answer of 10 for any 2 × 2 square chosen.

c The diagram shows a calendar square (where the numbers are arranged in rows of seven).

Prove that you always get a value of 7 if you repeat the procedure in part **a i**.

1	2	3	4	5	6	7
8	9	10	11	12	13	14
15	16	17	18	19	20	21
22	23	24	25	26	27	28
29	30	31				

d Prove that in a number square that is arranged in rows of n numbers then the difference is always n if you repeat the procedure in part **a i**.

AU 6 Prove that if you add any two-digit number from the 9 times table to the reverse of itself (that is, swap the tens digit and units digit), the result will always be 99.

A*

AU 7 Speed Cabs charges 45 pence per kilometre for each journey. Evans Taxis has a fixed charge of 90p plus 30p per kilometre.

a **i** Verify that Speed Cabs is cheaper for a journey of 5 km.

ii Verify that Evans Taxis is cheaper for a journey of 7 km.

b Show clearly why both companies charge the same for a journey of 6 km.

c Show that if Speed Cabs charges a pence per kilometre, and Evans Taxis has a fixed charge of £b plus a charge of c pence per kilometre, both companies charge the same for a journey of $\frac{4}{3x - 1}$ kilometres.

AU 8 You are given that:

$$(a + b)^2 + (a - b)^2 = 2(a^2 + b^2)$$

a Verify that this result is true for $a = 3$ and $b = 4$.

b Show that the LHS is the same as the RHS.

c Prove that the LHS can be simplified to the RHS.

AU 9 Prove that: $(a + b)^2 - (a - b)^2 = 4ab$.

AU 10 The rule for converting from degrees Fahrenheit to degrees Celsius is to subtract 32° and then to multiply by $\frac{5}{9}$.

Prove that the temperature that has the same value in both scales is –40°.

AU 11 The sum of the series $1 + 2 + 3 + 4 + \ldots + (n - 2) + (n - 1) + n$ is given by $\frac{1}{2}n(n + 1)$.

a Verify that this result is true for $n = 6$.

b Write down a simplified value, in terms of n, for the sum of these two series.

$$1 + 2 + 3 + \ldots + (n - 2) + (n - 1) + n$$

and $$n + (n - 1) + (n - 2) + \ldots + 3 + 2 + 1$$

c Prove that the sum of the first n integers is $\frac{1}{2}n(n + 1)$.

AU 12 The following is a 'think of a number' trick.

- Think of a number.
- Multiply it by 2.
- Add 10.
- Divide the result by 2.
- Subtract the original number.

The result is always 5.

a Verify that the trick works when you pick 7 as the original number.

b Prove why the trick always works.

AU 13 You are told that 'when two numbers have a difference of 2, the difference of their squares is twice the sum of the two numbers'.

a Verify that this is true for 5 and 7.

b Prove that the result is true.

c Prove that when two numbers have a difference of n, the difference of their squares is n times the sum of the two numbers.

AU 14 Four consecutive numbers are 4, 5, 6 and 7.

	n^2	$-n$	-1
n^2	n^4		$-n^2$
$-n$		n^2	
-1			

a Verify that their product plus 1 is a perfect square.

b Complete the multiplication square and use it to show that:

$$(n^2 - n - 1)^2 = n^4 - 2n^3 - n^2 + 2n + 1$$

c Let four consecutive numbers be $(n - 2)$, $(n - 1)$, n, $(n + 1)$. Prove that the product of four consecutive numbers plus 1 is a perfect square.

AU 15 Here is another mathematical trick to try on a friend.

- Think of two single-digit numbers.
- Multiply one number (your choice) by 2.
- Add 5 to this answer.
- Multiply this answer by 5.
- Add the second number.
- Subtract 4.
- Ask your friend to state the final answer.
- Mentally subtract 21 from this answer.

The two digits you get are the two digits your friend first thought of.

Prove why this works.

EXERCISE 13C

You may not be able algebraically to prove all of these results. Some of them can be disproved by a counter-example. You should first try to verify each result, then attempt to prove it — or at least try to demonstrate that the result is probably true by trying lots of examples.

AU 1 T represents any triangular number. Prove the following.

a $8T + 1$ is always a square number.

b $9T + 1$ is always another triangular number.

AU 2 Lewis Carroll, who wrote *Alice in Wonderland*, was also a mathematician. In 1890, he suggested the following results.

a For any pair of numbers, x and y, if $x^2 + y^2$ is even, then $\frac{1}{2}(x^2 + y^2)$ is the sum of two squares.

b For any pair of numbers, x and y, $2(x^2 + y^2)$ is always the sum of two squares.

c Any number of which the square is the sum of two squares is itself the sum of two squares.

Can you prove these statements to be true or false?

AU 3 For all values of n, $n^2 - n + 41$ gives a prime number. True or false?

AU 4 Pythagoras' theorem says that for a right-angled triangle with two short sides a and b and a long side c, $a^2 + b^2 = c^2$. For any integer n, $2n$, $n^2 - 1$ and $n^2 + 1$ form three numbers that obey Pythagoras' theorem. Can you prove this?

AU 5 Waring's theorem states that: "Any whole number can be written as the sum of not more than four square numbers".

For example, $27 = 3^2 + 3^2 + 3^2$ and $23 = 3^2 + 3^2 + 2^2 + 1^2$

Is this always true?

AU 6 Take a three-digit multiple of 37, for example, $7 \times 37 = 259$. Write these digits in a cycle.

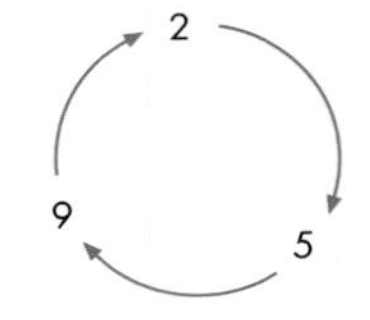

Take all possible three-digit numbers from the cycle, for example, 259, 592 and 925.

Divide each of these numbers by 37 to find that:

$259 = 7 \times 37 \qquad 592 = 16 \times 37 \qquad 925 = 25 \times 37$

Is this true for all three-digit multiples of 37?

Is it true for a five-digit multiple of 41?

AU 7 Prove that the sum of the squares of two consecutive integers is an odd number.

8 The difference of two squares is an identity, i.e.,

$$a^2 - b^2 \equiv (a + b)(a - b)$$

which means that it is true for all values of a and b whether they are numeric or algebraic.

Prove that $a^2 - b^2 \equiv (a + b)(a - b)$ is true when $a = 2x + 1$ and $b = x - 1$

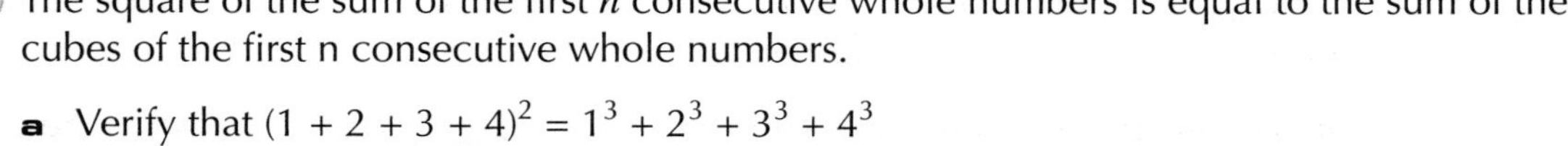

9 The square of the sum of the first n consecutive whole numbers is equal to the sum of the cubes of the first n consecutive whole numbers.

a Verify that $(1 + 2 + 3 + 4)^2 = 1^3 + 2^3 + 3^3 + 4^3$

b The sum of the first n consecutive whole numbers is $\frac{1}{2}n(n + 1)$

Write down a formula for the sum of the cubes of the first n whole numbers.

c Test your formula for $n = 6$

A*

GRADE BOOSTER

B You can solve linear equations involving algebraic fractions where the subject appears as the numerator

B You can understand the proofs of simple theorems, such as an exterior angle of a triangle is the sum of the two opposite interior angles

A You can combine fractions, using the four algebraic rules of addition, subtraction, multiplication and division

A You can show that an algebraic statement is true, using both sides of the statement to justify your answer

A* You can solve a quadratic equation obtained from algebraic fractions where the variable appears in the denominator

A* You can simplify algebraic fractions by factorisation and cancellation

A* You can prove algebraic results with rigorous and logical mathematical arguments

What you should know now

- How to manipulate algebraic fractions and solve equations resulting from the simplified fractions
- The meaning of the terms 'verify that', 'show that' and 'prove'
- How to prove some standard results in mathematics
- How to use your knowledge of proof to answer the questions throughout the book that are flagged with the proof icon

1 Simplify $\frac{3(x+2)}{(x+2)^2}$ *(1 mark)*

Edexcel, June 2008, Paper 10 Higher, Question 5(c)

2 Simplify fully $\frac{(x+3)^2}{x+3}$ *(1 mark)*

Edexcel, March 2008, Paper 11 Higher, Question 5(e)

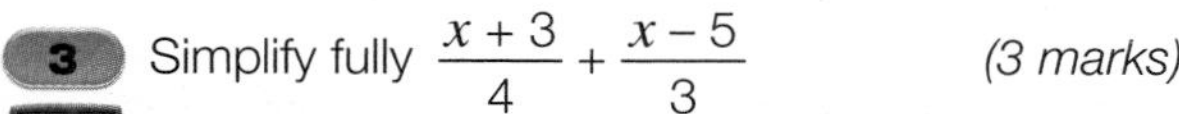

3 Simplify fully $\frac{x+3}{4} + \frac{x-5}{3}$ *(3 marks)*

Edexcel, November 2008, Paper 10 Higher, Question 8

4 Write as a single fraction:

$$\frac{4}{x(x+3)} + \frac{5}{(x+3)}$$ *(2 marks)*

Edexcel, March 2007, Paper 11 Higher, Question 9

5 Simplify fully $\frac{3x+6}{x^2-4}$ *(3 marks)*

Edexcel, March 2007, Paper 11 Higher, Question 8

6 Simplify $\frac{x^2+5x+6}{x+2}$ *(3 marks)*

Edexcel, June 2007, Paper 11 Higher, Question 8

7 Prove that $(n+2)^2 - (n-2)^2 = 8n$ for all values of n. *(2 marks)*

Edexcel, June 2007, Paper 11 Higher, Question 9

8 Write $\frac{x}{x-2} - \frac{3}{x(x-2)}$

as a single fraction in its simplest form. *(2 marks)*

Edexcel, June 2007, Paper 11 Higher, Question 9

9 Simplify fully $\frac{4a-20}{a^2-25}$ *(3 marks)*

Edexcel, November 2007, Paper 11 Higher, Question 11

10 Write as a single fraction in its simplest form

$$\frac{4}{x+5} + \frac{1}{x-3}$$ *(4 marks)*

Edexcel, November 2007, Paper 11 Higher, Question 8

11 Simplify fully $\frac{x}{x-3} - \frac{x}{x+3}$ *(3 marks)*

Edexcel, March 2008, Paper 11 Higher, Question 9(b)

12 Simplify $\frac{p^2-9}{2p+6}$ *(3 marks)*

Edexcel, March 2008, Paper 10 Higher, Question 8

13 Simplify fully $\frac{6x^2+3x}{4x^2-1}$ *(3 marks)*

Edexcel, June 2008, Paper 10 Higher, Question 9

14 The nth even number is $2n$.

The next even number after $2n$ is $2n + 2$.

a Explain why. *(1 mark)*

b Write down an expression, in terms of n, for the next even number after $2n + 2$. *(1 mark)*

c Show algebraically that the sum of any 3 consecutive even numbers is always a multiple of 6. *(3 marks)*

Edexcel, November 2008, Paper 4 Higher, Question 20

15 Simplify $\frac{x^2+2x+1}{x^2+3x+2}$ *(3 marks)*

Edexcel, November 2008, Paper 11 Higher, Question 10

16 Simplify fully $\frac{x^2+x-6}{x^2-7x+10}$ *(3 marks)*

Edexcel, May 2008, Paper 3 Higher, Question 28

17 Simplify $\frac{3(x-2)}{x^2-7x+10}$ *(2 marks)*

Edexcel, November 2008, Paper 10 Higher, Question 6(b)

Worked Examination Questions

1 Make g the subject of the following formula.

$$\frac{t(3 + g)}{8 - g} = 2$$

1 $t(3 + g) = 2(8 - g)$ — Cross multiply to get rid of the fraction. This gives 1 mark for method.

$3t + gt = 16 - 2g$ — Expand the brackets. This gives 1 mark for accuracy.

$gt + 2g = 16 - 3t$

$g(t + 2) = 16 - 3t$ — Collect all the g terms on the left-hand side and other terms on the right-hand side. This gives 1 mark for method.

$g = \frac{16 - 3t}{t + 2}$ — Simplify, $gt + 2g = g(t + 2)$, and divide by $(t + 2)$. This gives 1 mark for accuracy.

Total: 4 marks

Worked Examination Questions

PS **2** **a** n is a positive integer.

i Explain why $n(n + 1)$ must be an even number.

ii Explain why $2n + 1$ must be an odd number.

b Expand and simplify $(2n + 1)^2$.

c Prove that the square of any odd number is always 1 more than a multiple of 8.

2 a i If n is odd, $n + 1$ is even.
If n is even, $n + 1$ is odd.
Even times odd is always even.

> This is a lead-in to the rest of the task. An explanation in words is good enough. Keep the words to a minimum. This gives 1 mark.

ii $2n$ must be even, so $2n + 1$ must be odd.

> An explanation in words is good enough. This gives 1 mark.

b $(2n + 1)^2 = (2n + 1)(2n + 1) =$
$4n^2 + 2n + 2n + 1 = 4n^2 + 4n + 1$

> Always write down a squared bracket twice, then expand it by whichever method you prefer. This gives 1 mark.

c $(2n + 1)^2 = 4n^2 + 4n + 1$

> Use the fact that $2n + 1$ is odd, and it has been 'squared' in part **b**.

$4n^2 + 4n + 1 = 4n(n + 1) + 1$

> The 'one more than' is taken care of with the +1. This gives 1 mark for method.

$4 \times n(n + 1) + 1 = 4 \times$ even $+ 1$,
which must be a multiple of 8 plus 1.

> Show that the $4n^2 + 4n$ is a multiple of 8 using the result in part **a i**. This gives 1 mark for accuracy.

Total: 5 marks

13 Problem Solving
Picture proofs

People often use algebra as a tool to write proofs of mathematical results. Sometimes we can use a 'picture' or diagram to demonstrate a result or to indicate how it might be proved.

For example, you know that the sum of an odd number and another odd number is always an even number. We can illustrate this with a diagram.

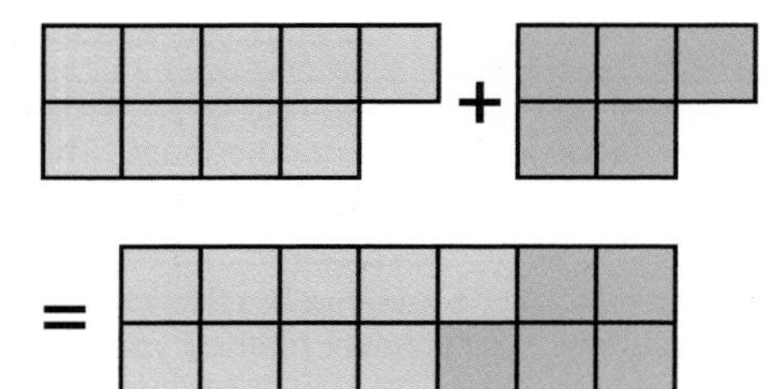

Although the numbers in the diagram are 9 and 5, you can see that it will work for any two odd numbers. The result will always be two rows of the same length, which means the sum of the two odd numbers is an even number.

Getting started

1 Draw a diagram to represent three consecutive numbers. Use the same idea as in the example above.

How can you adjust the diagram to show that the sum is a multiple of 3?

Hint: What would a multiple of 3 look like as a block of squares?
You can move the squares around.

2 The sum of four consecutive numbers is **not** a multiple of 4. Can you draw a diagram to illustrate this?

3 Is the sum of five consecutive numbers a multiple of 5? Draw a diagram to justify your conclusion.

Your task

Investigate how algebraic results, like those you have already considered, can be written and displayed. Draw your own conclusions about what mathematical proof really is.

1 On the right are some diagrams, with the algebraic results they illustrate. In each case, explain how the diagram illustrates the result.

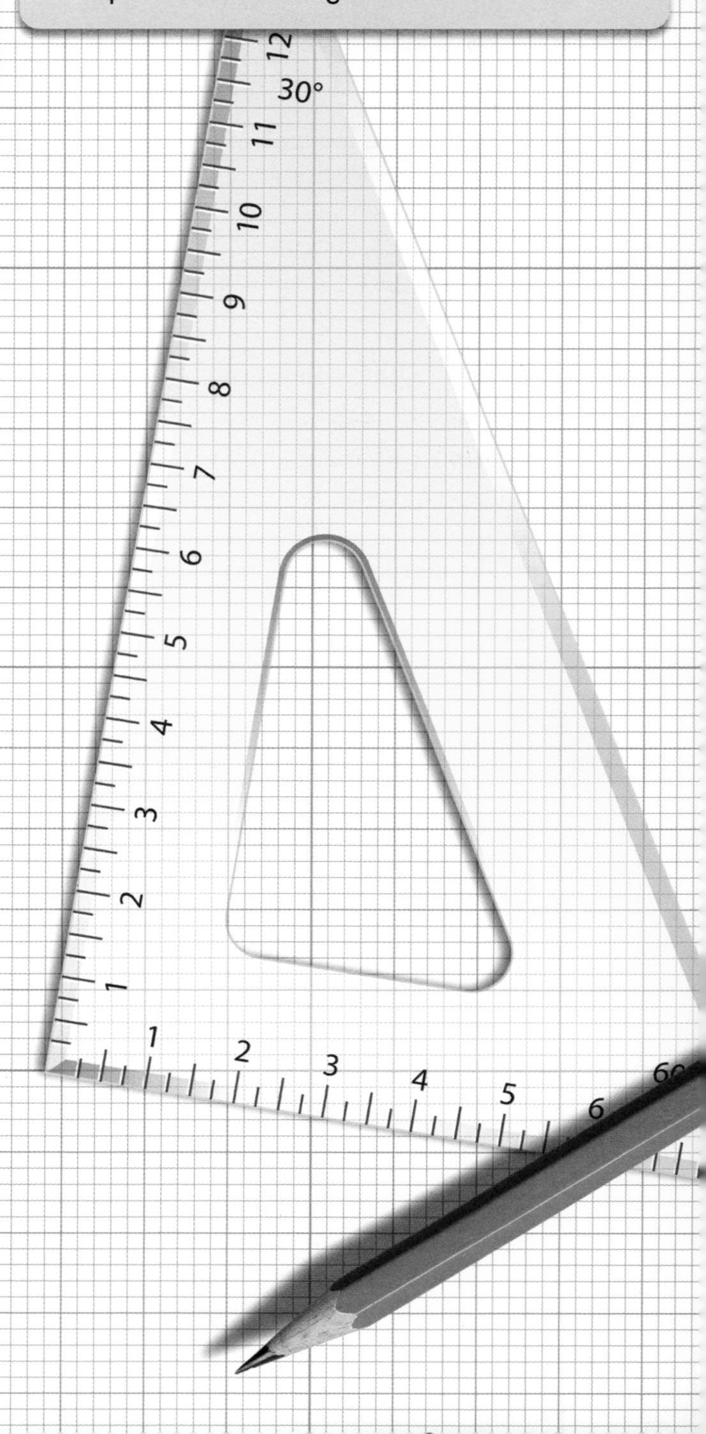

1a

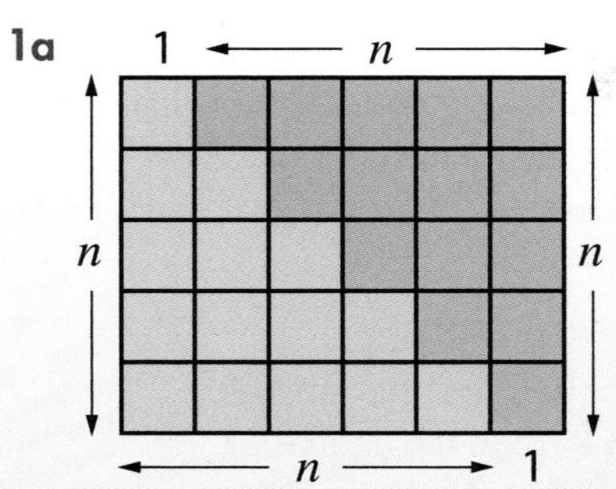

$1 + 2 + 3 + \ldots + n = \frac{1}{2}n(n + 1)$

1b

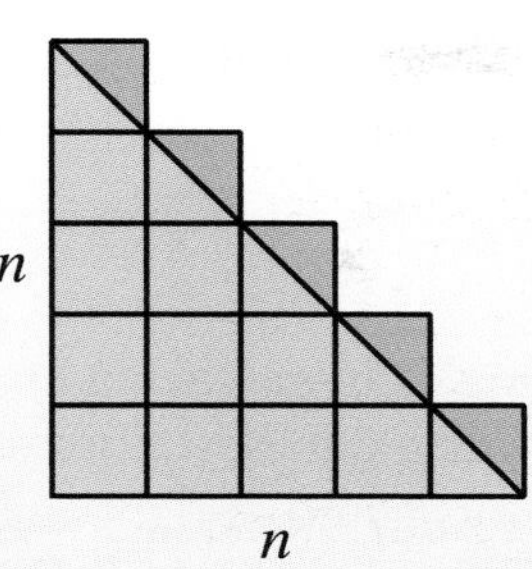

$1 + 2 + 3 + \ldots + n = \frac{1}{2}n^2 + \frac{1}{2}n$

1c

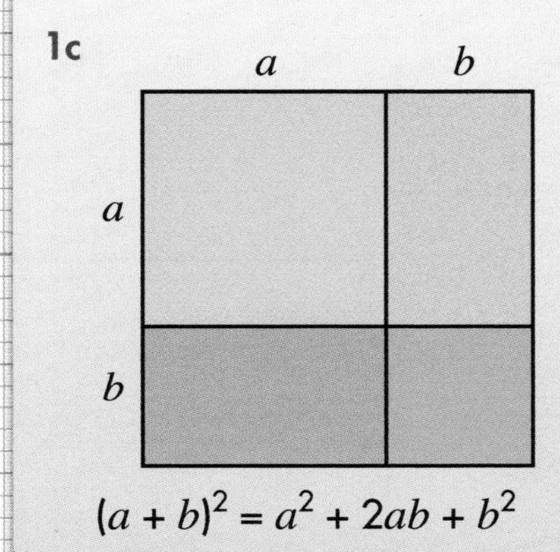

$(a + b)^2 = a^2 + 2ab + b^2$

1d

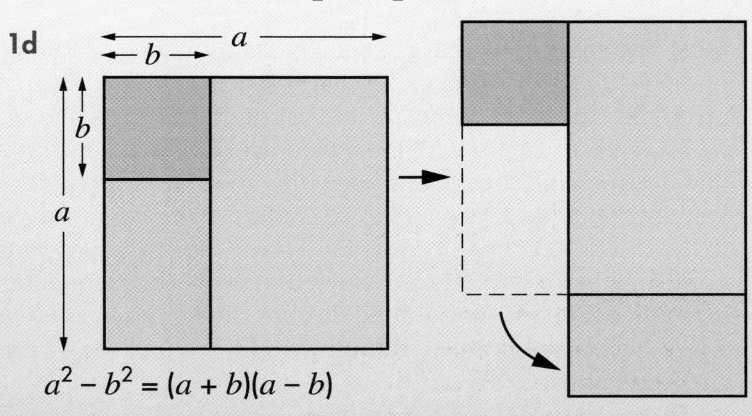

$a^2 - b^2 = (a + b)(a - b)$

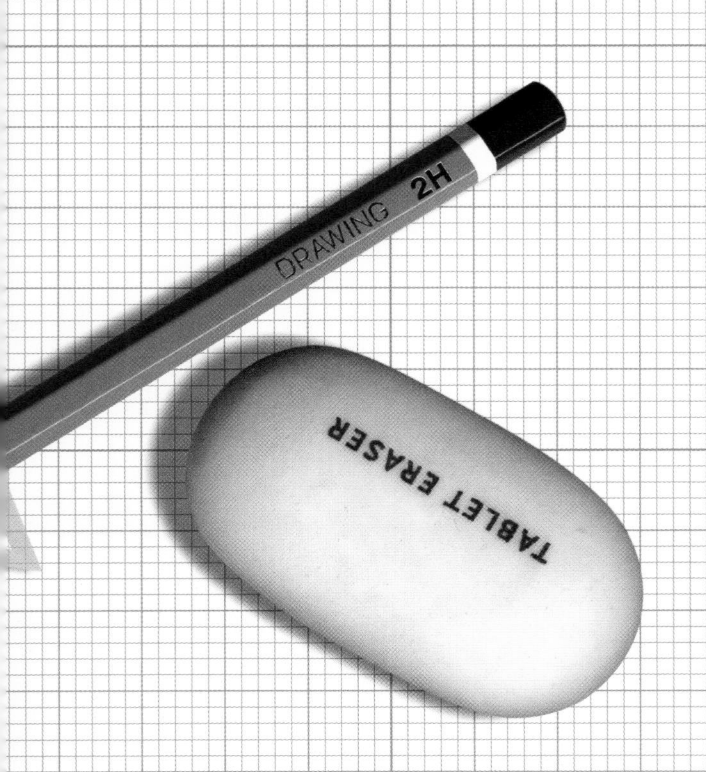

Your task (continued)

2 Now, look at these statements.

a The sum of the first n odd numbers is n^2.

b $(2a + 1)^2 = 4a^2 + 4a + 1$

c The sum of two consecutive triangular numbers is always a square number.

d $(a + b)^2 - (a - b)^2 = 4ab$

Draw diagrams to illustrate each statement. In each case, explain how your diagrams show that the statement is true.

3 Write a report of your investigation, setting out your explanations and diagrams clearly. You must:

- give algebraic proofs of some of the results from your investigation
- explain whether you think that a diagram on its own is a proof.

Why this chapter matters

It is essential to understand angles. They help us to construct everything, from a building to a table. So angles literally shape our world.

Ancient civilisations used **right angles** in surveying and in constructing buildings. The ancient Greeks used the right angle to describe relationships between other angles. However, not everything can be measured in right angles. There is a need for a smaller, more useful unit. The ancient Babylonians chose a unit angle that led to the development of the **degree**, which is what we still use today.

Most historians think that the ancient Babylonians thought of the 'circle' of the year as consisting of 360 days. This is not a bad approximation, given the crudeness of the ancient astronomical tools and often having to measure small angles with the naked eye. Mathematics historians believe that the ancient Babylonians knew that the side of a **regular hexagon** inscribed in a circle is equal to the **radius** of the circle. This may have led to the division of the full circle (360 days) into six equal parts, each part consisting of 60 days. They divided one angle of an **equilateral triangle** into 60 equal parts, now called degrees, then further subdivided a degree into 60 equal parts, called **minutes**, and a minute into 60 equal parts, called **seconds**.

Although many historians believe this is why 60 was the base of the Babylonian system of angle measurement, others think there was a different reason. The number 60 has many **factors**. Work with fractional parts of the whole (60) is greatly simplified, because 2, 3, 4, 5, 6, 10, 12, 15, 20 and 30 are all factors of 60.

Modern measurement of angles

Modern surveyors use theodolites for measuring angles.

A theodolite can be used for measuring both horizontal and vertical angles. It is a key tool in surveying and engineering work, particularly on inaccessible ground, but theodolites have been adapted for other specialised purposes in fields such as meteorology and rocket-launching technology. A modern theodolite comprises a movable telescope mounted within two perpendicular axes – the horizontal and the vertical axis. When the telescope is pointed at a desired object, the angle of each of these axes can be measured with great precision, typically on the scale of arcseconds. (There are 3600 **arcseconds** in 1°.)

Modern theodolite

Chapter

Geometry: Properties of circles

1. Circle theorems
2. Cyclic quadrilaterals
3. Tangents and chords
4. Alternate segment theorem

The grades given in this chapter are target grades.

This chapter will show you ...

B to A* how to find angles using circle theorems

B to A* how to find the sizes of angles in cyclic quadrilaterals

B to A* how to use tangents and chords to find the sizes of angles in circles

A to A* how to use the alternate segment theorem to find the sizes of angles in circles

Visual overview

What you should already know

- The three interior angles of a triangle add up to 180°. So, $a + b + c = 180°$
  (KS3 level 5, GCSE grade E)

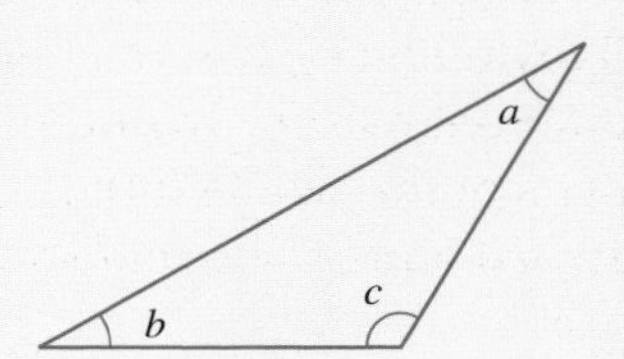

- The four interior angles of a quadrilateral quadrilateral add up to 360°.

  So, $a + b + c + d = 360°$
  (KS3 level 5, GCSE grade E)

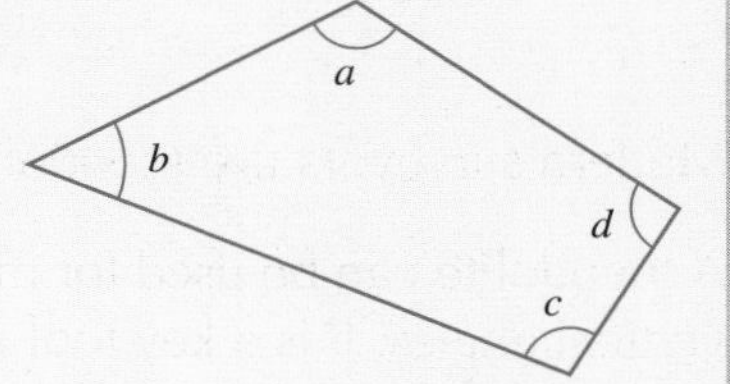

- Angles in parallel lines

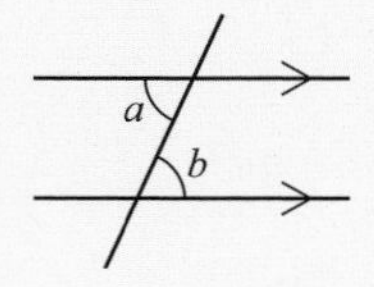

a and b are equal
a and b are alternate angles

a and b are equal
a and b are corresponding angles

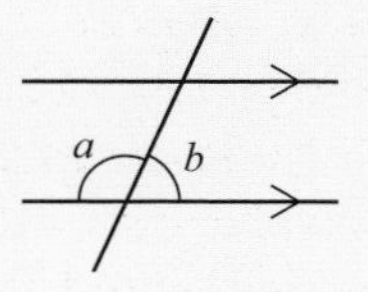

$a + b = 180°$
a and b are allied angles

(KS3 level 6, GCSE grade D)

continued

- Circle terms

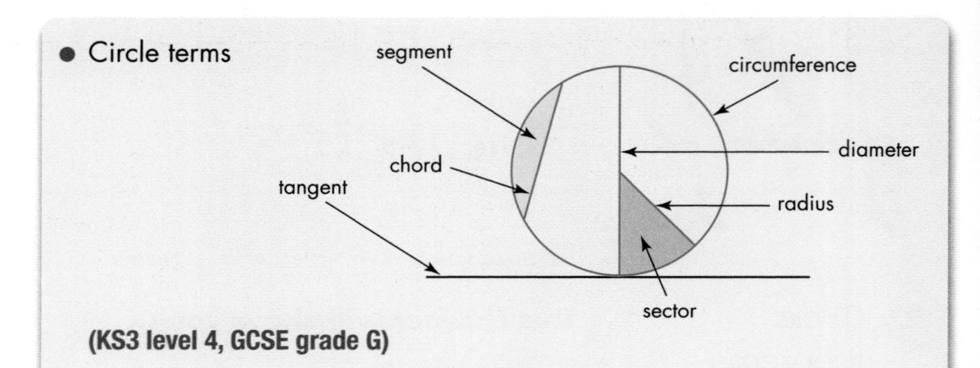

(KS3 level 4, GCSE grade G)

Quick check

Find the sizes of the lettered angles in these diagrams.

1

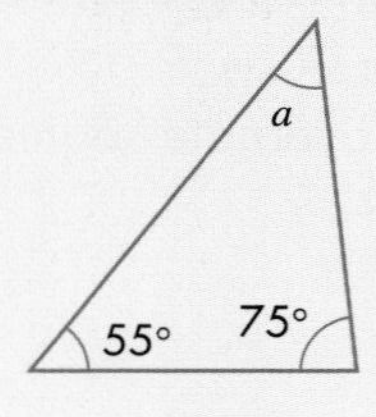

2

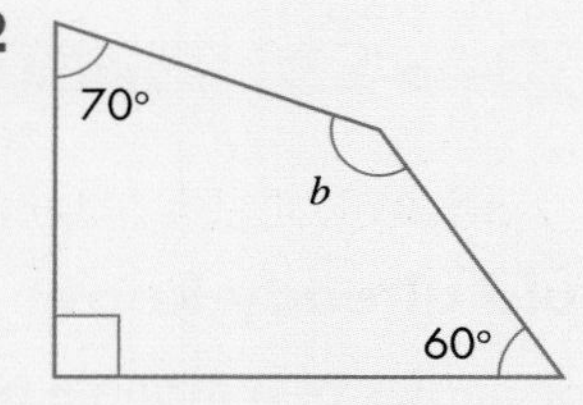

3

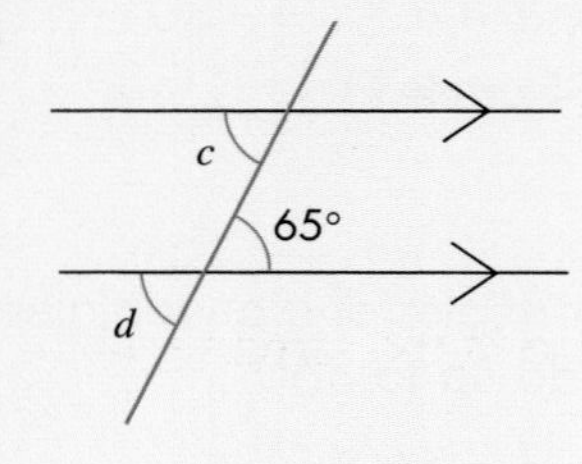

14.1 Circle theorems

This section will show you how to:

- work out the sizes of angles in circles

Key words

arc
circle
circumference
diameter
segment
semicircle
subtended

Here are three **circle** theorems you need to know.

- **Circle theorem 1**

  The angle at the centre of a circle is twice the angle at the **circumference** that is **subtended** by the same **arc**.

  $\angle AOB = 2 \times \angle ACB$

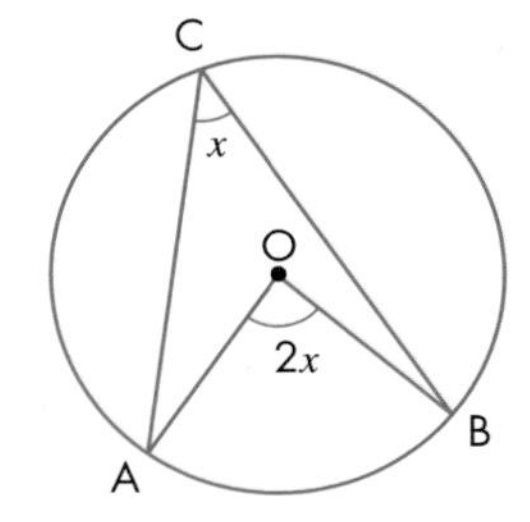

- **Circle theorem 2**

  Every angle at the circumference of a **semicircle** that is subtended by the **diameter** of the semicircle is a right angle.

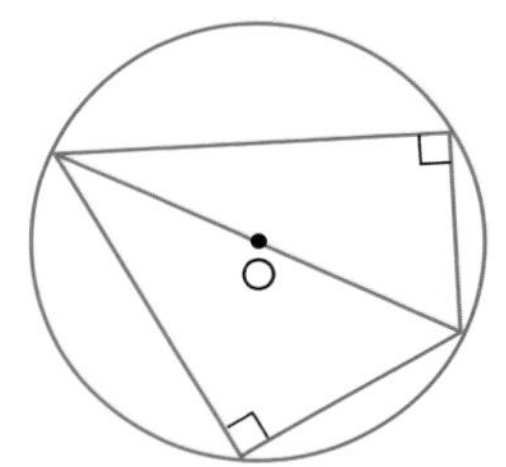

- **Circle theorem 3**

  Angles subtended at the circumference in the same **segment** of a circle are equal.

  Points C_1, C_2, C_3 and C_4 on the circumference are subtended by the same arc AB.

  So $\angle AC_1B = \angle AC_2B = \angle AC_3B = \angle AC_4B$

  Follow through Examples 1–3 to see how these theorems are applied.

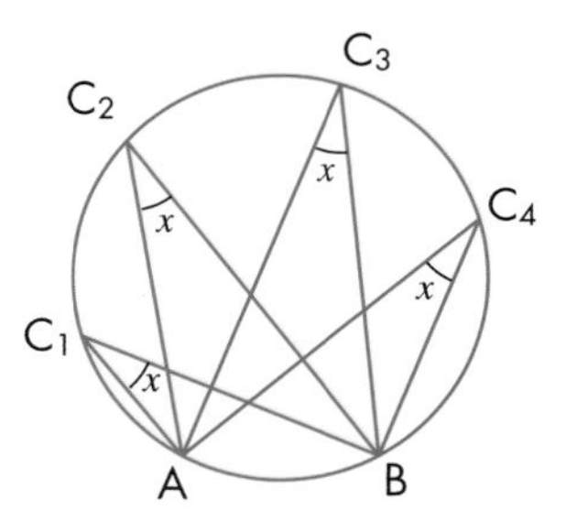

EXAMPLE 1

O is the centre of each circle. Find the angles marked a and b in each circle.

i

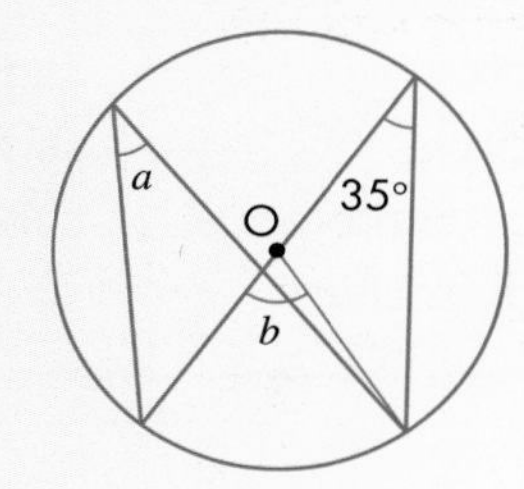

ii

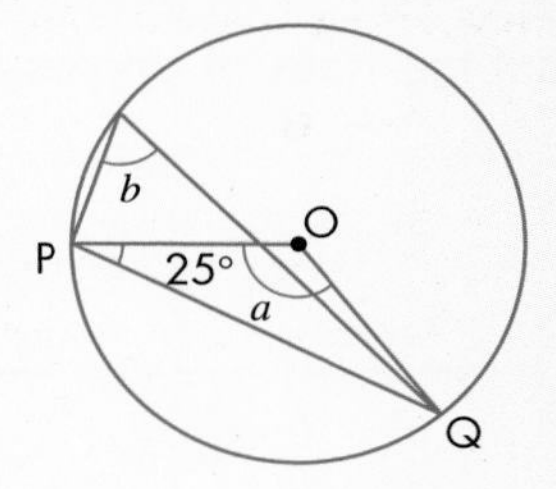

i $a = 35°$ (angles in same segment)

$b = 2 \times 35°$ (angle at centre = twice angle at circumference)

$= 70°$

ii With OP = OQ, triangle OPQ is isosceles and the sum of the angles in this triangle = 180°

So $a + (2 \times 25°) = 180°$

$a = 180° - (2 \times 25°)$

$= 130°$

$b = 130° \div 2$ (angle at centre = twice angle at circumference)

$= 65°$

EXAMPLE 2

O is the centre of the circle. PQR is a straight line.

Find the angle labelled a.

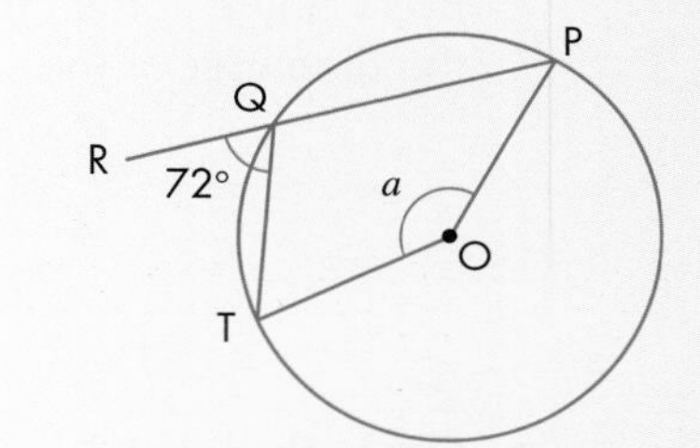

$\angle PQT = 180° - 72° = 108°$ (angles on straight line)

The reflex angle $\angle POT = 2 \times 108°$

(angle at centre = twice angle at circumference)

$= 216°$

$a + 216° = 360°$ (sum of angles around a point)

$a = 360° - 216°$

$a = 144°$

EXAMPLE 3

O is the centre of the circle. POQ is parallel to TR.

Find the angles labelled a and b.

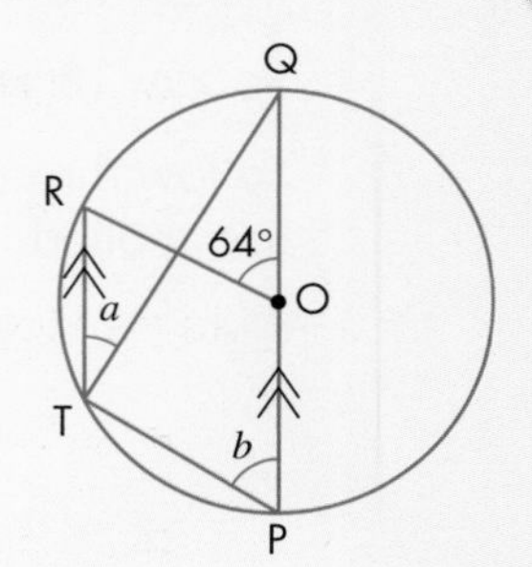

$a = 64° \div 2$ (angle at centre = twice angle at circumference)

$a = 32°$

$\angle TQP = a$ (alternate angles)

$= 32°$

$\angle PTQ = 90°$ (angle in a semicircle)

$b + 90° + 32° = 180°$ (sum of angles in $\triangle PQT$)

$b = 180° - 122°$

$b = 58°$

EXERCISE 14A

1 Find the angle marked x in each of these circles with centre O.

a

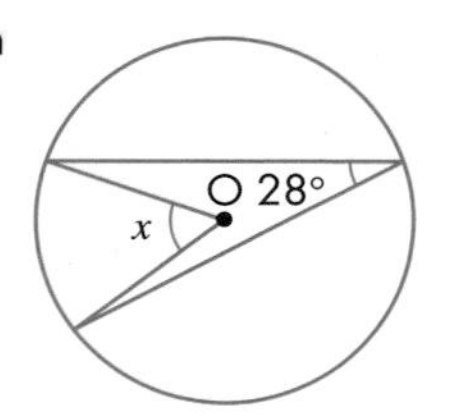

b

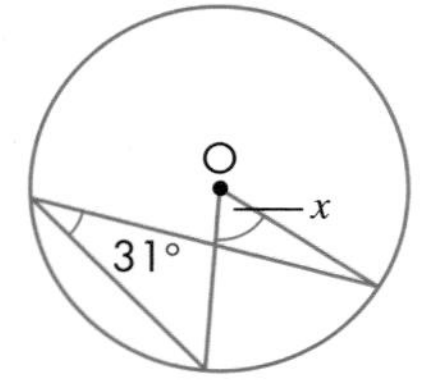

c

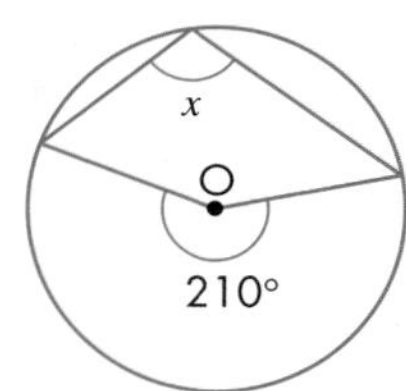

d

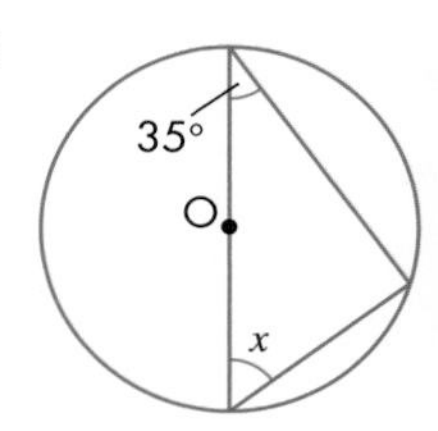

e

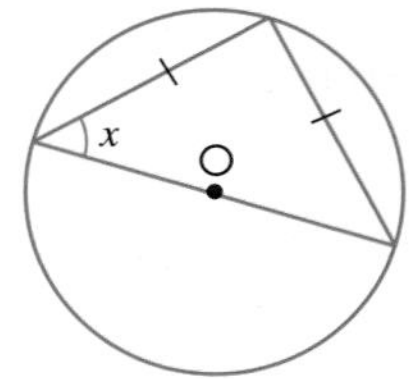

f

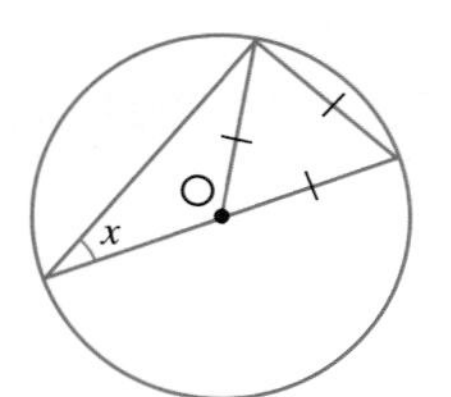

g

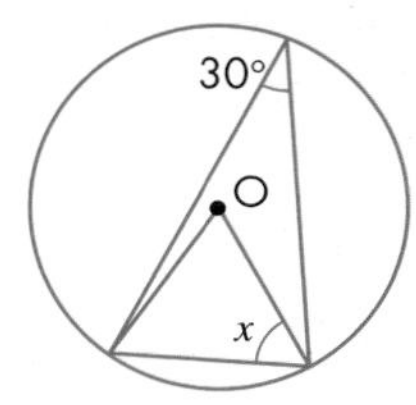

h

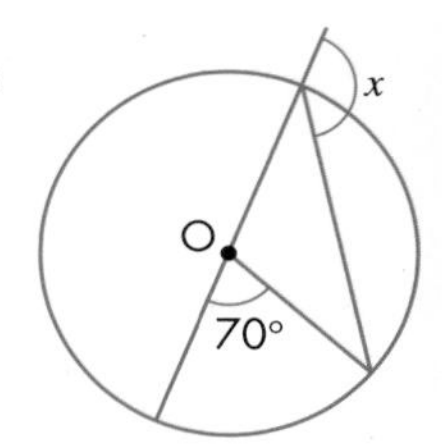

2 Find the angle marked x in each of these circles with centre O.

a

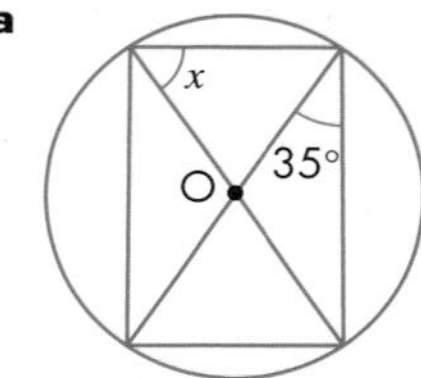

b

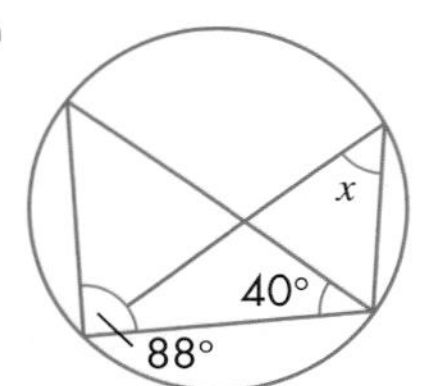

c

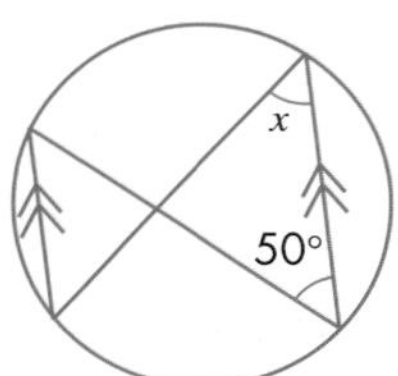

d

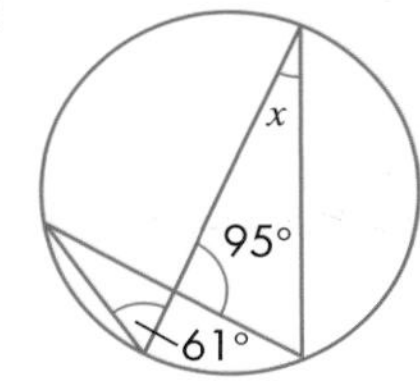

e

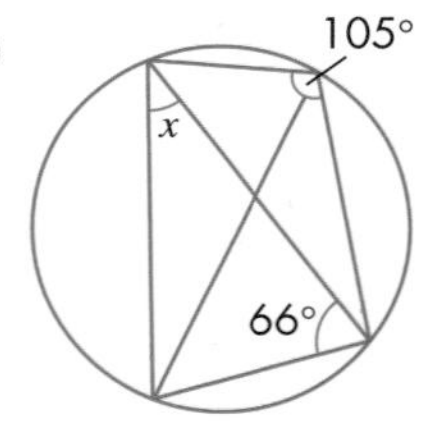

f

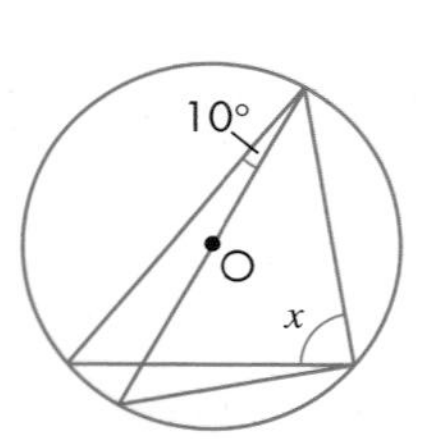

g

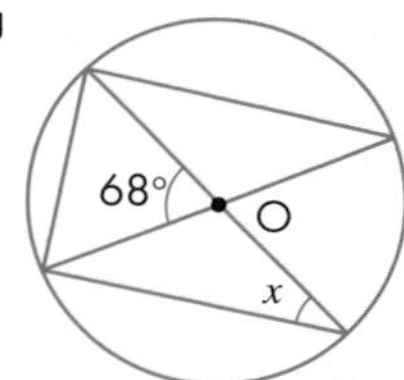

h

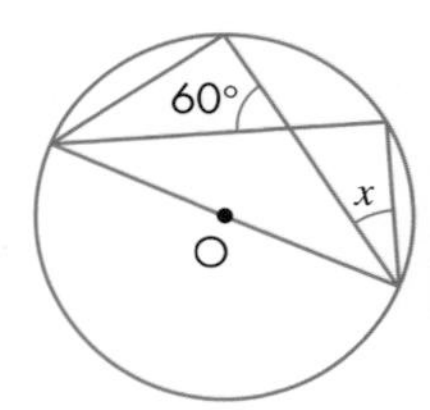

3 In the diagram, O is the centre of the circle. Find these angles.

a ∠ADB

b ∠DBA

c ∠CAD

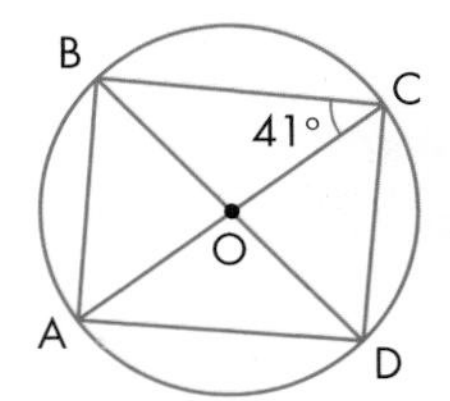

4 In the diagram, O is the centre of the circle. Find these angles.

a ∠EDF

b ∠DEG

c ∠EGF

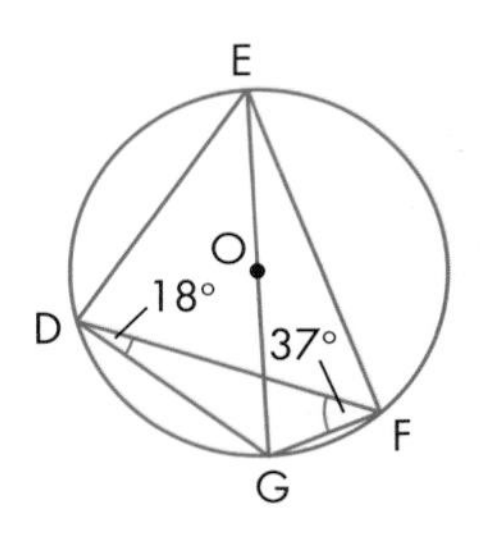

AU 5 In the diagram XY is a diameter of the circle and ∠AZX is a.

Ben says that the value of a is 55°.

Give reasons to explain why he is wrong.

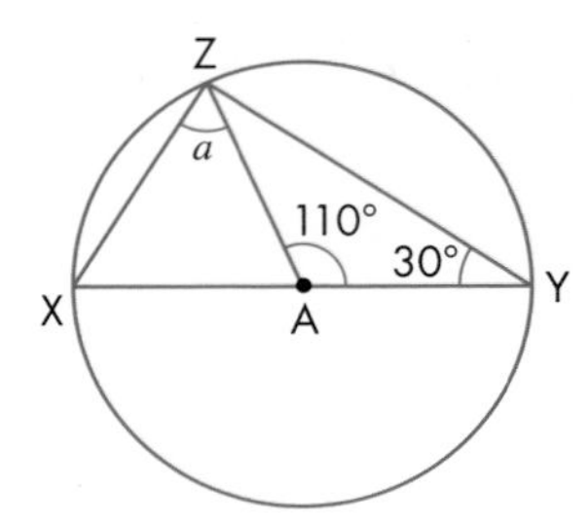

6 Find the angles marked x and y in each of these circles. O is the centre where shown.

a

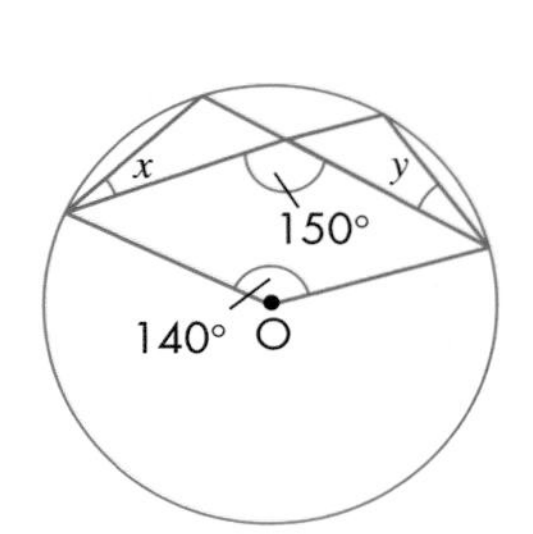

b

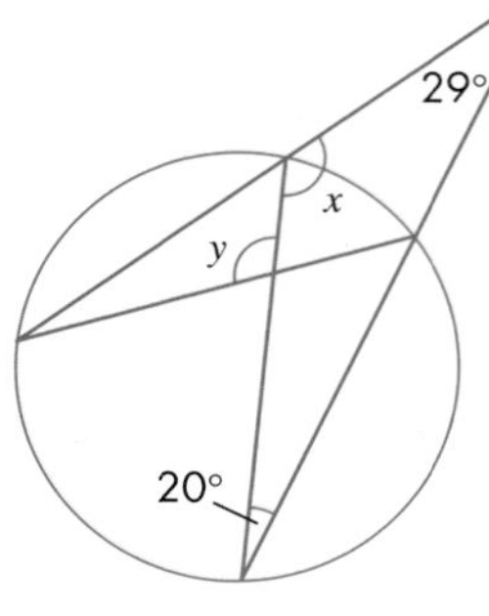

c

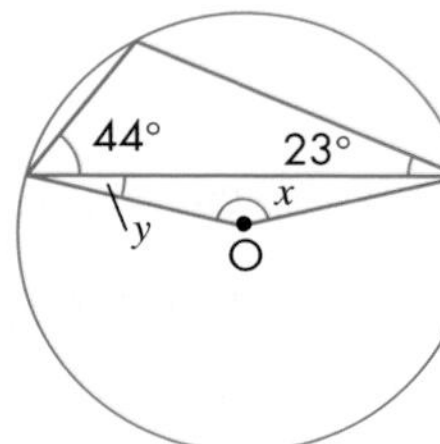

d

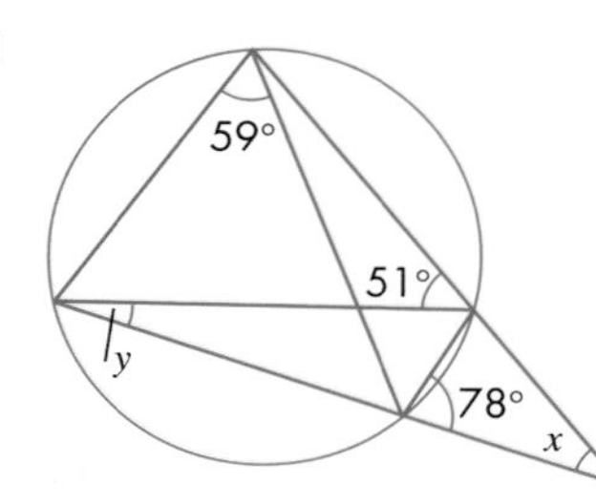

e

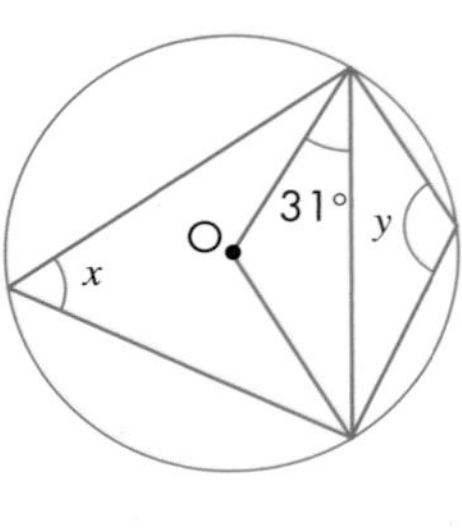

f

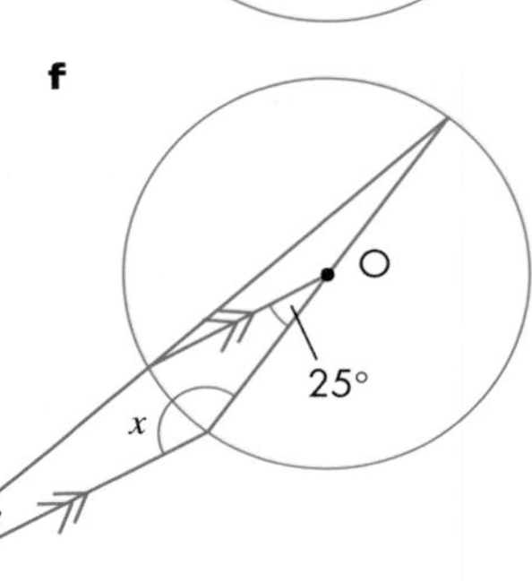

7 In the diagram, O is the centre and AD a diameter of the circle. Find x.

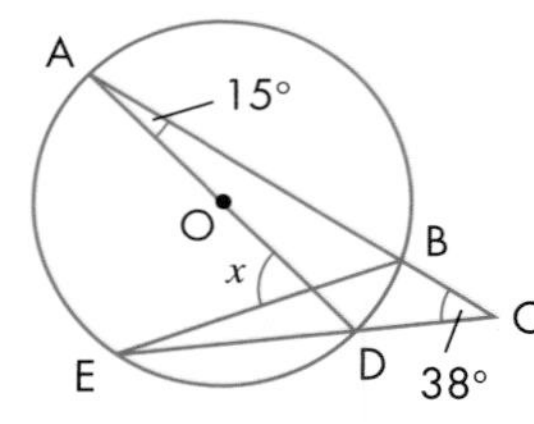

PS 8 In the diagram, O is the centre of the circle and ∠CBD is x.

Show that the reflex ∠AOC is $2x$, giving reasons to explain your answer.

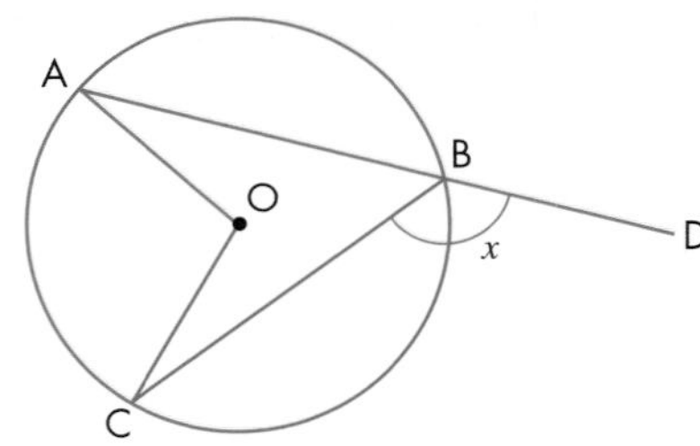

9 A, B, C and D are points on the circumference of a circle with centre O.
Angle ABO is x° and angle CBO is y°.

a State the value of angle BAO.

b State the value of angle AOD.

c Prove that the angle subtended by the chord AC at the centre of a circle is twice the angle subtended at the circumference.

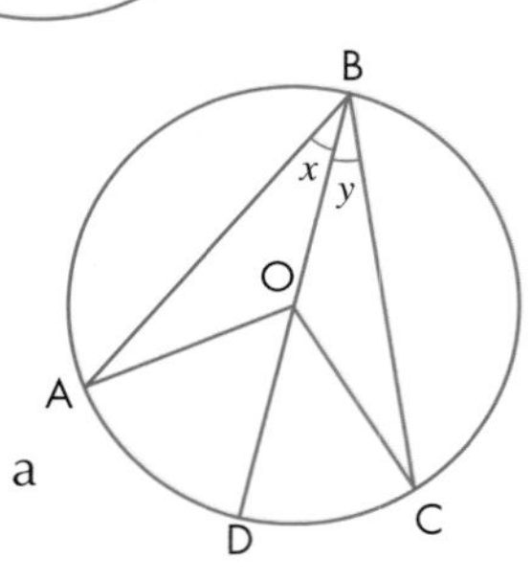

Cyclic quadrilaterals

This section will show you how to:
- find the sizes of angles in cyclic quadrilaterals

Key words
cyclic quadrilateral

A quadrilateral whose four vertices lie on the circumference of a circle is called a **cyclic quadrilateral**.

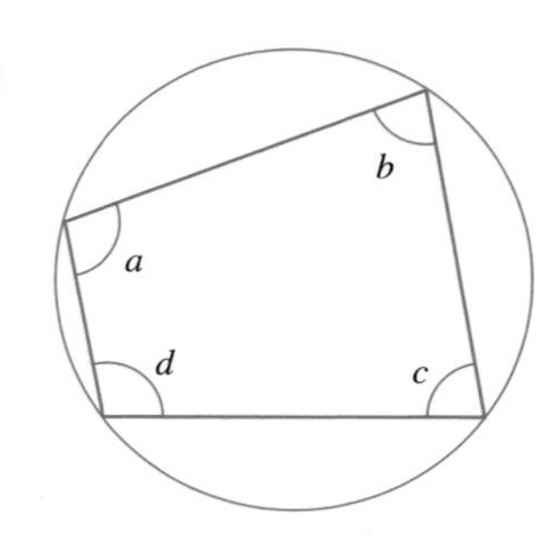

- **Circle theorem 4**

  The sum of the opposite angles of a cyclic quadrilateral is 180°.

  $a + c = 180°$ and $b + d = 180°$

EXAMPLE 4

Find the angles marked x and y in the diagram.

$x + 85° = 180°$ (angles in a cyclic quadrilateral)

So, $x = 95°$

$y + 108° = 180°$ (angles in a cyclic quadrilateral)

So, $y = 72°$

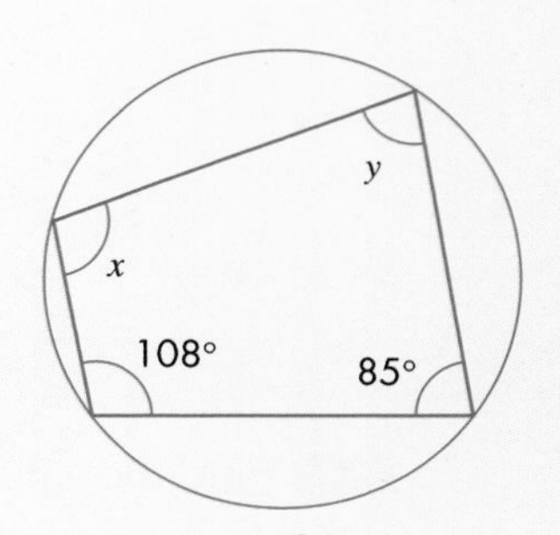

EXERCISE 14B

1 Find the sizes of the lettered angles in each of these circles.

a

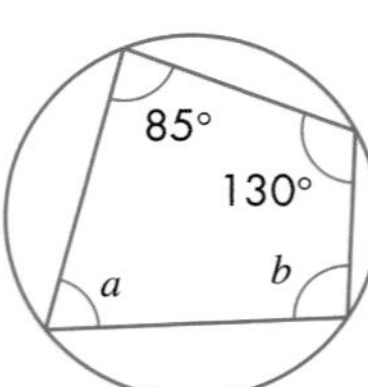

b

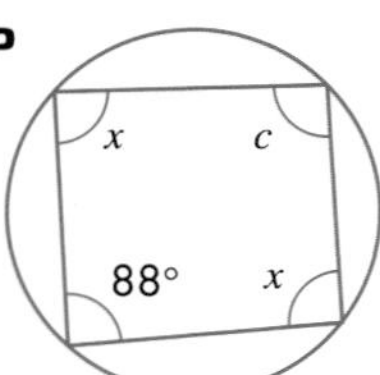

c

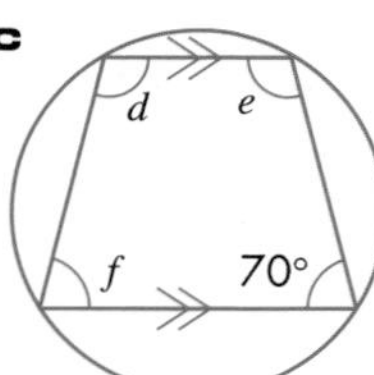

d

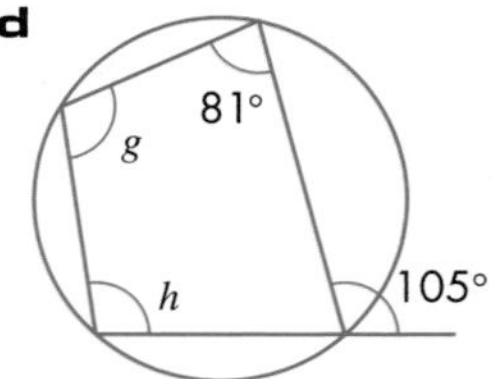

e

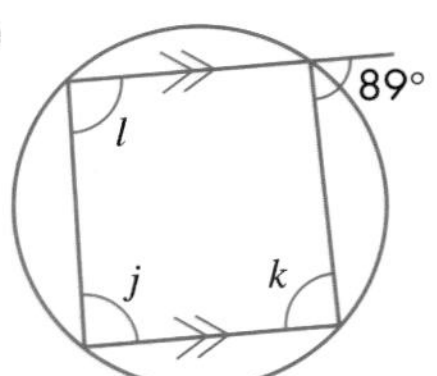

f

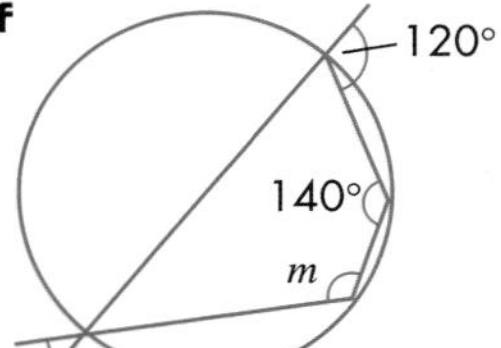

g

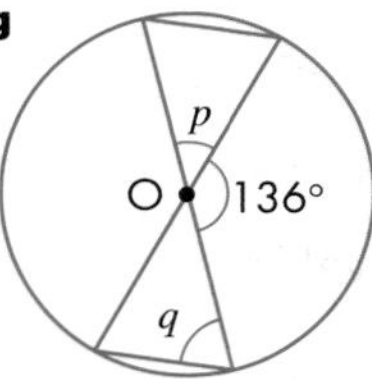

h

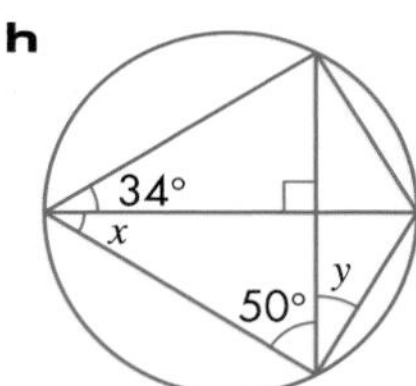

2 Find the values of x and y in each of these circles. Where shown, O marks the centre of the circle.

a

b

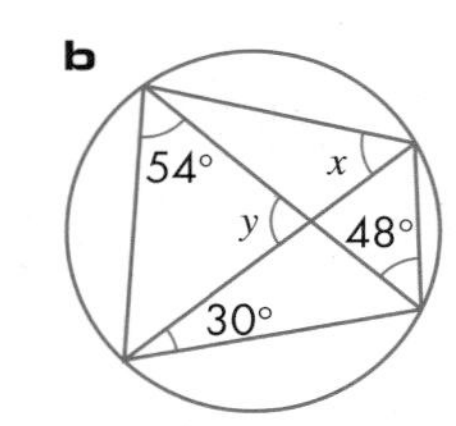

c

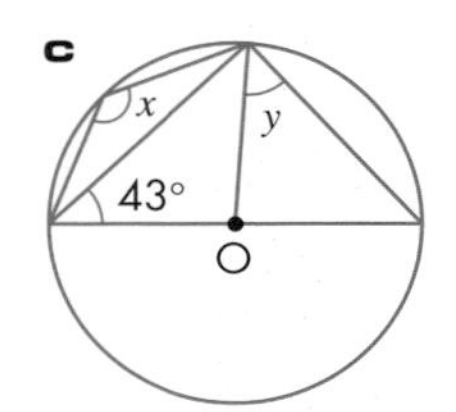

d

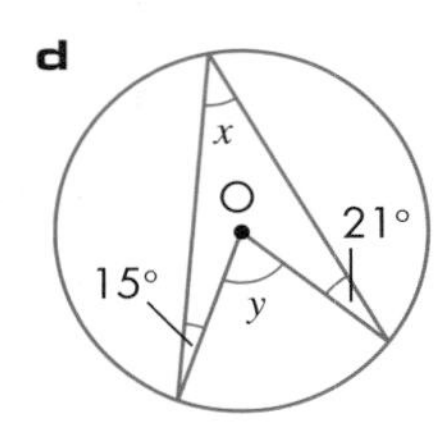

e

f

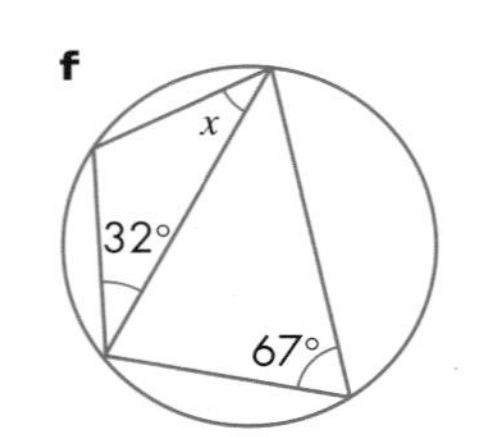

g

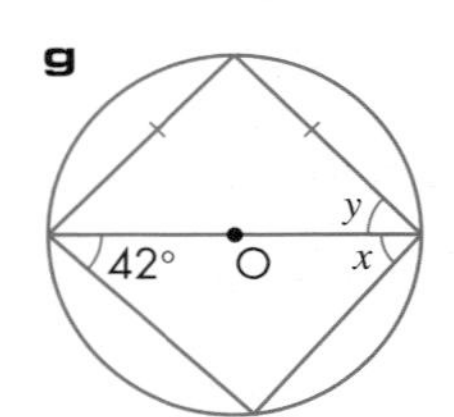

h

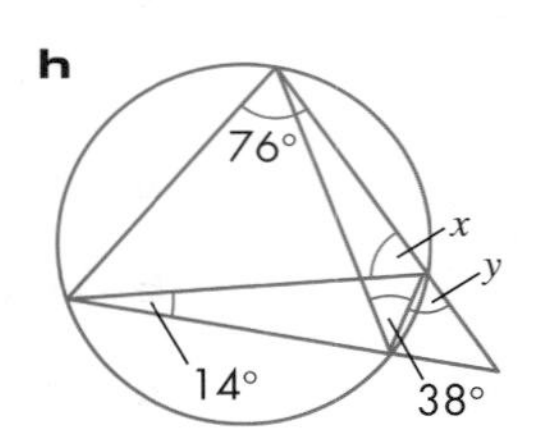

3 Find the values of x and y in each of these circles. Where shown, O marks the centre of the circle.

a

b

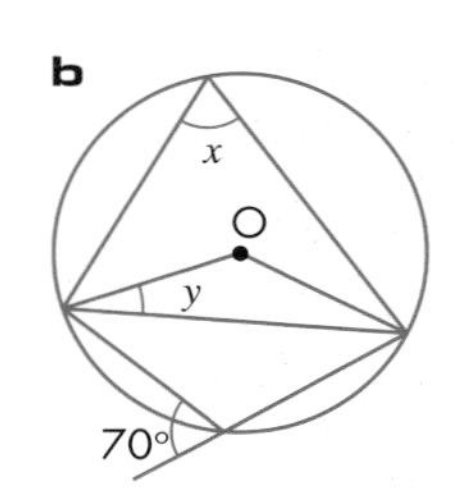

c

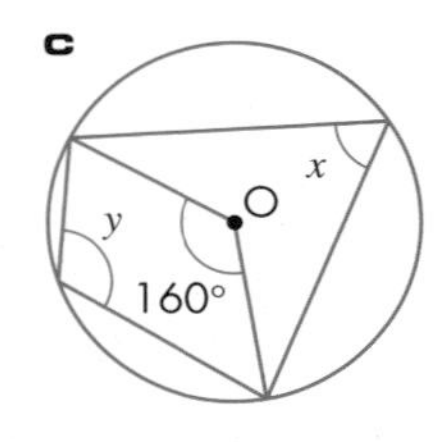

d

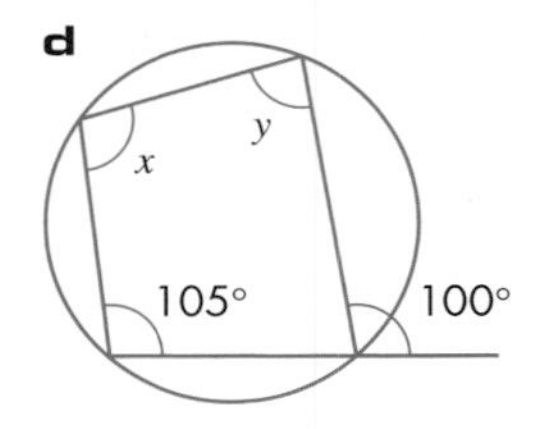

4 Find the values of x and y in each of these circles.

a

b

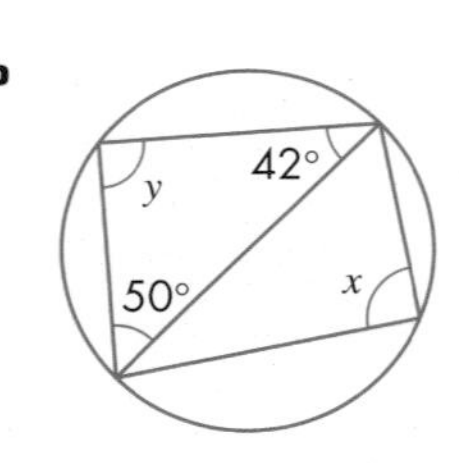

c

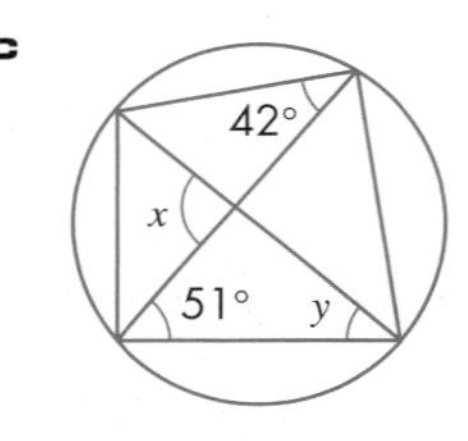

d

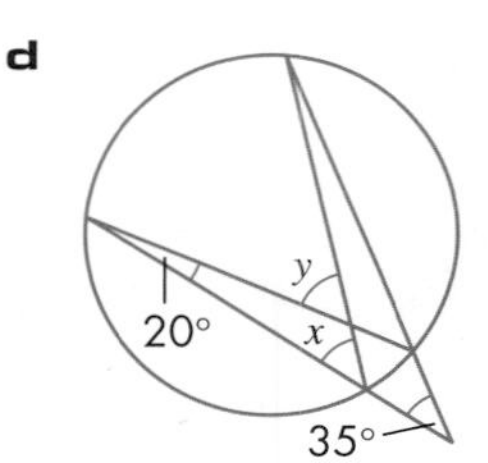

5 Find the values of x and y in each of these circles with centre O.

a

b

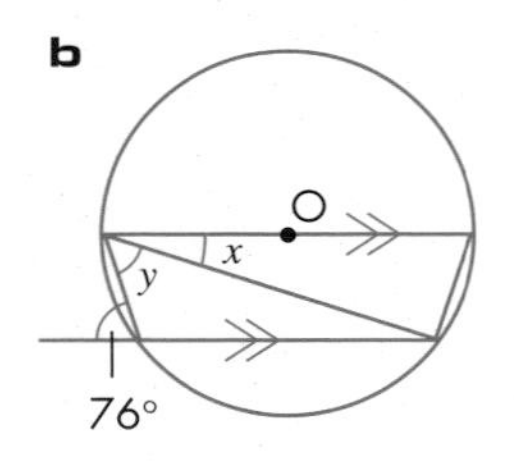

c

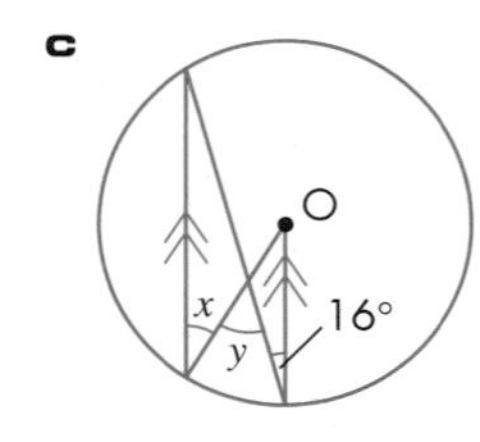

d

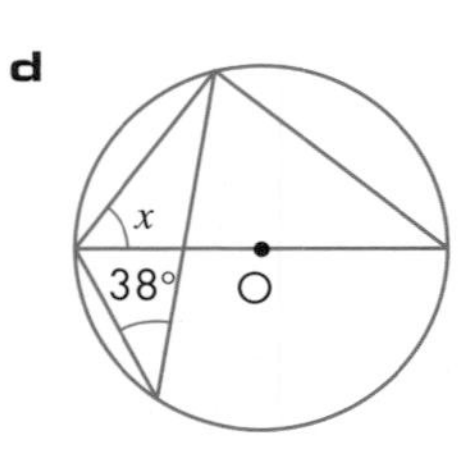

PS 6 The cyclic quadrilateral PQRT has ∠ROQ equal to 38° where O is the centre of the circle. POT is a diameter and parallel to QR. Calculate these angles.

a ∠ROT **b** ∠QRT **c** ∠QPT

AU **7** In the diagram, O is the centre of the circle.

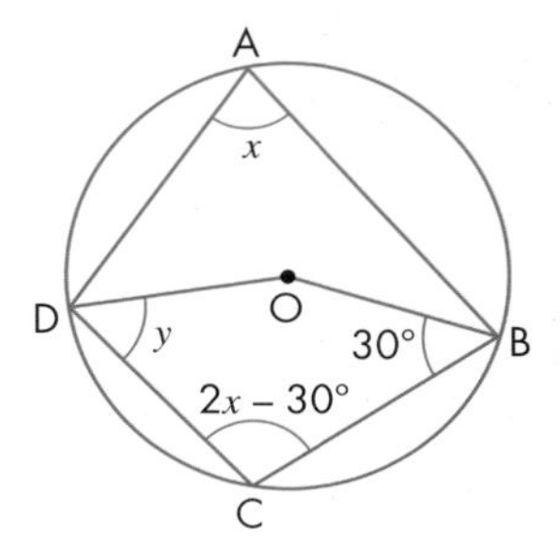

a Explain why $3x - 30° = 180°$.

b Work out the size of ∠CDO, marked y on the diagram.

Give reasons in your working.

8 ABCD is a cyclic quadrilateral within a circle centre O and ∠AOC is $2x°$.

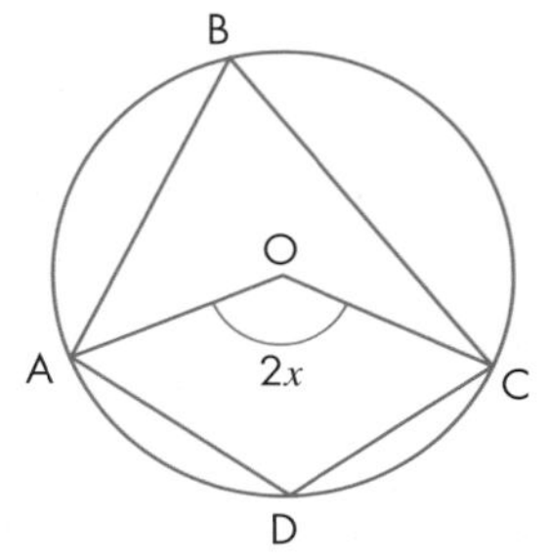

a Write down the value of ∠ABC.

b Write down the value of the reflex angle AOC.

c Prove that the sum of a pair of opposite angles of a cyclic quadrilateral is 180°.

PS **9** In the diagram, ABCE is a parallelogram.

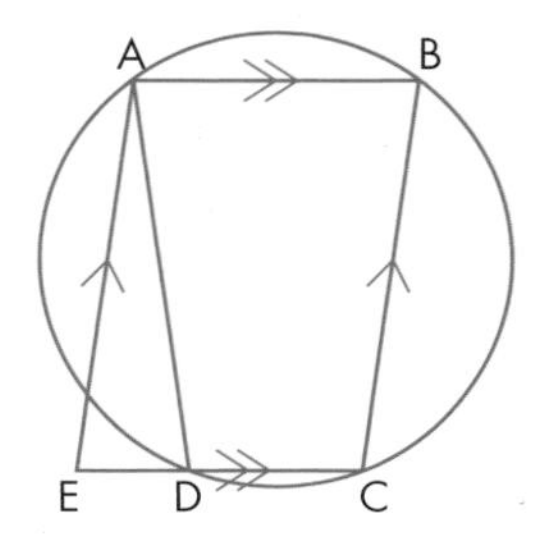

Prove ∠AED = ∠ADE.

Give reasons in your working.

14.3 Tangents and chords

This section will show you how to:

- use tangents and chords to find the sizes of angles in circles

Key words

chord
point of contact
radius
tangent

A **tangent** is a straight line that touches a circle at one point only. This point is called the **point of contact**. A **chord** is a line that joins two points on the circumference.

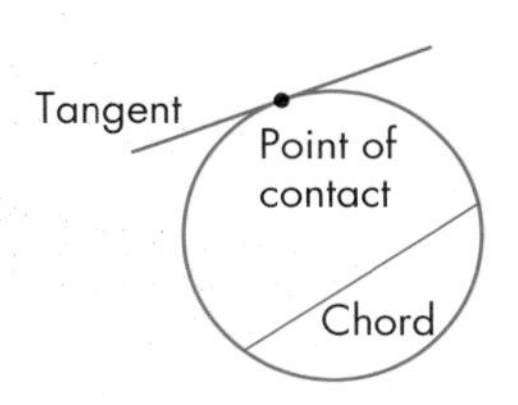

- **Circle theorem 5**

  A tangent to a circle is perpendicular to the **radius** drawn to the point of contact.

  The radius OX is perpendicular to the tangent AB.

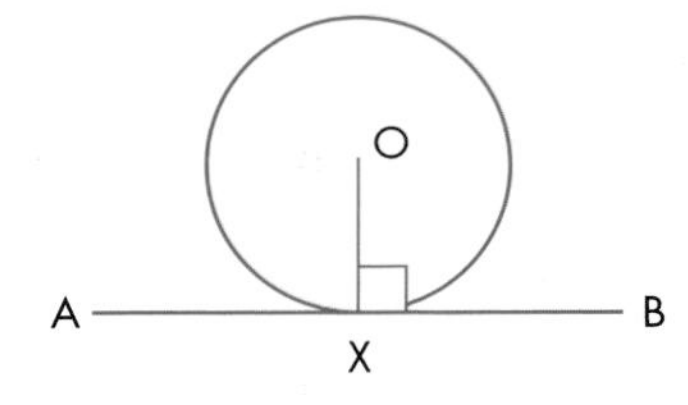

- **Circle theorem 6**

  Tangents to a circle from an external point to the points of contact are equal in length.

  AX = AY

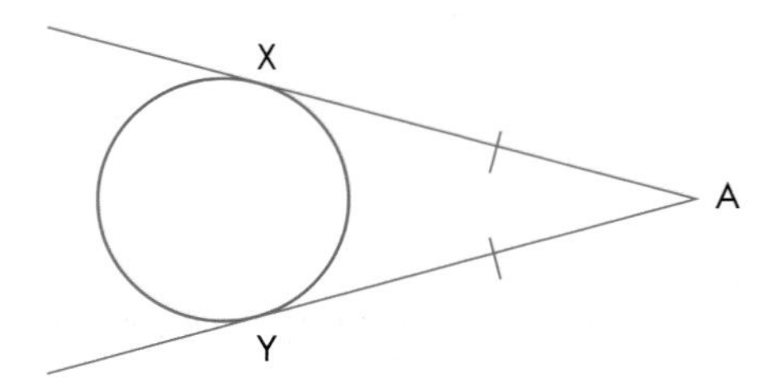

- **Circle theorem 7**

  The line joining an external point to the centre of the circle bisects the angle between the tangents.

  ∠OAX = ∠OAY

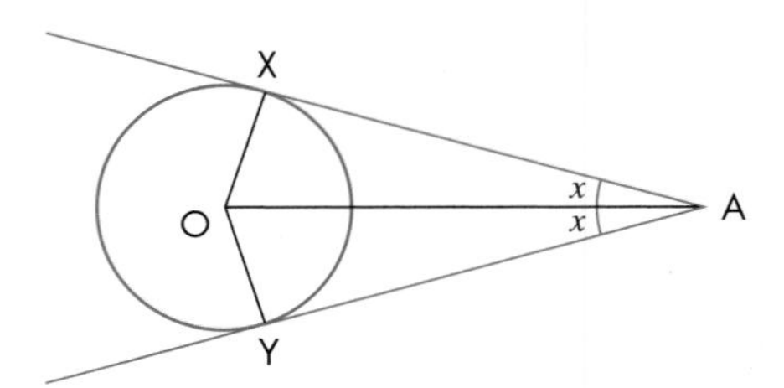

- **Circle theorem 8**

  A radius bisects a chord at 90°.

  If O is the centre of the circle,

  ∠BMO = 90° and BM = CM

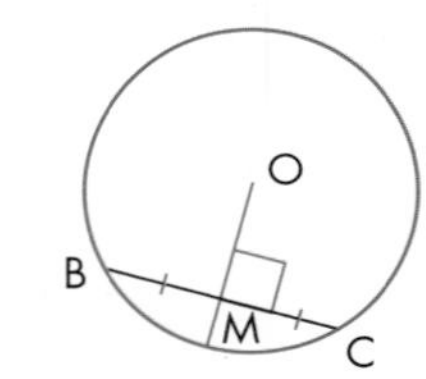

EXAMPLE 5

OA is the radius of the circle and AB is a tangent.

OA = 5 cm and AB = 12 cm

Calculate the length OB.

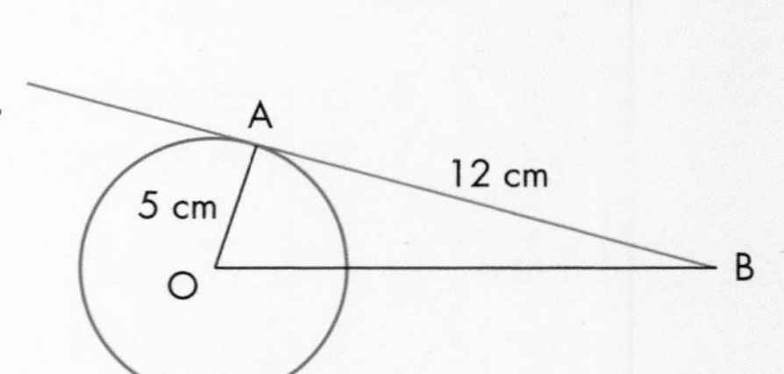

∠OAB = 90° (radius is perpendicular to a tangent)

Let OB = x

By Pythagoras' theorem,

$x^2 = 5^2 + 12^2 \text{ cm}^2$

$x^2 = 169 \text{ cm}^2$

So $x = \sqrt{169} = 13$ cm

EXERCISE 14C 

1 In each diagram, TP and TQ are tangents to a circle with centre O. Find each value of x.

a

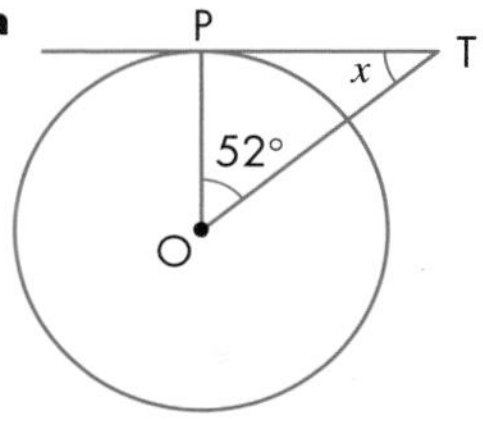

b 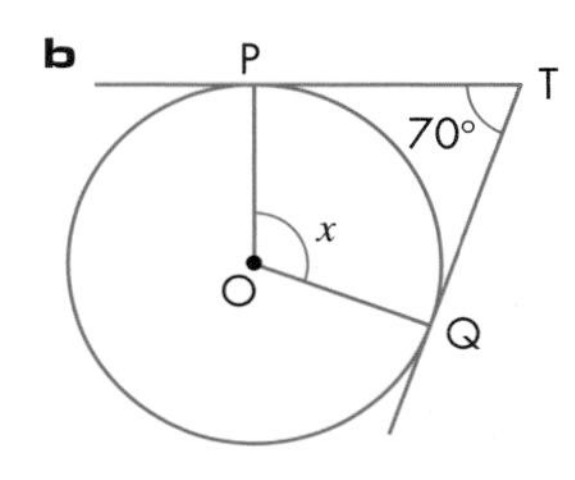

c 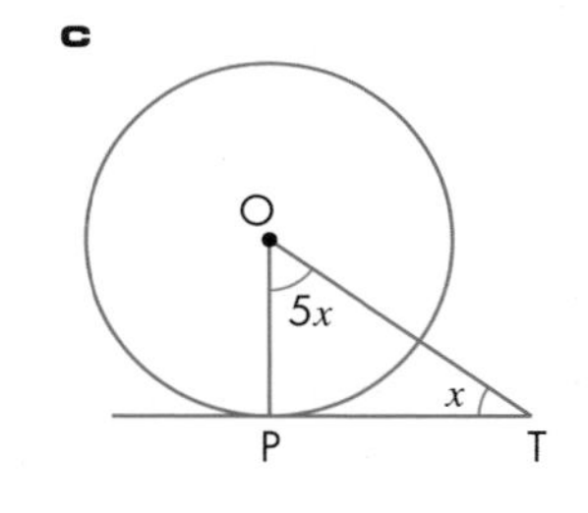

d 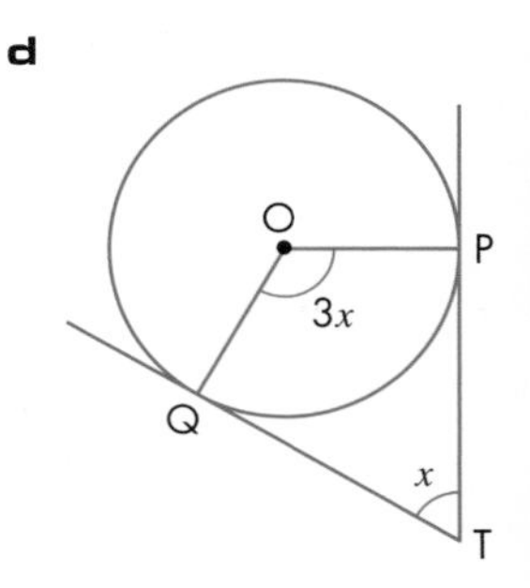

2 Each diagram shows tangents to a circle with centre O. Find each value of y.

a

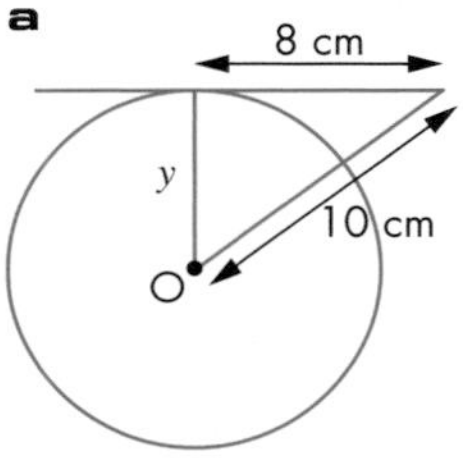

b 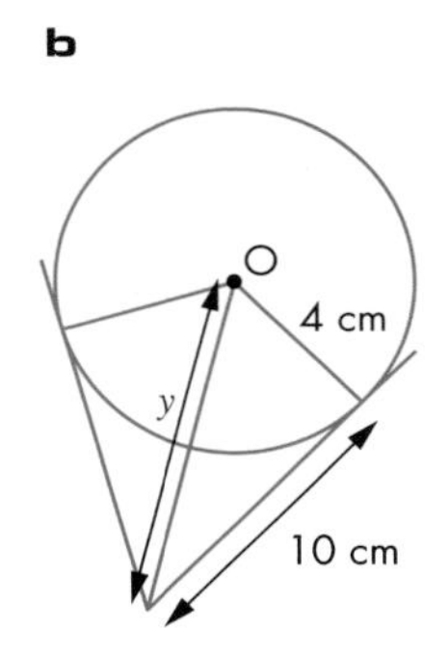

c 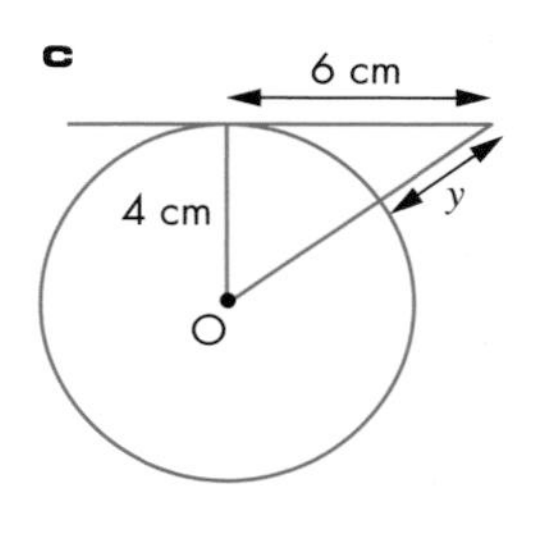

d 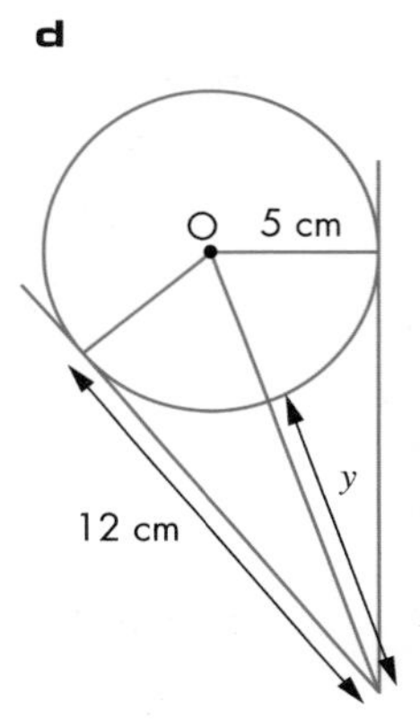

3 Each diagram shows a tangent to a circle with centre O. Find x and y in each case.

a

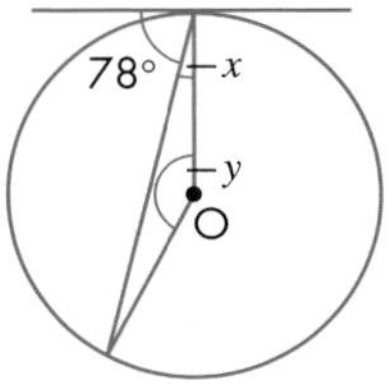

b 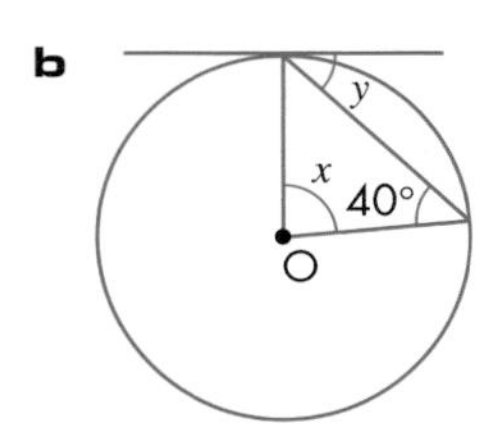

c 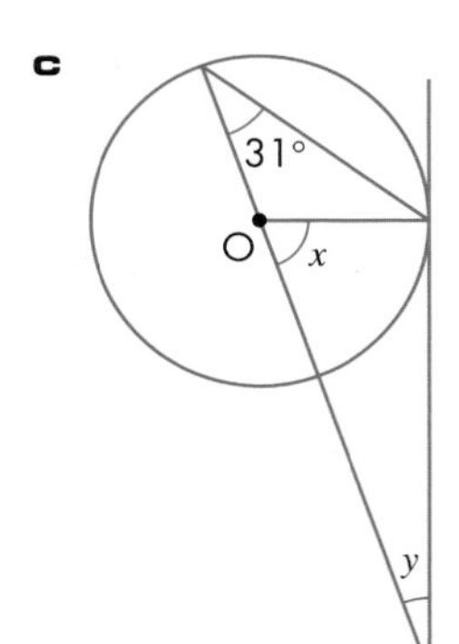

d 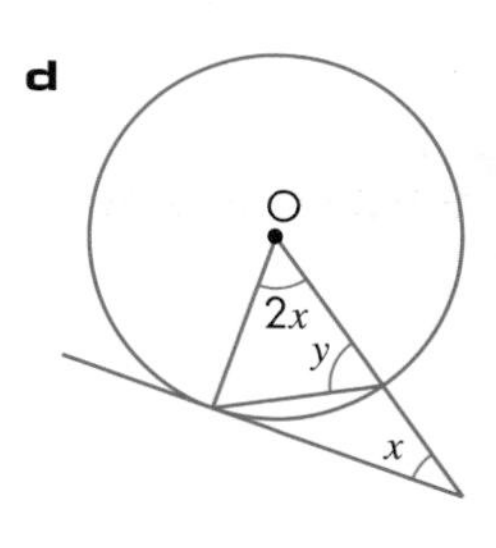

4 In each of the diagrams, TP and TQ are tangents to the circle with centre O. Find each value of x.

a

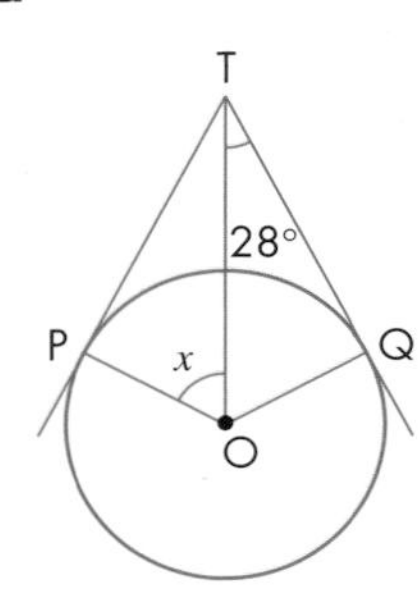

b

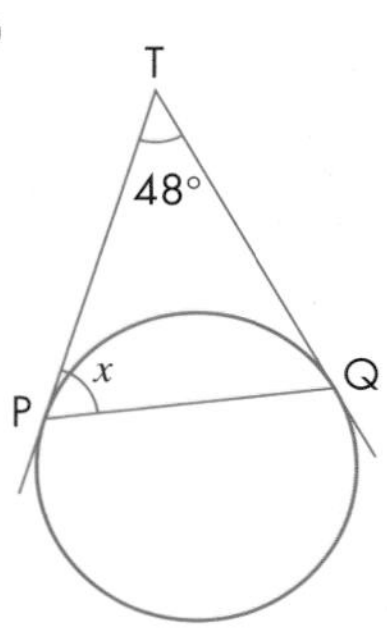

c

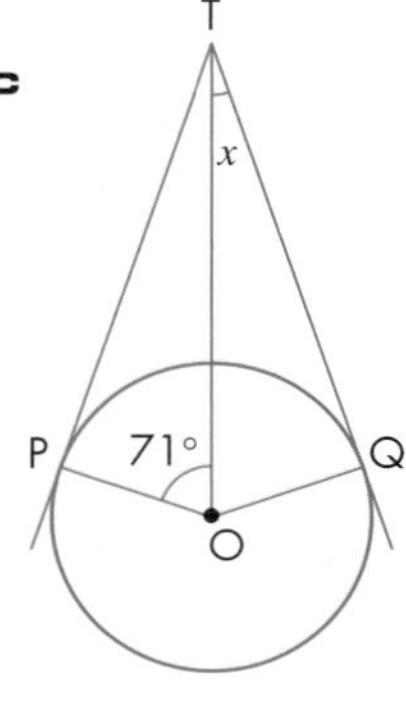

d

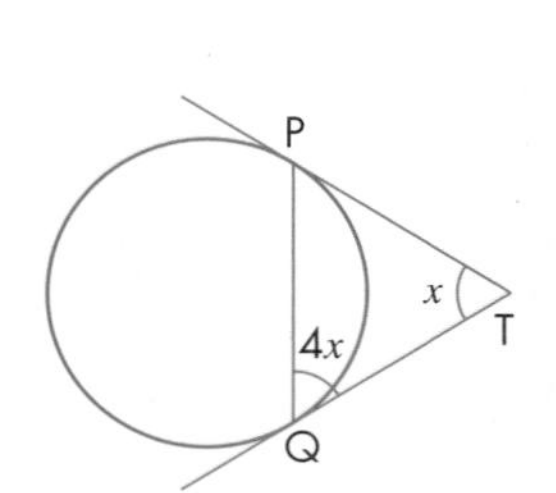

PS 5 Two circles with the same centre have radii of 7 cm and 12 cm respectively. A tangent to the inner circle cuts the outer circle at A and B. Find the length of AB.

PS 6 The diagram shows a circle with centre O.
The circle fits exactly inside an equilateral triangle XYZ.
The lengths of the sides of the triangle are 20 cm.

Work out the radius of the circle.

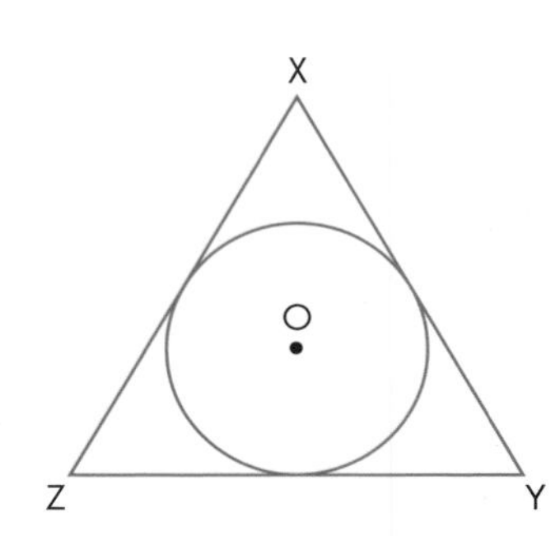

HINTS AND TIPS

Remember you can use Pythagoras' theorem and trigonometry to solve problems.

AU 7 In the diagram, O is the centre of the circle and AB is a tangent to the circle at C.

Explain why triangle BCD is isosceles.

Give reasons to justify your answer.

8 AB and CB are tangents from B to the circle with centre O. OA and OC are radii.

a Prove that angles AOB and COB are equal.

b Prove that OB bisects the angle ABC.

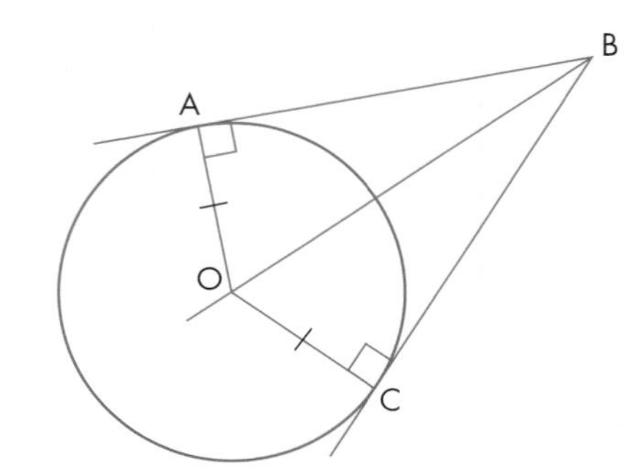

14.4 Alternate segment theorem

This section will show you how to:

- use the alternate segment theorem to find the sizes of angles in circles

Key words
alternate segment
chord
tangent

PTQ is the **tangent** to a circle at T. The segment containing ∠TBA is known as the **alternate segment** of ∠PTA, because it is on the other side of the **chord** AT from ∠PTA.

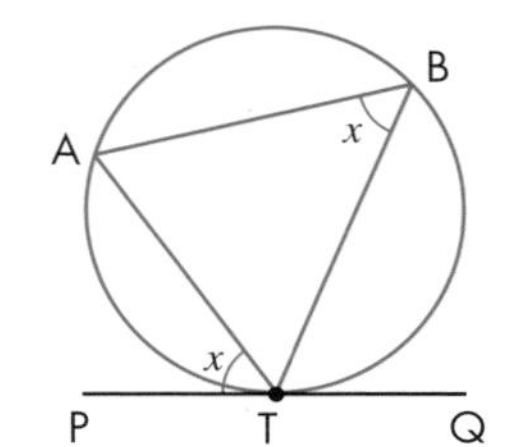

- **Circle theorem 9**

  The angle between a tangent and a chord through the point of contact is equal to the angle in the alternate segment.

  ∠PTA = ∠TBA

EXAMPLE 6

In the diagram, find **a** ∠ATS and **b** ∠TSR.

a ∠ATS = 80° (angle in alternate segment)

b ∠TSR = 70° (angle in alternate segment)

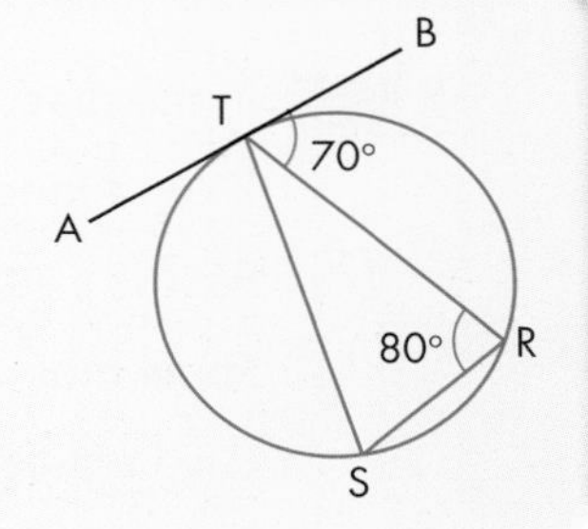

EXERCISE 14D

1 Find the size of each lettered angle.

a

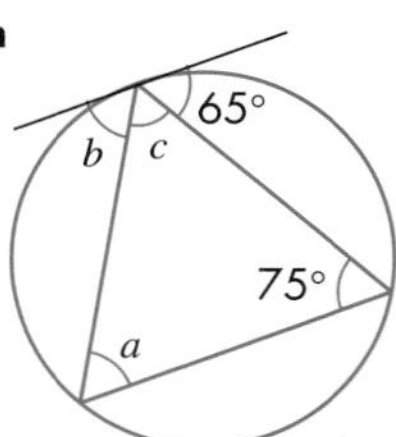

b

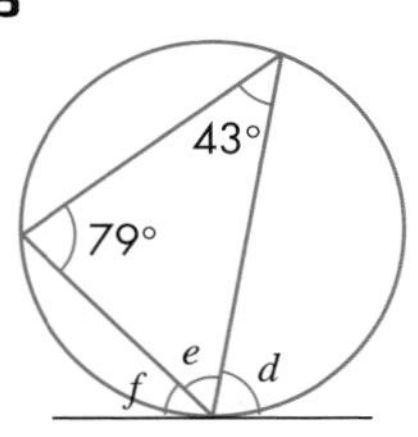

c

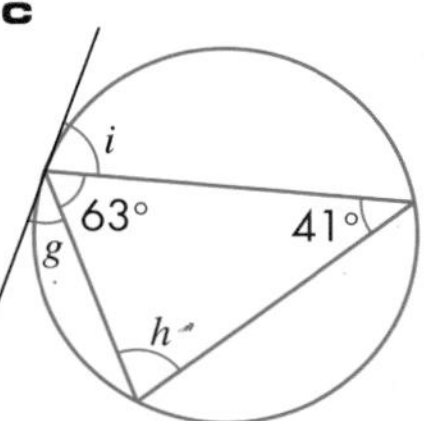

d

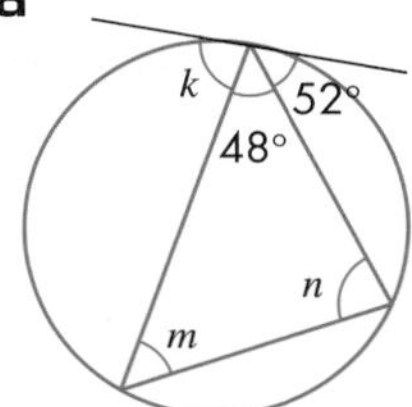

2 In each diagram, find the size of each lettered angle.

a

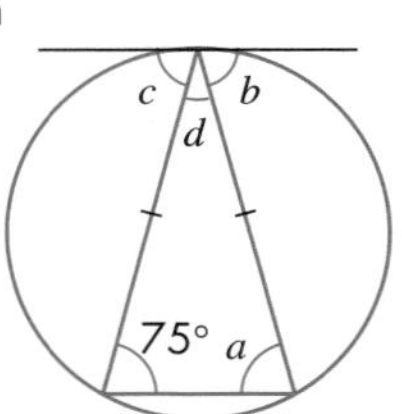

b

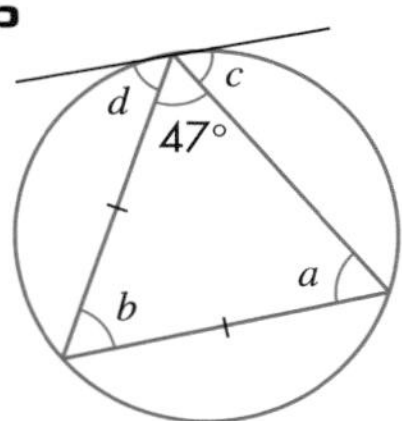

c

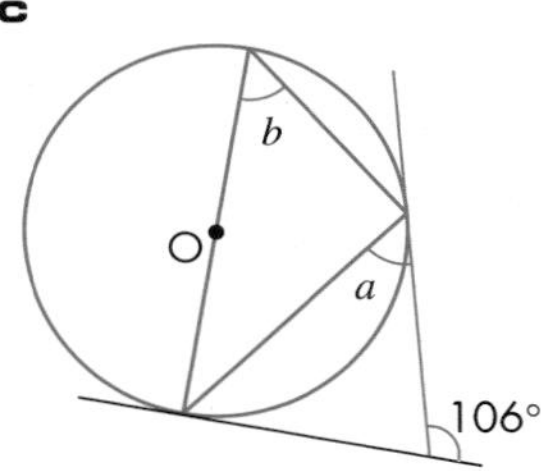

d

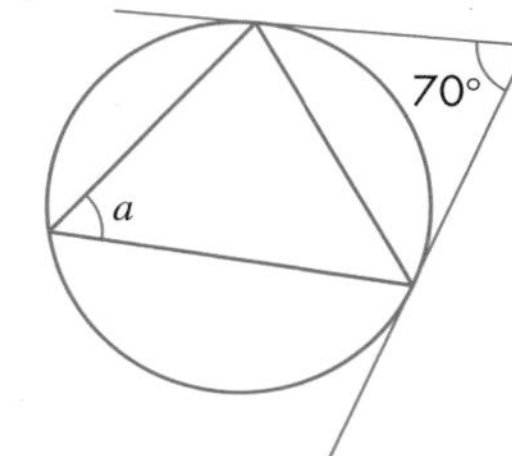

A

3 In each diagram, find the value of x.

a

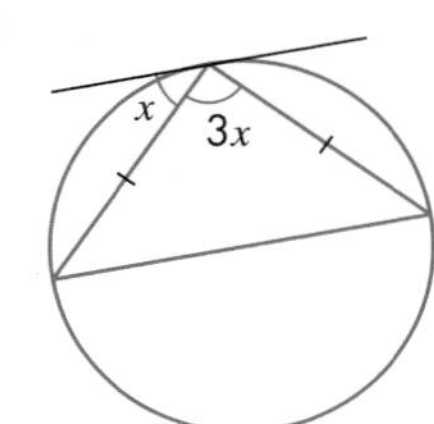

b

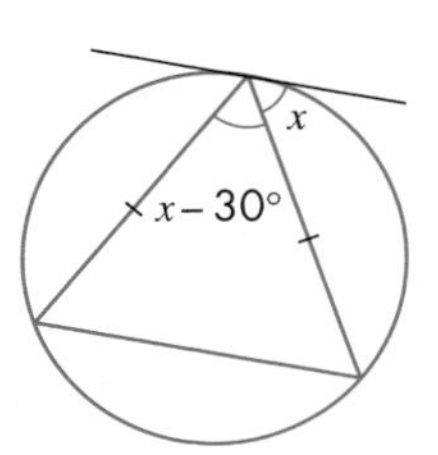

4 ATB is a tangent to each circle with centre O. Find the size of each lettered angle.

a

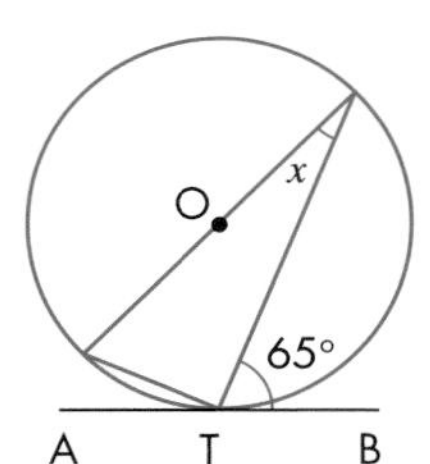

b

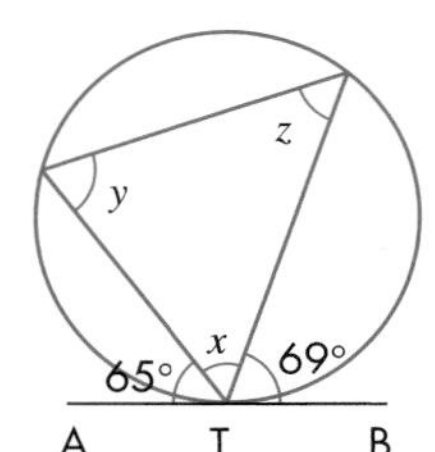

c

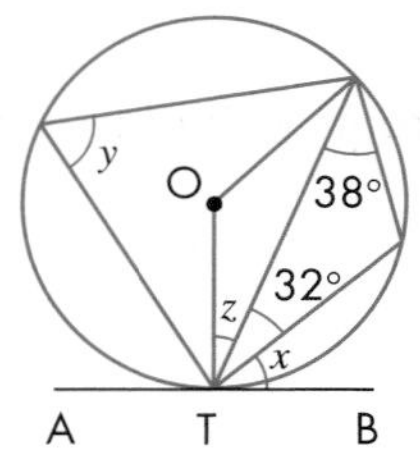

d

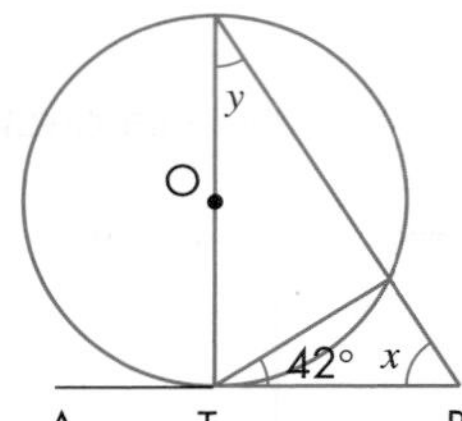

PS 5 In the diagram, O is the centre of the circle.

XY is a tangent to the circle at A.

BCX is a straight line.

Show that triangle ACX is isosceles.

Give reasons to justify your answer.

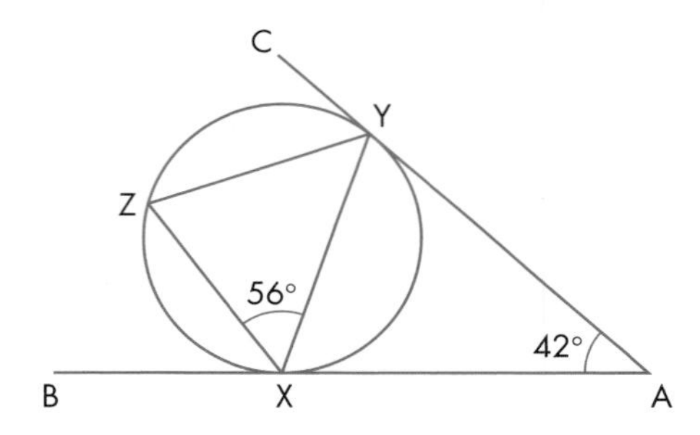

AU 6 AB and AC are tangents to the circle at X and Y.

Work out the size of ∠XYZ.

Give reasons to justify your answer.

A*

7 PT is a tangent to a circle with centre O.
AB are points on the circumference. Angle PBA is x°.

a Write down the value of angle AOP.

b Calculate the angle OPA in terms of x.

c Prove that the angle APT is equal to the angle PBA.

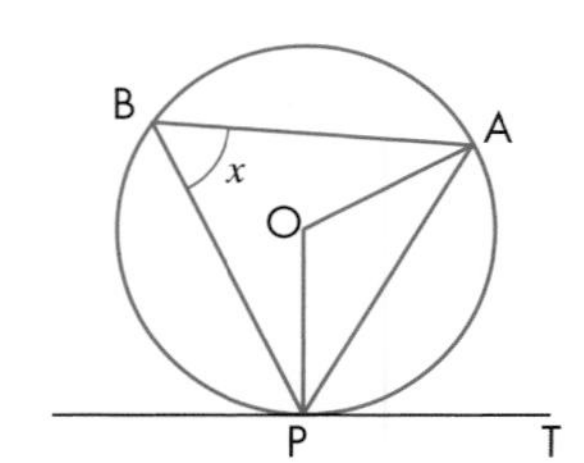

GRADE BOOSTER

B You can find angles in circles

A You can find angles in circles, using the alternate segment theorem

A* You can use circle theorems to prove geometrical results

What you should know now

- How to use circle theorems to find angles

EXAMINATION QUESTIONS

1 

A, B and C are points on the circumference of a circle, centre O.

AC is a diameter of the circle.

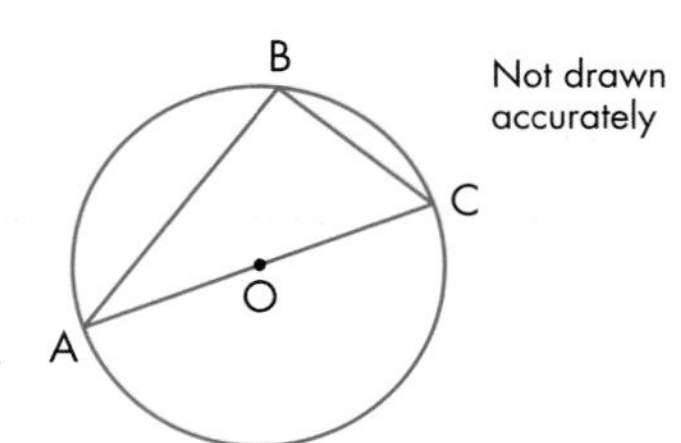

a **i** Write down the size of angle ABC.

ii Give a reason for your answer. *(2 marks)*

D, E and F are points on the points on the circumference of a circle, centre O.

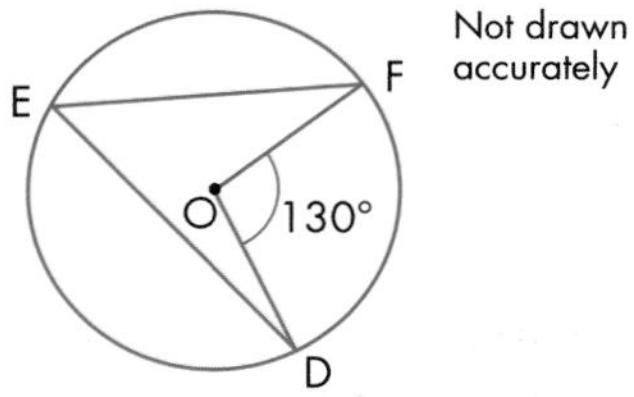

Angle DOF = 130°.

b **i** Work out the size of angle DEF.

ii Give a reason for your answer. *(2 marks)*

Edexcel, November 2008, Paper 14 Higher, Question 13(a, b)

2 In the diagram, A, B, C and D are points on the circumference of a circle, centre O.

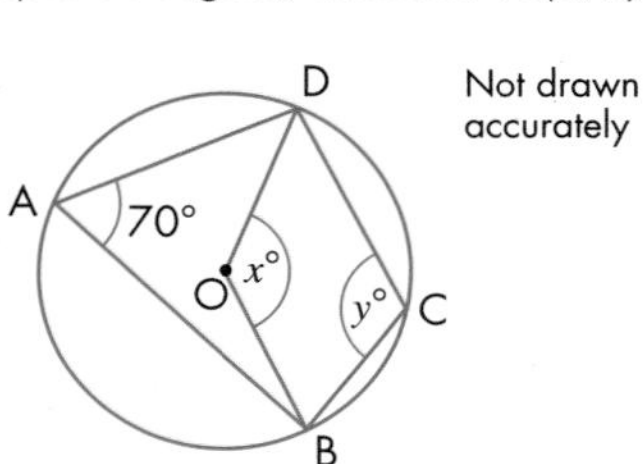

Angle BAD = 70°.
Angle BOD = x.
Angle BCD = y.

a **i** Work out the value of x.

ii Give a reason for your answer. *(2 marks)*

b **i** Work out the value of y.

ii Give a reason for your answer. *(2 marks)*

Edexcel, May 2008, Paper 14 Higher, Question 10(a, b)

3 P and Q are two points on a circle, centre O.

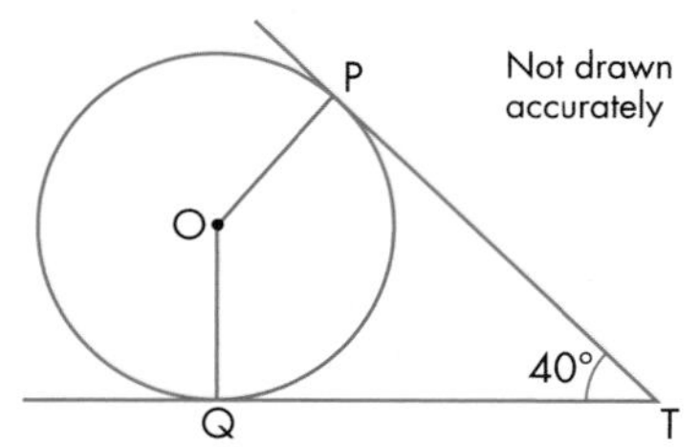

The tangents to the circle at P and Q intersect at the point T.

a Write down the size of angle OQT. *(1 mark)*

b Calculate the size of the obtuse angle POQ. *(2 marks)*

c Give reasons why angle PQT is 70°. *(2 marks)*

Edexcel, June 2007, Paper 11 Higher, Question 8

4 

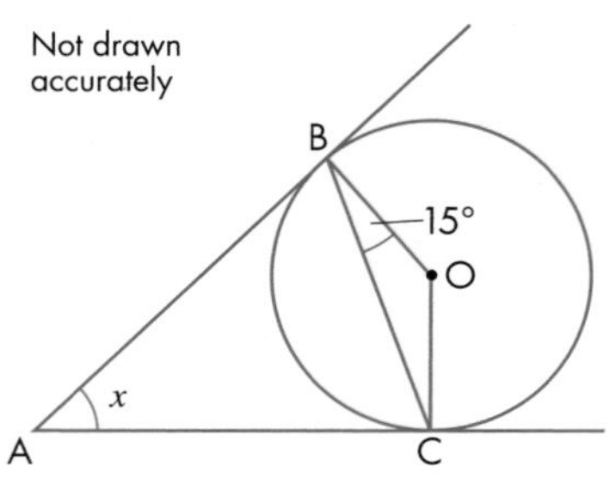

B and C are two points on a circle, centre O.

Angle OBC = 15°.

AB and AC are tangents to the circle.

a Calculate the size of the angle marked x. *(2 marks)*

b Give reasons for your answer. *(2 marks)*

Edexcel, June 2008, Paper 11 Higher, Question 8(a, b)

5 

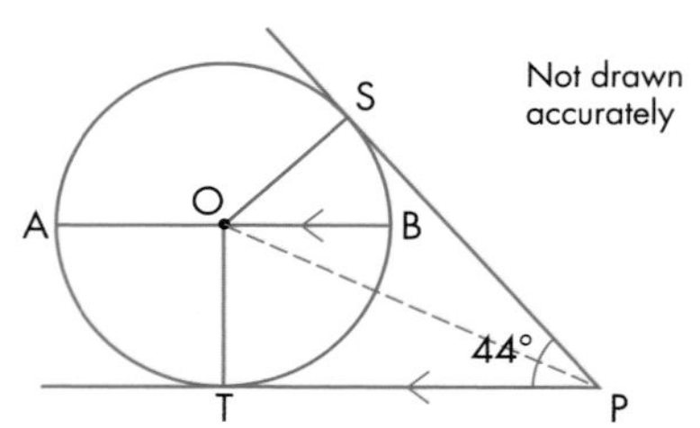

The diagram shows a circle, centre O.

A, S, B and T are points on the circumference of the circle.

PT and PS are tangents to the circle.

AB is parallel to TP.

Angle SPT = 44°.

Work out the size of angle SOB. *(4 marks)*

Edexcel, March 2009, Paper 10 Higher, Question 7

6

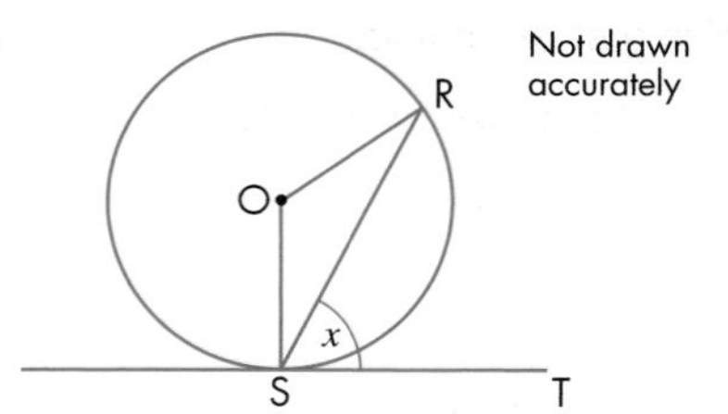

R and S are two points on a circle, centre O.

TS is a tangent to the circle.

Angle RST = x.

Prove that angle ROS = $2x$.

You must give reasons for each stage of your working. *(4 marks)*

Edexcel, March 2008, Paper 10 Higher, Question 9

A* A B

Worked Examination Questions

1 a

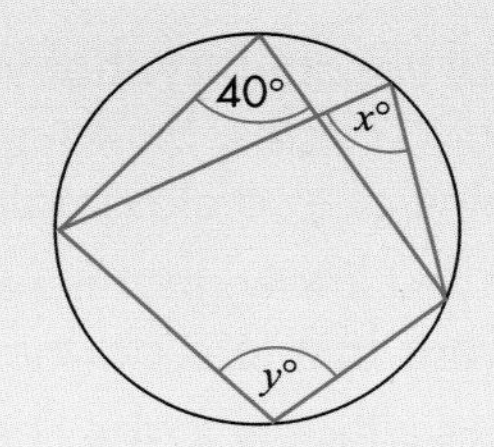

i Write down the value of x.

ii Calculate the value of y.

b A and C are points on the circumference of a circle centre B. AD and CD are tangents and ∠ADB = 40°.

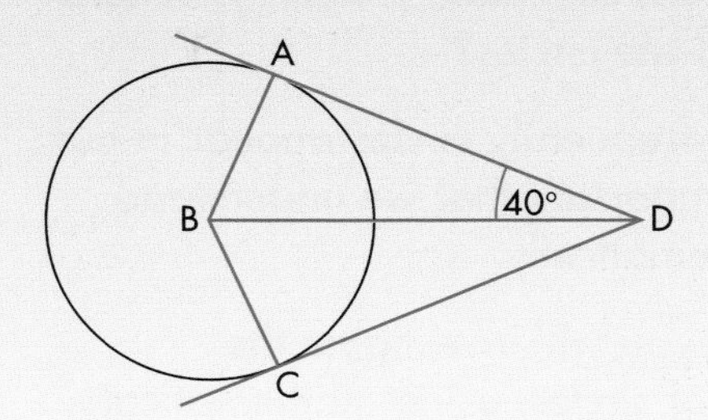

Explain why ABC is 100°. Give **reasons** for your answers.

1 a i $x = 40°$ (angle in same segment) — This gets 1 mark.

ii $x = 140°$ (opposite angles in cyclic quadrilateral = 180°) — This gets 1 mark.

b BAD = 90° (radius is perpendicular to tangent) — This gets 1 mark.

∠ABD = 50° (angles in a triangle) — This gets 1 mark for method.

Similarly:

∠BCD = 90° (radius is perpendicular to tangent)

∠CBD = 50° (angles in a triangle)

So ∠ABC = 2 x 50° = 100° — This gets 1 mark for accuracy.

Total: 5 marks

2 In the diagram, XY is a tangent to the circle at A.

BCY is a straight line.

Work out the size of ∠ABC.

Give reasons to justify your answer.

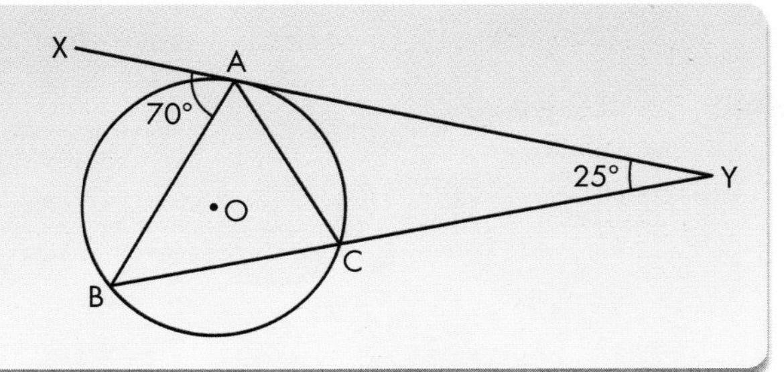

2 ∠ACB = 70° (angle in alternate segment) — This gets 1 mark.

∠ACY = 110° (angles on a line = 180°)

∠CAY = 45° (angles in a triangle = 180°) — This gets 1 method mark for using angles in a triangle.

so ∠ABC = 45° (angle in alternate segment) — This gets 1 accuracy mark.

Total: 3 marks

14 Problem Solving
Proving properties of circles

Circular shapes form the basis of many of the objects that we see and use every day. For example, we see circular shapes in DVDs, wheels, coins and jewellery. Where else do you see circles?

Given how frequently circles appear in our lives, it is important that we understand them mathematically.

In this task you will investigate the properties of circles, using mathematical theory and proof to help you understand this shape more fully.

Getting started

- List the mathematical vocabulary that you know, that is related to circles. Explain each of the words you think of to a classmate.
- Select one fact that you know, that is related to circles. Explain your fact to a partner.
- Select a real-life object that is in the shape of a circle.

  What mathematical questions could you ask about this object?

  How would you use mathematics to find the answers to your questions?

Your task

1 With a partner, develop a statement about the properties of circles. For example, "A radius that is perpendicular to a chord bisects that chord."

Think carefully about your statement: it will be the hypothesis that forms the basis of a mathematical investigation.

Alternatively, create a question to form the basis of your investigation. For example, "What is the relationship between an angle subtended at the centre of a circle and the angle subtended by the same two points at the circumference, opposite and on the same arc?"

2 Swap your hypothesis or question with that of another pair.

Now, use your problem-solving skills to investigate the hypothesis or question. You should create a presentation, using slides or an interactive whiteboard, to explain the mathematical process that you go through in your investigation, the mathematics that you used and the conclusion that you reach.

In your presentation you must:

- explain the overall approach you took in your investigation
- summarise each step taken during your investigation
- advance a solution to the hypothesis or problem with which you were presented
- find the most effective way of representing your solution
- give examples to support your conclusions
- show what you have now learnt about circles
- or any other aspect of geometry.

Why this chapter matters

The theory of linear programming has been used by many companies to reduce their costs and increase productivity.

The theory of linear programming, which uses inequalities in two dimensions, was developed at the start of the Second World War in 1939.

It was used to work out ways to get armaments as efficiently as possible and to increase the effectiveness of resources. It was so powerful an analytical tool that the Allies did not want the Germans to know about it, so it was not made public until 1947.

As a student, George Dantzig, who was one of the inventors of linear programming, came late to a lecture at University one day and saw two problems written on the blackboard. He copied them, thinking they were the homework assignment. He solved both problems, but had to apologise to the lecturer as he found them a little harder than the usual homework, so he took a few days to solve them and was late handing them in.

The lecturer was astonished. The problems he had written on the board were not homework but examples of 'impossible problems'. Not any more!

TOP SECRET

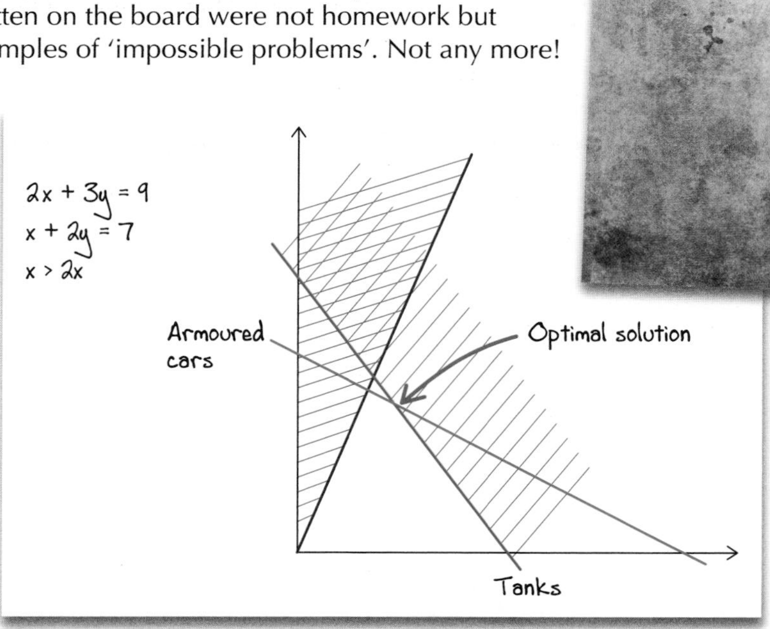

Chapter 15 Algebra: Inequalities and regions

The grades given in this chapter are target grades.

This chapter will show you ...

C how to solve a linear inequality

B how to find a region on a graph that obeys a linear inequality in two variables

Visual overview

What you should already know

- How to solve linear equations **(KS3 level 6, GCSE grade D)**
- How to draw linear graphs **(KS3 level 6, GCSE grade D)**

Quick check

1 Solve these equations.

a $\frac{2x + 5}{3} = 7$ **b** $2x - 7 = 13$

2 On a grid with x- and y-axes from 0 to 10, draw the graphs of these equations.

a $y = 3x + 1$ **b** $2x + 3y = 12$

15.1 Solving inequalities

This section will show you how to:
- solve a simple linear inequality

Key words
inclusive inequality
inequality
number line
strict inequality

Inequalities behave similarly to equations, which you have already met. In the case of linear inequalities, you use the same rules to solve them as you use for linear equations. There are four inequality signs, $<$ which means 'less than', $>$ which means 'greater than', $\leqslant$ which means 'less than or equal to' and $\geqslant$ which means 'greater than or equal to'.

Be careful. Never replace the inequality sign with an equals sign in an examination or you could end up getting no marks.

EXAMPLE 1

Solve $2x + 3 < 14$

Rewrite this as: $2x < 14 - 3$

$2x < 11$

Divide both sides by 2: $\frac{2x}{2} < \frac{11}{2}$

$\Rightarrow x < 5.5$

This means that x can take any value below 5.5 but *not* the value 5.5.

$<$ and $>$ are called **strict inequalities**.

Note: The inequality sign given in the problem is the sign to use in the answer.

EXAMPLE 2

Solve $\frac{x}{2} + 4 \geqslant 13$

Solve just like an equation but leave the inequality sign in place of the equals sign.

Subtract 4 from both sides: $\frac{x}{2} \geqslant 9$

Multiply both sides by 2: $x \geqslant 18$

This means that x can take any value above and including 18.

$\leqslant$ and $\geqslant$ are called **inclusive inequalities**.

EXAMPLE 3

Solve $\frac{3x + 7}{2} < 14$

Rewrite this as: $3x + 7 < 14 \times 2$

That is: $3x + 7 < 28$

$\Rightarrow \quad 3x < 28 - 7$

$\Rightarrow \quad 3x < 21$

$\Rightarrow \quad x < 21 \div 3$

$\Rightarrow \quad x < 7$

EXAMPLE 4

Solve $1 < 3x + 4 \leqslant 13$

Divide the inequality into two parts, and treat each part separately.

$1 < 3x + 4$	$3x + 4 \leqslant 13$
$\Rightarrow \quad 1 - 4 < 3x$	$\Rightarrow \quad 3x \leqslant 13 - 4$
$\Rightarrow \quad -3 < 3x$	$\Rightarrow \quad 3x \leqslant 9$
$\Rightarrow \quad -\frac{3}{3} < x$	$\Rightarrow \quad x \leqslant \frac{9}{3}$
$\Rightarrow \quad -1 < x$	$\Rightarrow \quad x \leqslant 3$

Hence, $-1 < x \leqslant 3$

EXERCISE 15A

1 Solve the following linear inequalities.

a $x + 4 < 7$ **b** $t - 3 > 5$ **c** $p + 2 \geqslant 12$

d $2x - 3 < 7$ **e** $4y + 5 \leqslant 17$ **f** $3t - 4 > 11$

g $\frac{x}{2} + 4 < 7$ **h** $\frac{y}{5} + 3 \leqslant 6$ **i** $\frac{t}{3} - 2 \geqslant 4$

j $3(x - 2) < 15$ **k** $5(2x + 1) \leqslant 35$ **l** $2(4t - 3) \geqslant 34$

2 Write down the largest integer value of x that satisfies each of the following.

a $x - 3 \leqslant 5$, where x is positive

b $x + 2 < 9$, where x is positive and even

c $3x - 11 < 40$, where x is a square number

d $5x - 8 \leqslant 15$, where x is positive and odd

e $2x + 1 < 19$, where x is positive and prime

C

C

3 Write down the smallest integer value of x that satisfies each of the following.

a $x - 2 \geqslant 9$, where x is positive

b $x - 2 > 13$, where x is positive and even

c $2x - 11 \geqslant 19$, where x is a square number

FM 4 Ahmed went to town with £20 to buy two CDs. His bus fare was £3. The CDs were both the same price. When he arrived home he still had some money in his pocket. What was the most each CD could cost?

> **HINTS AND TIPS**
>
> Set up an inequality and solve it.

AU 5 **a** Explain why you cannot make a triangle with three sticks of length 3 cm, 4 cm and 8 cm.

b Three sides of a triangle are x, $x + 2$ and 10 cm.

x is a whole number.

What is the smallest value x can take?

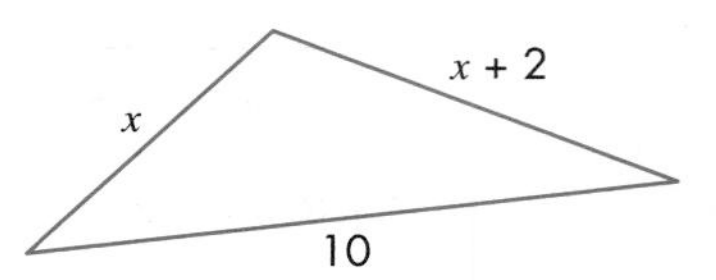

B

PS 6 Five cards have inequalities and equations marked on them.

$x > 0$ | $x < 3$ | $x \geqslant 4$ | $x = 2$ | $x = 6$

The cards are shuffled and then turned over, one at a time.

If two consecutive cards have any numbers in common, then a point is scored.

If they do not have any numbers in common, then a point is deducted.

a The first two cards below score –1 because $x = 6$ and $x < 3$ have no numbers in common.

Explain why the total for this combination scores 0.

$x = 6$ | $x < 3$ | $x > 0$ | $x = 2$ | $x \geqslant 4$

b What does this combination score?

$x > 0$ | $x = 6$ | $x \geqslant 4$ | $x = 2$ | $x < 3$

c Arrange the cards to give a maximum score of 4.

7 Solve the following linear inequalities.

a $4x + 1 \geqslant 3x - 5$

b $5t - 3 \leqslant 2t + 5$

c $3y - 12 \leqslant y - 4$

d $2x + 3 \geqslant x + 1$

e $5w - 7 \leqslant 3w + 4$

f $2(4x - 1) \leqslant 3(x + 4)$

8 Solve the following linear inequalities.

a $\dfrac{x + 4}{2} \leqslant 3$

b $\dfrac{x - 3}{5} > 7$

c $\dfrac{2x + 5}{3} < 6$

d $\dfrac{4x - 3}{5} \geqslant 5$

e $\dfrac{2t - 2}{7} > 4$

f $\dfrac{5y + 3}{5} \leqslant 2$

9 Solve the following linear inequalities.

a $7 < 2x + 1 < 13$

b $5 < 3x - 1 < 14$

c $-1 < 5x + 4 \leqslant 19$

d $1 \leqslant 4x - 3 < 13$

e $11 \leqslant 3x + 5 < 17$

f $-3 \leqslant 2x - 3 \leqslant 7$

The number line

The solution to a linear inequality can be shown on the **number line** by using the following conventions.

$x \leqslant$

$x \geqslant$

$x <$

$x >$

A strict inequality does not include the boundary point but an inclusive inequality does include the boundary point.

Below are five examples.

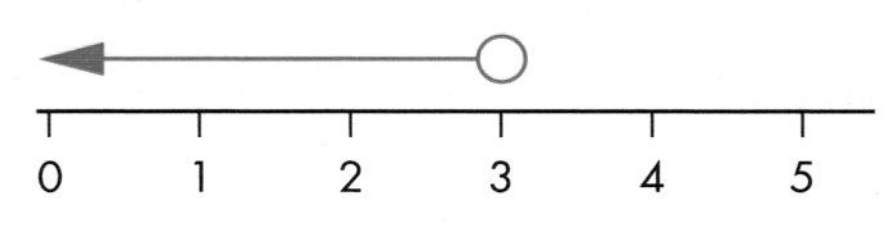

represents $x < 3$

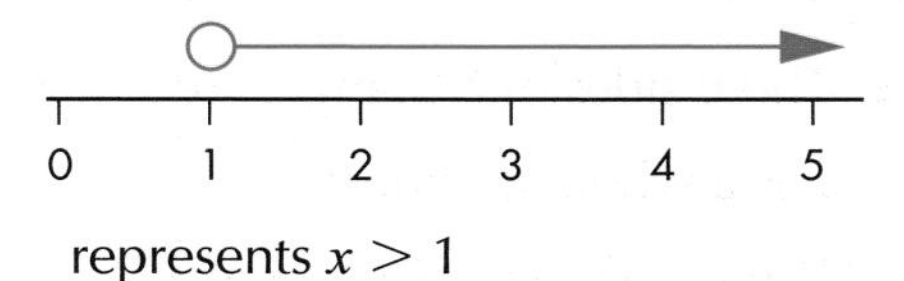

represents $x > 1$

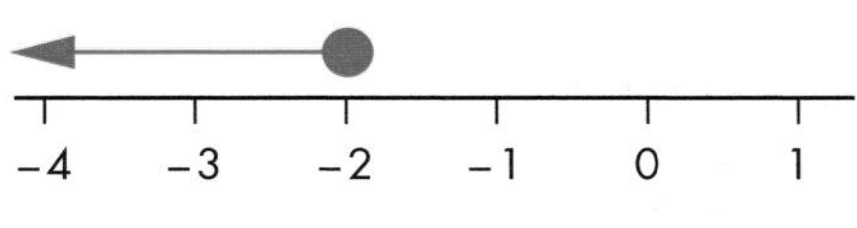

represents $x \leqslant -2$

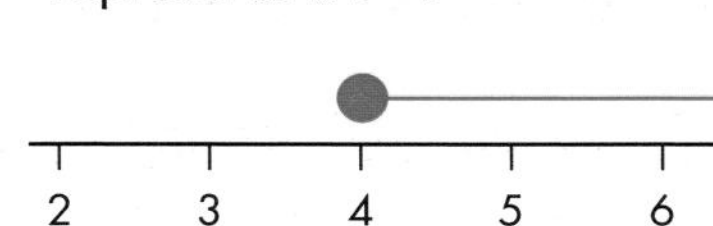

represents $x \geqslant 4$

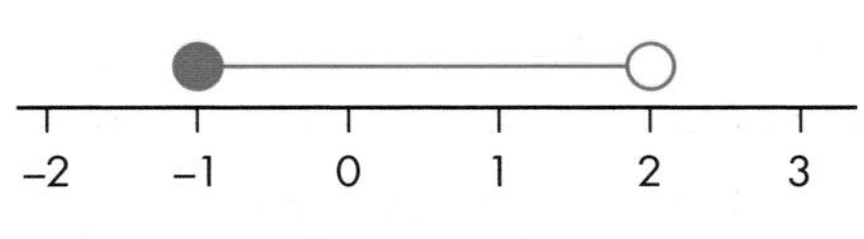

represents $-1 \leqslant x < 2$

This is a 'between' inequality. It can be written as $x \geqslant -1$ and $x < 2$, but the notation $-1 \leqslant x < 2$ is much neater.

EXAMPLE 5

a Write down the inequality shown by this diagram.

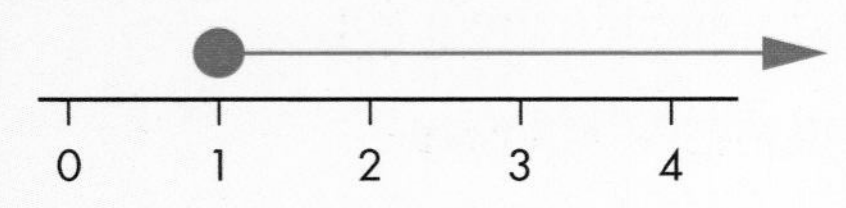

b **i** Solve the inequality $2x + 3 < 11$.

ii Mark the solution on a number line.

c Write down the integers that satisfy both the inequalities in **a** and **b**.

a The inequality shown is $x \geqslant 1$.

b **i** $2x + 3 < 11$

$\Rightarrow 2x < 8$

$\Rightarrow x < 4$

ii

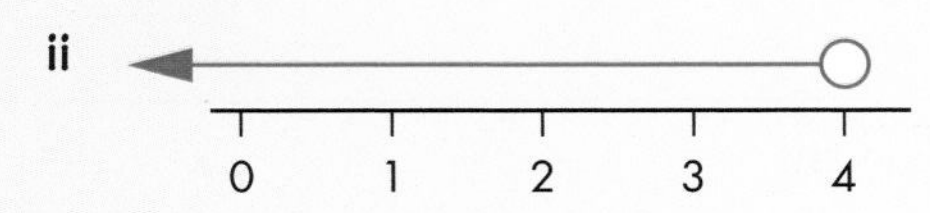

c The integers that satisfy both inequalities are 1, 2 and 3.

EXERCISE 15B

C

1 Write down the inequality that is represented by each diagram below.

a
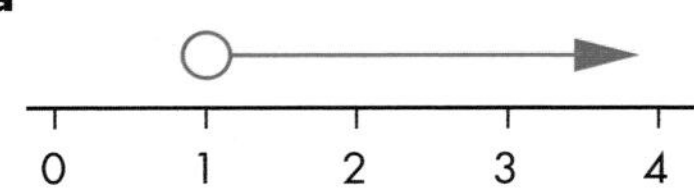

b
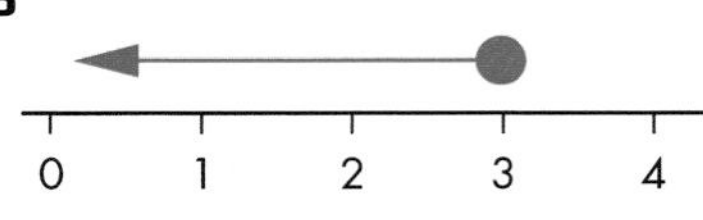

c
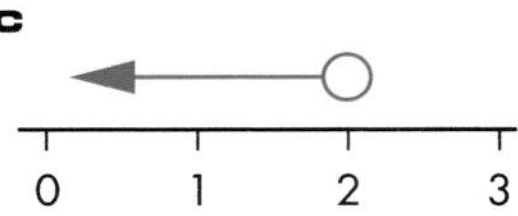

d
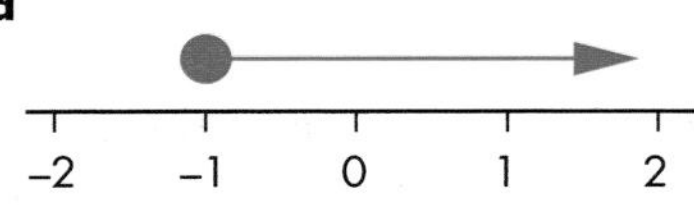

e
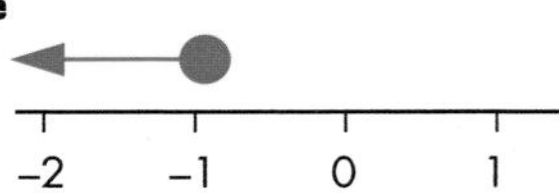

f
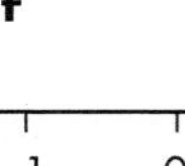
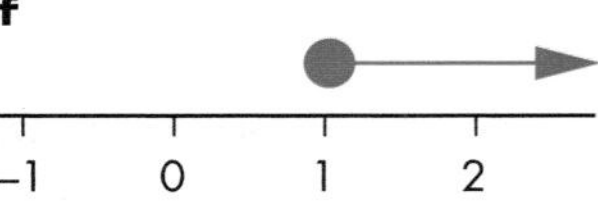

2 Draw diagrams to illustrate these inequalities.

a $x \leqslant 3$ **b** $x > -2$ **c** $x \geqslant 0$ **d** $x < 5$

e $x \geqslant -1$ **f** $2 < x \leqslant 5$ **g** $-1 \leqslant x \leqslant 3$ **h** $-3 < x < 4$

3 Solve the following inequalities and illustrate their solutions on number lines.

a $x + 4 \geqslant 8$ **b** $x + 5 < 3$ **c** $4x - 2 \geqslant 12$ **d** $2x + 5 < 3$

e $2(4x + 3) < 18$ **f** $\frac{x}{2} + 3 \leqslant 2$ **g** $\frac{x}{5} - 2 > 8$ **h** $\frac{x}{3} + 5 \geqslant 3$

FM 4 Max went to the supermarket with £1.20. He bought three apples costing x pence each and a chocolate bar costing 54p. When he got to the till, he found he didn't have enough money.

Max took one of the apples back and paid for two apples and the chocolate bar. He counted his change and found he had enough money to buy a 16p chew.

a Explain why $3x + 54 > 120$ and solve the inequality.

b Explain why $2x + 54 \leqslant 104$ and solve the inequality.

c Show the solution to both of these inequalities on a number line.

d What is the possible price of an apple?

AU 5 On copies of the number lines below, draw two inequalities so that only the integers {–1, 0, 1, 2} are common to both inequalities.

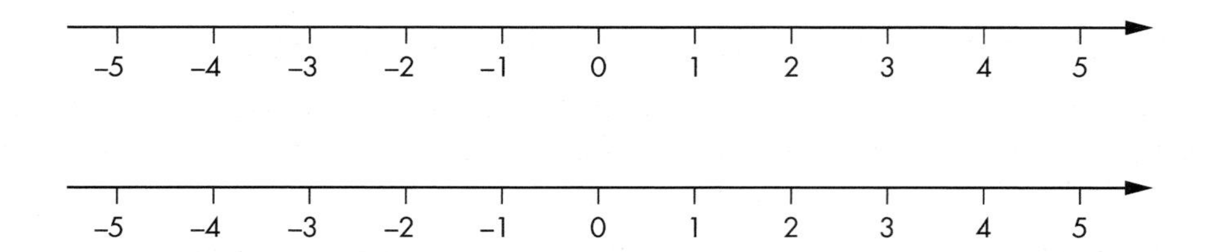

PS 6 What numbers are being described?

x is a square number.

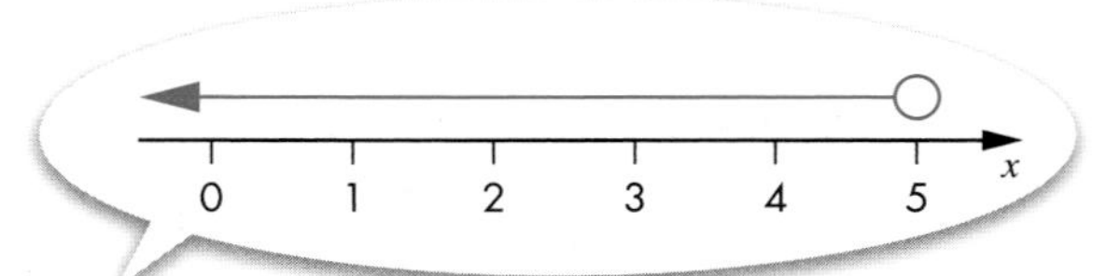

$2x + 3 > 5$

7 Solve the following inequalities and illustrate their solutions on number lines.

a $\frac{2x + 5}{3} > 3$ **b** $\frac{3x + 4}{2} \geqslant 11$ **c** $\frac{2x + 8}{3} \leqslant 2$ **d** $\frac{2x - 1}{3} \geqslant -3$

15.2 Graphical inequalities

This section will show you how to:
- show a graphical inequality
- find regions that satisfy more than one graphical inequality

Key words
boundary
included
origin
region

A linear inequality can be plotted on a graph. The result is a **region** that lies on one side or the other of a straight line. You will recognise an inequality by the fact that it looks like an equation but instead of the equals sign it has an inequality sign: $<$, $>$, $\leqslant$, or $\geqslant$.

The following are examples of linear inequalities that can be represented on a graph.

$y < 3$ $x > 7$ $-3 \leqslant y < 5$ $y \geqslant 2x + 3$ $2x + 3y < 6$ $y \leqslant x$

The method for graphing an inequality is to draw the **boundary** line that defines the inequality. This is found by replacing the inequality sign with an equals sign. When a strict inequality is stated ($<$ or $>$), the boundary line should be drawn as a *dashed* line to show that it is not included in the range of values. When $\leqslant$ or $\geqslant$ is used to state the inequality, the boundary line should be drawn as a *solid* line to show that the boundary is **included**.

After the boundary line has been drawn, shade the *required region*.

To confirm on which side of the line the region lies, choose any point that is not on the boundary line and test it in the inequality. If it satisfies the inequality, that is the side required. If it doesn't, the other side is required.

Work through the six inequalities in the following example to see how the procedure is applied.

EXAMPLE 6

Show each of the following inequalities on a graph.

a $y \leqslant 3$ **b** $x > 7$ **c** $-3 \leqslant y < 5$

d $y \leqslant 2x + 3$ **e** $2x + 3y < 6$ **f** $y \leqslant x$

a Draw the line $y = 3$. Since the inequality is stated as $\leqslant$, the line is *solid*. Test a point that is not on the line. The **origin** is always a good choice if possible, as 0 is easy to test.

Putting 0 into the inequality gives $0 \leqslant 3$. The inequality is satisfied and so the region containing the origin is the side we want.

Shade it in.

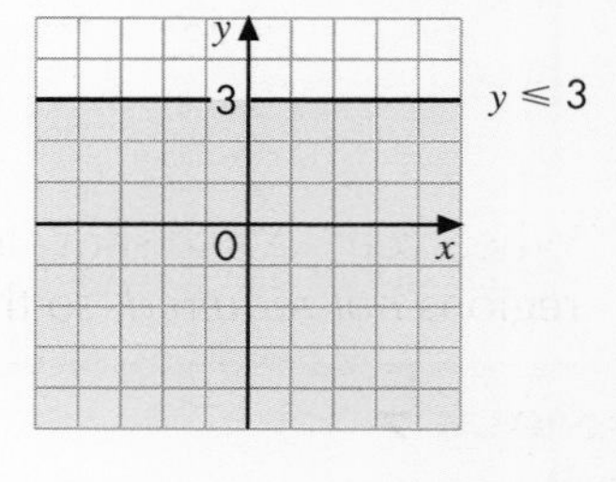

b Since the inequality is stated as $>$, the line is *dashed*. Draw the line $x = 7$.

Test the origin (0, 0), which gives $0 > 7$. This is not true, so you want the other side of the line from the origin.

Shade it in.

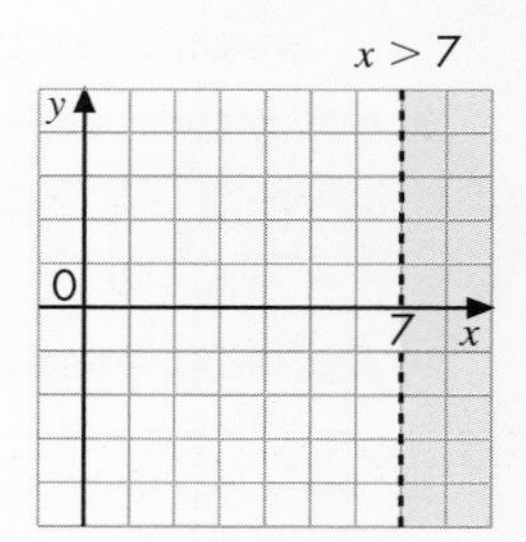

c Draw the lines $y = -3$ (solid for $\leqslant$) and $y = 5$ (dashed for $<$).

Test a point that is not on either line, say (0, 0). Zero is between –3 and 5, so the required region lies between the lines.

Shade it in.

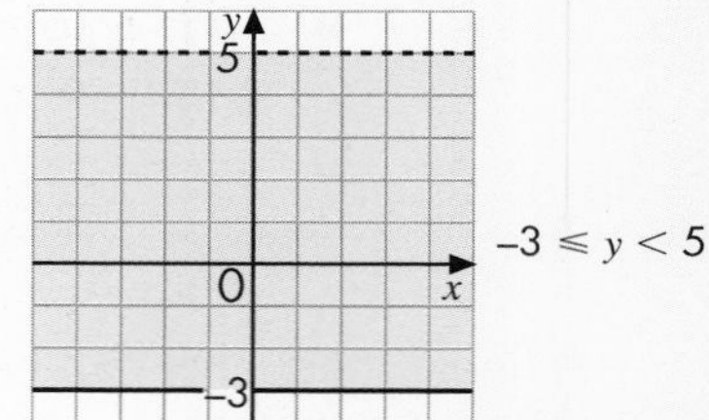

d Draw the line $y = 2x + 3$. Since the inequality is stated as $\leqslant$, the line is solid.

Test a point that is not on the line, (0, 0). Putting these x- and y-values in the inequality gives $0 \leqslant 2(0) + 3$, which is true. So the region that includes the origin is what you want.

Shade it in.

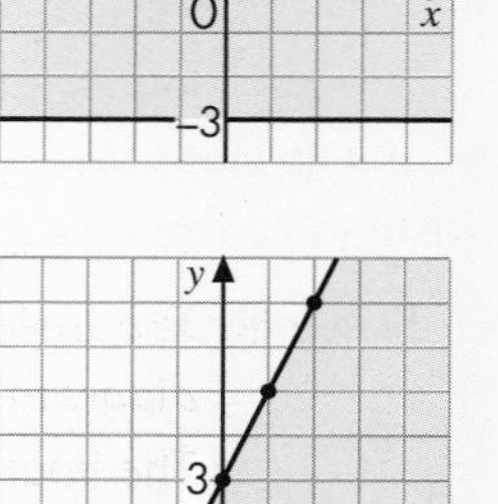

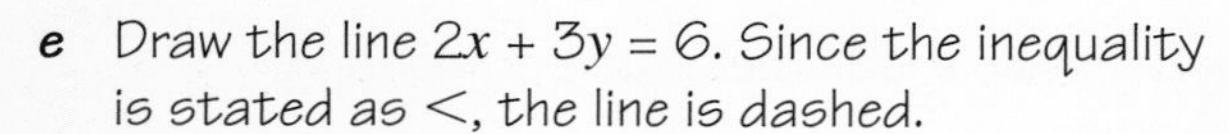

e Draw the line $2x + 3y = 6$. Since the inequality is stated as $<$, the line is dashed.

Test a point that is not on the line, say (0, 0). Is it true that $2(0) + 3(0) < 6$? The answer is yes, so the origin is in the region that you want.

Shade it in.

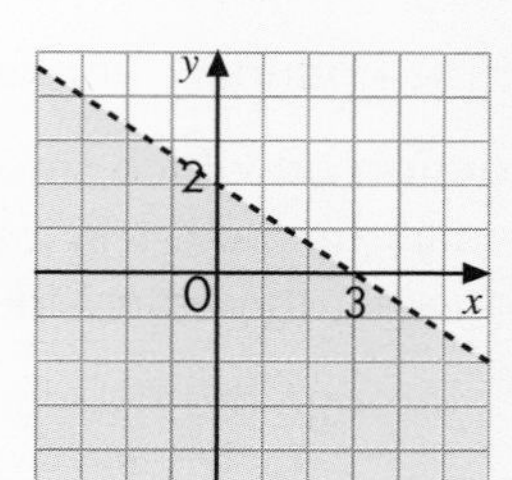

f Draw the line $y = x$. Since the inequality is stated as $\leqslant$, the line is solid.

This time the origin is on the line, so pick any other point, say (1, 3). Putting $x = 1$ and $y = 3$ in the inequality gives $3 \leqslant 1$. This is not true, so the point (1, 3) is not in the region you want.

Shade in the other side to (1, 3).

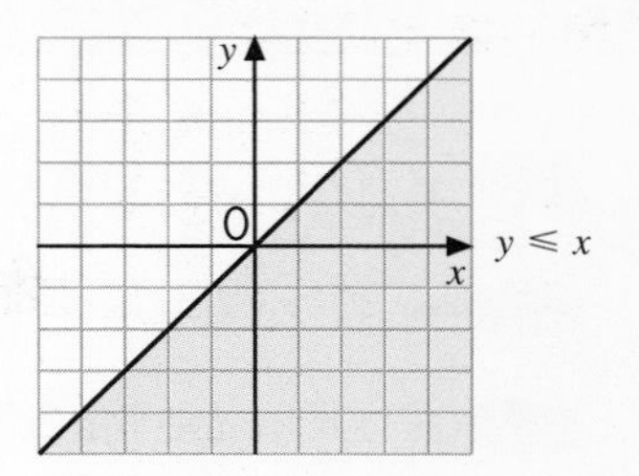

More than one inequality

When you have to show a region that satisfies more than one inequality, it is clearer to *shade* the regions *not required,* so that the *required region* is left *blank*.

EXAMPLE 7

a On the same grid, show the regions that represent the following inequalities by shading the unwanted regions.

i $x > 2$ **ii** $y \geqslant x$ **iii** $x + y < 8$

b Are these points

i (3, 4) **ii** (2, 6) **iii** (3, 3)

in the region that satisfies all three inequalities?

a **i**

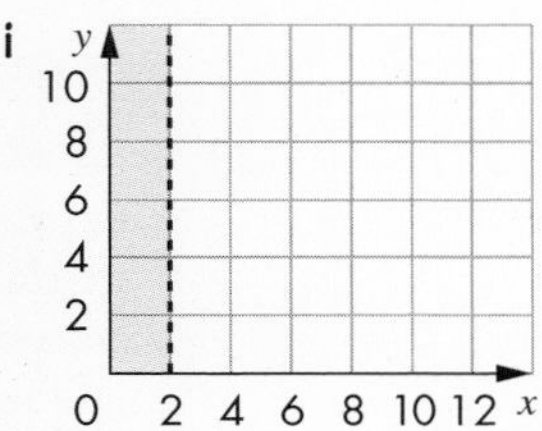

ii

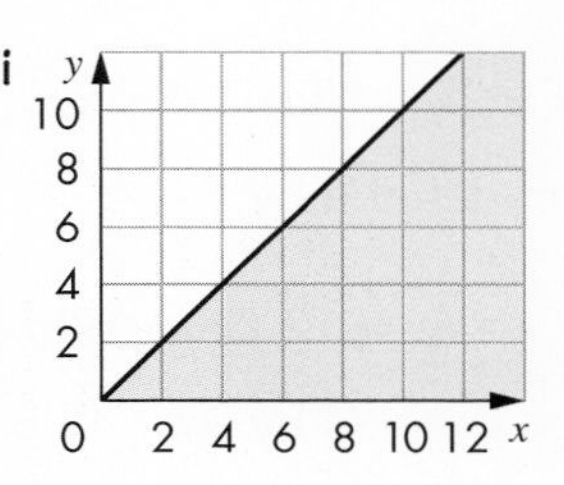

iii

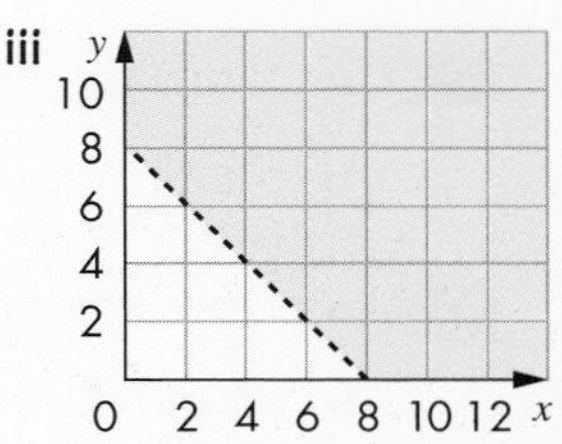

i The region $x > 2$ is shown unshaded in diagram **i**. The boundary line is $x = 2$ (dashed).

ii The region $y \geqslant x$ is shown unshaded in diagram **ii**. The boundary line is $y = x$ (solid).

iii The region $x + y < 8$ is shown unshaded in diagram **iii**.

The boundary line is $x + y = 8$ (dashed). The regions have first been drawn separately so that each may be clearly seen. The diagram on the right shows all three regions on the same grid. The white triangular area defines the region that satisfies all three inequalities.

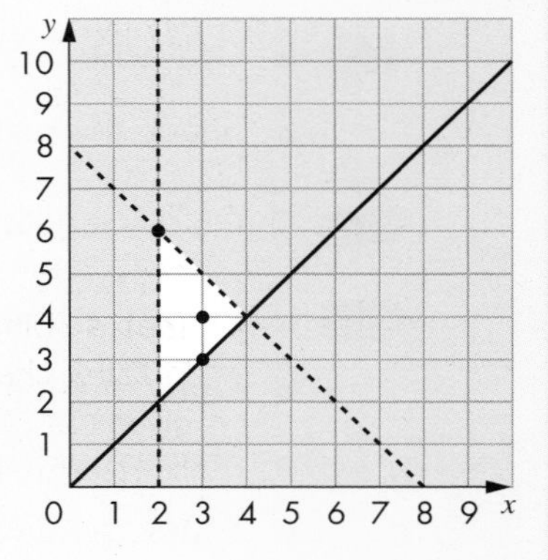

b **i** The point (3, 4) is clearly within the region that satisfies all three inequalities.

ii The point (2, 6) is on the boundary lines $x = 2$ and $x + y = 8$. As these are dashed lines, they are not included in the region defined by all three inequalities. So, the point (2, 6) is not in this region.

iii The point (3, 3) is on the boundary line $y = x$. As this is a solid line, it is included in the region defined by all three inequalities. So, the point (3, 3) is included in this region.

EXERCISE 15C

C

1 **a** Draw the line $x = 2$ (as a solid line). **b** Shade the region defined by $x \leqslant 2$.

2 **a** Draw the line $y = -3$ (as a dashed line). **b** Shade the region defined by $y > -3$.

3 **a** Draw the line $x = -2$ (as a solid line).

b Draw the line $x = 1$ (as a solid line) on the same grid.

c Shade the region defined by $-2 \leqslant x \leqslant 1$.

4 **a** Draw the line $y = -1$ (as a dashed line).

b Draw the line $y = 4$ (as a solid line) on the same grid.

c Shade the region defined by $-1 < y \leqslant 4$.

B

5 **a** On the same grid, draw the regions defined by these inequalities.

i $-3 \leqslant x \leqslant 6$ **ii** $-4 < y \leqslant 5$

b Are the following points in the region defined by both inequalities?

i (2, 2) **ii** (1, 5) **iii** (–2, –4)

6 **a** Draw the line $y = 2x - 1$ (as a dashed line).

b Shade the region defined by $y < 2x - 1$.

7 **a** Draw the line $3x - 4y = 12$ (as a solid line).

b Shade the region defined by $3x - 4y \leqslant 12$.

8 **a** Draw the line $y = \frac{1}{2}x + 3$ (as a solid line).

b Shade the region defined by $y \geqslant \frac{1}{2}x + 3$.

HINTS AND TIPS

In exams it is always made clear which region is to be labelled or shaded. Make sure you do as the question asks, and label or shade as required, otherwise you could lose a mark.

9 Shade the region defined by $y < -3$.

10 **a** Draw the line $y = 3x - 4$ (as a solid line).

b Draw the line $x + y = 10$ (as a solid line) on the same diagram.

c Shade the diagram so that the region defined by $y \geqslant 3x - 4$ is left *unshaded*.

d Shade the diagram so that the region defined by $x + y \leqslant 10$ is left *unshaded*.

e Are the following points in the region defined by both inequalities?

i (2, 1) **ii** (2, 2) **iii** (2, 3)

11 **a** Draw the line $y = x$ (as a solid line).

b Draw the line $2x + 5y = 10$ (as a solid line) on the same diagram.

c Draw the line $2x + y = 6$ (as a dashed line) on the same diagram.

d Shade the diagram so that the region defined by $y \geqslant x$ is left *unshaded*.

e Shade the diagram so that the region defined by $2x + 5y \geqslant 10$ is left *unshaded*.

f Shade the diagram so that the region defined by $2x + y < 6$ is left *unshaded*.

g Are the following points in the region defined by these inequalities?

i (1, 1) **ii** (2, 2) **iii** (1, 3)

12 **a** On the same grid, draw the regions defined by the following inequalities. (Shade the diagram so that the overlapping region is left blank.)

i $y > x - 3$ **ii** $3y + 4x \leqslant 24$ **iii** $x \geqslant 2$

b Are the following points in the region defined by all three inequalities?

i (1, 1) **ii** (2, 2) **iii** (3, 3) **iv** (4, 4)

AU 13 The graph shows three points (1, 2), (1, 3) and (2, 3).

Write down three inequalities that between them surround these three grid intersection points and *no others*.

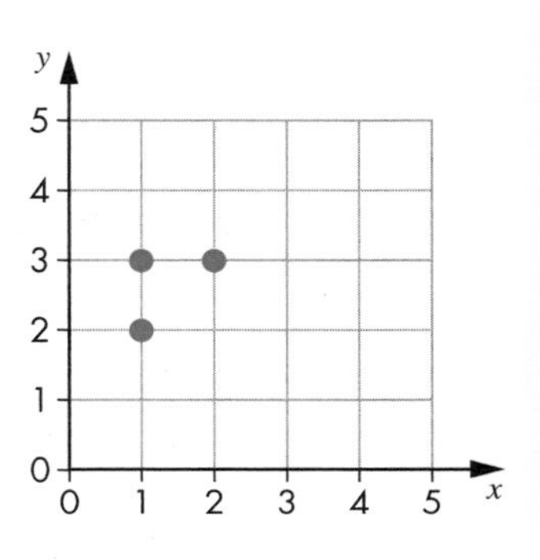

PS 14 If $x + y > 40$, which of the following may be true (M), must be false (F) or must be true (T)?

a $x > 40$ **b** $x + y \leqslant 20$ **c** $x - y = 10$

d $x \leqslant 5$ **e** $x + y = 40$ **f** $y > 40 - x$

g $y = 2x$ **h** $x + y \geqslant 39$

AU 15 Explain how you would find which side of the line represents the inequality $y < x + 2$.

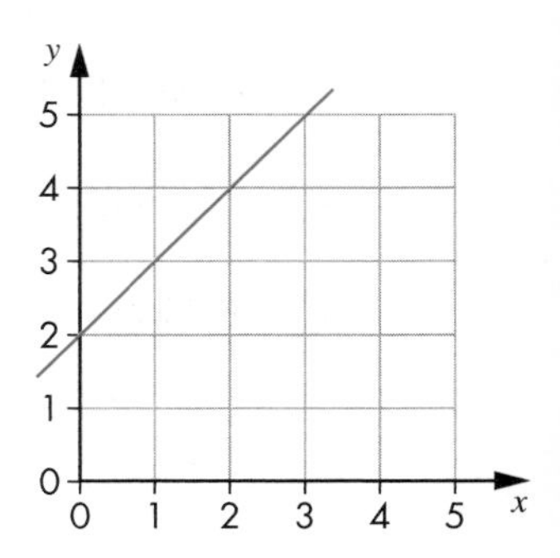

AU 16 The region marked R is the overlap of the inequalities:

$x + y \geqslant 3$ $y \leqslant \frac{1}{2}x + 3$ $y \geqslant 5x - 15$

a For which point in the region R is the value of the function $2x - y$ the greatest ?

b For which point in the region R is the value of the function $x - 3y$ the least?

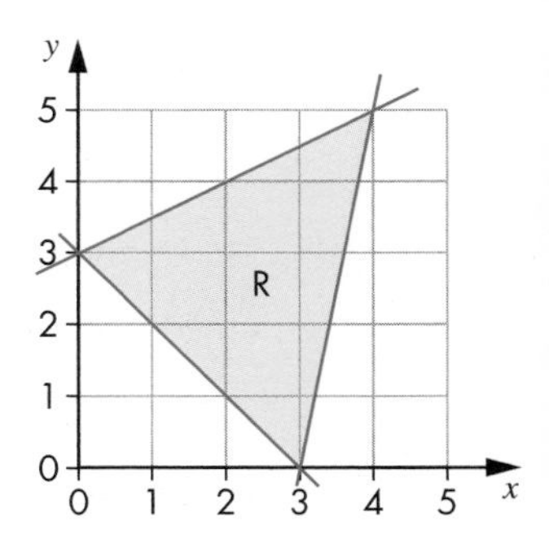

SUMMARY

GRADE BOOSTER

C You can solve inequalities such as $3x + 2 < 5$ and represent the solution on a number line

B You can represent a region that satisfies a linear inequality graphically, and solve more complex linear inequalities

B You can represent a region that simultaneously satisfies more than one linear inequality graphically

---

What you should know now

- How to solve simple inequalities
- How to create algebraic inequalities from verbal statements
- How to represent linear inequalities on a graph
- How to depict a region satisfying more than one linear inequality

1 **a** Solve the inequality $4x + 3 < 13$

b Write down the largest integer that is a solution of $4x + 3 < 13$

2 $-2 \leqslant x < 3$

x is an integer.

Write down all the possible values of x.

(2 marks)

Edexcel, May 2008, Paper 3 Higher, Question 12

3 $-2 \leqslant n < 4$

n is an integer.

a Write down all the possible values of n.

(2 marks)

b Solve the inequality $6x - 3 < 9$ *(2 marks)*

Edexcel, May 2008, Paper 3 Higher, Question 13

4 The region R satisfies the inequalities
$x \geqslant 2$, $y \geqslant 1$, $x + y \leqslant 6$

On a copy of the grid below, draw straight lines and use shading to show the region R.

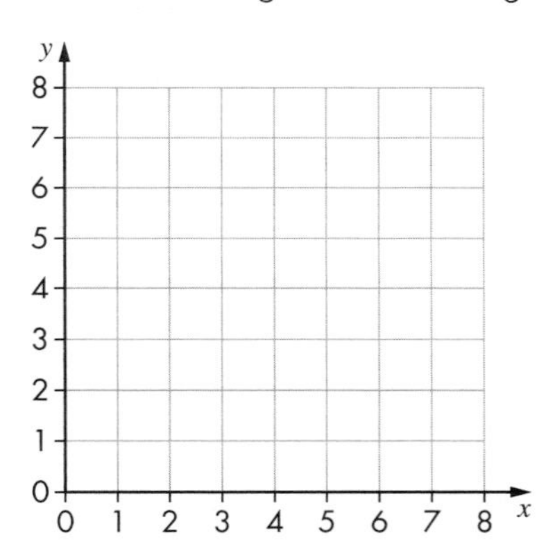

(3 marks)

Edexcel, June 2008, Paper 4 Higher, Question 18

5 **a** Solve the inequality $5x - 9 \geqslant 6$

b **i** Solve the inequality $2x + 7 > 15 - 3x$

ii If x is an integer, what is the smallest possible value of x?

6 Find all the integer values of n that satisfy the inequality

$-3 \leqslant 2n + 1 < 9$

7 **a** Solve the inequality $4x - 9 < x + 3$

b Solve the inequality $x^2 < 25$

8 On a grid show the region that is defined by the three inequalities

$x \geqslant 1$ $y \leqslant 3$ $x + y \leqslant 5$

Mark the region with an R.

9 On a grid show the region that is defined by the three inequalities

$x \geqslant 0$ $y \leqslant 2x + 1$ $x + y \leqslant 4$

Mark the region with an R.

10 Copy the grid below and indicate clearly on it the region defined by the three inequalities

$y \leqslant 5$

$x \geqslant -2$

$y \geqslant 2x - 1$

Mark the region with an R.

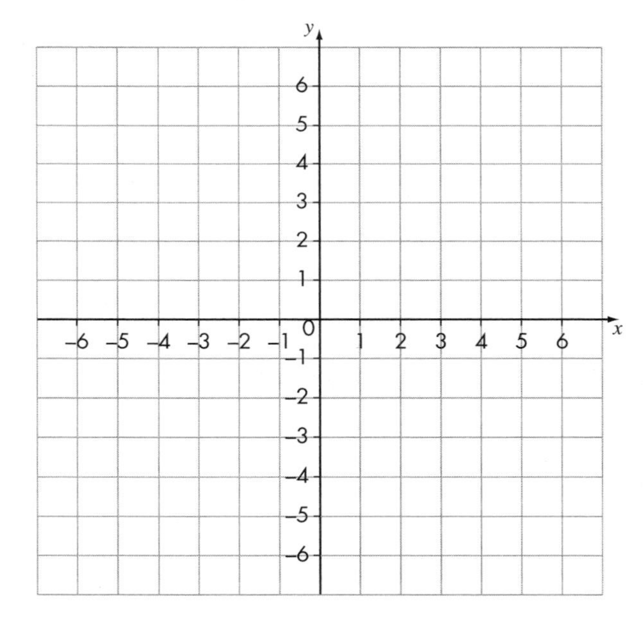

11 n is an integer that satisfies the inequality

$\frac{121}{n^2} \geqslant 8$

List all the possible values of n.

Worked Examination Questions

1 a On the number lines show these inequalities.

i $-2 \leqslant n < 4$

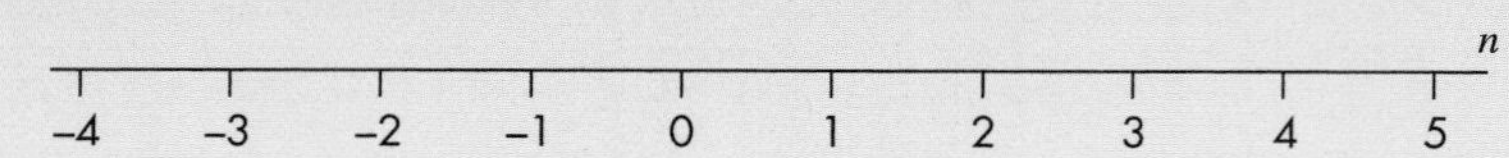

ii $n < 2$

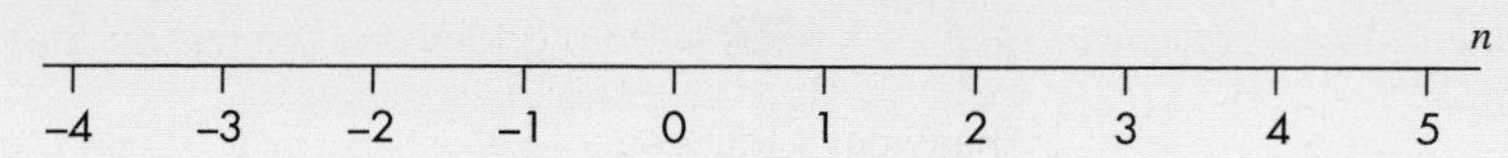

b n is an integer. Find the values of n that satisfy both inequalities in part **a**.

c Solve these inequalities.

i $3x + 8 > 2$

ii $3(x - 4) \leqslant \frac{1}{2}(x + 1)$

1 a i

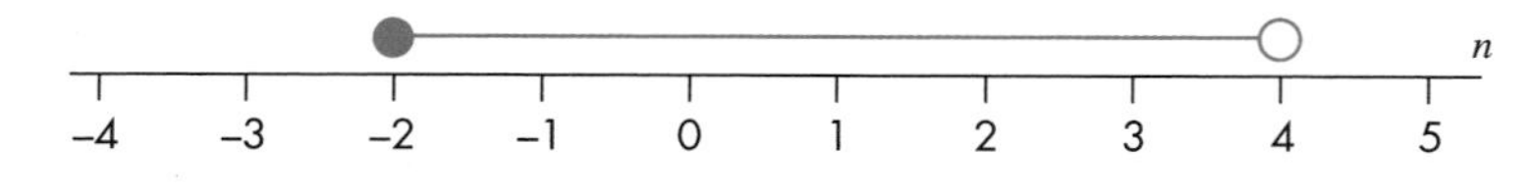

ii

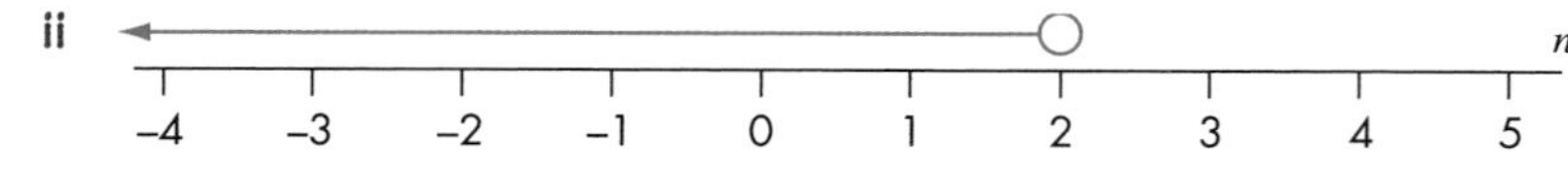

Remember that a strict inequality has an open circle to show the boundary and an inclusive inequality has a solid circle to show the boundary. These get 1 mark each.

b {–2, –1, 0, 1}

The integers that satisfy both inequalities are in the overlap of both lines. This gets 1 mark.

c i $3x + 8 > 2$

$3x > -6$

$x > -2$

As when solving an equation do the same thing to both sides. First subtract 8, then divide by 3. This gets 1 mark for method and 1 mark for accuracy.

ii $3(x - 4) \leqslant \frac{1}{2}(x + 1)$

$6(x - 4) \leqslant x + 1$

$6x - 24 < x + 1$

$5x < 25$

$x < 5$

First multiply by 2 to get rid of the fraction, then expand the brackets. This gets 1 mark for method. Then collect all the x-terms on the left-hand side and the number terms on the right-hand side. This gets 1 mark for method.
Then simplify and divide by 5.
This gets 1 mark for accuracy.

Total: 7 marks

FM 2 A school uses two coach firms, Excel and Storm, to take students home from school. An Excel coach holds 40 students and a Storm coach holds 50 students. 1500 students need to be taken home by coach. If E Excel coaches and S Storm coaches are used, explain why:

$4E + 5S \geqslant 150$

2 E Excel coaches take $40E$ and S Storm coaches take $50S$ students. Together they take $40E + 50S$ students.

Write down the number of students and the total carried by each company's coaches.
This gets 1 mark.

There are 1500 students to use the coaches, so $40E + 50S$ must be at least 1500.

$40E + 50S \geqslant 1500$, then cancel by 10.

Explain that this must be at least the total number of students to be carried, and that the equation will cancel by 10.
This gets 1 mark for method and 1 mark for accuracy.

Total: 3 marks

Worked Examination Questions

PS **3** A bookshelf holds P paperback and H hardback books. The bookshelf can hold a total of 400 books. Which of the following may be true?

a $P + H < 300$ **b** $P \geqslant H$ **c** $P + H > 500$

3 $P + H < 300$ and $P \geqslant H$.

Both of these inequalities could be true. The bookshelf doesn't have to be full and there could be more paperbacks than hardbacks. The final inequality cannot be true as there can only be a maximum of 400 books.
These get 1 mark each.

Total: 2 marks

AU **4** The region R is shown shaded below.

Write down three inequalities which together describe the shaded region.

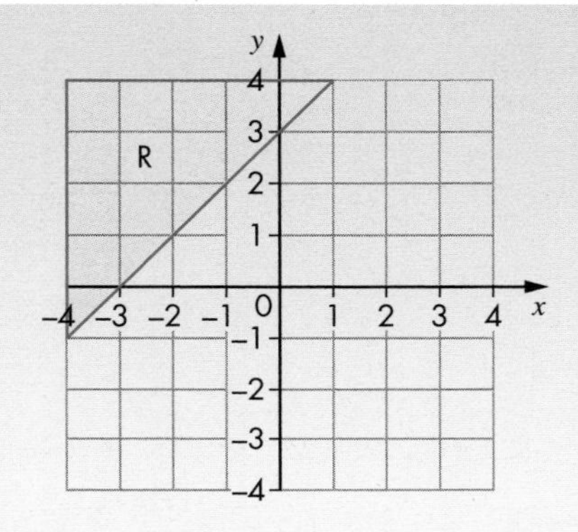

4 $y \leqslant 4, x \geqslant -4$ and $y \geqslant x + 3$

Work out the equation of each boundary line and then decide if points off the line are greater or less than the boundary.
These get 1 mark each.

Total: 3 marks

15 Problem Solving
Linear programming

Linear programming is a mathematical method that uses two-dimensional inequalities. It takes into account maximum and minimum values, and constraints to find the optimum solution to a problem. It is often used by shops to work out stock levels, and reduce cost to increase profit.

Getting started

A boy goes to the fair with £6.00 in his pocket. He only likes rides on the big wheel and eating hot-dogs. A big wheel ride costs £1.50 and a hot-dog costs £2.00. He has W big wheel rides and D hot-dogs.

a Explain why:

i $W \leq 4$

ii $D \leq 3$

iii $3W + 4D \leq 12$

b If he cannot eat more than two hot-dogs before feeling full, write down an inequality that must be true.

c Which of these combinations of big wheel rides and hot-dogs are possible if they obey all of the above conditions?

i two big wheel rides and one hot-dog

ii three big wheel rides and two hot-dogs

iii two big wheel rides and two hot-dogs

iv one big wheel ride and one hot-dog

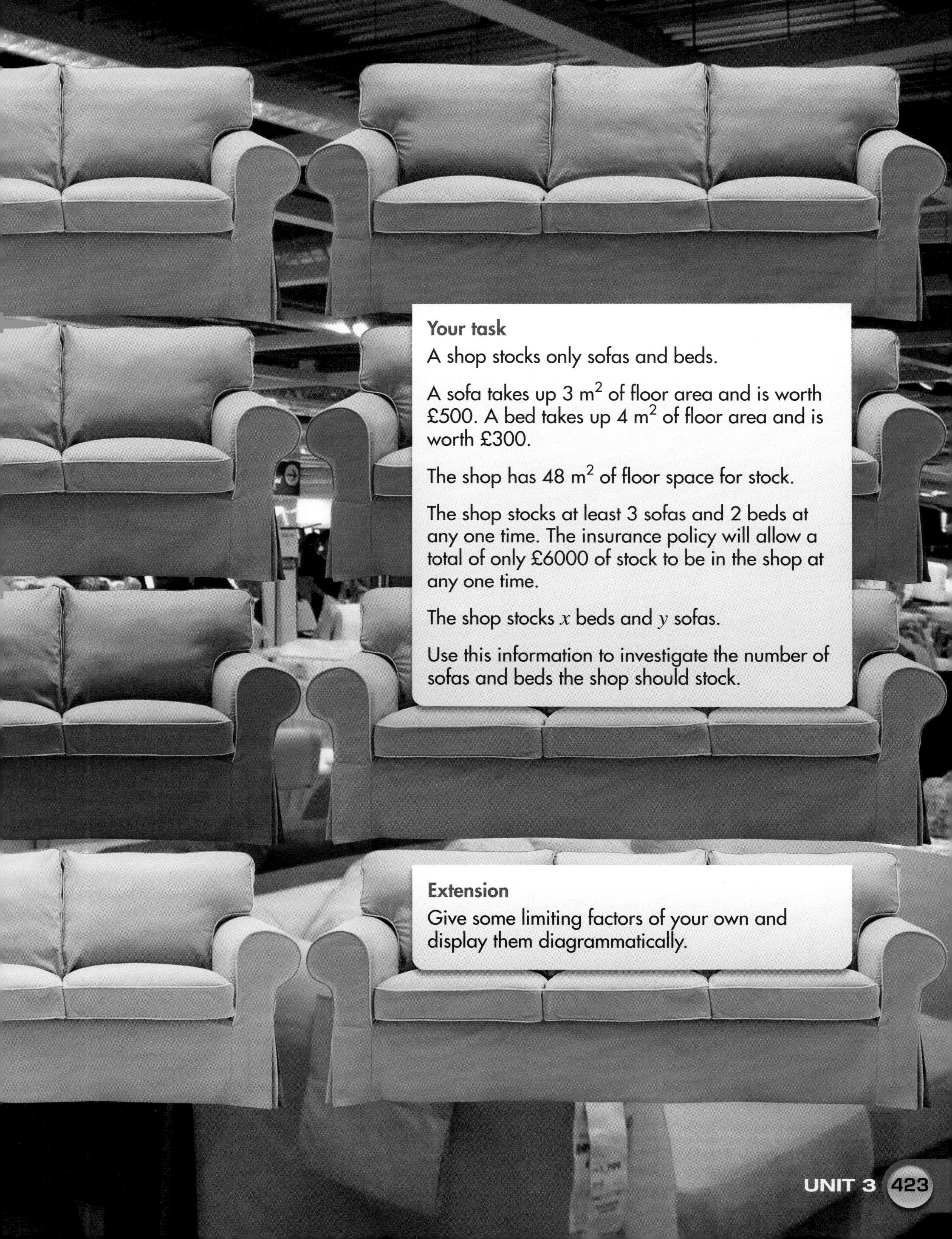

Your task

A shop stocks only sofas and beds.

A sofa takes up 3 m^2 of floor area and is worth £500. A bed takes up 4 m^2 of floor area and is worth £300.

The shop has 48 m^2 of floor space for stock.

The shop stocks at least 3 sofas and 2 beds at any one time. The insurance policy will allow a total of only £6000 of stock to be in the shop at any one time.

The shop stocks x beds and y sofas.

Use this information to investigate the number of sofas and beds the shop should stock.

Extension

Give some limiting factors of your own and display them diagrammatically.

Why this chapter matters

In many real-life situations, variables are connected by a rule or relationship. It may be that as one variable increases the other increases. Alternatively, it may be that as one variable increases the other decreases.

This chapter looks at how quantities vary when they are related in some way.

As this plant gets older it becomes taller.

As the storm increases the number of sunbathers decreases.

As this car gets older it is worth less (and eventually it is worthless!).

As more songs are downloaded, there is less money left on the voucher.

Try to think of other variables that are connected in this way.

Chapter

Number: Variation

The grades given in this chapter are target grades.

1 Direct variation

2 Inverse variation

This chapter will show you ...

A how to solve problems where two variables are connected by a relationship that varies in direct or inverse proportion

Visual overview

Direct proportion → Inverse proportion

What you should already know

- Squares, square roots, cubes and cube roots of integers **(KS3 level 4–5, GCSE grade G–E)**
- How to substitute values into algebraic expressions **(KS3 level 5, GCSE grade E)**
- How to solve simple algebraic equations **(KS3 level 6, GCSE grade D)**

Quick check

1 Write down the value of each of the following.

a 5^2

b $\sqrt{81}$

c 3^3

d $\sqrt[3]{64}$

2 Calculate the value of y if $x = 4$.

a $y = 3x^2$

b $y = \frac{1}{\sqrt{x}}$

16.1 Direct variation

This section will show you how to:
- solve problems where two variables have a directly proportional relationship (direct variation)
- work out the constant of proportionality

Key words
constant of proportionality, k
direct proportion
direct variation

The term **direct variation** means the same as as **direct proportion**.

There is direct variation (or direct proportion) between two variables when one variable is a simple multiple of the other. That is, their ratio is a constant.

For example:

1 kilogram = 2.2 pounds	There is a multiplying factor of 2.2 between kilograms and pounds.
Area of a circle = πr^2	There is a multiplying factor of π between the area of a circle and the square of its radius.

An examination question involving direct variation usually requires you first to find this multiplying factor (called the **constant of proportionality**), then to use it to solve a problem.

The symbol for variation or proportion is $\propto$.

So the statement 'Pay is directly proportional to time' can be mathematically written as:

$pay \propto time$

which implies that:

$pay = k \times time$

where k is the constant of proportionality.

There are four steps to be followed when you are using proportionality to solve problems.

Step 1: Set up the statement, using the proportionality symbol (you may use symbols to represent the variables).

Step 2: Set up the equation, using a constant of proportionality.

Step 3: Use given information to work out the value of the constant of proportionality.

Step 4: Substitute the value of the constant of proportionality into the equation and use this equation to find unknown values.

EXAMPLE 1

The cost of an article is directly proportional to the time spent making it. An article taking 6 hours to make costs £30. Find:

a the cost of an article that takes 5 hours to make

b the length of time it takes to make an article costing £40.

Step 1: Let C be the cost of making an article and t the time it takes.

$C \propto t$

Step 2: Setting up the equation gives:

$C = kt$

where k is the constant of proportionality.

Note that you can 'replace' the proportionality sign $\propto$ with $= k$ to obtain the proportionality equation.

Step 3: Since $C = £30$ when $t = 6$ hours, then $30 = 6k$

$\Rightarrow \frac{30}{60} = k$

$\Rightarrow k = 5$

Step 4: So the formula is $C = 5t$.

a When $t = 5$ hours $\quad C = 5 \times 5 = 25$

So the cost is £25.

b When $C = £40 \quad 40 = 5 \times t$

$\Rightarrow \frac{40}{5} = t \Rightarrow t = 8$

So the time spent making the article is 8 hours.

EXERCISE 16A

For questions **1** to **4**, first find k, the constant of proportionality, and then the formula connecting the variables.

1 T is directly proportional to M. If $T = 20$ when $M = 4$, find:

a T when $M = 3$ **b** M when $T = 10$.

2 W is directly proportional to F. If $W = 45$ when $F = 3$, find:

a W when $F = 5$ **b** F when $W = 90$.

3 Q varies directly with P. If $Q = 100$ when $P = 2$, find:

a Q when $P = 3$ **b** P when $Q = 300$.

4 X varies directly with Y. If $X = 17.5$ when $Y = 7$, find:

a X when $Y = 9$

b Y when $X = 30$.

5 The distance covered by a train is directly proportional to the time taken for the journey. The train travels 105 miles in 3 hours.

a What distance will the train cover in 5 hours?

b How much time will it take for the train to cover 280 miles?

6 The cost of fuel delivered to your door is directly proportional to the weight received. When 250 kg is delivered, it costs £47.50.

a How much will it cost to have 350 kg delivered?

b How much would be delivered if the cost were £33.25?

FM **7** The number of children who can play safely in a playground is directly proportional to the area of the playground. A playground with an area of 210 m^2 is safe for 60 children.

a How many children can safely play in a playground of area 154 m^2?

b A playgroup has 24 children. What is the smallest playground area in which they could safely play?

8 The number of spaces in a car park is directly proportional to the area of the car park.

FM **a** A car park has 300 parking spaces in an area of 4500 m^2.

It is decided to increase the area of the car park by 500 m^2 to make extra spaces.

How many extra spaces will be made?

PS **b** The old part of the car park is redesigned so that the original area has 10% more parking spaces.

How many more spaces than originally will there be altogether if the number of spaces in the new area is directly proportional to the number in the redesigned car park?

AU **9** The number of passengers in a bus queue is directly proportional to the time that the person at the front of the queue has spent waiting.

Karen is the first to arrive at a bus stop. When she has been waiting 5 minutes the queue has 20 passengers.

A bus has room for 70 passengers.

How long had Karen been in the queue if the bus fills up from empty when it arrives and all passengers get on?

Direct proportions involving squares, cubes, square roots and cube roots

The process is the same as for a linear direct variation, as the next example shows.

EXAMPLE 2

The cost of a circular badge is directly proportional to the square of its radius. The cost of a badge with a radius of 2 cm is 68p. Find:

a the cost of a badge of radius 2.4 cm **b** the radius of a badge costing £1.53.

Step 1: Let C be the cost and r the radius of a badge.

$C \propto r^2$

Step 2: Setting up the equation gives:

$C = kr^2$

where k is the constant of proportionality.

Step 3: $C = 68\text{p}$ when $r = 2$ cm. So:

$68 = 4k$

$\Rightarrow \frac{68}{4} = k \Rightarrow k = 17$

Step 4: So the formula is $C = 17r^2$.

a When $r = 2.4$ cm $\quad C = 17 \times 2.4^2\text{p} = 97.92\text{p}$

Rounding gives the cost as 98p.

b When $C = 153\text{p}$ $\quad 153 = 17r^2$

$\Rightarrow \frac{153}{7} = 9 = r^2$

$\Rightarrow r = \sqrt{9} = 3$

Hence, the radius is 3 cm.

EXERCISE 16B

For questions **1** to **6**, first find k, the constant of proportionality, and then the formula connecting the variables.

1 T is directly proportional to x^2. If $T = 36$ when $x = 3$, find:

a T when $x = 5$ **b** x when $T = 400$.

2 W is directly proportional to M^2. If $W = 12$ when $M = 2$, find:

a W when $M = 3$ **b** M when $W = 75$.

3 E varies directly with $\sqrt{C}$. If $E = 40$ when $C = 25$, find:

a E when $C = 49$ **b** C when $E = 10.4$.

4 X is directly proportional to $\sqrt{Y}$. If $X = 128$ when $Y = 16$, find:

a X when $Y = 36$ **b** Y when $X = 48$.

5 P is directly proportional to f^3. If $P = 400$ when $f = 10$, find:

a P when $f = 4$ **b** f when $P = 50$.

6 y is directly proportional to $\sqrt[3]{x}$. If $y = 100$ when $x = 125$, find:

a y when $x = 64$ **b** x when $y = 40$.

7 The cost of serving tea and biscuits varies directly with the square root of the number of people at the buffet. It costs £25 to serve tea and biscuits to 100 people.

a How much will it cost to serve tea and biscuits to 400 people?

b For a cost of £37.50, how many could be served tea and biscuits?

8 In an experiment, the temperature, in °C, varied directly with the square of the pressure, in atmospheres (atm). The temperature was 20 °C when the pressure was 5 atm.

a What will the temperature be at 2 atm?

b What will the pressure be at 80 °C?

9 The mass, in grams, of ball bearings varies directly with the cube of the radius, measured in millimetres. A ball bearing of radius 4 mm has a mass of 115.2 g.

a What will be the mass of a ball bearing of radius 6 mm?

b A ball bearing has a mass of 48.6 g. What is its radius?

10 The energy, in J, of a particle varies directly with the square of its speed, in m/s. A particle moving at 20 m/s has 50 J of energy.

a How much energy has a particle moving at 4 m/s?

b At what speed is a particle moving if it has 200 J of energy?

11 The cost, in £, of a trip varies directly with the square root of the number of miles travelled. The cost of a 100-mile trip is £35.

a What is the cost of a 500-mile trip (to the nearest £1)?

b What is the distance of a trip costing £70?

FM 12 A sculptor is making statues.

The amount of clay used is directly proportional to the cube of the height of the statue.

A statue is 10 cm tall and uses 500 cm^3 of clay.

How much clay will a similar statue use if it is twice as tall?

FM 13 The cost of making different-sized machines is proportional to the time taken.

A small machine costs £100 and takes two hours to make.

How much will a large machine cost that takes 5 hours to build?

PS 14 The sketch graphs show each of these proportion statements.

a $y \propto x^2$ **b** $y \propto x$ **c** $y \propto \sqrt{x}$

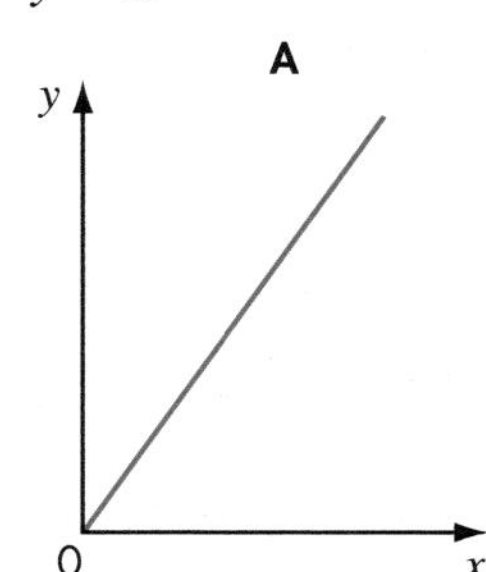

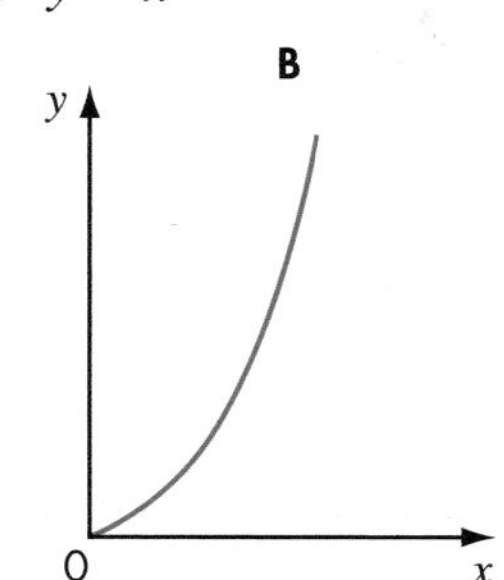

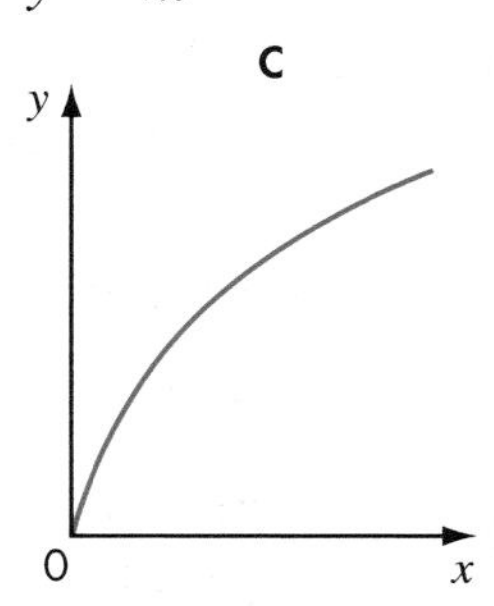

Match each statement to the correct sketch.

AU 15 Here are two tables.

Match each table to a graph in question **14**.

a

x	1	2	3
y	3	12	27

b

x	1	2	3
y	3	6	9

16.2 Inverse variation

This section will show you how to:

- solve problems where two variables have an inversely proportional relationship (inverse variation)
- work out the constant of proportionality

Key words

constant of proportionality, k
inverse proportion
inverse variation

The term **inverse variation** means the same as **inverse proportion**.

There is inverse variation between two variables when one variable is directly proportional to the *reciprocal* of the other. That is, the product of the two variables is constant. So, as one variable increases, the other decreases.

For example, the faster you travel over a given distance, the less time it takes. So there is an inverse variation between speed and time. Speed is inversely proportional to time.

$S \propto \frac{1}{T}$ and so $S = \frac{k}{T}$

which can be written as $ST = k$.

EXAMPLE 3

M is inversely proportional to R. If $M = 9$ when $R = 4$, find the value of:

a M when $R = 2$ **b** R when $M = 3$.

Step 1: $M \propto \frac{1}{R}$

Step 2: Setting up the equation gives:

$M = \frac{k}{R}$

where k is the **constant of proportionality**.

Step 3: $M = 9$ when $R = 4$. So $9 = \frac{k}{4}$

$\Rightarrow 9 \times 4 = k \Rightarrow k = 36$

Step 4: The formula is $M = \frac{36}{R}$

a When $R = 2$, then $M = \frac{36}{2} = 18$

b When $M = 3$, then $3 = \frac{36}{R} \Rightarrow 3R = 36 \Rightarrow R = 12$

EXERCISE 16C

For questions **1** to **6**, first find the formula connecting the variables.

A

1 T is inversely proportional to m. If $T = 6$ when $m = 2$, find:

a T when $m = 4$ **b** m when $T = 4.8$.

2 W is inversely proportional to x. If $W = 5$ when $x = 12$, find:

a W when $x = 3$ **b** x when $W = 10$.

3 Q varies inversely with $(5 - t)$. If $Q = 8$ when $t = 3$, find:

a Q when $t = 10$ **b** t when $Q = 16$.

4 M varies inversely with t^2. If $M = 9$ when $t = 2$, find:

a M when $t = 3$ **b** t when $M = 1.44$.

5 W is inversely proportional to $\sqrt{T}$. If $W = 6$ when $T = 16$, find:

a W when $T = 25$ **b** T when $W = 2.4$.

6 y is inversely proportional to the cube of x. If $y = 4$ when $x = 2$, find:

a y when $x = 1$ **b** x when $y = \frac{1}{2}$.

7 The grant available to a section of society was inversely proportional to the number of people needing the grant. When 30 people needed a grant, they received £60 each.

a What would the grant have been if 120 people had needed one?

b If the grant had been £50 each, how many people would have received it?

8 While doing underwater tests in one part of an ocean, a team of scientists noticed that the temperature, in °C, was inversely proportional to the depth, in kilometres. When the temperature was 6 °C, the scientists were at a depth of 4 km.

a What would the temperature have been at a depth of 8 km?

b To what depth would they have had to go to find the temperature at 2 °C?

9 A new engine was being tested, but it had serious problems. The distance it went, in kilometres, without breaking down was inversely proportional to the square of its speed in metres per second (m/s). When the speed was 12 m/s, the engine lasted 3 km.

a Find the distance covered before a breakdown, when the speed is 15 m/s.

b On one test, the engine broke down after 6.75 km. What was the speed?

10 In a balloon it was noticed that the pressure, in atmospheres (atm), was inversely proportional to the square root of the height, in metres. When the balloon was at a height of 25 m, the pressure was 1.44 atm.

a What was the pressure at a height of 9 m?

b What would the height have been if the pressure was 0.72 atm?

FM 11 The amount of waste which a firm produces, measured in tonnes per hour, is inversely proportional to the square root of the area of the filter beds, in square metres (m^2). The firm produces 1.25 tonnes of waste per hour, with filter beds of size 0.16 m^2.

a The filter beds used to be only 0.01 m^2. How much waste did the firm produce then?

b How much waste could be produced if the filter beds were 0.75 m^2?

PS 12 Which statement is represented by the graph?
Give a reason for your answer.

A $y \propto x$ B $y \propto \frac{1}{x}$ C $y \sqrt{x}$

y

x

AU 13 In the table, y is inversely proportional to the cube root of x.

Complete the table, leaving your answers as fractions.

x	8	27	
y	1		$\frac{1}{2}$

FM 14 The fuel consumption, in miles per gallon (mpg) of a car is inversely proportional to its speed, in miles per hour (mph). When the car is travelling at 30 mph the fuel consumption is 60 mpg.

How much further would the car travel on 1 gallon of fuel by travelling at 60 mph instead of 70 mph on a motorway?

GRADE BOOSTER

A You can find formulae describing direct or inverse variation and use them to solve problems

What you should know now

- How to recognise direct and inverse variation
- What a constant of proportionality is, and how to find it
- How to find formulae describing inverse or direct variation
- How to solve problems involving direct or inverse variation

1 y is proportional to $\sqrt{x}$. Complete the table.

x	25		400
y	10	20	

2 The energy, E, of an object moving horizontally is directly proportional to the speed, v, of the object. When the speed is 10 m/s the energy is 40 000 Joules.

a Find an equation connecting E and v.

b Find the speed of the object when the energy is 14 400 Joules.

3 y is inversely proportional to the cube root of x. When $y = 8$, $x = \frac{1}{8}$.

a Find an expression for y in terms of x,

b Calculate

i the value of y when $x = \frac{1}{125}$

ii the value of x when $y = 2$.

4 The mass of a cube is directly proportional to the cube of its side. A cube with a side of 4 cm has a mass of 320 grams. Calculate the side length of a cube made of the same material with a mass of 36 450 grams.

5 y is directly proportional to the cube of x. When $y = 16$, $x = 3$. Find the value of y when $x = 6$.

6 d is directly proportional to the square of t. $d = 80$ when $t = 4$.

a Express d in terms of t.

b Work out the value of d when $t = 7$.

c Work out the positive value of t when $d = 45$. *(? marks)*

Edexcel, June 2005, Paper 5 Higher, Question 16

7 M is directly proportional to L^3.

When $L = 2$, $M = 160$

Find the value of M when $L = 3$ *(4 marks)*

Edexcel, May 2009, Paper 3, Question 21

8 Two variables, x and y, are known to be proportional to each other. When $x = 10$, $y = 25$.

Find the constant of proportionality, k, if:

a $y \propto x$

b $y \propto x^2$

c $y \propto \frac{1}{x}$

d $\sqrt{y} \propto \frac{1}{x}$

9 y is directly proportional to the cube root of x. When $x = 27$, $y = 6$.

a Find the value of y when $x = 125$.

b Find the value of x when $y = 3$.

10 The surface area, A, of a solid is directly proportional to the square of the depth, d. When $d = 6$, $A = 12\pi$.

a Find the value of A when $d = 12$. Give your answer in terms of π.

b Find the value of d when $A = 27\pi$.

11 Q is inversely proportional to the square of t.

When $t = 4$, $q = 8.5$

a Find a formula for q in terms of t. *(3 marks)*

b Calculate the value of q when $t = 5$ *(1 mark)*

Edexcel, June 2008, Paper 4, Question 20

12 D is proportional to S^2.

$D = 900$ when $S = 20$

Calculate the value of D when $S = 25$ *(4 marks)*

Edexcel, November 2008, Paper 4, Question 22

13 The frequency, f, of sound is inversely proportional to the wavelength, w. A sound with a frequency of 36 hertz has a wavelength of 20.25 metres.

Calculate the frequency when the frequency and the wavelength have the same numerical value.

14 t is proportional to m^3.

a When $m = 6$, $t = 324$. Find the value of t when $m = 10$.

Also, m is inversely proportional to the square root of w.

b When $t = 12$, $w = 25$. Find the value of w when $m = 4$.

15 P and Q are positive quantities. P is inversely proportional to Q^2. When $P = 160$, $Q = 20$. Find the value of P when $P = Q$.

A* A

Worked Examination Questions

1 y is inversely proportional to the square of x. When y is 40, $x = 5$.

a Find an equation connecting x and y.

b Find the value of y when $x = 10$.

1 a $y \propto \frac{1}{x^2}$

$y = \frac{k}{x^2}$

First set up the proportionality relationship and replace the proportionality sign with $= k$.

This gets 1 method mark for stating first or second line or both.

$40 = \frac{k}{25}$

Substitute the given values of y and x into the proportionality equation to find the value of k.

This gets 1 accuracy mark for finding k.

$\Rightarrow k = 40 \times 25 = 1000$

$y = \frac{1000}{x^2}$

Substitute the value of k to get the final equation connecting y and x.

This gets 1 mark for accuracy.

or $yx^2 = 1000$

b When $x = 10$, $y = \frac{1000}{10^2} = \frac{1000}{100} = 10$

Substitute the value of x into the equation to find y.

This gets 1 method mark for substitution of $x = 10$ and 1 accuracy mark for correct answer.

Total: 5 marks

Worked Examination Questions

PS **2** The mass of a solid, M, is directly proportional to the cube of its height, h. When $h = 10$, $M = 4000$.

The surface area, A, of the solid is directly proportional to the square of the height, h. When $h = 10$, $A = 50$.

Find A, when $M = 32\,000$.

2 $M \propto h^3$ — First set up the proportionality statement. This gets 1 method mark for writing either the proportionality statement or the proportionality equation.

$M = kh^3$

$4000 = k \times 1000 \Rightarrow k = 4$ — First, find the relationship between M and h using the given information. This gets 1 accuracy mark for obtaining the correct value of k.

So, $M = 4h^3$ — This gets 1 accuracy mark for writing out the equation with the value of k substituted.

$A = ph^2$ — Be careful when using the second equation to use a different letter for the constant of proportionality to avoid confusion. As setting up the second equation is the same technique as in the first part of the question the marks for method are only awarded in one part.

$50 = p \times 100 \Rightarrow p = \frac{1}{2}$ — Next, find the relationship between A and h using the given information, $h = 10$ and $A = 50$

So, $A = \frac{1}{2}h^2$ — This gets 1 mark for accuracy.

$32\,000 = 4h^3$

$h^3 = 8000 \Rightarrow h = 20$ — Find the value of h when $M = 32\,000$.

$A = \frac{1}{2}(20)^2 = \frac{400}{2} = 200$ — Now find the value of A for that value of h. This gets 1 mark for accuracy.

Total: 5 marks

16 Functional Maths
Voting in the European Union

The Council of the European Union is the main decision-making body for Europe. One minister from each of the EU's national governments attends Council meetings and decisions are taken by voting. The bigger the country's population, the more votes it has, but numbers are currently weighted in favour of the less populous countries.

Getting started

In June 2007, Poland argued for a change to the rules for the voting in the Council of the European Union. The Polish suggested that each country's voting strength should be directly proportional to the square root of its population. This idea is known as **Pensore's rule**.

Let V be the voting strength (that is the number of votes a country gets) and P be the country's population.

- Write down a mathematical statement for Pensore's rule, using the symbol of variation, $\propto$.
- Write a proportionality equation for Pensore's rule, using a constant of proportionality, k.

Country	Population	Current number of votes in the Council of the European Union
UK	61 600 835	29
Poland	38 125 478	27
Romania	21 398 181	14
The Netherlands	16 518 199	13
Belgium	10 574 595	12
Sweden	9 290 113	10
Ireland	4 434 925	7
Luxembourg	472 569	4
Malta	408 009	3

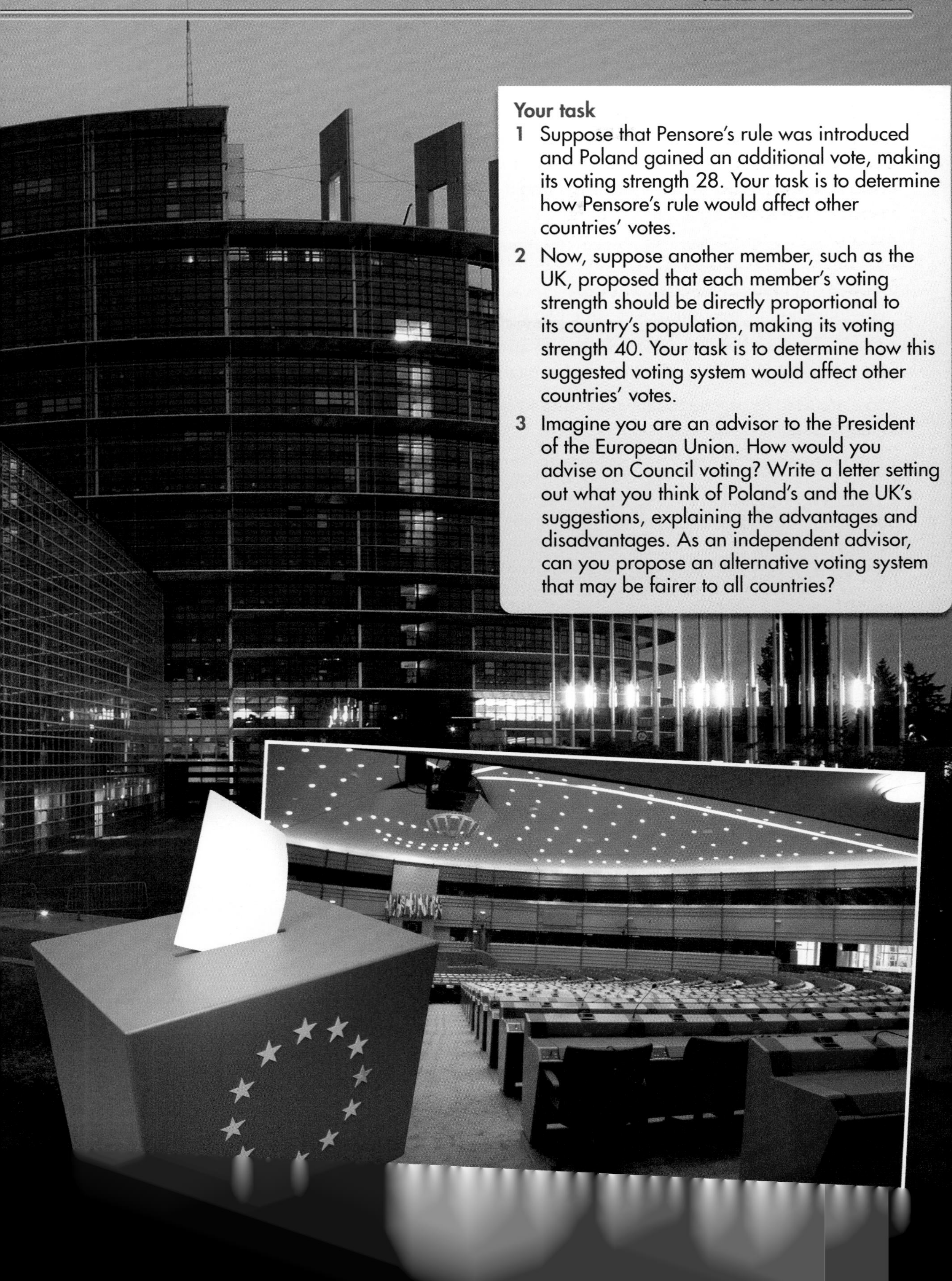

Your task

1. Suppose that Pensore's rule was introduced and Poland gained an additional vote, making its voting strength 28. Your task is to determine how Pensore's rule would affect other countries' votes.
2. Now, suppose another member, such as the UK, proposed that each member's voting strength should be directly proportional to its country's population, making its voting strength 40. Your task is to determine how this suggested voting system would affect other countries' votes.
3. Imagine you are an advisor to the President of the European Union. How would you advise on Council voting? Write a letter setting out what you think of Poland's and the UK's suggestions, explaining the advantages and disadvantages. As an independent advisor, can you propose an alternative voting system that may be fairer to all countries?

Why this chapter matters

Vectors are used to represent any quantity that has both magnitude and direction. The velocity of a speeding car may be described in terms of its direction and its speed. The speed is the magnitude, but when it has direction it becomes the velocity – a vector.

To understand how a force acts on an object, you need to know the magnitude of the force and the direction in which it moves – the two bits of information that define a vector.

And when you watch the weather report, you are told which way the wind will blow tomorrow, and how strongly – again, a direction and a magnitude together making up a vector.

Vectors are used to describe many quantities in science, such as displacement, acceleration and momentum.

But are vectors used in real life? Yes! Here are some examples.

In the 1950s, a group of talented Brazilian footballers invented the **swerving free kick**. By kicking the ball in just the right place, they managed to make it curl around the wall of defending players and, quite often, go straight into the back of the net. When a ball is in flight, it is acted upon by various forces. Some of these depend on the way the ball is spinning. The forces at work here can be described by vectors.

Formula One teams always employ physicists and mathematicians to help them build the perfect racing car. **Aerodynamics** is the study of how the air moves. Since vectors describe movements and forces, they are used as the basis of a car's design.

Pilots have to consider wind speed and direction when they plan to land an aircraft at an airport. Vectors are an integral part of the computerised landing system.

The science of aerodynamics is used in the design of aircraft; vectors play a key role in the design of wings, where an upward force or lift is needed to enable the aircraft to fly.

Meteorologists or weather forecasters use vectors to map out weather patterns. Wind speeds can be represented by vectors of different lengths to indicate the intensity of the wind.

Vectors are used extensively in computer graphics. Software designed to give the viewer the impression that an object or person is moving around a scene makes extensive use of the mathematics of vectors.

Chapter 17

Geometry: Vectors

The grades given in this chapter are target grades.

This chapter will show you ...

- **A** the properties of vectors
- **A** how to add and subtract vectors
- **A*** how to use vectors to solve geometrical problems
- **A*** how to prove geometric results using rigorous and logical mathematical arguments

Visual overview

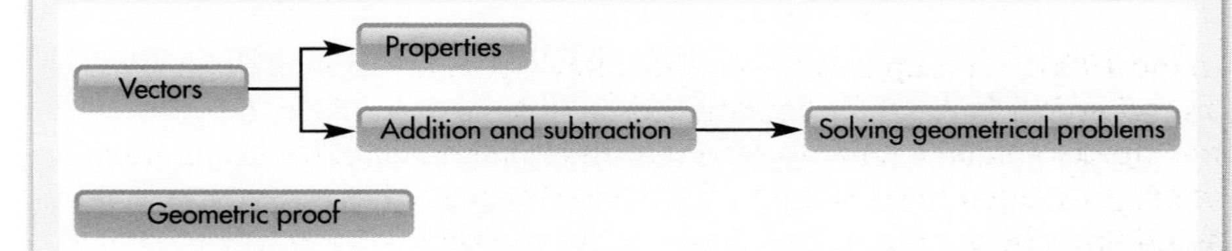

What you should already know

- Vectors are used to describe translations **(KS3 level 7, GCSE grade C)**

Quick check

Use column vectors to describe these translations.

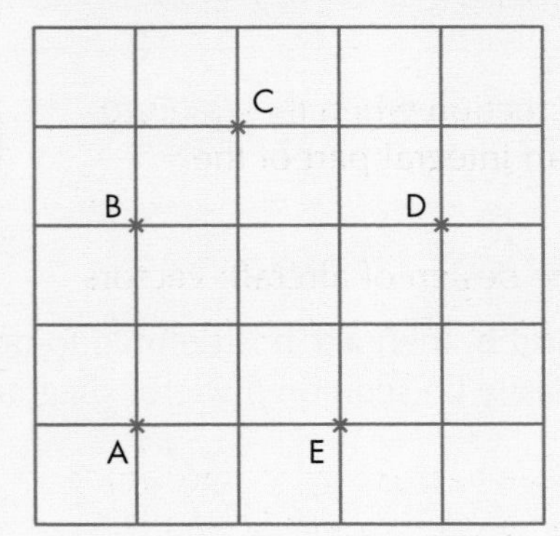

a A to C
b B to D
c C to D
d D to E

17.1 Properties of vectors

This section will show you how to:
- add and subtract vectors

Key words
direction
magnitude
vector

A **vector** is a quantity which has both **magnitude** and **direction**. It can be represented by a straight line which is drawn in the direction of the vector and whose length represents the magnitude of the vector. Usually, the line includes an arrowhead.

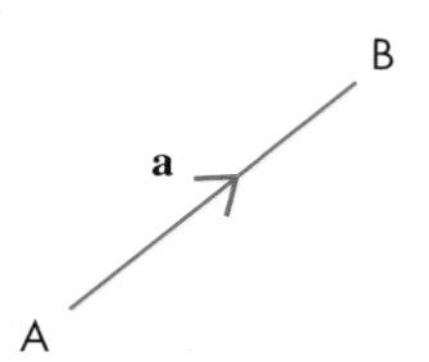

The translation or movement from A to B is represented by the vector **a**.

a is always printed in bold type, but is written as $\underline{a}$.

a can also be written as $\overrightarrow{AB}$.

A quantity which is completely described by its magnitude, and has no direction associated with it, is called a scalar. The mass of a bus (10 tonnes) is an example of a scalar. Another example is a linear measure, such as 25.4 mm.

Multiplying a vector by a number (scalar) alters its magnitude (length) but not its direction. For example, the vector 2**a** is twice as long as the vector **a**, but in the same direction.

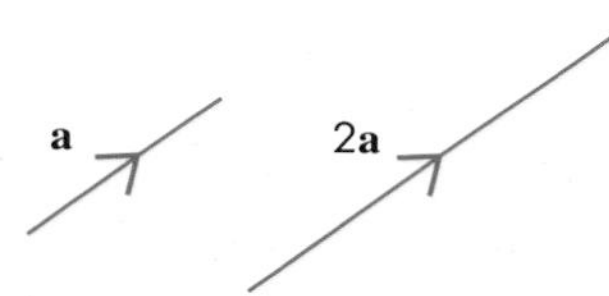

A negative vector, for example **–b**, has the same magnitude as the vector **b**, but is in the opposite direction.

Addition and subtraction of vectors

Take two non-parallel vectors **a** and **b**, then **a** + **b** is defined to be the translation of **a** followed by the translation of **b**. This can easily be seen on a vector diagram.

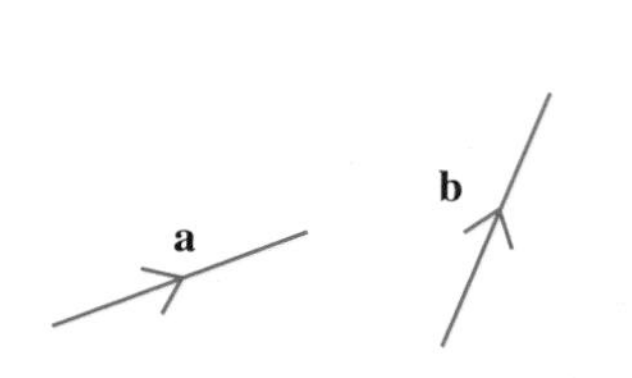

a + b
b
a

Similarly, $\mathbf{a} - \mathbf{b}$ is defined to be the translation of $\mathbf{a}$ followed by the translation of $-\mathbf{b}$.

Look at the parallelogram grid below. **a** and **b** are two independent vectors that form the basis of this grid. It is possible to define the position, with reference to O, of any point on this grid by a vector expressed in terms of **a** and **b**. Such a vector is called a position vector.

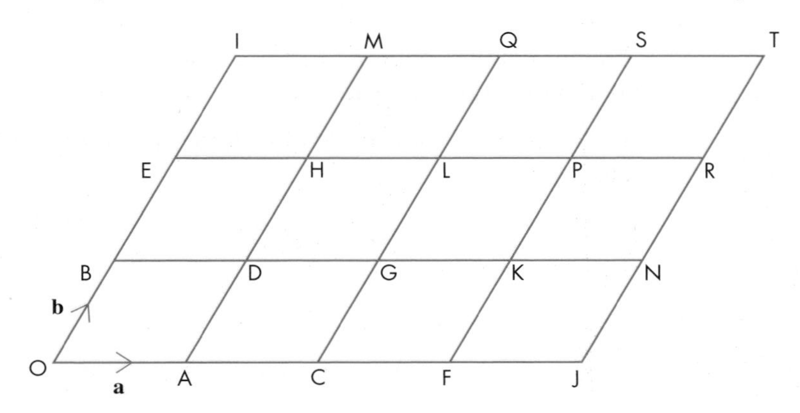

For example, the position vector of K is $\overrightarrow{OK}$ or $\mathbf{k} = 3\mathbf{a} + \mathbf{b}$, the position vector of E is $\overrightarrow{OE}$ or $\mathbf{e} = 2\mathbf{b}$. The vector $\overrightarrow{HT} = 3\mathbf{a} + \mathbf{b}$, the vector $\overrightarrow{PN} = \mathbf{a} - \mathbf{b}$, the vector $\overrightarrow{MK} = 2\mathbf{a} - 2\mathbf{b}$, and the vector $\overrightarrow{TP} = -\mathbf{a} - \mathbf{b}$.

Note $\overrightarrow{OK}$ and $\overrightarrow{HT}$ are called equal vectors because they have exactly the same length and are in the same direction. $\overrightarrow{MK}$ and $\overrightarrow{PN}$ are parallel vectors but $\overrightarrow{MK}$ is twice the magnitude of $\overrightarrow{PN}$.

EXAMPLE 1

a Using the grid above, write down the following vectors in terms of **a** and **b**.

i $\overrightarrow{BH}$ **ii** $\overrightarrow{HP}$ **iii** $\overrightarrow{GT}$

iv $\overrightarrow{TI}$ **v** $\overrightarrow{FH}$ **vi** $\overrightarrow{BQ}$

b What is the relationship between the following vectors?

i $\overrightarrow{BH}$ and $\overrightarrow{GT}$ **ii** $\overrightarrow{BQ}$ and $\overrightarrow{GT}$ **iii** $\overrightarrow{HP}$ and $\overrightarrow{TI}$

c Show that B, H and Q lie on the same straight line.

a **i** $\mathbf{a} + \mathbf{b}$ **ii** $2\mathbf{a}$ **iii** $2\mathbf{a} + 2\mathbf{b}$ **iv** $-4\mathbf{a}$ **v** $-2\mathbf{a} + 2\mathbf{b}$ **vi** $2\mathbf{a} + 2\mathbf{b}$

b **i** $\overrightarrow{BH}$ and $\overrightarrow{GT}$ are parallel and $\overrightarrow{GT}$ is twice the length of $\overrightarrow{BH}$.

ii $\overrightarrow{BQ}$ and $\overrightarrow{GT}$ are equal.

iii $\overrightarrow{HP}$ and $\overrightarrow{TI}$ are in opposite directions and $\overrightarrow{TI}$ is twice the length of $\overrightarrow{HP}$.

c $\overrightarrow{BH}$ and $\overrightarrow{BQ}$ are parallel and start at the same point B. Therefore, B, H and Q must lie on the same straight line.

EXAMPLE 2

Use a vector diagram to show that $\mathbf{a} + \mathbf{b} = \mathbf{b} + \mathbf{a}$.

Take two independent vectors **a** and **b**:

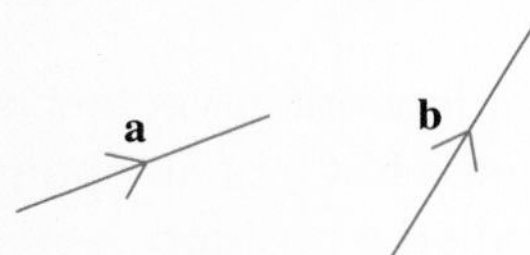

$\mathbf{a} + \mathbf{b}$ and $\mathbf{b} + \mathbf{a}$ have the same magnitude and direction and are therefore equal.

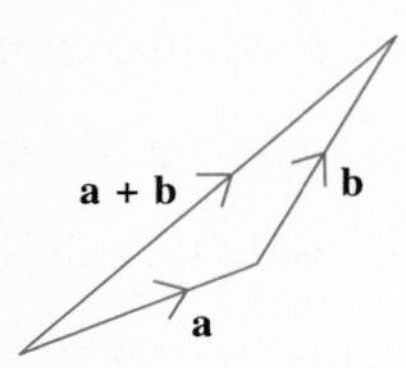

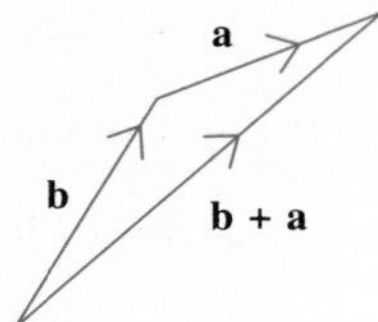

EXERCISE 17A

1 On this grid, $\overrightarrow{OA}$ is **a** and $\overrightarrow{OB}$ is **b**.

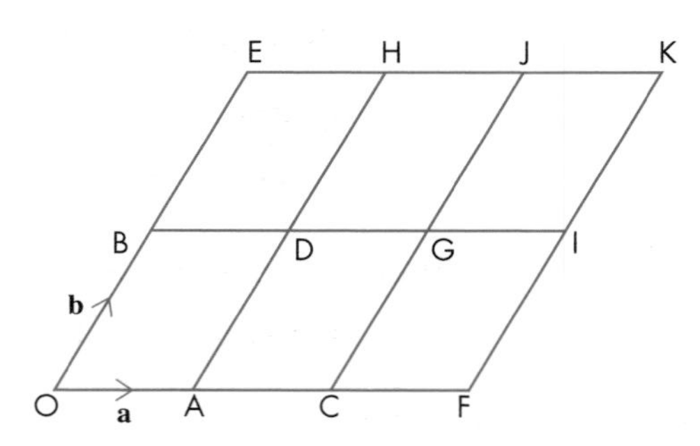

a Name three other vectors equivalent to **a**.

b Name three other vectors equivalent to **b**.

c Name three vectors equivalent to –**a**.

d Name three vectors equivalent to –**b**.

2 Using the same grid as in question **1**, give the following vectors in terms of **a** and **b**.

a $\overrightarrow{OC}$ **b** $\overrightarrow{OE}$ **c** $\overrightarrow{OD}$ **d** $\overrightarrow{OG}$ **e** $\overrightarrow{OJ}$

f $\overrightarrow{OH}$ **g** $\overrightarrow{AG}$ **h** $\overrightarrow{AK}$ **i** $\overrightarrow{BK}$ **j** $\overrightarrow{DI}$

k $\overrightarrow{GJ}$ **l** $\overrightarrow{DK}$

3 **a** What do the answers to parts **2c** and **2g** tell you about the vectors $\overrightarrow{OD}$ and $\overrightarrow{AG}$?

b On the grid in question **1**, there are three vectors equivalent to $\overrightarrow{OG}$. Name all three.

4 **a** What do the answers to parts **2c** and **2e** tell you about vectors $\overrightarrow{OD}$ and $\overrightarrow{OJ}$?

b On the grid in question **1**, there is one other vector that is twice the size of $\overrightarrow{OD}$. Which is it?

c On the grid in question **1**, there are three vectors that are three times the size of $\overrightarrow{OA}$. Name all three.

5 On a copy of this grid, mark on the points C to P to show the following.

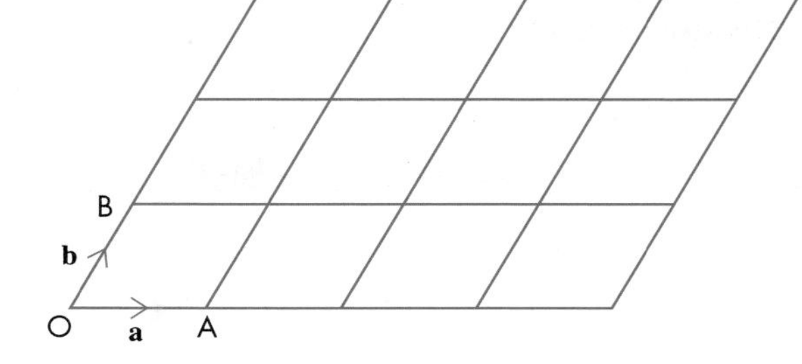

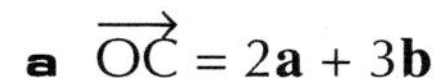

a $\overrightarrow{OC} = 2\mathbf{a} + 3\mathbf{b}$ **b** $\overrightarrow{OD} = 2\mathbf{a} + \mathbf{b}$

c $\overrightarrow{OE} = \mathbf{a} + 2\mathbf{b}$ **d** $\overrightarrow{OF} = 3\mathbf{b}$

e $\overrightarrow{OG} = 4\mathbf{a}$ **f** $\overrightarrow{OH} = 4\mathbf{a} + 2\mathbf{b}$

g $\overrightarrow{OI} = 3\mathbf{a} + 3\mathbf{b}$ **h** $\overrightarrow{OJ} = \mathbf{a} + \mathbf{b}$

i $\overrightarrow{OK} = 2\mathbf{a} + 2\mathbf{b}$ **j** $\overrightarrow{OM} = 2\mathbf{a} + \frac{3}{2}\mathbf{b}$ **k** $\overrightarrow{ON} = \frac{1}{2}\mathbf{a} + 2\mathbf{b}$ **l** $\overrightarrow{OP} = \frac{5}{2}\mathbf{a} + \frac{3}{2}\mathbf{b}$

6 **a** Look at the diagram in question **5**. What can you say about the points O, J, K and I?

b How could you tell this by looking at the vectors for parts **5g**, **5h** and **5i**?

c There is another point on the same straight line as O and D. Which is it?

d Copy and complete these statements and then mark the appropriate points on the diagram you drew for question **5**.

i The point Q is on the straight line ODH. The vector $\overrightarrow{OQ}$ is given by:

$\overrightarrow{OQ} = \mathbf{a} + \ldots\ldots\ \mathbf{b}$

ii The point R is on the straight line ODH. The vector $\overrightarrow{OR}$ is given by:

$\overrightarrow{OR} = 3\mathbf{a} + \ldots\ldots\ \mathbf{b}$

e Copy and complete the following statement.

Any point on the line ODH has a vector $n\mathbf{a} + \ldots\ldots\ \mathbf{b}$, where n is any number.

7 On this grid, $\overrightarrow{OA}$ is **a** and $\overrightarrow{OB}$ is **b**.

Give the following vectors in terms of **a** and **b**.

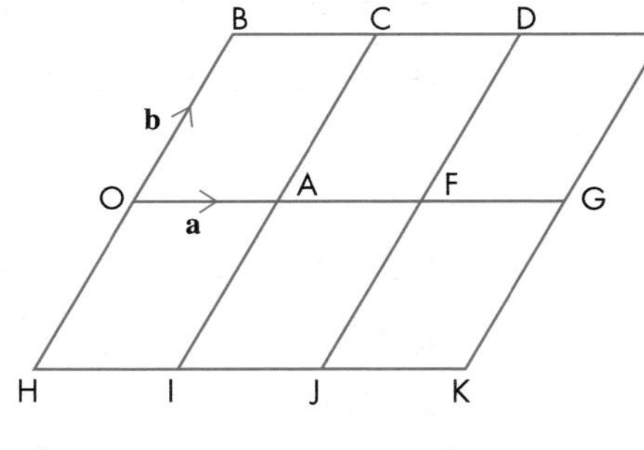

a $\overrightarrow{OH}$ **b** $\overrightarrow{OK}$

c $\overrightarrow{OJ}$ **d** $\overrightarrow{OI}$

e $\overrightarrow{OC}$ **f** $\overrightarrow{CO}$

g $\overrightarrow{AK}$ **h** $\overrightarrow{DI}$

i $\overrightarrow{JE}$ **j** $\overrightarrow{AB}$ **k** $\overrightarrow{CK}$ **l** $\overrightarrow{DK}$

8 **a** What do the answers to parts **7e** and **7f** tell you about the vectors $\overrightarrow{OC}$ and $\overrightarrow{CO}$?

b On the grid in question **7**, there are five other vectors opposite to $\overrightarrow{OC}$. Name at least three.

9 **a** What do the answers to parts **7j** and **7k** tell you about vectors $\overrightarrow{AB}$ and $\overrightarrow{CK}$?

b On the grid in question **7**, there are two vectors that are twice the size of $\overrightarrow{AB}$ and in the opposite direction. Name both of them.

c On the grid in question **7**, there are three vectors that are three times the size of $\overrightarrow{OA}$ and in the opposite direction. Name all three.

A

10 On a copy of this grid, mark on the points C to P to show the following.

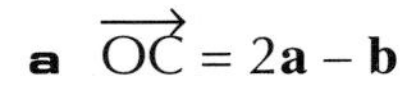

a $\overrightarrow{OC} = 2\mathbf{a} - \mathbf{b}$ **b** $\overrightarrow{OD} = 2\mathbf{a} + \mathbf{b}$

c $\overrightarrow{OE} = \mathbf{a} - 2\mathbf{b}$ **d** $\overrightarrow{OF} = \mathbf{b} - 2\mathbf{a}$

e $\overrightarrow{OG} = -\mathbf{a}$ **f** $\overrightarrow{OH} = -\mathbf{a} - 2\mathbf{b}$

g $\overrightarrow{OI} = 2\mathbf{a} - 2\mathbf{b}$ **h** $\overrightarrow{OJ} = -\mathbf{a} + \mathbf{b}$

i $\overrightarrow{OK} = -\mathbf{a} - \mathbf{b}$ **j** $\overrightarrow{OM} = -\mathbf{a} - \frac{3}{2}\mathbf{b}$ **k** $\overrightarrow{ON} = -\frac{1}{2}\mathbf{a} - 2\mathbf{b}$ **l** $\overrightarrow{OP} = \frac{3}{2}\mathbf{a} - \frac{3}{2}\mathbf{b}$

AU 11 The diagram shows two sets of parallel lines.

$\overrightarrow{OA} = \mathbf{a}$ and $\overrightarrow{OB} = \mathbf{b}$

$\overrightarrow{OC} = 3\overrightarrow{OA}$ and $\overrightarrow{OD} = 2\overrightarrow{OB}$

a Write down the following vectors in terms of **a** and **b**.

i $\overrightarrow{OF}$ **ii** $\overrightarrow{OG}$ **iii** $\overrightarrow{EG}$ **iv** $\overrightarrow{CE}$

b Write down two vectors that can be written as $3\mathbf{a} - \mathbf{b}$.

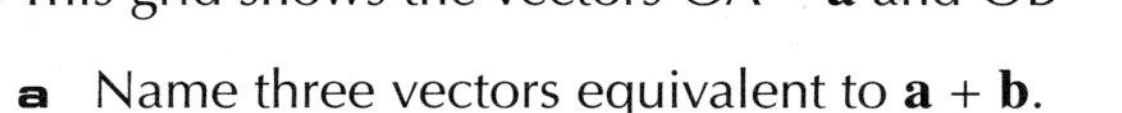

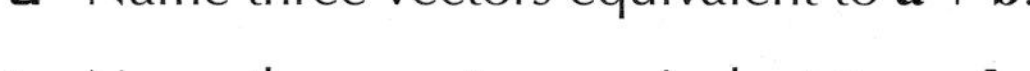

12 This grid shows the vectors $\overrightarrow{OA} = \mathbf{a}$ and $\overrightarrow{OB} = \mathbf{b}$.

a Name three vectors equivalent to $\mathbf{a} + \mathbf{b}$.

b Name three vectors equivalent to $\mathbf{a} - \mathbf{b}$.

c Name three vectors equivalent to $\mathbf{b} - \mathbf{a}$.

d Name three vectors equivalent to $-\mathbf{a} - \mathbf{b}$.

e Name three vectors equivalent to $2\mathbf{a} - \mathbf{b}$.

f Name three vectors equivalent to $2\mathbf{b} - \mathbf{a}$.

g For each of these, name one equivalent vector.

i $3\mathbf{a} - \mathbf{b}$ **ii** $2(\mathbf{a} + \mathbf{b})$ **iii** $3\mathbf{a} - 2\mathbf{b}$

iv $3(\mathbf{a} - \mathbf{b})$ **v** $3(\mathbf{b} - \mathbf{a})$ **vi** $3(\mathbf{a} + \mathbf{b})$

vii $-3(\mathbf{a} + \mathbf{b})$ **viii** $2\mathbf{a} + \mathbf{b} - 3\mathbf{a} - 2\mathbf{b}$ **ix** $2(2\mathbf{a} - \mathbf{b}) - 3(\mathbf{a} - \mathbf{b})$

13 The points P, Q and R lie on a straight line. The vector $\overrightarrow{PQ}$ is $2\mathbf{a} + \mathbf{b}$, where **a** and **b** are vectors. Which of the following vectors could be the vector $\overrightarrow{PR}$ and which could not be the vector $\overrightarrow{PR}$ (two of each).

a $2\mathbf{a} + 2\mathbf{b}$ **b** $4\mathbf{a} + 2\mathbf{b}$ **c** $2\mathbf{a} - \mathbf{b}$ **d** $-6\mathbf{a} - 3\mathbf{b}$

14 The points P, Q and R lie on a straight line. The vector $\overrightarrow{PQ}$ is $3\mathbf{a} - \mathbf{b}$, where **a** and **b** are vectors.

a Write down any other vector that could represent $\overrightarrow{PR}$.

b How can you tell from the vector $\overrightarrow{PS}$ that S lies on the same straight line as P, Q and R?

15 Use a vector diagram to prove that $\mathbf{a} + (\mathbf{b} + \mathbf{c}) = (\mathbf{a} + \mathbf{b}) + \mathbf{c}$.

 OABC is a quadrilateral.

P, Q, R and S are the midpoints of OA, AB, BC and OC respectively.

$\overrightarrow{OA} = 2\mathbf{a}$, $\overrightarrow{OB} = 2\mathbf{b}$ and $\overrightarrow{OC} = 2\mathbf{c}$

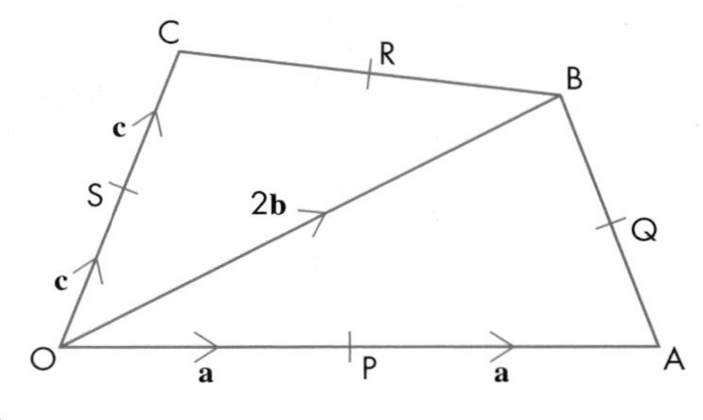

a Find the following vectors in terms of **a**, **b** and **c**.

Give your answers in their simplest form.

i $\overrightarrow{AB}$ **ii** $\overrightarrow{SP}$ **iii** $\overrightarrow{BC}$ **iv** $\overrightarrow{PR}$

b Use vectors to prove that PQRS is a parallelogram.

17.2 Vectors in geometry

This section will show you how to:
- use vectors to solve geometrical problems

Key words
vector

Vectors can be used to prove many results in geometry, as the following examples show.

EXAMPLE 3

In the diagram, $\overrightarrow{OA} = \mathbf{a}$, $\overrightarrow{OB} = \mathbf{b}$, and $\overrightarrow{BC} = 1.5\mathbf{a}$. M is the midpoint of BC, N is the midpoint of AC and P is the midpoint of OB.

a Find these vectors in terms of **a** and **b**.

i $\overrightarrow{AC}$ **ii** $\overrightarrow{OM}$ **iii** $\overrightarrow{BN}$

b Prove that $\overrightarrow{PN}$ is parallel to $\overrightarrow{OA}$.

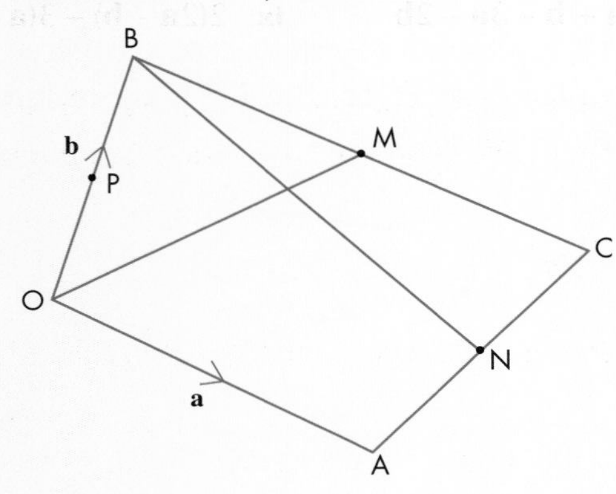

EXAMPLE 3 (continued)

a **i** You have to get from A to C in terms of vectors that you know.

$$\overrightarrow{AC} = \overrightarrow{AO} + \overrightarrow{OB} + \overrightarrow{BC}$$

Now $\overrightarrow{AO} = -\overrightarrow{OA}$, so you can write,

$$\overrightarrow{AC} = -\mathbf{a} + \mathbf{b} + \tfrac{3}{2}\mathbf{a}$$
$$= \tfrac{1}{2}\mathbf{a} + \mathbf{b}$$

Note that the letters 'connect up' as we go from A to C, and that the negative of a vector represented by any pair of letters is formed by reversing the letters.

ii In the same way:

$$\overrightarrow{OM} = \overrightarrow{OB} + \overrightarrow{BM} = \overrightarrow{OB} + \tfrac{1}{2}\overrightarrow{BC}$$
$$= \mathbf{b} + \tfrac{1}{2}(\tfrac{3}{2}\mathbf{a})$$
$$\overrightarrow{OM} = \tfrac{3}{4}\mathbf{a} + \mathbf{b}$$

iii $\overrightarrow{BN} = \overrightarrow{BC} + \overrightarrow{CN} = \overrightarrow{BC} - \tfrac{1}{2}\overrightarrow{AC}$

$$= \tfrac{3}{2}\mathbf{a} - \tfrac{1}{2}(\tfrac{1}{2}\mathbf{a} + \mathbf{b})$$
$$= \tfrac{3}{2}\mathbf{a} - \tfrac{1}{4}\mathbf{a} - \tfrac{1}{2}\mathbf{b}$$
$$= \tfrac{5}{4}\mathbf{a} - \tfrac{1}{2}\mathbf{b}$$

Note that if you did this as $\overrightarrow{BN} = \overrightarrow{BO} + \overrightarrow{OA} + \overrightarrow{AN}$, you would get the same result.

b $\overrightarrow{PN} = \overrightarrow{PO} + \overrightarrow{OA} + \overrightarrow{AN}$

$$= \tfrac{1}{2}(-\mathbf{b}) + \mathbf{a} + \tfrac{1}{2}(\tfrac{1}{2}\mathbf{a} + \mathbf{b})$$
$$= -\tfrac{1}{2}\mathbf{b} + \mathbf{a} + \tfrac{1}{4}\mathbf{a} + \tfrac{1}{2}\mathbf{b}$$
$$= \tfrac{5}{4}\mathbf{a}$$

$\overrightarrow{PN}$ is a multiple of $\mathbf{a}$ only, so must be parallel to $\overrightarrow{OA}$.

EXAMPLE 4

OACB is a parallelogram. $\overrightarrow{OA}$ is represented by the vector $\mathbf{a}$. $\overrightarrow{OB}$ is represented by the vector $\mathbf{b}$. P is a point $\tfrac{2}{3}$ the distance from O to C, and M is the midpoint of AC. Show that B, P and M lie on the same straight line.

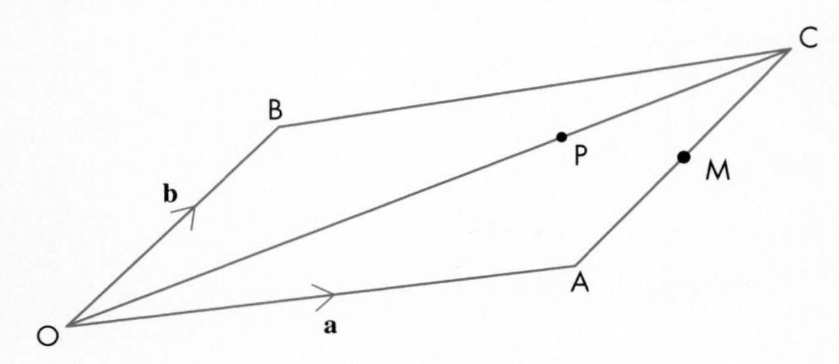

$\overrightarrow{OC} = \overrightarrow{OA} + \overrightarrow{AC} = \mathbf{a} + \mathbf{b}$

$\overrightarrow{OP} = \frac{2}{3}\overrightarrow{OC} = \frac{2}{3}\mathbf{a} + \frac{2}{3}\mathbf{b}$

$\overrightarrow{OM} = \overrightarrow{OA} + \overrightarrow{AM} = \overrightarrow{OA} + \frac{1}{2}\overrightarrow{AC} = \mathbf{a} + \frac{1}{2}\mathbf{b}$

$\overrightarrow{BP} = \overrightarrow{BO} + \overrightarrow{OP} = -\mathbf{b} + \frac{2}{3}\mathbf{a} + \frac{2}{3}\mathbf{b} = \frac{2}{3}\mathbf{a} - \frac{1}{3}\mathbf{b} = \frac{1}{3}(2\mathbf{a} - \mathbf{b})$

$\overrightarrow{BM} = \overrightarrow{BO} + \overrightarrow{OM} = -\mathbf{b} + \mathbf{a} + \frac{1}{2}\mathbf{b} = \mathbf{a} - \frac{1}{2}\mathbf{b} = \frac{1}{2}(2\mathbf{a} - \mathbf{b})$

Therefore, $\overrightarrow{BM}$ is a multiple of $\overrightarrow{BP}$ ($\overrightarrow{BM} = \frac{3}{2}\overrightarrow{BP}$).

Therefore, $\overrightarrow{BP}$ and $\overrightarrow{BM}$ are parallel and as they have a common point, B, they must lie on the same straight line.

EXERCISE 17B

1 The diagram shows the vectors $\overrightarrow{OA} = \mathbf{a}$ and $\overrightarrow{OB} = \mathbf{b}$. M is the midpoint of AB.

a **i** Work out the vector $\overrightarrow{AB}$.

ii Work out the vector $\overrightarrow{AM}$.

iii Explain why $\overrightarrow{OM} = \overrightarrow{OA} + \overrightarrow{AM}$.

iv Using your answers to parts **ii** and **iii**, work out $\overrightarrow{OM}$ in terms of **a** and **b**.

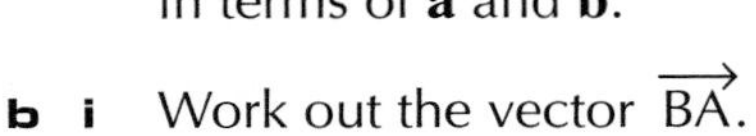

b **i** Work out the vector $\overrightarrow{BA}$.

ii Work out the vector $\overrightarrow{BM}$.

iii Explain why $\overrightarrow{OM} = \overrightarrow{OB} + \overrightarrow{BM}$.

iv Using your answers to parts **ii** and **iii**, work out $\overrightarrow{OM}$ in terms of **a** and **b**.

c Copy the diagram and show on it the vector $\overrightarrow{OC}$ which is equal to **a** + **b**.

d Describe in geometrical terms the position of M in relation to O, A, B and C.

2 The diagram shows the vectors $\overrightarrow{OA} = \mathbf{a}$ and $\overrightarrow{OC} = -\mathbf{b}$. N is the midpoint of AC.

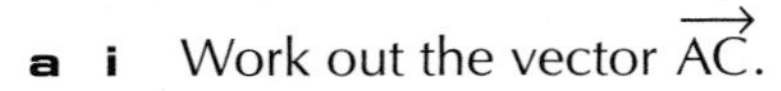

a **i** Work out the vector $\overrightarrow{AC}$.

ii Work out the vector $\overrightarrow{AN}$.

iii Explain why $\overrightarrow{ON} = \overrightarrow{OA} + \overrightarrow{AN}$.

iv Using your answers to parts **ii** and **iii**, work out $\overrightarrow{ON}$ in terms of **a** and **b**.

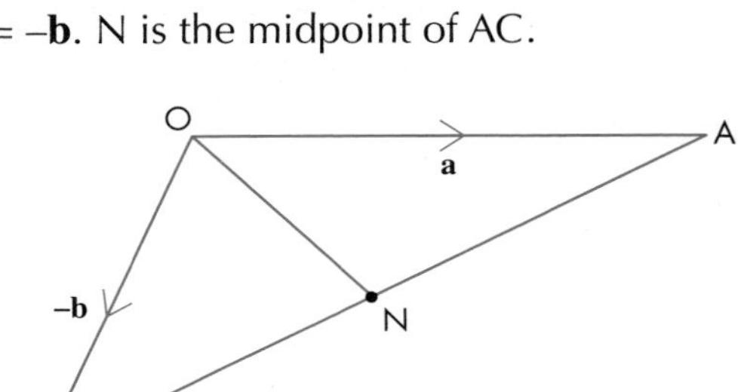

b **i** Work out the vector $\overrightarrow{CA}$.

ii Work out the vector $\overrightarrow{CN}$.

iii Explain why $\overrightarrow{ON} = \overrightarrow{OC} + \overrightarrow{CN}$.

iv Using your answers to parts **ii** and **iii**, work out $\overrightarrow{ON}$ in terms of **a** and **b**.

c Copy the diagram above and show on it the vector $\overrightarrow{OD}$ which is equal to **a** – **b**.

d Describe in geometrical terms the position of N in relation to O, A, C and D.

3 The diagram shows the vectors $\overrightarrow{OA} = \mathbf{a}$ and $\overrightarrow{OB} = \mathbf{b}$.
The point C divides the line AB in the ratio 1:2.

HINTS AND TIPS

AC is $\frac{1}{3}$ the distance from A to B.

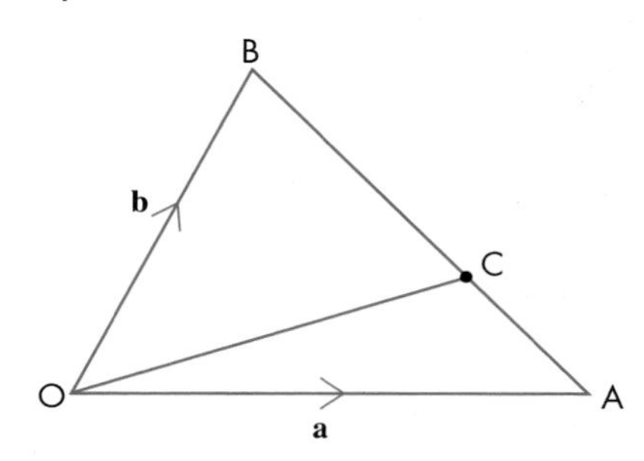

a **i** Work out the vector $\overrightarrow{AB}$.   **ii** Work out the vector $\overrightarrow{AC}$.

iii Work out the vector $\overrightarrow{OC}$ in terms of **a** and **b**.

b If C now divides the line AB in the ratio 1:3, write down the vector that represents $\overrightarrow{OC}$.

HINTS AND TIPS

AC is now $\frac{1}{4}$ the distance from A to B.

4 The diagram shows the vectors $\overrightarrow{OA} = \mathbf{a}$ and $\overrightarrow{OB} = \mathbf{b}$.

HINTS AND TIPS

OC is $\frac{2}{3}$ the distance from O to B.

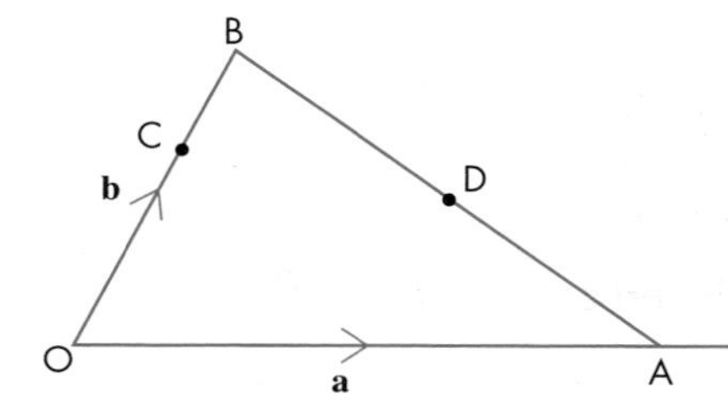

The point C divides OB in the ratio 2:1. The point E is such that $\overrightarrow{OE} = 2\overrightarrow{OA}$. D is the midpoint of AB.

a Write down (or work out) these vectors in terms of **a** and **b**.

i $\overrightarrow{OC}$ **ii** $\overrightarrow{OD}$ **iii** $\overrightarrow{CO}$

b The vector $\overrightarrow{CD}$ can be written as $\overrightarrow{CD} = \overrightarrow{CO} + \overrightarrow{OD}$. Use this fact to work out $\overrightarrow{CD}$ in terms of **a** and **b**.

c Write down a similar rule to that in part **b** for the vector $\overrightarrow{DE}$. Use this rule to work out $\overrightarrow{DE}$ in terms of **a** and **b**.

d Explain why C, D and E lie on the same straight line.

5 ABCDEF is a regular hexagon. $\overrightarrow{AB}$ is represented by the vector $\mathbf{a}$ and $\overrightarrow{BC}$ by the vector $\mathbf{b}$.

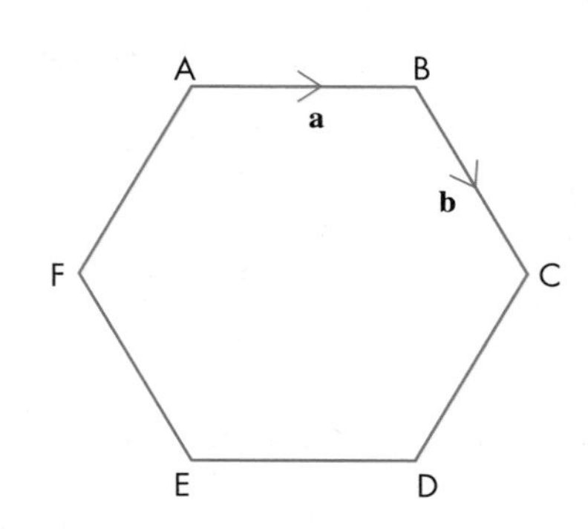

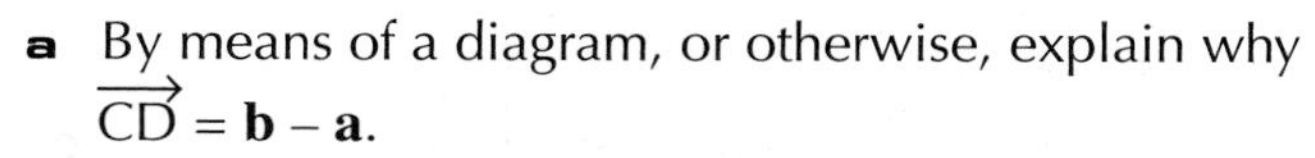

a By means of a diagram, or otherwise, explain why $\overrightarrow{CD} = \mathbf{b} - \mathbf{a}$.

b Express these vectors in terms of $\mathbf{a}$ and $\mathbf{b}$.

i $\overrightarrow{DE}$ **ii** $\overrightarrow{EF}$ **iii** $\overrightarrow{FA}$

c Work out the answer to:

$\overrightarrow{AB} + \overrightarrow{BC} + \overrightarrow{CD} + \overrightarrow{DE} + \overrightarrow{EF} + \overrightarrow{FA}$

Explain your answer.

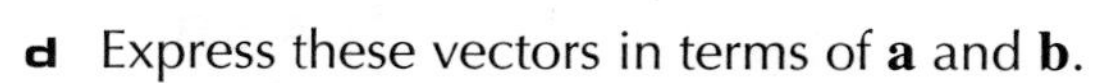

d Express these vectors in terms of $\mathbf{a}$ and $\mathbf{b}$.

i $\overrightarrow{AD}$ **ii** $\overrightarrow{BE}$ **iii** $\overrightarrow{CF}$ **iv** $\overrightarrow{AE}$ **v** $\overrightarrow{DF}$

6 ABCDEFGH is a regular octagon. $\overrightarrow{AB}$ is represented by the vector $\mathbf{a}$, and $\overrightarrow{BC}$ by the vector $\mathbf{b}$.

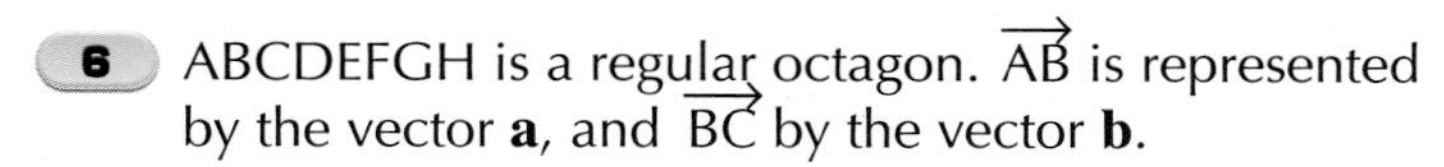

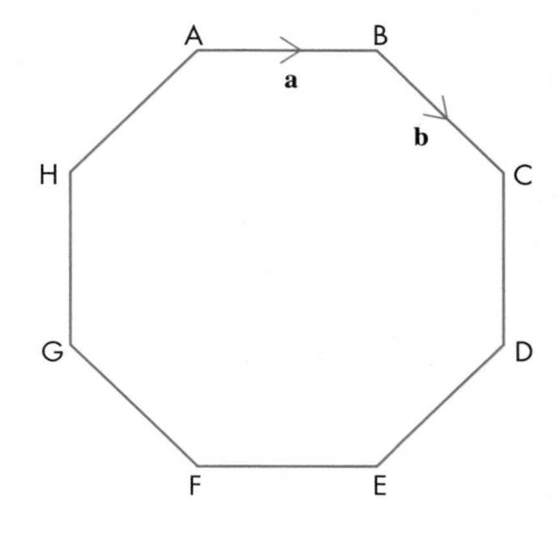

a By means of a diagram, or otherwise, explain why $\overrightarrow{CD} = \sqrt{2}\mathbf{b} - \mathbf{a}$.

b By means of a diagram, or otherwise, explain why $\overrightarrow{DE} = \mathbf{b} - \sqrt{2}\mathbf{a}$.

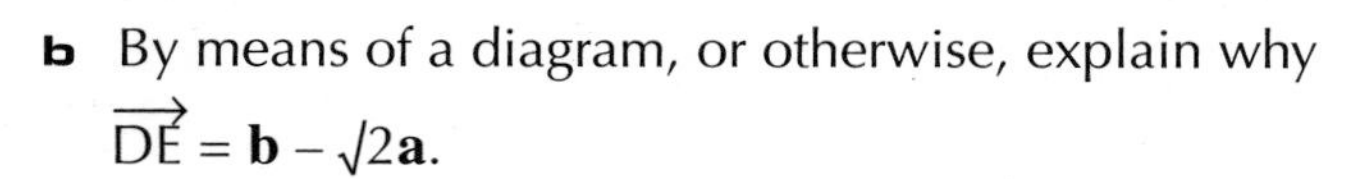

c Express the following vectors in terms of $\mathbf{a}$ and $\mathbf{b}$.

i $\overrightarrow{EF}$ **ii** $\overrightarrow{FG}$ **iii** $\overrightarrow{GH}$ **iv** $\overrightarrow{HA}$

v $\overrightarrow{HC}$ **vi** $\overrightarrow{AD}$ **vii** $\overrightarrow{BE}$ **viii** $\overrightarrow{BF}$

7 In the quadrilateral OABC, M, N, P and Q are the midpoints of the sides as shown. $\overrightarrow{OA}$ is represented by the vector $\mathbf{a}$, and $\overrightarrow{OC}$ by the vector $\mathbf{c}$. The diagonal $\overrightarrow{OB}$ is represented by the vector $\mathbf{b}$.

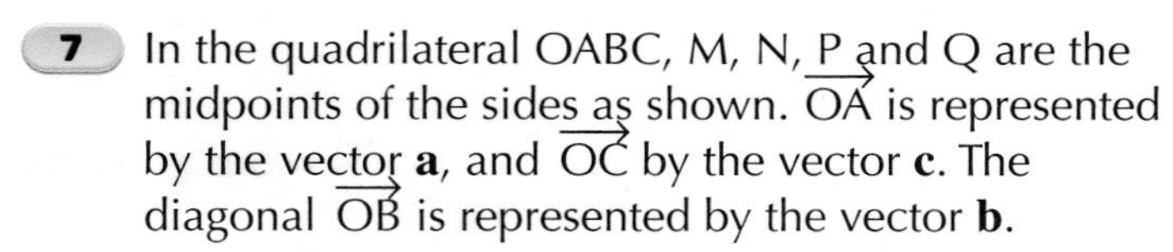

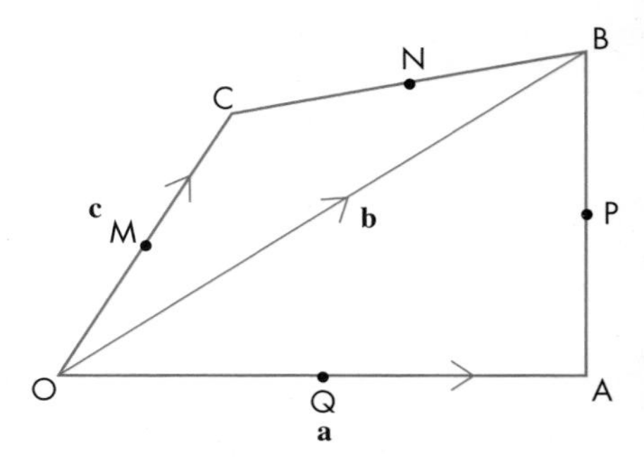

a Express these vectors in terms of $\mathbf{a}$, $\mathbf{b}$ and $\mathbf{c}$.

i $\overrightarrow{AB}$ **ii** $\overrightarrow{AP}$ **iii** $\overrightarrow{OP}$

Give your answers as simply as possible.

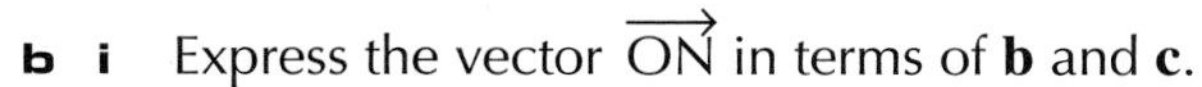

b **i** Express the vector $\overrightarrow{ON}$ in terms of $\mathbf{b}$ and $\mathbf{c}$.

ii Hence express the vector $\overrightarrow{PN}$ in terms of $\mathbf{a}$ and $\mathbf{c}$.

c **i** Express the vector $\overrightarrow{QM}$ in terms of $\mathbf{a}$ and $\mathbf{c}$.

ii What relationship is there between $\overrightarrow{PN}$ and $\overrightarrow{QM}$?

iii What sort of quadrilateral is PNMQ?

d Prove that $\overrightarrow{AC} = 2\overrightarrow{QM}$.

8 L, M, N, P, Q, R are the midpoints of the line segments, as shown.

$\overrightarrow{OA} = \mathbf{a}$, $\overrightarrow{OB} = \mathbf{b}$ and $\overrightarrow{OC} = \mathbf{c}$

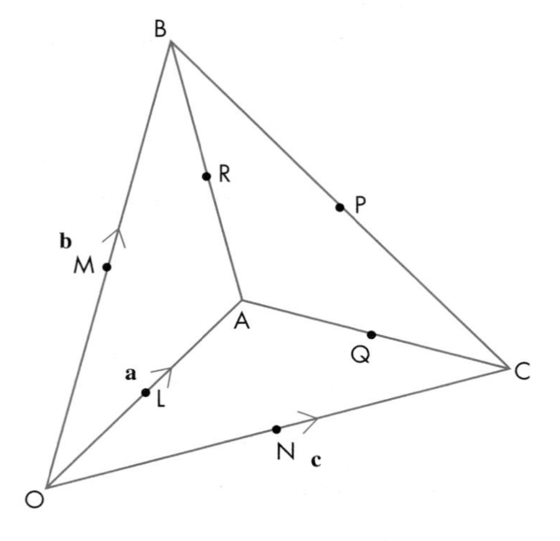

a Express these vectors in terms of **a** and **c**.

i $\overrightarrow{OL}$ **ii** $\overrightarrow{AC}$

iii $\overrightarrow{OQ}$ **iv** $\overrightarrow{LQ}$

b Express these vectors in terms of **a** and **b**.

i $\overrightarrow{LM}$ **ii** $\overrightarrow{QP}$

c Prove that the quadrilateral LMPQ is a parallelogram.

d Find two other sets of four points that form parallelograms.

AU 9 In the triangle OAB, M is the midpoint of AB.

$\overrightarrow{OA} = \mathbf{a}$ and $\overrightarrow{OB} = \mathbf{b}$

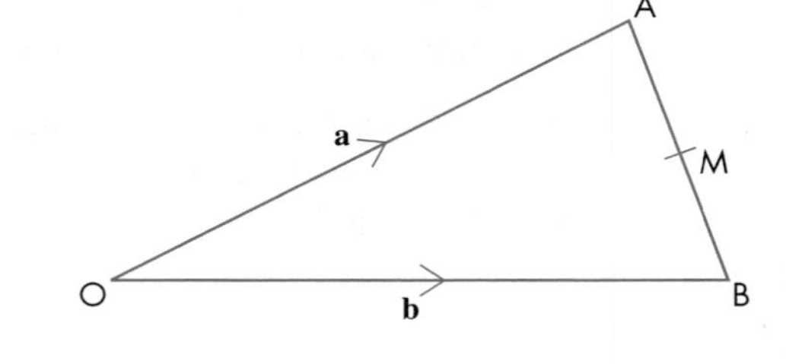

a Find $\overrightarrow{AM}$ in terms of **a** and **b**.

Give your answer in its simplest form.

b $\overrightarrow{OC} = \mathbf{a} + \mathbf{b}$

The length of OA is equal to the length of OB.

i Write down the name of the shape OACB.

ii Write down one fact about the points O, M and C.

Give a reason for your answer.

AU 10 ABCD is a trapezium with AB parallel to DC.

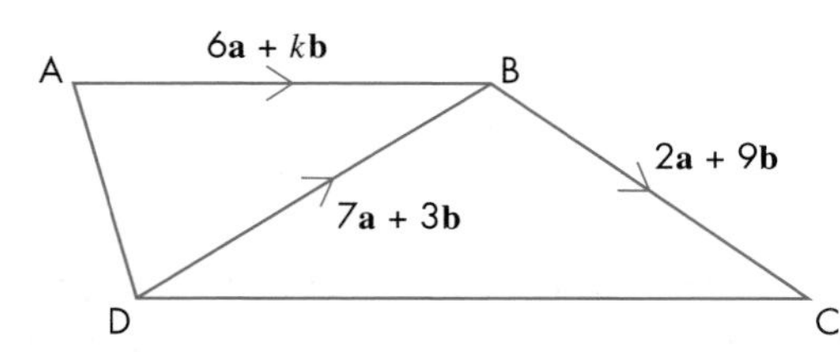

$\overrightarrow{AB} = 6\mathbf{a} + k\mathbf{b}$, $\overrightarrow{BC} = 2\mathbf{a} + 9\mathbf{b}$ and $\overrightarrow{DB} = 7\mathbf{a} + 3\mathbf{b}$, where k is a number.

Work out the value of k.

17.3 Geometric proof

This section will show you how to:

- understand the difference between a proof and a demonstration

Key words
demonstration
proof
prove

You should already know these.

- The angle sum of the interior angles in a triangle (180°)
- The circle theorems
- Pythagoras' theorem

Can you **prove** them?

For a mathematical **proof**, you must proceed in logical steps, establishing a series of mathematical statements by using facts that are already known to be true.

Below are three standard proofs: *the sum of the interior angles of a triangle is 180°*, *Pythagoras' theorem* and *congruency*. Read through them, following the arguments carefully. Make sure you understand each step in the process.

Proof that the sum of the interior angles of a triangle is 180°

One of your earlier activities in geometry may have been to draw a triangle, to cut off its corners and to stick them down to *show that* they make a straight line.

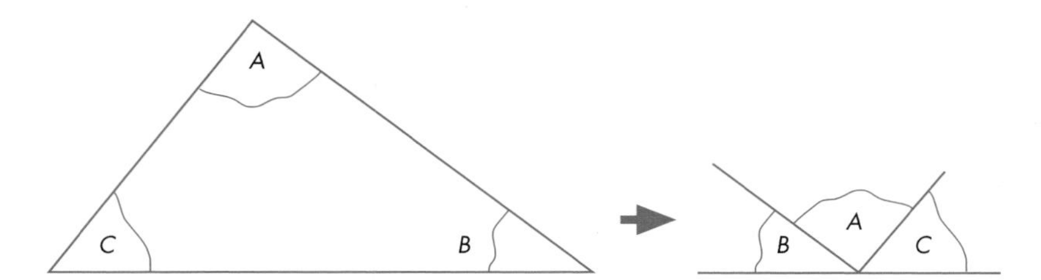

Does this prove that the interior angles make 180° or were you just lucky and picked a triangle that worked? Was the fact that everyone else in the class managed to pick a triangle that worked also a lucky coincidence?

Of course not! But this was a **demonstration**, not a proof. You would have to show that this method worked for *all* possible triangles (there is an infinite number!) to say that you have proved this result.

Your proof must establish that the result is true for *all* triangles.

Look at the following proof.

Start with triangle ABC with angles α, β and γ (figure **i**).

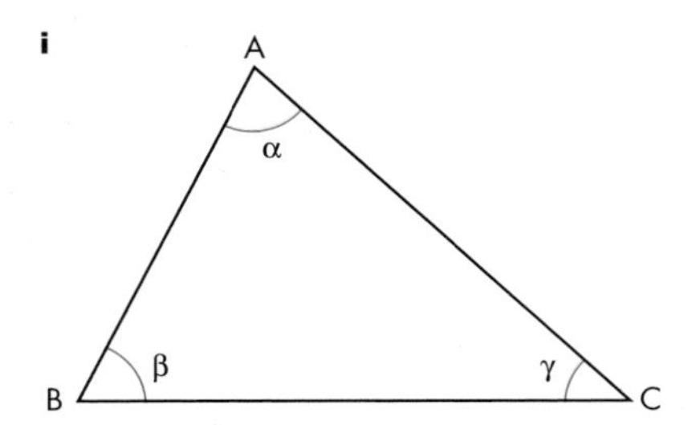

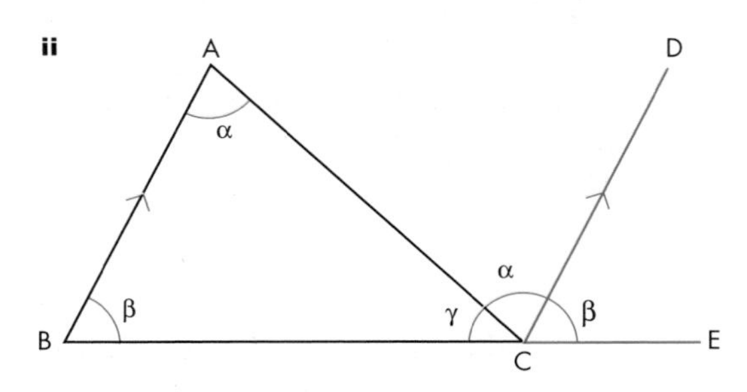

On figure **i** draw a line CD parallel to side AB and extend BC to E, to give figure **ii**.

Since AB is parallel to CD:

$\angle ACD = \angle BAC = \alpha$ (alternate angles) $\angle DCE = \angle ABC = \beta$ (corresponding angles)

BCE is a straight line, so $\gamma + \alpha + \beta = 180°$. Therefore the interior angles of a triangle = 180°.

This proof assumes that alternate angles are equal and that corresponding angles are equal. Strictly speaking, we should prove these results, but we have to accept certain results as true. These are based on Euclid's axioms from which all geometric proofs are derived.

Proof of Pythagoras' theorem

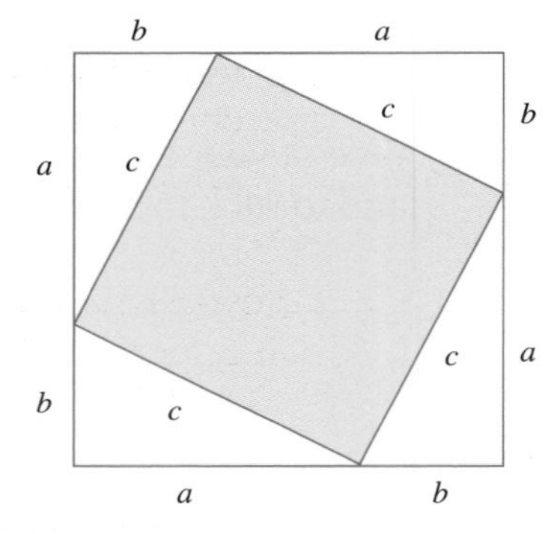

Draw a square of side c inside a square of side $(a + b)$, as shown.

The area of the exterior square is $(a + b)^2 = a^2 + 2ab + b^2$.

The area of each small triangle around the shaded square is $\frac{1}{2}ab$.

The total area of all four triangles is $4 \times \frac{1}{2}ab = 2ab$.

Subtracting the total area of the four triangles from the area of the large square gives the area of the shaded square:

$$a^2 + 2ab + b^2 - 2ab = a^2 + b^2$$

But the area of the shaded square is c^2, so

$$c^2 = a^2 + b^2$$

which is Pythagoras' theorem.

Congruency proof

There are four conditions to prove congruency. These are commonly known as SSS (three sides the same), SAS (two sides and the included angle the same), ASA (or AAS) (two angles and one side the same) and RHS (right-angled triangle, hypotenuse, and one short side the same). **Note:** AAA (three angles the same) is not a condition for congruency.

When you prove a result, you must explain or justify every statement or line. Proofs have to be rigorous and logical.

EXAMPLE 5

ABCD is a parallelogram. X is the point where the diagonals meet.

Prove that triangles AXB and CXD are congruent.

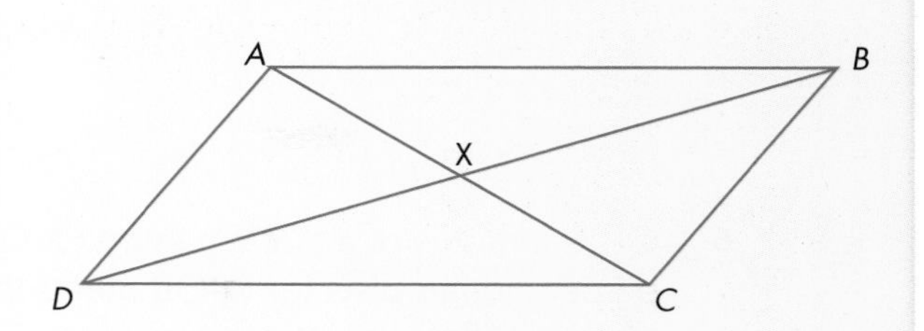

$\angle BAX = \angle DCX$ (alternate angles)

$\angle ABX = \angle CDX$ (alternate angles)

$AB = CD$ (opposite sides in a parallelogram)

Hence ΔAXB is congruent to ΔCXD (ASA).

Note that you could have used $\angle AXB = \angle CXD$ (vertically opposite angles) as the second line but whichever approach is used you *must* give a reason for each statement.

EXERCISE 17C

PS 1 **a** Show that the triangle ABC is isosceles.

b Prove that the triangle DEF with one angle of $x°$ and an exterior angle of $90° + \frac{x°}{2}$ is isosceles.

i

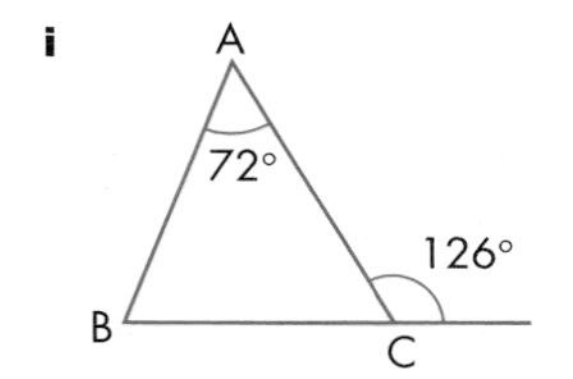

ii

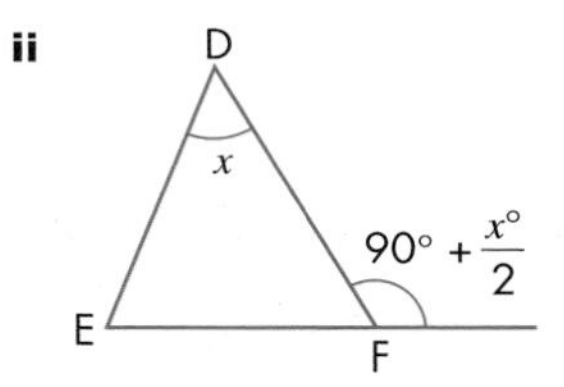

PS 2 Prove that a triangle with an interior angle of $\frac{x°}{2}$ and an exterior angle of $x°$ is isosceles.

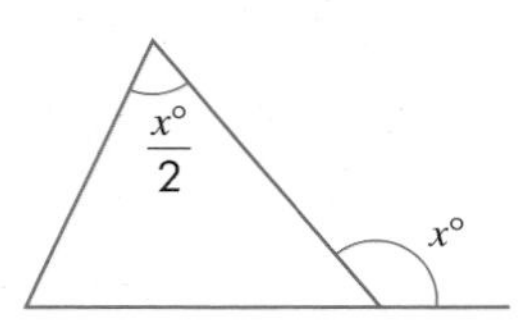

PS 3 **a** Using the theorem that the angle subtended by an arc at the centre of a circle is twice the angle subtended by the same arc at the circumference, find the values of angles DAB and DCB in the circle shown in figure **i**.

b Prove that the sum of the opposite angles of a cyclic quadrilateral is 180°. (You may find figure **ii** useful.)

i

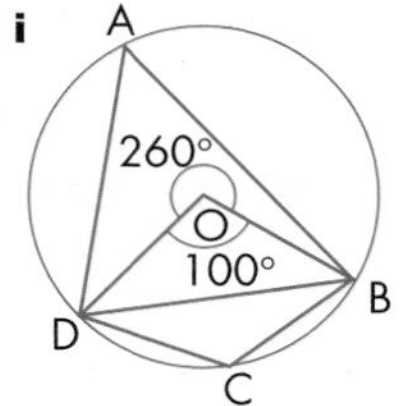

ii

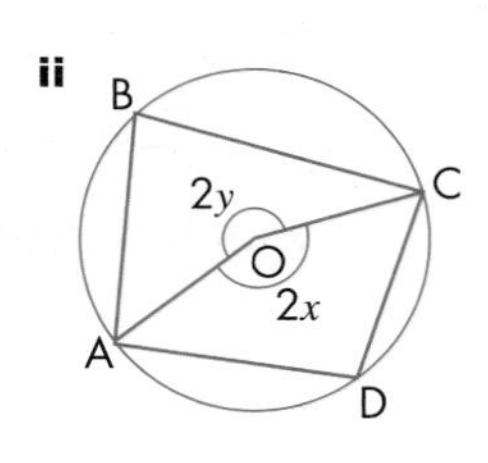

PS 4 **a** The triangle ABC is isosceles. BCD and AED are straight lines. Find the value of the angle CED, marked x, in figure **i**.

b Prove that angle ACB = angle CED in figure **ii**.

i

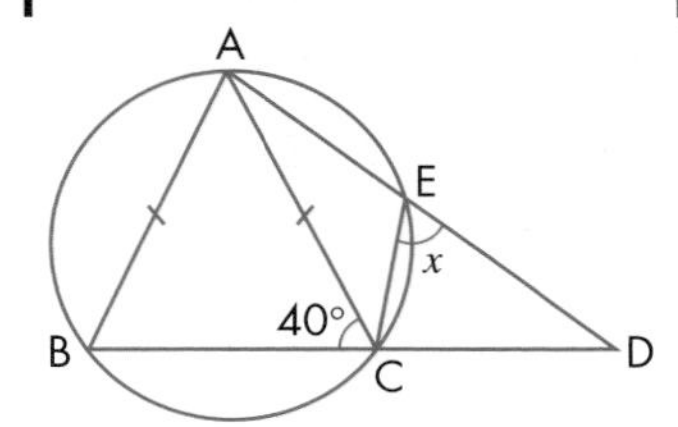

ii

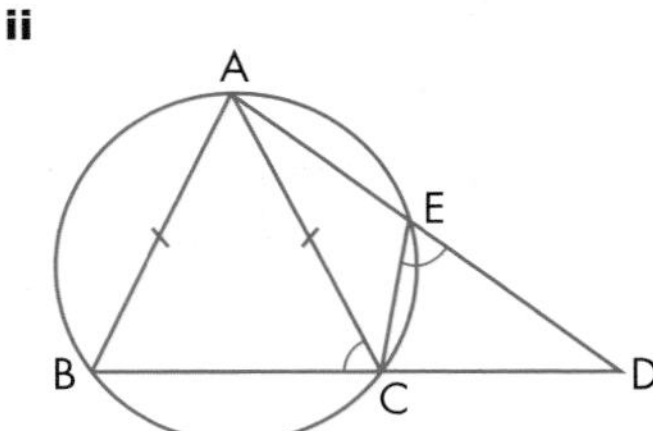

PS 5 PQRS is a parallelogram. Prove that triangles PQS and RQS are congruent.

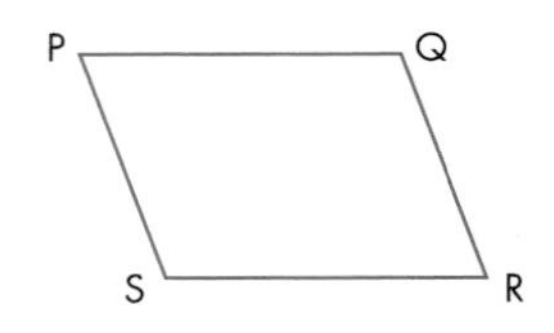

PS 6 OB is a radius of a circle, centre O. C is the point where the perpendicular bisector of OB meets the circumference. Prove that triangle OBC is equilateral.

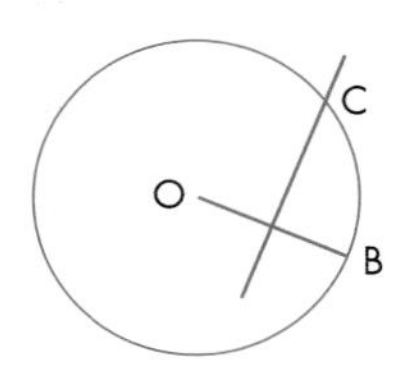

PS 7 The grid is made up of identical parallelograms.
In the grid, $\overrightarrow{OA} = \mathbf{a}$ and $\overrightarrow{OB} = \mathbf{b}$.
Prove that AB is parallel to EF.

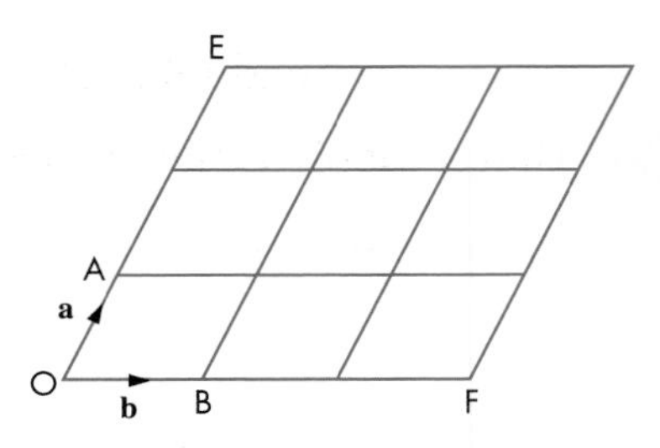

PS 8 **a** Prove the alternate segment theorem.

b Two circles touch internally at T. The common tangent at T is drawn. Two lines TAB and TXY are drawn from T. Prove that AX is parallel to BY.

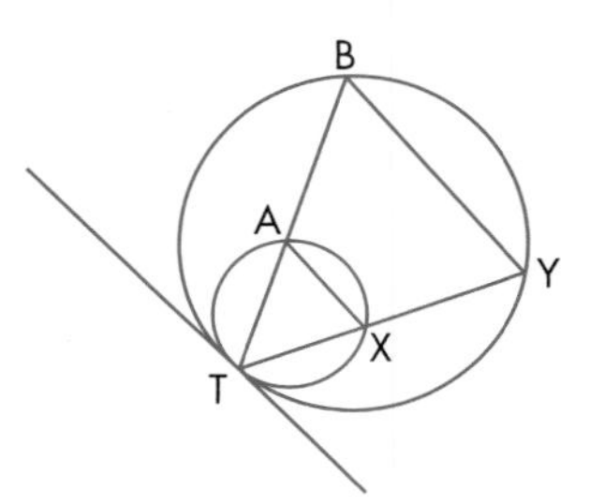

PS 9 Two circles touch externally at T. A line ATB is drawn through T. The common tangent at T and the tangents at A and B meet at P and Q. Prove that PB is parallel to AQ.

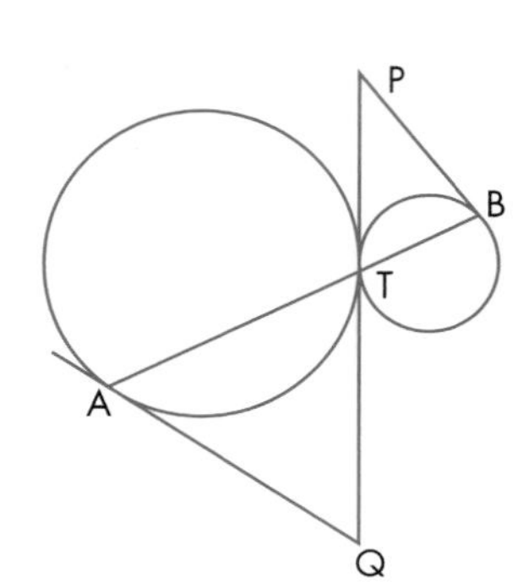

10 **a** and **b** are vectors.

$\overrightarrow{XY} = \mathbf{a} + \mathbf{b}$ $\quad\overrightarrow{YZ} = 2\mathbf{a} + \mathbf{b}$ $\quad\overrightarrow{ZW} = \mathbf{a} + 2\mathbf{b}$

a Show that $\overrightarrow{YW}$ is parallel to $\overrightarrow{XY}$.

b Write down the ratio YW : XY.

c What do your answers to **a** and **b** tell you about the points X, Y and W?

d O is the origin.

A, B and C are three points such that:

$\overrightarrow{OA} = \begin{pmatrix} 6 \\ 2 \end{pmatrix}$ $\quad\overrightarrow{OB} = \begin{pmatrix} 1 \\ 1 \end{pmatrix}$ $\quad\overrightarrow{OC} = \begin{pmatrix} 2 \\ -4 \end{pmatrix}$

Prove that angle ABC is a right angle.

GRADE BOOSTER

A You can solve problems, using addition and subtraction of vectors

A* You can solve complex geometrical problems, using vectors

A* You can use proof in geometrical problems

What you should know now

- How to add and subtract vectors
- How to apply vector methods to solve geometrical problems

1

A N B M a O c C Not drawn accurately

OABC is a parallelogram.

M is the midpoint of CB.

N is the midpoint of AB.

$\overrightarrow{OA} = \mathbf{a}$ $\quad\quad$ $\overrightarrow{OC} = \mathbf{c}$

a Find, in terms of **a** and/or **c**, the vectors:

i $\overrightarrow{MB}$ $\quad\quad$ **ii** $\overrightarrow{MN}$. *(2 marks)*

b Show that CA is parallel to MN. *(2 marks)*

Edexcel, May 2008, Paper 3, Question 25

2

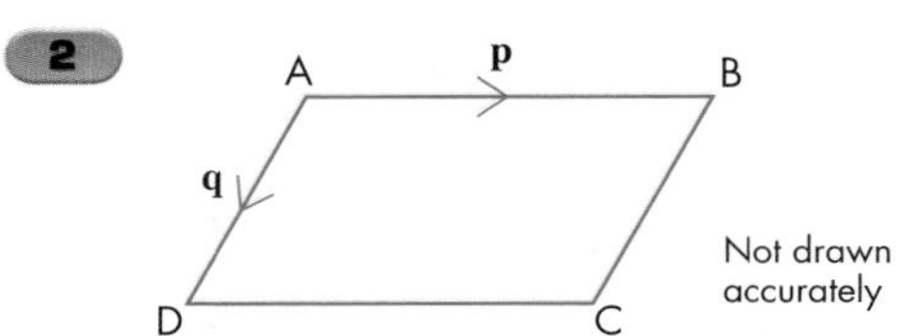

ABCD is a parallelogram.

AB is parallel to DC.

AD is parallel to BC.

$\overrightarrow{AB} = \mathbf{p}$

$\overrightarrow{AD} = \mathbf{q}$

a Express, in terms of **p** and **q**:

i $\overrightarrow{AC}$ $\quad\quad$ **ii** $\overrightarrow{BD}$ *(2 marks)*

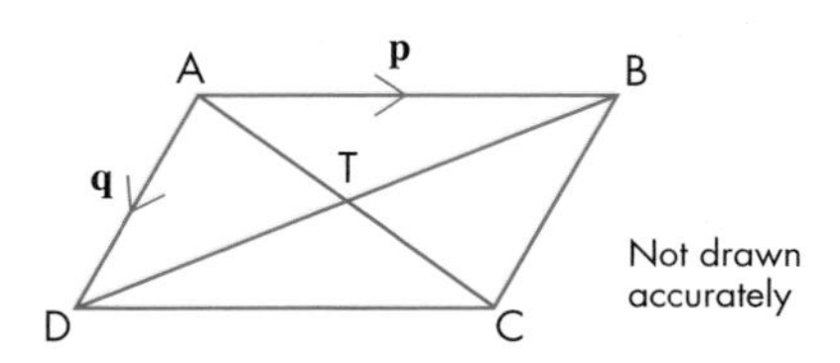

AC and BD are diagonals of parallelogram ABCD.

AC and BD intersect at T.

b Express AT in terms of **p** and **q**. *(1 mark)*

Edexcel, June 2006, Paper 6 Higher, Question 13

3 The diagram shows a parallelogram, ABCD.

M is the midpoint of BC.

N is the midpoint of AD.

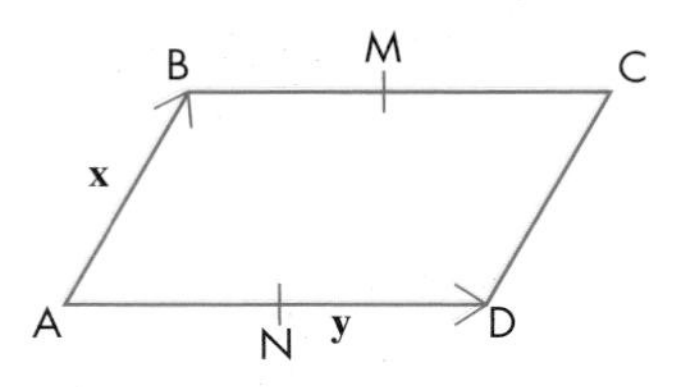

$\overrightarrow{AB} = x$

$\overrightarrow{AD} = y$

Find, in terms of x and/or y, the vectors:

a $\overrightarrow{MN}$ *(1 mark)*

b $\overrightarrow{AC}$ *(1 mark)*

P is the point such that $\overrightarrow{CP} = y - \frac{1}{2}x$.

c Find, in terms of x and/or y, the vector $\overrightarrow{PA}$.

Simplify your answer as much as possible. *(3 marks)*

Edexcel, June 2009, Paper 4H, Question 18

4

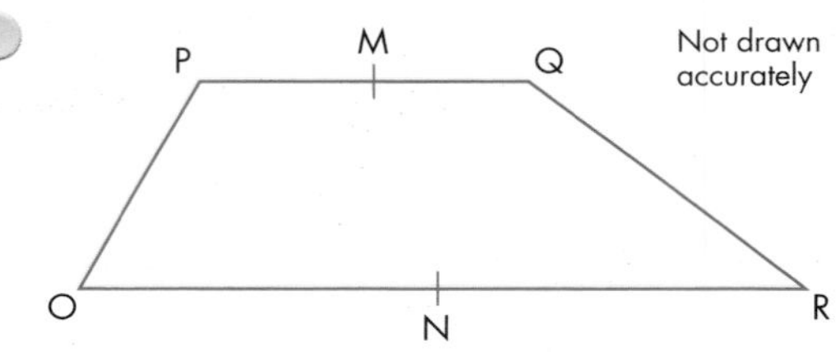

OPQR is a trapezium with PQ parallel to OR.

$\overrightarrow{OP} = 2\mathbf{b}$ $\quad\quad$ $\overrightarrow{PQ} = 2\mathbf{a}$ $\quad\quad$ $\overrightarrow{OR} = 6\mathbf{a}$

M is the midpoint of PQ and N is the midpoint of OR.

a Find the vector $\overrightarrow{MN}$ in terms of **a** and **b** *(2 marks)*

X is the midpoint of MN and Y is the midpoint of QR.

b Prove that XY is parallel to OR. *(2 marks)*

Edexcel, June 2005, Paper 5, Question 22

A* A

5

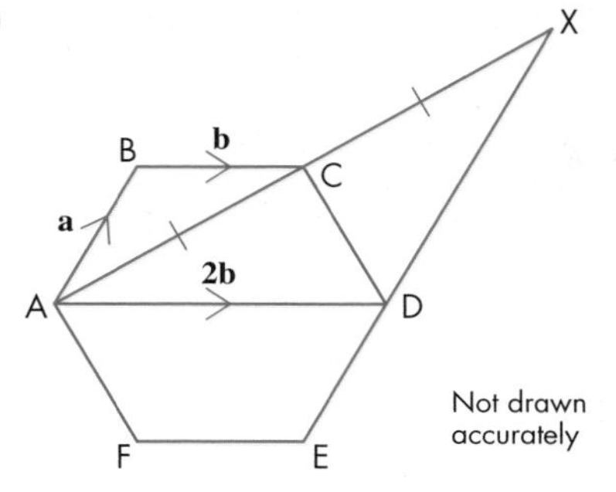

ABCDEF is a regular hexagon.

$\overrightarrow{AB} = \mathbf{a}$ $\overrightarrow{BC} = \mathbf{b}$ $\overrightarrow{AD} = 2\mathbf{b}$

a Find the vector $\overrightarrow{AC}$ in terms of **a** and **b**. *(1 mark)*

$\overrightarrow{AC} = \overrightarrow{CX}$

b Prove that AB is parallel to DX. *(3 marks)*

Edexcel, June 2007, Paper 5, Question 22

6 ABCD is a square.

BEC and DCF are equilateral triangles.

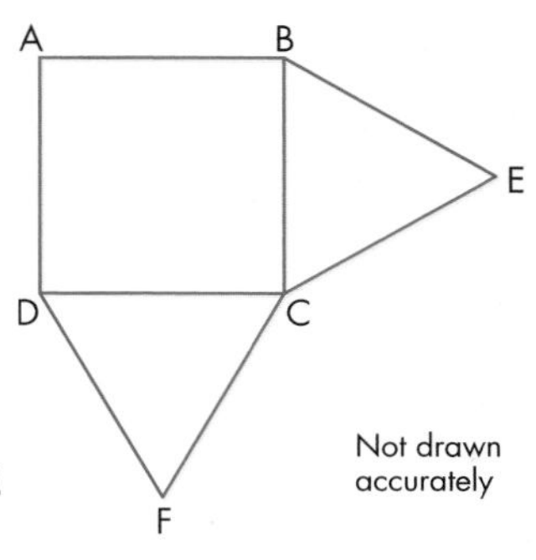

a Prove that triangle ECD is congruent to triangle BCF. *(3 marks)*

G is the point such that BEGF is a parallelogram.

b Prove that ED = EG. *(2 marks)*

Edexcel, June 2006, Paper 5, Question 21

Worked Examination Questions

1 The diagram shows triangle OAB. M is the midpoint of OA.
P lies on BM and BP = $\frac{2}{3}$BM.

$\overrightarrow{OA} = 2\mathbf{a}$ and $\overrightarrow{OB} = 2\mathbf{b}$

a Find expressions, in terms of $\mathbf{a}$ and $\mathbf{b}$, for **i** $\overrightarrow{BM}$ and **ii** $\overrightarrow{OP}$.
Write each answer in its simplest form.

b N is the midpoint of OB. Q lies on AN and AQ = $\frac{2}{3}$AN.

i Find an expression for $\overrightarrow{OQ}$, in terms of $\mathbf{a}$ and $\mathbf{b}$.
Write your answer in its simplest form.

ii What do your answers for $\overrightarrow{OP}$ and $\overrightarrow{OQ}$ tell you about the points P and Q?

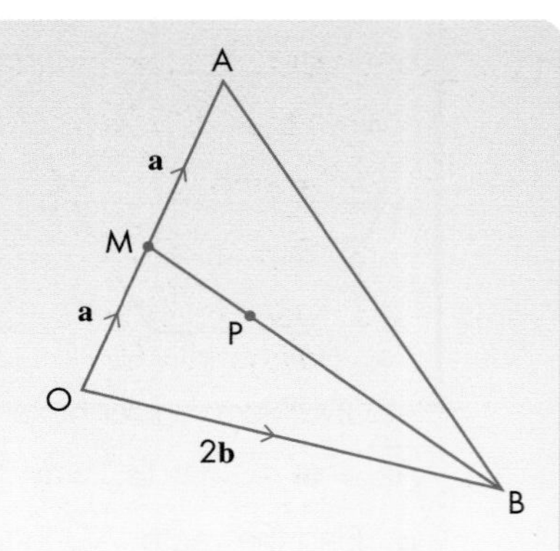

1 *a* i $\overrightarrow{BM} = \mathbf{a} - 2\mathbf{b}$

> Find a route from B to M in terms of known vectors. $\overrightarrow{BM} = \overrightarrow{BO} + \overrightarrow{OM}$
> This scores 1 mark for accuracy.

ii $\overrightarrow{OP} = \overrightarrow{OB} + \frac{2}{3}\overrightarrow{BM} = 2\mathbf{b} + \frac{2}{3}\mathbf{a} - \frac{4}{3}\mathbf{b} = \frac{2}{3}\mathbf{a} + \frac{2}{3}\mathbf{b}$

> Find a route from O to P in terms of known vectors. $\overrightarrow{OP} = \overrightarrow{OB} + \overrightarrow{BP}$
> This scores 1 mark each for accuracy and method.

b i $\overrightarrow{OQ} = \overrightarrow{OA} + \overrightarrow{AQ} = \overrightarrow{OA} + \frac{2}{3}\overrightarrow{AN}$ with $\overrightarrow{AN} = \mathbf{b} - 2\mathbf{a}$

> Find a route from O to Q in terms of known vectors. $\overrightarrow{OQ} = \overrightarrow{OA} + \overrightarrow{AQ}$
> This scores 1 mark each for accuracy and method mark.

So $\overrightarrow{OQ} = 2\mathbf{a} + \frac{2}{3}(\mathbf{b} - 2\mathbf{a}) = 2\mathbf{a} + \frac{2}{3}\mathbf{b} - \frac{4}{3}\mathbf{a} = \frac{2}{3}\mathbf{a} + \frac{2}{3}\mathbf{b}$

> This scores 1 mark for accuracy.

ii $\overrightarrow{OP} = \overrightarrow{OQ}$, so P and Q are the same point

> This statement scores 1 mark.

Total: 7 marks

Worked Examination Questions

AU **2** OABC is a parallelogram.

M is the midpoint of the diagonal OB.

$\overrightarrow{OA} = 2\mathbf{a}$ and $\overrightarrow{OC} = 2\mathbf{c}$

a Express $\overrightarrow{OM}$ in terms of $\mathbf{a}$ and $\mathbf{c}$.

b Use vectors to prove that M is also the midpoint of the diagonal AC.

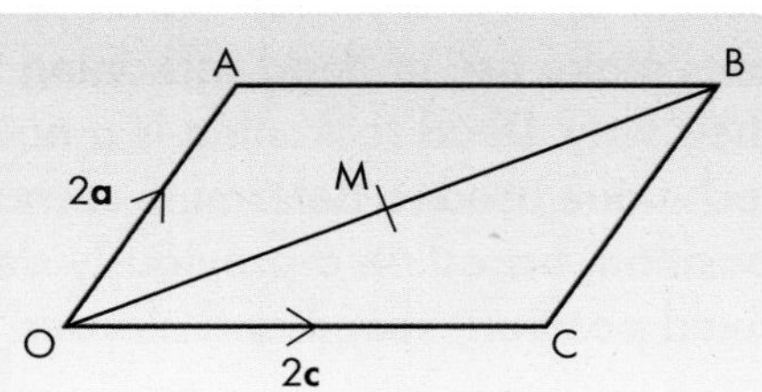

2 a $\overrightarrow{OB} = 2\mathbf{a} + 2\mathbf{c}$ and $\overrightarrow{OM} = \frac{1}{2}\overrightarrow{OB}$, so $\overrightarrow{OM} = \mathbf{a} + \mathbf{c}$.

The correct answer scores 1 mark for accuracy.

b $\overrightarrow{AC} = 2\mathbf{c} - 2\mathbf{a}$ and $\overrightarrow{AM} = \overrightarrow{AO} + \overrightarrow{OM} = -2\mathbf{a} + \mathbf{a} + \mathbf{c} = \mathbf{c} - \mathbf{a}$.

This scores 1 method mark for $2\mathbf{c} - 2\mathbf{a}$ and 1 method mark for $\mathbf{c} - \mathbf{a}$.

So $\overrightarrow{AM} = \frac{1}{2}\overrightarrow{AC}$, hence M is the midpoint of AC.

This statement scores 1 mark for accuracy.

Total: 4 marks

PS **3** In the diagram the lines VX and WY intersect at Z. VW and YX are parallel and VW = XY.

Prove that triangles VWZ and XYZ are congruent.

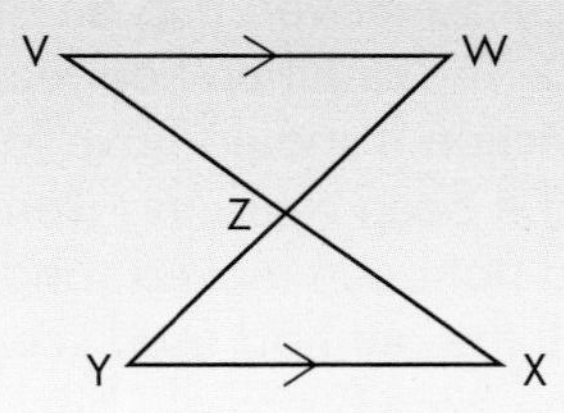

3 ∠VWZ = ∠ZYX (Alternate angles)
∠WVZ = ∠ZXY (Alternate angles)
∠VZW = ∠YZX (Vertically opposite)

State which angles are equal and give a reason. You only need two angles. It doesn't matter which two. You can state all three but there will still only be two marks available.
This scores 2 marks.

VW = YZ (Given)

Even though this information is given in the question, do not assume that the examiner knows you know it. If information given in the question is needed in the answer, restate it.
This scores 1 mark.

Hence the triangles are congruent, ASA.

State the reason for congruency.
This scores 1 mark.

Total: 4 marks

17

Functional Maths
Navigational techniques of ants

Scientists have discovered that some desert ants make use of dead reckoning to find their way. Dead reckoning is a navigational technique used to determine current position, based on a previously determined fixed position, speed and elapsed time.

Getting started

Scientists conducted an experiment to prove that Tunisian desert ants navigate using dead reckoning rather than other means such as laying down scents.

They conducted the following experiment:

The Tunisian desert ant set out from its nest at A. As soon as it found food at O, the scientists moved it to an alternative position at B. The ant then headed in exactly the direction it should have taken to find its nest, had it been returning from O to A, as if it had not been moved, and so ending up at C. If the ant had used scent it would have gone from B straight back to A.

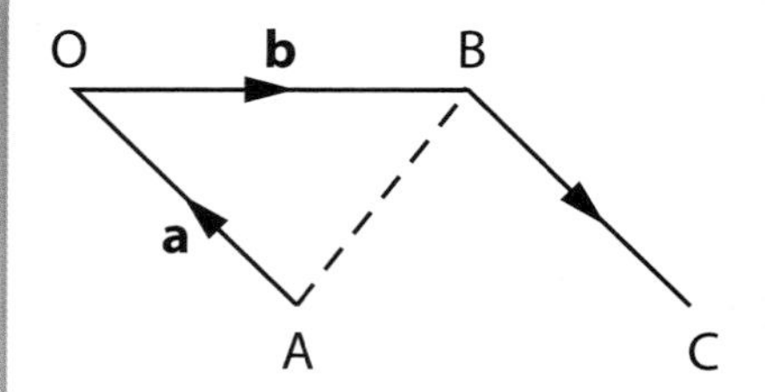

The diagram uses the vectors $\overrightarrow{OA} = -\mathbf{a}$ and $\overrightarrow{OB} = \mathbf{b}$.

Describe in terms of **a** and **b**:

- the route the ant took, from its nest to its end point after it had been moved, if it was using dead reckoning
- the route the ant would have taken from its nest to its end point, if it had navigated by scent rather than by dead reckoning.

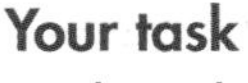

Your task

Look at the diagram below. An ant's nest is at O. Each time the ant sets off it finds food at a different location: D, E, F, G, H, I or J.

1 Suggest possible vectors for its route to food at each of these points, writing them in terms of **a** and **b**.

2 Now imagine you are a scientist researching colonies of ants in the jungle and the desert. The position of the ants' nest, in each case, is again described as O. Choose one location for food at K, L, M, N or P. As soon as an ant finds food, you move it to point C.

   Use vectors to describe the route the ant takes:

   - from its nest to the food
   - then when it is moved to C
   - back to its nest, assuming it is an ant that lives in the jungle and relies on scent
   - to its end point, assuming it is a desert ant and relies on dead reckoning.

   Write a scientific report, with three sections, describing the method, results and conclusion, and using a vector diagram to explain your theory.

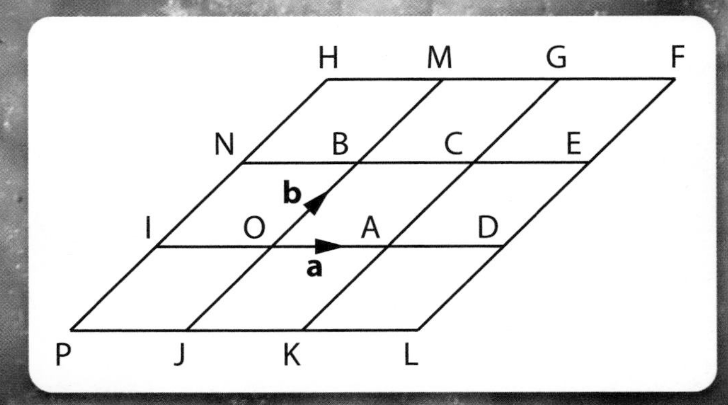

Why this chapter matters

By transforming graphs into other shapes, it is possible to change a circle into an aerofoil. This enables aeroplane engineers to do much simpler calculations when designing wings. This chapter will not help you to design aeroplane wings, but everyone has to start somewhere!

So far you have met four transformations:

- Translation
- Rotation
- Reflection
- Enlargement

Can you remember how to describe each of these?

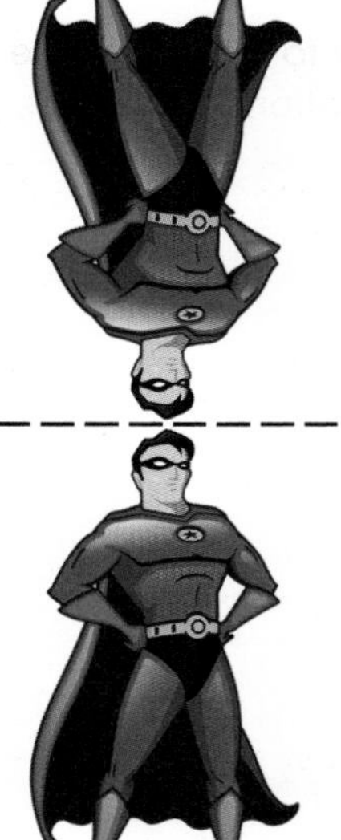

Reflection

Original

Enlargement

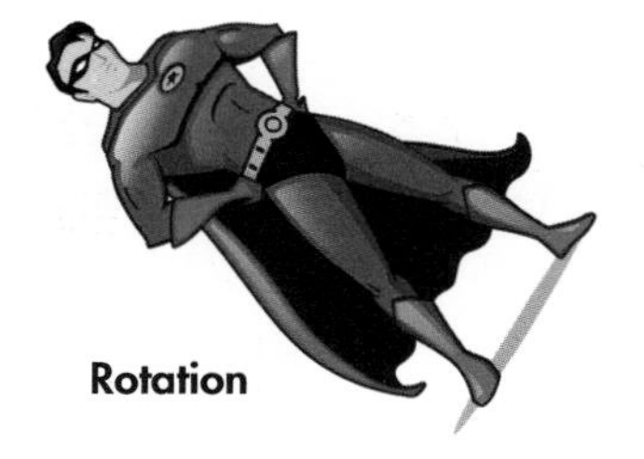

Rotation

Translation

There are many other transformations, but one that you will need here is the 'stretch'. It is exactly what it says it is.

The stretch can be in any direction, but in GCSE mathematics the stretch will be in the x- and y-directions.

A stretch is defined by a direction and a scale factor.

Original

Stretch scale factor 1.5 in the y-direction

Stretch scale factor 0.5 in the x-direction

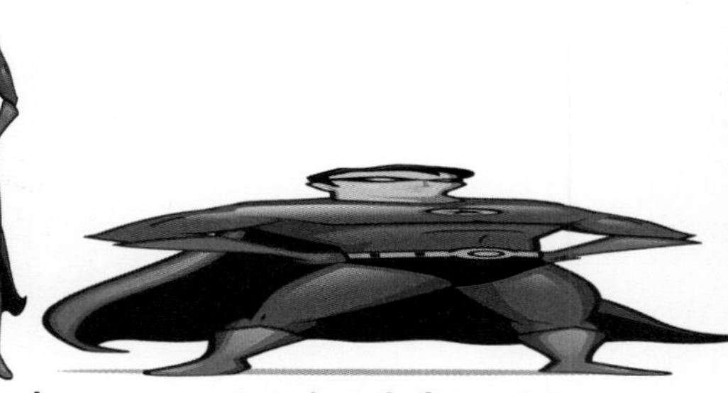

Stretch scale factor 3 in the x-direction and 0.5 in the y-direction

Chapter 18

Algebra: Transformation of graphs and other graphs

The grades given in this chapter are target grades.

1. Transformation of the graph $y = f(x)$
2. Cubic, exponential and reciprocal graphs

This chapter will show you ...

- **A** how to transform a graph
- **A*** how to recognise the relationships between graphs and their equations
- **B to A*** how to complete and interpret cubic, exponential and reciprocal graphs

Visual overview

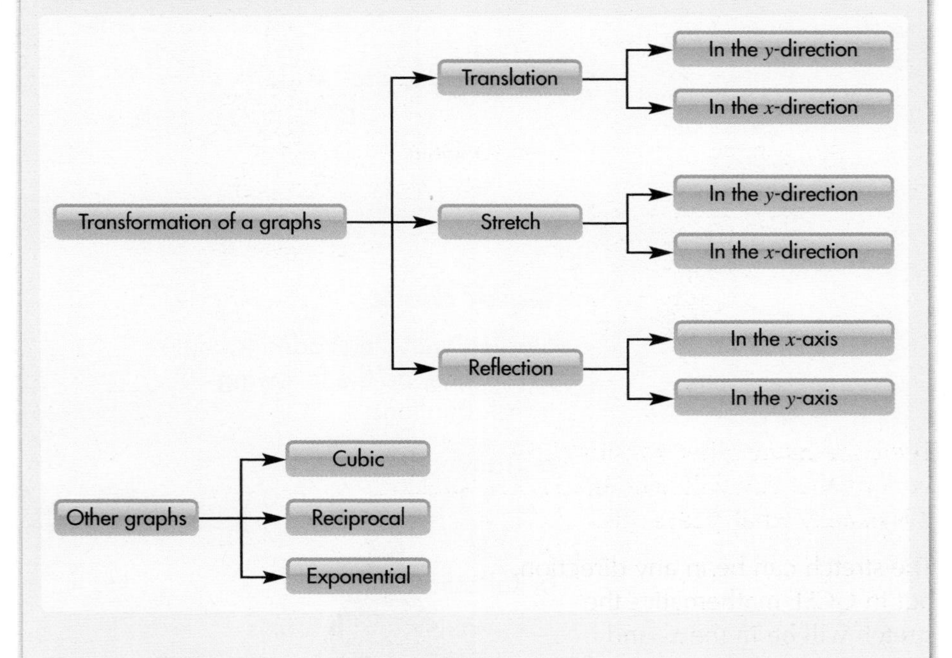

What you should already know

- How to transform a shape by a translation and a reflection **(KS3 level 6, GCSE grade D)**
- A translation is described by a column vector **(KS3 level 7, GCSE grade C)**
- A reflection is described by a mirror line **(KS3 level 5, GCSE grade E)**

continued

- The graphs of $y = x^2$, $y = x^3$, $y = \frac{1}{x}$, $y = \sin x$, $y = \cos x$ and $y = \tan x$

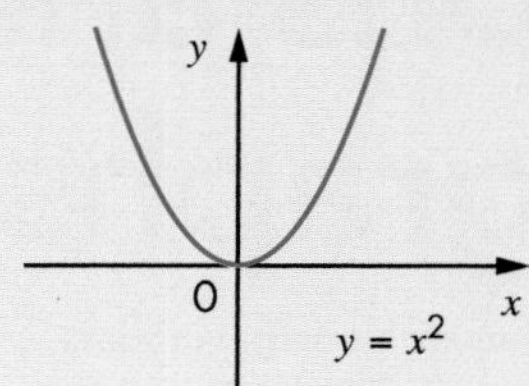

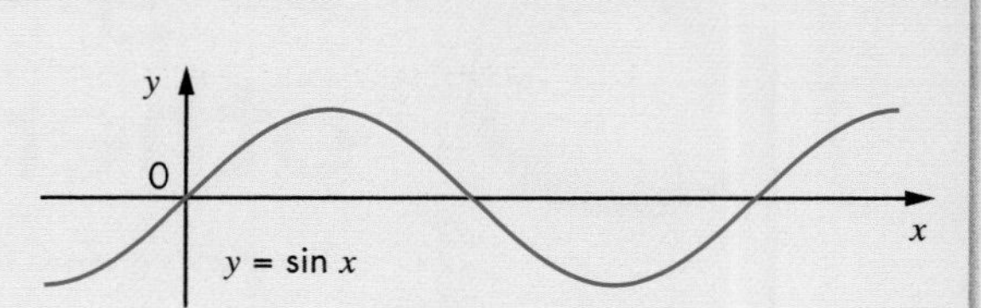

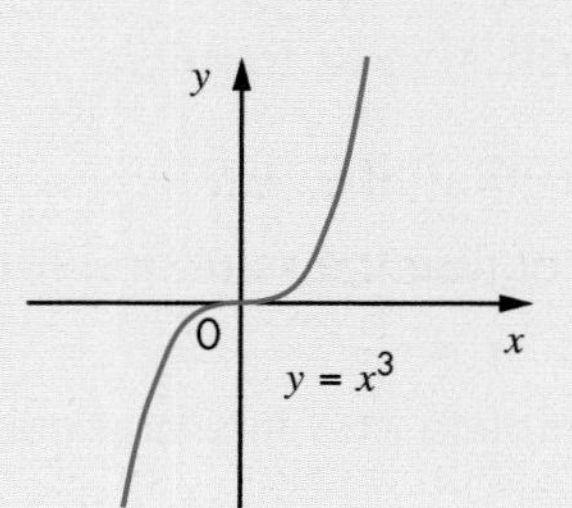

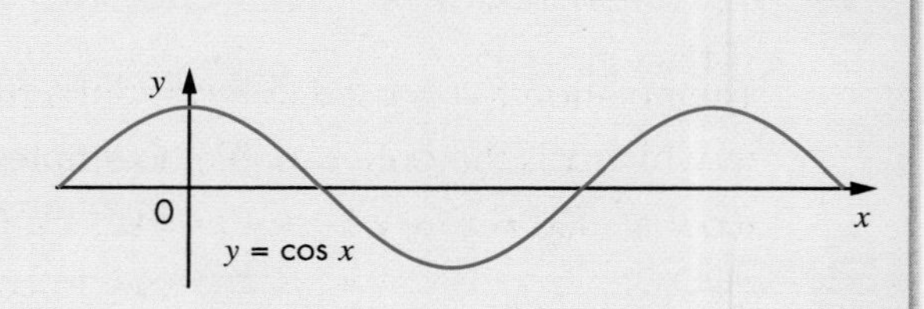

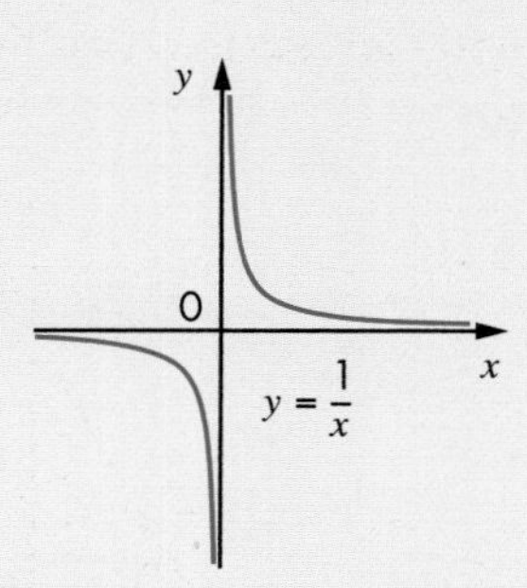

Quick check

Starting with the shaded triangle every time, do the following transformations.

a Translation

i $\begin{pmatrix} 3 \\ 0 \end{pmatrix}$ **ii** $\begin{pmatrix} 0 \\ -2 \end{pmatrix}$

b Reflection in the

i y-axis **ii** x-axis

c Rotation of 180° about the origin

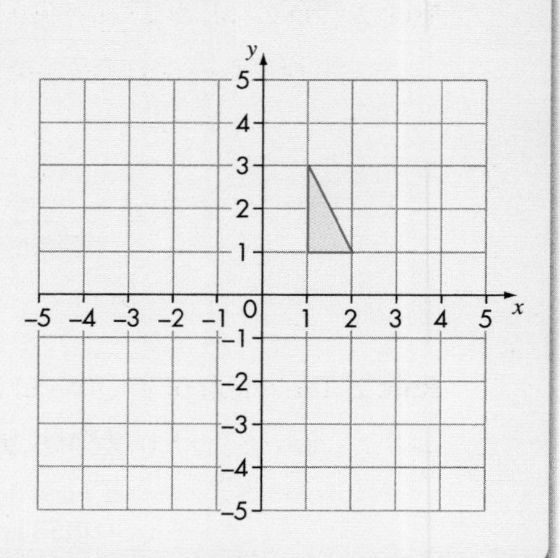

18.1 Transformation of the graph $y = f(x)$

This section will show you how to:
- transform a graph

Key words
function, transform, reflection, translation, scale factor, vector, stretch

The notation f(x) is used to represent a **function** of x. A function of x is any algebraic expression in which x is the only variable. Examples of functions are: $f(x) = x + 3$, $f(x) = 5x$, $f(x) = 2x - 7$, $f(x) = x^2$, $f(x) = x^3 + 2x - 1$, $f(x) = \sin x$ and $f(x) = \frac{1}{x}$.

On this and the next page are six general statements or rules about **transforming** graphs.

This work is much easier to understand if you can use to a graphics calculator or a graph-drawing computer program.

The graph on the right represents any function $y = f(x)$.

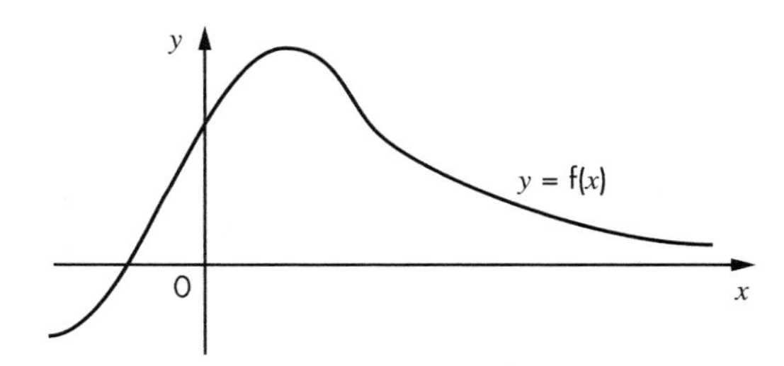

Rule 1 The graph of $y = f(x) + a$ is a **translation** of the graph of $y = f(x)$ by a **vector** $\begin{pmatrix} 0 \\ a \end{pmatrix}$.

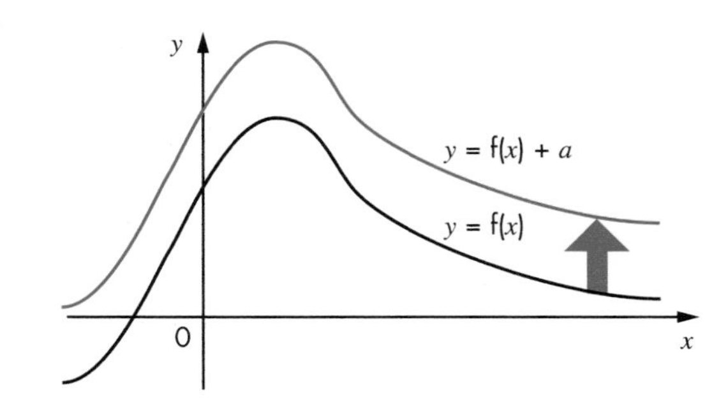

Rule 2 The graph of $y = f(x - a)$ is a translation of the graph of $y = f(x)$ by a vector $\begin{pmatrix} a \\ 0 \end{pmatrix}$.

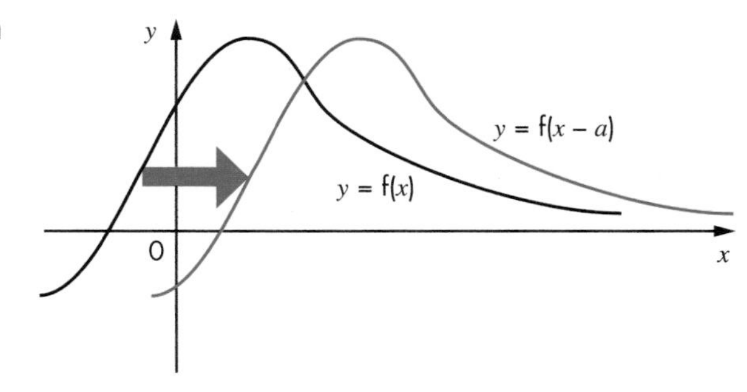

Note: The sign in front of a in the bracket is negative, but the translation is in the positive direction. $f(x + a)$ would translate $f(x)$ by the vector $\begin{pmatrix} -a \\ 0 \end{pmatrix}$.

A **stretch** is an enlargement that takes place in one direction only. It is described by a **scale factor** and the direction of the stretch.

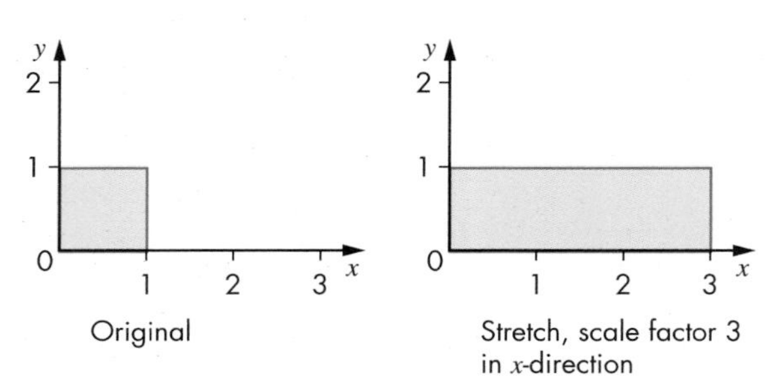

Rule 3 The graph of $y = af(x)$ is a stretch of the graph $y = f(x)$ by a scale factor of a in the y-direction.

Note: Points on the x-axis do not move. These are called invariant points.

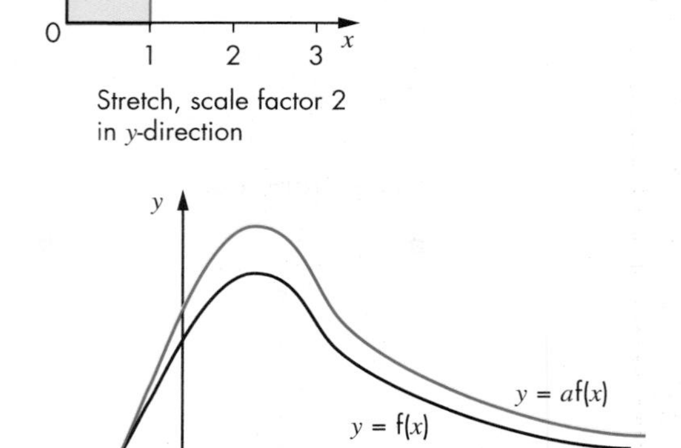

Rule 4 The graph of $y = f(ax)$ is a stretch of the graph $y = f(x)$ by a scale factor of $\frac{1}{a}$ in the x-direction.

Note: Points on the y-axis do not move and the scale factor of the stretch is the reciprocal of the constant multiplier inside the bracket.

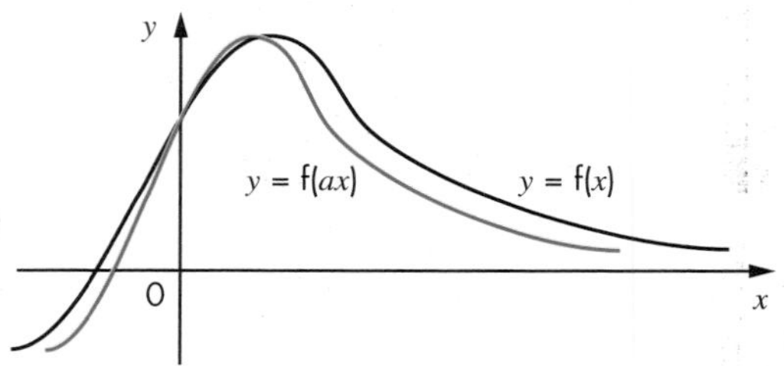

Rule 5 The graph of $y = -f(x)$ is the **reflection** of the graph $y = f(x)$ in the x-axis.

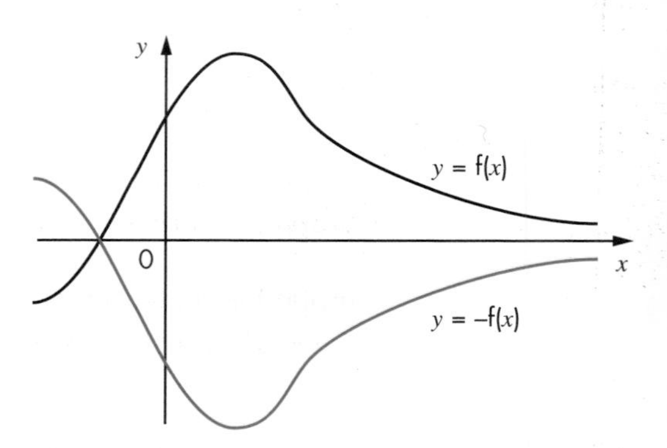

Rule 6 The graph of $y = f(-x)$ is the reflection of the graph $y = f(x)$ in the y-axis.

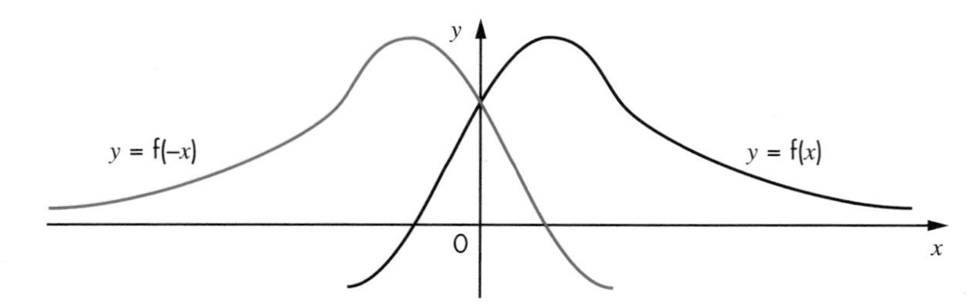

EXAMPLE 1

Sketch the following graphs.

a $y = x^2$ **b** $y = 5x^2$ **c** $y = x^2 - 5$

d $y = -x^2$ **e** $y = (x - 5)^2$ **f** $y = 2x^2 + 3$

Describe the transformation(s) that change(s) graph **a** to each of the other graphs.

Graph **a** is the basic graph to which you will apply the rules to make the necessary transformations: graph **b** uses Rule 3, graph **c** uses Rule 1, graph **d** uses Rule 5, graph **e** uses Rule 2, and graph **f** uses Rules 3 and 1.

The graphs are:

a

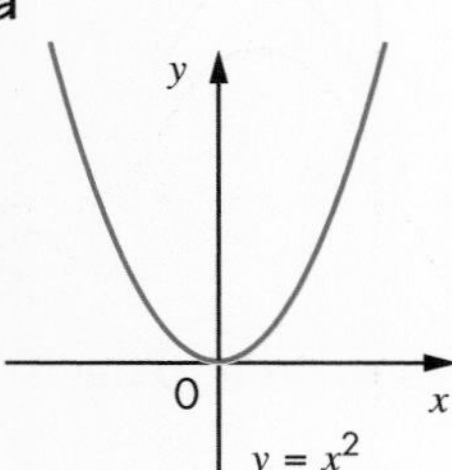

b

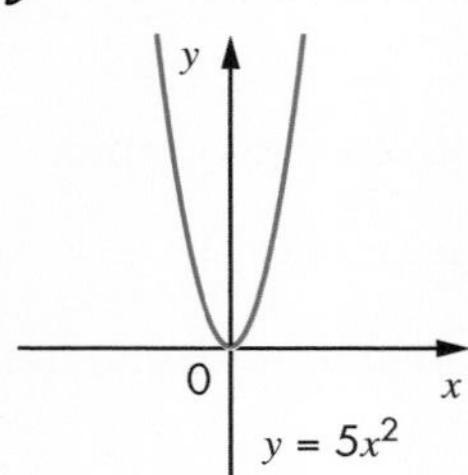

c

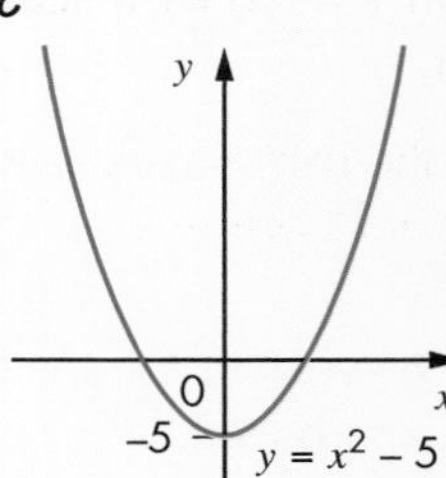

d

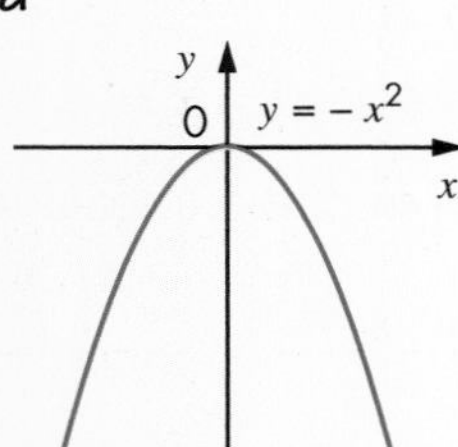

e

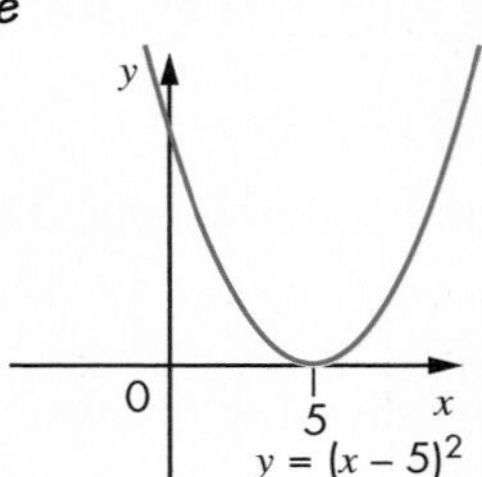

f

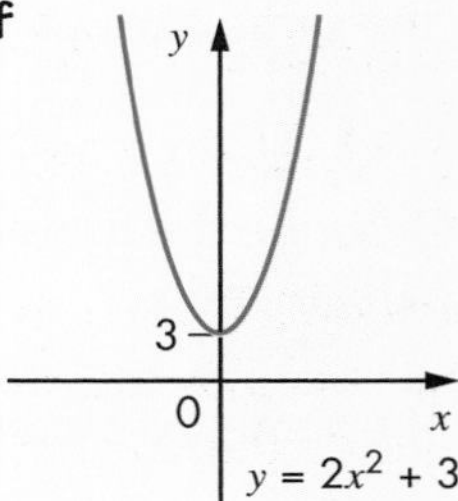

The transformations are:

graph **b** is a stretch of scale factor 5 in the y-direction,

graph **c** is a translation of $\begin{pmatrix} 0 \\ -5 \end{pmatrix}$,

graph **d** is a reflection in the x-axis,

graph **e** is a translation of $\begin{pmatrix} 5 \\ 0 \end{pmatrix}$,

graph **f** is a stretch of scale factor 2 in the y-direction, followed by a translation of $\begin{pmatrix} 0 \\ 3 \end{pmatrix}$.

Note that two of the transformations cause problems because they seem to do the opposite of what is expected. These are:

$y = f(x + a)$ (Rule 2)

The translation is $\begin{pmatrix} -a \\ 0 \end{pmatrix}$, so the sign of the constant inside the bracket changes in the vector (see part **e** in Example 1).

$y = f(ax)$ (Rule 4)

This does not look like a stretch. It actually closes the graph up. Just like an enlargement (see Chapter 6) can make something smaller, a stretch can make it squeeze closer to the axes.

EXERCISE 18A 

A*

1 On the same axes sketch the following graphs.

a $y = x^2$ **b** $y = 3x^2$ **c** $y = \frac{1}{2}x^2$ **d** $y = 10x^2$

e Describe the transformation(s) that take(s) the graph in part **a** to each of the graphs in parts **b** to **d**.

2 On the same axes sketch the following graphs.

a $y = x^2$ **b** $y = x^2 + 3$ **c** $y = x^2 - 1$ **d** $y = 2x^2 + 1$

e Describe the transformation(s) that take(s) the graph in part **a** to each of the graphs in parts **b** to **d**.

3 On the same axes sketch the following graphs.

a $y = x^2$ **b** $y = (x + 3)^2$ **c** $y = (x - 1)^2$ **d** $y = 2(x - 2)^2$

e Describe the transformation(s) that take(s) the graph in part **a** to each of the graphs in parts **b** to **d**.

4 On the same axes sketch the following graphs.

a $y = x^2$ **b** $y = (x + 3)^2 - 1$ **c** $y = 4(x - 1)^2 + 3$

d Describe the transformation(s) that take(s) the graph in part **a** to each of the graphs in parts **b** and **c**.

5 On the same axes sketch the following graphs.

a $y = x^2$ **b** $y = -x^2 + 3$ **c** $y = -3x^2$ **d** $y = -2x^2 + 1$

e Describe the transformation(s) that take(s) the graph in part **a** to each of the graphs in parts **b** to **d**.

6 On the same axes sketch the following graphs.

a $y = \sin x$ **b** $y = 2\sin x$ **c** $y = \frac{1}{2}\sin x$ **d** $y = 10\sin x$

e Describe the transformation(s) that take(s) the graph in part **a** to each of the graphs in parts **b** to **d**.

7 On the same axes sketch the following graphs.

a $y = \sin x$ **b** $y = \sin 3x$ **c** $y = \sin \frac{x}{2}$ **d** $y = 5\sin 2x$

e Describe the transformation(s) that take(s) the graph in part **a** to each of the graphs in parts **b** to **d**.

8 On the same axes sketch the following graphs.

a $y = \sin x$ **b** $y = \sin (x + 90°)$ **c** $y = \sin (x - 45°)$ **d** $y = 2\sin (x - 90°)$

e Describe the transformation(s) that take(s) the graph in part **a** to the graphs in parts **b** to **d**.

A*

9 On the same axes sketch the following graphs.

a $y = \sin x$ **b** $y = \sin x + 2$ **c** $y = \sin x - 3$ **d** $y = 2\sin x + 1$

e Describe the transformation(s) that take(s) the graph in part **a** to the graphs parts **b** to **d**.

10 On the same axes sketch the following graphs.

a $y = \sin x$ **b** $y = -\sin x$ **c** $y = \sin(-x)$ **d** $y = -\sin(-x)$

e Describe the transformation(s) that take(s) the graph in part **a** to the graphs in parts **b** to **d**.

11 On the same axes sketch the following graphs.

a $y = \cos x$ **b** $y = 2\cos x$ **c** $y = \cos(x - 60°)$ **d** $y = \cos x + 2$

e Describe the transformation(s) that take(s) the graph in part **a** to the graphs in parts **b** to **d**.

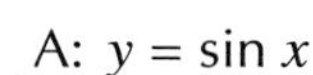

AU 12 Which of the equations below represents the graph shown?

A: $y = \sin x$

B: $y = \cos(x - 90°)$

C: $y = -\sin(-x)$

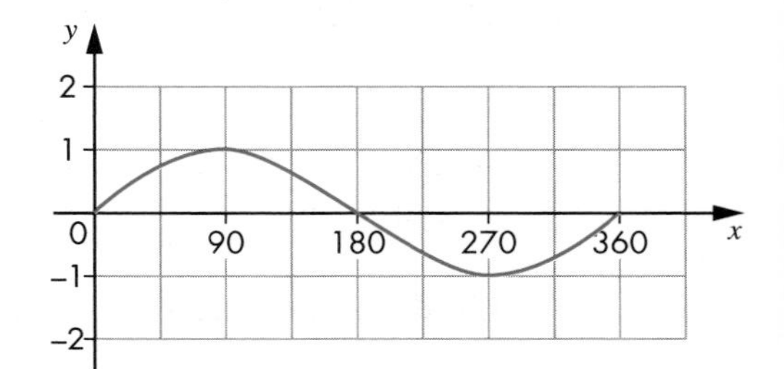

13 **a** Describe the transformations of the graph of $y = x^2$ needed to obtain these graphs.

i $y = 4x^2$ **ii** $y = 9x^2$ **iii** $y = 16x^2$

b Describe the transformations of the graph of $y = x^2$ needed to obtain these graphs.

i $y = (2x)^2$ **ii** $y = (3x)^2$ **iii** $y = (4x)^2$

c Describe two different transformations that take the graph of $y = x^2$ to the graph of $y = (ax)^2$, where a is a positive number.

14 On the right is a sketch of the function $y = f(x)$. Use this to sketch the following.

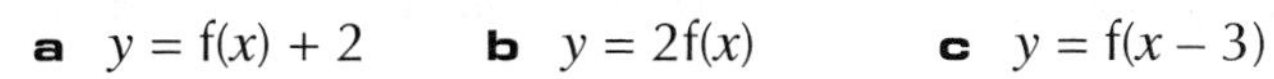

a $y = f(x) + 2$ **b** $y = 2f(x)$ **c** $y = f(x - 3)$

d $y = -f(x)$ **e** $y = 2f(x) + 3$ **f** $y = -f(x) - 2$

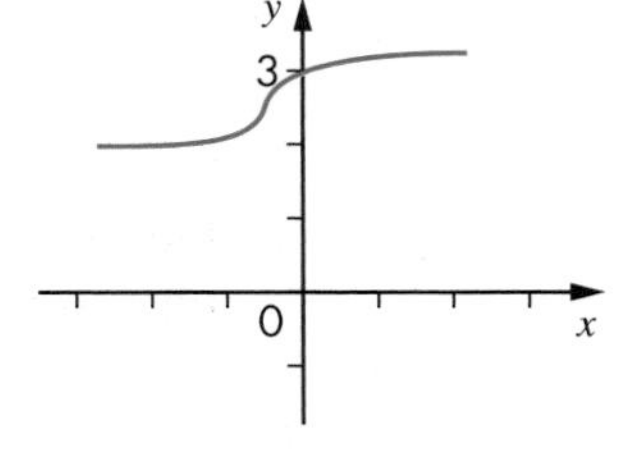

15 What is the equation of the graph obtained when the following transformations are performed on the graph of $y = x^2$?

a Stretch by a factor of 5 in the y-direction

b Translation of $\begin{pmatrix} 0 \\ 7 \end{pmatrix}$ **c** Translation of $\begin{pmatrix} -3 \\ 0 \end{pmatrix}$ **d** Translation of $\begin{pmatrix} -2 \\ -3 \end{pmatrix}$

e Stretch by a factor of 3 in the y-direction followed by a translation of $\begin{pmatrix} 0 \\ 4 \end{pmatrix}$

f Reflection in the x-axis, followed by a stretch, scale factor 3, in the y-direction

16 What is the equation of the graph obtained when the following transformations are performed on the graph of $y = \cos x$?

a Stretch by a factor of 6 in the y-direction **b** Translation of $\begin{pmatrix} 0 \\ 3 \end{pmatrix}$

c Translation of $\begin{pmatrix} -30 \\ 0 \end{pmatrix}$ **d** Translation of $\begin{pmatrix} 45 \\ -2 \end{pmatrix}$

e Stretch by a factor of 3 in the y-direction followed by a translation of $\begin{pmatrix} 0 \\ -2 \end{pmatrix}$

17 a Sketch the graph $y = x^3$.

b Use your sketch in part **a** to draw the graphs obtained after $y = x^3$ is transformed as follows.

i Reflection in the x-axis **ii** Translation of $\begin{pmatrix} 0 \\ -2 \end{pmatrix}$

iii Stretch by a scale factor of 3 in the y-direction **iv** Translation of $\begin{pmatrix} -2 \\ 0 \end{pmatrix}$

c Give the equation of each of the graphs obtained in part **b**.

18 a Sketch the graph of $y = \frac{1}{x}$.

b Use your sketch in part **a** to draw the graphs obtained after $y = \frac{1}{x}$ is transformed as follows.

i Translation of $\begin{pmatrix} 0 \\ 4 \end{pmatrix}$ **ii** Translation of $\begin{pmatrix} 4 \\ 0 \end{pmatrix}$

iii Stretch, scale factor 3 in the y-direction

iv Stretch, scale factor $\frac{1}{2}$ in the x-direction

c Give the equation of each of the graphs obtained in part **b**.

PS 19 A teacher asked her class to apply the following transformations to the function $f(x) = x^2$.

a $f(-x)$ **b** $-f(x)$

Martyn said that they must be the same as $-x^2 = x^2$.

Is Martyn correct? Explain your answer.

20 The graphs below are all transformations of $y = x^2$. Two points through which each graph passes are indicated. Use this information to work out the equation of each graph.

a

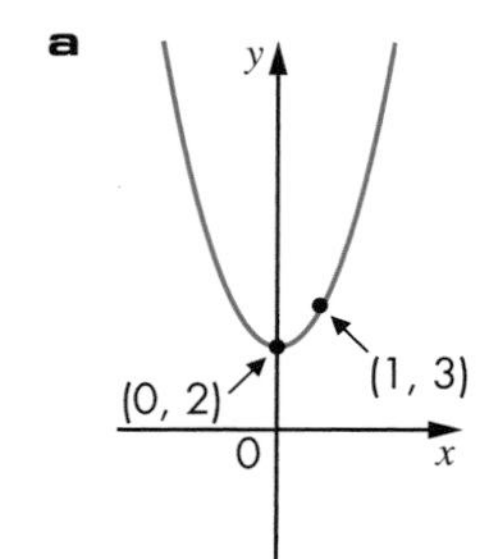

b

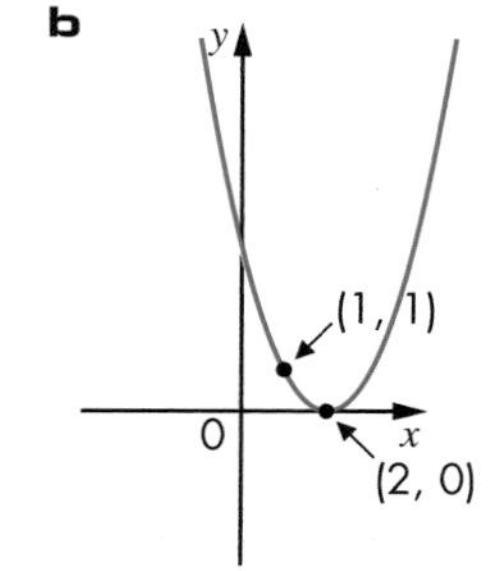

c

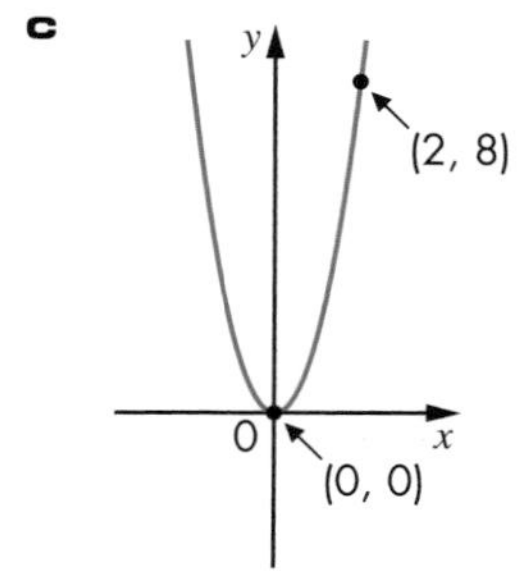

d

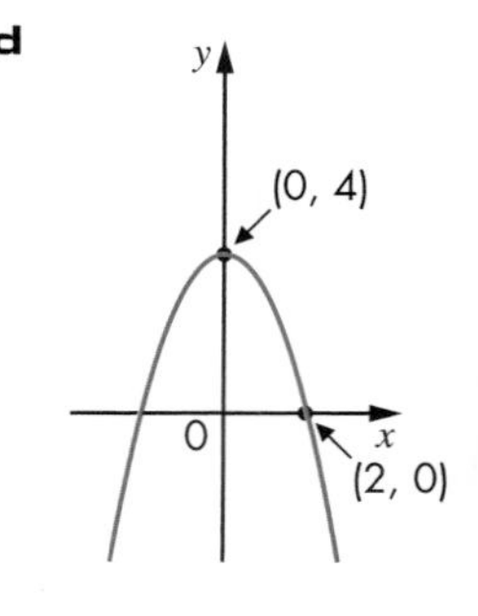

21 The graphs below are all transformations of $y = \sin x$. Two points through which each graph passes are indicated. Use this information to work out the equation of each graph.

a

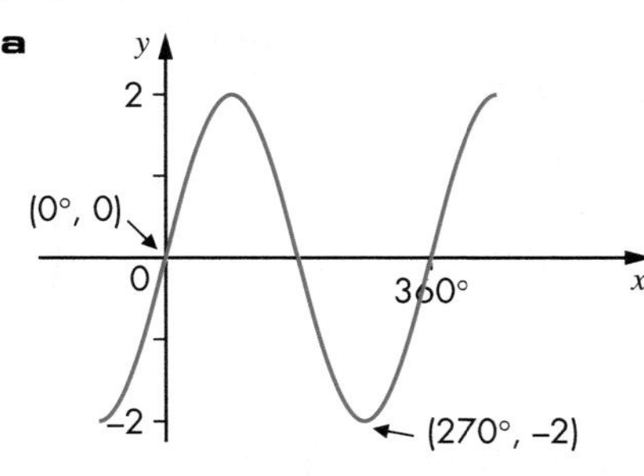

b

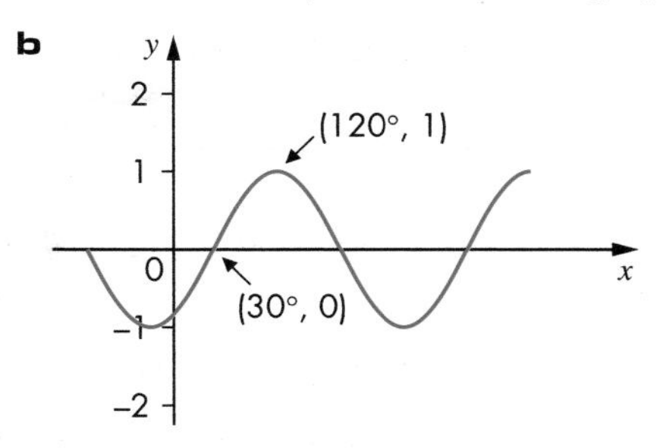

c

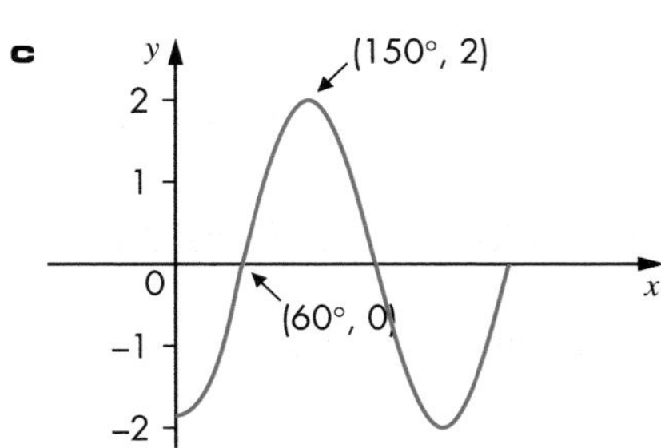

d

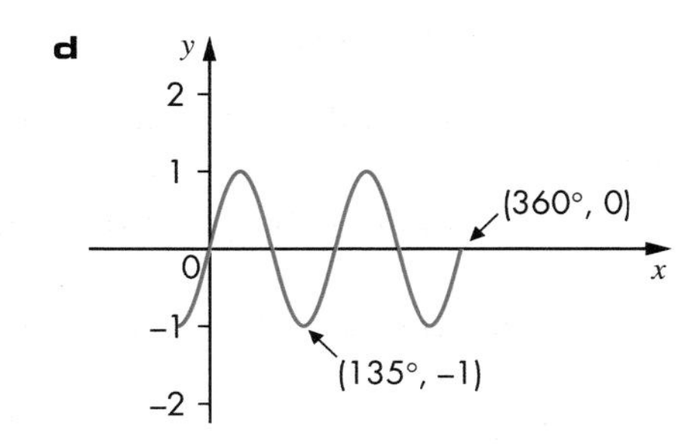

22 Below are the graphs of $y = \sin x$ and $y = \cos x$.

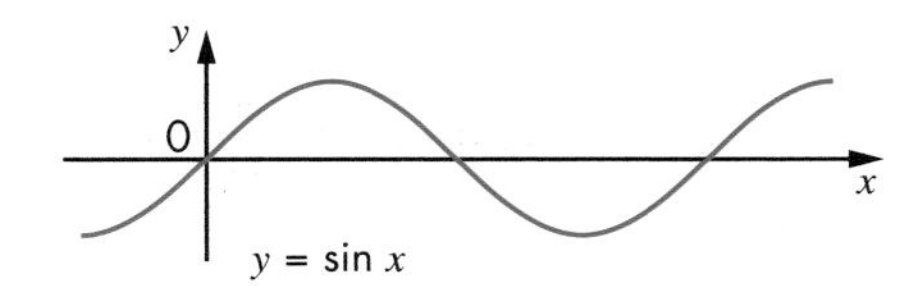

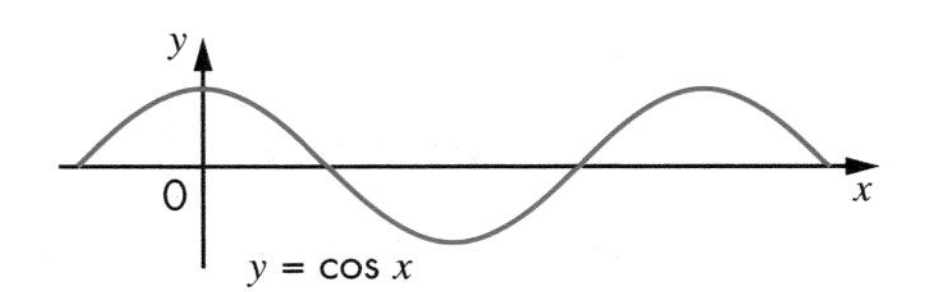

a Describe a series of transformations that would take the first graph to the second.

b Which of these is equivalent to $y = \cos x$?

i $y = \sin (x + 90°)$ **ii** $y = -\sin (x - 90°)$ **iii** $y = 2\cos \frac{x}{2}$

23 **A**

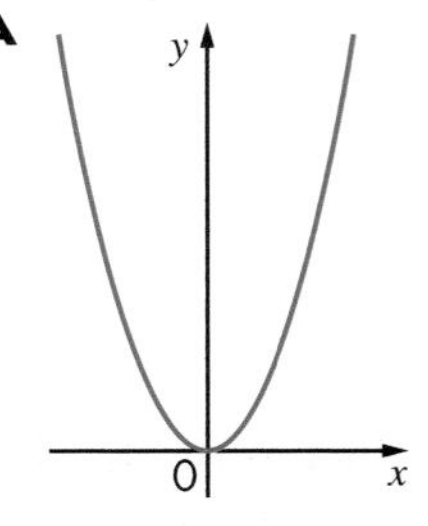

B

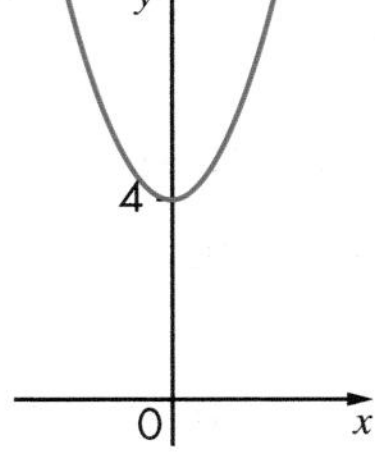

C

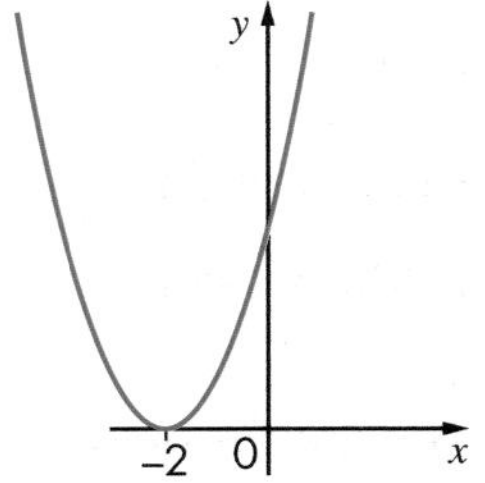

D

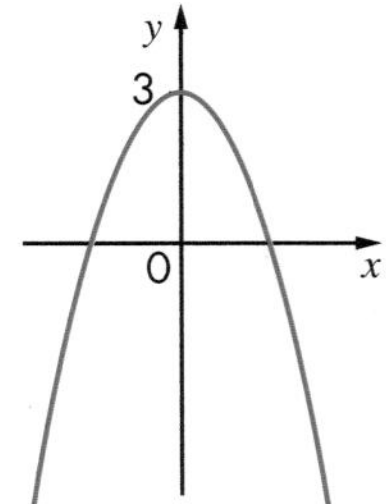

E

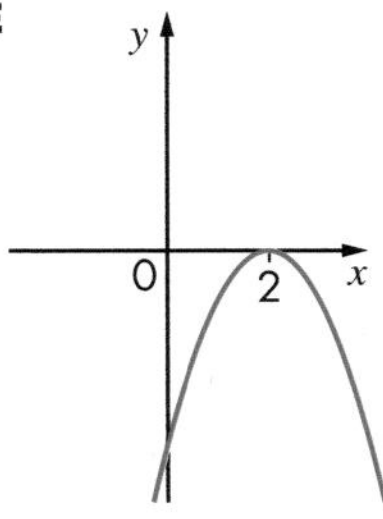

Match each of the graphs **A**, **B**, **C**, **D** and **E** to one of these equations.

i $y = x^2$ **ii** $y = -x^2 + 3$ **iii** $y = -(x - 2)^2$ **iv** $y = (x + 2)^2$ **v** $y = x^2 + 4$

18.2 Cubic, exponential and reciprocal graphs

This section will show you how to:
- recognise and plot cubic, exponential and reciprocal graphs

Key words
asymptote
cubic
exponential function
reciprocal

Cubic graphs

A **cubic** function or graph is one that contains a term in x^3. The following are examples of cubic graphs.

$y = x^3$ $\quad\quad$ $y = x^3 + 3x$ $\quad\quad$ $y = x^3 + x^2 + x + 1$

The techniques used to draw them are exactly the same as those for quadratic and reciprocal graphs.

Questions requiring an accurate drawing of a cubic graph are not very common in examinations, but questions asking if you can recognise a cubic graph occur quite often.

This is the basic graph $y = x^3$.

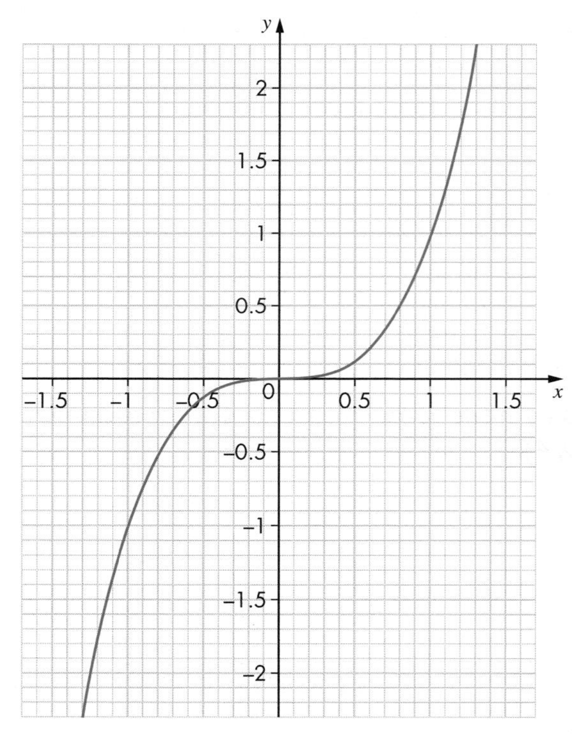

It has a characteristic shape which you should learn to recognise.

Example 5 shows you how to draw a cubic graph accurately.

You should use a calculator to work out the value of y and to round to 1 or 2 decimal places.

EXAMPLE 2

a Complete the table to draw the graph of $y = x^3 - x^2 - 4x + 4$ for $-3 \leqslant x \leqslant 3$.

x	−3	−2.5	−2	−1.5	−1	−0.5	0	0.5	1	1.5	2	2.5	3
y	−20.00		0.00		6.00		4.00	1.88				3.38	10.00

b Use your graph to give the roots of the equation $x^3 - x^2 - 4x + 4 = 0$.

c Write down the coordinates of:

i the minimum vertex **ii** the maximum vertex.

d Write down the coordinates of the point where the graph intersects the y-axis.

a The completed table (to 2 decimal places) is given below and the graph is shown below, right.

x	−3	−2.5	−2	−1.5	−1	−0.5	0	0.5	1	1.5	2	2.5	3
y	−20.00	−7.88	0.00	4.38	6.00	5.63	4.00	1.88	0.00	−0.88	0.00	3.38	10.00

b Just as in quadratic graphs, the roots are the points where the graph crosses the x-axis.

So the roots are $x = -2$, 1 and 2.

Note that, in the table, these are the x-values where the y-value is 0.

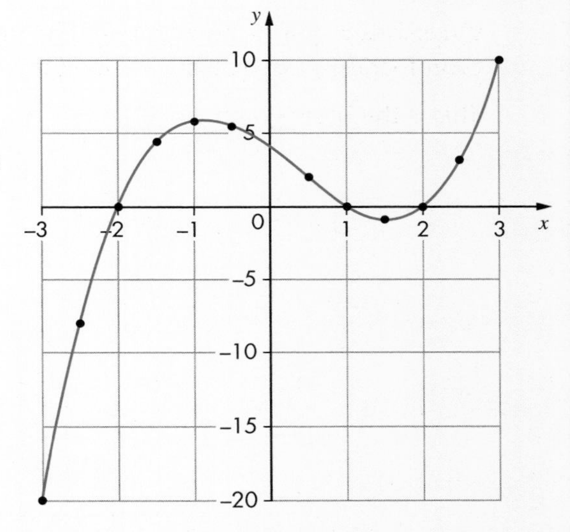

c **i** The minimum vertex is at the point (1.5, −0.88).

ii The maximum vertex is at the point (−1, 6).

Note that the minimum and maximum values of the function are ± infinity, as the arms of the curve continue forever.

d Just as in the quadratic, this is the constant term in the equation, so the point is (0, 4).

Reciprocal graphs

A **reciprocal** equation has the form $y = \frac{a}{x}$.

Examples of reciprocal equations are:

$$y = \frac{1}{x} \qquad y = \frac{4}{x} \qquad y = -\frac{3}{x}$$

All reciprocal graphs have a similar shape and some symmetry properties.

EXAMPLE 3

Complete the table to draw the graph of $y = \frac{1}{x}$ for $-4 \leqslant x \leqslant 4$.

x	–4	–3	–2	–1	1	2	3	4
y								

Values are rounded to two decimal places, as it is unlikely that you could plot a value more accurately than this. The completed table looks like this.

x	–4	–3	–2	–1	1	2	3	4
y	–0.25	–0.33	–0.5	–1	1	0.5	0.33	0.25

The graph plotted from these values is shown in **A**. This is not much of a graph and does not show the properties of the reciprocal function. If you take x-values from –0.8 to 0.8 in steps of 0.2, you get the next table.

Note that you cannot use $x = 0$ since $\frac{1}{0}$ is infinity.

x	–0.8	–0.6	–0.4	–0.2	0.2	0.4	0.6	0.8
y	–1.25	–1.67	–2.5	–5	5	2.5	1.67	1.25

Plotting these points as well gives the graph in **B**.

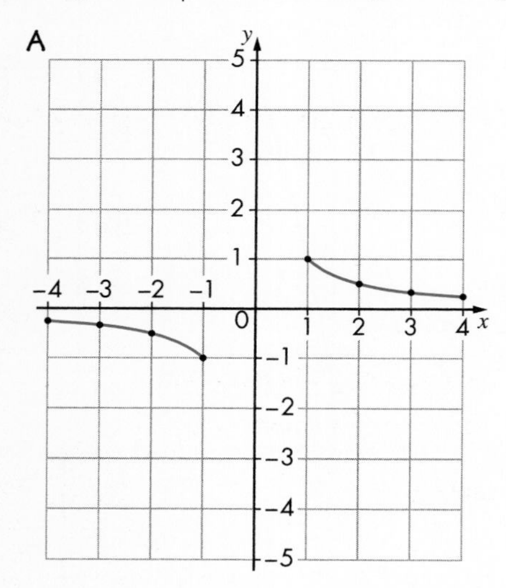

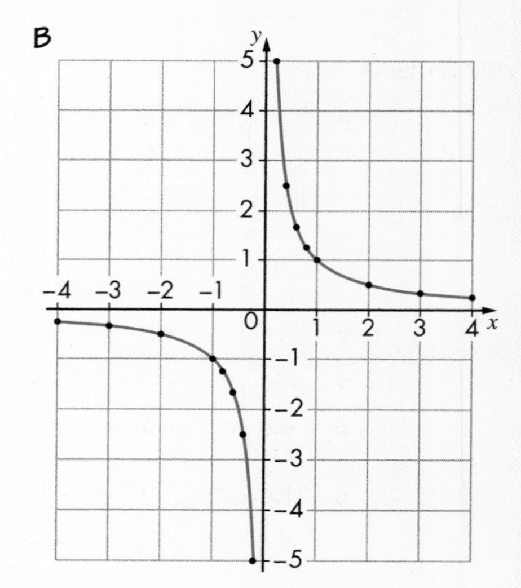

From the graph in **B**, the following properties can be seen.

- The lines $y = x$ and $y = -x$ are lines of symmetry.
- The closer x gets to zero, the nearer the graph gets to the y-axis.
- As x increases, the graph gets closer to the x-axis.

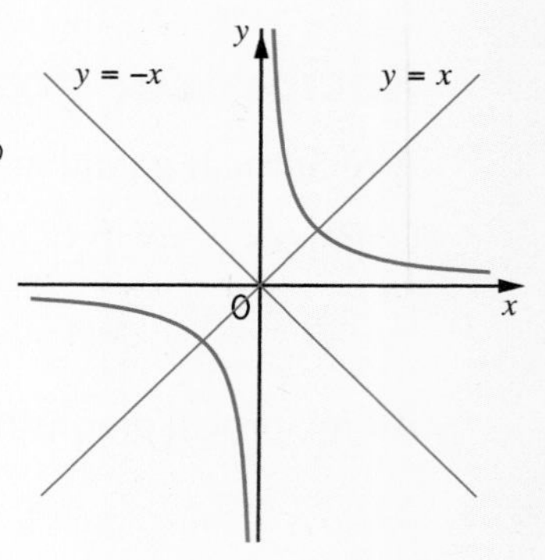

The graph never actually touches the axes, it just gets closer and closer to them. A line to which a graph gets closer but never touches or crosses is an **asymptote**.

These properties are true for *all reciprocal graphs*.

Exponential graphs

Equations that have the form $y = k^x$, where k is a positive number, are called **exponential functions**.

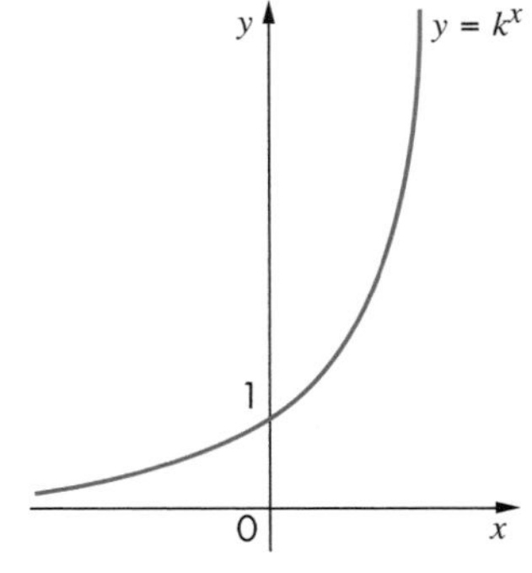

Exponential functions share the following properties.

- When k is greater than 1, the value of y increases steeply as x increases, which you can see from the graph on the right.
- Also when k is greater than 1, as x takes on increasingly large negative values, the closer y gets to zero, and so the graph gets nearer and nearer to the negative x-axis. y never actually becomes zero and so the graph never touches the negative x-axis. That is, the negative x-axis is an asymptote to the graph. (See also previous page.)

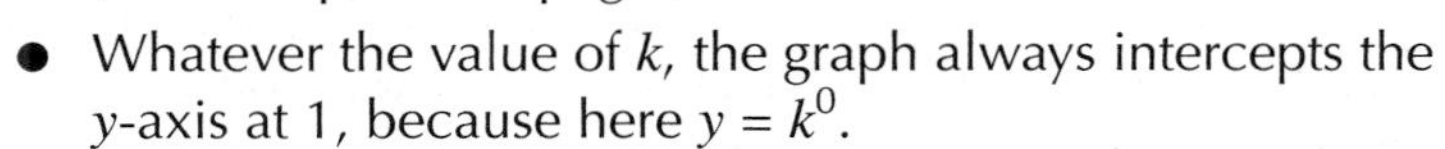

- Whatever the value of k, the graph always intercepts the y-axis at 1, because here $y = k^0$.
- The reciprocal graph, $y = k^{-x}$, is the reflection in the y-axis of the graph of $y = k^x$, as you can see from the graph on the right.

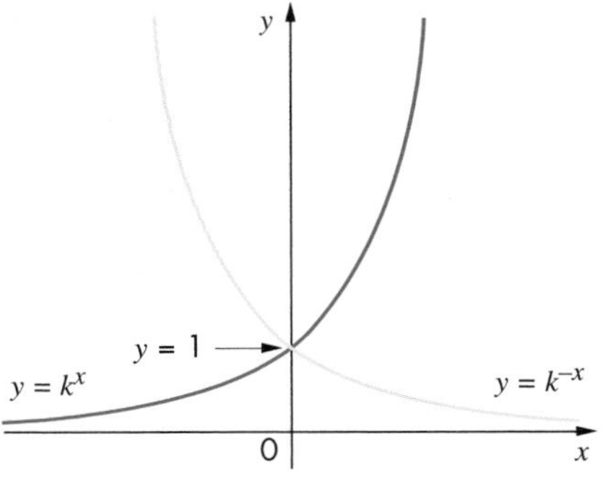

EXAMPLE 4

a Complete the table below for $y = 2^x$ for values of x from –5 to +5. (Values are rounded to 2 decimal places.)

x	–5	–4	–3	–2	–1	0	1	2	3	4	5
$y = 2^x$	0.03	0.06	0.13			1	2	4			32

b Plot the graph of $y = 2^x$ for $-5 \leqslant x \leqslant 5$.

c Use your graph to estimate the value of y when $x = 2.5$.

d Use your graph to estimate the value of x when $y = 0.75$.

a The values missing from the table are:

0.25, 0.5, 8 and 16

b Part of the graph (drawn to scale) is shown on the right.

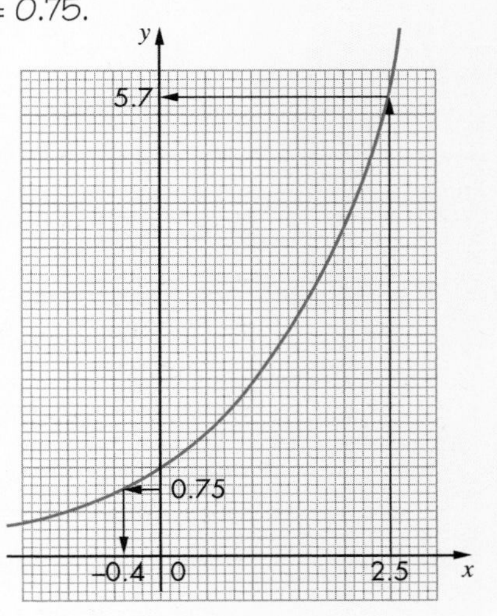

c Draw a line vertically from $x = 2.5$ until it meets the graph and then read across.
The y-value is 5.7.

d Draw a line horizontally from $y = 0.75$, the x-value is –0.4.

See exercise 12I on p383 Student Book 1 for extra practice

GRADE BOOSTER

B You can recognise the shapes of the graphs $y = x^3$ and $y = \frac{1}{x}$

A You can recognise a variety of graphs such as exponential graphs and reciprocal graphs using a table of values

A* You can transform the graph of a given function

A* You can identify the equation of a function from its graph, which has been formed by a transformation on a known function

What you should know now

- How to sketch the graphs of functions such as $y = \mathrm{f}(ax)$ and $y = \mathrm{f}(x + a)$ from the known graph of $y = \mathrm{f}(x)$
- How to describe from their graphs the transformation of one function into another
- How to identify equations from the graphs of transformations of known graphs

1 The diagram shows part of the curve with equation $y = \mathrm{f}(x)$.

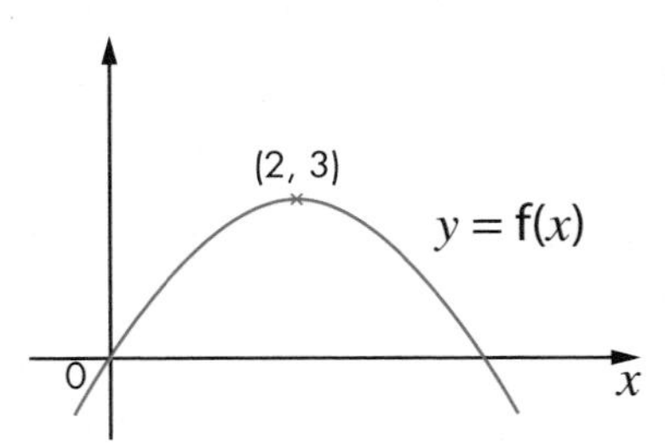

The coordinates of the maximum point of this curve are (2, 3).

Write down the coordinates of the maximum point of the curve with equation:

a $y = \mathrm{f}(x - 2)$ *(1 mark)*

b $y = 2\mathrm{f}(x)$ *(1 mark)*

Edexcel, May 2008, Paper 3 Higher, Question 27

2 The graph of $y = \sin x°$ for $0 \leqslant x \leqslant 360$ is drawn below.

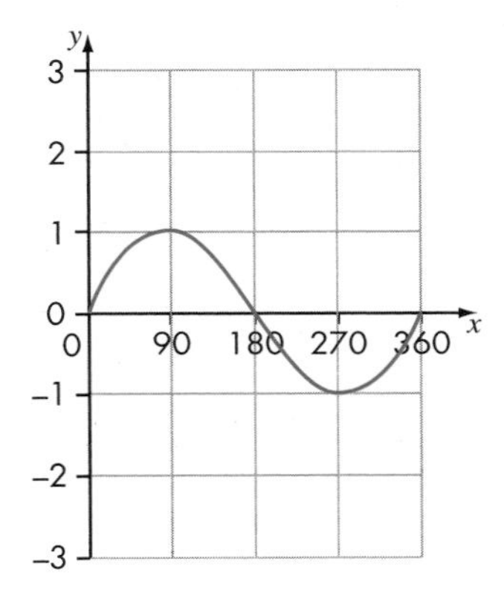

On copies of the same axes draw sketches of the following graphs.

a $y = 2 \sin x°$

b $y = \sin 2x°$

c $y = \sin x° - 2$

3 The graph of $y = \mathrm{f}(x)$ is shown below.

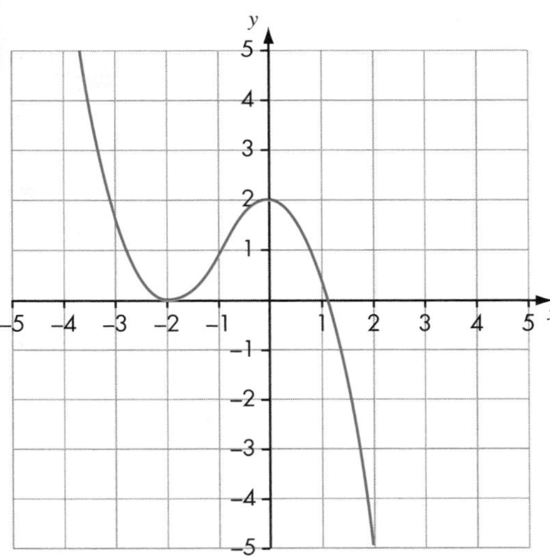

On copies of the grid, sketch the following graphs:

a $y = \mathrm{f}(x + 1)$ *(1 mark)*

b $y = 2\mathrm{f}(x)$ *(1 mark)*

Edexcel, June 2005, Paper 6 Higher, Question 20

Worked Examination Questions

1 The sketch shows the graph $y = x^3$.

Copy the axes below and sketch the graphs indicated.

p is a positive integer greater than 1.
(The graph $y = x^3$ is shown dotted to help you.)

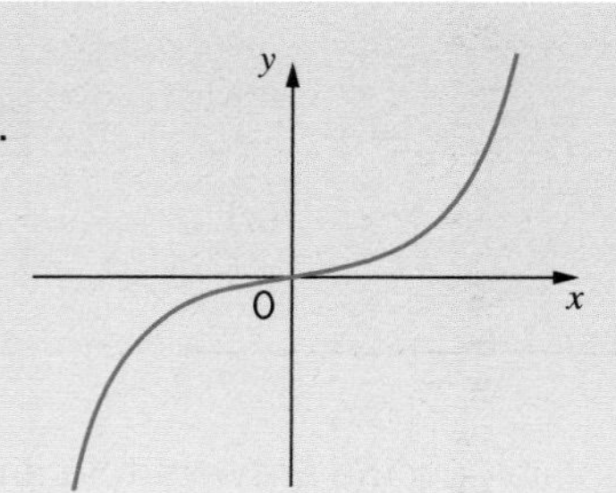

a $4 = x^3 - p$

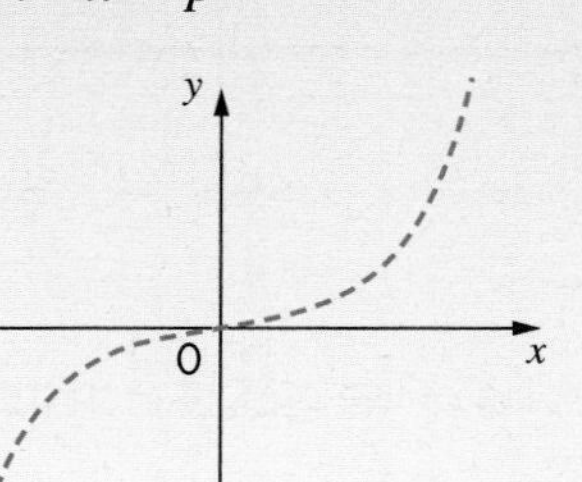

b $y = (x + p)^3$

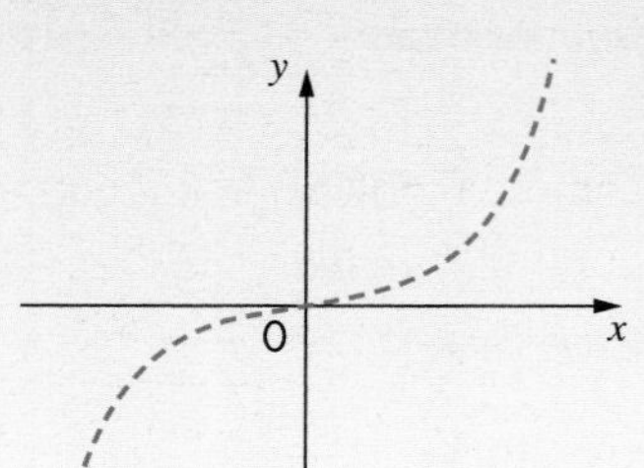

c $y = \frac{x^3}{p}$

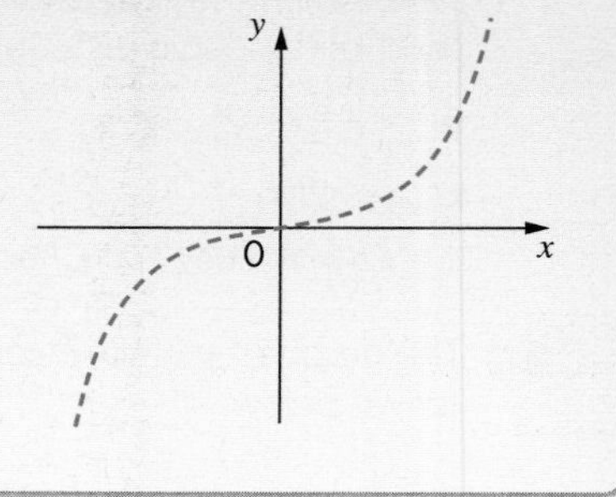

1 a This is a translation of $y = x^3$ by the vector $\begin{pmatrix} 0 \\ -p \end{pmatrix}$.

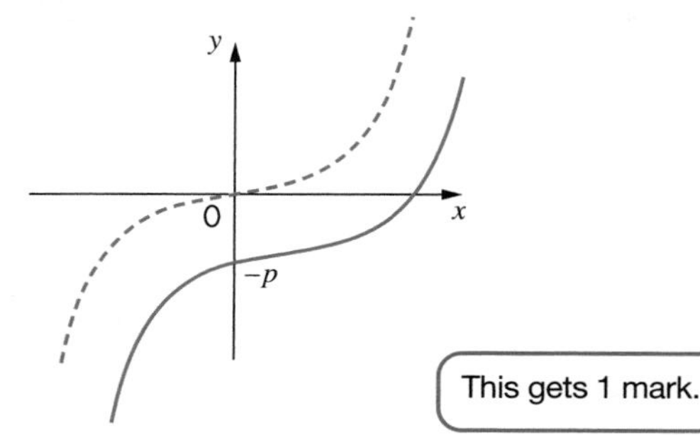

This gets 1 mark.

b This is a translation of $y = x^3$ by the vector $\begin{pmatrix} 0 \\ -p \end{pmatrix}$.

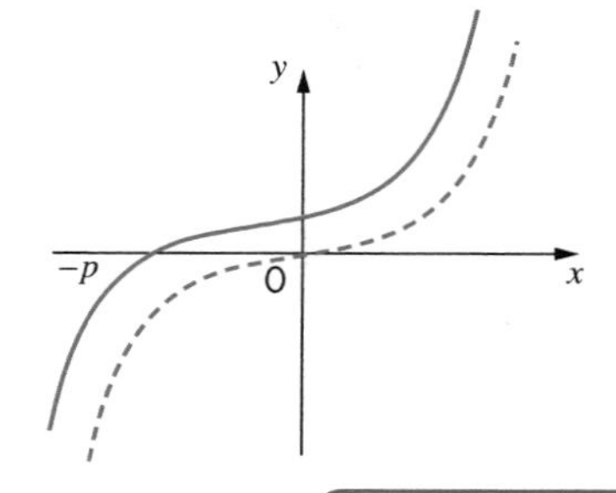

You do not know what the actual value of p is so make sure that the translation is clear. Alternatively, make a value for p up, say 2. This gets 1 mark.

c This is a stretch of $y = x^3$ by a scale factor of $\frac{1}{p}$ in the y-direction.

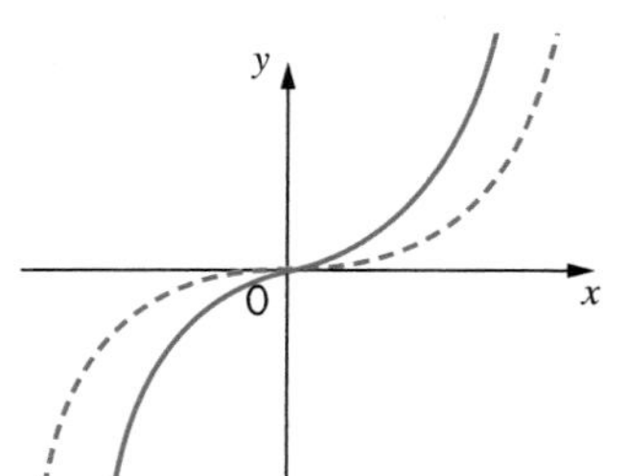

This gets 1 mark.

Total: 3 marks

Worked Examination Questions

PS **2** The sketch shows the graph of $y = x^2 - 6x + 5$.

The minimum point of the graph is (3, –4).

a Describe what happens to the graph of $f(x) = x^2$ under the transformation $f(x - 3)$.

b Describe what happens to the graph of $f(x) = x^2$ under the transformation $f(x) - 4$.

c Explain how the answers to **a** and **b** connect with the equation $y = x^2 - 6x + 5$ and the minimum point (3, –4).

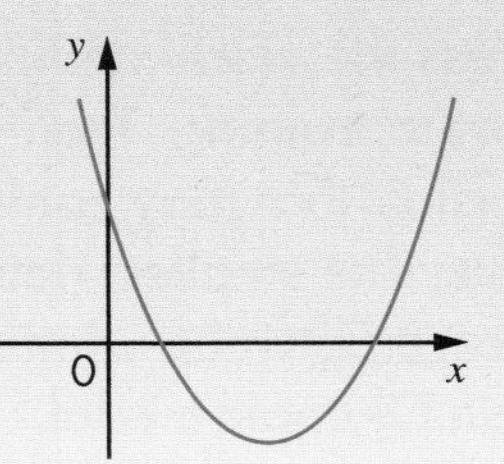

2 a A translation of $\begin{pmatrix} 3 \\ 0 \end{pmatrix}$

b A translation of $\begin{pmatrix} 0 \\ -4 \end{pmatrix}$

c $x^2 - 6x + 5 = (x - 3)^2 - 4$,

so the graph is a translation of $\begin{pmatrix} 3 \\ -4 \end{pmatrix}$,

which takes the original minimum point (0, 0) to (3, –4).

The answers to parts **b** and **c** give the clue to link with the minimum point. Remember that writing the equation in 'completing the square' form $(x - a)^2 - b$ gives the minimum point $(a, -b)$. This gets 1 mark for method and 1 mark for accuracy.

Total: 2 marks

18 Functional Maths
Diving at the Olympics

In preparation for the Olympics, sportsmen and women across the world dedicate a lot of time and effort to training. Their coaches analyse every action that may contribute to their performance. The smallest detail can make the difference between winning or missing out on an Olympic medal.

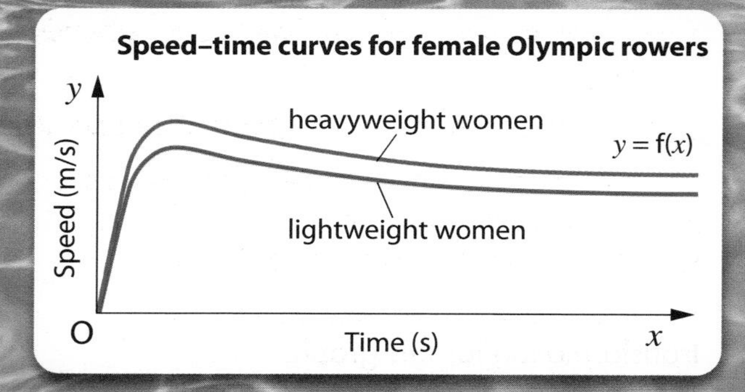

Getting started

The graph shows the mean boat speed against time at the beginning of a race for two female Olympic rowers: one a heavyweight and the other a lightweight.

Which of these statements describes the graphs?

- The blue graph is a translation of the red graph by a vector $\binom{0}{a}$.
- The blue graph is a translation of the red graph by a vector $\binom{a}{0}$.
- The blue graph is a stretch of the red graph in the vertical or y-direction.
- The blue graph is a reflection of the red graph in the horizontal or x-axis.

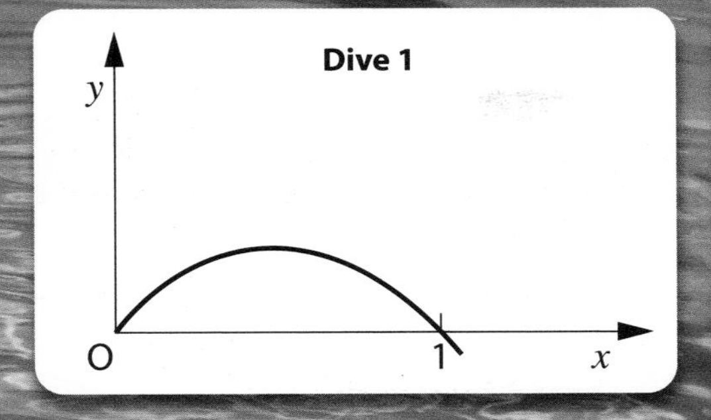

Your task

An Olympic diver begins her training for the day by performing a warm-up dive from the side of the pool. Her coach analyses the path of the dive and describes it as

$$y = -x^2 + x \qquad \text{(dive 1)}$$

where y represents the distance from the surface of the water and x represents the distance from the side of the pool. The graph to the left shows how this looks.

The diver performs a second dive from the side of the pool. Its graph is like the one for dive 1 but with a stretch in the y-direction (dive 2).

Then the diver climbs to the first diving platform, arches her back with arms outstretched and performs a swan dive (dive 3). The graph for this third dive is a transformation of the graph for dive 2.

Finally, the diver is joined by a team-mate to practise for the synchronised diving event. They dive, in unison, from two adjacent platforms along the side of the pool, both at the height of the second diving platform. The coach calls this dive 4.

Read the description of each dive above carefully.

Then write an appropriate equation and an accurate description for each of the transformations for dive 2 and dive 3.

Now consider dive 4. Describe the transformation for this graph.

Write the coach's end-of-day diving team report, sketching some graphs to show the shapes of the dives.

APPENDIX

Your Higher Edexcel Modular Specification

This table will show you exactly where to find all the content that you require to complete your Higher Edexcel Modular Specification.

Ref.	Descriptor	Unit 1	Unit 2	Unit 3
Statistics and Probability				
SPa	Understand and use statistical problem-solving process/handling data cycle	Book 1, lesson 4.8		
SPb	Identify possible sources of bias	Book 1, lesson 4.10		
SPc	Design an experiment or survey	Book 1, lessons 4.6–4.8, 4.10		
SPd	Design data collection sheets distinguishing between different types of data	Book 1, lessons 4.2–4.4, 4.6		
SPe	Extract data from printed tables and lists	Book 1, lessons 4.1, 4.2		
SPf	Design and use two-way tables for discrete and grouped tables	Book 1, lesson 6.4		
SPg	Produce charts and diagrams for various data types	Book 1, lessons 4.4, 4.5, 5.1–5.5		
SPh	Calculate median, mean, range, quartiles, interquartile range, mode, modal class	Book 1, lessons 4.1, 4.3, 5.4, 5.5		
SPi	Interpret a wide range of graphs and diagrams and draw conclusions	Book 1, lessons 4.2–4.4, 4.5, 5.1–5.5		
SPj	Look at data to find patterns and exceptions	Book 1, lessons 4.1–4.10, 5.1–5.5		
SPk	Recognise correlation and draw/use lines of best fit by eye	Book 1, lesson 5.3		
SPl	Compare distributions and make inferences	Book 1, lessons 4.1, 4.4, 5.1–5.5		
SPm	Understand and use the vocabulary of probability and probability scale	Book 1, lesson 6.1		
SPn	Understand and use estimates or measures of probability from theoretical models or from relative frequency	Book 1, lesson 6.1		
SPo	List all the outcomes for single events and two successive events	Book 1, lessons 6.1, 6.4, 6.6, 6.7		
SPp	Identify different mutually exclusive outcomes and know the sum is 1	Book 1, lesson 6.2		
SPq	Know when to add or multiply two probabilities	Book 1, lessons 6.2, 6.5, 6.7 6.8–6.10		
SPr	Use tree diagrams to represent outcomes of compound events	Book 1, lesson 6.7		
SPs	Compare experimental data and theoretical probabilities	Book 1, lessons 6.1, 6.3, 6.4		
SPt	Understand that if they repeat an experiment they will get different outcomes	Assumed knowledge		
SPu	Use calculators efficiently and effectively, including statistical functions	Book 1, lesson 4.2		
Algebra				
Aa	Distinguish the roles played by letter symbols in algebra	Prior knowledge	Prior knowledge	
Ab	Distinguish between 'equation', 'formula', 'identity' and 'expression'	Book 1, lesson 11.1	Book 1, lesson 11.1	
Ac	Manipulate algebraic expressions		Book 1, lessons 11.1, 11.2, 16.1–16.3	Book 2, lessons 11.1–11.3, 11.5
Ad	Set up and solve equations, including simultaneous equations			Book 2, lessons 7.2, 7.3, 7.5–7.7, 12.6
Ae	Solve quadratic equations			Book 2, lessons 11.3–11.6
Af	Derive a formula and substitute numbers into a formula		Book 1, lessons 11.1, 11.3	Book 2, lesson 7.1
Ag	Solve linear inequalities and represent the solution			Book 2, lesson 15.1, 15.2
Ah	Use systematic trial and improvement to find approximate solutions			Book 2, lesson 7.4
Ai	Generate terms of a sequence		Book 1, lessons 10.1, 10.3	
Aj	Use linear expressions to describe the nth term of a sequence		Book 1, lesson 10.2	
Ak	Use the conventions for coordinates in the plane		Book 1, lesson 17.1	Book 2, lesson 19.1
Al	Recognise and plot equations that correspond to straight-line graphs	Book 1, lessons 12.1, 12.3	Book 1, lessons 12.1, 12.3, 14.1	Book 2, lessons 12.1, 12.2
Am	Understand that $y = mx + c$ represents a straight line	Book 1, lessons 14.1, 14.2	Book 1, lessons 14.1, 14.2	
An	Understand the gradients of parallel lines		Book 1, lesson 12.5	
Ao	Find intersection points of the graphs of a linear and quadratic function			Book 2, lesson 12.6
Ap	Draw and recognise graphs of simple cubic, reciprocal and trigonometric functions	Book 1, lessons 12.6, 15.1		Book 2, lesson 18.2
Aq	Construct graphs of simple loci			Book 2, lesson 12.6
Ar	Construct linear, quadratic and other functions from real-life problems and plot their graphs		Book 1, lessons 12.1, 12.3	Book 2, lessons 12.3, 12.4, 12.6
As	Discuss, plot and interpret graphs modelling real-life situations	Book 1, lessons 12.1, 12.2, 12.4	Book 1, lessons 12.1, 12.2, 12.4	Book 2, lessons 12.1–12.7, 18.1, 18.2
At	Generate points and plot graphs of simple quadratic functions			Book 2, lessons 12.3, 12.4
Au	Direct and inverse proportion			Book 2, lessons 16.1, 16.2
Av	Transformation of functions			Book 2, lesson 18.1

Ref.	Descriptor	Unit 1	Unit 2	Unit 3
Number				
Na	Add, subtract, multiply and divide any number	Throughout Book 1	Throughout Book 1	Throughout Book 2
Nb	Order rational numbers		Prior knowledge	
Nc	Use the concepts and vocabulary of factor, multiple, common factor, HCF, LCM, prime number and prime factor decomposition		Book 1, lessons 7.2, 7.3	
Nd	Use the terms 'square', 'positive' and 'negative' 'square root', 'cube', 'cube root'		Book 1, lesson 7.2	
Ne	Use index notation for squares, cubes and powers of 10		Book 1, lessons 7.2, 9.1	
Nf	Use index laws, including for fractional and negative powers		Book 1, lesson 9.1	
Ng	Interpret, order and calculate with numbers written in standard index form		Book 1, lesson 9.2	Book 2, lesson 2.6
Nh	Understand equivalent fractions and simplify fractions	Prior knowledge	Prior knowledge	
Ni	Add and subtract fractions		Book 1, lesson 7.4	
Nj	Use decimal notation	Prior knowledge	Prior knowledge	
Nk	Recognise that recurring decimals are exact fractions		Book 1, lesson 9.3	Book 2, lesson 2.5
Nl	Understand what 'percentage' means and compare proportions	Prior knowledge	Prior knowledge	
Nm	Use percentage, repeated proportional change	Book 1, lesson 2.4		Book 2, lessons 2.2, 2.3
Nn	Understand and use direct and indirect proportion			Book 2, lessons 16.1, 16.2
No	Interpret fractions, decimals and percentages as operators	Book 1, lessons 2.1–2.3, 7.1, 7.4–7.6	Book 1, lessons 2.1–2.3, 7.1, 7.4–7.6	Book 1, lessons 2.2, 2.3, 7.1, 7.4–7.6, Book 2 lesson 2.2
Np	Use ratio notation	Book 1, lesson 2.6	Book 1, lesson 2.6	
Nq	Understand and use number operations, including inverse operations	Book 1, lesson 1.1	Book 1, lesson 1.1	Book 2, lessons 2.3, 2.5, 3.3
Nr	Use surds and π in exact calculations		Book 1, lesson 9.4	
Ns	Calculate upper and lower bounds			Book 2, lesson 3.1
Nt	Divide a quantity into a given ratio	Book 1, lesson 2.6	Book 1, lesson 2.6	
Nu	Approximate to specified or appropriate degrees of accuracy	Book 1, lesson 1.2	Book 1, lesson 1.2	
Nv	Use calculators effectively and efficiently	Throughout Book 1		Throughout Book 2
Geometry and Measures				
GMa	Recall and use properties of angles	Prior knowledge	Prior knowledge	
GMb	Understand and use the angle properties of intersecting lines, quadrilaterals and triangles		Book 1, lessons 8.4, 8.5	
GMc	Calculate and use the sums of interior and exterior angles of polygons		Book 1, lesson 8.5	
GMd	Recall the properties and definitions of special types of quadrilateral		Book 1, lesson 8.4	
GMe	Recognise reflection and rotation symmetry of 2D shapes		Prior knowledge	
GMf	Understand congruence and similarity			Book 2, lessons 6.1, 9.1
GMg	Use Pythagoras' theorem in 2D and 3D			Book 2, lessons 5.1–5.4
GMh	Use trigonometric ratios, sine and cosine rules to solve problems in 2D and 3D			Book 2, lessons 5.5–5.11, 10.1–10.6
GMi	Distinguish between 'centre', 'radius', 'chord', 'diameter', 'circumference', 'tangent', 'arc', 'sector', 'segment'		Prior knowledge	
GMj	Understand and construct geometrical proofs using circle theorems		Book 1, lesson 13.1–13.3	Book 2, lessons 14.1–14.3
GMk	Use 2D representations of 3D shapes		Prior knowledge	
GMl	Transform 2D shapes and distinguish properties preserved			Book 2, lessons 6.2–6.6
GMm	Use and interpret maps and scale drawings			Prior knowledge
GMn	Understand and use the effect of enlargement for perimeter, area and volume			Book 2, lessons 6.5, 9.2
GMo	Interpret scales on a range of measuring instruments and recognise inaccuracies	Prior knowledge	Prior knowledge	
GMp	Convert measurements from one unit to another	Prior knowledge	Prior knowledge	Prior knowledge
GMq	Make sensible estimates of a range of measures	Prior knowledge	Prior knowledge	
GMr	Understand and use bearings			Book 2, lesson 5.11
GMs	Understand and use compound measures		Book 1, lesson 2.8	Book 2, lessons 3.2, 3.4
GMt	Measure and draw lines and angles	Prior knowledge	Prior knowledge	
GMu	Draw triangles and other 2D shapes using ruler and protractor			Book 2, lesson 8.1
GMv	Use straight edge and a pair of compasses to do constructions			Book 2, lessons 8.1–8.4
GMw	Construct loci			Book 2, lessons 8.3, 8.4
GMx	Calculate perimeters and areas of shapes made from triangles and rectangles		Book 1, lesson 8.1	
GMy	Calculate the area of a triangle using $\frac{1}{2}ab$ sin C			Book 2, lesson 10.6
GMz	Find the circumferences, and areas of circles			Book 1, 8.2, Book 2, lesson 4.1
GMaa	Calculate volumes of right prisms and shapes made from cubes and cuboids		Book 1, lesson 8.3	Book 1, lesson 8.3
GMbb	Solve mensuration problems involving complex shapes/solids			Book 2, lessons 4.4, 4.5
GMcc	Use vectors to solve problems			Book 2, lessons 17.1–17.3

Answers to Chapter 1

Quick check

1 **a** 137 **b** 65 **c** 161 **d** 42
e 6.5 **f** 4.6 **g** 13.5 **h** 1.3

2 **a** **i** $\frac{12}{5}$ **ii** $\frac{13}{4}$ **iii** $\frac{16}{9}$
b **i** $1\frac{5}{6}$ **ii** $2\frac{1}{3}$ **iii** $3\frac{2}{7}$

3 **a** 14 **b** 20 **c** 1 **d** 7

4 **a** $1\frac{5}{12}$ **b** $\frac{17}{35}$ **c** $\frac{11}{20}$ **d** $\frac{14}{15}$

1.1 Basic calculations and using brackets

Exercise 1A

1 **a** 144 **b** 108
2 **a** 12.54 **b** 27.45
3 **a** 26.7 **b** 24.5 **c** 145.3 **d** 1.5
4 Sovereign is 102.47p per litre, Bridge is 102.73p per litre so Sovereign is cheaper.
5 Abby 1.247, Bobby 2.942, Col 5.333, Donna 6.538
Col is correct.
6 $31 \times 3600 \div 1610 = 69.31677 \approx 70$
7 **a** 167.552 **b** 196.48
8 **a** 2.77 **b** 6
9 **a** 497.952 **b** 110.978 625

1.2 Adding and subtracting fractions with a calculator

Exercise 1B

1 **a** $6\frac{11}{20}$ **b** $8\frac{8}{15}$ **c** $16\frac{1}{4}$ **d** $11\frac{147}{200}$
e $7\frac{43}{80}$ **f** $11\frac{63}{80}$ **g** $3\frac{11}{30}$ **h** $2\frac{29}{48}$
i $3\frac{17}{96}$ **j** $7\frac{167}{240}$ **k** $7\frac{61}{80}$ **l** $4\frac{277}{396}$

2 $\frac{1}{12}$

3 **a** $12\frac{1}{4}$ **b** $3\frac{1}{4}$

4 Use the fraction key [fraction key] to input $\frac{3}{25}$, then key in [+] and then use the fraction key again to input $\frac{7}{10}$.

[fraction key] [3] ▼ [2] [5] ▶ [+] [fraction key] [7] ▼ [1] [0] ▶ [=]

5 $\frac{47}{120}$

6 **a** $-\frac{77}{1591}$ **b** Answer negative

7 **a** $\frac{223}{224}$ **b** $\frac{97}{1248}$ **c** $-\frac{97}{273}$
d One negative and one positive so $\frac{5}{7} > \frac{14}{39} > \frac{9}{32}$.

8 **a** Answers will vary
b Yes, always true, unless fractions are equivalent then answer is also equivalent.

9 $18\frac{11}{12}$ cm

10 $\frac{5}{12}$ (anticlockwise) or $\frac{7}{12}$ (clockwise)

1.3 Multiplying and dividing fractions with a calculator

Exercise 1C

1 **a** $\frac{3}{5}$ **b** $\frac{7}{12}$ **c** $\frac{9}{25}$ **d** $\frac{27}{200}$
e $\frac{21}{320}$ **f** $\frac{27}{128}$ **g** $5\frac{2}{5}$ **h** $5\frac{1}{7}$
i $2\frac{1}{16}$ **j** $\frac{27}{40}$ **k** $3\frac{9}{32}$ **l** $\frac{11}{18}$

2 $\frac{1}{6}$ m^2

3 15

4 **a** $\frac{27}{64}$ **b** $\frac{27}{64}$

5 **a** $\frac{4}{5}$ **b** $\frac{4}{5}$ **c** $\frac{16}{21}$ **d** $\frac{16}{21}$

6 **a** $8\frac{11}{20}$ **b** $18\frac{1}{60}$ **c** $65\frac{91}{100}$ **d** $22\frac{1}{8}$
e $7\frac{173}{320}$ **f** $52\frac{59}{160}$ **g** $2\frac{17}{185}$ **h** $2\frac{22}{103}$
i $1\frac{305}{496}$ **j** $5\frac{17}{65}$ **k** $7\frac{881}{4512}$ **l** $5\frac{547}{1215}$

7 $18\frac{5}{12}$ m^2

8 $3\frac{11}{32}$ cm^3

9 $90\frac{5}{8}$ miles

10 3
11 3

Examination questions

1 4.8
2 172.425
3 11.94117647
4 **a** 27.0343336
 b 27.0
5 **a** 18.34747939
6 9.476841579
7 **a** 14.41666667
 b 14.42
8 0.310407623

Answers to Chapter 2

Quick check

1 **a** 3335 **b** 41 **c** 625
2 **a** 17 **b** 25 **c** 28
3 **a** £25 **b** £3.90 **c** 47p

2.1 Multiplication and division with decimals

Exercise 2A

1 **a** 0.028 **b** 0.09 **c** 0.192 **d** 3.0264 **e** 7.134 **f** 50.96 **g** 3.0625 **h** 46.512
2 **a** 35, 35.04, 0.04 **b** 16, 18.24, 2.24 **c** 60, 59.67, 0.33 **d** 180, 172.86, 7.14 **e** 12, 12.18, 0.18 **f** 24, 26.016, 2.016 **g** 40, 40.664, 0.664 **h** 140, 140.58, 0.58
3 **a** 572 **b i** 5.72 **ii** 1.43 **iii** 22.88
4 **a** Incorrect as should end in the digit 2
 b Incorrect since 9 × 5 = 45, so answer must be less than 45
5 26.66 ÷ 3.1 (answer 8.6) since approximately 27 ÷ 3 = 9
6 **a** 18 **b** 140 **c** 1.4 **d** 12 **e** 21.3 **f** 6.9 **g** 2790 **h** 12.1 **i** 18.9
7 **a** 280 **b** 12 **c** 0.18 **d** 450 **e** 0.62 **f** 380 **g** 0.26 **h** 240 **i** 12
8 750
9 300
10 **a** 27 **b i** 27 **ii** 0.027 **iii** 0.27
11 £54.20
12 Mark bought a DVD, some jeans and a pen.

2.2 Compound interest and repeated percentage change

Exercise 2B

1 **a i** 10.5 g **ii** 11.03 g **iii** 12.16 g **iv** 14.07 g
 b 9 days
2 12 years
3 **a** £14 272.27 **b** 20 years
4 **a i** 2550 **ii** 2168 **iii** 1331
 b 7 years
5 **a** £6800 **b** £5440 **c** £3481.60
6 **a i** 1.9 million litres **ii** 1.6 million litres **iii** 1.2 million litres
 b 10th August
7 **a i** 51 980 **ii** 84 752 **iii** 138 186
 b 2021
8 **a** 21 years **b** 21 years
9 3 years
10 30 years
11 1.1 × 1.1 = 1.21 (21% increase)
12 Bradley Bank account is worth £1032, Monastery Building Society account is worth £1031.30, so Bradley Bank by 70p
13 4 months: fish weighs 3 × 1.14 = 4.3923 kg; crab weighs 6 × 0.94 = 3.9366 kg
14 4 weeks
15 20

2.3 Reverse percentage (working out the original quantity)

Exercise 2C

1 **a** 800 g **b** 250 m **c** 60 cm **d** £3075 **e** £200 **f** £400
2 80
3 T shirt £8.40, Tights £1.20, Shorts £5.20, Sweater £10.74, Trainers £24.80, Boots £32.40
4 £833.33
5 £300
6 240

7 £350

8 4750 blue bottles

9 £22

10 **a** £1600
b With 10% cut each year he earns £1440 × 12 + £1296 × 12 = £17 280 + £15 552 = £32 832
With immediate 14% cut he earns £1376 × 24 = £33 024, so correct decision

11 **a** 30% **b** 15%

12 Less by $\frac{1}{4}$%

13 £900

14 Calculate the pre-VAT price for certain amounts, and $\frac{5}{6}$ of that amount.
Show the error grows as the amount increases. Up to £280 the error is less than £5.

15 £1250

16 £1267.35

17 Baz has assumed that 291.2 is 100% instead of 112%. He rounded his wrong answer to the correct answer of £260.

2.4 Powers (indices)

Exercise 2D

1 **a** 2^4 **b** 3^5 **c** 7^2 **d** 5^3
e 10^7 **f** 6^4 **g** 4^1 **h** 1^7
i 0.5^4 **j** 100^3

2 **a** 3 × 3 × 3 × 3
b 9 × 9 × 9
c 6 × 6
d 10 × 10 × 10 × 10 × 10
e 2 × 2 × 2 × 2 × 2 × 2 × 2 × 2 × 2 × 2
f 8
g 0.1 × 0.1 × 0.1
h 2.5 × 2.5
i 0.7 × 0.7 × 0.7
j 1000 × 1000

3 **a** 16 **b** 243
c 49 **d** 125
e 10 000 000 **f** 1296
g 4 **h** 1
i 0.0625 **j** 1 000 000

4 **a** 81 **b** 729
c 36 **d** 100 000
e 1024 **f** 8
g 0.001 **h** 6.25
i 0.343 **j** 1 000 000

5 125 m^3

6 **b** 10^2 **c** 2^3 **d** 5^2

7 **a** 1 **b** 4 **c** 1
d 1 **e** 1

8 Any power of 1 is equal to 1.

9 10^6

10 10^6

11 **a** 1 **b** −1 **c** 1
d 1 **e** −1

12 **a** 1 **b** −1 **c** −1
d 1 **e** 1

13 2^{24}, 4^{12}, 8^8, 16^6

Exercise 2E

1 **a** $\frac{1}{5^3}$ **b** $\frac{1}{6}$ **c** $\frac{1}{10^5}$ **d** $\frac{1}{3^2}$
e $\frac{1}{8^2}$ **f** $\frac{1}{9}$ **g** $\frac{1}{w^2}$ **h** $\frac{1}{t}$
i $\frac{1}{x^m}$ **j** $\frac{4}{m^3}$

2 **a** 3^{-2} **b** 5^{-1} **c** 10^{-3} **d** m^{-1}
e t^{-n}

3 **a** **i** 2^4 **ii** 2^{-1} **iii** 2^{-4} **iv** -2^3
b **i** 10^3 **ii** 10^{-1} **iii** 10^{-2} **iv** 10^6
c **i** 5^3 **ii** 5^{-1} **iii** 5^{-2} **iv** 5^{-4}
d **i** 3^2 **ii** 3^{-3} **iii** 3^{-4} **iv** -3^5

4 **a** $\frac{5}{x^3}$ **b** $\frac{6}{t}$ **c** $\frac{7}{m^2}$ **d** $\frac{4}{q^4}$
e $\frac{10}{y^5}$ **f** $\frac{1}{2x^3}$ **g** $\frac{1}{2m}$ **h** $\frac{3}{4t^4}$
i $\frac{4}{5y^3}$ **j** $\frac{7}{8x^5}$

5 **a** $7x^{-3}$ **b** $10p^{-1}$ **c** $5t^{-2}$
d $8m^{-5}$ **e** $3y^{-1}$

6 **a** **i** 25 **b** **i** 64
ii $\frac{1}{125}$ **ii** $\frac{1}{16}$
iii $\frac{4}{5}$ **iii** $\frac{5}{256}$
c **i** 8 **d** **i** 1 000 000
ii $\frac{1}{32}$ **ii** $\frac{1}{1000}$
iii $\frac{9}{2}$ or $4\frac{1}{2}$ **iii** $\frac{1}{4}$

7 24 (32 − 8)

8 $x = 8$ and $y = 4$ (or $x = y = 1$)

9 $\frac{1}{2097152}$

10 **a** x^{-5}, x^0, x^5 **b** x^5, x^0, x^{-5}
c x^5, x^{-5}, x^0

Exercise 2F

1 **a** 5^4 **b** 5^3 **c** 5^2
d 5^3 **e** 5^{-5}

2 **a** 6^3 **b** 6^0 **c** 6^6
d 6^{-7} **e** 6^2

3 **a** a^3 **b** a^5 **c** a^7
d a^4 **e** a^2 **f** a^1

4 **a** Any two values such that $x + y = 10$
b Any two values such that $x - y = 10$

5 **a** 4^6 **b** 4^{15} **c** 4^6
d 4^{-6} **e** 4^6 **f** 4^0

6 **a** $6a^5$ **b** $9a^2$ **c** $8a^6$
d $-6a^4$ **e** $8a^8$ **f** $-10a^{-3}$

7 **a** $3a$ **b** $4a^3$ **c** $3a^4$
d $6a^{-1}$ **e** $4a^7$ **f** $5a^{-4}$

8 **a** $8a^5b^4$ **b** $10a^3b$ **c** $30a^{-2}b^{-2}$
d $2ab^3$ **e** $8a^{-5}b^7$

9 **a** $3a^3b^2$ **b** $3a^2c^4$ **c** $8a^2b^2c^3$

10 **a** Possible answer: $6x^2 \times 2y^5$ and $3xy \times 4xy^4$
b Possible answer: $24x^2y^7 \div 2y^2$ and $12x^6y^8 \div x^4y^3$

11 12 ($a = 2$, $b = 1$, $c = 3$)

12 $1 = \; = a^x \div a^x = a^{x-x} = a^0$

Exercise 2G

1 **a** 5 **b** 10 **c** 8 **d** 9
e 25 **f** 3 **g** 4 **h** 10
i 5 **j** 8 **k** 12 **l** 20
m 5 **n** 3 **o** 10 **p** 3
q 2 **r** 24 **s** 6 **t** 6
u $\frac{1}{4}$ **v** $\frac{1}{2}$ **w** $\frac{1}{3}$ **x** $\frac{1}{5}$
y $\frac{1}{10}$

2 **a** $\frac{5}{6}$ **b** $1\frac{2}{3}$ **c** $\frac{8}{9}$ **d** $1\frac{4}{5}$
e $\frac{5}{8}$ **f** $\frac{3}{5}$ **g** $\frac{1}{4}$ **h** $2\frac{1}{2}$
i $\frac{4}{5}$ **j** $1\frac{1}{7}$

3 $(x^{\frac{1}{n}})^n = x^{\frac{1}{h \times h}} = x^1 = x$, but
$(\sqrt[n]{x})^n = \sqrt[n]{x} \times \sqrt[n]{x} \ldots n$ times $= x$,
so $x^{\frac{1}{n}} = \sqrt[n]{x}$

4 $64^{-\frac{1}{2}} = \frac{1}{8}$, others are both $\frac{1}{2}$

5 Possible answer: The negative power gives the reciprocal, so $27^{-\frac{1}{3}} = \frac{1}{27}^{\frac{1}{3}}$
The power one-third means cube root, so you need the cube root of 27 which is 3, so $27^{\frac{1}{3}} = 3$ and $\frac{1}{27}^{\frac{1}{3}} = \frac{1}{3}$

6 Possible answer: $x = 1$ and $y = 1$,
$x = 8$ and $y = \frac{1}{64}$

Exercise 2H

1 **a** 16 **b** 25 **c** 216 **d** 81
2 **a** $t^{\frac{2}{3}}$ **b** $m^{\frac{3}{4}}$ **c** $k^{\frac{2}{5}}$ **d** $x^{\frac{3}{2}}$
3 **a** 4 **b** 9 **c** 64 **d** 3125
4 **a** $\frac{1}{5}$ **b** $\frac{1}{6}$ **c** $\frac{1}{2}$ **d** $\frac{1}{3}$
e $\frac{1}{4}$ **f** $\frac{1}{2}$ **g** $\frac{1}{2}$ **h** $\frac{1}{3}$

5 **a** $\frac{1}{125}$ **b** $\frac{1}{216}$ **c** $\frac{1}{8}$ **d** $\frac{1}{27}$
e $\frac{1}{256}$ **f** $\frac{1}{4}$ **g** $\frac{1}{4}$ **h** $\frac{1}{9}$
6 **a** $\frac{1}{100000}$ **b** $\frac{1}{12}$ **c** $\frac{1}{25}$ **d** $\frac{1}{27}$
e $\frac{1}{32}$ **f** $\frac{1}{32}$ **g** $\frac{1}{81}$ **h** $\frac{1}{13}$
7 $8^{-\frac{2}{3}} = \frac{1}{4}$, others are both $\frac{1}{8}$

8 Possible answer: The negative power gives the reciprocal, so $27^{-\frac{2}{3}} = \frac{1}{27}^{\frac{2}{3}}$
The power one-third means cube root, so we need the cube root of 27 which is 3 and the power 2 means square, so $3^2 = 9$, so $27^{\frac{2}{3}} = 9$ and $\frac{1}{27}^{\frac{2}{3}} = \frac{1}{9}$

2.5 Reciprocals and rational numbers

Exercise 2I

1 **a** 0.5 **b** $0.\dot{3}$ **c** 0.25 **d** 0.2
e $0.1\dot{6}$ **f** $0.\dot{1}4285\dot{7}$
g 0.125 **h** $0.\dot{1}$ **i** 0.1
j $0.\dot{0}7692\dot{3}$
2 **b** They all contain the same pattern of digits, starting at a different point in the pattern.
3 $0.\dot{1}$, $0.\dot{2}$, $0.\dot{3}$, etc. Digit in decimal fraction same as numerator.
4 $0.\dot{0}\dot{9}$, $0.\dot{1}\dot{8}$, $0.\dot{2}\dot{7}$, etc. Sum of digits in recurring pattern = 9. First digit is one less than numerator.
5 0.444 ..., 0.454 ..., 0.428 ..., 0.409 ..., 0.432 ..., 0.461 ..., $\frac{9}{22}$, $\frac{3}{7}$, $\frac{16}{37}$, $\frac{4}{9}$, $\frac{5}{11}$, $\frac{6}{13}$
6 $\frac{38}{120}$, $\frac{35}{120}$, $\frac{36}{120}$, $\frac{48}{120}$, $\frac{50}{120}$
$\frac{7}{24}$, $\frac{3}{10}$, $\frac{19}{60}$, $\frac{2}{5}$, $\frac{5}{12}$

7 **a** $\frac{1}{8}$ **b** $\frac{17}{50}$ **c** $\frac{29}{40}$ **d** $\frac{5}{16}$
e $\frac{89}{100}$ **f** $\frac{1}{20}$ **g** $2\frac{7}{20}$ **h** $\frac{7}{32}$
8 **a** $0.08\dot{3}$ **b** 0.0625 **c** 0.05 **d** 0.04
e 0.02
9 **a** $\frac{4}{3}$ **b** $\frac{6}{5}$ **c** $\frac{5}{2}$ **d** $\frac{10}{7}$
e $\frac{20}{11}$ **f** $\frac{15}{4}$
10 **a** 0.75, $1.\dot{3}$; $0.8\dot{3}$, 1.2; 0.4, 2.5; 0.7, $1.\dot{4}28571\dot{1}$; 0.55, $1.\dot{8}\dot{1}$; $0.2\dot{6}$, 3.75
b Not always true, e.g. reciprocal of 0.4 $(\frac{2}{5})$ is $\frac{5}{2} = 2.5$
11 $1 \div 0$ is infinite, so there is no finite answer.
12 **a** 10
b 2
c The reciprocal of a reciprocal is always the original number.

13 The reciprocal of x is greater than the reciprocal of y. For example, $2 < 10$, reciprocal of 2 is 0.5, reciprocal of 10 is 0.1, and $0.5 > 0.1$
14 Possible answer: $-\frac{1}{2} \times -2 = 1$, $-\frac{1}{3} \times -3 = 1$
15 **a** 24.24242 ... **b** 24
c $\frac{24}{99} = \frac{8}{33}$
16 **a** $\frac{8}{9}$ **b** $\frac{34}{99}$ **c** $\frac{5}{11}$ **d** $\frac{21}{37}$
e $\frac{4}{9}$ **f** $\frac{2}{45}$ **g** $\frac{13}{90}$ **h** $\frac{1}{22}$
i $2\frac{7}{9}$ **j** $7\frac{7}{11}$ **k** $3\frac{1}{3}$ **l** $2\frac{2}{33}$
17 **a** true **b** true **c** recurring
18 **a** $\frac{9}{9}$ **b** $\frac{45}{90} = \frac{1}{2} = 0.5$

2.6 Standard form

Exercise 2J

1 **a** 31 **b** 310 **c** 3100
d 31 000
2 **a** 65 **b** 650 **c** 6500
d 65 000
3 **a** 0.31 **b** 0.031 **c** 0.0031
d 0.000 31
4 **a** 0.65 **b** 0.065 **c** 0.0065
d 0.000 65
5 **a** 250 **b** 34.5 **c** 4670
d 346 **e** 207.89 **f** 56 780
g 246 **h** 0.76 **i** 999 000
j 23 456 **k** 98 765.4
l 43 230 000 **m** 345.78
n 6000 **o** 56.7 **p** 560 045
6 **a** 0.025 **b** 0.345 **c** 0.004 67
d 3.46 **e** 0.207 89
f 0.056 78 **g** 0.0246
h 0.0076 **i** 0.000 000 999
j 2.3456 **k** 0.098 765 4
l 0.000 043 23
m 0.000 000 034 578
n 0.000 000 000 006
o 0.000 000 567 **p** 0.005 600 45
7 **a** 230 **b** 578 900
c 4790 **d** 57 000 000
e 216 **f** 10 500 **g** 0.000 32
h 9870
8 **a**, **b** and **c**
9 Power 24 means more digits in the answer.
10 6

Exercise 2K

1 **a** 0.31 **b** 0.031 **c** 0.0031
d 0.000 31
2 **a** 0.65 **b** 0.065 **c** 0.0065
d 0.000 65
3 **a** $9999999999 \times 10^{99}$
b $0.000000001 \times 10^{-99}$ (depending on number of digits displayed)
4 **a** 31 **b** 310 **c** 3100
d 31 000
5 **a** 65 **b** 650 **c** 6500
d 65 000
6 **a** 250 **b** 34.5 **c** 0.004 67
d 34.6 **e** 0.020 789
f 5678 **g** 246 **h** 7600
i 897 000 **j** 0.008 65
k 60 000 000 **l** 0.000 567
7 **a** 2.5×10^2 **b** 3.45×10^{-1}
c 4.67×10^4 **d** 3.4×10^9
e 2.078×10^{10} **f** 5.678×10^{-4}
g 2.46×10^3 **h** 7.6×10^{-2}
i 7.6×10^{-4} **j** 9.99×10^{-1}
k 2.3456×10^2 **l** 9.87654×10^1
m 6×10^{-4} **n** 5.67×10^{-3}
o 5.60045×10^1
8 2.7797×10^4
9 2.81581×10^5, 3×10^1, 1.382101×10^6
10 1.298×10^7, 2.997×10^9, 9.3×10^4
11 100
12 36 miles

Exercise 2L

1 **a** 5.67×10^3 **b** 6×10^2 **c** 3.46×10^{-1} **d** 7×10^{-4} **e** 5.6×10^2 **f** 6×10^5 **g** 7×10^3 **h** 1.6 **i** 2.3×10^7 **j** 3×10^{-6} **k** 2.56×10^6 **l** 4.8×10^2 **m** 1.12×10^2 **n** 6×10^{-1} **o** 2.8×10^6
2 **a** 1.08×10^8 **b** 4.8×10^6 **c** 1.2×10^9 **d** 1.08 **e** 6.4×10^2 **f** 1.2×10^1 **g** 2.88 **h** 2.5×10^7 **i** 8×10^{-6}
3 **a** 1.1×10^8 **b** 6.1×10^6 **c** 1.6×10^9 **d** 3.9×10^{-2} **e** 9.6×10^8 **f** 4.6×10^{-7} **g** 2.1×10^3 **h** 3.6×10^7 **i** 1.5×10^2 **j** 3.5×10^9 **k** 1.6×10^4
4 **a** 2.7×10 **b** 1.6×10^{-2} **c** 2×10^{-1} **d** 4×10^{-8} **e** 2×10^5 **f** 6×10^{-2} **g** 2×10^{-5} **h** 5×10^2 **i** 2×10
5 **a** 5.4×10 **b** 2.9×10^{-3} **c** 1.1 **d** 6.3×10^{-10} **e** 2.8×10^2 **f** 5.5×10^{-2} **g** 4.9×10^2 **h** 8.6×10^6
6 2×10^{13}, 1×10^{-10}, mass $= 2 \times 10^3$g (2 kg)
7 **a** (2^{63}) 9.2×10^{18} grains **b** $2^{64} - 1 = 1.8 \times 10^{19}$
8 **a** 6×10^7 sq miles **b** 30%
9 1.5×10^7 sq miles
10 5×10^4
11 2.3×10^5
12 455 070 000 kg or 455 070 tonnes
13 80 000 000 (80 million)
14 **a** 2.048×10^6 **b** 4.816×10^6
15 250
16 9.41×10^4
17 Any value from $1.000\,000\,01 \times 10^8$ to 1×10^9 (excluding 1×10^9), i.e. any value of the form $a \times 10^8$ where $1 \leqslant a < 10$

Examination questions

1 **a** 9.476 841 579 **b** 9.48
2 **a** 2.638 465 463 **b** 2.6
3 4576.876712
4 47.12 kg (2dp)
5 **a** 20% **b** 147
6 £7375.53
7 **a** £1815 **b** £780
8 £3244.80
9 **a** £4867.20 **b** 5 years
10 5.8×10^{-4}
11 **a** $3^4 = 81$ **b** 6 **c** -2
12 **a** $\frac{1}{9}$ **b** 343
13 **a** t^8 **b** m^5 **c** $8x^3$ **d** $12a^7h^5$
14 20%
15 **a** 4.5×10^4 **b** 0.06
16 3.02×10^{27}
17 **a** t^7 **b** m^{-5} **c** $12k^5m^3$
18 **a** $4xy^2$ **b** $8m^{12}p^6$

Answers to Chapter 3

Quick check

1 **a** 6370 **b** 6400 **c** 6000
2 **a** 2.4 **b** 2.39
3 **a** 50 **b** 47.3

3.1 Limits of accuracy

Exercise 3A

1 **a** $6.5 \leqslant 7 < 7.5$ **b** $115 \leqslant 120 < 125$ **c** $3350 \leqslant 3400 < 3450$ **d** $49.5 \leqslant 50 < 50.5$ **e** $5.50 \leqslant 6 < 6.49$ **f** $16.75 \leqslant 16.8 < 16.85$ **g** $15.5 \leqslant 16 < 16.5$ **h** $14\,450 \leqslant 14\,500 < 14\,549$ **i** $54.5 \leqslant 55 < 55.5$ **j** $52.5 \leqslant 55 < 57.5$
2 **a** $5.5 \leqslant 6 < 6.5$ **b** $16.5 \leqslant 17 < 17.5$ **c** $31.5 \leqslant 32 < 32.5$ **d** $237.5 \leqslant 238 < 238.5$ **e** $7.25 \leqslant 7.3 < 7.35$ **f** $25.75 \leqslant 25.8 < 25.85$ **g** $3.35 \leqslant 3.4 < 3.45$ **h** $86.5 \leqslant 87 < 87.5$ **i** $4.225 \leqslant 4.23 < 4.235$ **j** $2.185 \leqslant 2.19 < 2.195$ **k** $12.665 \leqslant 12.67 < 12.675$ **l** $24.5 \leqslant 25 < 25.5$ **m** $35 \leqslant 40 < 45$ **n** $595 \leqslant 600 < 605$ **o** $25 \leqslant 30 < 35$ **p** $995 \leqslant 1000 < 1050$ **q** $3.95 \leqslant 4.0 < 4.05$ **r** $7.035 \leqslant 7.04 < 7.045$ **s** $11.95 \leqslant 12.0 < 12.05$ **t** $6.995 \leqslant 7.00 < 7.005$
3 **a** 7.5, 8.5 **b** 25.5, 26.5 **c** 24.5, 25.5 **d** 84.5, 85.5 **e** 2.395, 2.405 **f** 0.15, 0.25 **g** 0.055, 0.065 **h** 250 g, 350 g **i** 0.65, 0.75 **j** 365.5, 366.5 **k** 165, 175 **l** 205, 215

4 There are 16 empty seats and the number getting on the bus is from 15 to 24 so it is possible if 15 or 16 get on.
5 C: The chain and distance are both any value between 29.5 and 30.5 metres, so there is no way of knowing if the chain is longer or shorter than the distance.
6 2 kg 450 grams
7 **a** <65.5 g **b** 64.5 g **c** <2620 g **d** 2580 g

3.2 Speed, time and distance

Exercise 3B

1 18 mph
2 280 miles
3 52.5 mph
4 11.50 am
5 500 seconds
6 **a** 75 mph **b** 6.5 h **c** 175 miles **d** 240 km **e** 64 km/h **f** 325 km **g** 4.3 h (4 h 18 min)
7 **a** 7.75 h **b** 52.9 mph
8 **a** 2.25 h **b** 99 miles
9 **a** 1.25 h **b** 1 h 15 min
10 **a** 48 mph **b** 6 h 40 min
11 **a** 120 km **b** 48 km/h
12 **a** 30 min **b** 6 mph
13 **a** 10 m/s **b** 3.3 m/s **c** 16.7 m/s **d** 41.7 m/s **e** 20.8 m/s
14 **a** 90 km/h **b** 43.2 km/h **c** 14.4 km/h **d** 108 km/h **e** 1.8 km/h
15 **a** 64.8 km/h **b** 28 s **c** 8.07
16 **a** 6.7 m/s **b** 66 km **c** 5 minutes **d** 133.3 metres
17 7 minutes
18 **a** 20 mph **b** 07.30

3.3 Direct proportion problems

Exercise 3C

1 60 g
2 £5.22
3 45
4 £6.72
5 **a** £312.50 **b** 8
6 **a** 56 litres **b** 350 miles
7 **a** 300 kg **b** 9 weeks
5 40 seconds
9 **a** **i** 100 g, 200 g, 250 g, 150 g
ii 150 g, 300 g, 375 g, 225 g
iii 250 g, 500 g, 625 g, 375 g
b 24
10 Peter: £2.30 ÷ 6 = 38.33p each; I can buy four packs (24 sausages) from him (£9.20)
Paul: £3.50 ÷ 10 = 35p each; I can only buy two packs (20 sausages) from him (£7)
I should use Peter's shop to get the most sausages for £10.
11 11 minutes 40 seconds + 12 minutes = 23 minutes 40 seconds
12 Possible answer:
30 g plain flour (rounding to nearest 10 g)
60 ml whole milk (rounding to nearest 10 ml)
1 egg (need an egg)
1 g salt (nearest whole number)
10 ml beef dripping or lard (rounding to nearest 10 ml)
13 30 litres

3.4 Density

Exercise 3D

1 **a** 0.75 g/cm^3
2 $8\frac{1}{3}$ g/cm^3
3 32 g
4 120 cm^3
5 156.8 g
6 3200 cm^3
7 2.72 g/cm^3
8 36 800 kg
9 1.79 g/cm^3 (3 sf)
10 1.6 g/cm^3
11 First statue is the fake as density is approximately 26 g/cm^3
12 Second piece by 1 cm^3
13 0.339 m^3

Examination questions

1 **a** 90 g
b 400 ml
2 £234.36
3 80 km/h
4 40 miles per hour
5 **a** 62.5 cm
b 63.499… cm
6 **a** 22.499… cm
b 21.5 cm
7 **a** 216 km
b 88 km/h
8 **a** 9.75 m/s^2
b 30.7 metres
9 8.75 miles per litre
10 **a** 59.5 cm
b 60.499… cm
11 2.92
12 7.39 m/s
13 1930 g

Answers to Chapter 4

Quick check

1 **a** 90 mm^2 **b** 40 cm^2 **c** 21 m^2 **2** 120 cm^3

4.1 Circumference and area of a circle

Exercise 4A

1 **a** 8 cm, 25.1 cm, 50.3 cm^2
b 5.2 m, 16.3 m, 21.2 m^2
c 6 cm, 37.7 cm, 113 cm^2
d 1.6 m, 10.1 m, 8.04 m^2
2 **a** 5π cm **b** 8π cm
c 18π m **d** 12π cm
3 **a** 25π cm^2 **b** 36π cm^2
c 100π cm^2 **d** 0.25π m^2
4 8.80 m
5 4 complete revolutions
6 1p : 3.1 cm^2, 2p : 5.3 cm^2, 5p : 2.3 cm^2, 10p : 4.5 cm^2
7 0.83 m
8 38.6 cm
9 Claim is correct (ratio of the areas is just over 1.5 : 1)
10 **a** 18π cm^2 **b** 4π cm^2
11 9π cm^2
12 28.3 m^2
13 Diameter of tree is 9.96 m

4.2 Cylinders

Exercise 4B

1 **a** **i** 226 cm^3
ii 207 cm^2
b **i** 14.9 cm^3
ii 61.3 cm^2
c **i** 346 cm^3
ii 275 cm^2
d **i** 1060 cm^3
ii 636 cm^2
2 **a** **i** 72π cm^3
ii 48π cm^2
b **i** 112π cm^3
ii 56π cm^2
c **i** 180π cm^3
ii 60π cm^2
d **i** 600π m^3
ii 120π m^2
3 £80
4 1.23 tonnes
5 665 cm^3
6 Label should be less than 10.5 cm wide so that it fits the can and does not overlap the rim and more than 23.3 cm long to allow an overlap.
7 332 litres
8 There is no right answer. Students could start with the dimensions of a real can. Often drinks cans are not exactly cylindrical. One possible answer is height of 6.6 cm and diameter of 8 cm.
9 1.71 g/cm^3
10 7.78 g/cm^3
11 About 127 cm
12 A diameter of 10 cm and a length of 5 cm give a volume close to 400 cm^3 (0.4 litres).

4.3 Volume of a pyramid

Exercise 4C

1 **a** 56 cm^3 **b** 168 cm^3
c 1040 cm^3 **d** 84 cm^3
e 160 cm^3
2 270 cm^3
3 **a** Put the apexes of the pyramids together. The 6 square bases will then form a cube.
b If the side of the base is a then the height will be $\frac{1}{2}a$.
Total volume of the 6 pyramids is a^3.
Volume of one pyramid is $\frac{1}{6}a^3 =$
$\frac{1}{3} \times \frac{1}{2} \times a \times a^2 =$
$\frac{1}{3} \times$ height $\times$ base area
4 6.9 m ($\frac{1}{3}$ height of pyramid)
5 **a** 73.3 m^3 **b** 45 m^3
c 3250 cm^3
6 208 g
7 1.5 g
8 6.0 cm
9 14.4 cm
10 260 cm^3

4.4 Cones

Exercise 4D

1 **a** **i** 3560 cm^3
ii 1430 cm^2
b **i** 314 cm^3
ii 283 cm^2
c **i** 1020 cm^3
ii 679 cm^2
2 935 g
3 24π cm^2
4 **a** 816π cm^3 **b** 720π mm^3
5 **a** 4 cm
b 6 cm
c Various answers, e.g. 60° gives 2 cm, 240° gives 8 cm
6 24π cm^2
7 If radius of base is r, slant height is $2r$. Area of curved surface $= \pi r \times 2r = 2\pi r^2$, area of base $= \pi r^2$
8 140 g
9 2.81 cm

4.5 Spheres

Exercise 4E

1 **a** 36π cm^3 **b** 288π cm^3 **c** 1330π cm^3
2 **a** 36π cm^2 **b** 100π cm^2 **c** 196π cm^2
3 65 400 cm^3, 7850 cm^2
4 **a** 1960 cm^2 **b** 8180 cm^3
5 125 cm
6 6231
7 **a** The surface area, because this is the amount of material (leather or plastic) needed to make the ball
b Surface area can vary from about 1470 cm^2 to 1560 cm^2, difference of about 90 cm^2. This seems surprisingly large.
8 7.8 cm
9 48%
10 Radius of sphere = base radius of cylinder = r, height of cylinder = $2r$
Curved surface area of cylinder = circumference × height = $2\pi r \times 2r = 4\pi r^2$ = surface area of sphere

Examination questions

1 $(5\pi + 10)$ cm
2 **a** 2260 cm^3
b 3390 g
3 **a** 503 cm^3
b 302 grams
4 96π cm^2
5 $6x$
6 **a** 905 m^3
b 4.92 m

Answers to Chapter 5

Quick check

1 5.3
2 246.5
3 0.6
4 2.8
5 16.1
6 0.7

5.1 Pythagoras' theorem

Exercise 5A

1 10.3 cm
2 5.9 cm
3 8.5 cm
4 20.6 cm
5 18.6 cm
6 17.5 cm
7 5 cm
8 13 cm
9 10 cm
10 The square in the first diagram and the two squares in the second have the same area.

5.2 Finding a shorter side

Exercise 5B

1 **a** 15 cm **b** 14.7 cm **c** 6.3 cm **d** 18.3 cm
2 **a** 20.8 m **b** 15.5 cm **c** 15.5 m **d** 12.4 cm
3 **a** 5 m **b** 6 m **c** 3 m **d** 50 cm
4 There are infinite possibilities, e.g. any multiple of 3, 4, 5 such as 6, 8, 10; 9, 12, 15; 12, 16, 20; multiples of 5, 12, 13 and of 8, 15, 17.
5 42.6 cm

5.3 Applying Pythagoras' theorem to real-life situations

Exercise 5C

1 No. The foot of the ladder is about 6.6 m from the wall.
2 2.06 m
3 11.3 m
4 About 17 minutes, assuming it travels at the same speed.
5 127 m − 99.6 m = 27.4 m
6 4.58 m
7 **a** 3.87 m
b 1.74 m
8 3.16 m
9 13 units
10 **a** 4.85 m
b 4.83 m (There is only a small difference.)
11 Yes, because $24^2 + 7^2 = 25^2$
12 6 cm
13 Greater than 20 cm (no width) and less than 28.3 cm (a square)

Exercise 5D

1 **a** 32.2 cm^2
b 2.83 cm^2
c 50.0 cm^2
2 22.2 cm^2
3 15.6 cm^2
4 **a**

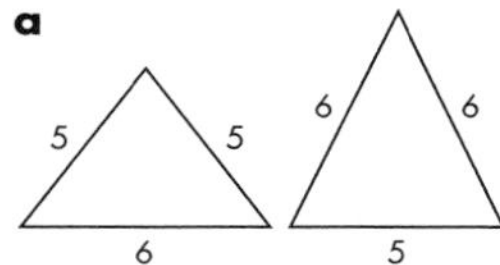

b The areas are 12 cm^2 and 13.6 cm^2 respectively, so triangle with 6 cm, 6 cm, 5 cm sides has the greater area
5 **a** **b** 166.3 cm^2
6 259.8 cm^2
7 **a** No, areas vary from 24.5 cm^2 to 27.7 cm^2
b No, equilateral triangle gives the largest area
c The closer the isosceles triangle gets to an equilateral triangle the larger its area becomes
8 19.8 or 20 m^2
9 48 cm^2
10 **a** 10 cm **b** 26 cm **c** 9.6 cm

5.4 Pythagoras' theorem in three dimensions

Exercise 5E

1 **a** **i** 14.4 cm **ii** 13 cm **iii** 9.4 cm
b 15.2 cm
2 No, 6.6 m is longest length
3 **a** 20.6 cm **b** 15.0 cm
4 21.3 cm
5 **a** 8.49 m **b** 9 m
6 17.3 cm
7 20.6 cm
8 **a** 11.3 cm **b** 7 cm
c 8.06 cm
9 **a** 50.0 cm **b** 54.8 cm
c 48.3 cm **d** 27.0 cm

5.5 Trigonometric ratios

Exercise 5F

1 **a** 0.682 **b** 0.829 **c** 0.922
d 1 **e** 0.707 **f** 0.342
g 0.375 **h** 0
2 **a** 0.731 **b** 0.559 **c** 0.388
d 0 **e** 0.707 **f** 0.940
g 0.927 **h** 1
3 45°
4 **a** **i** 0.574 **ii** 0.574
b **i** 0.208 **ii** 0.208
c **i** 0.391 **ii** 0.391
d Same
e **i** sin 15° is the same as cos 75°
ii cos 82° is the same as sin 8°
iii sin x is the same as cos $(90° - x)$
5 **a** 0.933 **b** 1.48 **c** 2.38
d Infinite **e** 1 **f** 0.364
g 0.404 **h** 0
6 **a** 0.956 **b** 0.899 **c** 2.16
d 0.999 **e** 0.819 **f** 0.577
g 0.469 **h** 0.996
7 Has values > 1
8 **a** 4.53 **b** 4.46 **c** 6 **d** 0
9 **a** 10.7 **b** 5.40 **c** Infinite **d** 0
10 **a** 3.56 **b** 8.96 **c** 28.4 **d** 8.91
11 **a** 5.61 **b** 11.3 **c** 6 **d** 10
12 **a** 1.46 **b** 7.77 **c** 0.087 **d** 7.15
13 **a** 7.73 **b** 48.6 **c** 2.28 **d** 15.2
14 **a** 29.9 **b** 44.8 **c** 20.3 **d** 2.38
15 **a** $\frac{4}{5}, \frac{3}{5}, \frac{4}{3}$
b $\frac{5}{13}, \frac{12}{13}, \frac{5}{12}$
c $\frac{7}{25}, \frac{24}{25}, \frac{7}{24}$

5.6 Calculating angles

Exercise 5G

1 **a** 30° **b** 51.7° **c** 39.8°
d 61.3° **e** 87.4° **f** 45.0°
2 **a** 60° **b** 50.2° **c** 2.6°
d 45.0 **e** 78.5° **f** 45.6°
3 **a** 31.0° **b** 20.8° **c** 41.8°
d 46.4° **e** 69.5° **f** 77.1°
4 **a** 53.1° **b** 41.8° **c** 44.4°
d 56.4° **e** 2.4° **f** 22.6°
5 **a** 36.9° **b** 48.2° **c** 45.6°
d 33.6° **e** 87.6° **f** 67.4°
6 **a** 31.0° **b** 37.9° **c** 15.9°
d 60.9° **e** 57.5° **f** 50.2°
7 Error message, largest value 1, smallest value -1
8 **a** **i** 17.5° **ii** 72.5° **iii** 90°
b Yes

5.7 Using the sine and cosine functions

Exercise 5H

1 **a** 17.5° **b** 22.0°
c 32.2°
2 **a** 5.29 cm **b** 5.75 cm
c 13.2 cm
3 **a** 4.57 cm **b** 6.86 cm
c 100 cm
4 **a** 5.12 cm **b** 9.77 cm
c 11.7 cm **d** 15.5 cm
5 **a** 47.2° **b** 5.42 cm
c 13.7 cm **d** 38.0°
6 **a** 6 **b** 15
c 30

Exercise 5I

1 **a** 51.3° **b** 75.5°
c 51.3°
2 **a** 6.47 cm **b** 32.6 cm
c 137 cm
3 **a** 7.32 cm **b** 39.1 cm
c 135 cm
4 **a** 5.35 cm **b** 14.8 cm
c 12.0 cm **d** 8.62 cm
5 **a** 5.59 cm **b** 46.6°
c 9.91 cm **d** 40.1°
6 **a** 10 **b** 39
c 2.5

5.8 Using the tangent function

Exercise 5J

1 **a** 33.7° **b** 36.9° **c** 52.1°
2 **a** 5.09 cm **b** 30.4 cm **c** 1120 cm
3 **a** 8.24 cm **b** 62.0 cm **c** 72.8 cm
4 **a** 9.02 cm **b** 7.51 cm **c** 7.14 cm **d** 8.90 cm
5 **a** 13.7 cm **b** 48.4° **c** 7.03 cm **d** 41.2°
6 **a** 12 **b** 12 **c** 2

5.9 Which ratio to use

Exercise 5K

1 **a** 12.6 **b** 59.6 **c** 74.7 **d** 16.0 **e** 67.9 **f** 20.1
2 **a** 44.4° **b** 39.8° **c** 44.4° **d** 49.5° **e** 58.7° **f** 38.7°
3 **a** 67.4° **b** 11.3 **c** 134 **d** 28.1° **e** 39.7 **f** 263 **g** 50.2° **h** 51.3° **i** 138 **j** 22.8
4 **a** Sides of right-hand triangle are sine and cosine **b** Pythagoras' theorem **c** Students should check the formulae

5.10 Solving problems using trigonometry 1

Exercise 5L

1 65°
2 The safe limits are between 1.04 m and 2.05 m. The ladder will reach between 5.63 m and 5.90 m up the wall.
3 44°
4 6.82 m
5 31°
6 **a** 25° **b** 2.10 m **c** Thickness of wood has been ignored
7 **a** 20° **b** 4.78 m
8 She would calculate 100 tan 23°. The answer is about 42.4 m
9 21.1 m
10 One way is stand opposite a feature, such as a tree, on the opposite bank, move a measured distance, x, along your bank and measure the angle, θ, between your bank and the feature. Width of river is $x \tan \theta$. This of course requires measuring equipment! An alternative is to walk along the bank until the angle is 45° (if that is possible). This angle is easily found by folding a sheet of paper. This way an angle measurer is not required.

Exercise 5M

1 10.1 km
2 22°
3 429 m
4 **a** 156 m **b** No. the new angle of depression is $\tan^{-1}(\frac{200}{312})$ $= 33°$ and half of 52° is 26°
5 **a** 222 m **b** 42°
6 **a** 21.5 m **b** 17.8 m
7 13.4 m
8 19°
9 The angle is 16° so Cara is not quite correct.

5.11 Solving problems using trigonometry 2

Exercise 5N

1 **a** 73.4 km **b** 15.6 km
2 **a** 14.7 miles **b** 8.5 miles
3 120°
4 **a** 59.4 km **b** 8.4 km
5 **a** 15.9 km **b** 24.1 km **c** 31.2 km **d** 052°
6 2.28 km
7 235°
8 **a** 66.2 km **b** 11.7 km **c** 13.1 km **d** 170°
9 48.4 km, 100°

Exercise 5O

1 **a** 5.79 cm **b** 48.2° **c** 7.42 cm **d** 81.6 cm
2 9.86 m
3 **a** 36.4 cm^2 **b** 115 cm^2 **c** 90.6 cm^2 **d** 160 cm^2
4 473 cm^2

Examination questions

1 150 m
2 4.1 m
3 13.4 cm
4 **a** 33.7° **b** 17.7 cm
5 **a** 28 cm^2 **b** 10.63 cm **c** 34.8°
6 21.8°
7 **a** 30.7° **b** 121°

Answers to Chapter 6

Quick check

Trace shape **a** and check whether it fits exactly on top of the others.
You should find that shape **b** is not congruent to the others.

6.1 Congruent triangles

Exercise 6A

1 **a** SAS **b** SSS **c** ASA
d RHS **e** SSS **f** ASA
2 **a** SSS. A to R, B to P, C to Q
b SAS. A to R, B to Q, C to P
3 **a** 60° **b** 80° **c** 40° **d** 5 cm
4 **a** 110° **b** 55° **c** 85° **d** 110° **e** 4 cm
5 SSS or RHS
6 SSS or SAS or RHS
7 For example, use △ADE and △CDG. AD = CD (sides of large square), DE = DG (sides of small square), ∠ADE = ∠CDG (angles sum to 90° with ∠ADG), so △ADE ≡ △CDG (SAS), so AE = CG
8 AB and PQ are the corresponding sides to the 42° angle, but they are not equal in length.

6.2 Translations

Exercise 6B

1 **a** **i** $\begin{pmatrix}1\\3\end{pmatrix}$ **ii** $\begin{pmatrix}4\\2\end{pmatrix}$ **iii** $\begin{pmatrix}2\\-1\end{pmatrix}$
iv $\begin{pmatrix}5\\1\end{pmatrix}$ **v** $\begin{pmatrix}-1\\6\end{pmatrix}$ **vi** $\begin{pmatrix}4\\6\end{pmatrix}$
b **i** $\begin{pmatrix}-1\\-3\end{pmatrix}$ **ii** $\begin{pmatrix}3\\-1\end{pmatrix}$ **iii** $\begin{pmatrix}1\\-4\end{pmatrix}$
iv $\begin{pmatrix}4\\-2\end{pmatrix}$ **v** $\begin{pmatrix}-2\\3\end{pmatrix}$ **vi** $\begin{pmatrix}3\\3\end{pmatrix}$
c **i** $\begin{pmatrix}-4\\-2\end{pmatrix}$ **ii** $\begin{pmatrix}-3\\1\end{pmatrix}$ **iii** $\begin{pmatrix}-2\\-3\end{pmatrix}$
iv $\begin{pmatrix}1\\-1\end{pmatrix}$ **v** $\begin{pmatrix}-5\\4\end{pmatrix}$ **vi** $\begin{pmatrix}0\\4\end{pmatrix}$
d **i** $\begin{pmatrix}3\\2\end{pmatrix}$ **ii** $\begin{pmatrix}-4\\2\end{pmatrix}$ **iii** $\begin{pmatrix}5\\-4\end{pmatrix}$
iv $\begin{pmatrix}-2\\-7\end{pmatrix}$ **v** $\begin{pmatrix}5\\0\end{pmatrix}$ **vi** $\begin{pmatrix}1\\-5\end{pmatrix}$

2

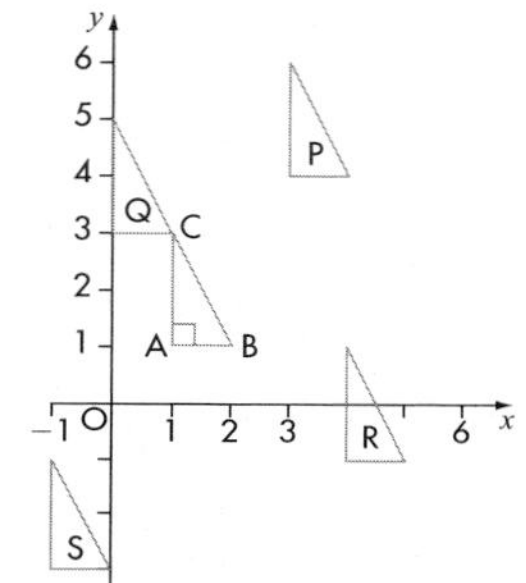

3 **a** $\begin{pmatrix}-3\\-1\end{pmatrix}$ **b** $\begin{pmatrix}4\\-4\end{pmatrix}$ **c** $\begin{pmatrix}-5\\-2\end{pmatrix}$
d $\begin{pmatrix}4\\7\end{pmatrix}$ **e** $\begin{pmatrix}-1\\5\end{pmatrix}$ **f** $\begin{pmatrix}1\\6\end{pmatrix}$
g $\begin{pmatrix}-4\\4\end{pmatrix}$ **h** $\begin{pmatrix}-4\\-7\end{pmatrix}$
4 10 × 10 = 100 (including $\begin{pmatrix}0\\0\end{pmatrix}$)
5 Check students' designs for a Snakes and ladders board.
6 $\begin{pmatrix}-x\\-y\end{pmatrix}$
7 $\begin{pmatrix}-300\\-500\end{pmatrix}$
8 $\begin{pmatrix}-1\\4\end{pmatrix}$

6.3 Reflections

Exercise 6C

1

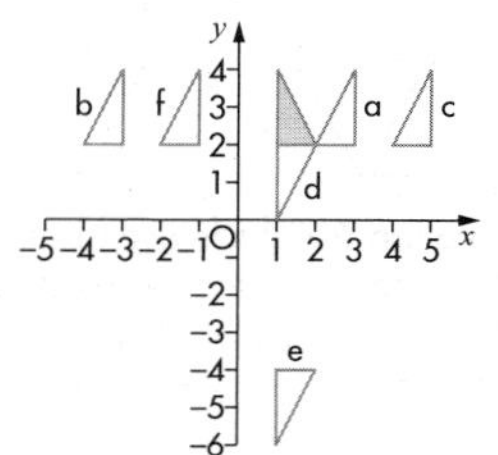

2 **a–e**

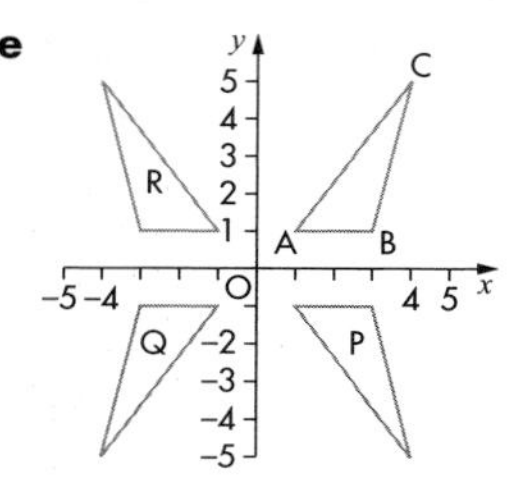

f Reflection in the y-axis

3 **a–b**

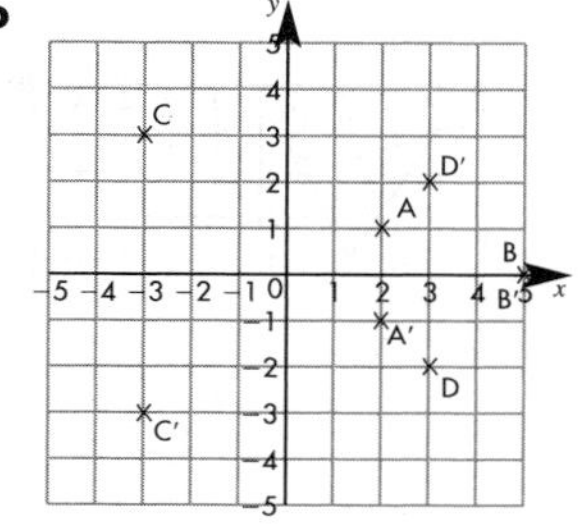

c y-value changes sign
d $(a, -b)$

4 **a–b**

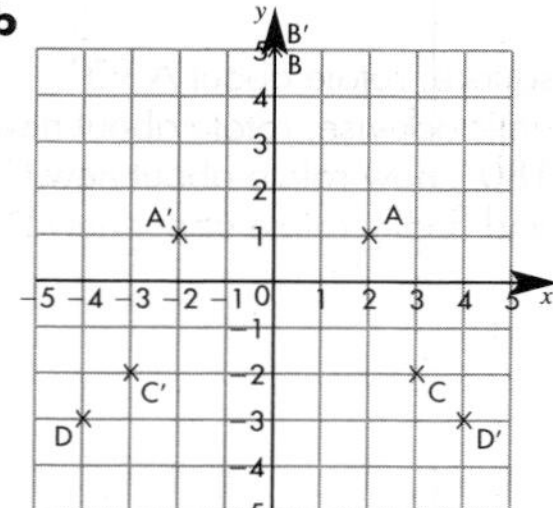

c x-value changes sign
d $(-a, b)$
5 Possible answer: Take the centre square as ABCD then reflect this square each time in the line, AB, then BC, then CD and finally AD.
6 $x = -1$

7 Possible answer:

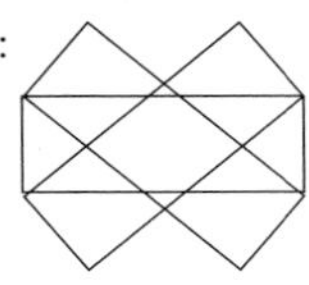

8

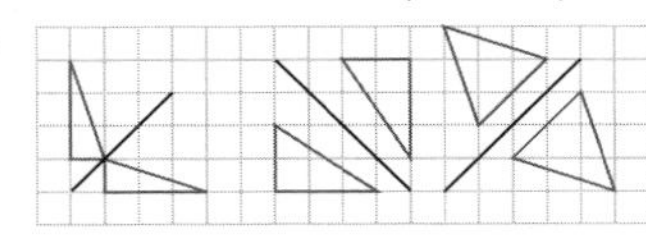

9 a–i

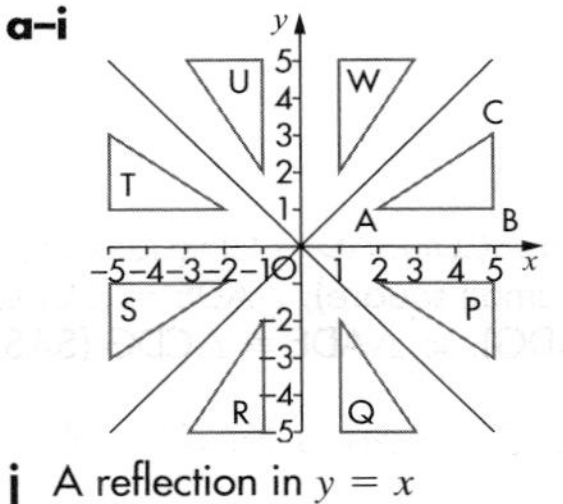

i A reflection in $y = x$

10

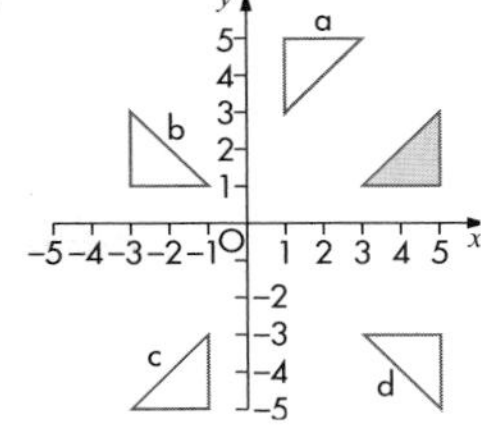

11 a–c

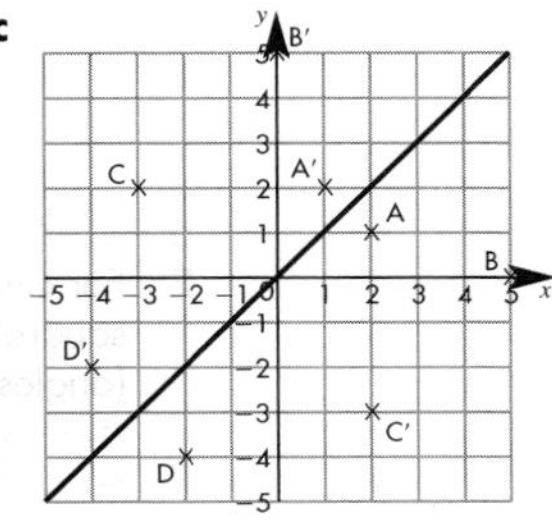

d Coordinates are reversed: x becomes y and y becomes x

e (b, a)

12 a–c

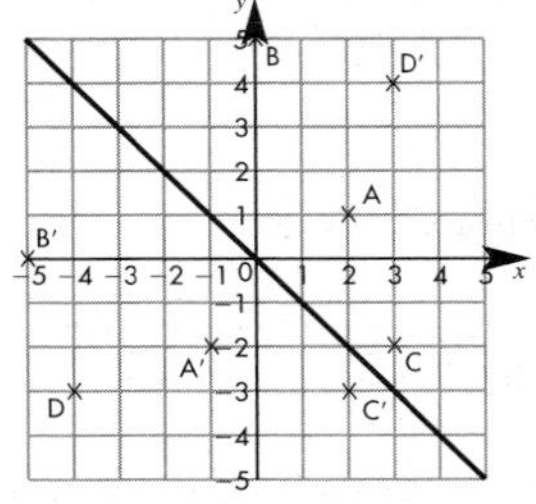

d Coordinates are reversed and change sign, x becomes $-y$ and y becomes $-x$

e $(-b, -a)$

6.4 Rotations

Exercise 6D

1 a

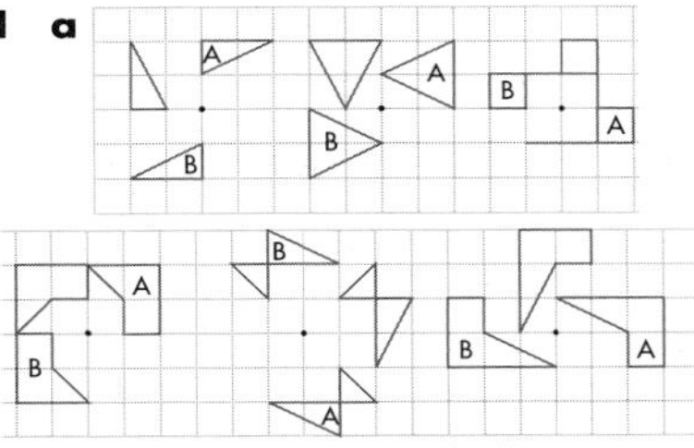

b **i** Rotation 90° anticlockwise

ii Rotation 180°

2

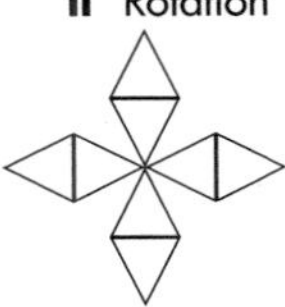

3 Possible answer: If ABCD is the centre square, rotate about A 90° anticlockwise, rotate about new B 180°, now rotate about new C 180°, and finally rotate about new D 180°.

4

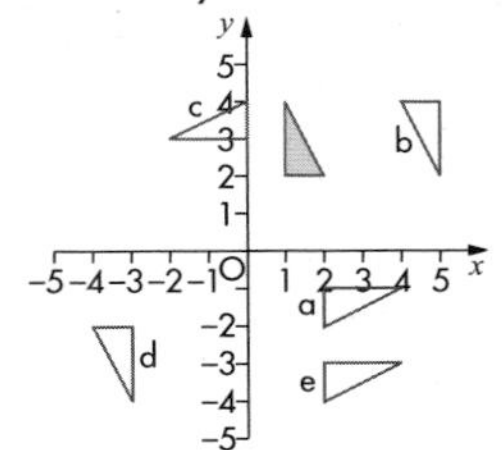

5 **a** 90° anticlockwise

b 270° anticlockwise

c 300° clockwise

d 260° clockwise

6 a b c i

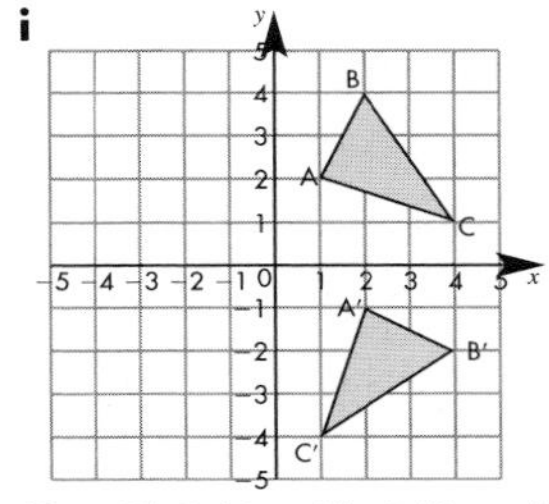

ii A′(2, −1), B′(4, −2), C′(1, −4)

iii Original coordinates (x, y) become $(y, -x)$

iv Yes

7 i

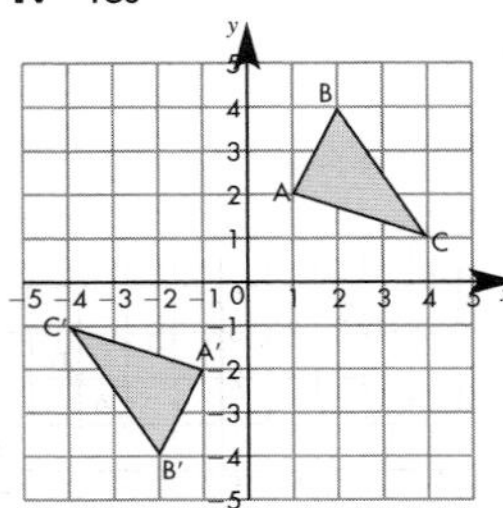

ii A′(−1, −2), B′(−2, −4), C′(−4, −1)

iii Original coordinates (x, y) become $(-x, -y)$

iv Yes

8 i

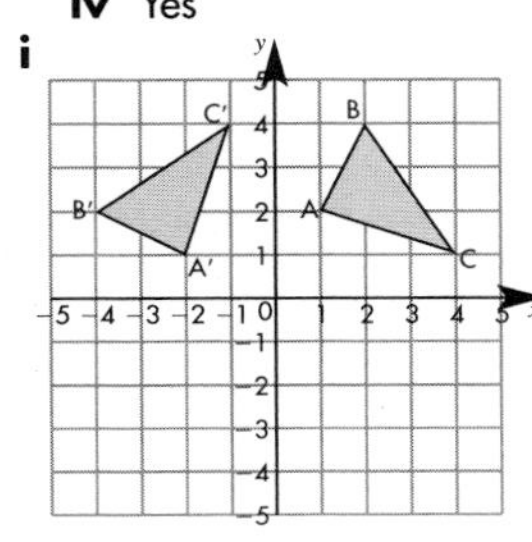

ii A′(−2, 1), B′(−4, 2), C′(−1, 4)

iii Original coordinates (x, y) become $(-y, x)$

iv Yes

9 Show by drawing a shape or use the fact that (a, b) becomes $(a, -b)$ after reflection in the x-axis, and $(a, -b)$ becomes $(-a, -b)$ after reflection in the y-axis, which is equivalent to a single rotation of 180°.

10 Show by drawing a shape or use the fact that (a, b) becomes (b, a) after reflection in the line $y = x$, and (b, a) becomes $(-a, -b)$ after reflection in the line $y = -x$, which is equivalent to a single rotation of 180°.

11 a

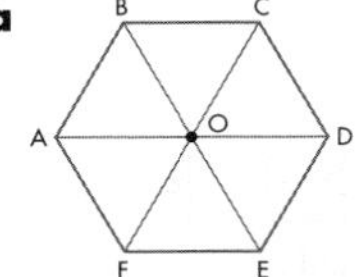

b **i** Rotation 60° clockwise about O

ii Rotation 120° clockwise about O

iii Rotation 180° about O

iv Rotation 240° clockwise about O

c **i** Rotation 60° clockwise about O

ii Rotation 180° about O

12 Rotation 90° anticlockwise about (2, −1)

6.5 Enlargement

Exercise 6E

1

2 **a** **b**

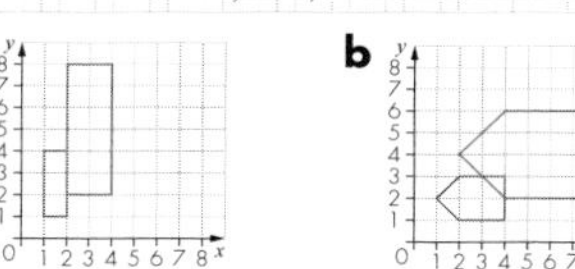

c

3 **a**

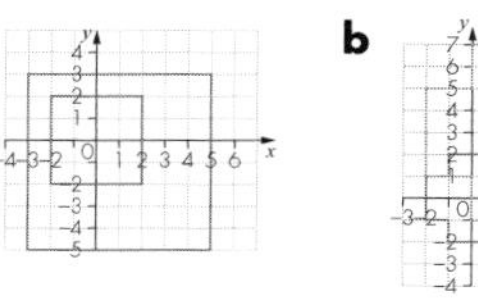

b

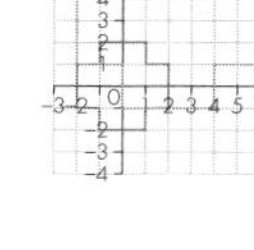

4

5

6 **a**

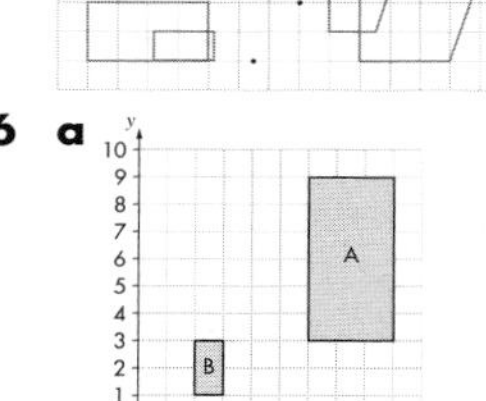

b 3 : 1
c 3 : 1
d 9 : 1

7

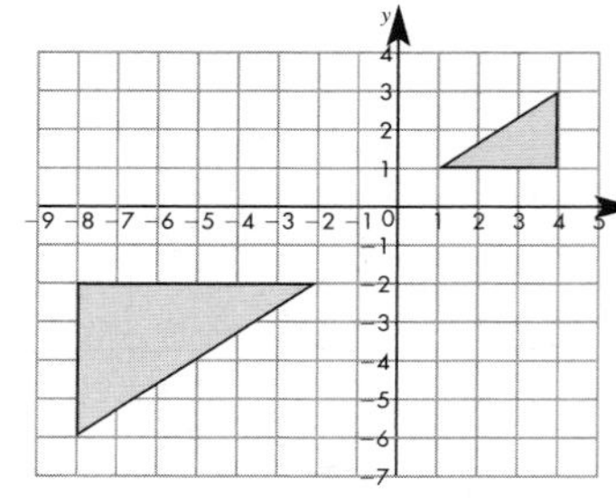

8 **a–c**

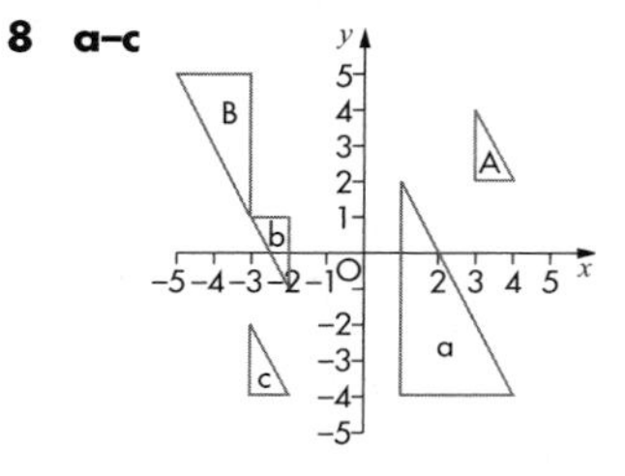

d Scale factor $-\frac{1}{2}$, centre (1, 3)
e Scale factor −2, centre (1, 3)
f Scale factor −1, centre (−2.5, −1.5)
g Scale factor −1, centre (−2.5, −1.5)
h Same centres, and the scale factors are reciprocals of each other

9 Enlargement, scale factor −2, about (1, 3)

6.6 Combined transformations

Exercise 6F

1 (−4, −3)
2 **a** (−5, 2)
b Reflection in y-axis
3 A: translation $\begin{pmatrix} 1 \\ -2 \end{pmatrix}$, B: reflection in y-axis, C: rotation 90°clockwise about (0, 0), D: reflection in $x = 3$, E: reflection in $y = 4$, F: enlargement by scale factor 2, centre (0, 1)
4 **a** T_1 to T_2: rotation 90°clockwise about (0, 0)
b T_1 to T_6: rotation 90°anticlockwise about (0, 0)
c T_2 to T_3: translation $\begin{pmatrix} 2 \\ 2 \end{pmatrix}$
d T_6 to T_2: rotation 180°about (0, 0)
e T_6 to T_5: reflection in y-axis
f T_5 to T_4: translation $\begin{pmatrix} 4 \\ 0 \end{pmatrix}$

5 **a–d**

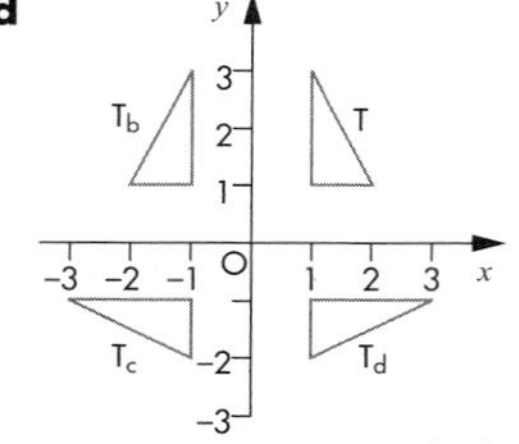

e T_d to T: rotation 90° anticlockwise about (0, 0)

6 Reflection in x-axis, translation $\begin{pmatrix} 0 \\ -5 \end{pmatrix}$, rotation 90°clockwise about (0, 0)

7 Translation $\begin{pmatrix} 0 \\ -8 \end{pmatrix}$, reflection in x-axis, rotation 90°clockwise about (0, 0)

8 **a**

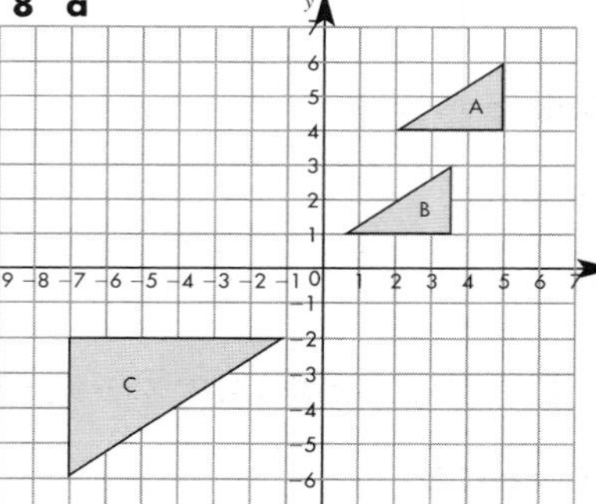

b enlargement of scale factor $-\frac{1}{2}$ about (1, 2)

Examination questions

1 a

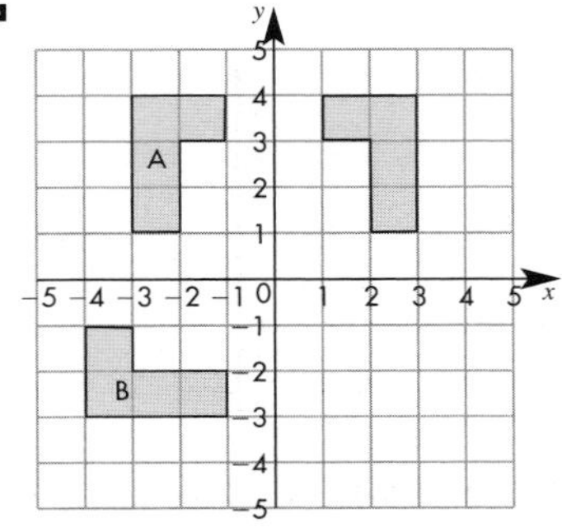

b rotation of 90° clockwise about (0, 0)

2

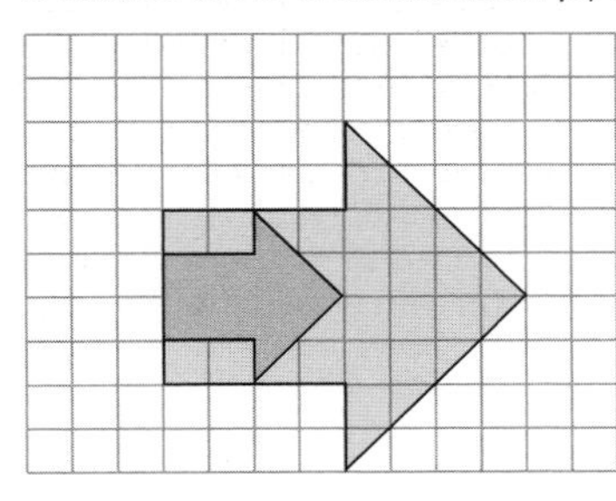

3 a
b

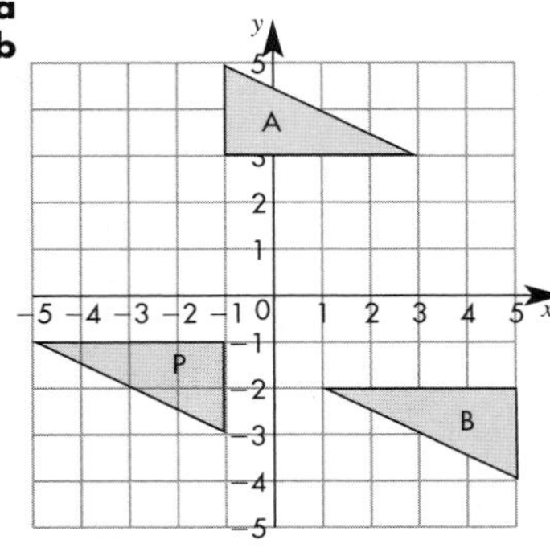

c

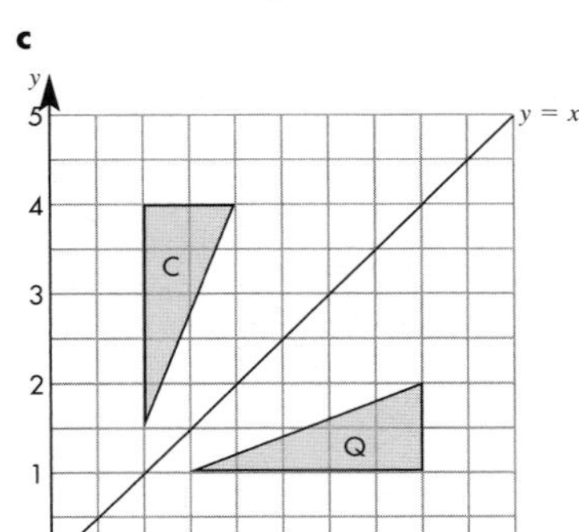

4 Translation by the vector $\begin{pmatrix}2\\0\end{pmatrix}$

5 AD = CD (given), AB = BC (given), BD (common), so triangles are congruent (SSS).

Answers to Chapter 7

Quick check

1 a $3x - 15$ **b** $2x + 14$ **c** $14x - 21$
2 a $6y$ **b** $7x - 3$ **c** $x + 5$
3 a $8x$ **b** $18y^2$ **c** $3x^3$
4 a $6y$ **b** $7x - 3$ **c** $5 + 1x$ **d** $6x$ **e** $8y$ **f** $2c^3$
5 a 4 **b** 6 **c** 40 **d** −1 **e** 4 **f** 33

7.1 Changing the subject of a formula

Exercise 7A

1 $-8y$
2 $-\frac{x}{8}$
3 $\frac{b + 5c}{4}$
4 $\frac{b(q + p)}{q - p}$
5 $\frac{a(q - p)}{q + p}$
6 $\frac{A}{\pi(2h + k)}$
7 $\frac{u}{\sqrt{(1 - a)}}$
8 $\frac{3 + st}{2 + s}$
9 $\frac{6 + st}{2 + s}$
10 $\frac{2R - 3}{R - 1}$
11 a $r = \frac{P}{\pi + 2k}$
b $r = \sqrt{\frac{2A}{\pi + \sqrt{(k^2 - 1)}}}$
12 $\frac{100A}{100 + RY}$
13 a $b = \frac{Ra}{a - R}$
b $a = \frac{Rb}{b - R}$
14 a $\frac{2 + 2y}{y - 1}$
b $y - 1 = \frac{4}{x - 2}$, $(x - 2)(y - 1) = 4$, $x - 2 = \frac{4}{y - 1}$, $x = 2 + \frac{4}{y - 1}$
c $y = 1 + \frac{4}{x - 2} = \frac{x - 2 + 4}{x - 2} = \frac{x + 2}{x - 2}$
d Same formula as in **a**
15 a Cannot factorise the expression
b $\frac{3V}{r^2(2r + 3h)}$
c Yes, $\sqrt[3]{\frac{3V}{5\pi}}$
16 $x = \frac{2W - 2zy}{z + y}$
17 $x = \frac{1 - 3y}{2y - 5}$

The first number at the top of the answer is the constant term on the top of the original. The coefficient of y at the top of the answer is the negative constant term on the bottom of the original. The coefficient of y at the bottom of the answer the coefficient of x on the bottom of the original. The constant term on the bottom is negative the coefficient of x on the top of the original.

18 a $b^2 = c^2 + a^2 - 2ac \cos B$
b $\cos C = \frac{a^2 + b^2 - c^2}{2ab}$

7.2 Solving linear equations

Exercise 7B

1 **a** 30 **b** 21 **c** 72 **d** 12
e 6 **f** $10\frac{1}{2}$ **g** −10 **h** 7
i 11 **j** −4 **k** 7 **l** $2\frac{4}{5}$
m 1 **n** $11\frac{1}{2}$ **o** $\frac{1}{5}$

2 Any valid equations such as $\frac{x+3}{2} = 6$ or $\frac{x-5}{4} = 1$

3 **a** Amanda
b First line: Betsy adds 4 instead of multiplying by 5.
Second line: Betsy adds 5 instead of multiplying by 5.
Fourth line: Betsy subtracts 2 instead of dividing by 2.

Exercise 7C

1 **a** 3 **b** 7 **c** 5 **d** 3
e 4 **f** 6 **g** 8 **h** 1
i $1\frac{1}{2}$ **j** $2\frac{1}{2}$ **k** $\frac{1}{2}$ **l** $1\frac{1}{5}$
m 2 **n** −2 **o** −1 **p** −2
q −2 **r** −1

2 Any values that work, e.g. $a = 2$, $b = 3$ and $c = 30$.

3 55

Exercise 7D

1 **a** $x = 2$ **b** $y = 1$ **c** $a = 7$
d $t = 4$ **e** $p = 2$ **f** $k = -1$
g $m = 3$ **h** $s = -2$

2 $3x - 2 = 2x + 5, x = 7$

3 **a** $d = 6$ **b** $x = 11$ **c** $y = 1$
d $h = 4$ **e** $b = 9$ **f** $c = 6$

4 $6x + 3 = 6x + 10$; $6x - 6x = 10 - 3$; $0 = 7$, which is obviously false. Both sides have $6x$, which cancels out.

5 $8x + 7 + x + 4 = 11x + 5 - x - 4$, $x = 10$

6 Because both sides expand to $12x + 18$. $12x + 18 = 12x + 18$ cannot be solved for a unique value of x.

7.3 Setting up equations

Exercise 7E

1 90p

2 **a** 1.5 **b** 2

3 **a** 1.5 cm **b** 6.75 cm^2

4 17

5 8

6 **a** $8c - 10 = 56$
b £8.25

7 **a** B: 450 cars, C: 450 cars, D: 300 cars
b 800
c 750

8 Length is 5.5 m, width is 2.5 m and area is 13.75 m^2. Carpet costs £123.75

9 3 years

10 9 years

11 3 cm

12 5

13 **a** $4x + 40 = 180$
b $x = 35°$

14 **a** $\frac{x+10}{5} = 9.50$
b £37.50

15 **a** 15
b −1
c $2(n + 3)$, $2(n + 3) - 5$
d $2(n + 3) - 5 = n$, $2n + 6 - 5 = n$, $2n + 1 = n$, $n = -1$

16 No, as $x + x + 2 + x + 4 + x + 6 = 360$ gives $x = 87°$ so the consecutive numbers (87, 89, 91, 93) are not even but odd

17 $4x + 18 = 3x + 1 + 50$, $x = 33$. Large bottle 1.5 litres, small bottle 1 litre

7.4 Trial and improvement

Exercise 7F

1 **a** 4 and 5 **b** 4 and 5
c 2 and 3

2 $x = 3.5$

3 $x = 3.7$

4 $x = 2.5$

5 $x = 1.5$

6 **a** $x = 2.4$ **b** $x = 2.8$
c $x = 3.2$

7 **a** $x(x + 5) = 100 \Rightarrow x^2 + 5x = 100$
b $x = 7.8$ cm, 12.8 cm

8 $x = 5.8$

9 Volume $= x \times 2x(x + 8) = 500$, $x^3 + 8x^2 = 250$, $4 \Rightarrow 192$, $5 \Rightarrow 325$, $4.4 \Rightarrow 240.064$, $4.5 \Rightarrow 253.125$, $4.45 \Rightarrow 246.541125$, so dimensions are 4.5 cm, 9 cm and 12.5 cm

10 **a** Cube is x^3, hole is $\frac{x}{2} \times \frac{x}{2} \times 8 = 2x^3$.
Cube minus hole is 1500
b $12 \Rightarrow 1440$, $13 \Rightarrow 1859$, $12.1 \Rightarrow 1478.741$, $12.2 \Rightarrow 1518.168$, $12.15 \Rightarrow 1498.368375$ so the value of $x = 12.2$ (to 1 dp)

11 2.76 and 7.24

7.5 Simultaneous equations

Exercise 7G

1 **a** $x = 4, y = 1$ **b** $x = 1, y = 4$ **c** $x = 3, y = 1$
d $x = 5, y = -2$ **e** $x = 7, y = 1$ **f** $x = 5, y = \frac{1}{2}$
g $x = 4\frac{1}{2}, y = 1\frac{1}{2}$ **h** $x = -2, y = 4$ **i** $x = 2\frac{1}{2}, y = -1\frac{1}{2}$
j $x = 2\frac{1}{4}, y = 6\frac{1}{2}$ **k** $x = 4, y = 3$ **l** $x = 5, y = 3$

2 **a** 3 is the first term. The next term is $3 \times a + b$, which equals 14.
b $14a + b = 47$ **c** $a = 3, b = 5$ **d** 146, 443

Exercise 7H

1 **a** $x = 2, y = -3$ **b** $x = 7, y = 3$ **c** $x = 4, y = 1$
d $x = 2, y = 5$ **e** $x = 4, y = -3$ **f** $x = 1, y = 7$
g $x = 2\frac{1}{2}, y = 1\frac{1}{2}$ **h** $x = -1, y = 2\frac{1}{2}$ **i** $x = 6, y = 3$
j $x = \frac{1}{2}, y = -\frac{3}{4}$ **k** $x = -1, y = 5$ **l** $x = 1\frac{1}{2}, y = \frac{3}{4}$

2 **a** They are the same equation. Divide the first by 2 and it is the second, so they have an infinite number of solutions.
b Double the second equation to get $6x + 2y = 14$ and subtract to get $9 = 14$. The left-hand sides are the same if the second is doubled so they cannot have different values.

Exercise 7I

1 **a** $x = 5, y = 1$ **b** $x = 3, y = 8$ **c** $x = 9, y = 1$
d $x = 7, y = 3$ **e** $x = 4, y = 2$ **f** $x = 6, y = 5$
g $x = 3, y = -2$ **h** $x = 2, y = \frac{1}{2}$ **i** $x = -2, y = -3$
j $x = -1, y = 2\frac{1}{2}$ **k** $x = 2\frac{1}{2}, y = -\frac{1}{2}$ **l** $x = -1\frac{1}{2}, y = 4\frac{1}{2}$
m $x = -\frac{1}{2}, y = -6\frac{1}{2}$ **n** $x = 3\frac{1}{2}, y = 1\frac{1}{2}$ **o** $x = -2\frac{1}{2}, y = -3\frac{1}{2}$

2 $(1, -2)$ is the solution to equations A and C; $(-1, 3)$ is the solution to equations A and D; $(2, 1)$ is the solution to B and C; $(3, -3)$ is the solution to C and D.

3 Intersection points are (0, 2), (6, 0) and (2, 4). Area is 8 square units.

4 Intersection points are (0, 3), (6, 0) and (4, −1). Area is 4.5 square units.

7.6 Solving problems using simultaneous equations

Exercise 7J

1 Amul £7.20, Kim £3.50

2 **a** $10x + 5y = 420$, $8x + 10y = 540$
b £2.11

3 **a** $6x + 3y = 435$, $11x + 7y = 880$
b £5.55

4 **a** My age minus 6 equals 2 × (my son's age minus 6)
b $x = 46$ and $y = 26$

5 **a** $3t + 5b = 810$, $3t + 3b = 630$
b £10.20

6 84p

7 10.3 kg

8 £4.40

9 £62

10 £195

11 2 hr 10 min

12 $p = 36$, $q = 22$. Total weight for Baz is 428 pounds so he can carry the load safely on his trailer.

13 $b = £3.50$, $p = £1.75$. Camilla needs £35 so she will not have enough money.

14 When Carmen worked out (2) − (3), she should have got $y = 6$
When Jeff rearranged $2x + 8 - x = 10$, he should have got $x = 2$
They also misunderstood 'two, six' as this is means $x = 2$ and $y = 6$, not the other way round.

7.7 Linear and non-linear simultaneous equations

Exercise 7K

1 **a** (5, −1)
b (4, 1)
c (8, −1)

2 **a** (1, 2) and (−2, −1)
b (−4, 1) and (−2, 2)

3 **a** (3, 4) and (4, 3)
b (0, 3) and (−3, 0)
c (3, 2) and (−2, 3)

4 **a** (2, 5) and (−2, −3)
b (−1, −2) and (4, 3)
c (3, 3) and (1, −1)

5 **a** (−3, −3), (1, 1)
b (3, −2), (−2, 3)
c (−2, 1), (1, 2)
d (2, −1), (3, 1)
e (−2, 1), (3, 6)
f (1, −4), (4, 2)
g (4, 5), (−5, −4)

6 **a** (1, 0)
b **iii** as the straight line just touches the curve

7 **a** (−2, 1)
b **i** (2, 1)
ii (−2, −1)
iii (2, −1)

8 **a** (2, 4)
b (1, 0)
c The line is a tangent to the curve.

Examination questions

1 $t = \frac{1}{2}(y + 10)$

2 $x = 8$

3 $x^3 + 2x = 26$

2	12	too small
3	33	too big
2.5	20.625	too small
2.8	27.552	too big by 1.552
2.7	25.083	too small by 0.917
2.75	26.2968	

Take the value as 2.7 to 1 dp.

4 **a** $5x + 60 = 360°$
b $5x = 300$, $x = 60°$

5 $6x + 30 = 180°$, $6x = 150°$, $x = 25°$

6 $x^3 + 4x = 26$

2	16	too small
3	39	too big
2.5	25.625	too small by 0.375
2.6	27.976	too big by 1.976
2.55	26.781375	

Take the value as 2.5 to 1 dp.

7 $x = +8$

8 $x = 3$, $y = -2$

9 **a** 9 or −9
b $s = \frac{v^2 - u^2}{2a}$

10 $b = \frac{5a + 2}{a + 7}$

11 $x = 0.4$, $y = 2.2$ and $x = -1$, $y = -2$

Answers to Chapter 8

Quick check

1 **a** 6 cm **b** 7.5 cm **c** 11 cm

2 **a** 30° **b** 135°

8.1 Constructing triangles

Exercise 8A

1 **a** BC = 2.9 cm, ∠B = 53°, ∠C = 92°
b EF = 7.4 cm, ED = 6.8 cm, ∠E = 50°
c ∠G = 105°, ∠H = 29°, ∠I = 46°
d ∠J = 48°, ∠L = 32°, JK = 4.3 cm
e ∠N = 55°, ON = OM = 7 cm
f ∠P = 51°, ∠R = 39°, QP = 5.7 cm

2 **a** Students can check one another's triangles.
b ∠ABC = 44°, ∠BCA = 79°, ∠CAB = 57°

3 **a** 5.9 cm
b 18.8 cm^2

4 BC = 2.6 cm, 7.8 cm

5 **a** 4.5
b 11.25 cm^2

6 **a** 4.3 cm
b 34.5 cm^2

7 **a** Right-angled triangle constructed with sides 3, 4, 5 and 4.5, 6 and 7.5, and scale marked 1 cm : 1 m
b Right-angled triangle constructed with 12 equally spaced dots

8 An equilateral triangle of side 4 cm

9 Even with all three angles, you need to know at least one length

8.2 Bisectors

Exercise 8B

1–9 Practical work; check students' constructions

10 Leicester

11 The centre of the circle

12 Start with a base line AB; then construct a perpendicular to the line from point A. At point B, construct an angle of 60°. Ensure that the line for this 60° angle crosses the perpendicular line; where they meet will be the final point C.

13–15 Practical work; check students' constructions

8.3 Defining a locus

Exercise 8C

1 Circle with radius:
a 2 cm **b** 4 cm **c** 5 cm

2 **a** **b** **c**

A • • B A • • B A • • B

3 **a** Circle with radius 4 m
b

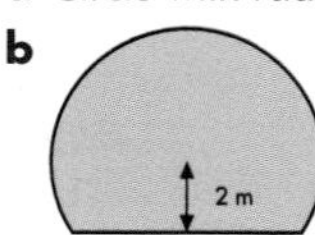

4

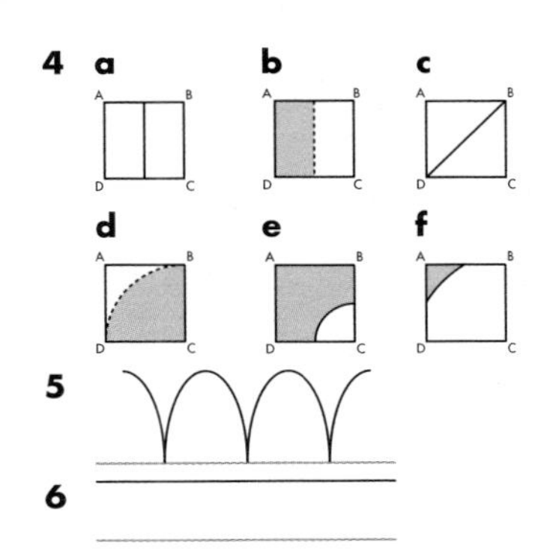

5

6

7 Construct the bisector of angle BAC and the perpendicular bisector of the line AC.

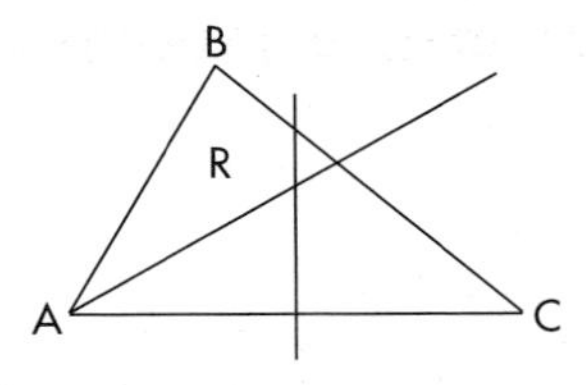

8

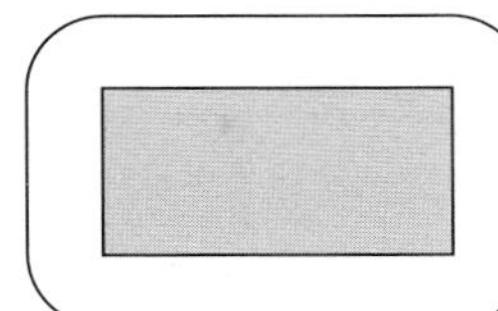

8.4 Loci problems

Exercise 8D

1

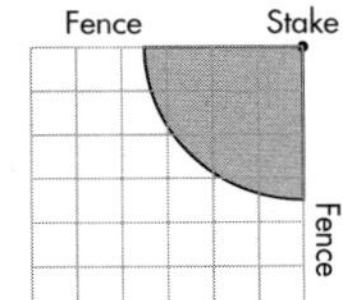

2

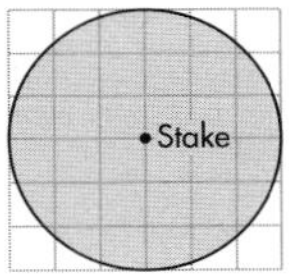

3

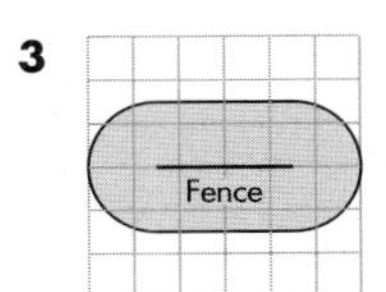

4

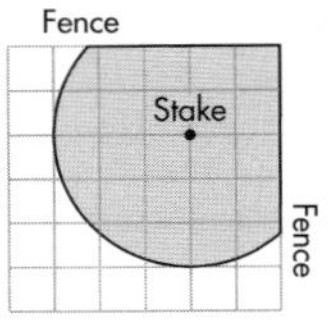

5

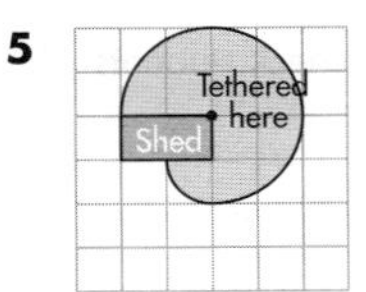

6

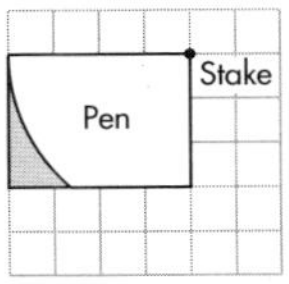

7

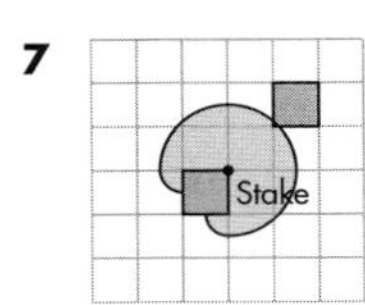

8 **a** Sketch should show a circle of radius 6 cm around London and one of radius 4 cm around Glasgow.
b No
c Yes

9 **a** Yes
b Sketch should show a circle of radius 4 cm around Leeds and one of radius 4 cm around Exeter. The area where they overlap should be shaded.

10 **a** This is the perpendicular bisector of the line from York to Birmingham. It should pass just below Manchester and just through the top of Norwich.
b Sketch should show a circle of radius 7 cm around Glasgow and one of radius 5 cm around London.
c The transmitter can be built anywhere on line constructed in part **a** that is within the area shown in part **b**.

11 Sketch should show two circles around Birmingham, one of radius 3 cm and one of radius 5 cm. The area of good reception is the area between the two circles.

12 Sketch should show a circle of radius 6 cm around Glasgow, two circles around York, one of radius 4 cm and one of radius 6 cm and a circle around London of radius 8 cm. The small area in the Irish Sea that is between the two circles around York and inside both the circle around Glasgow and the circle around London is where the boat can be.

13 Sketch should show two circles around Newcastle upon Tyne, one of radius 4 cm and one of radius 6 cm, and two circles around Bristol, one of radius 3 cm and one of radius 5 cm. The area that is between both pairs of circles is the area that should be shaded.

14 Sketch should show the perpendicular bisector of the line running from Newcastle upon Tyne to Manchester and that of the line running from Sheffield to Norwich. Where the lines cross is where the oil rig is located.

15 Sketch should show the perpendicular bisector of the line running from Glasgow to Norwich and that of the line running from Norwich to Exeter. Where the lines cross is where Fred's house is.

16 Sketch should show the bisectors of the angles made by the piers and the sea wall at points A and B. These are the paths of each boat.

17 Leeds

18 On a map, draw a straight line from Newcastle to Bristol, construct the line bisector, then the search will be anywhere on the sea along that line.

Examination questions

1

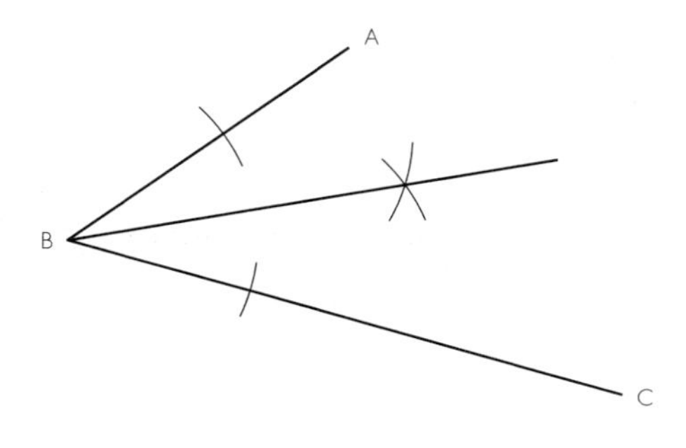

2

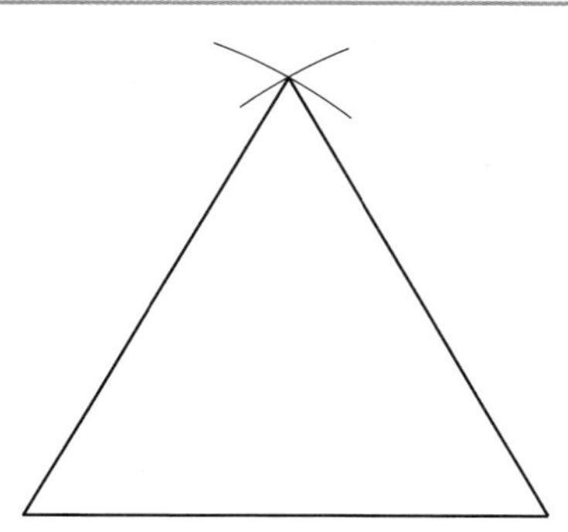

3

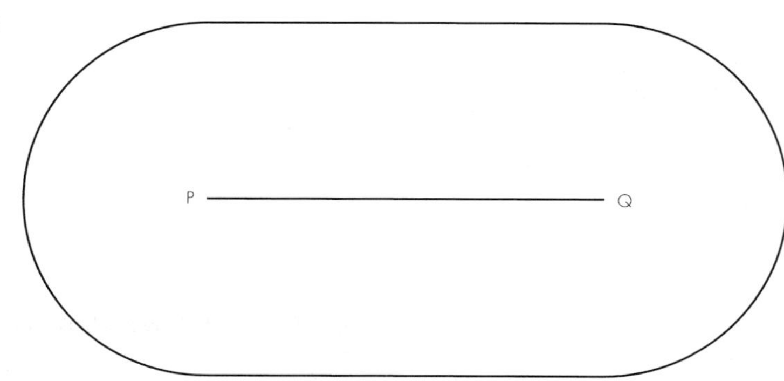

4

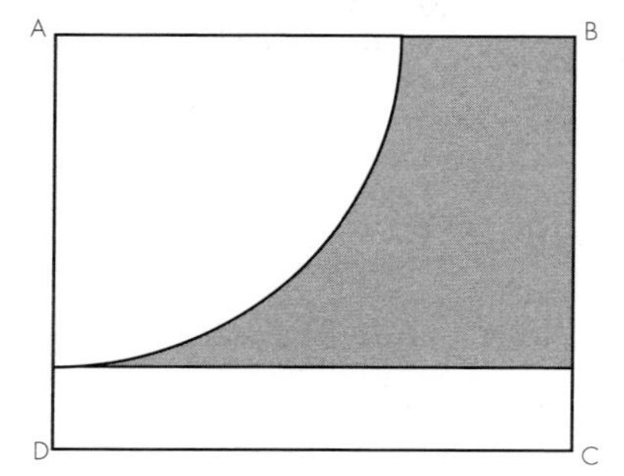

Answers to Chapter 9

Quick check

1 **a** 3 : 4 **b** 4 : 5 **c** 4 : 1 **d** 3 : 2
2 **a** 14 **b** 2.5 **c** 8 **d** 6

9.1 Similar triangles

Exercise 9A

1 2, 3
2 **a** Yes, 4
b No, corresponding sides have different ratios.

3 **a** PQR is an enlargement of ABC
b 1 : 3
c Angle R
d BA

4 **a** Sides in same ratio
b Angle P **c** PR
5 **a** Same angles **b** Angle Q
c AR

6 a 8 cm
b 7.5 cm
c $x = 6.67$ cm, $y = 13.5$ cm
d $x = 24$ cm, $y = 13$ cm
e AB = 10 cm, PQ = 6 cm
f 4.2 cm

7 a Sides in same ratio
b 1 : 3 c 13 cm d 39 cm

8 5.2 m

9 Corresponding sides are not in the same ratio, $12 : 15 \neq 16 : 19$.

10 DE = 17.5 cm; AC : EC = BA : DE, 5 : 12.5 = 7 : DE, DE = $7 \times 12.5 \div 5 = 17.5$ cm

Exercise 9B

1 a 9 cm
b 12 cm

2 a 5 cm
b 5 cm
c $x = 60$ cm, $y = 75$ cm
d $x = 45$ cm, $y = 60$ cm
e DC = 10 cm, EB = 8 cm

3 82 m

4 220 feet

5 15 m

6 3.3 m

7 1.8 m

8 13.5 cm

9 c In triangles AXY and ABC, $\frac{AX}{AB} = \frac{15}{10}$

so $AX = \frac{15 \times 8}{10} = 12$ cm.

$BX = AX - AB = 12 - 8 = 4$ cm

Exercise 9C

1 5 cm

2 6 cm

3 10 cm

4 $x = 6$ cm, $y = 7.5$ cm

5 $x = 15$ cm, $y = 21$ cm

6 $x = 3$ cm, $y = 2.4$ cm

9.2 Area and volume of similar shapes

Exercise 9D

1 a 4 : 25 b 8 : 125

2 a 16 : 49 b 64 : 343

3 Linear scale factor 2, 3, $\frac{1}{4}$, 5, $\frac{1}{10}$;

linear ratio 1 : 2, 1 : 3, 4 : 1, 1: 5, 10 : 1;

linear fraction $\frac{2}{1}$, $\frac{3}{1}$, $\frac{1}{4}$, $\frac{5}{1}$, $\frac{1}{10}$;

area scale factor 4, 9, $\frac{1}{16}$, 25, $\frac{1}{100}$;

volume scale factor 8, 27, $\frac{1}{64}$, $\frac{1}{125}$, $\frac{1}{1000}$

4 135 cm^2

5 a 56 cm^2
b 126 cm^2

6 a 48 m^2
b 3 m^2

7 a 2400 cm^3
b 8100 cm^3

8 4 litres

9 1.38 m^3

10 £6

11 4 cm

12 8×60p = £4.80 so it is better value

13 a 3 : 4
b 9 : 16
c 27 : 64

14 $720 \div 8 = 90$ cm^3

Exercise 9E

1 6.2 cm, 10.1 cm

2 4.26 cm, 6.74 cm

3 9.56 cm

4 3.38 m

5 8.39 cm

6 26.5 cm

7 16.9 cm

8 a 4.33 cm, 7.81 cm
b 143 g, 839 g

9 53.8 kg

10 1.73 kg

11 8.8 cm

12 7.9 cm and 12.6 cm

13 b

Examination questions

1 a 12 cm b 5 cm

2 a 39 cm b 30 cm

3 a 2 cm b 5.25 cm

4 a 8 cm b 96 cm^3

5 a 12 cm b 2700 π cm^3

Answers to Chapter 10

Quick check

1 8.60 cm
2 13.0 cm
3 21.6°
4 8.40 cm

10.1 Some 2D problems

Exercise 10A

1 13.1 cm
2 73.7°
3 9.81 cm
4 33.5 m
5 **a** 10.0 cm **b** 11.5° **c** 4.69 cm
6 63.0°
7 200 m
8 **a** $\sqrt{2}$ cm
b **i** $\frac{\sqrt{2}}{2}$ (an answer of $\frac{\sqrt{1}}{\sqrt{2}}$ would also be accepted)
ii $\frac{\sqrt{2}}{2}$ **iii** 1
9 14.1°

10.2 Some 3D problems

Exercise 10B

1 25.1°
2 **a** 58.6° **b** 20.5 cm
c 2049 cm^3 **d** 64.0°
3 **a** 3.46 m **b** 75.5°
c 73.2° **d** 60.3 m^2
e £1507.50, or £1525, if tiles must be bought in full square metres
4 **a** 24.0° **b** 48.0°
c 13.5 cm **d** 16.6°
5 **a** 3.46 m
b 70.5°
6 For example, the length of the diagonal of the base is $\sqrt{b^2 + c^2}$ and taking this as the base of the triangle with the height of the edge, then the hypotenuse is $\sqrt{(a^2 + (\sqrt{b^2 + c^2})^2)} = \sqrt{a^2 + b^2 + c^2}$
7 It is 44.6°; use triangle XDM where M is the midpoint of BD; triangle DXB is isosceles, as X is over the point where the diagonals of the base cross; the length of DB is $\sqrt{626}$, the cosine of the required angle is $0.5\sqrt{626} \div 18$.

10.3 Trigonometric ratios of angles between 90° and 360°

Exercise 10C

1 **a** 36.9°, 143.1° **b** 53.1°, 126.9°
c 48.6°, 131.4° **d** 224.4°, 315.6°
e 194.5°, 345.5° **f** 198.7°, 341.3°
g 190.1°, 349.9° **h** 234.5°, 305.5°
i 28.1°, 151.9° **j** 185.6°, 354.4°
k 33.6°, 146.4° **l** 210°, 330°
2 Sin 234°, as the others all have the same numerical value.
3 **a** 438° or 78° + 360n°
b −282° or 78° − 360n°
c Line symmetry about ±90n° where n is an odd integer.
Rotational symmetry about ±180n° where n is an integer.

Exercise 10D

1 **a** 53.1°, 306.9° **b** 54.5°, 305.5°
c 62.7°, 297.3° **d** 54.9°, 305.1°
e 79.3°, 280.7° **f** 143.1°, 216.9°
g 104.5°, 255.5° **h** 100.1°, 259.9°
i 111.2°, 248.8° **j** 166.9°, 193.1°
k 78.7°, 281.3° **l** 44.4°, 315.6°
2 Cos 58°, as the others are negative.
3 **a** 492° or 132° + 360n°
b −228° or 132° − 360n°
c Line symmetry about ±180n° where n is an integer.
Rotational symmetry about ±90n° where n is an odd integer.

Exercise 10E

1 **a** 0.707 **b** −1 (−0.9998)
c −0.819 **d** 0.731
2 **a** −0.629 **b** −0.875
c −0.087 **d** 0.999
3 **a** 21.2°, 158.8° **b** 209.1°, 330.9°
c 50.1°, 309.9° **d** 150.0°, 210.0°
e 60.9°, 119.1° **f** 29.1°, 330.9°
4 30°, 150°
5 −0.755
6 **a** 1.41 **b** −1.37
c −0.0367 **d** −0.138
e 1.41 **f** −0.492
7 True
8 **a** Cos 65° **b** Cos 40°
9 **a** 10°, 130° **b** 12.7°, 59.3°
10 38.2°, 141.8°

Exercise 10F

1 **a** 14.5°, 194.5° **b** 38.1°, 218.1° **c** 50.0°, 230.0° **d** 61.9°, 241.9° **e** 68.6°, 248.6° **f** 160.3°, 340.3° **g** 147.6°, 327.6° **h** 135.4°, 315.4° **i** 120.9°, 300.9° **j** 105.2°, 285.2° **k** 54.4°, 234.4° **l** 42.2°, 222.2° **m** 160.5°, 340.5° **n** 130.9°, 310.9° **o** 76.5°, 256.5° **p** 116.0°, 296.0° **q** 174.4°, 354.4° **r** 44.9°, 224.9° **s** 50.4°, 230.4° **t** 111.8°, 291.8°
2 Tan 235°, as the others have a numerical value of 1
3 **a** 425° or 65° + $180n°$, $n \geq 2$
b −115° or 65° − $180n°$
c No Line symmetry
Rotational symmetry about $\pm 180n°$ where n is an integer.

10.4 Solving any triangle

Exercise 10G

1 **a** 3.64 m **b** 8.05 cm **c** 19.4 cm
2 **a** 46.6° **b** 112.0° **c** 36.2°
3 50.3°, 129.7°
4 2.88 cm, 20.9 cm
5 **a** **i** 30° **ii** 40° **b** 19.4 m
6 36.5 m
7 22.2 m
8 3.47 m
9 767 m
10 26.8 km/h
11 64.6 km
12 Check students' answers.
13 134°
14 Using any triangle, find two expressions for the height in terms of lengths of sides and sines of angles and set them equal.

Exercise 10H

1 **a** 7.71 m **b** 29.1 cm **c** 27.4 cm
2 **a** 76.2° **b** 125.1° **c** 90° **d** Right-angled triangle
3 5.16 cm
4 65.5 cm
5 **a** 10.7 cm **b** 41.7° **c** 38.3° **d** 6.69 cm **e** 54.4 cm^2
6 72.3°
7 25.4 cm, 38.6 cm
8 58.4 km at 092.5°
9 21.8°
10 **a** 82.8° **b** 8.89 cm
11 42.5 km
12 Check students' answers.
13 111°; the largest angle is opposite the longest side

Exercise 10I

1 **a** 8.60 m **b** 90° **c** 27.2 cm **d** 26.9° **e** 41.0° **f** 62.4 cm **g** 90.0° **h** 866 cm **i** 86.6 cm
2 7 cm
3 11.1 km
4 19.9 knots
5 **a** 27.8 miles **b** 262°
6 **a** $A = 90°$; this is Pythagoras' theorem **b** A is acute **c** A is obtuse
7 142 m

10.5 Trigonometric ratios in surd form

Exercise 10J

1 $\frac{3}{5}$
2 $\frac{\sqrt{2}}{5}$
3 $\sqrt{19}$, $\sin x = \frac{\sqrt{6}}{\sqrt{19}}$, $\cos x = \frac{\sqrt{13}}{\sqrt{19}}$, $\tan x = \frac{\sqrt{6}}{\sqrt{13}}$
4 **a** $\sqrt{157}$ **b** $\sin A = \frac{\sqrt{6}}{\sqrt{157}}$, $\cos A = \frac{\sqrt{11}}{\sqrt{157}}$
5 $9\sqrt{3}$ cm^2
6 400 cm^2
7 $3\sqrt{6}$ cm

10.6 Using sine to find the area of a triangle

Exercise 10K

1 **a** 24.0 cm^2 **b** 26.7 cm^2 **c** 243 cm^2 **d** 21 097 cm^2 **e** 1224 cm^2
2 4.26 cm
3 **a** 42.3° **b** 49.6°
4 103 cm^2
5 2033 cm^2
6 21.0 cm^2
7 **a** 33.2° **b** 25.3 cm^2
8 Find an expression for a vertical height, in terms of the length of a side and the sine of an angle, then apply the usual formula.
9 **a** $\frac{1}{\sqrt{2}}$ **b** 21 cm^2
10 726 cm^2
11 $\frac{\sqrt{3}a^2}{4}$
12 c

Examination questions

1 85.5°
2 53.8 cm^2
3 5.89 cm
4 **a** **i** $\angle ABM = 60°$, BM = 1m, $\cos 60° = \frac{BM}{AB} = \frac{1}{2}$
ii $\frac{\sqrt{3}}{2}$
b $(\frac{1}{2})^2 + (\frac{\sqrt{3}}{2})^2 = \frac{1}{4} + \frac{3}{4} = 1$
5 21.7 m
6 **a** $\frac{\sqrt{3}}{2}$ **b** $-\frac{\sqrt{3}}{2}$

Answers to Chapter 11

Quick check

1 **a** $-3x$ **b** $2x$ **c** $-3x$ **d** $6m^2$ **e** $-6x^2$ **f** $-12p^2$ **2** **a** -6 **b** $-\frac{1}{2}$ **c** $\frac{2}{3}$

11.1 Expanding brackets

Exercise 11A

1 $x^2 + 5x + 6$
2 $t^2 + 7t + 12$
3 $w^2 + 4w + 3$
4 $m^2 + 6m + 5$
5 $k^2 + 8k + 15$
6 $a^2 + 5a + 4$
7 $x^2 + 2x - 8$
8 $t^2 + 2t - 15$
9 $w^2 + 2w - 3$
10 $f^2 - f - 6$
11 $g^2 - 3g - 4$
12 $y^2 + y - 12$
13 $x^2 + x - 12$
14 $p^2 - p - 2$
15 $k^2 - 2k - 8$
16 $y^2 + 3y - 10$
17 $a^2 + 2a - 3$
18 $x^2 - 9$
19 $t^2 - 25$
20 $m^2 - 16$
21 $t^2 - 4$
22 $y^2 - 64$
23 $p^2 - 1$
24 $25 - x^2$
25 $49 - g^2$
26 $x^2 - 36$
27 $(x + 2)$ and $(x + 3)$
28 **a** B: $1 \times (x - 2)$
C: 1×2
D: $2 \times (x - 1)$
b $(x - 2) + 2 + 2(x - 1)$
$= 3x - 2$
c Area A $= (x - 1)(x - 2)$
= area of square minus areas (B + C + D)
$= x^2 - (3x - 2) = x^2 - 3x + 2$
29 **a** $x^2 - 9$
b **i** 9991
ii 39991

Exercise 11B

1 $6x^2 + 11x + 3$
2 $12y^2 + 17y + 6$
3 $6t^2 + 17t + 5$
4 $8t^2 + 2t - 3$
5 $10m^2 - 11m - 6$
6 $12k^2 - 11k - 15$
7 $6p^2 + 11p - 10$
8 $10w^2 + 19w + 6$
9 $6a^2 - 7a - 3$
10 $8r^2 - 10r + 3$
11 $15g^2 - 16g + 4$
12 $12d^2 + 5d - 2$
13 $8p^2 + 26p + 15$
14 $6t^2 + 7t + 2$
15 $6p^2 + 11p + 4$
16 $6 - 7t - 10t^2$
17 $12 + n - 6n^2$
18 $6f^2 - 5f - 6$
19 $12 + 7q - 10q^2$
20 $3 - 7p - 6p^2$
21 $4 + 10t - 6t^2$
22 **a** $x^2 + 2x + 1$
b $x^2 - 2x + 1$
c $x^2 - 1$
d $p + q = (x + 1 + x - 1) = 2x$
$(p + q)^2 = (2x)^2 = 4x^2$
$p^2 + 2pq + q^2 = x^2 + 2x + 1 + 2(x^2 - 1) + x^2 - 2x + 1$
$= 4x^2 + 2x - 2x + 2 - 2 = 4x^2$
23 **a** $(3x - 2)(2x + 1) = 6x^2 - x - 2$, $(2x - 1)(2x - 1) = 4x^2 - 4x + 1$, $(6x - 3)(x + 1) = 6x^2 + 3x - 3$, $(3x + 2)(2x + 1) = 6x^2 + 7x + 2$
b Multiply the x terms to match the x^2 term and/or multiply the constant terms to get the constant term in the answer.

Exercise 11C

1 $4x^2 - 1$
2 $9t^2 - 4$
3 $25y^2 - 9$
4 $16m^2 - 9$
5 $4k^2 - 9$
6 $16h^2 - 1$
7 $4 - 9x^2$
8 $25 - 4t^2$
9 $36 - 25y^2$
10 $a^2 - b^2$
11 $9t^2 - k^2$
12 $4m^2 - 9p^2$
13 $25k^2 - g^2$
14 $a^2b^2 - c^2d^2$
15 $a^4 - b^4$
16 **a** $a^2 - b^2$
b Dimensions: $a + b$ by $a - b$; Area: $a^2 - b^2$
c Areas are the same, so $a^2 - b^2 = (a + b) \times (a - b)$
17 First shaded area is $(2k)^2 - 1^2 = 4k^2 - 1$
Second shaded area is $(2k + 1)(2k - 1) = 4k^2 - 1$

Exercise 11D

1 $x^2 + 10x + 25$
2 $m^2 + 8m + 16$
3 $t^2 + 12t + 36$
4 $p^2 + 6p + 9$
5 $m^2 - 6m + 9$
6 $t^2 - 10t + 25$
7 $m^2 - 8m + 16$
8 $k^2 - 14k + 49$
9 $9x^2 + 6x + 1$
10 $16t^2 + 24t + 9$
11 $25y^2 + 20y + 4$
12 $4m^2 + 12m + 9$
13 $16t^2 - 24t + 9$
14 $9x^2 - 12x + 4$
15 $25t^2 - 20t + 4$
16 $25r^2 - 60r + 36$
17 $x^2 + 2xy + y^2$
18 $m^2 - 2mn + n^2$
19 $4t^2 + 4ty + y^2$
20 $m^2 - 6mn + 9n^2$
21 $x^2 + 4x$
22 $x^2 - 10x$
23 $x^2 + 12x$
24 $x^2 - 4x$
25 **a** Bernice has just squared the first term and the second term. She hasn't written down the brackets twice.
b Pete has written down the brackets twice but has worked out $(3x)^2$ as $3x^2$ and not $9x^2$.
c $9x^2 + 6x + 1$
26 Whole square is $(2x)^2 = 4x^2$.
Three areas are $2x - 1$, $2x - 1$ and 1.
$4x^2 - (2x - 1 + 2x - 1 + 1) = 4x^2 - (4x - 1) = 4x^2 - 4x + 1$

11.2 Quadratic factorisation

Exercise 11E

1 $(x + 2)(x + 3)$
2 $(t + 1)(t + 4)$
3 $(m + 2)(m + 5)$
4 $(k + 4)(k + 6)$
5 $(p + 2)(p + 12)$
6 $(r + 3)(r + 6)$
7 $(w + 2)(w + 9)$
8 $(x + 3)(x + 4)$
9 $(a + 2)(a + 6)$
10 $(k + 3)(k + 7)$
11 $(f + 1)(f + 21)$
12 $(b + 8)(b + 12)$
13 $(t - 2)(t - 3)$
14 $(d - 4)(d - 1)$
15 $(g - 2)(g - 5)$
16 $(x - 3)(x - 12)$
17 $(c - 2)(c - 16)$
18 $(t - 4)(t - 9)$
19 $(y - 4)(y - 12)$
20 $(j - 6)(j - 8)$
21 $(p - 3)(p - 5)$
22 $(y + 6)(y - 1)$
23 $(t + 4)(t - 2)$
24 $(x + 5)(x - 2)$
25 $(m + 2)(m - 6)$
26 $(r + 1)(r - 7)$
27 $(n + 3)(n - 6)$
28 $(m + 4)(m - 11)$
29 $(w + 4)(w - 6)$
30 $(t + 9)(t - 10)$
31 $(h + 8)(h - 9)$
32 $(t + 7)(t - 9)$
33 $(d + 1)^2$
34 $(y + 10)^2$
35 $(t - 4)^2$
36 $(m - 9)^2$
37 $(x - 12)^2$
38 $(d + 3)(d - 4)$
39 $(t + 4)(t - 5)$
40 $(q + 7)(q - 8)$
41 $(x + 2)(x + 3)$, giving areas of $2x$ and $3x$, or $(x + 1)(x + 6)$, giving areas of x and $6x$.
42 **a** $x^2 + (a + b)x + ab$
b **i** $p + q = 7$
ii $pq = 12$
c 7 can only be 1×7 and $1 + 7 \neq 12$

Exercise 11F

1 $(x + 3)(x - 3)$
2 $(t + 5)(t - 5)$
3 $(m + 4)(m - 4)$
4 $(3 + x)(3 - x)$
5 $(7 + t)(7 - t)$
6 $(k + 10)(k - 10)$
7 $(2 + y)(2 - y)$
8 $(x + 8)(x - 8)$
9 $(t + 9)(t - 9)$
10 **a** x^2
b **i** $(x - 2)$ **ii** $(x + 2)$ **iii** x^2 **iv** 4
c $A + B - C = x^2 - 4$, which is the area of D, which is $(x + 2)(x - 2)$.
11 **a** $x^2 + 4x + 4 - (x^2 + 2x + 1) = 2x + 3$
b $(a + b)(a - b)$
c $(x + 2 + x + 1)(x + 2 - x - 1) = (2x + 3)(1) = 2x + 3$
d The answers are the same.
e $(x + 1 + x - 1)(x + 1 - x + 1) = (2x)(2) = 4x$
12 $(x + y)(x - y)$
13 $(x + 2y)(x - 2y)$
14 $(x + 3y)(x - 3y)$
15 $(3x + 1)(3x - 1)$
16 $(4x + 3)(4x - 3)$
17 $(5x + 8)(5x - 8)$
18 $(2x + 3y)(2x - 3y)$
19 $(3t + 2w)(3t - 2w)$
20 $(4y + 5x)(4y - 5x)$

Exercise 11G

1 $(2x + 1)(x + 2)$
2 $(7x + 1)(x + 1)$
3 $(4x + 7)(x - 1)$
4 $(3t + 2)(8t + 1)$
5 $(3t + 1)(5t - 1)$
6 $(4x - 1)^2$
7 $3(y + 7)(2y - 3)$
8 $4(y + 6)(y - 4)$
9 $(2x + 3)(4x - 1)$
10 $(2t + 1)(3t + 5)$
11 $(x - 6)(3x + 2)$
12 $(x - 5)(7x - 2)$
13 $4x + 1$ and $3x + 2$
14 **a** All the terms in the quadratic have a common factor of 6.
b $6(x + 2)(x + 3)$. This has the highest common factor taken out.

11.3 Solving quadratic equations by factorisation

Exercise 11H

1 −2, −5
2 −3, −1
3 −6, −4
4 −3, 2
5 −1, 3
6 −4, 5
7 1, −2
8 2, −5
9 7, −4
10 3, 2
11 1, 5
12 4, 3
13 −4, −1
14 −9, −2
15 2, 4
16 3, 5
17 −2, 5
18 −3, 5
19 −6, 2
20 −6, 3
21 −1, 2
22 −2
23 −5
24 4
25 −2, −6
26 7
27 **a** $x(x - 3) = 550$, $x^2 - 3x - 550 = 0$
b $(x - 25)(x + 22) = 0$, $x = 25$
28 $x(x + 40) = 48000$, $x^2 + 40x - 48000 = 0$, $(x + 240)(x - 200) = 0$
Fence is $2 \times 200 + 2 \times 240 = 880$ m.
29 −6, −4
30 2, 16
31 −6, 4
32 −9, 6
33 −10, 3
34 −4, 11
35 −8, 9
36 8, 9
37 1
38 Mario was correct. Sylvan did not make it into a standard quadratic and only factorised the x terms. She also incorrectly solved the equation $x - 3 = 4$.

Exercise 11I

1 **a** $\frac{1}{3}, -3$ **b** $1\frac{1}{3}, -\frac{1}{2}$ **c** $-\frac{1}{5}, 2$ **d** $-2\frac{1}{2}, 3\frac{1}{2}$ **e** $-\frac{1}{6}, -\frac{1}{3}$ **f** $\frac{2}{3}, 4$ **g** $\frac{1}{2}, -3$ **h** $\frac{5}{2}, -\frac{7}{6}$ **i** $-1\frac{2}{3}, 1\frac{2}{5}$ **j** $1\frac{3}{4}, 1\frac{2}{7}$ **k** $\frac{2}{3}, \frac{1}{8}$ **l** $\pm\frac{1}{4}$ **m** $-2\frac{1}{4}, 0$ **n** $\pm 1\frac{2}{5}$ **o** $-\frac{1}{3}, 3$

2 **a** 7, −6 **b** $-2\frac{1}{2}, 1\frac{1}{2}$ **c** 7, −6 **d** $-1, \frac{11}{13}$ **e** 3, −2 **f** $-\frac{2}{5}, \frac{1}{2}$ **g** $-\frac{1}{3}, -\frac{1}{2}$ **h** $\frac{1}{5}, -2$ **i** 4 **j** $-2, \frac{1}{8}$ **k** $-\frac{1}{3}, 0$ **l** ±5 **m** $-1\frac{2}{3}$ **n** $\pm 3\frac{1}{2}$ **o** $-2\frac{1}{2}, 3$

3 **a** Both only have one solution: $x = 1$.
b B is a linear equation, but A and C are quadratic equations.

4 **a** $(5x - 1)^2 = (2x + 3)^2 + (x + 1)^2$, when expanded and collected into the general quadratics, gives the required equation.
b $(10x + 3)(2x - 3)$, $x = 1.5$; area = 7.5 cm^2.

11.4 Solving a quadratic equation by the quadratic formula

Exercise 11J

1 1.77, −2.27
2 −0.23, −1.43
3 3.70, −2.70
4 0.29, −0.69
5 −0.19, −1.53
6 −1.23, −2.43
7 −0.20, −0.92
8 −1.39, −2.27
9 1.37, −4.37
10 2.18, 0.15
11 −0.39, −5.11
12 0.44, −1.69
13 1.64, 0.61
14 0.36, −0.79
15 1.89, 0.11
16 13
17 $x^2 + 3x - 7 = 0$
18 Terry gets $x = \frac{4 + \sqrt{0}}{8}$ and June gets $(2x - 1)^2 = 0$ which only give one solution $x = \frac{1}{2}$

11.5 Solving a quadratic equation by completing the square

Exercise 11K

1 **a** $(x + 2)^2 - 4$ **b** $(x + 7)^2 - 49$ **c** $(x - 3)^2 - 9$ **d** $(x + 3)^2 - 9$ **e** $(x - 2)^2 - 4$ **f** $(x + 3)^2 - 9$ **g** $(x - 5)^2 - 25$ **h** $(x + 10)^2 - 100$ **i** $(x + 5)^2 - 25$ **j** $(x + 4)^2 - 16$ **k** $(x - 1)^2 - 1$ **l** $(x + 1)^2 - 1$

2 **a** $(x + 2)^2 - 5$ **b** $(x + 7)^2 - 54$ **c** $(x - 3)^2 - 6$ **d** $(x + 3)^2 - 2$ **e** $(x - 2)^2 - 5$ **f** $(x + 3)^2 - 6$ **g** $(x - 5)^2 - 30$ **h** $(x + 10)^2 - 101$ **i** $(x + 4)^2 - 22$ **j** $(x + 1)^2 - 2$ **k** $(x - 1)^2 - 8$ **l** $(x + 1)^2 - 10$

3 **a** $-2 \pm \sqrt{5}$ **b** $-7 \pm 3\sqrt{6}$ **c** $3 \pm \sqrt{6}$ **d** $-3 \pm \sqrt{2}$ **e** $2 \pm \sqrt{5}$ **f** $-3 \pm \sqrt{6}$ **g** $5 \pm \sqrt{30}$ **h** $-10 \pm \sqrt{101}$ **i** $-4 \pm \sqrt{22}$ **j** $-1 \pm \sqrt{2}$ **k** $1 \pm 2\sqrt{2}$ **l** $-1 \pm \sqrt{10}$

4 **a** 1.45, −3.45 **b** 5.32, −1.32 **c** −4.16, 2.16

5 Check or correct proof.

6 $p = -14, q = -3$

7 **a** 3rd, 1st, 4th and 2nd – in that order
b Rewrite $x^2 - 4x - 3 = 0$ as $(x - 2)^2 - 7 = 0$, take − 7 over the equals sign, square root both sides, take − 2 over the equals sign
c **i** $x = -3 \pm \sqrt{2}$ **ii** $x = 2 \pm \sqrt{7}$

8 H, C, B, E, D, J, A, F, G, I

11.6 Problems involving quadratic equations

Exercise 11L

1 52, two
2 65, two
3 24, two
4 85, two
5 145, two
6 68, two
7 −35, none
8 −23, none
9 41, two
10 40, two
11 −135, none
12 37, two
13 $x^2 + 3x - 1 = 0$; $x^2 - 3x - 1 = 0$; $x^2 + x - 3 = 0$; $x^2 - x - 3 = 0$

Exercise 11M

1 15 m by 20 m
2 29
3 6.54, 0.46
4 5, 0.5
5 16 m by 14 m
6 48 km/h
7 45, 47
8 2.54 m by 3.54 m
9 30 km/h
10 10p
11 1.25, 0.8
12 10
13 5 h
14 0.75 m
15 Area = 22.75, width = 3.5 m

Examination questions

1 $e^2 + 7e + 12$
2 $x^2 - 2x - 15$
3 $x^2 - x - 2$
4 $2x^2 + x - 10$
5 $y^2 + 5y + 6$
6 $(t + 4)(t - 4)$
7 $x^2 + x - 12$
8 $x^2 + 13x + 40$
9 $3x^2 - 2x - 5$
10 **a** $x^2 + 7x + 12$
b $(y + 3)(y + 5)$
11 $(a - 8)(a - 8)$ or $(a - 8)^2$
12 $(x - 3)(x + 5)$
13 $15x^2 + 17x - 4$
14 $(x - 1)(x - 5)$
15 $(x + 6)(x - 6)$
16 $10x^2 + x - 3$
17 $6x^2 - 5x - 6$
18 **a** $(x + y)(x - y)$
b $(x + y + 2)(x - y)$
19 **a** $5(x - 1) = (4 - 3x)(x + 2) \Rightarrow 5x - 5 = 4x + 8 - 3x^2 - 6x \Rightarrow 5x - 5 = 8 - 3x^2 - 2x \Rightarrow 3x^2 + 7x - 13 = 0$
b $x = 1.22$ or $x = -3.55$
20 **a** Combining the areas of the two parts of the shape: $5x + x(2x + 1) = 95 \Rightarrow 5x + 2x^2 + x = 95 \Rightarrow 2x^2 + 6x = 95 \Rightarrow 2x^2 + 6x - 95 = 0$
b $x = 5.39$ (3 sf) or $x = -8.39$ (3 sf)

Answers to Chapter 12

Quick check

1 **a** 13 **b**

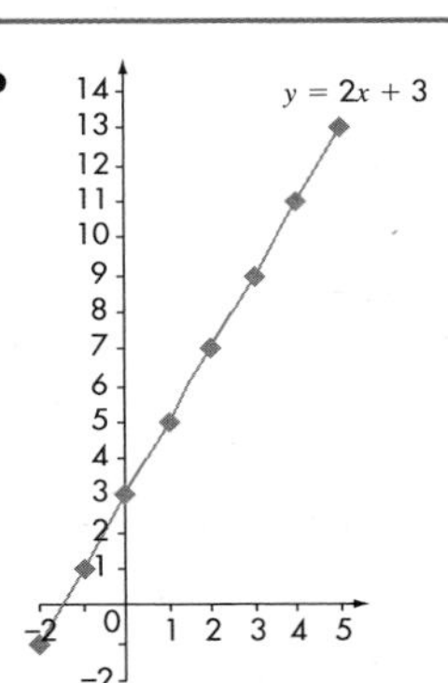

12.1 Drawing graphs by the gradient-intercept method

Exercise 12A

1 **a, b, c, d**

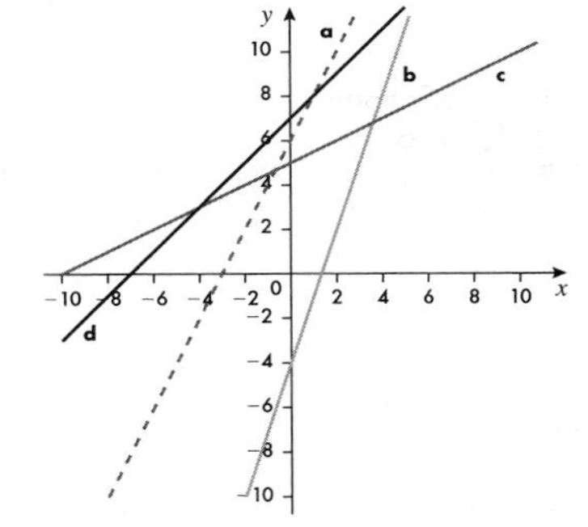

e, f, g, h

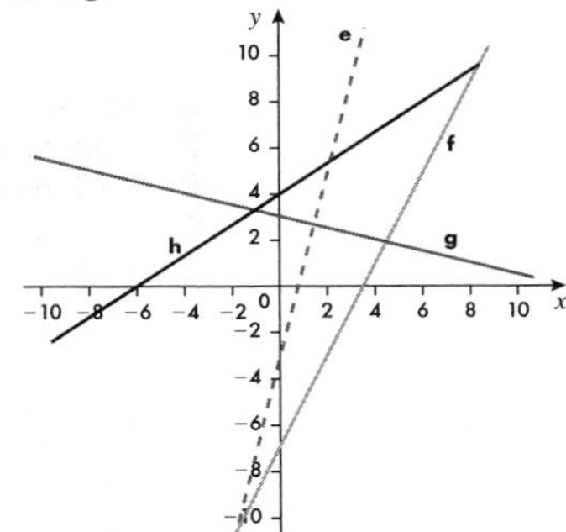

i, j, k, l

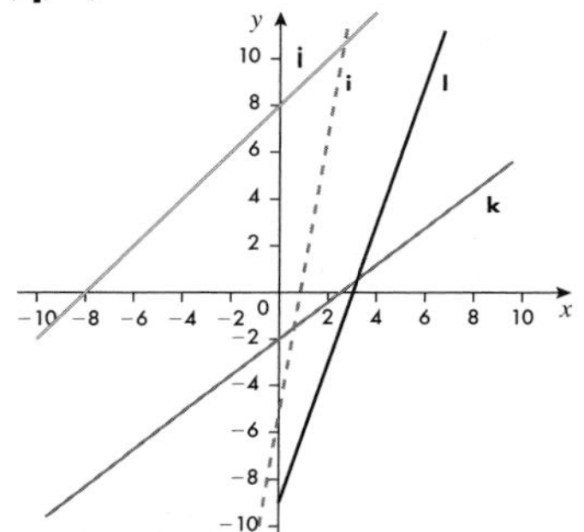

2 a

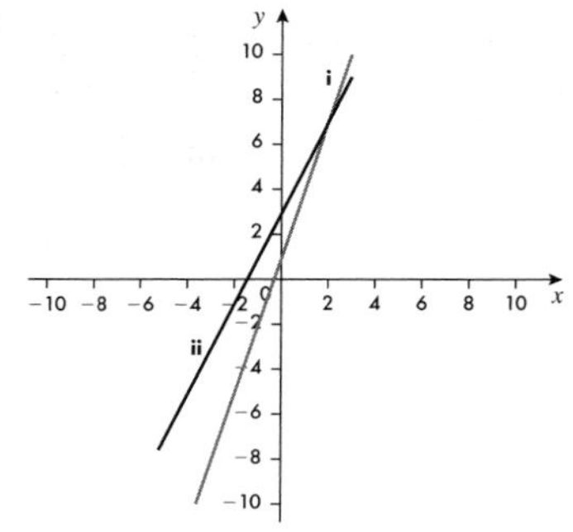

b (2, 7)

3 a

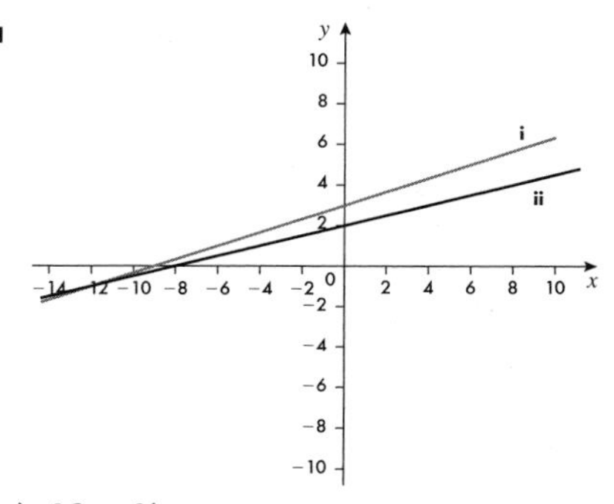

b (−12, −1)

4 a

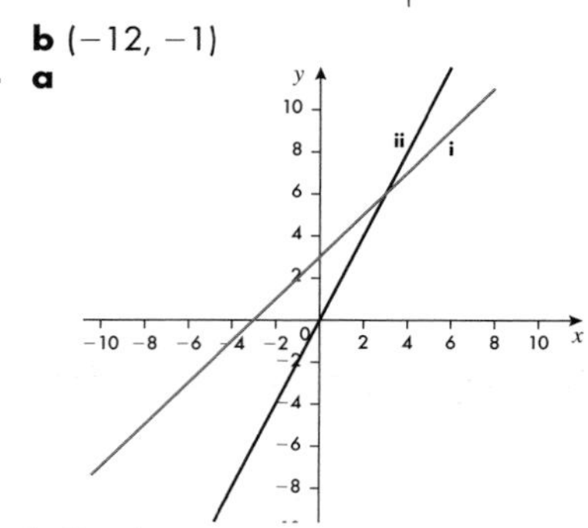

b (3, 6)

5 **a** They have the same gradient (2).
b They intercept the y-axis at the same point (0, −2).
c (−1, −4)

6 **a** −2 **b** $\frac{1}{2}$ **c** 90°
d Negative reciprocal
e $-\frac{1}{3}$

Exercise 12B

1 a, b, c, d

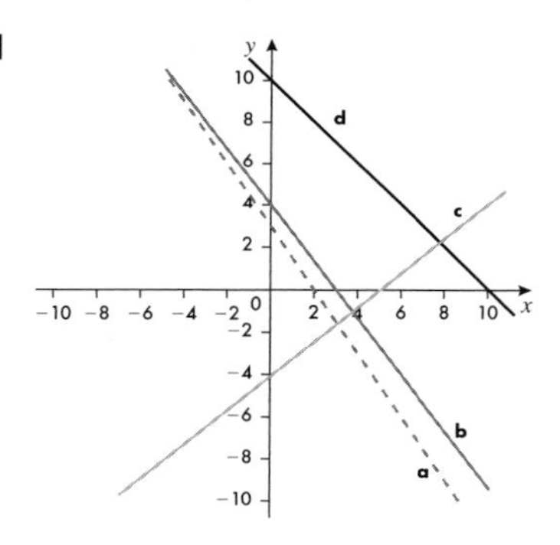

e, f, g, h

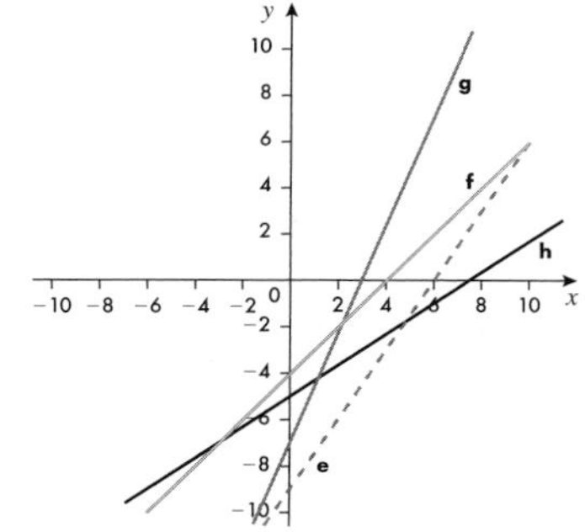

i, j, k, l

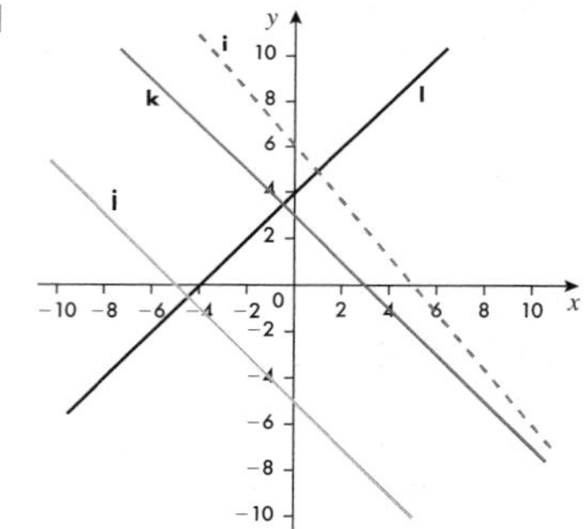

2 a

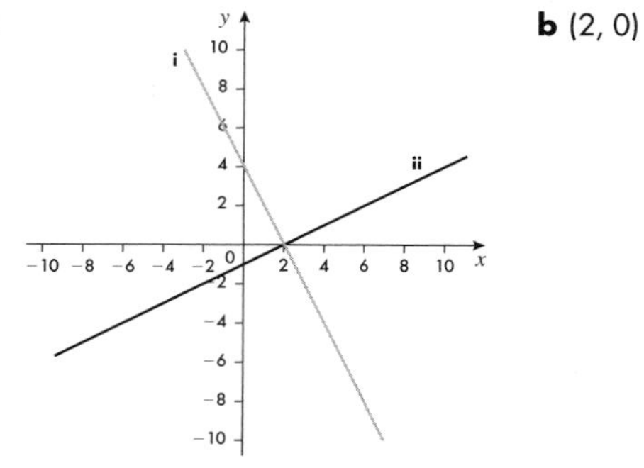

b (2, 0)

3 a

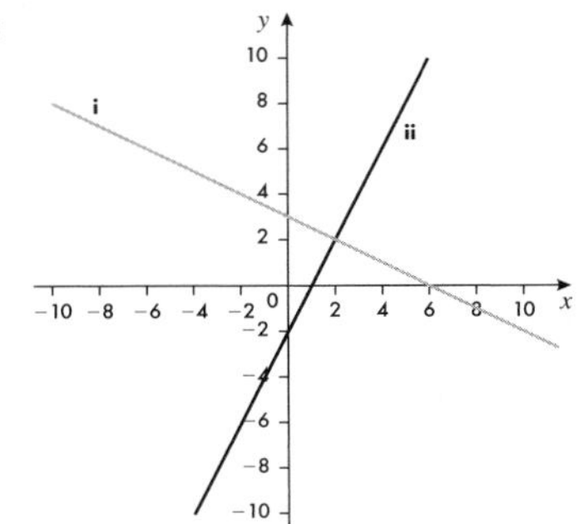

b (2, 2)

4 a

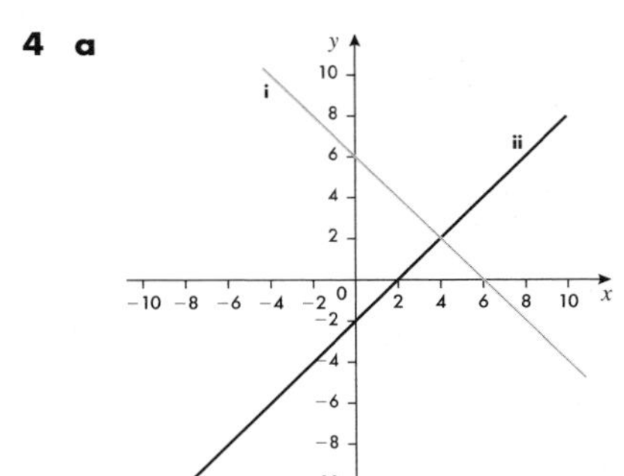

b (4, 2)

5 a Intersect at (6, 0) **b** Intersect at (0, −3)
c Parallel **d** $-2x + 9y = 18$

6 a i $x = 3$ **ii** $x - y = 4$ **iii** $y = -3$
iv $x + y = -4$ **v** $x = -3$ **vi** $y = x + 4$

b **i** -3 **ii** $\frac{1}{3}$ **iii** $-\frac{1}{3}$

12.2 Finding the equation of a line from its graph

Exercise 12C

1 a $y = \frac{4}{3}x - 2$ or $3y = 4x - 6$ **b** $y = x + 1$
c $y = 2x - 3$ **d** $2y = x + 6$
e $y = x$ **f** $y = 2x$

2 a i $y = 2x + 1, y = -2x + 1$
ii Reflection in y-axis (and $y = 1$)
iii Different sign
b i $5y = 2x - 5, 5y = -2x - 5$
ii Reflection in y-axis (and $y = -1$)
iii Different sign
c i $y = x + 1, y = -x + 1$
ii Reflection in y-axis (and $y = 1$)
iii Different sign

3 a x-coordinates go from $2 \rightarrow 1 \rightarrow 0$ and y-coordinates go from $5 \rightarrow 3 \rightarrow 1$.
b x-step between the points is 1 and y-step is 2.
c $y = 3x + 2$

4 a $y = -2x + 1$
b $2y = -x$
c $y = -x + 1$
d $5y = -2x - 5$
e $y = \frac{3}{2}x - 3$ or $2y = -3x - 6$

5 a i $2y = -x + 1, y = -2x + 1$
ii Reflection in $x = y$
iii Reciprocal of each other
b i $2y = 5x + 5, 5y = 2x - 5$
ii Reflection in $x = y$
iii Reciprocal of each other
c i $y = 2, x = 2$
ii Reflection in $x = y$
iii Reciprocal of each other (reciprocal of zero is infinity)

12.3 Quadratic graphs

Exercise 12D

1 **a** Values of y: 27, 12, 3, 0, 3, 12, 27
b 6.8 **c** 1.8 or −1.8
2 **a** Values of y: 27, 18, 11, 6, 3, 2, 3, 6, 11, 18, 27
b 8.3 **c** 3.5 or −3.5
3 **a** Values of y: 27, 16, 7, 0, −5, −8, −9, −8, −5, 0, 7
b −8.8 **c** 3.4 or −1.4
4 **a** Values of y: 2, −1, −2, −1, 2, 7, 14
b 0.25 **c** 0.7 or −2.7
d

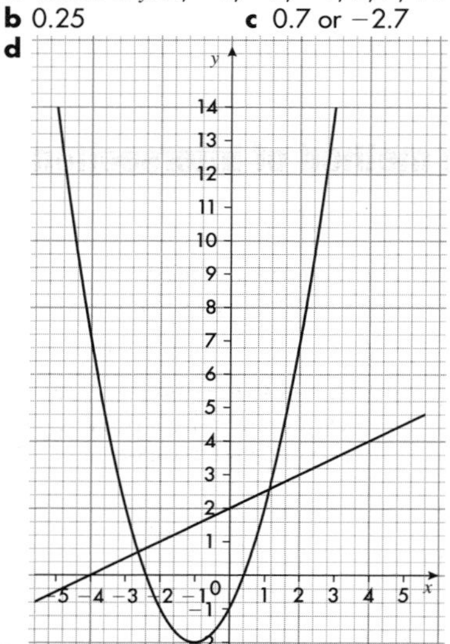

e (1.1, 2.6) and (−2.6, 0.7)
5 **a** Values of y: 18, 12, 8, 6, 6, 8, 12
b 9.75 **c** 2 or −1
d Values of y: 14, 9, 6, 5, 6, 9, 14
e (1, 6)
6 **a** Values of y: 4, 1, 0, 1, 4, 9, 16
b 7.3 **c** 0.4 or −2.4
d

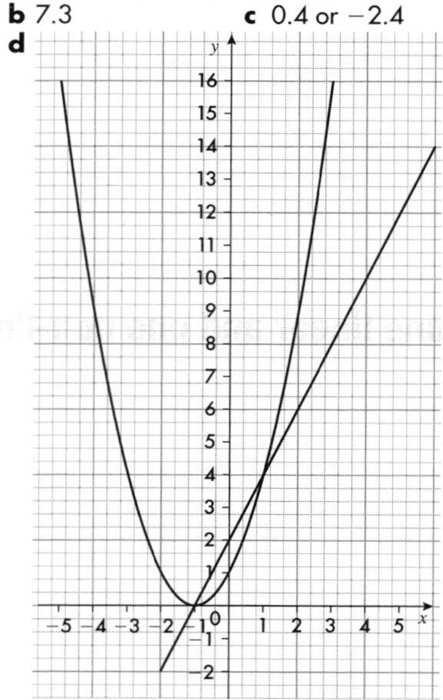

e (1, 4) and (−1, 0)
7 **a** Values of y: 15, 9, 4, 0, −3, −5, −6, −6, −5, −3, 0, 4, 9
b −0.5 and 3
8 Points plotted and joined should give parabolas.
9 Points plotted and joined should give a parabola.
10 Line A has a constant in front, so is 'thinner' than the rest.
Line B has a negative in front, so is 'upside down'.
Line C does not pass through the origin.

12.4 The significant points of a quadratic graph

Exercise 12E

1 **a** Values of y: 12, 5, 0, −3, −4, −3, 0, 5, 12
b 2 and −2
2 **a** Values of y: 7, 0, −5, −8, −9, −8, −5, 0, 7
b 3 and −3
3 **a** The roots are positive and negative square roots of the constant term.
b Check predictions
c Values of y: 15, 8, 3, 0, −1, 0, 3, 8, 15
d Values of y: 11, 4, −1, −4, −5, −4, −1, 4, 11
e 1 and −1, 2.2 and −2.2
4 **a** Values of y: 5, 0, −3, −4, −3, 0, 5, 12
b −4 and 0
5 **a** Values of y: 16, 7, 0, −5, −8, −9, −8, −5, 0, 7, 16
b 0 and 6
6 **a** Values of y: 10, 4, 0, −2, −2, 0, 4, 10, 18
b −3 and 0
7 **a** The roots are 0 and the negative of the coefficient of x.
b Check predictions
c Values of y: 10, 4, 0, −2, −2, 0, 4, 10
d Values of y: 6, 0, −4, −6, −6, −4, 0, 6, 14
e 0 and 3, −5 and 0
8 **a** Values of y: 9, 4, 1, 0, 1, 4, 9
b −2
c Only 1 root
9 **a** Values of y: 10, 3, −2, −5, −6, −5, −2, 3, 10
b 0.6 and 5.4
10 **a** Values of y: 19, 6, −3, −8, −9, −6, 1, 12
b 0.9 and −3.4
11 **a** (0, −4), (0, −9), (0, −1), (0, −5), (0, 0), (0, 0), (0, 0), (0, 0), (0, 0)
b (0, −4), (0, −9), (0, −1), (0, −5), (−2, −4), (3, −9), (−1.5, −2.25), (1.5, −2.25), (−2.5, −6.25)
c The y-intercept; the point where the x-value is the mean of the roots.
12 **a** $y = (x - 2)^2$ **b** 0
13 **a** $y = (x - 3)^2 - 6$ **b** −6
14 **a** $y = (x - 4)^2 - 14$ **b** −14
15 **a** $y = -(x - 1)^2 - 5$ **b** −5
16 **a** The minimum point is (a, b) **b** (−5, −28)
17 $y = (x - 3)^2 - 7$, $y = x^2 - 6x + 9 - 7$, $y = x^2 - 6x + 2$
18 **a** (−2, −7)
b **i** $(a, 2b - a^2)$
ii $(2a, b - 4a^2)$

12.5 The circular function graphs

Exercise 12F

1 115°
2 327°
3 324°
4 195°
5 210°, 330°
6 135°, 225°
7 **a** Say 32°, sin 32° = 0.53, cos 58° = 0.53
b Say 70°, sin 70° = 0.94, cos 20° = 0.94
c Sin x = cos (90 − x)°
d Cos x = sin (90 − x)°
8 **a** 64° **b** 206°, 334°
c 116°, 244°
9 **a** −0.384 **b** 113°
10 **a** 0.822 **b** 55.3
c No
d The calculator has given the value of the acute angle but the angle 124.7° has the same positive sign.
11 **a** 1.1307
b Error
c If you tried to draw this triangle accurately then you would see that the line that is 12 long does not intersect with the base.
12 **a to e** All true

12.6 Solving one linear and one non-linear equation by the method of intersection

Exercise 12G

1 **a** (0.7, 0.7), (−2.7, −2.7)
b (6, 12), (−1, −2)
c (4, −3), (−3, 4)
d (0.8, 1.8), (−1.8, −0.8)
e (4.6, 8.2), (0.5, 0)
f (3, 6), (−2, 1)
g (4.8, 6.6), (0.2, −2.6)
h (2.6, 1.6), (−1.6, −2.6)
2 **a** (1, 0)
b Only one intersection point
c $x^2 + x(3 - 5) + (-4 + 5) = 0$
d $(x - 1)^2 = 0 \Rightarrow x = 1$
e Only one solution as line is a tangent to curve.
3 **a** There is no solution.
b The graphs do not intersect.
c $x^2 + x + 4 = 0$
d $b^2 - 4ac = -15$
e No solution as the discriminant is negative and there is no square root of a negative number.

12.7 Solving equations by the method of intersection

Exercise 12H

1 **a** **i** −1.4, 4.4
ii −2, 5
iii −0.6, 3.6
b 2.6, 0.4
2 **a** −5, 1
b **i** −5.3, 1.3
ii −4.8, 0.8
iii −3.4, −0.6
3 **a** **i** 0, 6
ii 4.3, 0.7
b **i** 4.8, 0.2
ii 5.4, −0.4
4 **a** **i** −1.6, 2.6
ii 1.4, −1.4
b **i** 2.3, −2.3
ii 2, −2
5 **a** 0, 2 **b** 2.5
c −0.6, 1, 1.6 **d** 2.8
e −0.8, 0.6, 2.2
6 **a** −0.4, 4.4
b −1, 5
7

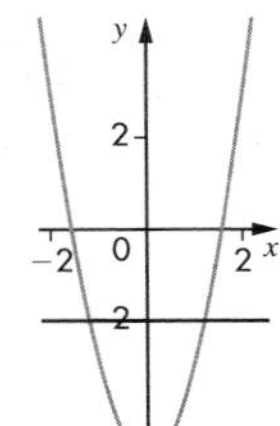

a 1.6, −1.6
b −1.2, 1.2
8

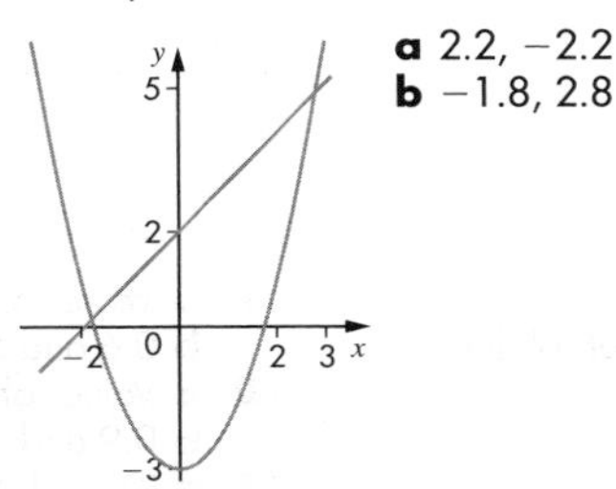

a 2.2, −2.2
b −1.8, 2.8
9

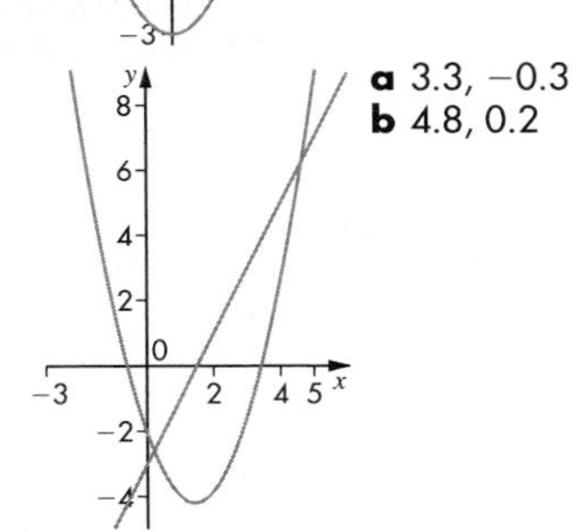

a 3.3, −0.3
b 4.8, 0.2
10

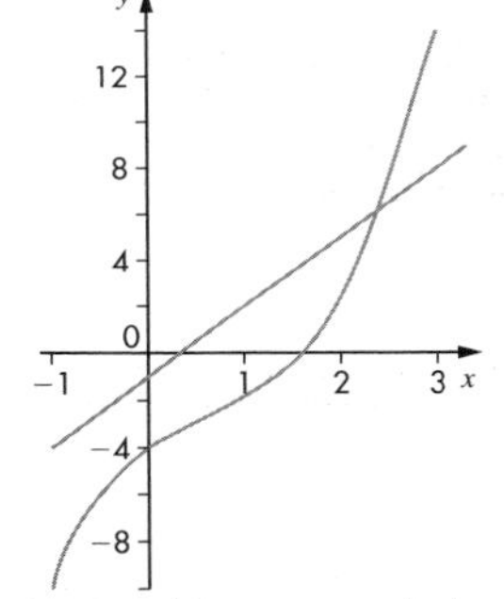

a 2
b 2.5
11 **a** C and D **b** A and D
c $x^2 + 4x - 1 = 0$
d (−1.5, −10.25)
12 **a** $(x + 2)(x - 1) = 0$
b 5 − −2 = + 7, not − 7
c $y = 2x + 7$

Examination questions

1

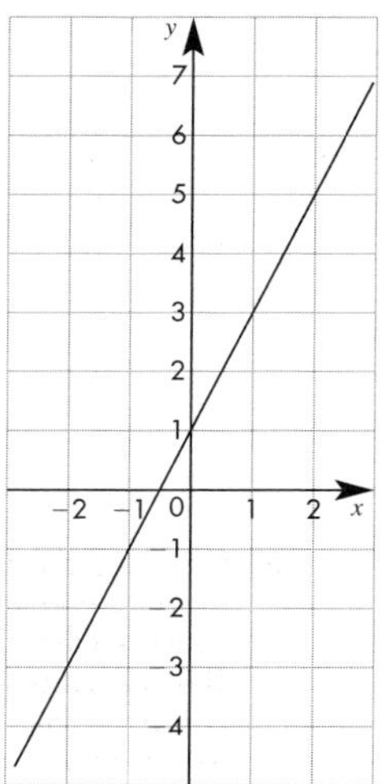

2

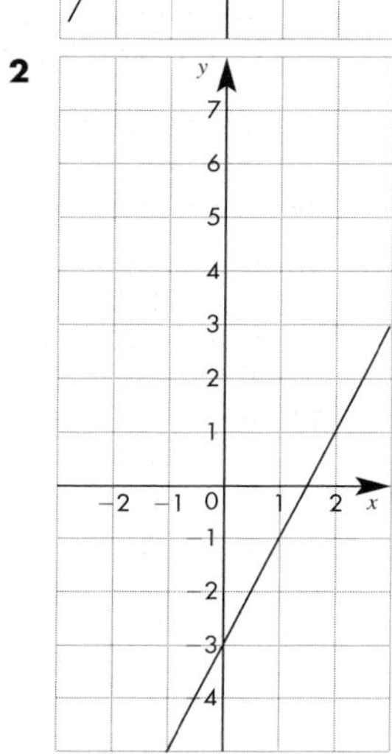

3

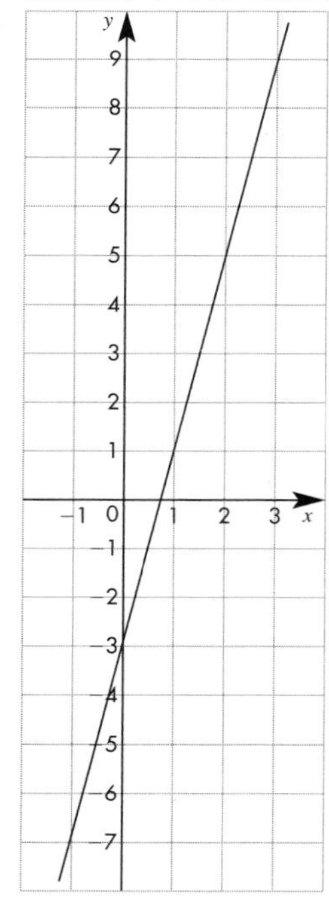

4 $\frac{7-3}{2-0} = \frac{4}{2} = 2$

5

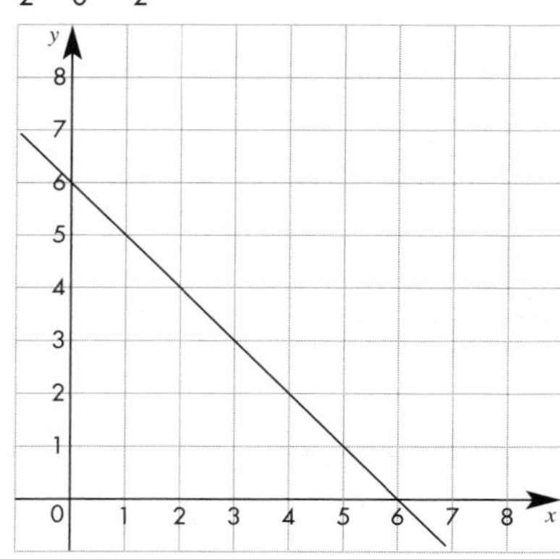

6 **a** $y = x^2 - 4x + 2$

x	−1	0	1	2	3	4	5
y	7	2	−1	−2	−1	2	7

b

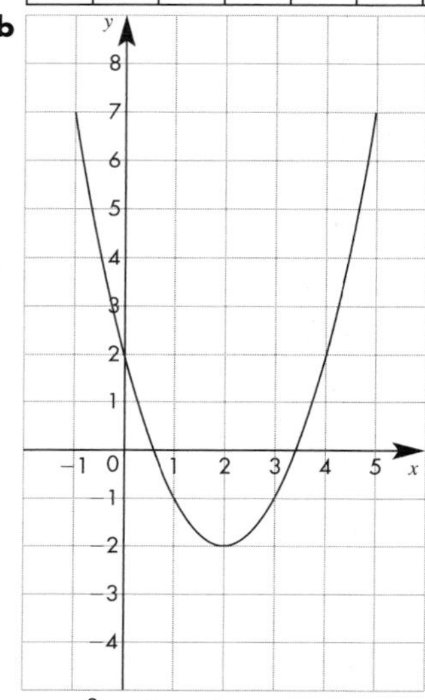

7 **a** $y = x^2 - 4x - 2$

x	−1	0	1	2	3	4	5
y	3	−2	−5	−6	−5	−2	3

b

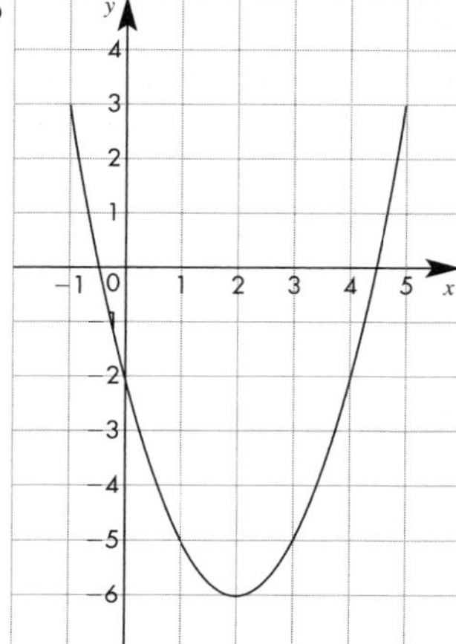

c 0.3, 3.7

Answers to Chapter 13

Quick check

1 **a** 17, 20, 23 **b** 28, 36, 45
2 **a** 1 **b** 4 **c** 7
3 **a** $2(x + 3)$ **b** $x(x - 1)$ **c** $2x(5x + 1)$
4 **a** $x^2 + 8x + 12$ **b** $2x^2 - 5x - 3$ **c** $x^2 - 4x + 4$
5 **a** $x = 3 - 2y$ **b** $x = 4 + 3y$ **c** $x = 4y - 3$

13.1 Algebraic fractions

Exercise 13A

1 **a** $\frac{5x}{6}$ **b** $\frac{19x}{20}$ **c** $\frac{23x}{20}$ **d** $\frac{3x + 2y}{6}$ **e** $\frac{x^2y + 8}{4x}$ **f** $\frac{5x + 7}{6}$ **g** $\frac{7x - 3}{4}$ **h** $\frac{13x - 5}{15}$ **i** $\frac{3x - 1}{4}$ **j** $\frac{5x - 10}{4}$

2 **a** $\frac{x}{6}$ **b** $\frac{11x}{20}$ **c** $\frac{7x}{20}$ **d** $\frac{3x - 2y}{6}$ **e** $\frac{xy^2 - 8}{4y}$ **f** $\frac{x - 1}{6}$ **g** $\frac{x + 1}{4}$ **h** $\frac{-7x - 5}{15}$ **i** $\frac{x - 1}{4}$ **j** $\frac{2 - 3x}{4}$

3 **a** 3 **b** 6 **c** 2 **d** 5 **e** 0.75 **f** 3

4 **a** $\frac{x^2}{6}$ **b** $\frac{3xy}{14}$ **c** $\frac{8}{3}$ **d** $\frac{2xy}{3}$ **e** $\frac{x^2 - 2x}{10}$ **f** $\frac{1}{6}$ **g** $\frac{6x^2 + 5x + 1}{8}$ **h** $\frac{2x^2 + x}{15}$ **i** $\frac{2x - 4}{x - 3}$ **j** $\frac{1}{2x}$

5 **a** $\frac{3}{2}$ **b** $\frac{x}{y}$ **c** $\frac{8}{3}$ **d** $\frac{2xy}{3}$ **e** $\frac{5x}{2x - 4}$ **f** $\frac{2x^2 - 12x + 18}{75}$ **g** 1 **h** $\frac{1}{4x + 2}$ **i** $\frac{x^2 - 5x + 6}{48}$ **j** $\frac{1}{2x}$

6 **a** x **b** $\frac{x}{2}$ **c** $\frac{3x^2}{16}$ **d** 3 **e** $\frac{17x + 1}{10}$ **f** $\frac{13x + 9}{10}$ **g** $\frac{3x^2 - 5x - 2}{10}$ **h** $\frac{x + 3}{2}$ **i** $\frac{2}{3}$ **j** $\frac{2x^2 - 6y^2}{9}$

7 All parts: student's own working

8 First, he did not factorise and just cancelled the x^2s. Then he cancelled 2 and 6 with the wrong signs. Then he said two minuses make a plus when adding, which is not true.

9 $\frac{2x^2 + x - 3}{4x^2 - 9}$

10 **a** 3, −1.5 **b** 4, −1.25 **c** 3, −2.5 **d** 0, 1

11 **a** $\frac{x - 1}{2x + 1}$ **b** $\frac{2x + 1}{x + 3}$ **c** $\frac{2x - 1}{3x - 2}$ **d** $\frac{x + 1}{x - 1}$

13.2 Algebraic proof

Exercise 13B

1 Student's own proof
2 Student's own proof
3 **a** 3, 5, 8, 13, 21, 34, 55
b $3a + 5b$, $5a + 8b$, $8a + 13b$, $13a + 21b$, $21a + 34b$
c The eighth term will be $8a + 13b$, the fifth will be $2a + 3b$, the difference is $6a + 10b = 2(3a + 5b)$ which is twice the sixth term.
4 Student's own proof
5 **a** **i** 10
b Student's own proof
c Student's own proof
d Student's own proof
6–15 Student's own proof

Exercise 13C

All answers student's own proof

Examination questions

1 $\frac{3}{x-2}$

2 $x + 3$

3 $\frac{7x - 11}{12}$

4 $\frac{4 + 5x}{x(x + 3)}$

5 $\frac{3}{x-2}$

6 $x + 3$

7 $(n + 2)^2 - (n - 2)^2 = n^2 + 4n + 4 - (n^2 - 4n + 4) = 8n$

8 $\frac{x^2 - 3}{x(x - 2)}$

9 $\frac{4}{a + 5}$

10 $\frac{5x - 7}{(x + 5)(x - 3)}$

11 $\frac{6x}{(x - 3)(x + 3)}$

12 $\frac{p - 3}{2}$

13 $\frac{3x}{2x - 1}$

14 **a** An even number add 2 gives the next even number
b $2n + 4$
c $2n + 2n + 2 + 2n + 4 = 6n + 6 = 6(n + 1)$, hence a multiple of 6.

15 $\frac{x + 1}{x + 2}$

16 $\frac{x + 3}{x - 5}$

17 $\frac{3}{x - 5}$

Answers to Chapter 14

Quick check

1 $a = 50°$

2 $b = 140°$

3 $c = d = 65°$

14.1 Circle theorems

Exercise 14A

1 **a** 56° **b** 62° **c** 105° **d** 55° **e** 45° **f** 30° **g** 60° **h** 145°

2 **a** 55° **b** 52° **c** 50° **d** 24° **e** 39° **f** 80° **g** 34° **h** 30°

3 **a** 41° **b** 49° **c** 41°

4 **a** 72° **b** 37° **c** 72°

5 ∠AZY = 40° (angles in a triangle), $a = 50°$ (angle in a semicircle = 90°)

6 **a** $x = y = 40°$
b $x = 131°, y = 111°$
c $x = 134°, y = 23°$
d $x = 32°, y = 19°$
e $x = 59°, y = 121°$
f $x = 155°, y = 12.5°$

7 68°

8 ∠ABC = 180° − x (angles on a line), ∠AOC = 360° − $2x$ (angle at centre is twice angle at circumference), reflex ∠AOC = 360° − (360° − $2x$) = $2x$ (angles at a point)

9 **a** x
b $2x$
c ∠ABC = $(x + y)$ and ∠AOC = $2(x + y)$

14.2 Cyclic quadrilaterals

Exercise 14B

1 **a** $a = 50°, b = 95°$
b $c = 92°, x = 90°$
c $d = 110°, e = 110°, f = 70°$
d $g = 105°, h = 99°$
e $j = 89°, k = 89°, l = 91°$
f $m = 120°, n = 40°$
g $p = 44°, q = 68°$
h $x = 40°, y = 34°$

2 **a** $x = 26°, y = 128°$
b $x = 48°, y = 78°$
c $x = 133°, y = 47°$
d $x = 36°, y = 72°$
e $x = 55°, y = 125°$
f $x = 35°$
g $x = 48°, y = 45°$
h $x = 66°, y = 52°$

3 **a** $x = 49°, y = 49°$
b $x = 70°, y = 20°$
c $x = 80°, y = 100°$
d $x = 100°, y = 75°$

4 **a** $x = 50°, y = 62°$
b $x = 92°, y = 88°$
c $x = 93°, y = 42°$
d $x = 55°, y = 75°$

5 **a** $x = 95°, y = 138°$
b $x = 14°, y = 62°$
c $x = 32°, y = 48°$
d 52°

6 **a** 71° **b** 125.5° **c** 54.5°

7 **a** $x + 2x - 30° = 180°$ (opposite angles in a cyclic quadrilateral), so $3x - 30° = 180°$
b $x = 70°$, so $2x - 30° = 110°$ ∠DOB = 140° (angle at centre equals twice angle at circumference), $y = 60°$ (angles in a quadrilateral)

8 **a** x
b $360° - 2x$
c ∠ADC = $\frac{1}{2}$ reflex ∠AOC = 180° − x, so ∠ADC + ∠ABC = 180°

9 Let ∠AED = x, then ∠ABC = x (opposite angles are equal in a parallelogram), ∠ADC = 180° − x (opposite angles in a cyclic quadrilateral), so ∠ADE = x (angles on a line)

14.3 Tangents and chords

Exercise 14C

1 **a** 38° **b** 110°
c 15° **d** 45°
2 **a** 6 cm **b** 10.8 cm
c 3.21 cm **d** 8 cm
3 **a** $x = 12°$, $y = 156°$
b $x = 100°$, $y = 50°$
c $x = 62°$, $y = 28°$
d $x = 30°$, $y = 60°$
4 **a** 62° **b** 66° **c** 19° **d** 20°
5 19.5 cm
6 5.77 cm
7 ∠OCD = 58° (triangle OCD is isosceles), ∠OCB = 90° (tangent/radius theorem), so ∠DCB = 32°, hence triangle BCD is isosceles (2 equal angles)
8 **a** $\angle AOB = \cos^{-1}\frac{OA}{OB} = \cos^{-1}\frac{OC}{OB} = \angle COB$
b As ∠AOB = ∠COB, so ∠ABO = ∠CBO, so OB bisects ∠ABC

14.4 Alternate segment theorem

Exercise 14D

1 **a** $a = 65°$, $b = 75°$, $c = 40°$
b $d = 79°$, $e = 58°$, $f = 43°$
c $g = 41°$, $h = 76°$, $i = 76°$
d $k = 80°$, $m = 52°$, $n = 80°$
2 **a** $a = 75°$, $b = 75°$, $c = 75°$, $d = 30°$
b $a = 47°$, $b = 86°$, $c = 86°$, $d = 47°$
c $a = 53°$, $b = 53°$
d $a = 55°$
3 **a** 36° **b** 70°
4 **a** $x = 25°$
b $x = 46°$, $y = 69°$, $z = 65°$
c $x = 38°$, $y = 70°$, $z = 20°$
d $x = 48°$, $y = 42°$
5 ∠ACB = 64° (angle in alternate segment), ∠ACX = 116° (angles on a line),
∠CAX = 32° (angles in a triangle), so triangle ACX is isosceles (two equal angles)
6 ∠AXY = 69° (tangents equal and so triangle AXY is isosceles), ∠XZY = 69° (alternate segment), ∠XYZ = 55° (angles in a triangle)
7 **a** $2x$
b $90° - x$
c OPT = 90°, so APT = x

Examination questions

1 **a** **i** 90°
ii angle in a semicircle
b **i** 65°
ii angle at centre = twice angle at circumference
2 **a** **i** 140°
ii angle at centre = twice angle at circumference
b **i** 110°
ii opposite angles in a cyclic quadrilateral add up to 180°
3 **a** 90° **b** 140°
b angle OQP = 20° (isosceles triangle),
angle PQT = 90° − 20°
4 **a** 30°
b angle ABC = 75° (angle ABO = 90°, radius/tangent), triangle ABC is isosceles (equal tangents)
5 46°
6 angle OST = 90° − x (radius meets tangent at 90°),
angle ORS = 90° − x (triangle OSR is isosceles),
angle ROS = 180° − (90° − x) − (90° − x) = $2x$

Answers to Chapter 15

Quick check

1 **a** 8 **b** 10

2 **a**

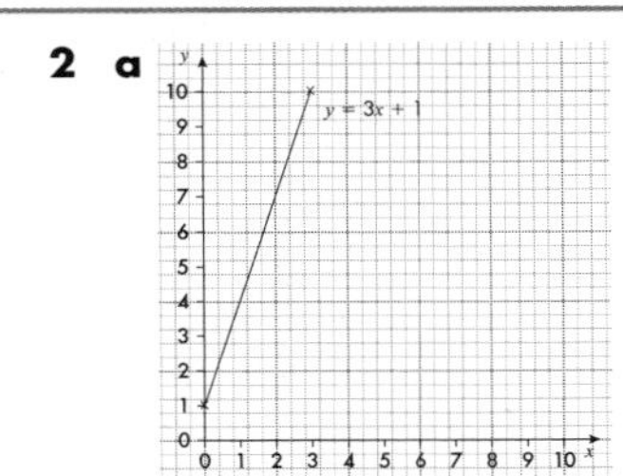

b

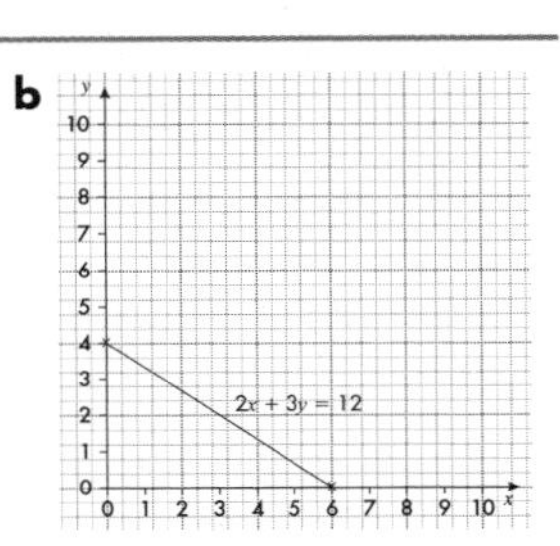

15.1 Solving inequalities

Exercise 15A

1 **a** $x < 3$ **b** $t > 8$ **c** $p \geqslant 10$ **d** $x < 5$ **e** $y \leqslant 3$ **f** $t > 5$ **g** $x < 6$ **h** $y \leqslant 15$ **i** $t \geqslant 18$ **j** $x < 7$ **k** $x \leqslant 3$ **l** $t \geqslant 5$

2 **a** 8 **b** 6 **c** 16 **d** 3 **e** 7

3 **a** 11 **b** 16 **c** 16

4 $2x + 3 < 20$, $x < 8.50$, so the most each could cost is £8.49

5 **a** Because $3 + 4 = 7$, which is less than the third side of length 8
b $x + x + 2 > 10$, $2x + 2 > 10$, $2x > 8$, $x > 4$, so smallest value of x is 5

6 **a** $x = 6$ and $x < 3$ scores -1 (nothing in common), $x < 3$ and $x > 0$ scores $+1$ (1 in common for example), $x > 0$ and $x = 2$ scores $+1$ (2 in common), $x = 2$ and $x \geqslant 4$ scores -1 (nothing in common), so we get $-1 + 1 + 1 - 1 = 0$
b $x > 0$ and $x = 6$ scores $+1$ (6 in common), $x = 6$ and $x \geqslant 4$ scores $+1$ (6 in common), $x \geqslant 4$ and $x = 2$ scores -1 (nothing in common), $x = 2$ and $x < 3$ scores $+1$ (2 in common). $+1 + 1 - 1 + 1 = 2$
c Any acceptable combination, e.g. $x = 2$, $x < 3$, $x > 0$, $x \geqslant 4$, $x = 6$

7 **a** $x \geqslant -6$ **b** $t \leqslant \frac{8}{3}$ **c** $y \leqslant 4$ **d** $x \geqslant -2$ **e** $w \leqslant 5.5$ **f** $x \leqslant \frac{14}{5}$

8 **a** $x \leqslant 2$ **b** $x > 38$ **c** $x < 6\frac{1}{2}$ **d** $x \geqslant 7$ **e** $t \leqslant 15$ **f** $y \leqslant \frac{7}{5}$

9 **a** $3 < x < 6$ **b** $2 < x < 5$ **c** $-1 < x \leqslant 3$ **d** $1 \leqslant x < 4$ **e** $2 \leqslant x < 4$ **f** $0 \leqslant x \leqslant 5$

Exercise 15B

1 **a** $x > 1$ **b** $x \leqslant 3$ **c** $x < 2$ **d** $x \geqslant -1$ **e** $x \leqslant -1$ **f** $x \geqslant 1$

2 **a** **b** 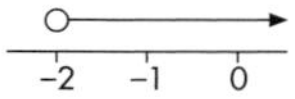 **c** 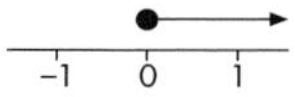 **d** 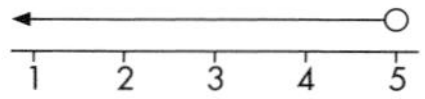 **e** 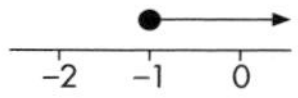 **f** 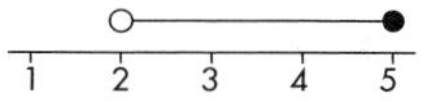 **g** 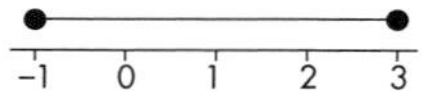 **h**

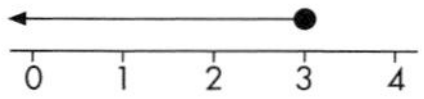

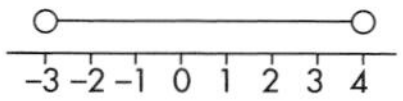

3 **a** $x \geqslant 4$ 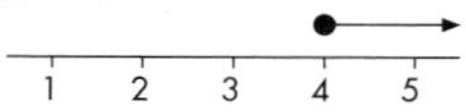
b $x < -2$
c $x \geqslant 3.5$ 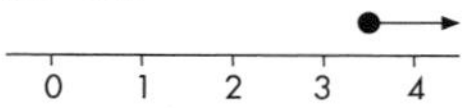
d $x < -1$
e $x < 1.5$
f $x \leqslant -2$ 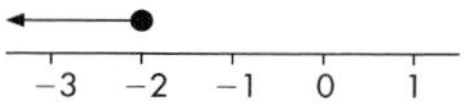
g $x > 50$ 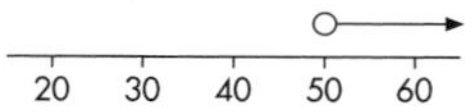
h $x \geqslant -6$

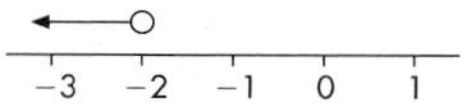

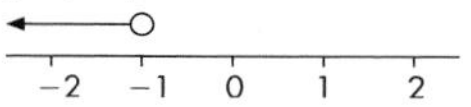

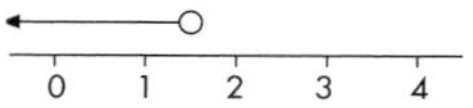

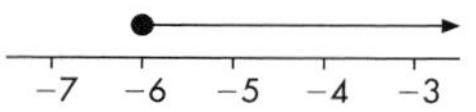

4 **a** Because 3 apples plus the chocolate bar cost more that £1.20: $x > 22$
b Because 2 apples plus the chocolate bar left Max with at least 16p change: $x \leqslant 25$
c 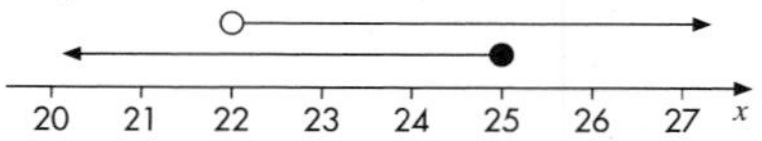
d Apples could cost 23p, 24p or 25p.

5 Any two inequalities that overlap only on the integers -1, 0, 1 and 2 – for example, $x \geqslant -1$ and $x < 3$

6 1 and 4

7 **a** $x > 2$ 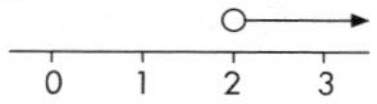
b $x \geqslant 6$ 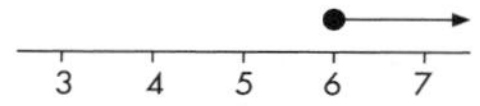
c $x \leqslant -1$ 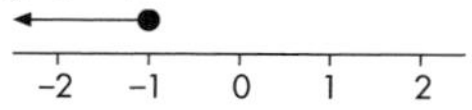
d $x \geqslant -4$

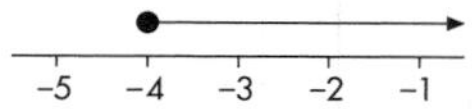

15.2 Graphical inequalities

Exercise 15C

1

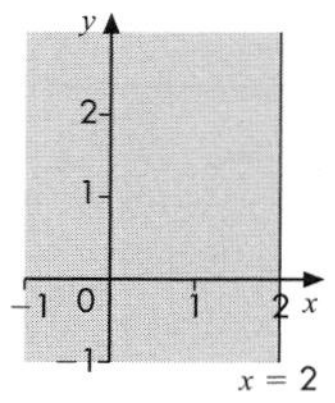

2

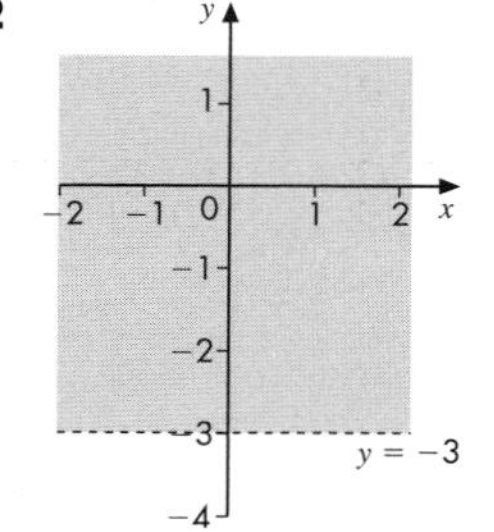

3

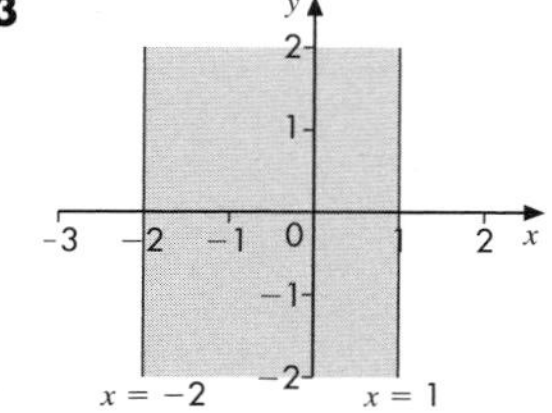

4

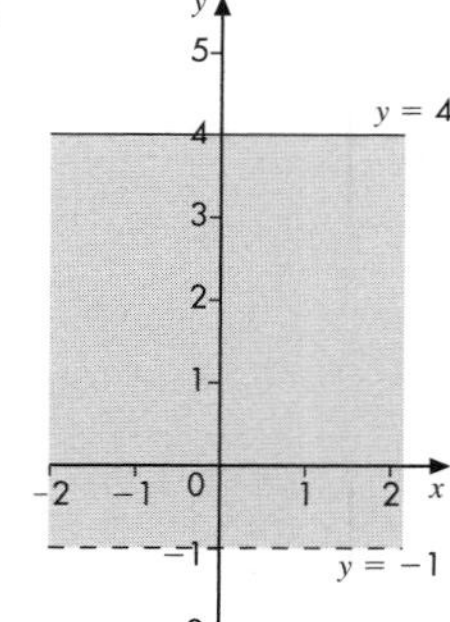

5 **a**

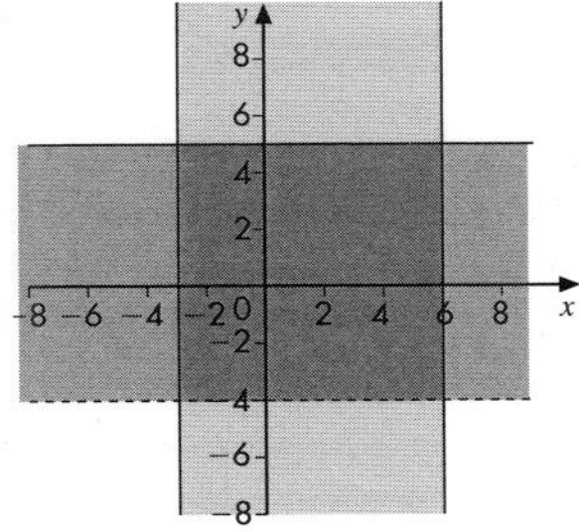

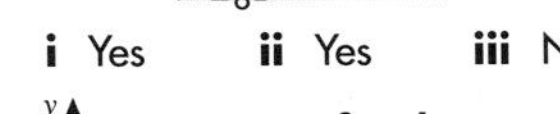

b **i** Yes **ii** Yes **iii** No

6

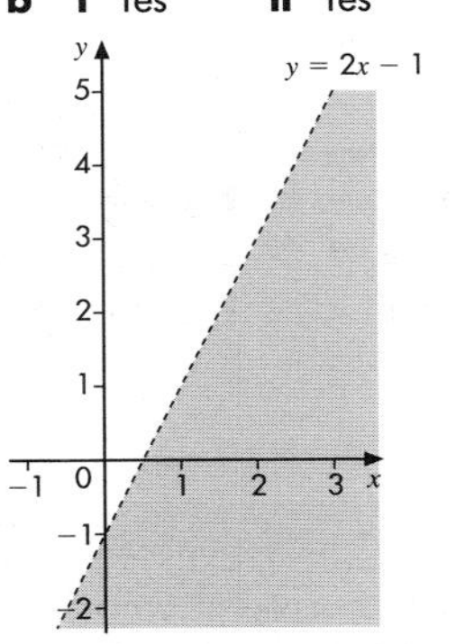

7

8

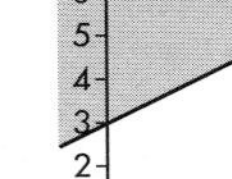

9

10 **a–d**

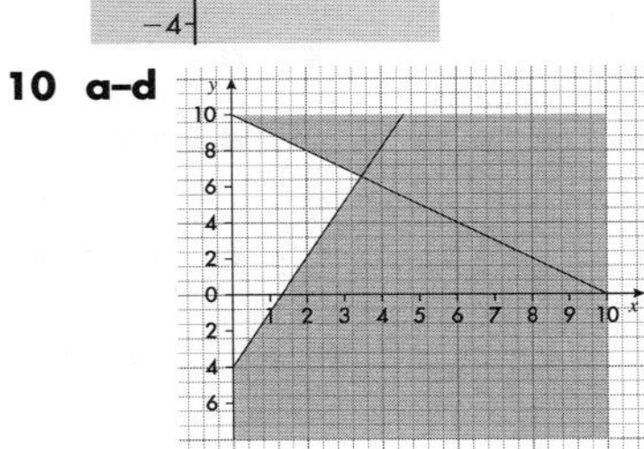

e **i** No **ii** No **iii** Yes

11 **a–f**

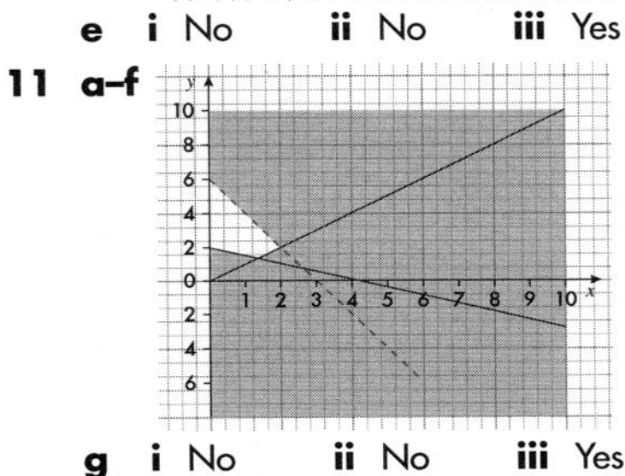

g **i** No **ii** No **iii** Yes

12 **a**

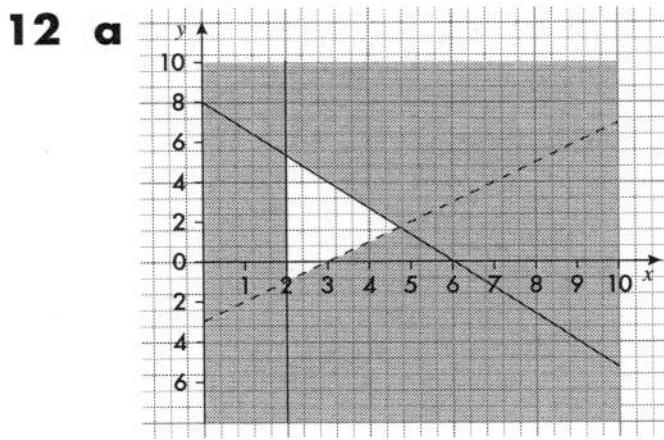

b **i** No **ii** Yes **iii** Yes **iv** No

13 For example, $x \geqslant 1$, $y \leqslant 3$ and $y \geqslant x + 1$. There are many other valid answers.

14 May be true: a, c, d, g
Must be false: b, e
Must be true: f, h

15 Substitute the co-ordinates of a point, such as (0, 0), that satifies the inequality and check which side of the line it is.

16 **a** (3, 1)
b (3, 4)

Examination questions

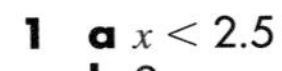

1 **a** $x < 2.5$
b 2

2 −2, −1, 0, 1, 2

3 **a** −2, −1, 0, 1, 2, 3
b $x < 2$

4

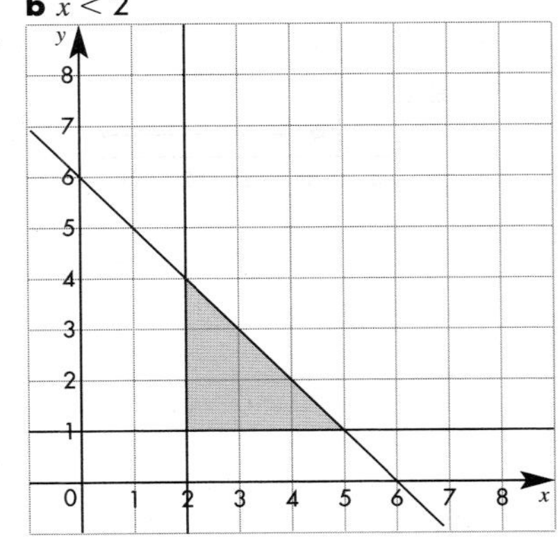

5 **a** $x \geqslant 3$
b **i** $x > 1.6$ **ii** 2

6 −2, −1, 0, 1, 2, 3

7 **a** $x < 4$
b $-5 < x < 5$

8

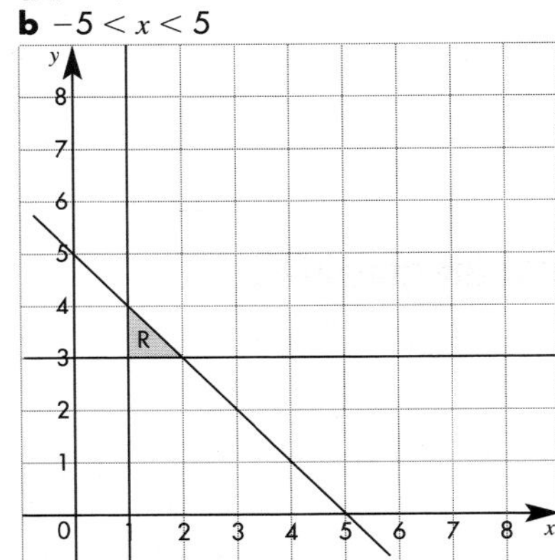

9

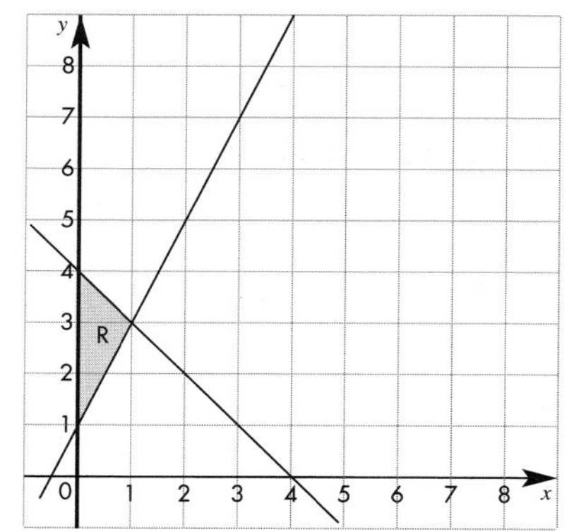

10

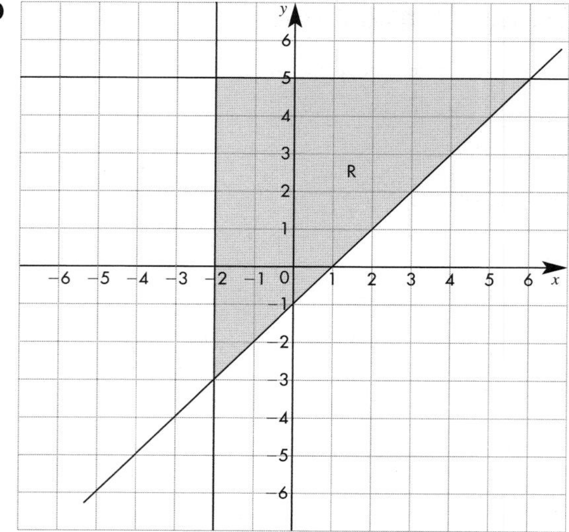

11 −3, −2, −1, 0, 1, 2, 3

Answers to Chapter 16

Quick check

1 a 25 b 9 c 27 d 4

2 a 48
b $\frac{1}{2}$

16.1 Direct variation

Exercise 16A

1 a 15 b 2
2 a 75 b 6
3 a 150 b 6
4 a 22.5 b 12
5 a 175 miles b 8 hours
6 a £66.50 b 175 kg
7 a 44 b 84 m^2
8 a 33 spaces
b 66 spaces since new car park has 366 spaces
9 17 minutes 30 seconds

Exercise 16B

1 a 100 b 10
2 a 27 b 5
3 a 56 b 1.69
4 a 192 b 2.25
5 a 25.6 b 5
6 a 80 b 8
7 a £50 b 225
8 a 3.2 °C b 10 atm
9 a 388.8 g b 3 mm
10 a 2 J b 40 m/s
11 a £78 b 400 miles
12 4000 cm^3
13 £250
14 a B b A c C
15 a B b A

16.2 Inverse variation

Exercise 16C

1 $Tm = 12$ a 3 b 2.5
2 $Wx = 60$ a 20 b 6
3 $Q(5 - t) = 16$ a −3.2 b 4
4 $Mt^2 = 36$ a 4 b 5
5 $W\sqrt{T} = 24$ a 4.8 b 100
6 a 32 b 4
7 $gp = 1800$ a £15 b 36
8 $td = 24$ a 3 °C b 12 km
9 $ds^2 = 432$ a 1.92 km b 8 m/s
10 $p\sqrt{h} = 7.2$ c 2.4 atm b 100 m
11 $W\sqrt{F} = 0.5$ a $5\frac{t}{h}$ b 0.58 t/h
12 B – This is inverse proportion, as x increases y decreases.
13

x	8	27	64
y	1	$\frac{2}{3}$	$\frac{1}{2}$

14 4.3 miles

Examination questions

1

x	25	100	400
y	10	20	40

2 **a** E _ $4000v$ **b** 3.6 m/s

3 **a** $y = 4x^{-\frac{1}{3}}$ or $y = \frac{4}{\sqrt[3]{x}}$

b **i** $\frac{4}{5}$ **ii** 8

4 19.4 cm

5 128

6 **a** $d = 5t^2$ **b** 245 **c** 3

7 540

8 **a** 2.5 **b** 0.25 **c** 250 **d** 50, −50

9 **a** 10 **b** 3.375

10 **a** 48π **b** 9

11 **a** $q = \frac{136}{t^2}$ **b** 5.44

12 1406.25

13 27 hertz

14 **a** 1500 **b** 6.25

15 40

Answers to Chapter 17

Quick check

a $\begin{pmatrix}1\\3\end{pmatrix}$ **b** $\begin{pmatrix}3\\0\end{pmatrix}$ **c** $\begin{pmatrix}2\\-1\end{pmatrix}$ **d** $\begin{pmatrix}-1\\-2\end{pmatrix}$

17.1 Properties of vectors

Exercise 17A

1 **a** Any three of: $\overrightarrow{AC}, \overrightarrow{CF}, \overrightarrow{BD}, \overrightarrow{DG}, \overrightarrow{GI}, \overrightarrow{EH}, \overrightarrow{HJ}, \overrightarrow{JK}$

b Any three of: $\overrightarrow{BE}, \overrightarrow{AD}, \overrightarrow{DH}, \overrightarrow{CG}, \overrightarrow{GJ}, \overrightarrow{FI}, \overrightarrow{IK}$

c Any three of: $\overrightarrow{AO}, \overrightarrow{CA}, \overrightarrow{FC}, \overrightarrow{IG}, \overrightarrow{GD}, \overrightarrow{DB}, \overrightarrow{KJ}, \overrightarrow{JH}, \overrightarrow{HE}$

d Any three of: $\overrightarrow{BO}, \overrightarrow{EB}, \overrightarrow{HD}, \overrightarrow{DA}, \overrightarrow{JG}, \overrightarrow{GC}, \overrightarrow{KI}, \overrightarrow{IF}$

2 **a** 2**a** **b** 2**b** **c** **a** + **b**
d 2**a** + **b** **e** 2**a** + 2**b** **f** **a** + 2**b**
g **a** + **b** **h** 2**a** + 2**b** **i** 3**a** + **b**
j 2**a** **k** **b** **l** 2**a** + **b**

3 **a** Equal **b** $\overrightarrow{AI}, \overrightarrow{BJ}, \overrightarrow{DK}$

4 **a** $\overrightarrow{OJ} = 2\overrightarrow{OD}$ and parallel **b** $\overrightarrow{AK}$ **c** $\overrightarrow{OF}, \overrightarrow{BI}, \overrightarrow{EK}$

5

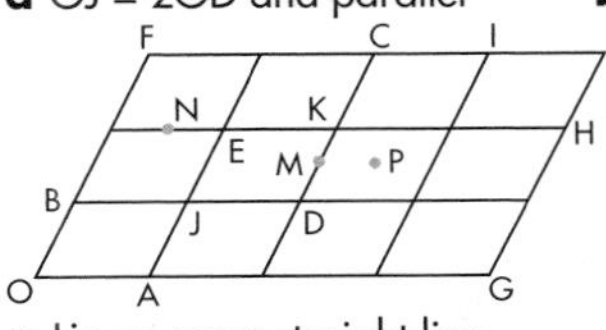

6 **a** Lie on same straight line
b All multiples of **a** + **b** and start at O
c H
d **i** $\overrightarrow{OQ} = \mathbf{a} + \frac{1}{2}\mathbf{b}$ **ii** $\overrightarrow{OR} = 3\mathbf{a} + \frac{3}{2}\mathbf{b}$
e $n\mathbf{a} + \frac{n}{2}\mathbf{b}$

7 **a** −**b** **b** 3**a** − **b** **c** 2**a** − **b**
d **a** − **b** **e** **a** + **b** **f** −**a** − **b**
g 2**a** − **b** **h** −**a** − 2**b** **i** **a** + 2**b**
j −**a** + **b** **k** 2**a** − 2**b** **l** **a** − 2**b**

8 **a** Equal but in opposite directions
b Any three of: $\overrightarrow{DA}, \overrightarrow{EF}, \overrightarrow{GJ}, \overrightarrow{FI}, \overrightarrow{AH}$

9 **a** Opposite direction and $\overrightarrow{AB} = -\frac{1}{2}\overrightarrow{CK}$
b $\overrightarrow{BJ}, \overrightarrow{CK}$
c $\overrightarrow{EB}, \overrightarrow{GO}, \overrightarrow{KH}$

10

11 **a** **i** 3**a** + 2**b** **ii** 3**a** + **b**
iii 2**a** − **b** **iv** 2**b** − 2**a**
b $\overrightarrow{DG}$ and $\overrightarrow{BC}$

12 **a** Any three of: $\overrightarrow{MJ}, \overrightarrow{AG}, \overrightarrow{HC}, \overrightarrow{BD}, \overrightarrow{OH}, \overrightarrow{NA}, \overrightarrow{PO}, \overrightarrow{KB}, \overrightarrow{IE}$
b Any three of: $\overrightarrow{DG}, \overrightarrow{HJ}, \overrightarrow{AL}, \overrightarrow{EH}, \overrightarrow{BA}, \overrightarrow{OM}, \overrightarrow{FB}, \overrightarrow{IO}, \overrightarrow{KN}$
c Any three of: $\overrightarrow{GD}, \overrightarrow{HE}, \overrightarrow{BF}, \overrightarrow{JH}, \overrightarrow{AB}, \overrightarrow{OI}, \overrightarrow{LA}, \overrightarrow{MO}, \overrightarrow{NK}$
d Any three of: $\overrightarrow{CH}, \overrightarrow{DB}, \overrightarrow{EI}, \overrightarrow{GA}, \overrightarrow{HO}, \overrightarrow{BK}, \overrightarrow{JM}, \overrightarrow{AN}, \overrightarrow{OP}$
e Any three of: $\overrightarrow{FH}, \overrightarrow{EG}, \overrightarrow{IA}, \overrightarrow{BJ}, \overrightarrow{KM}, \overrightarrow{OL}$
f Any three of: $\overrightarrow{JD}, \overrightarrow{AE}, \overrightarrow{OF}, \overrightarrow{LH}, \overrightarrow{MB}, \overrightarrow{NI}$
g **i** $\overrightarrow{FG}$, $\overrightarrow{IJ}$ or $\overrightarrow{KL}$ **ii** $\overrightarrow{OC}, \overrightarrow{KD}, \overrightarrow{NG}, \overrightarrow{PH}$
iii $\overrightarrow{FJ}$ or $\overrightarrow{IL}$ **iv** $\overrightarrow{FL}$
v $\overrightarrow{LF}$ **vi** $\overrightarrow{PC}$
vii $\overrightarrow{CP}$ **viii** Same as part **d** of this question
ix Same as part **a** of this question

13 Parts **b** and **d** could be, parts **a** and **c** could not be

14 **a** Any multiple (positive or negative) of 3**a** − **b**
b Will be a multiple of 3**a** − **b**

15 Check student's diagram

16 **a** **i** 2**b** − 2**a** **ii** **a** − **c**
iii 2**c** − 2**b** **iv** **b** + **c** − **a**
b $\overrightarrow{RQ}$ = **a** − **c** = $\overrightarrow{SP}$, so two opposite sides are equal and parallel, hence PQRS is a parallelogram

17.2 Vectors in geometry

Exercise 17B

1 a i $-\mathbf{a} + \mathbf{b}$

ii $\frac{1}{2}(-\mathbf{a} + \mathbf{b})$

iii

iv $\frac{1}{2}\mathbf{a} + \frac{1}{2}\mathbf{b}$

b i $\mathbf{a} - \mathbf{b}$

ii $\frac{1}{2}\mathbf{a} - \frac{1}{2}\mathbf{b}$

iii

iv $\frac{1}{2}\mathbf{a} + \frac{1}{2}\mathbf{b}$

c

d M is midpoint of parallelogram of which OA and OB are two sides.

2 a i $-\mathbf{a} - \mathbf{b}$

ii $-\frac{1}{2}\mathbf{a} - \frac{1}{2}\mathbf{b}$

iii

iv $\frac{1}{2}\mathbf{a} - \frac{1}{2}\mathbf{b}$

b i $\mathbf{b} + \mathbf{a}$

ii $\frac{1}{2}\mathbf{b} + \frac{1}{2}\mathbf{a}$

iii

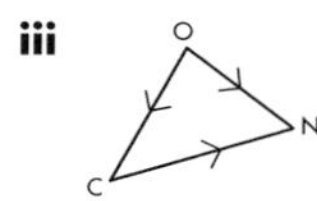

iv $\frac{1}{2}\mathbf{a} - \frac{1}{2}\mathbf{b}$

c

d N is midpoint of parallelogram of which OA and OC are two sides

3 a i $-\mathbf{a} + \mathbf{b}$

ii $\frac{1}{3}(-\mathbf{a} + \mathbf{b})$

iii $\frac{2}{3}\mathbf{a} + \frac{1}{3}\mathbf{b}$

b $\frac{3}{4}\mathbf{a} + \frac{1}{4}\mathbf{b}$

4 a i $\frac{2}{3}\mathbf{b}$

ii $\frac{1}{2}\mathbf{a} + \frac{1}{2}\mathbf{b}$

iii $-2.3\mathbf{b}$

b $\frac{1}{2}\mathbf{a} - \frac{1}{6}\mathbf{b}$

c $\overrightarrow{DE} = \overrightarrow{DO} + \overrightarrow{OE}$

$\rightarrow = \frac{3}{2}\mathbf{a} - \frac{1}{2}\mathbf{b}$

d $\overrightarrow{DE}$ parallel to $\overrightarrow{CD}$ (multiple of $\overrightarrow{CD}$) and D is a common point

5 a

$\overrightarrow{CD} = -\mathbf{a} + \mathbf{b} = \mathbf{b} - \mathbf{a}$

b i $-\mathbf{a}$

ii $-\mathbf{b}$

iii $\mathbf{a} - \mathbf{b}$

c 0, vectors return to starting point

d i $2\mathbf{b}$

ii $2\mathbf{b} - 2\mathbf{a}$

iii $-2\mathbf{a}$

iv $2\mathbf{b} - \mathbf{a}$

v $-\mathbf{a} - \mathbf{b}$

6 a

$\overrightarrow{CX} = \sqrt{1^2 + 1^2}\mathbf{b} = \sqrt{2}\mathbf{b}$

$\overrightarrow{CD} = \overrightarrow{CX} + \overrightarrow{XD}$

$= \sqrt{2}\mathbf{b} - \mathbf{a}$

b

$\overrightarrow{YE} = \sqrt{1^2 + 1^2}\mathbf{a} - \sqrt{2}\mathbf{a}$

$\overrightarrow{DE} = \overrightarrow{DY} + \overrightarrow{YE}$

$= \mathbf{b} - \sqrt{2}\mathbf{a}$

c i $-\mathbf{a}$

ii $-\mathbf{b}$

iii $\mathbf{a} - \sqrt{2}\mathbf{b}$

iv $\sqrt{2}\mathbf{a} - \mathbf{b}$

v $\sqrt{2}\mathbf{a} + \mathbf{a}$

vi $\sqrt{2}\mathbf{b} + \mathbf{b}$

vii $2\mathbf{b} + \sqrt{2}\mathbf{b} - \mathbf{a} - \sqrt{2}\mathbf{a}$

viii $2\mathbf{b} + \sqrt{2}\mathbf{b} - 2\mathbf{a} - \sqrt{2}\mathbf{a}$

d Student's own proof

7 a i $-\mathbf{a} + \mathbf{b}$

ii $\frac{1}{2}(-\mathbf{a} + \mathbf{b}) = -\frac{1}{2}\mathbf{a} + \frac{1}{2}\mathbf{b}$

iii $\frac{1}{2}\mathbf{a} + \frac{1}{2}\mathbf{b}$

b i $\frac{1}{2}\mathbf{b} + \frac{1}{2}\mathbf{c}$

ii $-\frac{1}{2}\mathbf{a} + \frac{1}{2}\mathbf{c}$

c i $-\frac{1}{2}\mathbf{a} + \frac{1}{2}\mathbf{c}$

ii Equal

iii Parallelogram

8 a i $\frac{1}{2}\mathbf{a}$

ii $\mathbf{c} - \mathbf{a}$

iii $\frac{1}{2}\mathbf{a} + \frac{1}{2}\mathbf{c}$

iv $\frac{1}{2}\mathbf{c}$

b i $-\frac{1}{2}\mathbf{a} + \frac{1}{2}\mathbf{b}$

ii $-\frac{1}{2}\mathbf{a} + \frac{1}{2}\mathbf{b}$

c Opposite sides are equal and parallel

d NMRQ and PNLR

9 a $\frac{1}{2}\mathbf{a} + \frac{1}{2}\mathbf{b}$

b i Rhombus

ii They lie on a straight line, $\overrightarrow{OM} = \frac{1}{2}\overrightarrow{OC}$

10 $k = 8$

17.3 Geometric proof

Exercise 17C

1–9 Check students' proofs.

10 a $\overrightarrow{YW} = \overrightarrow{YZ} + \overrightarrow{ZW} = 2\mathbf{a} + \mathbf{b} + \mathbf{a} + 2\mathbf{b} = 3\mathbf{a} + 3\mathbf{b}$

$= 3(\mathbf{a} + \mathbf{b}) = 3\overrightarrow{XY}$

b 3 : 1

c They lie on a straight line.

d Points are A(6, 2), B(1, 1) and C(2, −5). Using Pythagoras' theorem, $AB^2 = 26$, $BC^2 = 26$ and $AC^2 = 52$ so $AB^2 + BC^2 = AC^2$ hence $\angle ABC$ must be a right angle.

Examination questions

1 a i $\frac{1}{2}\mathbf{a}$ **ii** $\frac{1}{2}\mathbf{a} - \frac{1}{2}\mathbf{c}$

b $\overrightarrow{CA} = \mathbf{a} - \mathbf{c} = 2\overrightarrow{MN}$, so parallel

2 a i $\mathbf{p} + \mathbf{q}$ **ii** $\mathbf{q} - \mathbf{p}$

b $\frac{1}{2}\mathbf{p} + \frac{1}{2}\mathbf{q}$

3 a $-\mathbf{x}$ **b** $\mathbf{x} + \mathbf{y}$ **c** $-\frac{1}{2}\mathbf{x} - 2\mathbf{y}$

4 a $2\mathbf{a} - 2\mathbf{b}$ **b** $\overrightarrow{XY} = 2\mathbf{a}$ and $\overrightarrow{OR} = 6\mathbf{a}$, so parallel

5 a $\mathbf{a} + \mathbf{b}$ **b** $\overrightarrow{DX} = 2\mathbf{a}$ and $\overrightarrow{AB} = \mathbf{a}$, so parallel

6 a CD = BC, CE = CF, angle DCE = angle BCF, so congruent (SAS)

b ED = BF from **a** and BF = EG as parallelogram, so ED = EG

Answers to Chapter 18

Quick check

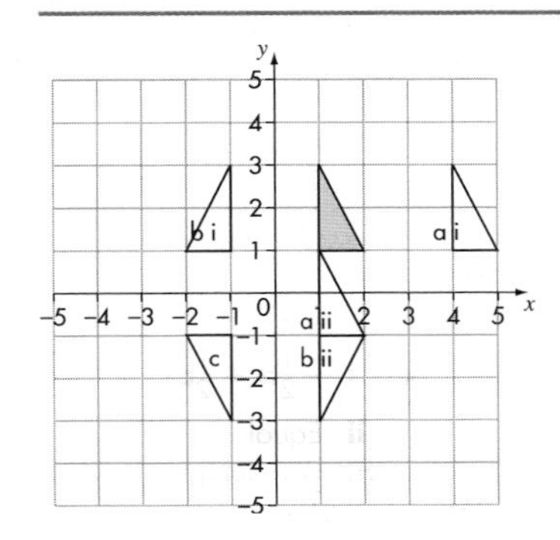

18.1 Transformation of the graph $y = f(x)$

Exercise 18A

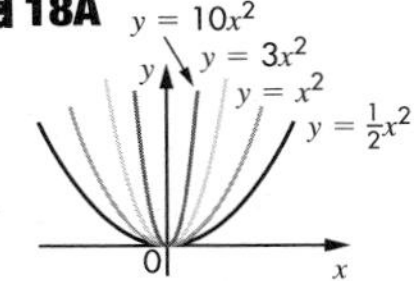

e Stretch sf in y-direction: 3, $\frac{1}{2}$, 10

2 a–d

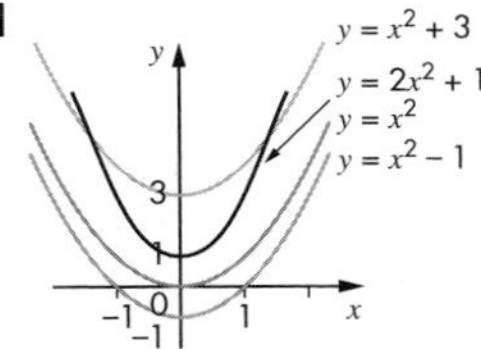

e *b* Translation $\begin{pmatrix} 0 \\ 3 \end{pmatrix}$

c Translation $\begin{pmatrix} 0 \\ -1 \end{pmatrix}$

d Stretch sf 2 in y-direction, followed by translation $\begin{pmatrix} 0 \\ 1 \end{pmatrix}$

3 a–d

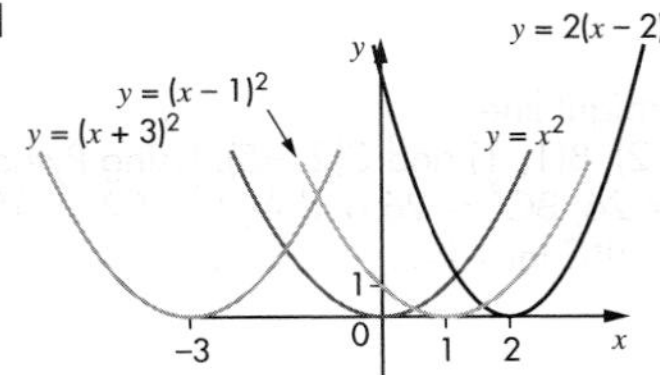

e *b* Translation $\begin{pmatrix} -3 \\ 0 \end{pmatrix}$

c Translation $\begin{pmatrix} 1 \\ 0 \end{pmatrix}$

d Stretch sf 2 in y-direction, followed by translation $\begin{pmatrix} 2 \\ 0 \end{pmatrix}$

4 a–c

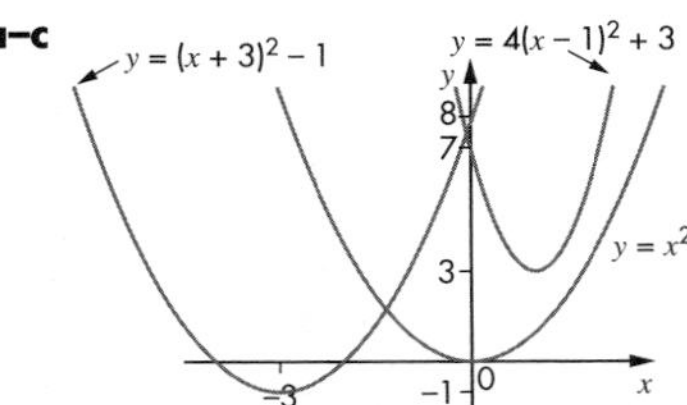

d *b* Translation $\begin{pmatrix} -3 \\ -1 \end{pmatrix}$

c Translation $\begin{pmatrix} 1 \\ 3 \end{pmatrix}$ followed by stretch sf 4 in y-direction

5 a–d

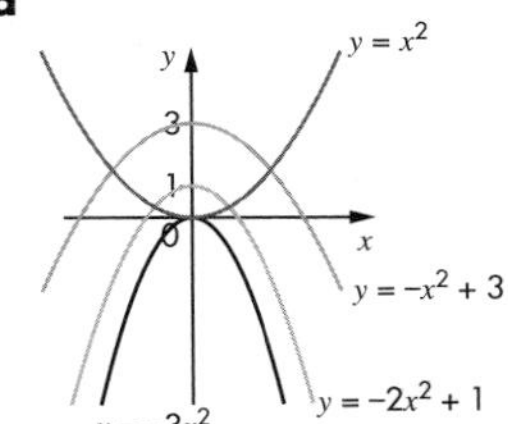

e *b* Reflection in x-axis, followed by translation $\begin{pmatrix} 0 \\ 3 \end{pmatrix}$

c Reflection in the x-axis, followed by stretch sf 3 in y-direction

d Reflection in x-axis, followed by stretch sf 2 in y-direction and translation $\begin{pmatrix} 0 \\ 1 \end{pmatrix}$

6 a–d

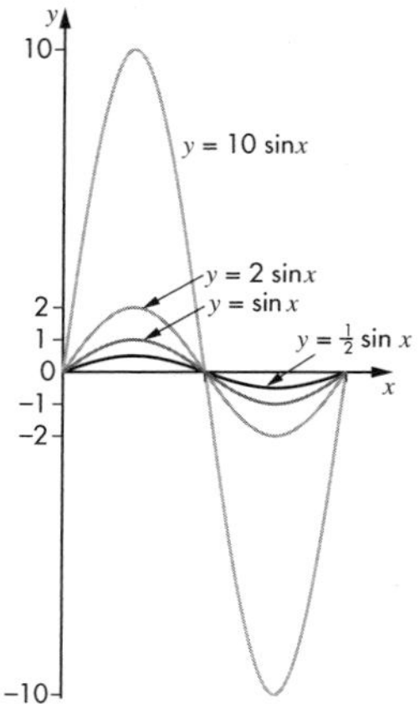

e Stretch sf in y-direction: 2, $\frac{1}{2}$, 10

7 a–d

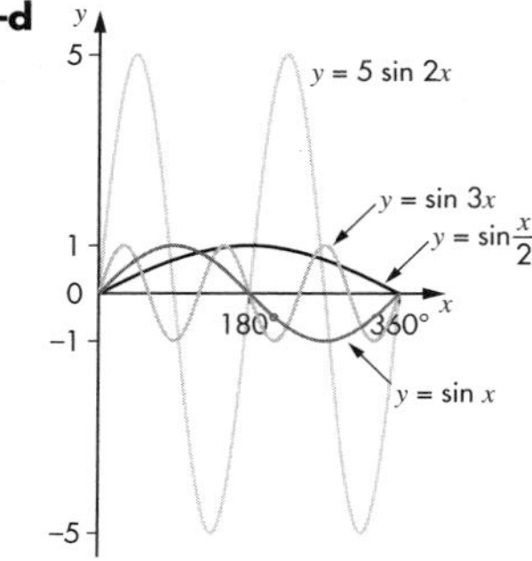

e *b* Stretch sf $\frac{1}{3}$ in x-direction

c Stretch sf 2 in x-direction

d Stretch sf 5 in y-direction, followed by stretch sf $\frac{1}{2}$ in x-direction

8 a–d

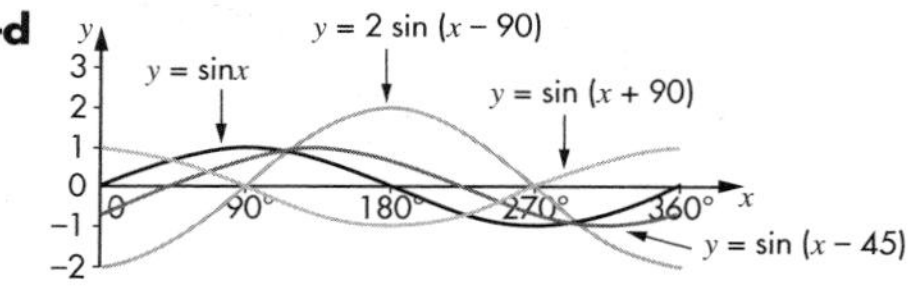

e *b* Translation $\begin{pmatrix} -90 \\ 0 \end{pmatrix}$

c Translation $\begin{pmatrix} 40 \\ 0 \end{pmatrix}$

d Stretch sf 2 in y-direction followed by translation $\begin{pmatrix} 90 \\ 0 \end{pmatrix}$

9 a–d

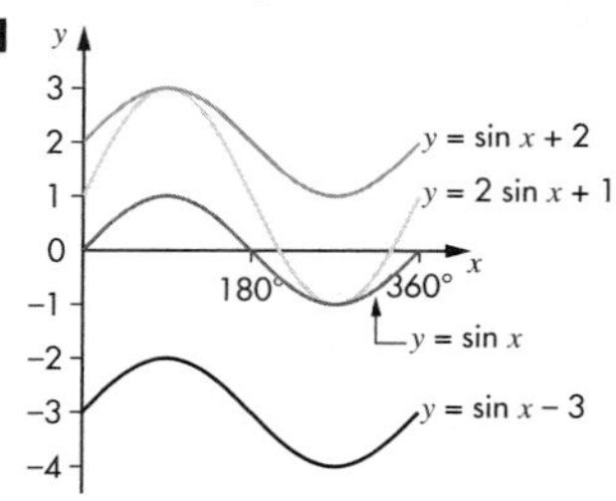

e *b* Translation $\begin{pmatrix} 0 \\ 2 \end{pmatrix}$

c Translation $\begin{pmatrix} 0 \\ -3 \end{pmatrix}$

d Stretch sf 2 in y-direction followed by translation $\begin{pmatrix} 0 \\ 1 \end{pmatrix}$

10 a–d

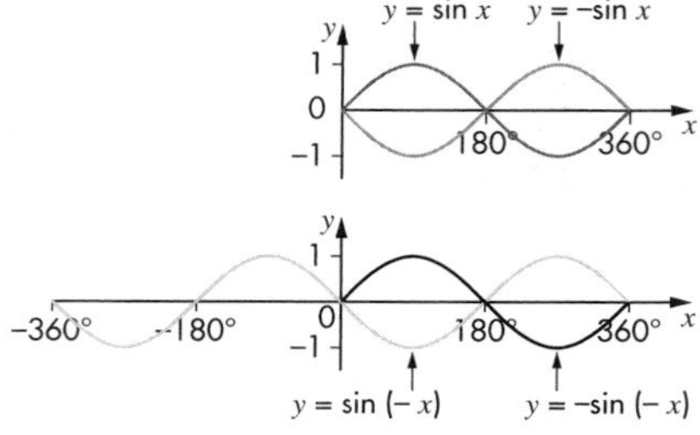

e *b* Reflection in x-axis

c Reflection in y-axis

d This leaves the graph in the same place and is the identity transformation

11 a–d

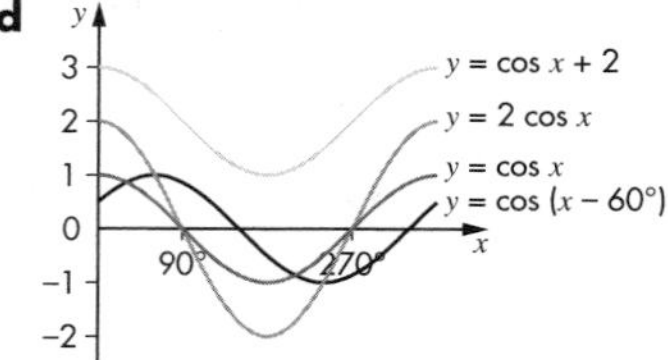

e *b* Stretch sf 2 in y-direction

c Translation $\begin{pmatrix} 60 \\ 0 \end{pmatrix}$

d Translation $\begin{pmatrix} 0 \\ 2 \end{pmatrix}$

12 All of them.

13 a **i** Stretch sf 4 in y-direction

ii Stretch sf 9 in y-direction

iii Stretch sf 16 in y-direction

b **i** Stretch sf $\frac{1}{2}$ in x-direction

ii Stretch sf $\frac{1}{3}$ in x-direction

iii Stretch sf $\frac{1}{4}$ in x-direction

c Stretch sf a^2 in y-direction, or stretch sf $\frac{1}{a}$ in x-direction

14 a **d**

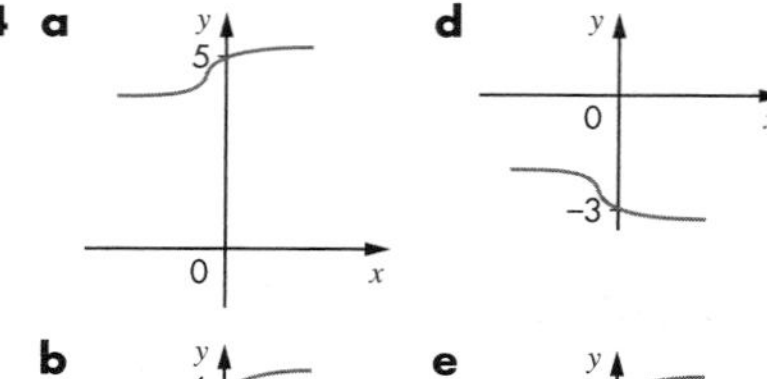

b **e**

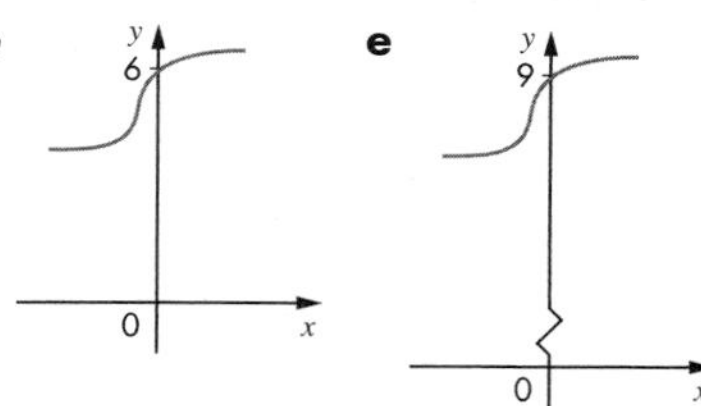

c **f**

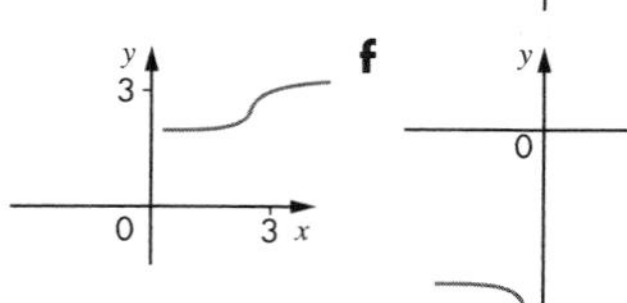

15 a $y = 5x^2$

b $y = x^2 + 7$

c $y = (x + 3)^2$

d $y = 3x^2 + 4$

e $y = (x + 2)^2 - 3$

f $y = -3x^2$

16 **a** $y = 6 \cos x$
b $y = \cos x + 3$
c $y = \cos (x + 30°)$
d $y = 3 \cos x - 2$
e $y = \cos (x - 45°) - 2$

17 **a**

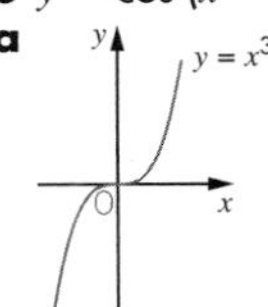

b **i**

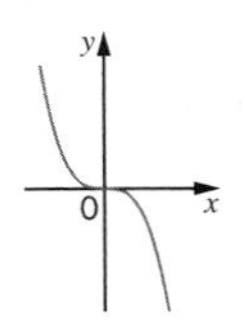

iii

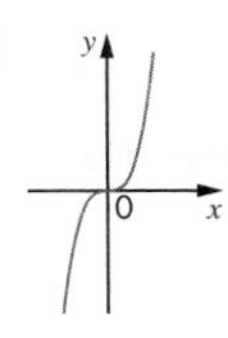

ii

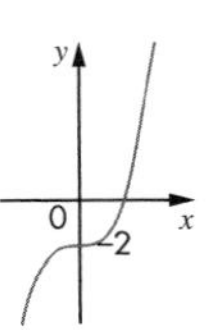

iv

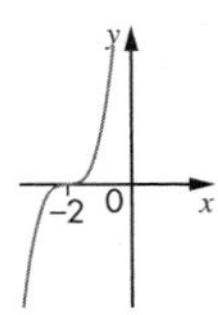

c **i** $y = -x^3$
ii $y = x^3 - 2$
iii $y = 3x^3$
iv $y = (x + 2)^3$

18 **a**

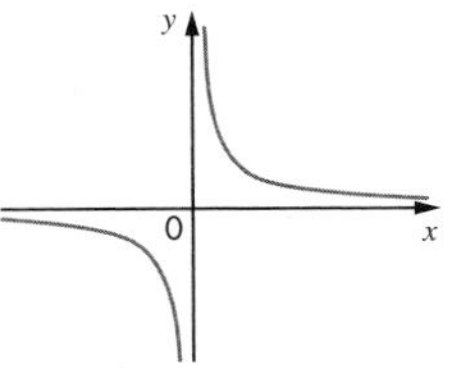

b **i**

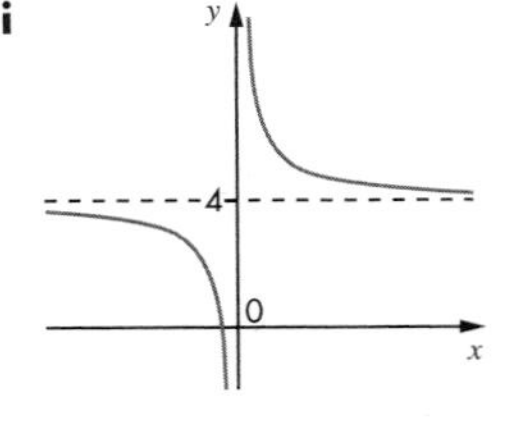

ii

iii

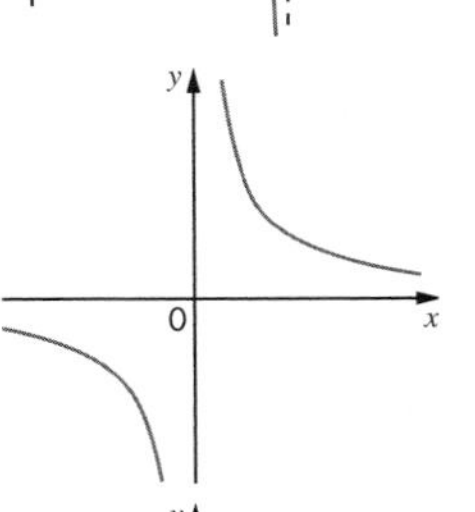

iv

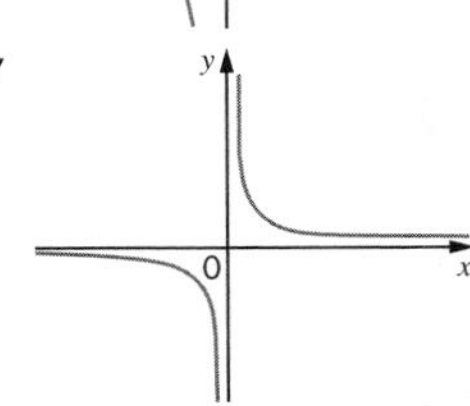

c **i** $y = \frac{1}{x} + 4$
ii $y = \frac{1}{x - 4}$
iii $y = \frac{3}{x}$
iv $y = \frac{1}{2x}$

19 No, as $f(-x) = (-x)^2 = x^2$, and $-f(x) = -(x)^2 = -x^2$

20 **a** $y = x^2 + 2$ **b** $y = (x - 2)^2$
c $y = 2x^2$ **d** $y = -x^2 + 4$

21 **a** $y = 2 \sin x$ **b** $y = \sin(x - 30°)$
c $y = 2 \sin(x - 60°)$ **d** $y = \sin 2x$

22 **a** Translation
b **i** Equivalent **ii** Equivalent
iii Not equivalent

23 **i** A **ii** D
iii E **iv** C

Examination questions

1 **a** (4, 3) **b** (2, 6)

2

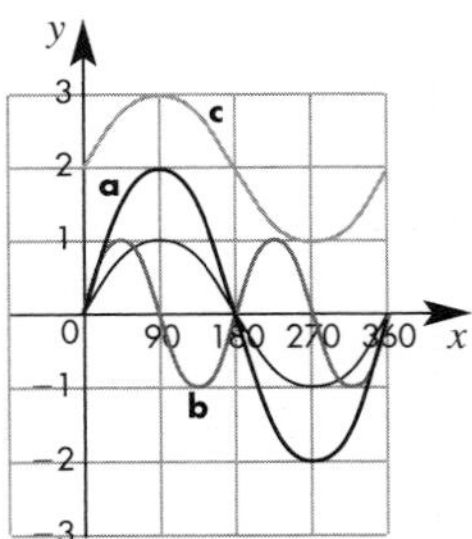

a $y = 2 \sin x°$
b $y = \sin 2x°$
c $y = \sin x° + 2$

3

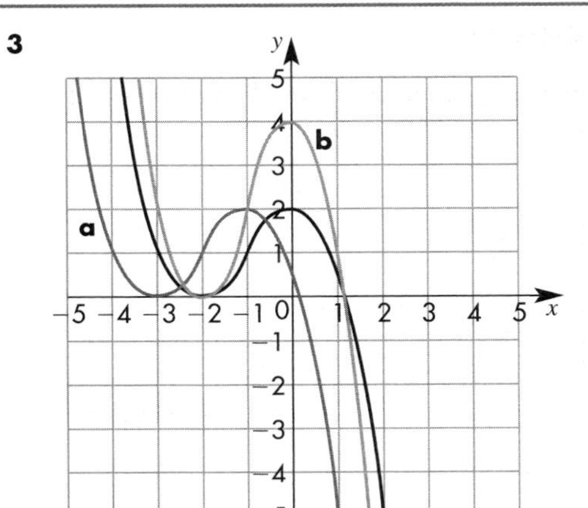

a $y = f(x + 1)$
b $y = 2f(x)$

GLOSSARY

3D A solid shape that has three dimensions (height, width, depth).

adjacent side In a triangle, rectangle or square, the side adjacent to (next to) the angle or side being worked on.

alternate segment The 'other' segment. In a circle divided by a chord, the alternate segment lies on the other side of the chord.

angle The space, usually measured in degrees (°), between two intersecting lines or surfaces (planes). The amount of turn needed to move from one line or plane to the other.

angle bisector A straight line or plane that divides an angle in half.

angle of depression The angle you have to turn downwards from looking along the horizontal to look at the ground or sea from the top of a tower, tree, or cliff, etc.

angle of elevation The angle you have to turn upwards from looking along the horizontal to look at the top of a tree, cliff, flagpole, etc.

angle of rotation The angle turned through to move from one direction to another.

annual rate The annual rate of interest is the percentage interest received or charged for one year.

anticlockwise Turning in the opposite direction to the movement of the hands of a clock. (Opposite of *clockwise*)

apex The highest vertex in the given orientation of a polygon such as a triangle or a 3D shape such as a pyramid or cone.

arc A curve forming part of the circumference of a circle.

area Measurement of the flat space a shape occupies. Usually measured in square units or hectares. (See also *surface area*.)

area ratio The ratio of the areas of two similar shapes is equal to the ratio of the squares of their corresponding lengths. The area ratio, or area scale factor, is the square of the length ratio.

area scale factor See *area ratio*.

area sine rule The area of a triangle is given by $\frac{1}{2}ab \sin C$.

asymptote A straight line whose perpendicular distance from a curve decreases to zero as the distance from the origin increases without limit.

average A single number that represents or typifies a collection of values. The three commonly used averages are mode, mean and median.

balance Equality on either side of an equation.

balance the coefficients The first step to solving simultaneous equations is to make the coefficients of one of the variables the same. This is called balancing the coefficients.

bearing The direction relative to a fixed point.

bisect Divide in half.

boundary When drawing a graph of an inequality first consider the equality. This line forms the limit or boundary of the inequality. When a strict inequality is stated ($<$ or $>$), the boundary line should be drawn as a dashed line to show that it is not included in the range of values. When $\leqslant$ or $\geqslant$ are used, the boundary line should be drawn as a solid line to show that the boundary is included.

brackets The symbols '(' and ')' which are used to separate part of an expression. This may be for clarity or to indicate a part to be worked out individually. When a number and/or value is placed immediately before an expression or value inside a pair of brackets, the two are to be multiplied together. For example, $6a(5b + c) = 30ab + 6ac$.

cancel A fraction can be simplified to an equivalent fraction by dividing the numerator and denominator by a common factor. This is called cancelling.

centre of enlargement The fixed point of an enlargement. The distance of each image point from the centre of enlargement is the distance of object point from centre of enlargement × scale factor. In simple terms, the centre of enlargement is used to change the size of an object without changing its shape, and can be compared to a projector magnifying an image. (See also *scale factor* and *image*.)

centre of rotation The fixed point around which a shape is rotated or turned.

check Calculations can be checked by carrying out the inverse operation. Solutions to equations can be checked by substituting values of the variable(s).

chord A line joining two points on the circumference of a circle. (See also *circumference*.)

circle A circle is the path of a point that is always equidistant from another point (the centre).

circular function A function with a repeating set of values, such as sine and cosine, where the values cycle through from 0 to 1 to 0 to –1 to 0, and repeat.

circumference The outline of a circle. The distance all the way around this outline.

clockwise Turning in the same direction as the movement of the hands of a clock. (Opposite of *anticlockwise*)

coefficient The number in front of an unknown quantity (the letter) in an algebraic term. For example, in $8x$, 8 is the coefficient of x.

comment A note whether a result in a trial and improvement problem is too high or too low.

compass Also called a pair of compasses, an instrument with two arms (one is sharp, and a pencil is attached to the other one), used for drawing circles or arcs.

completing the square A method of solving quadratic equations which involves rewriting the equation $x^2 + px + q$ in the form $(x + a)^2 + b$.

compound interest Instead of interest being calculated once at the end of a term (simple interest), it is calculated annually (or monthly or daily) and added to the principal before the next period starts. This has the effect of increasing the amount of interest earned or owed above that calculated by using simple interest.

congruent Exactly alike in shape and size.

constant of proportionality, k This describes the situation when the probability of an event is dependent upon the outcome of another event. For example, the probability of the colour of a second ball drawn from a bag is conditional to the colour of the first ball drawn from the bag – if the first ball is not replaced.

constant term A term in an algebraic expression that does not change because it does not contain a variable, the number term. For example, in $6x^2 + 5x + 7$, 7 is the constant term.

construct To draw angles, lines or shapes accurately, according to given requirements.

continuous data Data that can be measured rather than counted, such as weight and height. It can have an infinite number of values within a range.

cosine The ratio of the adjacent side to the hypotenuse in a right-angled triangle.

cosine rule A formula used to find the lengths of sides or the size of an angle in a triangle. For example, $a^2 = b^2 + c^2 - 2bc \cos A$.

cover-up method To plot a straight line graph, only two points need to be known. (Finding a third point acts as a check.) The easiest points to find are those when $x = 0$ and $y = 0$. By covering the x and y terms, you can easily see the associated values of y and x.

cross-multiply A method for taking the denominator from one side of an equals sign to the numerator on the other side of the equals sign. You are actually multiplying both sides of the equation by the denominator, but it cancels out on the original side. (Take care that you multiply all the terms by the denominator.)

cube A 3D shape with six identical square faces.

cubic A cubic expression or equation contains an 'x^3' term.

cyclic Arranged as if on a circle.

cyclic quadrilateral A quadrilateral whose vertices lie on a circle.

cylinder A solid or hollow prism with circular ends and uniform (unchanging) cross-section. For example, the shape of a can baked beans or length of drainpipe.

decimal place Every digit in a number has a place value (hundreds, tens, ones, etc.). The places after (to the right of) the decimal point have place values of tenths, hundredths, etc. These place values are called the decimal places.

decimal point A dot which is placed between the units and tenths column in a decimal number. Each column which is to the right of the decimal point is a decimal place.

demonstration Logically presented proof of how a theory generates a certain result.

density The ratio of the mass of an object to its volume. The mass per unit volume.

diameter A straight line across a circle, from circumference to circumference and passing through the centre. It is the longest chord of a circle and two radii long. (See also *radius*.)

difference of two squares The result $a^2 - b^2 = (a - b)(a + b)$ is called the difference of two squares.

direct proportion Two values or measurements may vary in direct proportion. That is, if one increases, then so does the other.

direct variation Another name for direct proportion.

direction The way something is facing or pointing. Direction can be described using the compass points (north, south, south-east, etc.) or using bearings (the clockwise angle turned from facing north). The direction of a vector is given by the angle it makes with a line of reference. Two vectors that differ only in magnitude will be parallel.

discrete data Data that is counted, rather than measured, such as favourite colour or a measurement that occurs in integer values, such as a number of days. Data can only have specific values within a range.

discriminant The quantity $(b^2 - 4ac)$ in the quadratic formula is called the discriminant.

distance The separation (usually along a straight line) of two points.

do the same to both sides To keep an equation balanced, you must do the same thing to both sides. For example, if you add something to one side, you must add the same thing to the other side. If you double one side, you must double the other side. If you manipulate a fraction, you must do the same thing to the numerator and the denominator to keep the value of fraction unchanged. However you can only multiply or divide the numbers. Adding or subtracting will alter the value of the fraction.

eliminate To remove a quantity such as a variable from an equation.

enlargement A transformation of a plane figure or solid object that increases the size of the figure or object by a scale factor but leaves it the same shape (i.e. the figure or object remains in the same ratio). (See also *scale factor*.)

equation A number sentence where one side is equal to the other. An equation always contains an equals sign (=).

equidistant The same distance apart.

exponential functions Equations which have the form $y = kx$, where k is a positive number, are called exponential functions.

expression Collection of symbols representing a number. These can include numbers, variables (x, y, etc.), operations (+, ×, etc.), functions (squaring, cosine, etc.), but there will be no equals sign (=).

factorisation (noun) Finding one or more factors of a given number or expression. (verb: factorise)

factorise See *factorisation*.

factors A whole number that divides exactly into a given number. For example, the factors of 16 are 1, 2, 4, 8, 16.

final amount The amount after working out, for example, the percentage.

frustum The base of a cone or pyramid. The shape obtained by removing the top of a cone or pyramid.

function A function of x is any algebraic expression in which x is the only variable. This is often represented by the function notation f(x) or 'function of x'.

gradient How steep a hill or the line of a graph is. The steeper the slope, the larger the value of the gradient. A horizontal line has a gradient of zero.

gradient-intercept The point at which the gradient of a curve or line crosses an axis.

guess Using your mathematical knowledge, you can make an estimate of an answer and use this as a starting point in a trial and improvement problem.

hypotenuse The longest side of a right-angled triangle. The side opposite the right angle.

image In geometry the 'image' is the result of a transformation. A good way to remember this is to relate it to the 'cut and paste' function on your computer.

included A point is included in a region on a graph if it satisfies certain criteria. A boundary is included if it is defined by an equality (=) or a 'weak' inequality ($\leqslant$ or $\geqslant$).

included angle The angle between two lines or sides of a polygon.

inclusive inequality Inequality in which the boundary value is included, such as $x \leqslant 5$, $y \geqslant 10$.

index (plural: indices) A power or exponent. For example, in the expression 3^4, 4 is the index, power or exponent.

indices See *index*.

inequality A relation that shows that two numbers or quantities are or may be unequal. The symbol $\neq$ is used.

integer A whole number. Integers include all the positive whole numbers, negative whole numbers and zero.

intercept The point where a line or graph crosses an axis.

inverse Inverse operations cancel each other out or reverse the effect of each other. The inverse of a number is the reciprocal of that number.

inverse cosine The reverse of the cosine function. It tells you the value of the angle with that cosine. For example, $\cos^{-1}A$.

inverse sine The reverse of the sine function. It tells you the value of the angle with that sine. For example, $\sin^{-1}A$.

isosceles triangle A triangle with two sides that are equal, and it has two equal angles.

length ratio The ratio of lengths in similar figures.

limit of accuracy No measurement is entirely accurate. The accuracy depends on the tool used to measure it. The value of every measurement will be rounded to within certain limits. For example, you can use a ruler to measure the nearest half-centimetre or millimeter, but any measurement you take could be inaccurate by up to half a centimeter or one millimetre. This is your limit of accuracy. (See also *lower bound* and *upper bound*.)

line bisector A point, a straight line or a plane that divides a line in half.

line symmetry Symmetry that uses a line to divide a figure or shape into halves, such that both halves are an exact mirror image of each other. Many shapes have more than one line of symmetry.

linear Forming a line.

linear scale factor Also called the *scale factor* or *length ratio*. The ratio of corresponding lengths in two similar shapes is constant.

loci See *locus*.

locus (plural: loci) The locus of a point is the path taken by the point following a rule or rules. For example, the locus of a point that is always the same distance from another point is the shape of a circle.

lower bound The lower limit of a measurement. (See also *limit of accuracy*.)

magnitude Size. Magnitude is always a positive value.

mass The mass is the amount of 'stuff' an object consists of. It does not vary if the object is moved somewhere else. The weight of an object is closely related to mass but depends on the effect of gravity. For example, the weight of an object on Earth will be different to that on the Moon. Its mass will remain constant.

maximum The greatest value of something. The turning point or point at which the graph of a parabola $y = -ax^2 + bx + c$ is at its highest.

minimum The smallest value of something. The turning point or point at which the graph of a parabola $y = -ax^2 + bx + c$ is at its lowest.

mirror line A line where a shape is reflected exactly on the other side.

multiplier The number used to multiply by.

negative reciprocal The reciprocal multiplied by –1. The gradients of perpendicular lines are the negative reciprocal of each other.

non-linear An expression or equation that does not form a straight line. The highest power of x is greater than 1.

number line A continuous line on which all the numbers (whole numbers and fractions, and can also include negative numbers) can be shown by points at distances from zero.

object (in maths) You carry out a transformation on an object to form an image. The object is the original or starting shape, line or point. (See also *enlargement*.)

opposite side In a triangle, rectangle or square, the side opposite (or, on the other side of) the angle or side being worked on.

origin The point (0, 0) on Cartesian coordinate axes.

original amount The amount you start with. For example, the amount before an increase or decrease is applied to it.

perpendicular bisector A line drawn at a right angle to a line segment which also divides it into two equal parts.

point of contact The point where a tangent touches a circle.

power/powers When a number or expression is multiplied by itself, the power is how many 'lots' are multiplied together. For example, $2^3 = 2 \times 2 \times 2$ (note that it is not the number of times that 2 is multiplied by itself; it is one more than the number of times it is multiplied by itself.) The name given to the symbol to indicate this, such as 2. (See also *square* and *cube*.)

principal The original amount of a sum of money.

proof An argument that establishes a fact about numbers or geometry for all cases. Showing that the fact is true for specific cases is a demonstration.

prove The process of explaining a proof. (See also *proof*.)

Pythagoras' theorem The theorem states that the square on the hypotenuse of a right-angled triangle is equal to the sum of the squares on the other two sides.

quadratic expression An expression involving an x^2 term.

quadratic formula A formula for solving quadratic equations. The solution of the equation

$ax^2 + bx + c = 0$ is given by: $x = \frac{-b \pm \sqrt{b^2 - 4ac}}{2a}$

radius (plural: radii) The distance from the centre of a circle to its circumference.

ratio The ratio of A to B is a number found by dividing A by B. It is written as A : B. For example, the ratio of 1 m to 1 cm is written as 1 m : 1 cm = 100 : 1. Notice that the two quantities must both be in the same units if they are to be compared in this way.

rational number A rational number is a number that can be written as a fraction. For example, $\frac{1}{4}$ or $\frac{10}{3}$.

rearrange See *rearrangement*.

rearrangement To change the arrangement of something. An equation can be rearranged using the rules of algebra to help you solve it. Data can be rearranged to help you analyse it. (See also *data*.)

reciprocal The reciprocal of any number is 1 divided by the number. The effect of finding the reciprocal of a fraction is to turn it upside down. For example, the reciprocal of 3 is $\frac{1}{3}$, the reciprocal of $\frac{1}{4}$ is 4, the reciprocal of $\frac{10}{3}$ is $\frac{3}{10}$.

recurring decimal A decimal number that repeats forever with a repeating pattern of digits

reflection The image formed after being reflected, for example, in a mirror. The process of reflecting an object.

region An area on a graph defined by certain rules or parameters.

roots The roots of a quadratic equation are the values of x when $y = 0$. (They are the solution to the equation.) They can be seen on a graph where the parabola crosses the x-axis.

rotation Turning. A geometrical transformation in which every point on a figure is rotated through the same angle.

rotational symmetry A shape which can be turned about a point so that it coincides exactly with its original position at least twice in a complete rotation.

scale A scale on a diagram shows the scale factor used to make the drawing. The axes on a graph or chart will use a scale depending on the space available to display the data. For example, each division on the axis may represent 1, 2, 5, 10 or 100, etc. units. (See also *scale factor*.)

scale factor The ratio by which a length or other measurement is increased or deceased.

sector A region of a circle, like a slice of a pie, bounded by an arc and two radii.

segment A part of a circle between a chord and the circumference.

semicircle Half a circle.

show The middle level in an algebraic proof. You are required to show that both sides of the result are the same algebraically. (See also *proof* and *verify*.)

side 1. A straight line forming part of the perimeter of a polygon. For example, a triangle has three sides.
2. A face (usually a vertical face) of a 3D object, such as the side of a box.

similar The same shape but a different size.

similar triangles Triangles with the same size angles. The lengths of the sides of the triangles are different but vary in a constant proportion

simultaneous equations Two or more equations that are true at the same time.

sine The ratio of the opposite side to the hypotenuse in a right-angled triangle

sine rule In a triangle, the ratio of the length of a side to the sine of the opposite angle is constant, hence $\frac{a}{\text{Sin}A} = \frac{b}{\text{Sin}B} = \frac{c}{\text{Sin}C}$.

slant height The distance along the sloping edge of a cone or pyramid.

soluble Something that can be solved or explained.

solution The result of solving a mathematical problem. Solutions are often given in equation form.

solve Finding the value or values of a variable (x) that satisfy the given equation or problem.

speed How fast something moves.

sphere A three-dimensional round body. All points of its surface are equidistant from its centre.

square 1. A polygon with four equal sides and all the interior angles equal to 90°.
2. The result of multiplying a number by itself. For example, 5^2 or 5 squared is equal to $5 \times 5 = 25$.

square root The square of a square root of a number gives you the number. The square root of 9 (or $\sqrt{9}$) is 3, $3 \times 3 = 3^2 = 9$.

standard form Also called standard index form. Standard form is a way of writing very large and very small numbers using powers of 10. A number is written as a value between 1 and 10 multiplied by a power of 10. That is, $a \times 10^n$ where $1 \leqslant a < 10$, and n is a whole number.

standard index form See *standard form.*

stretch An enlargement that takes place in one direction only. It is described by a scale factor and the direction of the stretch.

strict inequality An inequality, $<$ or $>$, that does not allow for equality.

subject The subject of a formula is the letter on its own on one side of the equals sign. For example, t is the subject of this formula: $t = 3f + 7$.

substitute When a letter in an equation, expression or formula is replaced by a number, we have substituted the number for the letter. For example, if $a = b + 2x$, and we know $b = 9$ and $x = 6$, we can write $a = 9 + 2 \times 6$. So $a = 9 + 12 = 17$.

subtended Standing on. An angle made by two radii at the centre of a circle is the angle subtended by the arc which joins the points on the circumference at the ends of the radii. (See also *radius*.)

surd A number written as $\sqrt{x}$. For example, $\sqrt{7}$.

surd form The square root sign is left in the final expression when $\sqrt{x}$ is an irrational number.

surface area The area of the surface of a 3D shape, such as a prism. The area of a net will be the same as the surface area of the shape.

tangent 1. A straight line that touches the circumference of a circle at one point only.
2. The ratio of the opposite side to the adjacent side in a right-angled triangle.

terminating decimal A terminating decimal can be written down exactly. $\frac{33}{100}$ can be written as 0.33, but $\frac{3}{10}$ is 0.3333… with the 3s recurring forever.

three-figure bearing The angle of a bearing is given with three digits. The angle is less than 100°, a zero (or zeros) is placed in front, such as 045° for north-east.

time How long something takes. Time is measured in days, hours, seconds, etc.

transform To change.

transformation An action such as translation, reflection, or rotation.

translation A transformation in which all points of a plane figure are moved by the same amount and in the same direction.

trial and improvement Some problems require knowledge of mathematics beyond GCSE level but sometimes these can be solved or estimated by making an educated guess and then refining this to a more accurate answer. This is known as trial and improvement.

trigonometry The branch of mathematics that shows how to explain and calculate the relationships between the sides and angles of triangles.

unit cost The cost of one unit of a commodity, such as the cost per kilogram or per pound.

unitary method A method of calculation where the value for item is found before finding the value for several items.

upper bound The higher limit of a measurement. (See also *limit of accuracy*.)

variable A quantity that can have many values. These values may be discrete or continuous. They are often represented by x and y in an expression. (See also *expression*.)

vector A quantity with magnitude and direction. (See also *magnitude* and *direction*.)

verify The first of the three levels in an algebraic proof. Each step must be shown clearly and at this level you need to substitute numbers into the result to show how it works. (See also *proof* and *show*.)

vertex
1. The points at which the sides of a polygon or the edges of a polyhedron meet.
2. The turning point (maximum or minimum) of a graph.

vertical height The perpendicular height from the base to the apex (or tip) of a triangle, cone, or pyramid.

volume The amount of space occupied by a substance or object or enclosed within a container.

volume ratio The ratio of the volumes of two similar shapes is equal to the ratio of the cubes of their corresponding lengths. The volume ratio, or volume scale factor, is the cube of the length ratio.

volume scale factor See *volume ratio*.

$y = mx + c$ The general equation of a straight line: m is the gradient and c is the y-intercept.

π Pronounced 'pie', the numerical value of the ratio of the circumference of a circle to its diameter (approximately 3.14159).

INDEX

William Collins' dream of knowledge for all began with the publication of his first book in 1819. A self-educated mill worker, he not only enriched millions of lives, but also founded a flourishing publishing house. Today, staying true to this spirit, Collins books are packed with inspiration, innovation and practical expertise. They place you at the centre of a world of possibility and give you exactly what you need to explore it.

Collins. Freedom to teach.

Published by Collins
An imprint of HarperCollins*Publishers*
77–85 Fulham Palace Road
Hammersmith
London
W6 8JB

Browse the complete Collins catalogue at
www.collinseducation.com

10 9 8 7 6 5 4 3 2 1

ISBN-13 978-0-00-733997-6

British Library Cataloguing in Publication Data

A Catalogue record for this publication is available from the British Library.

Commissioned by Katie Sergeant
Project managed by Priya Govindan and Hugo Wilkinson
FM and PS pages project-managed by Alexandra Riley
Edited and proofread by Joan Miller, Karen Westall and Brian Asbury
Answers checked by Amanda Dickson
Glossary by Marian Bond
Indexed by Esther Burd
Cover design by Angela English
Content design by Nigel Jordan
Typesetting by Jordan Publishing Design
FM and PS pages designed by EMC Design and Jerry Fowler
Production by Arjen Jansen and Leonie Kellman
Printed and bound by L.E.G.O. S.p.A. Italy

Acknowledgements

The publishers have sought permission from Edexcel to reproduce questions from past GCSE Mathematics papers.

The publishers wish to thank the following for permission to reproduce photographs. Every effort has been made to trace copyright holders and to obtain their permission for the use of copyright material. The publishers will gladly receive any information enabling them to rectify any error or omission at the first opportunity.

p. 6: A/W Nigel Jordan © HarperCollins Publishers, © iStockphoto.com/Patricia Burch, A/W Nigel Jordan © HarperCollins Publishers, © iStockphoto.com/Terry Wilson, © iStockphoto.com/Jaroslaw Wojcik, © Robert Gray; pp. 20–21: © iStockphoto.com/kavu, p21 © Dreamstime.com/Viktor Zhugin; p22: © iStockphoto.com/Karen Mower, © iStockphoto.com/D.Huss; p. 58: © BG Skutvik (Dr), © TR Harland & Wolff, A/W Jerry Fowler © HarperCollins Publishers; p. 59 © TL Tebnad (Dr); p. 60: © iStockphoto.com/Graeme Purdy, © iStockphoto.com/Darren Pearson, © iStockphoto.com/Andrew Howe; p. 78: © Dreamstime.com/Jörg Beuge, © Shutterstock/marco mayer, © Shutterstock/Martin Turzak, © Shutterstock/ Joe Gough, pp. 78–79: © Dreamstime.com/Saida Huseynova, p. 79: © Dreamstime.com/Jörg Beuge; p. 80: Wikipedia Commons, © iStockphoto.com/Vladimir Sinenko, © iStockphoto.com/Auke Holwerda; p. 98: © Dreamstime.com /Chris Turner, © Dreamstime.com/Ffolas, pp. 98–9: © Dreamstime.com/Milacroft, p. 99: © Dreamstime.com/Drbouz; p. 100: © iStockphoto.com/fotoVoyager, © iStockphoto.com/Matthias Weinrich; pp. 142–3: © Dreamstime.com/Davidmartyn, p. 143: © Dreamstime.com/ Eric Simard, © Dreamstime.com/Jun Mu; p. 144: © iStockphoto.com/Andrey Prokhorov, © iStockphoto.com/Roberto Gennaro; p. 174: A/W Nigel Jordan © HarperCollins Publishers, A/W Jerry Fowler © HarperCollins Publishers; pp. 206–7: © iStockphoto.com/Graeme Purdy, © iStockphoto.com/Darren Pearson, © iStockphoto.com/Andrew Howe, A/W Lesley Gray; p. 208: © iStockphoto.com/Branko Miokovic, © iStockphoto.com/Sam Valtenbergs; Wikipedia Commons; p. 234: © iStockphoto.com/Karimhesham, © iStockphoto.com/Luis Carlos Torres, © iStockphoto.com/magaliB, Wikipedia Commons; pp. 254–5: © Frank Boston/Dreamstime.com, p. 255: © Spooky2006/Dreamstime.com; p. 256: © iStockphoto.com/Lukasz Laska, A/W Nigel Jordan © HarperCollins Publishers, © iStockphoto.com/Paul Cowan, Wikipedia Commons; pp. 290–91: © Dreamstime.com/Gail Johnson, p. 291: © iStockphoto.com/Claudiad, © Dreamstime.com/Pei Lin; p. 292: © iStockphoto.com/Berenike, A/W Nigel Jordan © HarperCollins Publishers, © Mark A. Wilson; pp. 324–5: © Dreamstime.com/Slallison, p.325: © Dreamstime.com/Panagiotis Risvas, © Dreamstime.com/Konstantin Sutyagin; p. 326: © iStockphoto.com/Andreas Weber, © iStockphoto.com/ Philip Beasley; p. 364: © Dreamstime.com/Stasvolik, © Dreamstime.com/ Bagwold, pp. 364–5: © Dreamstime.com/Designerpix; p. 386: © iStockphoto.com/Henryk Sadura, © iStockphoto.com/mbbirdy; p. 406: © iStockphoto.com/narvikk, © iStockphoto.com/Goktug Gurellier; p. 422: © Dreamstime.com/Samum, © Dreamstime.com/Garuti, © Dreamstime.com/ Stephen Mcsweeny; p. 424: © iStockphoto.com/Ivan Kmit, © iStockphoto.com/ Scott Leigh, © iStockphoto.com/zxvisual, © iStockphoto.com/Dan Barnes, © iStockphoto.com/Gary Martin, © iStockphoto.com/Ivan Stevanovic; p. 438: © Dreamstime.com/Vladimir Wrangel, pp. 438–9: © Dreamstime.com/ Newphotoservice, p. 439: © Dreamstime.com/Mopic, © Dreamstime.com/ Claudio Monni; p. 440: A/W Jerry Fowler © HarperCollins Publishers, © iStockphoto.com/Mark Evans, © iStockphoto.com/David Joyner, © iStockphoto.com/Stefan Weichelt; pp. 462–3: © Dreamstime.com/Andrey Pavlov; p.464: © iStockphoto.com/marc brown; p. 482: © Shutterstock/ Anton Hlushchenko, © Dreamstime.com/Armando Iozzi, pp. 482–3: © Shutterstock/Stacie Stauff Smith Photography.

With thanks to Chris Pearce, Samantha Burns, Naomi Norman, Claire Beckett, Andy Edmonds, Anton Bush (Gloucester High School for Girls), Matthew Pennington (Wirral Grammar School for Girls), James Toyer (The Buckingham School), Gordon Starkey (Brockhill Park Performing Arts College), Laura Radford and Alan Rees (Wolfreton School) and Mark Foster (Sedgefield Community College).